Lecture Notes in Computer Sci

T0237775

Commenced Publication in 1973
Founding and Former Series Editors:
Gerhard Goos, Juris Hartmanis, and Jan van Leeuwen

Alfred J. van der Poorten   Andreas Stein (Eds.)

Algorithmic Number Theory

8th International Symposium, ANTS-VIII
Banff, Canada, May 17-22, 2008
Proceedings

 Springer

Volume Editors

Alfred J. van der Poorten
ceNTRe for Number Theory Research
1 Bimbil Place, Killara, NSW 2071, Australia
E-mail: alf@math.mq.edu.au

Andreas Stein
Carl von Ossietzky Universität Oldenburg
Institut für Mathematik
26111 Oldenburg, Germany
E-mail: andreas.stein1@uni-oldenburg.de

Library of Congress Control Number: 2008925108

CR Subject Classification (1998): F.2, G.2, E.3, I.1

LNCS Sublibrary: SL 1 – Theoretical Computer Science and General Issues

ISSN   0302-9743
ISBN-10   3-540-79455-7 Springer Berlin Heidelberg New York
ISBN-13   978-3-540-79455-4 Springer Berlin Heidelberg New York

Springer is a part of Springer Science+Business Media

springer.com

© Springer-Verlag Berlin Heidelberg 2008
Printed in Germany

Typesetting: Camera-ready by author, data conversion by Scientific Publishing Services, Chennai, India
Printed on acid-free paper   SPIN: 12262908   06/3180   5 4 3 2 1 0

Preface

The first Algorithmic Number Theory Symposium took place in May 1994 at Cornell University. The preface to its proceedings has the organizers expressing the hope that the meeting would be "the first in a long series of international conferences on the algorithmic, computational, and complexity theoretic aspects of number theory." ANTS VIII was held May 17–22, 2008 at the Banff Centre in Banff, Alberta, Canada. It was the eighth in this lengthening series.

The conference included four invited talks, by Johannes Buchmann (TU Darmstadt), Andrew Granville (Université de Montréal), François Morain (École Polytechnique), and Hugh Williams (University of Calgary), a poster session, and 28 contributed talks in appropriate areas of number theory.

Each submitted paper was reviewed by at least two experts external to the Program Committee; the selection was made by the committee on the basis of those recommendations. The Selfridge Prize in computational number theory was awarded to the authors of the best contributed paper presented at the conference.

The participants in the conference gratefully acknowledge the contribution made by the sponsors of the meeting.

May 2008   Alf van der Poorten and Andreas Stein (Editors)
                                          Renate Scheidler (Organizing Committee Chair)
                                          Igor Shparlinski (Program Committee Chair)

Conference Website

The names of the winners of the Selfridge Prize, material supplementing the contributed papers, and errata for the proceedings, as well as the abstracts of the posters and the posters presented at ANTS VIII, can be found at:
http://ants.math.ucalgary.ca .

I	Cornell University (Ithaca, NY, USA)	May 1994	LNCS 877
II	Université Bordeaux 1 (Talence, France)	May 1996	LNCS 1122
III	Reed College (Portland, Oregon, USA)	June 1998	LNCS 1423
IV	Universiteit Leiden (The Netherlands)	July 2000	LNCS 1838
V	University of Sydney (Australia)	July 2002	LNCS 2369
VI	University of Vermont (Burlington, VT, USA)	May 2004	LNCS 3076
VII	Technische Universität Berlin (Germany)	July 2006	LNCS 4076
VIII	Banff Centre (Banff, Alberta, Canada)	May 2008	LNCS 5011

Organization

Organizing Committee

Mark Bauer, University of Calgary, Canada
Joshua Holden, Rose-Hulman Institute of Technology, USA
Michael Jacobson Jr., University of Calgary, Canada
Renate Scheidler, University of Calgary, Canada (Chair)
Jonathan Sorenson, Butler University, USA

Program Committee

Dan Bernstein, University of Illinois at Chicago, USA
Nils Bruin, Simon Fraser University, Canada
Ernie Croot, Georgia Institute of Technology, USA
Andrej Dujella, University of Zagreb, Croatia
Steven Galbraith, Royal Holloway University of London, UK
Florian Heß, Technische Universität Berlin, Germany
Ming-Deh Huang, University of Southern California, USA
Jürgen Klüners, Heinrich-Heine-Universität Düsseldorf, Germany
Kristin Lauter, Microsoft Research, USA
Stéphane Louboutin, IML, France
Florian Luca, UNAM, Mexico
Daniele Micciancio, University of California at San Diego, USA
Victor Miller, IDA, USA
Oded Regev, Tel-Aviv University, Israel
Igor Shparlinski, Macquarie University, Australia (Chair)
Francesco Sica, Mount Allison University, USA
Andreas Stein, Carl-von-Ossietzky Universität Oldenburg, Germany
Arne Storjohann, University of Waterloo, Canada
Tsuyoshi Takagi, Future University – Hakodate, Japan
Edlyn Teske, University of Waterloo, Canada
Felipe Voloch, University of Texas, USA

Sponsoring Institutions

The Pacific Institute for the Mathematical Sciences (PIMS)
The Fields Institute
The Alberta Informatics Circle of Research Excellence (iCORE)
The Centre for Information Security and Cryptography (CISaC)
Microsoft Research
The Number Theory Foundation
The University of Calgary
Butler University

Table of Contents

Running Time Predictions
for Factoring Algorithms

Ernie Croot[1], Andrew Granville[2], Robin Pemantle[3], and Prasad Tetali[4,*]

[1] School of Mathematics, Georgia Tech, Atlanta, GA 30332-0160, USA
ecroot@math.gatech.edu
[2] Département de mathématiques et de statistique, Université de Montréal,
Montréal QC H3C 3J7, Canada
andrew@dms.umontreal.ca
[3] Department of Mathematics, University of Pennsylvania, 209 S. 33rd Street,
Philadelphia, Pennsylvania 19104, USA
pemantle@math.upenn.edu
[4] School of Mathematics and College of Computing, Georgia Tech, Atlanta,
GA 30332-0160, USA
tetali@math.gatech.edu

In 1994, Carl Pomerance proposed the following problem:

Select integers $a_1, a_2, \ldots, a_J$ at random from the interval $[1, x]$, stopping when some (non-empty) subsequence, $\{a_i : i \in I\}$ where $I \subseteq \{1, 2, \ldots, J\}$, has a square product (that is $\prod_{i \in I} a_i \in \mathbb{Z}^2$). *What can we say about the possible stopping times, J?*

A 1985 algorithm of Schroeppel can be used to show that this process stops after selecting $(1+\epsilon)J_0(x)$ integers a_j with probability $1-o(1)$ (where the function $J_0(x)$ is given explicitly in (1) below. Schroeppel's algorithm actually finds the square product, and this has subsequently been adopted, with relatively minor modifications, by all factorers. In 1994 Pomerance showed that, with probability $1 - o(1)$, the process will run through at least $J_0(x)^{1-o(1)}$ integers a_j, and asked for a more precise estimate of the stopping time. We conjecture that there is a "sharp threshold" for this stopping time, that is, with probability $1 - o(1)$ one will first obtain a square product when (precisely) $\{e^{-\gamma} + o(1)\}J_0(x)$ integers have been selected. Herein we will give a heuristic to justify our belief in this sharp transition.

In our paper [4] we prove, with probability $1 - o(1)$, that the first square product appears in time

$$[(\pi/4)(e^{-\gamma} - o(1))J_0(x), \ (e^{-\gamma} + o(1))J_0(x)],$$

where $\gamma = 0.577\ldots$ is the Euler-Mascheroni constant, improving both Schroeppel and Pomerance's results. In this article we will prove a weak version of this theorem (though still improving on the results of both Schroeppel and Pomerance).

* The first author is supported in part by an NSF grant. Le deuxième auteur est partiellement soutenu par une bourse de la Conseil de recherches en sciences naturelles et en génie du Canada. The third author is supported in part by NSF Grant DMS-01-03635.

A.J. van der Poorten and A. Stein (Eds.): ANTS-VIII 2008, LNCS 5011, pp. 1–36, 2008.

We also confirm the well established belief that, typically, none of the integers in the square product have large prime factors.

Our methods provide an appropriate combinatorial framework for studying the large prime variations associated with the quadratic sieve and other factoring algorithms. This allows us to analyze what factorers have discovered in practice.

1 Introduction

Most factoring algorithms (including Dixon's random squares algorithm [5], the quadratic sieve [14], the multiple polynomial quadratic sieve [19], and the number field sieve [2] – see [18] for a nice expository article on factoring algorithms) work by generating a pseudorandom sequence of integers $a_1, a_2, ...$, with each

$$a_i \equiv b_i^2 \pmod{n},$$

for some known integer b_i (where n is the number to be factored), until some subsequence of the a_i's has product equal to a square, say

$$Y^2 = a_{i_1} \cdots a_{i_k},$$

and set

$$X^2 = (b_{i_1} \cdots b_{i_k})^2.$$

Then

$$n \mid Y^2 - X^2 = (Y - X)(Y + X),$$

and there is a good chance that $\gcd(n, Y - X)$ is a non-trivial factor of n. If so, we have factored n.

In his lecture at the 1994 International Congress of Mathematicians, Pomerance [16,17] observed that in the (heuristic) analysis of such factoring algorithms one assumes that the pseudo-random sequence $a_1, a_2, ...$ is close enough to random that we can make predictions based on this assumption. Hence it makes sense to formulate this question in its own right.

Pomerance's Problem. Select positive integers $a_1, a_2, \cdots \leq x$ independently at random (that is, $a_j = m$ with probability $1/x$ for each integer m, $1 \leq m \leq x$), stopping when some subsequence of the a_i's has product equal to a square (a *square product*). What is the expected stopping time of this process ?

There are several feasible positive practical consequences of resolving this question:

- It may be that the expected stopping time is far less than what is obtained by the algorithms currently used. Hence such an answer might point the way to speeding up factoring algorithms.
- Even if this part of the process can not be easily sped up, a good understanding of this stopping time might help us better determine the optimal choice of parameters for most factoring algorithms.

Let $\pi(y)$ denote the number of primes up to y. Call n a *y-smooth integer* if all of its prime factors are $\leq y$, and let $\Psi(x, y)$ denote the number of y-smooth integers up to x. Let $y_0 = y_0(x)$ be a value of y which maximizes $\Psi(x, y)/y$, and

$$J_0(x) := \frac{\pi(y_0)}{\Psi(x, y_0)} \cdot x. \tag{1}$$

In Pomerance's problem, let T be the smallest integer t for which $a_1, ..., a_t$ has a square dependence (note that T is itself a random variable). As we will see in section 4.1, Schroeppel's 1985 algorithm can be formalized to prove that for any $\epsilon > 0$ we have

$$\text{Prob}(T < (1 + \epsilon)J_0(x)) = 1 - o_\epsilon(1)$$

as $x \to \infty$. In 1994 Pomerance showed that

$$\text{Prob}(T > J_0(x)^{1-\epsilon}) = 1 - o_\epsilon(1).$$

as $x \to \infty$. Therefore there is a transition from "unlikely to have a square product" to "almost certain to have a square product" at $T = J_0(x)^{1+o(1)}$. Pomerance asked in [3] whether there is a sharper transition, and we conjecture that T has a *sharp threshold*:

Conjecture. For every $\epsilon > 0$ we have

$$\text{Prob}(T \in [(e^{-\gamma} - \epsilon)J_0(x), \ (e^{-\gamma} + \epsilon)J_0(x)]) = 1 - o_\epsilon(1), \tag{2}$$

as $x \to \infty$, where $\gamma = 0.577...$ is the Euler-Mascheroni constant.

The bulk of this article will be devoted to explaining how we arrived at this conjecture. In [4] we prove the upper bound in this conjecture using deep probabilistic methods in an associated random graph. Here we discuss a quite different approach which justifies the upper bound in this conjecture, but we have not been able to make all steps of the proof rigorous.

The constant $e^{-\gamma}$ in this conjecture is well-known to number theorists. It appears as the ratio of the proportion of integers free of prime divisors smaller than y, to the proportion of integers up to y that are prime, but this is not how it appears in our discusssion. Indeed herein it emerges from some complicated combinatorial identities, which have little to do with number theory, and we have failed to find a more direct route to this prediction.

Herein we will prove something a little weaker than the above conjecture (though stronger than the previously known results) using methods from combinatorics, analytic and probabilistic number theory:

Theorem 1. *We have*

$$\text{Prob}(T \in [(\pi/4)(e^{-\gamma} - o(1))J_0(x), \ (3/4)J_0(x)]) = 1 - o(1),$$

as $x \to \infty$.

To obtain the lower bound in our theorem, we obtain a good upper bound on the expected number of sub-products of the large prime factors of the a_i's that equal a square, which allows us to bound the probability that such a sub-product exists, for $T < (\pi/4)(e^{-\gamma} - o(1))J_0(x)$. This is the "first moment method". Moreover the proof gives us some idea of what the set I looks like: In the unlikely event that $T < (\pi/4)(e^{-\gamma}-o(1))J_0(x)$, with probability $1-o(1)$, the set I consists of a single number a_T, which is therefore a square. If T lies in the interval given in Theorem 1 (which happens with probability $1-o(1)$), then the square product I is composed of $y_0^{1+o(1)} = J_0(x)^{1/2+o(1)}$ numbers a_i (which will be made more precise in [4]).

Schroeppel's upper bound, $T \leq (1+o(1))J_0(x)$ follows by showing that one expects to have more than $\pi(y_0)$ y_0-smooth integers amongst $a_1, a_2, \ldots, a_T$, which guarantees a square product. To see this, create a matrix over $\mathbb{F}_2$ whose columns are indexed by the primes up to y_0, and whose (i, p)-th entry is given by the exponent on p in the factorization of a_i, for each y_0-smooth a_i. Then a square product is equivalent to a linear dependence over $\mathbb{F}_2$ amongst the corresponding rows of our matrix: we are guaranteed such a linear dependence once the matrix has more than $\pi(y_0)$ rows. Of course it might be that we obtain a linear dependence when there are far fewer rows; however, in section 3.1, we give a crude model for this process which suggests that we should not expect there to be a linear dependence until we have very close to $\pi(y_0)$ rows.

Schroeppel's approach is not only good for theoretical analysis, in practice one searches among the a_i for y_0-smooth integers and hunts amongst these for a square product, using linear algebra in $\mathbb{F}_2$ on the primes' exponents. Computing specialists have also found that it is easy and profitable to keep track of a_i of the form $s_i q_i$, where s_i is y_0-smooth and q_i is a prime exceeding y_0; if both a_i and a_j have exactly the same large prime factor $q_i = q_j$ then their product is a y_0-smooth integer times a square, and so can be used in our matrix as an extra smooth number. This is called the *large prime variation*, and the upper bound in Theorem 1 is obtained in section 4 by computing the limit of this method. (The best possible constant is actually a tiny bit smaller than 3/4.)

One can also consider the *double large prime variation* in which one allows two largish prime factors so that, for example, the product of three a_is of the form pqs_1, prs_2, qrs_3 can be used as an extra smooth number. Experience has shown that each of these variations has allowed a small speed up of various factoring algorithms (though at the cost of some non-trivial extra programming), and a long open question has been to formulate all of the possibilities for multi-large prime variations and to analyze how they affect the running time. Sorting out this combinatorial mess is the most difficult part of our paper. To our surprise we found that it can be described in terms of the theory of Huisimi cacti graphs (see section 6). In attempting to count the number of such smooth numbers (including those created as products of smooths times a few large primes) we run into a subtle convergence issue. We believe that we have a power series which yields the number of smooth numbers, created independently from $a_1, \ldots, a_J$, simply as a function of J/J_0; if it is correct then we obtain the upper bound in our conjecture.

In the graphs constructed here (which lead to the Husimi graphs), the vertices correspond to the a_j's, and the edges to common prime factors which are $> y_0$. In the random hypergraphs considered in [4] the vertices correspond to the prime factors which are $> y_0$ and the hyperedges, which are presented as subsets of the set of vertices, correspond to the prime factors of each a_j, which divide a_j to an odd power.

In [4] we were able to understand the speed up in running time using the k-large prime variation for each $k \geq 1$. We discuss the details of the main results of this work, along with some numerics, in section 8. We also compare, there, these theoretical findings, with the speed-ups obtained using large prime variations by the researchers who actually factor numbers. Their findings and our predictions differ significantly and we discuss what might contribute to this.

When our process terminates (at time T) we have some subset I of $a_1, ..., a_T$, including a_T, whose product equals a square.[1] If Schroeppel's argument comes close to reflecting the right answer then one would guess that a_i's in the square product are typically "smooth". In section 3.2 we prove that they will all be J_0^2-smooth with probability $1 - o(1)$, which we improve to

$$y_0^2 \exp((2 + \epsilon)\sqrt{\log y_0 \log \log y_0}) \quad - \text{ smooth.}$$

in [4], Theorem 2. We guess that this may be improvable to $y_0^{1+\epsilon}$-smooth for any fixed $\epsilon > 0$.

Pomerance's main goal in enunciating the random squares problem was to provide a model that would prove useful in analyzing the running time of factoring algorithms, such as the quadratic sieve. In section 7 we will analyze the running time of Pomerance's random squares problem showing that the running time will be inevitably dominated by finding the actual square product once we have enough integers. Hence to optimize the running time of the quadratic sieve we look for a square dependence among the y-smooth integers with y significantly smaller than y_0, so that Pomerance's problem is not quite so germane to factoring questions as it had at first appeared.

This article uses methods from several different areas not usually associated with factoring questions: the first and second moment methods from probabilistic combinatorics, Husimi graphs from statistical physics, Lagrange inversion from algebraic combinatorics, as well as comparative estimates on smooth numbers using precise information on saddle points.

2   Smooth Numbers

In this technical section we state some sharp results comparing the number of smooth numbers up to two different points. The key idea, which we took from

---

[1] Note that I is unique, else if we have two such subsets I and J then $(I \cup J) \setminus (I \cap J)$ is also a set whose product equals a square, but does not contain a_T, and so the process would have stopped earlier than at time T.

[10], is that such ratios are easily determined because one can compare very precisely associated saddle points – this seems to be the first time this idea has been used in the context of analyzing factoring algorithms.

2.1   Classical Smooth Number Estimates

From [10] we have that the estimate

$$\Psi(x,y) = x\rho(u)\left\{1+O\left(\frac{\log(u+1)}{\log y}\right)\right\} \quad \text{as} \quad x \to \infty \quad \text{where} \quad x = y^u, \quad (3)$$

holds in the range

$$\exp\left((\log\log x)^2\right) \le y \le x, \tag{4}$$

where $\rho(u) = 1$ for $0 \le u \le 1$, and where

$$\rho(u) = \frac{1}{u}\int_{u-1}^{u}\rho(t)\,dt \quad \text{for all } u > 1.$$

This function $\rho(u)$ satisfies

$$\rho(u) = \left(\frac{e+o(1)}{u\log u}\right)^u = \exp(-(u+o(u))\log u); \tag{5}$$

and so

$$\Psi(x,y) = x\exp(-(u+o(u))\log u). \tag{6}$$

Now let

$$L := L(x) = \exp\left(\sqrt{\frac{1}{2}\log x\log\log x}\right).$$

Then, using (6) we deduce that for $\beta > 0$,

$$\Psi(x, L(x)^{\beta+o(1)}) = xL(x)^{-1/\beta+o(1)}. \tag{7}$$

From this one can easily deduce that

$$y_0(x) = L(x)^{1+o(1)}, \quad \text{and} \quad J_0(x) = y_0^{2-\{1+o(1)\}/\log\log y_0} = L(x)^{2+o(1)}, \tag{8}$$

where y_0 and J_0 are as in the introduction (see (1)). From these last two equations we deduce that if $y = y_0^{\beta+o(1)}$, where $\beta > 0$, then

$$\frac{\Psi(x,y)/y}{\Psi(x,y_0)/y_0} = y_0^{2-\beta-\beta^{-1}+o(1)}.$$

For any $\alpha > 0$, one has

$$\Psi(x,y) \le \sum_{\substack{n\le x \\ P(n)\le y}} (x/n)^{\alpha} \le x^{\alpha}\prod_{p\le y}\left(1-\frac{1}{p^{\alpha}}\right)^{-1}, \tag{9}$$

which is minimized by selecting $\alpha = \alpha(x, y)$ to be the solution to

$$\log x = \sum_{p \leq y} \frac{\log p}{p^\alpha - 1}. \tag{10}$$

We show in [4] that for $y = L(x)^{\beta + o(1)} = y_0^{\beta + o(1)}$ we have

$$y^{1-\alpha} \sim \beta^{-2} \log y \sim \beta^{-1} \log y_0. \tag{11}$$

Moreover, by [10, Theorem 3], when $1 \leq d \leq y \leq x/d$ we have

$$\Psi\left(\frac{x}{d}, y\right) = \frac{1}{d^{\alpha(x,y)}} \Psi(x, y) \left(1 + O\left(\frac{1}{u} + \frac{\log y}{y}\right)\right). \tag{12}$$

By iterating this result we can deduce (see [4]) the following:

Proposition 1. *Throughout the range* (4), *for any* $1 \leq d \leq x$, *we have*

$$\Psi\left(\frac{x}{d}, y\right) \leq \frac{1}{d^{\alpha(x,y)}} \Psi(x, y)\{1 + o(1)\},$$

where α *is the solution to* (10).

Now Lemma 2.4 of [4] gives the following more accurate value for y_0:

$$\log y_0 = \log L(x) \left(1 + \frac{\log_3 x - \log 2}{2 \log_2 x} + O\left(\left(\frac{\log_3 x}{\log_2 x}\right)^2\right)\right). \tag{13}$$

It is usual in factoring algorithms to optimize by taking $\psi(x, y)$ to be roughly x/y:

Lemma 1. *If* $\psi(x, y) = x/y^{1 + o(1/\log \log y)}$ *then*

$$\log y = \log y_0 \left(1 - \frac{1 + o(1)}{\log_2 x}\right).$$

Proof. By (3) and (5) we have

$$u(\log u + \log \log u - 1 + o(1)) = \log y \left(1 + o\left(\frac{1}{\log \log y}\right)\right),$$

and from here it is a simple exercise to show that

$$u = \frac{\log y}{\log \log y} \left(1 + \frac{1 + o(1)}{\log \log y}\right).$$

Substituting $u = (\log x)/(\log y)$ and solving we obtain

$$\log y = \log L(x) \left(1 + \frac{\log_3 x - \log 2 - 2 + o(1)}{2 \log_2 x}\right),$$

from which our result follows using (13).   □

3   Some Simple Observations

3.1   A Heuristic Analysis

Let $M = \pi(y)$ and
$$p_1 = 2 < p_2 = 3 < \ldots < p_M$$
be the primes up to y. Any y-smooth integer
$$p_1^{e_1} p_2^{e_2} \ldots p_M^{e_M}$$
gives rise to the element $(e_1, e_2, \ldots e_M)$ of the vector space $\mathbb{F}_2^M$. The probability that any given element of $\mathbb{F}_2^M$ arises from Pomerance's problem (corresponding to a y-smooth value of a_i) varies depending on the entries in that element. Pomerance's problem can be rephrased as: Let $y = x$. Select elements $v_1, v_2, \ldots$ of $\mathbb{F}_2^M$, each with some specific probability (as above), and stop at v_T as soon as $v_1, v_2, \ldots, v_T$ are linearly dependent. The difficulty in this version is in quantifying the probabilities that each different $v \in \mathbb{F}_2^M$ occurs, and then manipulating those probabilities in a proof since they are so basis dependent.

As a first model we will work with an approximation to this question that avoids these difficulties: Now our problem will be to determine the distribution of T when each element of $\mathbb{F}_2^M$ is selected with probability $1/2^M$. We hope that this model will help us gain some insight into Pomerance's question.

If $v_1, v_2, \ldots, v_{\ell-1}$ are linearly independent they generate a subspace S_ℓ of dimension $\ell - 1$, which contains $2^{\ell-1}$ elements (if $1 \le \ell \le M + 1$). Then the probability that $v_1, v_2, \ldots, v_\ell$ are linearly dependent is the same as the probability that v_ℓ belongs to S_ℓ, which is $2^{\ell-1}/2^M$. Thus the expectation of T is

$$\sum_{\ell=1}^{M+1} \ell \frac{2^{\ell-1}}{2^M} \prod_{i=1}^{\ell-1} \left(1 - \frac{2^{i-1}}{2^M}\right) \longrightarrow$$

$$\prod_{i \ge 1} \left(1 - \frac{1}{2^i}\right) \left(\sum_{j=0}^{M} \frac{(M+1-j)}{2^j} \prod_{i=1}^{j} \left(1 - \frac{1}{2^i}\right)^{-1}\right)$$

$$= M - .60669515 \ldots \quad \text{as } M \to \infty.$$

(By convention, empty products have value 1.) Therefore $|T - M|$ has expected value $O(1)$. Furthermore,

$$\text{Prob}(|T - M| > n) = \sum_{\ell \ge n+1} \text{Prob}(T = M - \ell) < \sum_{\ell \ge n+1} 2^{-\ell-1} = 2^{-n-1},$$

for each $n \ge 1$, so that if $\phi(t) \to \infty$ as $t \to \infty$ then

$$\text{Prob}(T \in [M - \phi(M), M]) = 1 - o(1).$$

Hence this simplified problem has a very sharp transition function, suggesting that this might be so in Pomerance's problem.

3.2   No Large Primes, I

Suppose that we have selected integers $a_1, a_2, ..., a_T$ at random from $[1, x]$, stopping at T since there is a non-empty subset of these integers whose product is a square. Let q be the largest prime that divides this square. Then either q^2 divides one of $a_1, a_2, ..., a_T$, or q divides at least two of them. The probability that p^2 divides at least one of $a_1, a_2, ..., a_T$, for a given prime p, is $\leq T/p^2$; and the probability that p divides at least two of $a_1, a_2, ..., a_T$ is $\leq \binom{T}{2}/p^2$. Thus

$$\text{Prob}(q > T^2) \ll T^2 \sum_{p > T^2} \frac{1}{p^2} \ll \frac{1}{\log T},$$

by the Prime Number Theorem.

By Pomerance's result we know that $T \to \infty$ with probability $1 + o(1)$; and so the largest prime that divides the square product is $\leq T^2$ with probability $1 - o(1)$. We will improve this result later by more involved arguments.

4   Proof of the Upper Bound on T in Theorem 1

Our goal in this section is to prove that

$$\text{Prob}(T < (3/4)J_0(x)) = 1 - o(1),$$

as $x \to \infty$.

We use the following notation throughout. Given a sequence

$$a_1, \ldots, a_J \leq x$$

of randomly chosen positive integers, let

$$p_1 = 2 < p_2 = 3 < \cdots < p_{\pi(x)}$$

denote the primes up to x, and construct the J-by-$\pi(x)$ matrix A, which we take mod 2, where

$$a_i = \prod_{1 \leq j \leq \pi(x)} p_j^{A_{i,j}}.$$

Then, a given subsequence of the a_i has square product if the corresponding row vectors of A sum to the 0 vector modulo 2; and, this happens if and only if $\text{rank}(A) < J$. Here, and henceforth, the rank is always the $\mathbb{F}_2$-rank.

4.1   Schroeppel's Argument

Schroeppel's idea was to focus only on those rows corresponding to y_0-smooth integers so that they have no 1's beyond the $\pi(y_0)$-th column. If we let $S(y_0)$ denote the set of all such rows, then Schroeppel's approach amounts to showing that

$$|S(y_0)| > \pi(y_0)$$

holds with probability $1 - o(1)$ for $J = (1+\epsilon)J_0$, where J_0 and y_0 are as defined in (1). If this inequality holds, then the $|S(y_0)|$ rows are linearly dependent mod 2, and therefore some subset of them sums to the 0 vector mod 2.

Although Pomerance [15] gave a complete proof that Schroeppel's idea works, it does not seem to be flexible enough to be easily modified when we alter Schroeppel's argument, so we will give our own proof, seemingly more complicated but actually requiring less depth: Define the independent random variables $Y_1, Y_2, \ldots$ so that $Y_j = 1$ if a_j is y-smooth, and $Y_j = 0$ otherwise, where y will be chosen later. Let

$$N = Y_1 + \cdots + Y_J,$$

which is the number of y-smooth integers amongst $a_1, \ldots, a_J$. The probability that any such integer is y-smooth, that is that $Y_j = 1$, is $\Psi(x, y)/x$; and so,

$$\mathbb{E}(N) = \frac{J\psi(x, y)}{x}.$$

Since the Y_i are independent, we also have

$$V(N) = \sum_i (\mathbb{E}(Y_i^2) - \mathbb{E}(Y_i)^2) = \sum_i (\mathbb{E}(Y_i) - \mathbb{E}(Y_i)^2) \leq \frac{J\psi(x, y)}{x}.$$

Thus, selecting $J = (1 + \epsilon)x\pi(y)/\Psi(x, y)$, we have, with probability $1 + o_\epsilon(1)$, that

$$N = (1 + \epsilon + o(1))\pi(y) > \pi(y).$$

Therefore, there must be some non-empty subset of the a_i's whose product is a square. Taking $y = y_0$ we deduce that

$$\mathrm{Prob}(T < (1+\epsilon)J_0(x)) = 1 - o_\epsilon(1).$$

Remark. One might alter Schroeppel's construction to focus on those rows having only entries that are 0 (mod 2) beyond the $\pi(y_0)$-th column. These rows all correspond to integers that are a y_0-smooth integer times a square. The number of additional such rows equals

$$\sum_{\substack{d>1 \\ p(d)>y_0}} \Psi\left(\frac{x}{d^2}, y_0\right) \leq \sum_{y_0 < d \leq y_0^2} \Psi\left(\frac{x}{d^2}, y_0\right) + \sum_{d > y_0^2} \frac{x}{d^2} \ll \frac{\Psi(x, y_0)}{y_0^{1+o(1)}}$$

by Proposition 1, the prime number theorem, (11) and (7), respectively, which one readily sees are too few to significantly affect the above analysis. Here and henceforth, $p(n)$ denotes the smallest prime factor of n, and later on we will use $P(n)$ to denote the largest prime factor of n.

4.2   The Single Large Prime Variation

If, for some prime $p > y$, we have $ps_1, ps_2, \ldots, ps_r$ amongst the a_i, where each s_j is y-smooth, then this provides us with precisely $r - 1$ multiplicatively independent pseudo-smooths, $(ps_1)(ps_2), (ps_1)(ps_3), \ldots, (ps_1)(ps_r)$. We will count these using the combinatorial identity

$$r - 1 = \sum_{\substack{I \subset \{1,\dots,r\} \\ |I| \geq 2}} (-1)^{|I|},$$

which fits well with our argument. Hence the expected number of smooths and pseudo-smooths amongst $a_1, \dots, a_J$ equals

$$\frac{J\Psi(x,y)}{x} + \sum_{\substack{I \subset \{1,\dots,r\} \\ |I| \geq 2}} (-1)^{|I|} \text{Prob}(a_i = ps_i : \ i \in I, \ P(s_i) \leq y < p, \ p \text{ prime})$$

$$= \frac{J\Psi(x,y)}{x} + \sum_{k \geq 2} \binom{J}{k} (-1)^k \sum_{p > y} \left(\frac{\Psi(x/p, y)}{x} \right)^k. \quad (14)$$

Using (12) we have, by the prime number theorem, that

$$\sum_{p > y} \left(\frac{\Psi(x/p, y)}{\Psi(x, y)} \right)^k \sim \sum_{p > y} \frac{1}{p^{\alpha k}} \sim \frac{y^{1 - \alpha k}}{(\alpha k - 1) \log y} \sim \frac{1}{(k-1)\pi(y)^{k-1}} \ ;$$

using (11) for $y \asymp y_0$. Hence the above becomes, taking $J = \eta x \pi(y)/\Psi(x,y)$,

$$\sim \left(\eta + \sum_{k \geq 2} \frac{(-\eta)^k}{k!(k-1)} \right) \pi(y). \quad (15)$$

One needs to be a little careful here since the accumulated error terms might get large as $k \to \infty$. To avoid this problem we can replace the identity (14) by the usual inclusion-exclusion inequalities; that is the partial sum up to k even is an upper bound, and the partial sum up to k odd is a lower bound. Since these converge as $k \to \infty$, independently of x, we recover (15). One can compute that the constant in (15) equals 1 for $\eta = .74997591747934498263\dots$; or one might observe that this expression is $> 1.00003\pi(y)$ when $\eta = 3/4$.

4.3   From Expectation to Probability

Proposition 2. *The number of smooth and pseudosmooth integers (that is, integers that are a y_0-smooth number times at most one prime factor $> y_0$) amongst $a_1, a_2, \dots, a_J$ with $J = \eta J_0$ is given by (15), with probability $1 - o(1)$, as $x \to \infty$.*

Hence, with probability $1 - o(1)$, we have that the number of linear dependencies arising from the single large prime variation is (15) for $J = \eta J_0(x)$ with $y = y_0$ as $x \to \infty$. This is $> (1 + \epsilon)\pi(y_0)$ for $J = (3/4)J_0(x)$ with probability $1 - o(1)$, as $x \to \infty$, implying the upper bound on T in Theorem 1.

Proof (of Proposition 2). Suppose that $a_1, \dots, a_J \leq x$ have been chosen randomly. For each integer $r \geq 2$ and subset S of $\{1, \dots, J\}$ we define a random variable

$X_r(S)$ as follows: Let $X_r(S) = 1$ if each a_s, $s \in S$ equals p times a y-smooth for the same prime $p > y$, and let $X_r(S) = 0$ otherwise. Therefore if

$$Y_r = \sum_{\substack{S \subset \{1,...,J\} \\ |S|=r}} X_r(S),$$

then we have seen that

$$\mathbb{E}(Y_r) \sim \frac{\eta^r}{r!(r-1)} \pi(y).$$

Hence each

$$\mathbb{E}(X_r(S)) \sim \binom{J}{r}^{-1} \frac{\eta^r}{r!(r-1)} \pi(y)$$

for every $S \subset \{1, ..., J\}$, since each of the $X_r(S)$ have the same probability distribution.

Now, if S_1 and S_2 are disjoint, then $X_r(S_1)$ and $X_r(S_2)$ are independent, so that

$$\mathbb{E}(X_r(S_1)X_r(S_2)) = \mathbb{E}(X_r(S_1))\mathbb{E}(X_r(S_2)).$$

If S_1 and S_2 are not disjoint and both $X_r(S_1)$ and $X_r(S_2)$ equal 1, then $X_R(S) = 1$ where $S = S_1 \cup S_2$ and $R = |S|$. We just saw that

$$\mathbb{E}(X_R(S)) \sim \binom{J}{R}^{-1} \frac{\eta^R}{R!(R-1)} \pi(y).$$

Hence if $|S_1 \cap S_2| = j$ then

$$\mathbb{E}(X_r(S_1)X_r(S_2)) \sim \binom{J}{2r-j}^{-1} \frac{\eta^{2r-j}}{(2r-j)!(2r-j-1)} \pi(y).$$

Therefore

$$\mathbb{E}(Y_r^2) - \mathbb{E}(Y_r)^2 = \sum_{\substack{S_1,S_2 \subset \{1,...,J\} \\ |S_1|=|S_2|=r}} \mathbb{E}(X_r(S_1)X_r(S_2)) - \mathbb{E}(X_r(S_1))\mathbb{E}(X_r(S_2))$$

$$\lesssim \pi(y) \sum_{j=1}^{r} \binom{J}{2r-j}^{-1} \frac{\eta^{2r-j}}{(2r-j)!(2r-j-1)} \sum_{\substack{S_1,S_2 \subset \{1,...,J\} \\ |S_1|=|S_2|=r \\ |S_1 \cap S_2|=j}} 1$$

$$= \pi(y) \sum_{j=1}^{r} \frac{\eta^{2r-j}}{(2r-j-1)j!(r-j)!^2} \leq (1 + \eta^{2r-1})\pi(y).$$

Hence by Tchebychev's inequality we deduce that

$$\mathrm{Prob}(|Y_r - \mathbb{E}(Y_r)| > \epsilon\mathbb{E}(Y_r)) \ll_r \frac{\mathbb{E}(Y_r^2) - \mathbb{E}(Y_r)^2}{\epsilon^2\mathbb{E}(Y_r)^2} \ll_r \frac{1}{\epsilon^2\pi(y)},$$

so that $Y_r \sim \mathbb{E}(Y_r)$ with probability $1 - o(1)$.   □

5   The Lower Bound on T; a Sketch

We prove that

$$\text{Prob}(T > (\pi/4)(e^{-\gamma} - o(1))J_0(x)) = 1 - o(1),$$

in [4], by showing that for $J(x) = (\pi/4)(e^{-\gamma} - o(1))J_0(x)$ the expected number of square products among $a_1, \ldots, a_J$ is $o(1)$.

By considering the common divisors of all pairs of integers from $a_1, \ldots, a_J$ we begin by showing that the probability that a square product has size k, with $2 \leq k \leq \log x / 2 \log \log x$, is $O(J^2 \log x / x)$ provided $J < x^{o(1)}$.

Next we shall write $a_i = b_i d_i$ where $P(b_i) \leq y$ and where either $d_i = 1$ or $p(d_i) > y$ (here, $p(n)$ denotes the smallest prime divisor of n), for $1 \leq i \leq k$. If $a_1, \ldots, a_k$ are chosen at random from $[1, x]$ then

$$\text{Prob}(a_1 \ldots a_k \in \mathbb{Z}^2) \leq \text{Prob}(d_1 \ldots d_k \in \mathbb{Z}^2)$$

$$= \sum_{\substack{d_1, \ldots, d_k \geq 1 \\ d_1 \ldots d_k \in \mathbb{Z}^2 \\ d_i = 1 \text{ or } p(d_i) > y}} \prod_{i=1}^{k} \frac{\Psi(x/d_i, y)}{x}$$

$$\leq \left(\{1 + o(1)\} \frac{\Psi(x, y)}{x} \right)^k \sum_{n=1 \text{ or } p(n) > y} \frac{\tau_k(n^2)}{n^{2\alpha}}, \quad (16)$$

by Proposition 1. Out of $J = \eta J_0$ integers, the number of k-tuples is

$$\binom{J}{k} \leq (eJ/k)^k;$$

and so the expected number of k-tuples whose product is a square is at most

$$\left((e + o(1)) \frac{\eta y}{k \log y_0} \frac{\Psi(x, y)/y}{\Psi(x, y_0)/y_0} \right)^k \prod_{p > y} \left(1 + \frac{\tau_k(p^2)}{p^{2\alpha}} + \frac{\tau_k(p^4)}{p^{4\alpha}} + \cdots \right). \quad (17)$$

For $\log x / 2 \log \log x < k \leq y_0^{1/4}$ we take $y = y_0^{1/3}$ and show that the quantity in (17) is $< 1/x^2$.

For $y_0^{1/4} \leq k = y_0^{\beta} \leq J = \eta J_0 \leq J_0$ we choose y so that $[k/C] = \pi(y)$, with C sufficiently large. One can show that the quantity in (17) is $< ((1 + \epsilon)4\eta e^{\gamma}/\pi)^k$ and is significantly smaller unless $\beta = 1 + o(1)$. This quantity is $< 1/x^2$ since $\eta < 4\pi e^{-\gamma} - \epsilon$ and the result follows.

This proof yields further useful information: If either $J < (\pi/4)(e^{-\gamma} - o(1))J_0(x)$, or if $k < y_0^{1-o(1)}$ or $k > y_0^{1+o(1)}$, then the expected number of square products with $k > 1$ is $O(J_0(x)^2 \log x / x)$, whereas the expected number of squares in our sequence is $\sim J/\sqrt{x}$. This justifies the remarks immediately after the statement of Theorem 1.

Moreover with only minor modifications we showed the following in [4]: Let $y_1 = y_0 \exp((1 + \epsilon)\sqrt{\log y_0 \log \log y_0})$ and write each $a_i = b_i d_i$ where $P(b_i) \leq$

$y = y_1 < p(d_i)$. If $d_{i_1} \ldots d_{i_l}$ is a subproduct which equals a square n^2, but such that no subproduct of this is a square, then, with probability $1 - o(1)$, we have $l = o(\log y_0)$ and n is a squarefree integer composed of precisely $l - 1$ prime factors, each $\leq y^2$, where $n \leq y^{2l}$.

6   A Method to Examine All Smooth Products

In proving his upper bound on T, Schroeppel worked with the y_0-smooth integers amongst $a_1, \ldots, a_T$ (which correspond to rows of A with no 1's in any column that represents a prime $> y_0$), and in our improvement in section 4.2 we worked with integers that have no more than one prime factor $> y_0$ (which correspond to rows of A with at most one 1 in the set of columns representing primes $> y_0$). We now work with all of the rows of A, at the cost of significant complications.

Let A_{y_0} be the matrix obtained by deleting the first $\pi(y_0)$ columns of A. Note that the row vectors corresponding to y_0-smooth numbers will be 0 in this new matrix. If

$$\mathrm{rank}(A_{y_0}) \; < \; J - \pi(y_0), \tag{18}$$

then

$$\mathrm{rank}(A) \leq \mathrm{rank}(A_{y_0}) + \pi(y_0) \; < \; J,$$

which therefore means that the rows of A are dependent over $\mathbb{F}_2$, and thus the sequence $a_1, ..., a_J$ contains a square dependence.

So let us suppose we are given a matrix A corresponding to a sequence of a_j's. We begin by removing (extraneous) rows from A_{y_0}, one at a time: that is, we remove a row containing a 1 in its l-th column if there are no other 1's in the l-th column of the matrix (since this row cannot participate in a linear dependence). This way we end up with a matrix B in which no column contains exactly one 1, and for which

$$r(A_{y_0}) - \mathrm{rank}(A_{y_0}) = r(B) - \mathrm{rank}(B)$$

(since we reduce the rank by one each time we remove a row). Next we partition the rows of B into minimal subsets, in which the primes involved in each subset are disjoint from the primes involved in the other subsets (in other words, if two rows have a 1 in the same column then they must belong to the same subset). The i-th subset forms a submatrix, S_i, of rank ℓ_i, containing r_i rows, and then

$$r(B) - \mathrm{rank}(B) = \sum_i (r_i - \ell_i).$$

We will define a power series $f(\eta)$ for which we believe that

$$\mathbb{E}\left(\sum_i (r_i - \ell_i) \right) \sim f(\eta)\pi(y_0) \tag{19}$$

when $J = (\eta + o(1))J_0$, and we can show that

$$\lim_{\eta \to \eta_0^-} f(\eta) = 1, \tag{20}$$

where $\eta_0 := e^{-\gamma}$. Using the idea of section 4.3, we will deduce in section 6.9 that if (19) holds then

$$\sum_i (r_i - \ell_i) \sim f(\eta)\pi(y_0) \tag{21}$$

holds with probability $1 - o(1)$, and hence (18) holds with probability $1 - o(1)$ for $J = (\eta_0 + o(1))J_0$. That is we can replace the upper bound $3/4$ in Theorem 1 by $e^{-\gamma}$.

The simple model of section 3.1 suggests that A will not contain a square dependence until we have $\sim \pi(y_0)$ smooth or pseudo-smooth numbers; hence we believe that one can replace the lower bound $(\pi/4)e^{-\gamma}$ in Theorem 1 by $e^{-\gamma}$. This is our heuristic in support of the Conjecture.

6.1   The Submatrices

Let M_R denote the matrix composed of the set R of rows (allowing multiplicity), removing columns of 0's. We now describe the matrices M_{S_i} for the submatrices S_i of B from the previous subsection.

For an $r(M)$-by-$\ell(M)$ matrix M we denote the (i,j)th entry $e_{i,j} \in \mathbb{F}_2$ for $1 \le i \le r$, $1 \le j \le \ell$. We let

$$N(M) = \sum_{i,j} e_{i,j}$$

denote the number of 1's in M, and

$$\Delta(M) := N(M) - r(M) - \ell(M) + 1.$$

We denote the number of 1's in column j by

$$m_j = \sum_i e_{i,j},$$

and require each $m_j \ge 2$.[2] We also require that M is *transitive*. That is, for any j, $2 \le j \le \ell$ there exists a sequence of row indices $i_1, \ldots, i_g$, and column indices $j_1, \ldots, j_{g-1}$, such that

$$e_{i_1,1} = e_{i_g,j} = 1; \text{ and, } e_{i_h,j_h} = e_{i_{h+1},j_h} = 1 \text{ for } 1 \le h \le g - 1.$$

In other words we do not study M if, after a permutation, it can be split into a block diagonal matrix with more than one block, since this would correspond to independent squares.

---

[2] Else the prime corresponding to that column cannot participate in a square product.

It is convenient to keep in mind two reformulations:

Integer version: Given primes $p_1 < p_2 < \cdots < p_\ell$, we assign, to each row, a squarefree integer

$$n_i = \prod_{1 \leq j \leq \ell} p_j^{e_{i,j}}, \text{ for } 1 \leq i \leq r.$$

Graph version: Take a graph $G(M)$ with r vertices, where v_i is adjacent to v_I with an edge of colour p_j if p_j divides both n_i and n_I (or, equivalently, $e_{i,j} = e_{I,j} = 1$). Notice that M being *transitive* is equivalent to the graph $G(M)$ being *connected*, which is much easier to visualize.

Now we define a class of matrices $\mathcal{M}_k$, where $M \in \mathcal{M}_k$ if M is as above, is transitive and $\Delta(M) = k$. Note that the "matrix" with one row and no columns is in $\mathcal{M}_0$ (in the "integer version" this corresponds to the set with just the one element, 1, and in the graph version to the graph with a single vertex and no edges).

6.2   Isomorphism Classes of Submatrices

Let us re-order the rows of M so that, in the graph theory version, each new vertex connects to the graph that we already have, which is always possible as the overall graph is connected. Let

$$\ell_I = \#\{j \colon \text{there is an } i \leq I \text{ with } e_{i,j} = 1\},$$

the number of columns with a 1 in or before the I-th row, and

$$N_I := \sum_{i \leq I, \, j \leq \ell} e_{i,j},$$

the number of 1's up to, and including in, the I-th row. Define

$$\Delta_I = N_I - I - \ell_I + 1,$$

so that $\Delta_r = \Delta(M)$.

Now $N_1 = \ell_1$ and therefore $\Delta_1 = 0$. Let us consider the transition when we add in the $(I+1)$-th row. The condition that each new row connects to what we already have means that the number of new colours (that is, columns with a non-zero entry) is less than the number of 1's in the new row, that is

$$\ell_{I+1} - \ell_I \leq N_{I+1} - N_I - 1;$$

and so

$$\begin{aligned} \Delta_{I+1} &= N_{I+1} - I - \ell_{I+1} \\ &= N_I - I - \ell_I + (N_{I+1} - N_I) - (\ell_{I+1} - \ell_I) \geq N_I - I - \ell_I + 1 = \Delta_I. \end{aligned}$$

Therefore

$$\Delta(M) = \Delta_r \geq \Delta_{r-1} \geq \cdots \geq \Delta_2 \geq \Delta_1 = 0. \tag{22}$$

6.3 Restricting to the Key Class of Submatrices

Two matrices are said to be "isomorphic" if one can be obtained from the other by permuting rows and columns. In this subsection we estimate how many submatrices of A_{y_0} are isomorphic to a given matrix M, in order to exclude from our considerations all those M that occur infrequently.

Proposition 3. *Fix $M \in \mathcal{M}_k$. The expected number of submatrices S of A_{y_0} for which M_S is isomorphic to M is*

$$\sim \frac{\eta^r \pi(y_0)^{1-k}}{|\mathrm{Aut}_{\mathrm{Rows}}(M)|} \prod_{1 \leq j \leq \ell} \frac{1}{\nu_j}, \tag{23}$$

where $\nu_j := \sum_{i=j}^{\ell} (m_i - 1)$.

Note that we are not counting here the number of times a component S_i is isomorphic to M, but rather how many submatrices of A_{y_0} are isomorphic to M.

Since $\eta \leq 1$, the quantity in (23) is bounded if $k \geq 1$, but is a constant times $\pi(y_0)$ if $k = 0$. This is why we will restrict our attention to $M \in \mathcal{M}_0$, and our goal becomes to prove that

$$\mathbb{E}\left(\sum_{i:\, S_i \in \mathcal{M}} (r_i - \ell_i) \right) > \pi(y_0) \tag{24}$$

in place of (19), where henceforth we write $\mathcal{M} = \mathcal{M}_0$.

Proof. The expected number of times we get a set of integers of the form $\prod_{1 \leq j \leq \ell} p_j^{e_{i,j}}$ times a y_0-smooth times a square, for $i = 1, ..., r$, within our sequence of integers $a_1, ..., a_J$ is

$$\sim \binom{J}{r} |\mathrm{Orbit}_{\mathrm{Rows}}(M)| \prod_{1 \leq i \leq r} \frac{\Psi^*(x/\prod_{1 \leq j \leq \ell} p_j^{e_{i,j}}, y_0)}{x}, \tag{25}$$

where by $\mathrm{Orbit}_{\mathrm{Rows}}(M)$ we mean the set of distinct matrices produced by permuting the rows of M, and $\Psi^*(X, y) := \#\{n = mr^2 : P(m) \leq y < p(r)\}$ which is insignificantly larger than $\Psi(X, y)$ (as we saw at the end of section 4.1). Since r is fixed and J tends to infinity, we have

$$\binom{J}{r} \sim \frac{J^r}{r!};$$

and we know that[3]

$$r! = |\mathrm{Orbit}_{\mathrm{Rows}}(M)| \cdot |\mathrm{Aut}_{\mathrm{Rows}}(M)|$$

---

[3] This is a consequence of the "Orbit-Stabilizer Theorem" from elementary group theory. It follows from the fact that the cosets of $\mathrm{Aut}_{\mathrm{Rows}}(M)$ in the permutation group on the r rows of M, correspond to the distinct matrices (orbit elements) obtained by performing row interchanges on M.

where $\text{Aut}_{\text{Rows}}(M)$ denotes the number of ways to obtain exactly the same matrix by permuting the rows (this corresponds to permuting identical integers that occur). Therefore (25) is

$$\sim \frac{J^r}{|\text{Aut}_{\text{Rows}}(M)|} \prod_{1 \le i \le r} \frac{\Psi(x/\prod_{1 \le j \le \ell} p_j^{e_{i,j}}, y_0)}{x}$$

$$\sim \frac{1}{|\text{Aut}_{\text{Rows}}(M)|} \left(\frac{J\Psi(x, y_0)}{x}\right)^r \prod_{1 \le j \le \ell} \frac{1}{p_j^{m_j \alpha}}, \quad (26)$$

where $m_j = \sum_i e_{i,j} \ge 2$, by (12). Summing the last quantity in (26) over all $y_0 < p_1 < p_2 < \cdots < p_\ell$, we obtain, by the prime number theorem,

$$\sim \frac{(\eta\pi(y_0))^r}{|\text{Aut}_{\text{Rows}}(M)|} \int_{y_0 < v_1 < v_2 < \cdots < v_\ell} \prod_{1 \le j \le \ell} \frac{dv_j}{v_j^{m_j \alpha} \log v_j}$$

$$\sim \frac{\eta^r \pi(y_0)^{r+\ell-\sum_j m_j}}{|\text{Aut}_{\text{Rows}}(M)|} \int_{1 < t_1 < t_2 < \cdots < t_\ell} \prod_{1 \le j \le \ell} \frac{dt_j}{t_j^{m_j}}$$

using the approximation $\log v_j \sim \log y_0$ (because this range of values of v_j gives the main contribution to the integral), and the fact that $v_j^\alpha \sim v_j/\log y_0$ for v_j in this range. The result follows by making the substitution $t_j = v_j/y_0$. $\square$

6.4   Properties of $M \in \mathcal{M} := \mathcal{M}_I$

Lemma 2. *Suppose that $M \in \mathcal{M} := \mathcal{M}_I$. For each row of M, other than the first, there exists a unique column which has a 1 in that row as well as an earlier row. The last row of M contains exactly one 1.*

Proof. For each $M \in \mathcal{M}$, we have $\Delta_j = 0$ for each $j \ge 0$ by (22) so that

$$\ell_{j+1} - \ell_j = N_{j+1} - N_j - 1.$$

That is, each new vertex connects with a unique colour to the set of previous vertices, which is the first part of our result.[4] The second part comes from noting that the last row cannot have a 1 in a column that has not contained a 1 in an earlier row of M. $\square$

Lemma 3. *If $M \in \mathcal{M}$ then all cycles in its graph, $G(M)$, are monochromatic.*

Proof. If not, then consider a minimal cycle in the graph, where not all the edges are of the same color. We first show that, in fact, each edge in the cycle has a different color. To see this, we start with a cycle where not all edges are of the same color, but where at least two edges have the same color. Say we arrange

---

[4] Hence we confirm that $\ell = N - (r-1)$, since the number of primes involved is the total number of 1's less the unique "old prime" in each row after the first.

the vertices $v_1, ..., v_k$ of this cycle so that the edge (v_1, v_2) has the same color as (v_j, v_{j+1}), for some $2 \leq j \leq k-1$, or the same color as (v_k, v_1), and there are no two edges of the same colour in-between. If we are in the former case, then we reduce to the smaller cycle $v_2, v_3, ..., v_j$, where not all edges have the same color; and, if we are in the latter case, we reduce to the smaller cycle $v_2, v_3, ..., v_k$, where again not all the edges have the same color. Thus, if not all of the edges of the cycle have the same color, but the cycle does contain more than one edge of the same color, then it cannot be a minimal cycle.

Now let I be the number of vertices in our minimal cycle of different colored edges, and reorder the rows of M so that this cycle appears as the first I rows.[5] Then

$$N_I \geq 2I + (\ell_I - I) = \ell_I + I.$$

The term $2I$ accounts for the fact that each prime corresponding to a different colored edge in the cycle must divide at least two members of the cycle, and the $\ell_I - I$ accounts for the remaining primes that divide members of the cycle (that don't correspond to the different colored edges). This then gives $\Delta_I \geq 1$; and thus by (22) we have $\Delta(M) \geq 1$, a contradiction. We conclude that every cycle in our graph is monochromatic.   □

Lemma 4. *Every $M \in \mathcal{M}$ has rank $\ell(M)$.*

Proof (by induction on ℓ). For $\ell = 0, 1$ this is trivial. Otherwise, as there are no cycles the graph must end in a "leaf"; that is a vertex of degree one. Suppose this corresponds to row r and color ℓ. We now construct a new matrix M' which is matrix M less column ℓ, and any rows that only contained a 1 in the ℓ-th column. The new graph now consists of $m_\ell - 1$ disjoint subgraphs, each of which corresponds to an element of $\mathcal{M}$. Thus the rank of M is given by 1 (corresponding to the r-th row, which acts as a pivot element in Gaussian elimination on the ℓ-th column) plus the sum of the ranks of new connected subgraphs. By the induction hypothesis, they each have rank equal to the number of their primes, thus in total we have $1 + (\ell - 1) = \ell$, as claimed.   □

6.5   An Identity, and Inclusion-Exclusion Inequalities, for $\mathcal{M}$

Proposition 4. *If $M_R \in \mathcal{M}$ then*

$$\sum_{\substack{S \subseteq R \\ M_S \in \mathcal{M}}} (-1)^{N(S)} = r(M) - \text{rank}(M). \qquad (27)$$

Furthermore, if $N \geq 2$ is an even integer then

$$\sum_{\substack{S \subset R, N(S) \leq N \\ M_S \in \mathcal{M}}} (-1)^{N(S)} \geq r(M) - \text{rank}(M), \qquad (28)$$

---

[5] This we are allowed to do, because the connectivity of successive rows can be maintained, and because we will still have $\Delta(M) = 0$ after this permutation of rows.

and if $N \geq 3$ is odd then

$$\sum_{\substack{S \subset R, N(S) \leq N \\ M_S \in \mathcal{M}}} (-1)^{N(S)} \leq r(M) - \text{rank}(M). \tag{29}$$

Proof (by induction on $|R|$). It is easy to show when R has just one row and that has no 1's, and when $|R| = 2$, so we will assume that it holds for all R satisfying $|R| \leq r - 1$, and prove the result for $|R| = r$.

Let $\mathcal{N}$ be the set of integers that correspond to the rows of R. By Lemma 3 we know that the integer in $\mathcal{N}$ which corresponds to the last row of M must be a prime, which we will call p_ℓ. Note that p_ℓ must divide at least one other integer in $\mathcal{N}$, since $M_R \in \mathcal{M}$.

Case 1: p_ℓ Divides at Least Three Elements from our Set

We partition R into three subsets: R_0, the rows without a 1 in the ℓ-th column; R_1, the rows with a 1 in the ℓth column, but no other 1's (that is, rows which correspond to the prime p_ℓ); and R_2, the rows with a 1 in the ℓth column, as well as other 1's. Note that $|R_1| \geq 1$ and $|R_1| + |R_2| \geq 3$ by hypothesis.

Write each $S \subset R$ with $M_S \in \mathcal{M}$ as $S_0 \cup S_1 \cup S_2$ where $S_i \subset R_i$. If we fix S_0 and S_2 with $|S_2| \geq 2$ then $S_0 \cup S_2 \in \mathcal{M}$ if and only if $S_0 \cup S_1 \cup S_2 \in \mathcal{M}$ for any $S_1 \subset R_1$. Therefore the contribution of all of these S to the sum in (27) is

$$(-1)^{N(S_0)+N(S_2)} \sum_{S_1 \subset R_1} (-1)^{|S_1|} = (-1)^{N(S_0)+N(S_2)}(1-1)^{|R_1|} = 0 \tag{30}$$

Now consider those sets S with $|S_2| = 1$. In this case we must have $|S_1| \geq 1$ and equally we have $S_0 \cup \{p_\ell\} \cup S_2 \in \mathcal{M}$ if and only if $S_0 \cup S_1 \cup S_2 \in \mathcal{M}$ for any $S_1 \subset R_1$ with $|S_1| \geq 1$. Therefore the contribution of all of these S to the sum in (27) is

$$(-1)^{N(S_0)+N(S_2)} \sum_{\substack{S_1 \subset R_1 \\ |S_1| \geq 1}} (-1)^{|S_1|} = (-1)^{N(S_0)+N(S_2)}((1-1)^{|R_1|} - 1)$$

$$= (-1)^{N(S_0 \cup \{p_\ell\} \cup S_2)}. \tag{31}$$

Regardless of whether $|S_2| = 1$ or $|S_2| \geq 2$, we get that if we truncate the sums (30) or (31) to all those $S_1 \subset R_1$ with

$$N(S_1) = |S_1| \leq N - N(S_0) - N(S_2),$$

then the total sum is ≤ 0 if N is odd, and is ≥ 0 if N is even; furthermore, note that we get that these truncations are 0 in two cases: If $N - N(S_0) - N(S_2) \leq 0$ (which means that the above sums are empty, and therefore 0 by convention), or if $N - N(S_0) - N(S_2) \geq N(R_1)$ (which means that we have the complete sum over all $S_1 \subset R_1$).

It remains to handle those S where $|S_2| = 0$. We begin by defining certain sets H_j and T_j: If the elements of R_2 correspond to the integers $h_1, \ldots, h_k$ then let H_j be the connected component of the subgraph containing h_j, of the graph obtained by removing all rows divisible by p_ℓ except h_j. Let $T_j = H_j \cup \{p_\ell\}$. Note that if $S_2 = \{h_j\}$ then $S_0 \cup \{p_\ell\} \cup S_2 \subset T_j$ (in the paragraph immediately above).

Note that if S has $|S_2| = 0$, then $S = S_0 \subset T_j$ for some j (since the graph of S is connected), or $S = S_1$ with $|S| \geq 2$. The contribution of those $S = S_1$ with $|S| \geq 2$ to the sum in (27) is

$$\sum_{\substack{S_1 \subset R_1 \\ |S_1| \geq 2}} (-1)^{|S_1|} = (1-1)^{|R_1|} - (1 - |R_1|) = |R_1| - 1.$$

Furthermore, if we truncate this sum to all those S_1 satisfying

$$N(S_1) = |S_1| \leq N,$$

then the sum is $\geq |R_1| - 1$ if $N \geq 2$ is even, and the sum is $\leq |R_1| - 1$ if $N \geq 3$ is odd.

Finally note that if $S \subset T_j$ with $M_S \in \mathcal{M}$ then either $|S_2| = 0$ or $S = S_0 \cup \{p_\ell, h_j\}$ and therefore, combining all of this information,

$$\sum_{\substack{S \subset R \\ M_S \in \mathcal{M}}} (-1)^{N(S)} = |R_1| - 1 + \sum_{j=1}^{k} \sum_{\substack{S \subset T_j \\ M_S \in \mathcal{M}}} (-1)^{N(S)} = |R_1| - 1 + \sum_{j=1}^{k} (r(T_j) - \ell(T_j))$$

by the induction hypothesis (as each $|T_j| < |M|$). Also by the induction hypothesis, along with what we worked out above for N even and odd, in all possibilities for $|S_2|$ (i.e. $|S_2| = 0, 1$ or exceeds 1), we have that for $N \geq 3$ odd,

$$\sum_{\substack{S \subset R, \ N(S) \leq N \\ M_S \in \mathcal{M}}} (-1)^{N(S)} \leq |R_1| - 1 + \sum_{j=1}^{k} (r(T_j) - \ell(T_j));$$

and for $N \geq 2$ even,

$$\sum_{\substack{S \subset R, \ N(S) \leq N \\ M_S \in \mathcal{M}}} (-1)^{N(S)} \geq |R_1| - 1 + \sum_{j=1}^{k} (r(T_j) - \ell(T_j)).$$

The T_j less the rows $\{p_\ell\}$ is a partition of the rows of M less the rows $\{p_\ell\}$, and so

$$\sum_{j} (r(T_j) - 1) = r(M) - |R_1|.$$

The primes in T_j other than p_ℓ is a partition of the primes in M other than p_ℓ, and so

$$\sum_{j} (\ell(T_j) - 1) = \ell(M) - 1.$$

Combining this information gives (27), (28), and (29).

Case 2 : p_ℓ Divides Exactly Two Elements from our Set

Suppose these two elements are $n_r = p_\ell$ and $n_{r-1} = p_\ell q$ for some integer q. If $q = 1$ this is our whole graph and (27), (28) and (29) all hold, so we may assume $q > 1$. If $n_j \neq q$ for all j, then we create $M_1 \in \mathcal{M}$ with $r - 1$ rows, the first $r - 2$ the same, and with $n_{r-1} = q$. We have

$$N(M_1) = N(M) - 2, \ r(M_1) = r(M) - 1, \text{ and } \ell(M_1) = \ell(M) - 1.$$

We claim that there is a 1-1 correspondence between the subsets $S \subset \mathcal{R}(M)$ with $M_S \in \mathcal{M}$ and the subsets $T \subset \mathcal{R}(M_1)$ with $(M_1)_T \in \mathcal{M}$. The key observation to make is that $p_\ell \in S$ (ie row r) if and only if $p_\ell q \in S$ (ie row $r - 1$), since $M_S \in \mathcal{M}$. Thus if rows $r - 1$ and r are in S then S corresponds to T (ie $T = S_1$), which we obtain by replacing rows $r - 1$ and r of S by row $r - 1$ of T which corresponds to q. Otherwise we let $S = T$. Either way $(-1)^{N(S)} = (-1)^{N(T)}$ and so

$$\sum_{\substack{S \subset R \\ M_S \in \mathcal{M}}} (-1)^{N(S)} = \sum_{\substack{T \subset \mathcal{R}(M_1) \\ (M_1)_T \in \mathcal{M}}} (-1)^{N(T)} = r(M_1) - \ell(M_1) = r(M) - \ell(M),$$

by the induction hypothesis. Further, we have that for N even,

$$\sum_{\substack{S \subset R, N(S) \leq N \\ M_S \in \mathcal{M}}} (-1)^{N(S)} = \sum_{\substack{T \subset \mathcal{R}(M_1), N(T) \leq N-2 \\ (M_1)_T \in \mathcal{M}}} (-1)^{N(T)} \geq r(M) - \ell(M).$$

The analogous inequality holds in the case where N is odd. Thus, we have that (27), (28) and (29) all hold.

Finally, suppose that $n_j = q$ for some j, say $n_{r-2} = q$. Then q must be prime else there would be a non-monochromatic cycle in $M \in \mathcal{M}$. But since prime q is in our set it can only divide two of the integers of the set (by our previous deductions) and these are n_{r-2} and n_{r-1}. However this is then the whole graph and we observe that (27), (28), and (29) all hold. □

6.6 Counting Configurations

We partitioned B into connected components $S_1, \ldots, S_h$. Now we form the matrices B_k, the union of the $S_j \in \mathcal{M}_k$, for each $k \geq 0$, so that

$$r(B) - \operatorname{rank}(B) = \sum_{k \geq 0} r(B_k) - \operatorname{rank}(B_k), \qquad (32)$$

and

$$r(B_k) - \operatorname{rank}(B_k) = \sum_{j:\ S_j \in \mathcal{M}_k} r(S_j) - \operatorname{rank}(S_j).$$

More importantly

$$\sum_{j:\, M_j \in \mathcal{M}_0} r(M_j) - \mathrm{rank}(M_j)$$

$$= \sum_{\substack{j:\, M_j \in \mathcal{M}_0}} \sum_{\substack{S \subset \mathcal{R}(M_j) \\ M_S \in \mathcal{M}}} (-1)^{N(S)} = \sum_{\substack{S \subset \mathcal{R}(B_0) \\ M_S \in \mathcal{M}}} (-1)^{N(S)}, \quad (33)$$

by Proposition 4. If $k \geq 1$ then there are a bounded number of S_j isomorphic to any given matrix $M \in \mathcal{M}_k$, by Proposition 3, and so we believe that these contribute little to our sum (32). In particular we conjecture that

$$\sum_{k \geq 1} \sum_{j:\, M_j \in \mathcal{M}_k} \left(r(M_j) - \mathrm{rank}(M_j) - \sum_{\substack{S \subset \mathcal{R}(M_j) \\ M_S \in \mathcal{M}}} (-1)^{N(S)} \right) = o(\pi(y_0))$$

with probability $1 - o(1)$. Hence the last few equations combine to give what will now be our assumption.

Assumption

$$r(B) - \mathrm{rank}(B) = \sum_{\substack{S \subset \mathcal{R}(B) \\ M_S \in \mathcal{M}}} (-1)^{N(S)} + o(\pi(y_0)). \quad (34)$$

By combining (23), (34), and the identity

$$\sum_{\sigma \in S_\ell} \prod_{j=1}^{\ell} \frac{1}{\sum_{i=j}^{\ell} c_{\sigma(i)}} = \prod_{j=1}^{\ell} \frac{1}{c_i},$$

(here S_ℓ is the symmetric group on $1, ..., \ell$, and taking $c_i = m_i - 1$) we obtain, by summing over all orderings of the primes,

$$\mathbb{E}(r(B) - \mathrm{rank}(B)) \sim f(\eta)\pi(y_0) \quad (35)$$

where

$$f(\eta) := \sum_{M \in \mathcal{M}^*} \frac{(-1)^{N(M)}}{|\mathrm{Aut}_{\mathrm{Cols}}(M)| \cdot |\mathrm{Aut}_{\mathrm{Rows}}(M)|} \cdot \frac{\eta^{r(M)}}{\prod_{j=1}^{\ell}(m_j - 1)}, \quad (36)$$

assuming that when we sum and re-order our initial series, we do not change the value of the sum. Here $\mathrm{Aut}_{\mathrm{Cols}}(M)$ denotes the number of ways to obtain exactly the same matrix M when permuting the columns, and $\mathcal{M}^* = \mathcal{M}/\sim$ where two matrices are considered to be "equivalent" if they are isomorphic.

6.7 Husimi Graphs

All of the graphs $G(M), M \in \mathcal{M}$ are simple graphs, and have only monochromatic cycles: notice that these cycles are subsets of the complete graph formed by the edges of a particular colour (corresponding to the integers divisible by a particular prime). Hence any two-connected subgraph of $G(M)$ is actually a complete graph: This is precisely the definition of a *Husimi* graph (see [11]), and so the isomorphism classes of Husimi graphs are in one-to-one correspondence with the matrices in $\mathcal{M}^*$.

Husimi graphs have a rich history, inspiring the combinatorial theory of species, and are central to the thermodynamical study of imperfect gases (see [11] for references and discussion).

Lemma 5. *If G is a Husimi graph then*

$$\mathrm{Aut}(G) \cong \mathrm{Aut}_{\mathrm{Rows}}(M) \times \mathrm{Aut}_{\mathrm{Cols}}(M). \tag{37}$$

Proof. If $\sigma \in \mathrm{Aut}(G)$ then it must define a permutation of the colors of G; that is an element $\tau \in \mathrm{Aut}_{\mathrm{Cols}}(M)$. Then $\tau^{-1}\sigma \in \mathrm{Aut}(G)$ is an automorphism of G that leaves the colors alone; and therefore must permute the elements of each given color. However if two vertices of the same color in G are each adjacent to an edge of another color then permuting them would permute those colors which is impossible. Therefore $\tau^{-1}\sigma$ only permutes the vertices of a given color which are not adjacent to edges of any other color; and these correspond to automorphisms of the rows of M containing just one 1. However this is all of $\mathrm{Aut}_{\mathrm{Rows}}(M)$ since if two rows of M are identical then they must contain a single element, else G would contain a non-monochromatic cycle.   □

Let $\mathrm{Hu}(j_2, j_3, \dots)$ denote the set of Husimi graphs with j_i blocks of size i for each i, on

$$r = 1 + \sum_{i \geq 2}(i-1)j_i \tag{38}$$

vertices, with $\ell = \sum_i j_i$ and $N(M) = \sum_i i j_i$. (This corresponds to a matrix M in which exactly j_i columns contain precisely i 1's.) In this definition we count all distinct labellings, so that

$$\mathrm{Hu}(j_2, j_3, \dots) = \sum_G \frac{r!}{|\mathrm{Aut}(G)|} \ ,$$

where the sum is over all isomorphism classes of Husimi graphs G with exactly j_i blocks of size i for each i. The Mayer-Husimi formula (which is (42) in [11]) gives

$$\mathrm{Hu}(j_2, j_3, \dots) = \frac{(r-1)!}{\prod_{i \geq 2}((i-1)!^{j_i} j_i!)} \cdot r^{\ell-1}, \tag{39}$$

and so, by (36), (37) and the last two displayed equations we obtain

$$f(\eta) = \sum_{\substack{j_2, j_3, \dots \geq 0 \\ j_2 + j_3 + \dots < \infty}} (-1)^{r+\ell-1} \frac{r^{\ell-2}}{\prod_{i \geq 2}((i-1)!^{j_i}(i-1)^{j_i} j_i!)} \cdot \eta^r. \tag{40}$$

6.8   Convergence of $f(\eta)$

In this section we prove the following result under an appropriate (analytic) assumption.

"Theorem". *The function $f(\eta)$ has radius of convergence $e^{-\gamma}$, is increasing in $[0, e^{-\gamma})$, and $\lim_{\eta \to (e^{-\gamma})^-} f(\eta) = 1$.*

So far we have paid scant attention to necessary convergence issues. First note the identity

$$\exp\left(\sum_{i=1}^{\infty} c_i\right) = \sum_{\substack{k_1, k_2, \ldots \geq 0 \\ k_1 + k_2 + \cdots < \infty}} \prod_{i \geq 1} \frac{c_i^{k_i}}{k_i!}, \tag{41}$$

which converges absolutely for any sequence of numbers $c_1, c_2, \ldots$ for which $|c_1| + |c_2| + \cdots$ converges, so that the terms in the series on the right-hand-side can be summed in any order we please.

The summands of $f(\eta)$, for given values of r and ℓ, equal $(-1)^{r+\ell-1} r^{\ell-2} \eta^r$ times

$$\sum_{\substack{j_2, j_3, \cdots \geq 0 \\ \sum_{i \geq 2} j_i = \ell, \ \sum_{i \geq 2}(i-1)j_i = r-1}} \frac{1}{\prod_{i \geq 2}((i-1)!^{j_i} (i-1)^{j_i} j_i!)}, \tag{42}$$

which is exactly the coefficient of t^{r-1} in

$$\frac{1}{\ell!}\left(t + \frac{t^2}{2 \cdot 2!} + \frac{t^3}{3 \cdot 3!} + \cdots\right)^{\ell},$$

and so is less than $\tau^{\ell}/\ell!$ where $\tau = \sum_{j \geq 1} 1/(j \cdot j!) \approx 1.317902152$. Note that if $r \geq 2$ then $1 \leq \ell \leq r - 1$. Therefore the sum of the absolute values of all of the coefficients of η^r in $f(\eta)$ is less than

$$\sum_{2 \leq \ell \leq r-1} r^{\ell-2} \frac{\tau^{\ell}}{\ell!} \ll r^{r-2} \frac{\tau^r}{r!} \ll \frac{(e\tau)^r}{r^{5/2}}.$$

The first inequality holds since $\tau > 1$, the second by Stirling's formula. Thus $f(\eta)$ is absolutely convergent for $|\eta| \leq \rho_0 := 1/(e\tau) \approx 0.2791401779$. We can therefore manipulate the power series for f, as we wish, inside the ball $|\eta| \leq \rho_0$, and we want to extend this range.

Let

$$A(T) := -\sum_{j \geq 1} \frac{(-1)^j T^j}{j \cdot j!} = \int_0^T \frac{1 - e^{-t}}{t} dt.$$

The identity (41) implies that the coefficient of t^{r-1} in $\exp(rA(\eta t))$ is

$$\sum_{\substack{j_2, j_3, \ldots \\ j_2 + 2j_3 + 3j_4 + \cdots = r-1}} \frac{(-1)^{r+\ell-1} r^{\ell} \eta^{r-1}}{\prod_{i \geq 2}((i-1)!^{j_i} (i-1)^{j_i} j_i!)},$$

so that

$$f'(\eta) = \sum_{r \geq 1} \frac{\text{coeff of } t^{r-1} \text{ in } \exp(rA(\eta t))}{r}. \tag{43}$$

We will now obtain a functional equation for f' using Lagrange's Inversion formula:

Lagrange's Inversion Formula. *If $g(w)$ is analytic at $w = 0$, with $g(0) = a$ and $g'(0) \neq 0$, then*

$$h(z) = \sum_{r=1}^{\infty} \left(\frac{d}{dw}\right)^{r-1} \left(\frac{w}{g(w) - a}\right)^r \Bigg|_{w=0} \frac{(z-a)^r}{r!}$$

is the inverse of $g(w)$ in some neighbourhood around a (thus, $h(g(w)) = 1$).

If $g(w) = w/\varphi(w)$, where $\varphi(w)$ is analytic and non-zero in some neighbourhood of 0, then

$$h(z) = \sum_{r=1}^{\infty} \frac{c_{r-1} z^r}{r}$$

is the inverse of $g(w)$ in some neighbourhood around 0, where c_j is the coefficient of w^j in $\varphi(w)^{j+1}$. Applying this with $\varphi(w) = e^{A(\eta w)}$ we find that $g(w) = we^{-A(\eta w)}$ has an inverse $h(z)$ in a neighbourhood, Γ, around 0 where

$$h(1) = \sum_{r \geq 1} \frac{\text{coeff. of } z^{r-1} \text{ in } \exp(rA(\eta z))}{r} = f'(\eta).$$

We will assume that the neighbourhood Γ includes 1. Therefore, since

$$1 = g(h(1)) = h(1)e^{-A(\eta h(1))} = f'(\eta)e^{-A(\eta f'(\eta))},$$

we deduce that

$$f'(\eta) = e^{A(\eta f'(\eta))}. \tag{44}$$

(Note that this can only hold for η in some neighborhood of 0 in which the power series for $f'(\eta)$ converges.) Taking the logarithm of (44) and differentiating we get, using the formula $A'(T) = \frac{1-e^{-T}}{T}$,

$$\frac{f''(\eta)}{f'(\eta)} = (\eta f'(\eta))' \frac{1 - e^{-\eta f'(\eta)}}{\eta f'(\eta)}$$

so that $f'(\eta) = (\eta f'(\eta))' - \eta f''(\eta) = (\eta f'(\eta))' \, e^{-\eta f'(\eta)}$. Integrating and using the facts that $f(0) = 0$ and $f'(0) = 1$, we have

$$f(\eta) = 1 - e^{-\eta f'(\eta)}. \tag{45}$$

We therefore deduce that

$$\eta f'(\eta) = -\log(1 - f(\eta)) = \sum_{k \geq 1} \frac{f(\eta)^k}{k}. \tag{46}$$

Lemma 6. *The coefficients of $f(\eta)$ are all non-negative. Therefore $|f(z)| \leq f(|z|)$ so that $f(z)$ is absolutely convergent for $|z| < R$ if $f(\eta)$ converges for $0 \leq \eta < R$. Also all of the coefficients of $f'(\eta)$ are non-negative and $f'(0) = 1$ so that $f'(\eta) > 1$ for $0 \leq \eta < R$.*

Proof. Write $f(\eta) = \sum_{r \geq 0} a_r \eta^r$. We prove that $a_r > 0$ for each $r \geq 1$, by induction. We already know that $a_1 = 1$ so suppose $r \geq 2$. We will compare the coefficient of η^r on both sides of (46). On the left side this is obviously ra_r. For the right side, note that the coefficient of η^r in $f(\eta)^k$ is a polynomial, with positive integer coefficients (by the multinomial theorem), in variables $a_1, \ldots, a_{r+1-k}$ for each $k \geq 1$. This is 0 for $k > r$, and is positive for $2 \leq k \leq r$ by the induction hypothesis. Finally, for $r = 1$, the coefficient is a_r. Therefore we have that $ra_r > a_r$ which implies that $a_r > 0$ as desired. $\square$

Our plan is to determine R, the radius of convergence of $f(\eta)$, by determining the largest possible R_1 for which $f'(\eta)$ is convergent for $0 \leq \eta < R_1$. Then $R = R_1$.

Since f' is monotone increasing (as all the coefficients of f' are positive), we can define an inverse on the reals $\geq f'(0) = 1$. That is, for any given $y \geq 1$, let η_y be the (unique) value of $\eta \geq 0$ for which $f'(\eta) = y$. Therefore $R_1 = \lim_{y \to \infty} \eta_y$.

We claim that the value of $f'(\eta)$ is that unique real number y for which $B_\eta(y) := A(\eta y) - \log y = 0$. By (44) we do have that $B_\eta(f'(\eta)) = 0$, and this value is unique if it exists since $B_\eta(y)$ is monotone decreasing, as

$$B'_\eta(y) = \eta A'(\eta y) - 1/y = -e^{-\eta y}/y < 0.$$

This last equality follows since $A'(T) = (1 - e^{-T})/T$. Now $A'(T) > 0$ for $T > 0$, and so $A(t) > 0$ for all $t > 0$ as $A(0) = 0$. Therefore $B_\eta(1) = A(\eta) > 0$, and so, remembering that $B_\eta(y)$ is monotone decreasing, we have that a solution y exists to $B_\eta(y) := A(\eta y) - \log y = 0$ if and only if $B_\eta(\infty) < 0$. Therefore R_1 is precisely that value of $\eta = \eta_1$ for which $B_{\eta_1}(\infty) = 0$. Now

$$B_\eta(y) = B_\eta(1) + \int_1^y B'_\eta(t)dt = A(\eta) - \int_1^y \frac{e^{-\eta t}}{t}dt.$$

so that

$$B_\eta(\infty) = A(\eta) - \int_1^\infty \frac{e^{-\eta y}}{y}dy.$$

Therefore

$$\int_1^\infty \frac{e^{-\eta_1 y}}{y}dy = A(\eta_1) = A(0) + \int_0^{\eta_1} A'(v)dv = \int_0^{\eta_1} \frac{(1 - e^{-v})}{v}dv,$$

so that

$$\int_1^{\eta_1} \frac{dv}{v} = \int_1^\infty \frac{e^{-v}}{v}dv - \int_0^1 \frac{(1 - e^{-v})}{v}dv = -\gamma$$

(as is easily deduced from the third line of (6.3.22) in [1]). Exponentiating we find that $R_1 = \eta_1 = e^{-\gamma} = .561459\ldots$.

Finally by (45) we see that $f(\eta) < 1$ when $f'(\eta)$ converges, that is when $0 \leq \eta < \eta_0$, and $f(\eta) \to 1$ as $\eta \to \eta_0^-$.

6.9   From Expectation to Probability

One can easily generalize Proposition 2 to prove the following result, which implies that if $\mathbb{E}(r(B) - \mathrm{rank}(B)) > (1 + 2\epsilon)\pi(y_0)$ then

$$r(B) - \mathrm{rank}(B) > (1 + \epsilon)\pi(y_0) \text{ with probability } 1 - o_\epsilon(1).$$

Proposition 5. *If $M \in \mathcal{M}$ then*

$$\#\{S \subseteq A_{y_0} : M_S \simeq M\} \sim \mathbb{E}(\#\{S \subseteq A_{y_0} : M_S \simeq M\})$$

with probability $1 - o(1)$, as $x \to \infty$.

Hence, with probability $1 - o(1)$ we have, assuming (34) is true, that

$$\sum_{j:\ M_j \in \mathcal{M}} r(M_j) - \mathrm{rank}(M_j) \sim \mathbb{E}\left(\sum_{j:\ M_j \in \mathcal{M}} r(M_j) - \mathrm{rank}(M_j) \right)$$

as $x \to \infty$, which is why we believe that one can take $J = (e^{-\gamma} + o(1))J_0(x)$ with probability $1 - o(1)$.

7   Algorithms

7.1   The Running Time for Pomerance's Problem

We will show that, with current methods, the running time in the hunt for the first square product is dominated by the speed of finding a linear dependence in our matrix of exponents:

Let us suppose that we select a sequence of integers $a_1, a_2, \ldots, a_J$ in $[1, n]$ that appear to be random, as in Pomerance's problem, with $J \asymp J_0$. We will suppose that the time taken to determine each a_j, and then to decide whether a_j is y_0-smooth and, if so, to factor it, is $\ll y_0^{(1-\epsilon)/\log\log y_0}$ steps (note that the factoring can easily be done in $\exp(O(\sqrt{\log y_0 \log\log y_0}))$ steps by the elliptic curve method, according to [3], section 7.4.1).

Therefore, with probability $1 - o(1)$, the time taken to obtain the factored integers in the square dependence is $\ll y_0^{2-\epsilon/\log\log y_0}$ by (8).

In order to determine the square product we need to find a linear dependence mod 2 in the matrix of exponents. Using the Wiedemann or Lanczos methods (see section 6.1.3 of [3]) this takes time $O(\pi(y_0)^2\mu)$, where μ is the average number of prime factors of an a_i which has been kept, so this is by far the lengthiest part of the running time.

7.2   Improving the Running Time for Pomerance's Problem

If instead of wanting to simply find the first square dependence, we require an algorithm that proceeds as quickly as possible to find any square dependence then we should select our parameters so as to make the matrix smaller. Indeed if we simply create the matrix of y-smooths (without worrying about large prime variations) then we will optimize by taking

$$\frac{\pi(y)}{\Psi(x,y)/x} \asymp \pi(y)^2 \mu, \tag{47}$$

that is the expected number of a_j's selected should be taken to be roughly the running time of the matrix setp, in order to determine the square product. Here μ, is as in the previous section, and so we expect that μ is roughly

$$\frac{1}{\psi(x,y)} \sum_{n \le x, P(n) \le y} \sum_{p \le y: \, p|n} 1 = \sum_{p \le y} \frac{\psi(x/p, y)}{\psi(x,y)}$$

$$\sim \sum_{p \le y} \frac{1}{p^\alpha} \sim \frac{y^{1-\alpha}}{(1-\alpha)\log y} \sim \frac{\log y}{\log \log y}$$

by (12), the prime number theorem and (11). Hence we optimize by selecting $y = y_1$ so that $\rho(u_1) \asymp (\log \log y_1)/y_1$, which implies that

$$y_1 = y_0^{1-(1+o(1))/\log \log x},$$

by Lemma 1, which is considerably smaller than y_0. On the other hand, if J_1 is the expected running time, $\pi(y_1)/(\Psi(x,y_1)/x)$ then

$$J_1/J_0 \sim \frac{y_1/\rho(u_1)}{y_0/\rho(u_0)} = \exp\left(\{1+o(1)\}\frac{u_0 \log u_0}{(\log \log x)^2}\right) = y_0^{(1+o(1))/(\log \log x)^2}$$

by the prime number theorem, (3), and (22) in the proof of Lemma 2.3 in [4].

7.3   Smooth Squares

In factoring algorithms, the a_i are squares mod n (as explained at the beginning of section 1), which is not taken into account in Pomerance's problem. For instance, in Dixon's random squares algorithm one selects $b_1, b_2, \ldots, b_J \in [1, n]$ at random and lets a_i be the least residue of b_i^2 (mod n). We keep only those a_i that are y-smooth, and so to complete the analysis we need some idea of the probability that a y-smooth integer is also a square mod n. Dixon [5] gives an (unconditionally proven) lower bound for this probability which is too small by a non-trivial factor. We shall estimate this probability much more accurately though under the assumption of the Generalized Riemann Hypothesis.

Theorem 2. *Assume the Generalized Riemann Hypothesis and let n be an integer with smallest prime factor $> y$, which is $> 2^{3\omega(n)} L^\epsilon$ (where $\omega(n)$ denotes the number of distinct prime factors of n). For any $n \ge x \ge n^{1/4+\delta}$, the proportion of the positive integers $a \le x$ where a is a square mod n and coprime to n, which are y-smooth, is $\sim \Psi(x,y)/x$.*

We use the following result which is easily deduced from the remark following Theorem 2 of [9]:

Lemma 7. *Assume the Generalized Riemann Hypothesis. For any non-principal character χ (mod n), and $1 \leq x \leq n$ we have, uniformly,*

$$\left| \sum_{\substack{a \leq x \\ a\ y-\text{smooth}}} \chi(a) - \sum_{a \leq x} \chi(a) \right| \ll \frac{\Psi(x,y)(\log n)^3}{\sqrt{y}}.$$

Proof (of Theorem 2). Let $M(x)$ be the number of $a \leq x$ which are coprime with n, let $N(x)$ be the number of these a which are a square mod n, and let $N(x,y)$ be the number of these a which are also y-smooth. Then

$$\left(N(x,y) - \frac{\Psi(x,y)}{2^{\omega(n)}} \right) - \left(N(x) - \frac{M(x)}{2^{\omega(n)}} \right) =$$

$$= \left(\sum_{\substack{a \leq x, (a,n)=1 \\ a\ y-\text{smooth}}} - \sum_{\substack{a \leq x \\ (a,n)=1}} \right) \left(\prod_{p|n} \frac{1}{2} \left\{ 1 + \left(\frac{a}{p} \right) \right\} - \frac{1}{2^{\omega(n)}} \right)$$

$$= \frac{1}{2^{\omega(n)}} \sum_{\substack{d|n \\ d \neq 1}} \mu^2(d) \left(\sum_{\substack{a \leq x, (a,n)=1 \\ a\ y-\text{smooth}}} \left(\frac{a}{d} \right) - \sum_{\substack{a \leq x \\ (a,n)=1}} \left(\frac{a}{d} \right) \right) \ll \frac{\Psi(x,y)(\log n)^3}{\sqrt{y}}$$

by Lemma 7. Now Burgess's theorem tells us that $N(x) - M(x)/2^{\omega(n)} \ll x^{1-\epsilon}$ if $x \geq n^{1/4+\delta}$, the prime number theorem that $\omega(n) \leq \log n / \log y = o(\log x)$, and (7) that $\Psi(x,y) \geq x^{1-\epsilon/2}$ as $y > L^\epsilon$. Hence $N(x,y) \sim \Psi(n,y)/2^{\omega(n)}$. The number of integers $a \leq x$ which are coprime to n and a square mod n is $\sim (\phi(n)/n)(x/2^{\omega(n)})$, and $\phi(n) = n(1 + O(1/y))^{\omega(n)} \sim n$, so the result follows. □

7.4 Making the Numbers Smaller

In Pomerance's quadratic sieve the factoring stage of the algorithm is sped up by having the a_i be the reduced values of a polynomial, so that every p-th a_i is divisible by p, if any a_j is. This regularity means that we can proceed quite rapidly, algorithmically in factoring the a_i's. In addition, by an astute choice of polynomials, the values of a_i are guaranteed to be not much bigger than $\sqrt{n}$, which gives a big saving, and one can do a little better (though still bigger than $\sqrt{n}$) with Peter Montgomery's "multiple polynomial variation". For all this see section 6 of [3].

8 Large Prime Variations

8.1 A Discussion of Theorem 4.2 in [4] and Its Consequences

Define $\exp_k(z) := \sum_{j=0}^{k-1} z^j/j!$ so that $\lim_{k \to \infty} \exp_k(z) = \exp(z)$, and

$$A_M(z) := \int_{1/M}^{1} \frac{1 - e^{-zt}}{t}\, dt \text{ so that } \lim_{M \to \infty} A_M(z) = A(z) = \int_0^1 \frac{1 - e^{-zt}}{t}\, dt.$$

Recursively, define functions $\gamma_{m,M,k}$ by $\gamma_{0,M,k}(u) := u$ and

$$\gamma_{m+1,M,k}(u) := u \, \exp_k \left[A_M(\gamma_{m,M,k}(u)) \right]$$

for $m = 0, 1, 2, \ldots$. Note that $\gamma_{m,M,k}(u)$ is increasing in all four arguments. From this it follows that $\gamma_{m,M,k}(u)$ increases to $\gamma_{M,k}(u)$ as $m \to \infty$, a fixed point of the map $z \mapsto u \exp_k(A_m(z))$, so that

$$\gamma_{M,k}(u) := u \, \exp_k \left[A_M(\gamma_{M,k}(u)) \right]. \tag{48}$$

We now establish that $\gamma_{M,k}(u) < \infty$ except perhaps when $M = k = \infty$: We have $0 \le A_M(z) \le \log M$ for all z, so that $u < \gamma_{M,k}(u) \le Mu$ for all u; in particular $\gamma_{M,k}(u) < \infty$ if $M < \infty$. We have $A(z) = \log z + O(1)$ so that if $\gamma_{\infty,k}(u)$ is sufficiently large, we deduce from (48) that $\gamma_{\infty,k}(u) \sim u(\log u)^{k-1}/(k-1)!$; in particular $\gamma_{\infty,k}(u) < \infty$. As $M, k \to \infty$, the fixed point $\gamma_{M,k}(u)$ increases to the fixed point $\gamma(u)$ of the map $z \mapsto ue^{A(z)}$, or to ∞ if there is no such fixed point, in which case we write $\gamma(u) = \infty$. By comparing this with (44) we see that $\gamma(u) = uf'(u)$. In [4] we show that this map has a fixed point if and only if $u \le e^{-\gamma}$. Otherwise $\gamma(u) = \infty$ for $u > e^{-\gamma}$ so that $\int_0^\eta \frac{\gamma(u)}{u} \, du = \infty > 1$ for any $\eta > e^{-\gamma}$.

One might ask how the variables m, M, k, u relate to our problem? We are looking at the possible pseudosmooths (that is integers which are a y_0-smooth times a square) composed of products of a_j with $j \le uJ_0$. We restrict our attention to a_j that are My_0-smooth, and which have at most k prime factors $\ge y_0$. In the construction of our hypergraph we examine the a_j selecting only those with certain (convenient) properties, which corresponds to $m = 0$. Then we pass through the a_j again, selecting only those with convenient properties given the a_j already selected at the $m = 0$ stage: this corresponds to $m = 1$. We iterate this procedure which is how the variable m arises. The advantage in this rather complicated construction is that the count of the number of pseudosmooths created, namely

$$\sim \pi(y_0) \cdot \int_0^\eta \frac{\gamma_{m,M,k}(u)}{u} \, du \, ,$$

increases as we increase any of the variables so that it is relatively easy to deal with convergence issues (this is Theorem 2 in [4]). This technique is more amenable to analysis than the construction that we give in section 6, because here we use the inclusion-exclusion type formula (36) to determine $f(\eta)$, which has both positive and negative summands, and it has proved to be beyond us to establish unconditionally that this sum converges.

Note that as $m \to \infty$ we have that the number of pseudosmooths created is

$$\sim \pi(y_0) \cdot \int_0^\eta \frac{\gamma_{M,k}(u)}{u} \, du \, ; \tag{49}$$

hence if the value of this integral is > 1 then we are guaranteed that there is a square product. If we let M and k go to ∞ then the number of pseudosmooths created is

$$\sim \pi(y_0) \cdot \int_0^\eta \frac{\gamma(u)}{u} \, du \, .$$

The upper bound in the Conjecture follows. In terms of what we have proposed in section 6, we have now shown that the number of pseudosmooths created is indeed $\sim f(\eta)\pi(y_0)$.

We remarked above that this integral is an increasing function of η and, in fact, equals 1 for $\eta = e^{-\gamma}$. Hence if $\eta > e^{-\gamma}$ then we are guaranteed that there is a square product. One might expect that if $\eta = e^{-\gamma} + \epsilon$ then we are guaranteed $C(\epsilon)\pi(y_0)$ square products for some $C(\epsilon) > 0$. However we get rather more than that: if $\eta > e^{-\gamma}$ then $\int_0^\eta \frac{\gamma(u)}{u}\, du = \infty$ (that is $f(\eta)$ diverges) and hence the number of square products is bigger than any fixed multiple of $\pi(y_0)$ (we are unable to be more precise than this).

8.2   Speed-Ups

From what we have discussed above we know that we will find a square product amongst the y_0-smooth a_j's once $J = \{1 + o(1)\}J_0$, with probability $1 - o(1)$. When we allow the a_j's that are either y_0-smooth, or y_0-smooth times a single larger prime then we get a stopping time of $\{c_1 + o(1)\}J_0$ with probability $1 - o(1)$ where c_1 is close to $3/4$. When we allow any of the a_j's in our square product then we get a stopping time of $\{e^{-\gamma} + o(1)\}J_0$ with probability $1 - o(1)$ where $e^{-\gamma} = .561459\ldots$. It is also of interest to get some idea of the stopping time for the k-large primes variations, for values of k other than $0, 1$ and ∞. In practice we cannot store arbitrarily large primes in the large prime variation, but rather keep only those a_j where all of the prime factors are $\leq My_0$ for a suitable value of M – it would be good to understand the stopping time with the feasible prime factors restricted in this way. We have prepared a table of such values using the result from [4] as explained in section 8.1: First we determined a Taylor series for $\gamma_{M,k}(u)$ by solving for it in the equation (48). Next we found the appropriate multiple of $\pi(y_0)$, a Taylor series in the variable η, by substituting our Taylor series for $\gamma_{M,k}(u)$ into (49). Finally, by setting this multiple equal to 1, we determined the value of η for which the stopping time is $\{\eta + o(1)\}J_0$ with probability $1 - o(1)$, when we only use the a_j allowed by this choice of k and M to make square products.

k	$M = \infty$	$M = 100$	$M = 10$
0	1	1	1
1	.7499	.7517	.7677
2	.6415	.6448	.6745
3	.5962	.6011	.6422
4	.5764	.5823	.6324
5	.567	.575	.630

The expected stopping time, as a multiple of J_0.

What we have given here is the speed-up in Pomerance's problem; we also want to use our work to understand the speed-up of multiple prime variations in actual factoring algorithms. As dicussed in section 7 we optimize the running time by taking y_1 to be a solution to (47): If we include the implicit constant c on the left side of (47), then this is tantamount to a solution of $h(u_c) = \log(c \log \log y)$ where $h(u) := \frac{1}{u} \log x + \log \rho(u)$. For $u \approx u_c$ we have

$$h'(u) = -\frac{h(u)}{u} - \left(\frac{\log \rho(u)}{u} - \frac{\rho'(u)}{\rho(u)} \right) = -1 + o(1)$$

by (51), (56) and (42) of section III.5 of [20]. One can show that the arguments in [4] which lead to the speed-ups in the table above, work for y_1 just as for y_0; so if we use a multiprime variation to reduce the number of a_j's required by a factor η (taken from the table above), then we change the value of $h(u)$ by $\log \eta$, and hence we must change u to $u' := u - \{1 + o(1)\} \log \eta$. The change in our running time (as given by (47)) will therefore be by a factor of

$$\sim x^{\frac{2}{u'} - \frac{2}{u}} = \exp\left(\frac{2(u - u') \log x}{uu'} \right)$$

$$= \exp\left(\frac{\{2 + o(1)\} \log \eta \log x}{u^2} \right) = \frac{1}{(\log x)^{\{1 + o(1)\} \log(1/\eta)}};$$

with a little more care, one can show that this speed-up is actually a factor

$$\sim \left(\frac{2e^4 + o(1)}{\log x \log \log x} \right)^{\log(1/\eta)}.$$

8.3 A Practical Perspective

One approaches Pomerance's question, in practice, as part of an implementation of a factoring algorithm. The design of the computer, the language and the implementation of the algorithm, all affect the running time of each particular step. Optimally balancing the relative costs of the various steps of an algorithm (like the quadratic sieve) may be substantially different as these environmental factors change. This all makes it difficult to analyze the overall algorithm and to give one definitive answer.

The key parameter in Pomerance's problem and its use in factoring algorithms is the smoothness parameter $y = y_1$: We completely factor that part of a_j which is y-smooth. Given the origin of the a_j's it may be possible to do this very efficiently using a sieve method. One may obtain a significant speed-up by employing an "early abort" strategy for the a_j that have a particularly small y_2-smooth part, where y_2 is substantially smaller than $y = y_1$. The size of y also determines the size of the matrix in which we need to find a linear dependence – note though that the possible size of the matrix may be limited by the size of memory, and by the computer's ability to handle arrays above a certain size.

Suppose that a_j equals its y-smooth part times b_j, so that b_j is what is left after the initial sieving. We only intend to retain a_j if $b_j = 1$, or if b_j has no more than k prime factors, all of which are $\leq My$. Hence the variables M and

k are also key parameters. If M is large then we retain more a_j's, and thus the chance of obtaining more pseudosmooths. However this also slows down the sieving, as one must test for divisibility by more primes. Once we have obtained the b_j by dividing out of the a_j all of their prime factors $\leq y$ we must retain all of those $b_j \leq (My)^k$. If we allow k to be large then this means that only a very small proportion of the b_j that are retained at this stage will turn out to be My-smooth (as desired), so we will have wasted a lot of machine cycles on useless a_j. A recent successful idea to overcome this problem is to keep only those a_j where at most one of the prime factors is $> M'y$ for some M' that is not much bigger than 1 — this means that little time is wasted on a_j with two "large" prime factors. The resulting choice of parameters varies from program to program, depending on how reports are handled etc. etc., and on the prejudices and prior experiences of the programmers. Again, it is hard to make this an exact science.

Arjen Lenstra told us, in a private communication, that in his experience of practical implementations of the quadratic sieve, once n and y are large enough, the single large prime variation speeds things up by a factor between 2 and 2.5, and the double large prime variation by another factor between 2 and 2.5 (see, e.g. [13]), for a total speed-up of a factor between 4 and 6. An experiment with the triple large prime variation [12] seemed to speed things up by another factor of around 1.7.

Factorers had believed (see, e.g. [13] and [3]) that, in the quadratic sieve, there would be little profit in trying the triple large prime variation, postulating that the speed-up due to the extra pseudosmooths obtained had little chance of compensating for the slowdown due to the large number of superfluous a_j s considered, that is those for which $b_j \leq (My)^3$ but turned out to not be My-smooth. On the other hand, in practical implementations of the number field sieve, one obtains a_j with more than two large prime factors relatively cheaply and, after a slow start, the number of pseudosmooths obtained suddenly increases very rapidly (see [6]). This is what led the authors of [12] to their recent surprising and successful experiment with the triple large prime variation for the quadratic sieve (see Willemien Ekkelkamp's contribution to these proceedings [7] for further discussion of multiple prime variation speed-ups to the number field sieve).

This practical data is quite different from what we have obtained, theoretically, at the end of the previous section. One reason for this is that, in our analysis of Pomerance's problem, the variations in M and k simply affect the number of a_j being considered, whereas here these affect not only the number of a_j being considered, but also several other important quantities. For instance, the amount of sieving that needs to be done, and also the amount of data that needs to be "swapped" (typically one saves the a_j with several large prime factors to the disk, or somewhere else suitable for a lot of data). It would certainly be interesting to run experiments on Pomerance's problem directly to see whether our predicted speed comparisons are close to correct for numbers within computational range.

Acknowledgements

We thank François Bergeron for pointing out the connection with Husimi graphs, for providing mathematical insight and for citing references such as [11]. Thanks to Carl Pomerance for useful remarks, which helped us develop our analysis of the random squares algorithm in section 7, and to Arjen Lenstra for discussing with us a more practical perspective which helped us to formulate many of the remarks given in section 8.3.

References

1. Abramowitz, M., Stegun, I.: Handbook of mathematical functions. Dover Publications, New York (1965)
2. Buhler, J., Lenstra Jr., H.W., Pomerance, C.: Factoring integers with the number field sieve. Lecture Notes in Math., vol. 1554. Springer, Berlin (1993)
3. Crandall, R., Pomerance, C.: Prime numbers; A computational perspective. Springer, New York (2005)
4. Croot, E., Granville, A., Pemantle, R., Tetali, P.: Sharp transitions in making squares (to appear)
5. Dixon, J.D.: Asymptotically fast factorization of integers. Math. Comp. 36, 255–260 (1981)
6. Dodson, B., Lenstra, A.K.: NFS with four large primes: an explosive experiment. In: Coppersmith, D. (ed.) CRYPTO 1995. LNCS, vol. 963, pp. 372–385. Springer, Heidelberg (1995)
7. Ekkelkamp, W.: Predicting the sieving effort for the number field sieve. In: van der Poorten, A.J., Stein, A. (eds.) ANTS 2008. LNCS, vol. 5011, pp. 167–179. Springer, Heidelberg (2008)
8. Friedgut, E.: Sharp thresholds of graph properties, and the k-SAT problem. J. Amer. Math. Soc. 12, 1017–1054 (1999)
9. Granville, A., Soundararajan, K.: Large Character Sums. J. Amer. Math. Soc. 14, 365–397 (2001)
10. Hildebrand, A., Tenenbaum, G.: On integers free of large prime factors. Trans. Amer. Math. Soc. 296, 265–290 (1986)
11. Leroux, P.: Enumerative problems inspired by Mayer's theory of cluster integrals. Electronic Journal of Combinatorics. Paper R32, May 14 (2004)
12. Leyland, P., Lenstra, A., Dodson, B., Muffett, A., Wagstaff, S.: MPQS with three large primes. In: Fieker, C., Kohel, D.R. (eds.) ANTS 2002. LNCS, vol. 2369, pp. 446–460. Springer, Heidelberg (2002)
13. Lenstra, A.K., Manasse, M.S.: Factoring with two large primes. Math. Comp. 63, 785–798 (1994)
14. Pomerance, C.: The quadratic sieve factoring algorithm. Advances in Cryptology, Paris, pp. 169–182 (1984)
15. Pomerance, C.: The number field sieve. In: Gautschi, W. (ed.) Mathematics of Computation 1943–1993: a half century of computational mathematics. Proc. Symp. Appl. Math. 48, pp. 465–480. Amer. Math. Soc., Providence (1994)
16. Pomerance, C.: The role of smooth numbers in number theoretic algorithms. In: Proc. International Congress of Mathematicians (Zurich, 1994), Birhäuser, vol. 1, pp. 411–422 (1995)

17. Pomerance, C.: Multiplicative independence for random integers. In: Berndt, B.C., Diamond, H.G., Hildebrand, A.J. (eds.) Analytic Number Theory: Proc. Conf. in Honor of Heini Halberstam, Birhäuser, vol. 2, pp. 703–711 (1996)
18. Pomerance, C.: Smooth numbers and the quadratic sieve. In: Buhler, J.P., Stevenhagen, P. (eds.) Algorithmic Number Theory: Lattices, Number Fields, Curves and Cryptography, Mathematical Sciences Research Institute Publications 44 (to appear, 2007)
19. Silverman, R.: The multiple polynomial quadratic sieve. Math. Comp. 48, 329–339 (1987)
20. Tenenbaum, G.: Introduction to the analytic and probabilistic theory of numbers. Cambridge Univ. Press, Cambridge (1995)

A New Look at an Old Equation

R.E. Sawilla[1], A.K. Silvester[2], and H.C. Williams[2,*]

[1] Defense Research & Development Canada
3701 Carling Ave., Ottawa ON, K1A 0Z4, Canada
`reg.sawilla@drdc-rddc.gc.ca`
[2] Dept. of Mathematics and Statistics, University of Calgary,
2500 University Drive NW, Calgary AB, T2N 1N4, Canada
{aksilves,williams}@math.ucalgary.ca

Abstract. The general binary quadratic Diophantine equation

$$ax^2 + bxy + cy^2 + dx + ey + f = 0$$

was first solved by Lagrange over 200 years ago. Since that time little improvement has been made to Lagrange's technique. In this paper we show how to reduce this problem to that of determining whether or not an ideal of a certain quadratic order is principal and if so exhibiting a generator of that ideal. In the difficult case of the discriminant Δ of this order being positive, we develop a Las Vegas algorithm for solving the principal ideal problem that executes in expected time bounded by $O(\Delta^{1/6+\epsilon})$, whereas the complexity of Lagrange's (unconditional) technique for solving this problem is $O(\Delta^{1/2+\epsilon})$.

1 Introduction

We will be concerned with the Diophantine equation

$$ax^2 + bxy + cy^2 + dx + ey + f = 0, \tag{1.1}$$

where it is required to find integral values of x and y, given $a, b, c, d, e, f \in \mathbb{Z}$. A method for solving this equation was given over 200 years ago by Lagrange [17], and this method has not been improved significantly since that time. The reason for this is that Lagrange's method works perfectly well as long as the coefficients in (1.1) do not get very large. However, if we put $H = \max\{|a|, |b|, |c|, |d|, |e|, |f|\}$, Kornhauser [16] has shown that there is an infinite collection of equations of the form (1.1) having integer solutions, but none with $\max\{|x|, |y|\} \leq 2^{H/5}$. Thus it is possible for solutions of (1.1) to be very large, even when H is only moderately large. For such cases Lagrange's method will likely be far too slow to produce the solutions of (1.1). The purpose of this paper is to develop a faster method for dealing with this equation; in the process of doing this it will be necessary to investigate techniques for performing arithmetic efficiently in real quadratic fields.

* Research supported by NSERC of Canada and iCORE of Alberta.

A.J. van der Poorten and A. Stein (Eds.): ANTS-VIII 2008, LNCS 5011, pp. 37–59, 2008.

If we put $D = b^2 - 4ac$, $E = bd - 2ae$, $F = d^2 - 4af$, Lagrange realized that (1.1) could be written as

$$DY^2 = (Dy + E)^2 + DF - E^2, \tag{1.2}$$

where $Y = 2ax + by + d$. Clearly, if we put $N = E^2 - DF = -4a(ebd + 4acf - ae^2 - fb^2 - cd^2)$, then (1.2) can be written as

$$X^2 - DY^2 = N, \tag{1.3}$$

where $X = Dy + E$. Thus, if we have any solution X, Y of (1.3) such that there are integers x, y for which

$$X = Dy + E \quad \text{and} \quad Y = 2ax + by + d, \tag{1.4}$$

we get a solution x, y of (1.1).

Before proceeding any further, we will examine several cases of (1.3). If $D < 0$, then (1.3) can only have a finite number of solutions, and these can be determined by making use of the algorithm of Cornacchia (see, for example, Nitaj [21]). If $D = 0$ or $D > 0$ is a perfect integral square, the problem of solving (1.3) reduces to that of factoring N, and once again there can only be a finite number of solutions of (1.3). Also, if $D > 0$ and $N = 0$, (1.3) can only have a solution if D is a perfect integral square. In this case we get an infinitude of solutions of (1.3), but they are very easily characterized.

There remains, then, the case of $N \neq 0$, $D > 0$ and D not a perfect integral square. In this case (1.1) has an infinitude of solutions, if it has at least one. If (1.3) has a solution X, Y and $G = \gcd(X, Y)$, we must have $G^2 \mid N$ and (1.3) reduces to

$$X'^2 - DY'^2 = N', \tag{1.5}$$

where $X' = X/G$, $Y' = Y/G$, $N' = N/G^2$. Thus, in order to solve (1.3) we can find all the possible square divisors G^2 of N and solve (1.5) for each value of $N' = N/G^2$. We may, therefore, with no loss of generality assume that $\gcd(X, Y) = 1$ in (1.3). Such solutions are called *primitive*. Now suppose that X, Y is any primitive solution of (1.3). Let t, u denote the fundamental solution of the Pell equation

$$T^2 - DU^2 = 1.$$

If

$$X_n + Y_n \sqrt{D} = (X + Y\sqrt{D})(t + u\sqrt{D})^n \quad (n \in \mathbb{Z}), \tag{1.6}$$

we see that X_n, Y_n is also a primitive solution of (1.3). Indeed, as Lagrange was well aware, there exists a finite set $\mathcal{S}$ made up of ordered pairs (X, Y) of solutions X, Y of (1.3) such that if X', Y' is any solution of (1.3), then $X' = X_n$, $Y' = Y_n$ for some $n \in \mathbb{Z}$ and some $(X, Y) \in \mathcal{S}$. Thus, after having found $\mathcal{S}$, the problem reduces to that of identifying for each $(X, Y) \in \mathcal{S}$ those values of n for which

$$\begin{cases} X_n \equiv E \pmod{D}, \\ Y_n \equiv b(X_n - E)/D + d \pmod{2a}. \end{cases} \tag{1.7}$$

We will now rewrite (1.7) as

$$\begin{cases} X_n \equiv E \pmod{D}, \\ DY_n \equiv bX_n - bE + Dd \pmod{2aD}. \end{cases} \tag{1.8}$$

Lagrange noted that as there are only a finite number of possible values of

$$(t + u\sqrt{D})^n \pmod{2aD},$$

there must be a least positive integer r for which

$$(t + u\sqrt{D})^r \equiv 1 \pmod{2aD}.$$

Thus, (1.6) yields values of X_n and Y_n satisfying (1.8) if and only if it does so when n is replaced by $n + r$. It follows that in order to test all the solutions of (1.3) produced by (1.6) to see if they satisfy (1.8), it suffices to examine only those for which $0 \leq n \leq r - 1$.

Lagrange's method compels us to test up to r values of n to determine those congruence classes of $n \pmod{r}$ for which we produce solutions of (1.1) from (1.6). Unfortunately, this could be a very inefficient process when r is large, which is frequently the case when aD is.

2   Another Approach

If we define

$$T_n + U_n\sqrt{D} = (t + u\sqrt{D})^n \quad (n \in \mathbb{Z}), \tag{2.1}$$

we see from (1.6) that

$$X_n = XT_n + DYU_n, \quad Y_n = YT_n + XU_n.$$

Since we require that $X_n \equiv E \pmod{D}$, we must have $T_nX \equiv E \pmod{D}$. By (2.1) it is clear that $T_n \equiv t^n \pmod{D}$ and since $t^2 \equiv 1 \pmod{D}$, we get

$$T_n \equiv t^\epsilon \pmod{D}$$

when $n \equiv \epsilon \pmod{2}$, $\epsilon \in \{0, 1\}$. Thus, if neither $X \equiv E \pmod{D}$ nor $tX \equiv E \pmod{D}$ holds, then (1.6) will yield no solutions of (1.1).

Suppose that $T_nX \equiv E \pmod{D}$. By (1.8) we also require that

$$dD - bE \equiv (DY - bX)T_n + (DX - bDY)U_n \pmod{2aD}. \tag{2.2}$$

From (1.4) we can deduce that

$$\begin{aligned} X - bY &= Dy + E - 2abx - b^2y - bd \\ &= -2a(cy + e + bx). \end{aligned}$$

Thus, another necessary condition for (1.6) to produce solutions to (1.1) is that

$$2a \mid X - bY. \tag{2.3}$$

Since $b^2 \equiv D \pmod{2a}$, this means that $2a \mid DY - bX$. We next observe that

$$dD - bE = 2a(eb - 2dc).$$

Hence, we can now put (2.2) in the form

$$\frac{dD - bE}{2a} \equiv \left(\frac{DY - bX}{2a}\right) T_n + \left(\frac{DX - bDY}{2a}\right) U_n \pmod{D}.$$

By (2.1) we have $U_n \equiv nut^{n-1} \pmod{D}$; hence, (2.2) can be rewritten as

$$\frac{dD - bE}{2a} \equiv \left(\frac{DY - bX}{2a}\right) t^n + nut^{n-1}\left(\frac{DX - bDY}{2a}\right) \pmod{D}. \tag{2.4}$$

Since $t^2 \equiv 1 \pmod{D}$, this becomes a linear congruence in the unknown n.

However, by (2.3) we have $D \mid (DX - bDY)/2a$. Thus (2.4) can hold for all even n only if

$$D \mid \frac{dD - bE - DY + bX}{2a} \tag{2.5}$$

and for all odd n only if

$$D \mid \frac{dD - bE - DYt + bXt}{2a}. \tag{2.6}$$

Thus, it is no longer necessary to search for all possible values of n up to r. We need only check to see that (2.3) holds. If so and (2.5) holds, then (1.6) produces solutions of (1.1) for any even n and if (2.6) holds then (1.6) produces solutions of (1.1) for any odd n. If none of these conditions holds, then (1.6) produces no solutions of (1.1). This approach is another version of an idea of Legendre as modified by Dujardin (see Dickson [6, p. 416]).

3   Solutions of $X^2 - DY^2 = N$

We next turn our attention to the problem of finding all the primitive pairs (X, Y) for which (1.6) will yield all of the solutions of

$$X^2 - DY^2 = N.$$

As we have already mentioned there are only a finite number of such pairs. There may be none at all. We first notice that if $S^2 \mid \gcd(D, N)$, then $S \mid X$; thus, if we put $X = X/S$, $D' = D/S^2$, $N' = N/S^2$, then (1.3) becomes

$$X'^2 - D'Y^2 = N'.$$

We may therefore assume with no loss of generality that $\gcd(D, N)$ is squarefree.

In order to proceed further we will make use of some results from the theory of real quadratic number fields and some associated algorithms. Much of this material can be found in Williams and Wunderlich [28], Jacobson et al. [12,13,11] and de Haan et al. [8]. Let $\mathcal{O}$ be the order $\mathbb{Z}[\sqrt{D}]$ and $\mathbb{K}$ be the quadratic number field $\mathbb{Q}(\sqrt{D})$. The discriminant of $\mathcal{O}$ is $\Delta = 4D$ and we put $\omega = \sqrt{D}$. Suppose X, Y is any primitive solution of (1.3) and consider the principal $\mathcal{O}$-ideal $\mathfrak{a} = (X + Y\sqrt{D})$ generated by $X + Y\sqrt{D}$. If by $[\alpha, \beta]$ we denote the $\mathbb{Z}$-module $\{x\alpha + y\beta : x, y \in \mathbb{Z}\}$, where $\alpha, \beta \in \mathcal{O}$, then because $\gcd(X, Y) = 1$ we may write

$$\mathfrak{a} = [a, b + \omega], \tag{3.1}$$

where $a, b \in \mathbb{Z}$. (These integers a, b should not be confused with those in (1.1).) It is well known that $\mathfrak{a}$ can be an ideal of $\mathcal{O}$ if and only if $a \mid N(b + \omega)$. Also, we may assume that $a > 0$ and $0 \leq b < a$. Now $a = N(\mathfrak{a}) = |N(X + Y\sqrt{D})| = |N|$ and since $a \mid b^2 - D$, we get $b \equiv XY^{-1} \pmod{a}$. (We observe that since $\gcd(X, Y) = 1$, we must have $\gcd(Y, N) = 1$; hence, Y^{-1} exists modulo a.) It follows that even if we do not know a primitive solution of (1.3) a priori, we can find candidates for b by solving the simple quadratic congruence

$$Z^2 \equiv D \pmod{N}. \tag{3.2}$$

One of the solutions Z of (3.2) with $0 \leq Z < |N|$ must be b. For some such solution Z of (3.2), then, we can put $a = |N|$, $b = Z$ in (3.1). Also, since $\mathfrak{a}$ is principal, it must be invertible, which means that Z must be such that $\gcd(N, 2Z, (D - Z^2)/N) = 1$. If this is not the case, we must exclude the corresponding ideal $\mathfrak{a}$ from consideration.

Let ϵ_Δ (> 1) denote the fundamental unit, R $(= \log \epsilon_\Delta)$ the regulator and h the ideal class number of $\mathcal{O}$. If γ and μ are two generators of a principal $\mathcal{O}$-ideal $\mathfrak{a}$, then

$$\mu = \pm \epsilon_\Delta^n \gamma \quad (n \in \mathbb{Z}), \tag{3.3}$$

and

$$N(\mu) = N(\epsilon_\Delta)^n N(\gamma). \tag{3.4}$$

Having selected our candidate for $\mathfrak{a}$, we may now perform the following steps.

1. Determine whether or not $\mathfrak{a}$ is principal. If $\mathfrak{a}$ is not principal, then there can be no solutions of (1.3) corresponding to our selected value of Z.
2. If $\mathfrak{a}$ is principal, solve the discrete logarithm problem (DLP) for $\mathfrak{a}$ in $\mathcal{O}$ to produce a generator γ of $\mathfrak{a}$.
3. If $N(\gamma) = N$, we have a solution X, Y of (1.3) when $\gamma = X + Y\sqrt{D}$. If $N(\gamma) = -N$ and $N(\epsilon_\Delta) = -1$, we have a solution X, Y of (1.3) when $X + Y\sqrt{D} = \gamma \epsilon_\Delta$. If $N(\gamma) = -N$ and $N(\epsilon_\Delta) = 1$, we see from (3.4) that there can be no solution of (1.3) corresponding to our selected value of Z.

We see, then, that for each possible distinct solution Z_i of (3.2) we will either find a distinct value for λ_i such that $N(\lambda_i) = N$ or no such λ_i can exist. If we put

$$\eta = \begin{cases} \epsilon_\Delta & \text{when } N(\epsilon_\Delta) = 1, \\ \epsilon_\Delta^2 & \text{when } N(\epsilon_\Delta) = -1, \end{cases}$$

then $\eta = t + u\sqrt{D}$ and by (3.3)

$$\pm \lambda_i \eta^n \quad (n \in \mathbb{Z}) \qquad (3.5)$$

represents all the solutions of (1.3) that correspond to Z_i.

Let

$$|N| = 2^\alpha \prod_{i=1}^{k} p_i^{\alpha_i} \quad (\alpha \geq 0; \alpha_i \geq 0, i = 1, 2, \ldots, k),$$

where p_i $(i = 1, 2, \ldots, k)$ are distinct odd primes. Suppose that p is any of the primes that divide N and that $p^2 \mid N$ and $p \mid D$. If (3.2) has a solution, then $p \mid Z$ which means that $p^2 \mid D$, a case we have excluded. Thus, if $p^2 \mid N$, then $p \nmid D$. Denote by $\nu(D, p^\alpha)$ the number of distinct (modulo p^α) solutions of the congruence

$$Z^2 \equiv D \pmod{p^\alpha}.$$

It is well known that if $p = 2$, then

$$\nu(D, 2^\alpha) = \begin{cases} 1 \text{ when } \alpha = 1, \\ 2 \text{ when } \alpha = 2 \text{ and } D \equiv 1 \pmod 4, \\ 4 \text{ when } \alpha \geq 3 \text{ and } D \equiv 1 \pmod 8, \\ 0 \text{ otherwise.} \end{cases}$$

Also, if p is odd, then

$$\nu(D, p^\alpha) = 1 + \left(\frac{D}{p}\right),$$

where (D/p) is the Legendre symbol. Thus, if $\nu(D, N)$ denotes the number of distinct solutions modulo $|N|$ of (3.2), we see by the Chinese remainder theorem that

$$\nu(D, N) = \nu(D, 2^\alpha) \prod_{i=1}^{k} \nu(D, p^{\alpha_i}) = \nu(D, 2^\alpha) \prod_{i=1}^{k} \left(1 + \left(\frac{D}{p_i}\right)\right)$$

$$\leq 2^{\omega(N)+1},$$

where $\omega(N)$ is the number of distinct prime divisors of N. Notice that if $(D/p_i) = -1$ for any p_i, then $\nu(D, N) = 0$. The behaviour of $\omega(N)$ is quite irregular, but its average value (see, for example, Cojocaru and Murty [4, pp. 32–35]) is known to be $\log \log |N|$; hence, we expect that the usual value of $\nu(D, N)$ is bounded by a function of order $\log |N|$. This means that in most cases it is only necessary to solve for all the solutions of (1.3) by using only relatively few values of Z_i. The resulting number of classes of solutions as represented by (3.5) will therefore not likely be very large. For another approach to the problem of identifying the classes of solutions of (1.3), the reader is referred to Nagell [20] and Stolt [27].

4   The Principal Ideal Problem

As we have seen, once we can solve (3.2), the problem of solving (1.1) is reduced
to that of determining whether the ideal $\mathfrak{a}$ is principal and then finding a generator λ for $\mathfrak{a}$ such that $N(\lambda) = N$. For large values of D this problem is best
approached by making use of the index calculus method described by Jacobson
[9,10]. Under the assumption of a certain Riemann hypothesis, the complete version of this method, which requires the class group structure of $\mathcal{O}$, has expected
running time bounded by $L_\Delta[1/2, \sqrt{2} + o(1)]$, where

$$L_x[a, b] = \exp\left(b(\log x)^a (\log \log x)^{1-a}\right).$$

(See §11.5 of Buchmann and Vollmer [3].)

In order to contrast this attack on (1.3) with that of Lagrange, we must
introduce the concept of ideal reduction. A primitive ideal $\mathfrak{a}$ (an ideal with no
rational integer divisors except ± 1) is said to be *reduced* if $\mathfrak{a}$ does not contain
any nonzero α such that both $|\alpha| < N(\mathfrak{a})$ and $|\overline{\alpha}| < N(\mathfrak{a})$ hold. By referring to
Theorem 3.5 of [28], it is easy to derive a simple criterion for determining when
the primitive $\mathcal{O}$-ideal $\mathfrak{a} = [a, b + \omega]$ is reduced.

Theorem 4.1. *Let* $\mathfrak{a} = [a, b + \omega]$ *be any primitive $\mathcal{O}$-ideal, where $\Delta > 0$. Put*
$\beta = k|a| + b + \omega$, *where* $k = \lfloor -(b + \overline{\omega})/|a| \rfloor$; *then $\mathfrak{a}$ is reduced if and only if*
$\beta > |a|$.

Furthermore, if $\mathfrak{a}$ is reduced we must have $N(\mathfrak{a}) < \sqrt{\Delta}$ and if $\mathfrak{a}$ is any primitive
ideal such that $N(\mathfrak{a}) < \sqrt{\Delta}/2$, then $\mathfrak{a}$ must be reduced. We next point out that
given any $\mathcal{O}$-ideal $\mathfrak{a}$, we can always find a reduced $\mathcal{O}$-ideal $\mathfrak{b}$ such that $\mathfrak{b} \sim \mathfrak{a}$.
There are several algorithms (see [12]) for finding $\theta \in \mathbb{K}$ and $\mathfrak{b}$ such that $\mathfrak{a} = \theta\mathfrak{b}$.
Also, if for $\alpha \in \mathbb{K}$ we define $H(\alpha) = \max\{|\alpha|, |\overline{\alpha}|\}$, then these algorithms produce
θ such that $H(\theta) = O(N(\mathfrak{a}))$.

If $\mathfrak{a} = [a, b + \omega]$ is any $\mathcal{O}$-ideal, we define the $\mathcal{O}$-ideal $\rho(\mathfrak{a})$ to be $[a', b' + \omega]$,
where $q = \lfloor (b + \omega)/|a| \rfloor$, $b' = q|a| - b$, and $a' = -N(b' + \omega)/|a|$. If $\mathfrak{a}$ is a reduced
ideal, it is easy to show that $a' > 0$ and $\rho(\mathfrak{a})$ is also a reduced ideal. If $a > 0$ and
$\mathfrak{a}$ is reduced, then ρ is the same operation as that mentioned in [13, p. 214] and
ρ can be inverted. Note that $\rho(\mathfrak{a}) = \gamma\mathfrak{a}$, where $\gamma = (b' + \omega)/a$.

Since the norms of all reduced ideals are bounded above by $\sqrt{\Delta}$, there can
only be a finite number of them in $\mathcal{O}$. Indeed, if we begin with a reduced ideal
$\mathfrak{a}_1$ and compute $\mathfrak{a}_2 = \rho(\mathfrak{a}_1)$, $\mathfrak{a}_3 = \rho(\mathfrak{a}_2)$, $\ldots$, $\mathfrak{a}_{i+1} = \rho(\mathfrak{a}_i) = \rho^i(\mathfrak{a}_1)$, it turns out
that there is some minimal $l > 0$ such that $\mathfrak{a}_{l+1} = \mathfrak{a}_1$. In addition, if $\mathfrak{b}$ is any
reduced ideal such that $\mathfrak{b} \sim \mathfrak{a}_1$, then $\mathfrak{b} \in \mathcal{C} = \{\mathfrak{a}_1, \mathfrak{a}_2, \ldots, \mathfrak{a}_l\}$. The ordered set $\mathcal{C}$
is called the *cycle of reduced ideals* equivalent to $\mathfrak{a}_1$.

Put in a more modern setting, Lagrange's method for solving (1.3) essentially
takes each candidate ideal $\mathfrak{a}$ and finds a reduced ideal $\mathfrak{b} \sim \mathfrak{a}$ with $\mathfrak{a} = \theta\mathfrak{b}$. If
$\mathfrak{a}$ is principal, then $\mathfrak{b}$ must be principal, say $\mathfrak{b} = (\mu)$, and $\lambda = \theta\mu$. Since $\mathfrak{b}$
is reduced, $\mathfrak{b}$ must be in the cycle $\mathcal{C}$ of reduced ideals equivalent to $\mathcal{O} = (1)$;
thus, in order to determine whether $\mathfrak{a}$ is principal, we must search for $\mathfrak{b}$ among
the l ideals in $\mathcal{C}$. If $\mathfrak{b} \in \mathcal{C}$, we get a value for λ, if $\mathfrak{b} \notin \mathcal{C}$, then there is no

solution of (1.3) corresponding to our selected ideal $\mathfrak{a}$. Difficulties in applying this technique occur when D is large; this is because l is often of order $\sqrt{D}$, which means that the creation of and search through $\mathcal{C}$ can be very time-consuming. Thus for large values of D it seems that the use of the index calculus method is the better technique for finding solutions to (1.3). There are, however, two significant problems that could arise on employing this algorithm.

1. Even the smallest possible values of X and Y satisfying (1.3) when D is large can be absolutely enormous. Indeed, it is often not even possible to write them down in standard decimal notation. However, in order to solve (1.1) we need to have the values of X and Y (and t) modulo $2aD$; thus, we need to have a method of finding X and Y that allows for this. Fortunately, this problem is easy to solve because the index calculus methods can furnish us with an approximation to $\log \lambda$ and this can be used to produce a compact representation (see Buchmann et al. [2] and [3, §11.5.3]) to express λ (= $X + Y\sqrt{D}$). From this, we can then find X and Y modulo $2aD$ by using the process described by Jacobson and Williams [15].

2. We mentioned that the subexponential complexity of the index calculus procedure is dependent on the truth of a generalized Riemann hypothesis (GRH). This does not really cause a problem if the process yields a generator for $\mathfrak{a}$, but it can be a real difficulty if it doesn't or declares $\mathfrak{b}$ to be non-principal. In this case, we cannot rigourously prove that $\mathfrak{a}$ is not principal because this is dependent on the truth of the GRH. As it might be required to prove that (1.1) has no solutions, this could be a substantial problem.

For the remainder of this paper we will concentrate our efforts on how to deal with problem 2 above. We will develop a Las Vegas algorithm for solving this problem which executes in time $O(h\Delta^\epsilon)$. Our inputs to this process, besides D, N and the candidate ideal $\mathfrak{a}$ are R'_Δ and h, where $R'_\Delta \in \mathbb{Q}$, $|R'_\Delta - \log_2 \epsilon_\Delta| < 1$. We can produce values for R'_Δ and h by using the index calculus techniques described by Jacobson [9] and Maurer [19] in time bounded above by $L_\Delta[1/2, \sqrt{2} + o(1)]$, but we don't have a proof that they are correct because of the assumption of the GRH. Previous to the development of these techniques, the best method available for computing R was the $O(\Delta^{1/5+\epsilon})$ Las Vegas algorithm of Lenstra [18] and the best methods for computing h were the conditional algorithms in [18] and Srinivasan [25]. The correctness of R'_Δ can be verified deterministically in time $O(R^{1/3+\epsilon})$ by using the method described in [8] and de Haan [7]. Once R'_Δ has been determined we can establish a possible value for h by invoking the extended Riemann hypothesis on $L(1, \chi)$, where $\chi(n) = (\Delta/n)$ (see [10, p. 33]). Indeed, it is even possible to use Booker's technique [1] to verify the value of h unconditionally in time $O(\Delta^{1/4+\epsilon})$, but we will not require this here. We can also produce a compact representation of η and from this determine $t \pmod{2aD}$ for use in dealing with problem 1.

As it is well known that $R = O(\Delta^{1/2+\epsilon})$, we see that the complexity of our process, like Lagrange's, is still exponential, but it is much faster because $l = O(R)$ and Lagrange's search executes in $O(l)$ operations. As both the verification process for R'_Δ and our principal ideal testing technique require the algorithm AX mentioned in [8] and described in [7, pp. 44–46], it is useful to discuss an improved version of this in the next two sections.

5   The Algorithms NUCOMP and WNEAR

One of the most important concepts that we will require in the development of our techniques is that of the infrastructure of ideal classes, discovered by Shanks [23]. As this is discussed at some length in [18], [28], [11], and [13], we will simply assume here that the basic ideas behind it are known to the reader. Making use of the infrastructure, however, requires that we compute distances, and as such quantities are logarithms of quadratic irrationals, they must be transcendental numbers. This means, of course, that we cannot compute them to full accuracy, but must instead be content with approximations to a fixed number of figures. When Δ is small, this is not likely to cause many difficulties, but when Δ becomes large, we have no real handle on how much round-off or truncation error might accumulate. Numerical analysts pay a great deal of attention to this problem, but frequently computational number theorists ignore it, hoping or believing that their techniques are sufficiently robust that serious deviations of their results from the truth will not occur. It must be admitted that this is usually what happens, but if a computational algorithm is to produce a numerical answer that is to be formally accepted as correct, it must contain within it the same aspects of rigour that one would expect within any mathematical proof. This means that we must provide provable bounds on the possible errors in our results.

In the procedures that we will describe below, we will deal with this problem of error accumulation by making use of what we call (f,p) representations of ideals.

Definition 5.1. *Let* $p \in \mathbb{Z}^+$, $f \in \mathbb{R}$ *with* $1 \leq f < 2^p$ *and let* $\mathfrak{a}$ *be an* $\mathcal{O}$-*ideal. An* (f,p) *representation of* $\mathfrak{a}$ *is a triple* $(\mathfrak{b}, d, k)$ *where*

1. $\mathfrak{b}$ *is an* $\mathcal{O}$-*ideal equivalent to* $\mathfrak{a}$, $d \in \mathbb{N}$ *with* $2^p < d \leq 2^{p+1}$, $k \in \mathbb{Z}$;
2. *there exists a* $\theta \in \mathbb{K}$ *with* $\mathfrak{b} = \theta\mathfrak{a}$ *and*

$$\left| \frac{2^{p-k}\theta}{d} - 1 \right| < \frac{f}{2^p}.$$

Note that $(\mathfrak{a}, 2^{p+1}, -1)$ or $(\mathfrak{a}, 2^p + 1, 0)$ is an (f,p) representation of $\mathfrak{a}$ $(\theta = 1)$. Note also that $k \approx \log_2 \theta$.

An (f,p) representation of $\mathfrak{a}$ is said to be *reduced* if $\mathfrak{b}$ is a reduced $\mathcal{O}$-ideal. It is said to be *w-near* for some $w \in \mathbb{Z}^{\geq 0}$ if it is reduced and two additional conditions hold:

1. $k < w$,
2. If $\mathfrak{b}_1 = \mathfrak{b}$ and $\mathfrak{b}_2 = \rho(\mathfrak{b}_1) = \psi\mathfrak{b}$, then there exist integers d', k' with $k' \geq w$, $2^p < d' \leq 2^{p+1}$ such that

$$\left| \frac{2^{p-k'}\theta\psi}{d'} - 1 \right| < \frac{f}{2^p}.$$

If $(\mathfrak{b}, d, k)$ is a w-near (f, p) representation of some $\mathcal{O}$-ideal $\mathfrak{a}$ and f is not too large, then the parameters θ and k will not be far from 2^w and w, respectively. We can be more precise about this in the following lemma, which can be proved by using the same technique as that used in the proof of Corollary 4.1 of [13].

Lemma 5.1. *Let $(\mathfrak{b}, d, k)$ be a w-near (f, p) representation of some $\mathcal{O}$-ideal $\mathfrak{a}$ with $p > 4$ and $f < 2^{p-4}$. If θ and ψ have the meaning assigned to them above, then*

$$\frac{15N(\mathfrak{b})}{16\sqrt{\Delta}} < \frac{15}{16\psi} < \frac{\theta}{2^w} < \frac{17}{16}$$

and

$$0 > k - w > -\log_2\left(\frac{34\psi}{15}\right) > -\log_2\left(\frac{34\sqrt{\Delta}}{15N(\mathfrak{b})}\right).$$

From this result it easily follows that if $(\mathfrak{b}, d, k)$ is a w-near (f, p) representation with $f < 2^{p-4}$, then $0 < w - k = O(\log \Delta)$.

Suppose we are given p and f with $f < 2^{p-4}$. Let $\mathfrak{a}$ ($= \mathfrak{a}_1$) be any reduced $\mathcal{O}$-ideal. By our results in §4 and [28], we can use the simple continued fraction expansion of $(P + \sqrt{D})/Q$, where $\mathfrak{a} = [Q, P + \sqrt{D}]$, to produce a sequence of reduced ideals

$$\mathfrak{a}_1, \mathfrak{a}_2, \mathfrak{a}_3, \ldots, \mathfrak{a}_j, \ldots \tag{5.1}$$

with $\mathfrak{a}_j = \theta_j \mathfrak{a}_1$ ($j = 1, 2, \ldots$). We may also assume that for each $\mathfrak{a}_j$ we have $d_j, k_j \in \mathbb{Z}$ such that $(\mathfrak{a}_j, d_j, k_j)$ is a reduced (f, p) representation of $\mathfrak{a}$. Since

$$\left|\frac{2^p \theta_j}{2^{k_j} d_j} - 1\right| < \frac{1}{16}$$

and $2^p < d_j \leq 2^{p+1}$, we get

$$\frac{15}{16} 2^{k_j} < \theta_j < \frac{17}{8} 2^{k_j}.$$

By Theorem 2.1 of [8], we have $\theta_{j+2} > 2\theta_j$, $\theta_{j+i} > 3\theta_j$ ($i \geq 3$). Thus, if $i \geq 3$, then

$$2^{k_{j+i}} > \frac{8}{17}\theta_{j+i} > \frac{8 \cdot 3}{17}\theta_j > 2^{k_j}.$$

Hence $k_{j+i} > k_j$ when $j \geq 3$. If $j = 2$, then

$$2^{k_{j+2}} > \frac{15}{17} 2^{k_j} > 2^{k_j - 1};$$

consequently, $k_{j+2} \geq k_j$.

Now suppose that $(\mathfrak{a}_j, d_j, k_j)$ and $(\mathfrak{a}_h, d_h, k_h)$ are both w-near (f, p) representations of $\mathfrak{a}$. Since $\mathfrak{a}_{j+1} = \rho(\mathfrak{a}_j)$ and $\mathfrak{a}_{h+1} = \rho(\mathfrak{a}_h)$, we must have $k_j < w$, $k_{j+1} \geq w$, $k_h < w$, $k_{h+1} \geq w$. We will assume with no loss of generality that $h > j$. Clearly, we cannot have $h = j + 1$. If $h = j + i$, where $i \geq 3$, then

$$k_h = k_{j+1+i-1} \geq k_{j+1} \geq w,$$

a contradiction. Thus, if we have distinct $\mathcal{O}$-ideals $\mathfrak{a}_j$ and $\mathfrak{a}_h$ such that both $(\mathfrak{a}_j, d_j, k_j)$ and $(\mathfrak{a}_h, d_h, k_h)$ are w-near (f, p) representations of $\mathfrak{a}$, then $|h - j| = 2$. It follows that there can be at most two distinct $\mathcal{O}$-ideals which can occur in any w-near (f, p) representation of $\mathfrak{a}$. We will use the notation $\mathfrak{a}[w]$ to denote any one of these ideals, if there are two; certainly there must be at least one such ideal. That $\mathfrak{a}[w]$ needn't be unique will not be a problem in our applications of this concept.[1]

An algorithm for computing $\mathfrak{a}[x]$ when $\mathfrak{a} = \mathcal{O} = (1)$ was given as Algorithm 3.17 of [7]. In what follows we will provide an improved version of this algorithm. An essential ingredient in this investigation is the NUCOMP algorithm of Shanks [24]. Given two $\mathcal{O}$-ideals $\mathfrak{a}'$ and $\mathfrak{a}''$, Shanks discovered that there is a more efficient technique for finding a reduced ideal equivalent to $\mathfrak{a}'\mathfrak{a}''$ than first multiplying $\mathfrak{a}'$ by $\mathfrak{a}''$ and then using a reduction algorithm on their product $\mathfrak{a}$. He was guided in searching for such an algorithm by his need to keep the numbers involved in the calculations as small as possible. Since Q could be as large as about the size of D, and he wanted to keep all the values computed by his algorithm to be of size roughly $\sqrt{D}$, the technique of first multiplying $\mathfrak{a}'$ and $\mathfrak{a}''$ and then carrying out the reduction phase was not acceptable. Instead, he developed a new technique which he called NUCOMP. We will not discuss Shanks' version of this algorithm or its later improvements by Atkin, van der Poorten and Jacobson [22], [14] here. Instead we will consider the version of NUCOMP given in [13]. It is important to bear in mind that the operation of finding a reduced ideal equivalent to the product of two given ideals is of fundamental significance in performing arithmetic in $\mathcal{O}$. Thus, any improvement in this procedure is most desirable.

We begin by discussing some simple results from the theory of continued fractions. Let $q_0, q_1, q_2, \ldots, q_i, \ldots$ be any given sequence of integers (*partial quotients*). Let ϕ $(= \phi_0)$ be any given real number. If we define

$$\phi_{j+1} = \frac{1}{\phi_j - q_j} \quad (j = 0, 1, 2, \ldots, i),$$

then we can express ϕ_0 as the *continued fraction*

$$\phi_0 = q_0 + \cfrac{1}{q_1 + \cfrac{1}{q_2 + \cfrac{1}{\ddots \quad q_{i-1} + \cfrac{1}{q_i + \cfrac{1}{\phi_{i+1}.}}}}}$$

We denote this by

$$\phi_0 = \langle q_0, q_1, q_2, \ldots, q_i, \phi_{i+1} \rangle,$$

---

[1] The use of the notation $\mathfrak{a}(x)$ (instead of $\mathfrak{a}[x]$) was introduced in [8], but we have adopted the notation $\mathfrak{a}[x]$ here instead of the $\mathfrak{a}(x)$ used there in order to avoid functional notation which would imply a unique $\mathfrak{a}(x)$.

where ϕ_{i+1} is called a *complete quotient*. In the special case that $q_1, q_2, \ldots, q_i \geq 1$ and $\phi_{i+1} > 1$, we say that the continued fraction is *simple* (SCF) and denote this by

$$\phi_0 = [q_0, q_1, q_2, \ldots, q_i, \phi_{i+1}].$$

When ϕ_0 is a rational number K/L, where $K, L \in \mathbb{Z}$ and $L > 0$, we can produce the SCF expansion of ϕ_0 by simply employing the Euclidean algorithm. We put $R_{-2} = K$, $R_{-1} = L$ and define the sequence of remainders $\{R_i\}$ recursively by

$$R_{j-2} = q_j R_{j-1} + R_j \quad (0 < R_j < R_{j-1}; \ j = 0, 1, \ldots, n-1).$$

We must ultimately find some n such that $R_n = 0$ and then

$$K/L = [q_0, q_1, q_2, \ldots, q_n].$$

If we define the sequence $\{C_i\}$ by $C_{-2} = 0$, $C_{-1} = -1$ and

$$C_j = C_{j-2} - q_j C_{j-1},$$

it is an exercise in mathematical induction to prove the following theorem.

Theorem 5.1. *Suppose* $Q, D, P, N, L, K, P', P''$ *are integers such that* $D > 0$, $\sqrt{D} \notin \mathbb{Q}$, $Q \mid D - P^2$ *and*

$$P = P'' + NK, \quad Q = NL, \quad P \equiv P' \pmod{L}.$$

If $K/L = [q_0, q_1, q_2, \ldots, q_n]$ *and we put*

$$\frac{P + \sqrt{D}}{Q} = \langle q_0, q_1, \ldots, q_i, \phi_{i+1} \rangle \quad (i < n),$$

then $\phi_{i+1} = (P_{i+1} + \sqrt{D})/Q_{i+1}$, *where*

$$Q_{i+1} = (-1)^{i-1}(R_i M_1 - C_i M_2),$$
$$M_1 = (N R_i + (P' - P'')C_i)/L \in \mathbb{Z},$$
$$M_2 = (R_i(P' + P'') + T C_i)/L \in \mathbb{Z},$$
$$P_{i+1} = (N R_i + Q_{i+1} C_{i-1})/C_i - P'',$$
$$T = (D - P''^2)/N.$$

If $\mathfrak{a}' = [Q', P' + \sqrt{D}]$, $\mathfrak{a}'' = [Q'', P'' + \sqrt{D}]$ $(Q' > Q'' > 0)$, we can use Theorem 5.1 to produce a modification of the version of NUCOMP given in [13]. We begin with $R_{-2} = Q'/S$, $R_{-1} = U$ (In [13] we used b_i $(= R_{i+1})$.) and we search for that value of R_i such that

$$R_i < \sqrt{Q'/Q''} D^{1/4} < R_{i-1}.$$

We then produce the ideal $\mathfrak{a}_{i+2} = [Q_{i+1}, P_{i+1} + \sqrt{D}] \sim \mathfrak{a}'\mathfrak{a}''$ by using

$$Q_{i+1} = (-1)^{i+1}(R_i M_1 - C_i M_2),$$
$$P_{i+1} = ((Q''/S)R_i + Q_{i+1} C_{i-1})/C_i - P'',$$

where

$$M_1 = ((Q''/S)R_i + (P' - P'')C_i)/(Q'/S),$$
$$M_2 = ((P' + P'')R_i + SR''C_i)/(Q'/S),$$
$$R'' = (D - P''^2)/Q''.$$

It is not difficult to show that the value of Q_{i+1} found above must satisfy $|Q_{i+1}| < 3\sqrt{D}$ and from this it is a relatively simple matter to prove that either $\mathfrak{a}_{i+2}$ or $\rho(\mathfrak{a}_{i+2})$ must be reduced. Indeed, empirical studies suggest that $\mathfrak{a}_{i+2}$ is reduced about 98% of the time. We provide the pseudocode for our new version of NUCOMP in the Appendix.

At the conclusion of this version of NUCOMP we will have a reduced ideal $\mathfrak{b}$ such that

$$\mu\mathfrak{b} = \mathfrak{a}'\mathfrak{a}''.$$

Furthermore, it can be shown that $1 < \mu < \Delta^{3/4}$; indeed, $|\log \mu - \log \Delta^{1/4}|$ tends to be small, particularly when Δ is fairly large ($\Delta > 10^{10}$). Thus, at the end of executing NUCOMP we get $k \geq k' + k'' - t$, where $t = O(\log \mu) = O(\log \Delta)$. That is, $k' + k'' - k = O(\log \Delta)$.

We can also modify Algorithm NEAR in [13] to produce WNEAR. This algorithm will on input $(\mathfrak{b}, d, k)$, p, w, where $k < w$ and $(\mathfrak{b}, d, k)$ is a reduced (f, p) representation of some $\mathcal{O}$-ideal $\mathfrak{a}$, find a w-near $(f + 9/8, p)$ representation of $\mathfrak{a}$. Notice that NEAR is WNEAR with $w = 0$. As $w - k$ tends to be small in our application of WNEAR, we can dispense with some of the procedures used in NEAR. We provide the pseudocode for WNEAR in the Appendix. The method of proof of correctness of WNEAR is essentially that used to prove the correctness of NEAR used in [13] and the number of steps necessary to execute WNEAR is $O(w - k)$.

6   Algorithm AX

We will now develop an algorithm that can be used to compute an $\mathcal{O}$-ideal $\mathfrak{a}[x]$ in the important special case when $\mathfrak{a} = (1)$ and x is a positive integer. Our first algorithm ADDXY gives us the ability to determine, given $\mathcal{O}$-ideals $\mathfrak{a}[x]$ and $\mathfrak{a}[y]$, an $\mathcal{O}$-ideal $\mathfrak{a}[x + y]$. This will enable us to jump quickly through the cycle of reduced principal ideals in $\mathcal{O}$.

Algorithm 1.1. ADDXY

Input: $(\mathfrak{a}[x], d', k')$, $(\mathfrak{a}[y], d'', k'')$, p, x, y, where $(\mathfrak{a}[x], d', k')$ and $(\mathfrak{a}[y], d'', k'')$ are respectively x- and y-near (f', p) and (f'', p) representations of the $\mathcal{O}$-ideal $\mathfrak{a} = (1)$.

Output: $(\mathfrak{a}[x + y], d, k)$, an $(x + y)$-near (f, p) representation of $\mathfrak{a}$, where $f = 13/4 + f' + f'' + f'f''/2^p$.

1: Put $(\mathfrak{c}, g, h) = \text{NUCOMP}((\mathfrak{a}[x], d', k'), (\mathfrak{a}[y], d'', k''), p)$.
2: Put $(\mathfrak{c}', g', h') = \text{WNEAR}((\mathfrak{c}, g, h), p, x + y)$.
3: Put $\mathfrak{a}[x + y] = \mathfrak{c}'$, $d = g'$, $k = h'$.

We remark here that after step 1 has executed, we have $h \leq k' + k'' + 1 \leq x + y - 1$. Also, ADDXY will execute in $O(\log \Delta)$ elementary operations. This is because $k' + k'' - h = O(\log \Delta)$, $x - k' = O(\log \Delta)$ and $y - k'' = O(\log \Delta)$; hence, $x + y - h = O(\log \Delta)$ and $h < x + y$. Finally it is important to observe that since $\mathfrak{a} = (1)$, we have $\mathfrak{a}^2 = \mathfrak{a} = (1)$, and $\mathfrak{a}[x + y]$ as determined in the algorithm is principal.

The next algorithm, AX, finds for a given x and the $\mathcal{O}$-ideal $\mathfrak{a} = (1)$ an x-near (f, p) representation of $\mathfrak{a}$ for a certain value of f.

Algorithm 1.2. AX

Input: $x \in \mathbb{Z}^+$ and $p \in \mathbb{Z}^+$.
Output: $(\mathfrak{a}[x], d, k)$ an x-near (f, p) representation of $\mathfrak{a} = (1)$ for a suitable $f \in [1, 2^p)$.
1: Put $l = \lfloor \log_2 x \rfloor$ and compute the binary representation of x, say

$$x = \sum_{i=0}^{l} b_i 2^{l-i}$$

$(b_0 = 1, b_i \in \{0, 1\}$ for $1 \leq i \leq l)$.
2: Let $Q = 1$, $P = 0$, $\mathfrak{b} = [1, \sqrt{D}]$, $d = 2^p + 1$, $k = 0$, $i = 0$, $s_0 = 1$.
3: Put $(\mathfrak{b}_0, d_0, k_0) = \text{WNEAR}((\mathfrak{b}, d, k), p, 1)$
4: **while** $i < l$ **do**
5:     Put $(\mathfrak{b}_{i+1}, d_{i+1}, k_{i+1}) = \text{ADDXY}((\mathfrak{b}_i, d_i, k_i), (\mathfrak{b}_i, d_i, k_i), p, s_i, s_i)$.
6:     Put $s_{i+1} = 2s_i$
7:     **if** $b_{i+1} = 1$ **then**
8:         Put $s_{i+1} = 2s_i + 1$ and

$$(\mathfrak{b}_{i+1}, d_{i+1}, k_{i+1}) \leftarrow \text{WNEAR}((\mathfrak{b}_{i+1}, d_{i+1}, k_{i+1}), p, s_{i+1}).$$

9:     **end if**
10:    $i \leftarrow i + 1$.
11: **end while**
12: Put $\mathfrak{a}[x] = \mathfrak{b}_l$ $d = d_l$, $k = k_l$.

Clearly, Algorithm AX will execute in $O(\log x \log \Delta)$ elementary operations. That the algorithm is correct follows easily by observing that

$$\mathfrak{b}_j = \mathfrak{a}[s_j] \sim \mathfrak{a} \quad (j = 0, 1, 2, \ldots, l).$$

We now find an upper bound on f.

Theorem 6.1. *Suppose $p \geq 8$ and $h \in \mathbb{R}^+$ with $h \geq \log_2 x$. Put $m = 11.2$. If $hmx < 2^p$, then the value of f after AX has executed satisfies $f < mx$ and therefore $f < 2^p/h$.*

Proof. After Step 8, we see that $(b_{i+1}, d_{i+1}, k_{i+1})$ is an s_{i+1}-near (f_{i+1}, p) representation of $\mathfrak{a}$, where

$$f_{i+1} = \frac{9}{8} + \frac{13}{4} + 2f_i + \frac{f_i^2}{2^p} \quad (1 \leq i+1 \leq l) \tag{6.1}$$

and $f_0 = 1 + 9/8 = 17/8$. We put $f = f_l$. Since $s_l = x$, algorithm AX produces an x-near (f, p) representation (b_l, d_l, k_l) of $\mathfrak{a}$. We now define $a_0 = f_0$, $c = 37/8$ and

$$a_{i+1} = \left(2 + \frac{1}{h}\right) a_i + c.$$

A closed form representation for a_i is given by $a_i = g^i a_0 + c(g^i - 1)/(g - 1)$; hence, an analysis similar to that employed in the proof of Lemma 3.8 of [11] yields

$$a_l < g^l(a_0 + c) < 2^l e^{1/2}(a_0 + c) < 2^l m \leq mx,$$

where $m = 11.2$. As in the proof of Theorem 3.9 of [11] we have $h a_i < 2^p$ $(i = 0, 1, 2, \ldots, l)$ and $h f_0 < 2^p$. Thus, by using induction on (6.1), we can show that $f_i \leq a_i$ $(i = 0, 1, 2, \ldots, l)$. It follows that $f < mn$ and $hf < 2^p$.

Suppose now that we are given some $x \in \mathbb{R}$ and $a \in \mathbb{R}^{\geq 0}$ such that

$$|x - \log_2 \theta_j| \leq a,$$

where $\mathfrak{a}_j = \theta_j \mathfrak{a}_1$ in (5.1). If a is not too large, we would expect that if $\mathfrak{a}_j = \mathfrak{a}[x]$ in (5.1), then i and j should be close in value. However, just how close would they be? In order to answer this question we will begin by defining $c(m)$.

Definition 6.1. *Let $\{F_i\}$ be the sequence of Fibonacci numbers with $F_0 = 0$, $F_1 = 1$. For a fixed $m \in \mathbb{R}$, we define $c(m) = \max\{m_1, m_2\}$ where m_1 and m_2 are respectively the largest integers such that*

$$F_{m_1} < \frac{16}{15} 2^m \quad and \quad F_{m_2+1} < \frac{17}{16} 2^{m+1}.$$

Notice that $m_1 \geq 0$, $m_2 \geq -1$. For example, if $m = -3/2$ then $m_1 = 0$, $m_2 = -1$ and $c(-3/2) = 0$. A short table of values for $c(m)$ is given in Table 6.1.

It is easy to show that if $m' \leq m$, then $c(m') \leq c(m)$. It is also easy to deduce an upper bound on $c(m)$.

Table 6.1. Some values of $c(m)$

m	$c(m)$	m	$c(m)$
≤ -2	0	2	5
-1	1	3	6
0	2	4	8
1	3	5	9

Proposition 6.1. *If $m \geq 1$, then*

$$c(m) < 3 + \frac{3m}{2}.$$

We can now use $c(m)$ to bound the value of $|i - j|$.

Theorem 6.2. *Let $x, k \in \mathbb{Z}$, where $x \geq 1$. Suppose $a, b \in \mathbb{R}$ and*

$$a < \log_2 \theta_j - x < b.$$

If $\mathfrak{a}_i = \mathfrak{a}[x]$, then

$$j - c(b) \leq i \leq j + c(-a - 1).$$

Proof. We must have

$$2^{x+a} < \theta_j < 2^{x+b}. \tag{6.2}$$

By Lemma 5.1 we know that

$$\theta_i < \frac{17}{16} 2^x, \quad \theta_{i+1} > \frac{15}{16} 2^x. \tag{6.3}$$

By Theorem 2.1 of [8] when $n \geq 1$, $i \geq 0$, we have

$$\theta_{i+n} \geq F_n \theta_{i+1} > F_n \left(\frac{15}{16}\right) 2^x.$$

By (6.3) and Definition 6.1 we find that for $m = b$

$$\theta_{i+m_1+1} > F_{m_1+1} \theta_{i+1} > F_{m_1+1} \left(\frac{15}{16}\right) 2^x > 2^{m+x} = 2^{b+x} > \theta_j.$$

It follows that $j < i + m_1 + 1 \leq i + 1 + c(m)$. Hence $i \geq j - c(b)$.
  Also, if $i > n$, by (6.2) and (6.3) we get for $n \geq 0$ and $m = -a - 1$

$$F_{n+1} \theta_{i-n} \leq \theta_i < \left(\frac{17}{16}\right) 2^x = \left(\frac{17}{16}\right) 2^{-a} 2^{x+a} < \left(\frac{17}{16}\right) 2^{m+1} \theta_j.$$

Putting $n = m_2 + 1$ and noting that $F_{m_2+2} > (17/16)2^{m+1}$, we get

$$\theta_{i-m_2-1} < \theta_j.$$

Thus, $j > i - m_2 - 1 \geq i - c(m) - 1$ and $i \leq j + c(-a - 1)$. If $i \leq n = m_2 + 1$, then $i \leq c(m) + 1 \leq j + c(-a - 1)$. ∎

7   Verifying That an Ideal Is Or Is Not Principal

In this section we will only outline a Las Vegas process for determining whether a reduced ideal $\mathfrak{b}$ is (or more importantly is not) principal. A more detailed version of this process can be found in Silvester [26]. As mentioned in §4 we will assume we have been given R'_Δ and h produced by the index calculus algorithm. We now provide the steps needed. By using methods based on the infrastructure, it is possible to verify deterministically whether or not $\mathfrak{b}$ is principal in time complexity $O(R^{1/2+\epsilon})$. Since

$$hR = O(\Delta^{1/2+\epsilon}), \tag{7.1}$$

this means that we could certainly solve problem 2 in §4 unconditionally in time bounded by $O(\Delta^{1/4+\epsilon})$. The procedure that we will describe here, while conditional, should accomplish this in time bounded by $O(\Delta^{1/6+\epsilon})$.

1. We first execute algorithm $\mathrm{EXP}((\mathfrak{b}, 2^{p+1}, -1), h, p)$ of [13] to find a near reduced (f, p) representation $(\mathfrak{c}, d, k)$ of $\mathfrak{b}^h$, where $f < 2^{p-4}$. It is not difficult to show that $|\log_2 \phi - k| < 3/2$, where $\mathfrak{c} = \phi\mathfrak{b}^h$.
2. We next make use of the index calculus algorithm to solve the DLP for $\mathfrak{c}$ to obtain some $g \in \mathbb{Q}$ such that

$$|\log_2 \gamma - g| < 1,$$

   where $\mathfrak{c} = (\gamma)$ and $1 < \gamma < \epsilon_\Delta$. It this case we certainly expect this process to be successful because $\mathfrak{b}^h$ must be principal if h is really the class number. It is this aspect of our technique that renders it a Las Vegas algorithm, as we cannot be certain that this part of it will execute in subexponential time.
3. We put $\mathfrak{a} = (1)$ and use AX to compute $\mathfrak{d} = \mathfrak{a}[[g]]$. By Theorem 6.2, we must be able to find some $i \in \{\pm 3, \pm 2, \pm 1, 0\}$ such that $\rho^i(\mathfrak{d}) = \mathfrak{c}$. If we do not, then $\mathfrak{c}$ cannot be principal.
4. We next compute d', k' such that $(\mathfrak{c}, d', k')$ is a reduced (f, p) representation of $\mathfrak{a}$. This is very simple because in order to compute $\mathfrak{d}$, we had to produce an (f, p) representation of $\mathfrak{a}$. We also must have $|\log_2 \gamma - k'| < 3/2$, where $\mathfrak{c} = (\gamma)$. Thus

$$-3 + k' - k < \log_2 \gamma - \log_2 \phi < 3 + k' - k. \tag{7.2}$$

Before continuing to produce the next steps needed in this process, we must make a few observations. If $\mathfrak{b}$ is principal, then we may assume that $\mathfrak{b} = (\beta)$, where $\beta \in \mathcal{O}$ and $1 \le \beta < \epsilon_\Delta$. Also,

$$\beta^h = \gamma\phi^{-1}\lambda,$$

where $\lambda = \epsilon_\Delta^r$. Hence

$$h \log_2 \beta = \log_2 \gamma - \log_2 \phi + rR_\Delta \quad (R_\Delta = \log_2 \epsilon_\Delta). \tag{7.3}$$

By making use of this equation, we can prove two results.

Theorem 7.1. *If $R_\Delta > 9/2 + \log_2(34\sqrt{\Delta}/15)$, then r in (7.3) must satisfy*

$$-1 \le r < h. \tag{7.4}$$

Theorem 7.2. *If (7.3) holds and $b(r) = \lceil (rR'_\Delta + k' - k)/h \rceil$, then*

$$| \log_2 \beta - b(r)| < 2 + \frac{3}{h} \leq 5.$$

By Theorem 6.2, we see that if we put $S = \{\rho^i(\mathfrak{b}) : |i| \leq 9\}$, then $\mathfrak{b}$ will be principal if and only if $\mathfrak{a}[b(r)] \in S$ for some r satisfying (7.4). Thus, our final step is

5. For $r = -1, 0, 1, \ldots, h$ test to determine whether $\mathfrak{a}[b(r)] \in S$. $\mathfrak{b}$ is principal if and only if this happens for some r in the given range.

If h is large we can improve the execution of Step 5 by observing that

$$b(r + 1) = b(r) + \left\lceil \frac{R'_\Delta}{h} \right\rceil + k(r), \tag{7.5}$$

where $k(r) \in \{0, -1\}$. We can precompute $\mathfrak{a}[\lceil R'_\Delta/h \rceil]$ and $\mathfrak{a}[\lceil R'_\Delta/h \rceil - 1]$ and then we have

$$\mathfrak{a}[b(r + 1)] = \begin{cases} \text{ADDXY}(\mathfrak{a}[b(r)], \mathfrak{a}[\lceil R'_\Delta/h \rceil]) & \text{when } k(r) = 0, \\ \text{ADDXY}(\mathfrak{a}[b(r)], \mathfrak{a}[\lceil R'_\Delta/h \rceil - 1]) & \text{when } k(r) = -1, \end{cases}$$

The value of $k(r)$ is easily computed from (7.5) and the formula for $b(r + 1)$. Clearly, Step 5 executes in time complexity $O(h\Delta^\epsilon)$.

If we take into consideration that we must verify R'_Δ, a process that requires $O(R^{1/3+\epsilon})$ elementary operations, this together with Steps 1-5 will execute in expected time complexity

$$O(R^{1/3+\epsilon}) + O(h\Delta^\epsilon). \tag{7.6}$$

If $h > \Delta^{1/6}$, an unusual circumstance since h tends to be small (see Cohen and Lenstra [5]), then by (7.1) $R = O(\Delta^{1/3+\epsilon})$ and we can solve the principal ideal problem in time complexity $O(R^{1/2+\epsilon}) = O(\Delta^{1/6+\epsilon})$ by using infrastructure methods. If $h < \Delta^{1/6}$, then by (7.6) we can solve this problem in $O(\Delta^{1/6+\epsilon})$ operations by using the new procedure.

We conclude this section with a simple example left over from [15]. Let

$$d_1 = 187060083,$$
$$d_3 = 1311942540724389723505929002667880175005208,$$
$$j_1 = 2,$$
$$j_2 = 2104044625155634711504852164533487.$$

In [15] it was necessary to show that

$$d_1 x_3^2 - d_3 x_2^2 = \frac{d_3 j_1 - d_1 j_2}{j_2} = c = 88081306349606091164364 5 \tag{7.7}$$

has no integer solutions. Since $4 \mid d_3$, it is sufficient to show that all ideals of norm cd_1 in $\mathcal{O} = [1, \sqrt{D}]$, where $D = d_1 d_3 / 4$, are not principal. In this case Δ, the discriminant of $\mathcal{O}$ is the 51 digit number

$$\Delta = d_1 d_3 = 245412080559135221803366130231160886970528733912264$$

and, using the subexponential algorithm, we found that

$$h = 1024 \text{ and } R' = 6851106675369184895740.24677.$$

Here R' is an approximation to the regulator R of $\mathcal{O}$. Looking at the prime factors of cd_1, we see that

$$cd_1 = \underbrace{5 \cdot 769 \cdot 33809 \cdot 6775714175075849}_{\text{factors of } c} \cdot \underbrace{3 \cdot 7 \cdot 8907623}_{\text{factors of } d_1}$$

and since the prime factors of d_1 ramify in $\mathcal{O}$, we found a total of 16 ideals of norm cd_1. By excluding ideal conjugates, we can reduce this to only 8 candidates. By invoking the ERH it was possible to show that (7.7) had no solutions. However, by using the method described here we were able to show unconditionally that this equation has no solutions. Most (87%) of the time needed to perform this algorithm was required to verify R'_Δ.

References

1. Booker, A.: Quadratic class numbers and character sums. Math. Comp. 75, 1481–1492 (2006)
2. Buchmann, J., Thiel, C., Williams, H.C.: Short representation of quadratic integers. In: Mathematics and its Applications, vol. 325, pp. 159–185. Kluwer Academic Publishers, Dordrecht (1995)
3. Buchmann, J., Vollmer, U.: Binary Quadratic Forms: An Algorithmic Approach. Algorithms and Computation in Mathematics, vol. 20. Springer, Berlin (2007)
4. Cojocaru, A.C., Murty, M.R.: An Introduction to Sieve Methods and their Application. Cambridge University Press, Cambridge (2005)
5. Cohen, H., Lenstra Jr., H.W.: Heuristics on class groups of number fields. In: Number Theory. Lecture Notes in Math., vol. 1068, pp. 33–62. Springer, New York (1983)
6. Dickson, L.E.: History of the Theory of Numbers, Carnegie Institution of Washington, Publication No. 256 (1919), vol. 2. Dover Publications, New York (2005)
7. de Haan, R.: A fast, rigorous technique for verifying the regulator of a real quadratic field. Master's thesis, University of Amsterdam (2004)
8. de Haan, R., Jacobson Jr., M.J., Williams, H.C.: A fast, rigorous technique for computing the regulator of a real quadratic field. Math. Comp. 76, 2139–2160 (2007)
9. Jacobson Jr., M.J.: Subexponential Class Group Computation in Quadratic Orders. PhD thesis, Technische Universität Darmstadt, Darmstadt, Germany (1999)
10. Jacobson Jr., M.J.: Computing discrete logarithms in quadratic orders. Journal of Cryptology 13, 473–492 (2000)

11. Jacobson Jr., M.J., Scheidler, R., Williams, H.C.: The efficiency and security of a real quadratic field based key exchange protocol, Walter de Gruyter, Berlin, pp. 89–112 (2001)
12. Jacobson Jr., M.J., Sawilla, R.E., Williams, H.C.: Efficient ideal reduction in quadratic fields. International Journal of Mathematics and Computer Science 1, 83–116 (2006)
13. Jacobson Jr., M.J., Scheidler, R., Williams, H.C.: An improved real quadratic field based key-exchange procedure. J. Cryptology 19, 211–239 (2006)
14. Jacobson Jr., M.J., van der Poorten, A.J.: Computational aspects of NUCOMP. In: Fieker, C., Kohel, D.R. (eds.) ANTS 2002. LNCS, vol. 2369, pp. 120–133. Springer, Heidelberg (2002)
15. Jacobson Jr., M.J., Williams, H.C.: Modular arithmetic on elements of small norm in quadratic fields. Designs, Codes and Cryptography 27, 93–110 (2002)
16. Kornhauser, D.M.: On the smallest solution to the general binary quadratic Diophantine equation. Acta Arith. 55, 83–94 (1990)
17. Lagrange, J.L.: Sur la solution des problèmes indéterminés du second degré. In: Oeuvres, Gauthier-Villars, Paris, vol. II, pp. 377–535 (1868)
18. Lenstra Jr., H.W.: On the calculation of regulators and class numbers of quadratic fields. London Math. Soc. Lecture Notes Series 56, 123–150 (1982)
19. Maurer, M.H.: Regulator Approximation and Fundamental Unit Computation for Real-Quadratic Orders. PhD thesis, Technische Universität Darmstadt, Darmstadt, Germany (2000)
20. Nagell, T.: Introduction to Number Theory, Chelsea, NY (1964)
21. Nitaj, A.: L'algorithme de Cornacchia. Expositiones Math. 13, 358–365 (1995)
22. van der Poorten, A.J.: A note on NUCOMP. Math. Comp. 72, 1935–1946 (2003)
23. Shanks, D.: The infrastructure of real quadratic fields and its applications. In: Proc. 1972 Number Theory Conf., Boulder, Colorado, pp. 217–224 (1972)
24. Shanks, D.: On Gauss and composition I, II. NATO ASI, Series C, vol. 265, pp. 163–204. Kluwer, Dordrecht (1989)
25. Srinivasan, A.: Computations of class numbers of real quadratic fields. Math. Comp. 67, 1285–1308 (1998)
26. Silvester, A.K.: Fast and unconditional principal ideal testing. Master's thesis, University of Calgary (2006)., http://math.ucalgary.ca/~aksilves/papers/msc-thesis.pdf
27. Stolt, B.: On the Diophantine equation $u^2 - Dv^2 = \pm 4N$, Parts I, II, III. Ark. Mat. 2, 1–23, 251–268 (1952); 3, 117–132 (1955)
28. Williams, H.C., Wunderlich, M.C.: On the parallel generation of the residues for the continued fraction factoring algorithm. Math. Comp. 48, 405–423 (1987)

Appendix

In this brief appendix we provide the pseudocode for our versions of NUCOMP and WNEAR. Note that in NUCOMP we make use of the following theorem which can be proved in the same manner as Theorem 5.1 of [13].

Theorem 7.3. *Let $(\mathfrak{b}', d', k')$ be an (f', p) representation of an $\mathcal{O}$-ideal $\mathfrak{a}'$ and let $(\mathfrak{b}'', d'', k'')$ be an (f'', p) representation of an $\mathcal{O}$-ideal $\mathfrak{a}''$. If $d'd'' \leq 2^{2p+1}$, put $d = \lceil d'd''/2^p \rceil$, $k = k' + k''$. If $d'd'' > 2^{2p+1}$, put $d = \lceil d'd''/2^{p+1} \rceil$, $k = k' + k'' + 1$. Then $(\mathfrak{b}'\mathfrak{b}'', d, k)$ is an (f, p) representation of the product ideal $\mathfrak{a}'\mathfrak{a}''$, where $f = 1 + f' + f'' + 2^{-p}f'f''$.*

Algorithm 1.3. NUCOMP

Input: $(\mathfrak{b}', d', k')$, $(\mathfrak{b}'', d'', k'')$, p, where $(\mathfrak{b}', d', k')$ is a reduced (f', p) representation of an invertible $\mathcal{O}$-ideal $\mathfrak{a}'$ and $(\mathfrak{b}'', d'', k'')$ is reduced (f'', p) representation of an invertible $\mathcal{O}$-ideal $\mathfrak{a}''$. Here,

$$\mathfrak{b}' = \left[Q', P' + \sqrt{D}\right], \quad \mathfrak{b}'' = \left[Q'', P'' + \sqrt{D}\right], \quad Q' \geq Q'' > 0.$$

Output: A reduced (f, p) representation $(\mathfrak{b}, d, k)$ of $\mathfrak{a}'\mathfrak{a}''$, where $\mathfrak{b} = [Q, P + \sqrt{D}]$, $(P+\sqrt{D})/Q > 1$, $-1 < (P-\sqrt{D})/Q < 0$, $k \leq k'+k''+1$, $f = f^*+17/8$ with $f^* = f' + f'' + 2^{-p} f' f''$.

1: Compute $G = (Q', Q'')$ and solve $Q'' X \equiv G \pmod{Q'}$ for $X \in \mathbb{Z}$, $0 \leq X < Q'$.
2: Compute $S = (P' + P'', G)$ and solve $Y(P' + P'') + ZG = S$ for $Y, Z \in \mathbb{Z}$.
3: Put $R'' = (D - P''^2)/Q''$, $U \equiv XZ(P' - P'') + YR'' \pmod{Q'/S}$, where $0 \leq U < Q'/S$.
4: Put $R_{-1} = Q'/S$, $R_0 = U$, $C_{-1} = 0$, $C_0 = -1$, $i = -1$.
5: **if** $d'd'' \leq 2^{2p+1}$ **then**
6:     Put $d = \lfloor d'd''/2^p \rfloor$, $k = k' + k''$
7: **else**
8:     Put $d = \lfloor d'd''/2^{p+1} \rfloor$, $k = k' + k'' + 1$
9: **end if**
10: **if** $R_{-1} \leq \lfloor \sqrt{Q'/Q''} D^{1/4} \rfloor$ **then**
11:     Put

$$Q_{i+1} = Q'Q''/S^2,$$
$$P_{i+1} \equiv P'' + UQ''/S \pmod{Q_{i+1}}.$$

12:     Go to 21.
13: **end if**
14: **while** $R_i > \lfloor \sqrt{Q'/Q''} D^{1/4} \rfloor$ **do**
15:     $i \leftarrow i + 1$
16:     $q_i = \lfloor R_{i-2}/R_{i-1} \rfloor$
17:     $C_i = C_{i-2} - q_i C_{i-1}$
18:     $R_i = R_{i-2} - q_i R_{i-1}$
19: **end while**
20: Put

$$M_1 = ((Q''/rs)R_i + (P' - P'')C_i)/(Q'/S),$$
$$M_2 = ((P' + P'')R_i + rSR''C_i)/(Q'/S),$$
$$Q_{i+1} = (-1)^{i+1}(R_i M_1 - C_i M_2),$$
$$P_{i+1} = ((Q''/rS)R_i + Q_{i+1}C_{i-1})/C_i - P''.$$

21: Put $j = 1$,

$$Q'_{i+1} = |Q_{i+1}|,$$

$$k_{i+1} = \left\lfloor \frac{\lfloor \sqrt{D} \rfloor - P_{i+1}}{Q'_{i+1}} \right\rfloor,$$

$$P'_{i+1} = k_{i+1}Q'_{i+1} + P_{i+1},$$

and $\sigma = \text{sign}(Q_{i+1})$, $B_{i-1} = \sigma|C_{i-1}|$, $B_{i-2} = |C_{i-2}|$.

22: **if** $P'_{i+1} + \lfloor \sqrt{D} \rfloor \geq Q'_{i+1}$ **then**

23:     Go to 27.

24: **else**

25:     Put $j = 2$ and

$$q_{i+1} = \left\lfloor \frac{P_{i+1} + \lfloor \sqrt{D} \rfloor}{Q'_{i+1}} \right\rfloor,$$

$$P_{i+2} = q_{i+1}Q'_{i+1} - P_{i+1},$$

$$Q_{i+2} = \frac{D - P_{i+2}^2}{Q'_{i+1}},$$

$$Q'_{i+2} = |Q_{i+2}|,$$

$$k_{i+2} = \left\lfloor \frac{\lfloor \sqrt{D} \rfloor - P_{i+2}}{Q'_{i+2}} \right\rfloor,$$

$$P'_{i+2} = k_{i+2}Q'_{i+1} + P_{i+2}.$$

$$B_{i+1} = q_{i+1}B_i + B_{i-1}.$$

26: **end if**

27: Find $s \geq 0$ such that $2^s Q'_{i+j} > 2^{p+4} S B_{i+j-1}$.

28: Put $\mathfrak{b} = [Q'_{i+j}, P'_{i+j} + \sqrt{D}]$ and

$$T_{i+j} = 2^s Q_{i+j} B_{i+j-2} + B_{i+j-1}(2^s P_{i+j} - \lfloor 2^s \sqrt{D} \rfloor).$$

29: $(\mathfrak{b}, d, k) = \text{DIV}((\mathfrak{b}, e, h), ST_{i+j}, Q'_{i+j}, s, p)$.

Algorithm 1.4. WNEAR

Input: $(\mathfrak{b}, d, k), p, w$, where $k < w$ and $(\mathfrak{b}, d, k)$ is a reduced (f, p) representation of some $\mathcal{O}$-ideal $\mathfrak{a}$. Here $\mathfrak{b} = [Q, P + \sqrt{D}]$, where $P + \lfloor \sqrt{D} \rfloor \geq Q$, $0 \leq \lfloor \sqrt{D} \rfloor - P \leq Q$.

Output: $(\mathfrak{c}, g, h)$ a w-near $(f + 9/8, p)$ representation of $\mathfrak{a}$.

1: Find $s \in \mathbb{Z}^{\geq 0}$ such that $2^s Q \geq 2^{p+4}$. Put $Q_0 = Q$, $P_0 = P$, $Q_{-1} = (D - P^2)/Q$, $M = \lceil 2^{p+s-k+w} Q_0/d \rceil$, $T_{-2} = -2^s P_0 + \lfloor 2^s \sqrt{D} \rfloor$, $T_{-1} = 2^s Q_0$, $i = 1$.

2: **while** $T_{i-2} \leq M$ **do**

3:     $q_{i-1} = \lfloor (P_{i-1} + \lfloor \sqrt{D} \rfloor)/Q_{i-1} \rfloor$

4:     $P_i = q_{i-1}Q_{i-1} - P_{i-1}$

5:     $Q_i = Q_{i-2} - q_{i-1}(P_i - P_{i-1})$

6:     $T_{i-1} = q_{i-1}T_{i-2} + T_{i-3}$

7:     $i \leftarrow i + 1$

8: **end while**

9: Put $e_{i-1} = \lceil 2^{p-s+3}T_{i-3}/Q_0 \rceil$

10: **if** $de_{i-1} \leq 2^{2p-k+w+3}$ **then**

11:     Put $\mathfrak{c} = [Q_{i-2}, P_{i-2} + \sqrt{D}], \ e = e_{i-1}$.

12: **else**

13:     Put $\mathfrak{c} = [Q_{i-3}, P_{i-3} + \sqrt{D}], \ e = \lceil 2^{p-s+3}T_{i-4}/Q_0 \rceil$.

14: **end if**

15: Find t such that

$$2^t < \frac{ed}{2^{2p+3}} \leq 2^{t+1}.$$

16: Put

$$g = \left\lceil \frac{ed}{2^{p+t+3}} \right\rceil, \quad h = k + t.$$

Abelian Varieties with Prescribed Embedding Degree

David Freeman[1,*], Peter Stevenhagen[2], and Marco Streng[2]

[1] University of California, Berkeley
dfreeman@math.berkeley.edu
[2] Mathematisch Instituut, Universiteit Leiden
{psh,streng}@math.leidenuniv.nl

Abstract. We present an algorithm that, on input of a CM-field K, an integer $k \geq 1$, and a prime $r \equiv 1 \bmod k$, constructs a q-Weil number $\pi \in \mathcal{O}_K$ corresponding to an ordinary, simple abelian variety A over the field $\mathbf{F}$ of q elements that has an $\mathbf{F}$-rational point of order r and embedding degree k with respect to r. We then discuss how CM-methods over K can be used to explicitly construct A.

1 Introduction

Let A be an abelian variety defined over a finite field $\mathbf{F}$, and $r \neq \mathrm{char}(\mathbf{F})$ a prime number dividing the order of the group $A(\mathbf{F})$. Then the *embedding degree* of A with respect to r is the degree of the field extension $\mathbf{F} \subset \mathbf{F}(\zeta_r)$ obtained by adjoining a primitive r-th root of unity ζ_r to $\mathbf{F}$.

The embedding degree is a natural notion in pairing-based cryptography, where A is taken to be the Jacobian of a curve defined over $\mathbf{F}$. In this case, A is principally polarized and we have the non-degenerate *Weil pairing*

$$e_r : A[r] \times A[r] \longrightarrow \mu_r$$

on the subgroup scheme $A[r]$ of r-torsion points of A with values in the r-th roots of unity. If $\mathbf{F}$ contains ζ_r, we also have the non-trivial *Tate pairing*

$$t_r : A[r](\mathbf{F}) \times A(\mathbf{F})/rA(\mathbf{F}) \to \mathbf{F}^*/(\mathbf{F}^*)^r.$$

The Weil and Tate pairings can be used to 'embed' r-torsion subgroups of $A(\mathbf{F})$ into the multiplicative group $\mathbf{F}(\zeta_r)^*$, and thus the discrete logarithm problem in $A(\mathbf{F})[r]$ can be 'reduced' to the same problem in $\mathbf{F}(\zeta_r)^*$ [6,3]. In pairing-based cryptographic protocols [7], one chooses the prime r and the embedding degree k such that the discrete logarithm problems in $A(\mathbf{F})[r]$ and $\mathbf{F}(\zeta_r)^*$ are computationally infeasible, and of roughly equal difficulty. This means that r is typically large, whereas k is small. Jacobians of curves meeting such requirements are often said to be *pairing-friendly*.

---

* The first author is supported by a National Defense Science and Engineering Graduate Fellowship.

A.J. van der Poorten and A. Stein (Eds.): ANTS-VIII 2008, LNCS 5011, pp. 60–73, 2008.
© Springer-Verlag Berlin Heidelberg 2008

If $\mathbf{F}$ has order q, the embedding degree $k = [\mathbf{F}(\zeta_r) : \mathbf{F}]$ is simply the multiplicative order of q in $(\mathbf{Z}/r\mathbf{Z})^*$. As 'most' elements in $(\mathbf{Z}/r\mathbf{Z})^*$ have large order, the embedding degree of A with respect to a large prime divisor r of $\#A(\mathbf{F})$ will usually be of the same size as r, and A will not be pairing-friendly. One is therefore led to the question of how to efficiently construct A and $\mathbf{F}$ such that $A(\mathbf{F})$ has a (large) prime factor r and the embedding degree of A with respect to r has a prescribed (small) value k. The current paper addresses this question on two levels: the *existence* and the actual *construction* of A and $\mathbf{F}$.

Section 2 focuses on the question whether, for given r and k, there exist abelian varieties A that are defined over a finite field $\mathbf{F}$, have an $\mathbf{F}$-rational point of order r, and have embedding degree k with respect to r. We consider only abelian varieties A that are *simple*, that is, not isogenous (over $\mathbf{F}$) to a product of lower-dimensional varieties, as we can always reduce to this case. By Honda-Tate theory [10], isogeny classes of simple abelian varieties A over the field $\mathbf{F}$ of q elements are in one-to-one correspondence with $\mathrm{Gal}(\overline{\mathbf{Q}}/\mathbf{Q})$-conjugacy classes of q-*Weil numbers*, which are algebraic integers π with the property that all embeddings of π into $\mathbf{C}$ have absolute value $\sqrt{q}$. This correspondence is given by the map sending A to its q-th power Frobenius endomorphism π inside the number field $\mathbf{Q}(\pi) \subset \mathrm{End}(A) \otimes \mathbf{Q}$. The existence of abelian varieties with the properties we want is thus tantamount to the existence of suitable Weil numbers.

Our main result, Algorithm 2.12, constructs suitable q-Weil numbers π in a given *CM-field* K. It exhibits π as a *type norm* of an element in a *reflex field* of K satisfying certain congruences modulo r. The abelian varieties A in the isogeny classes over $\mathbf{F}$ that correspond to these Weil numbers have an $\mathbf{F}$-rational point of order r and embedding degree k with respect to r. Moreover, they are *ordinary*, i.e., $\#A(\overline{\mathbf{F}})[p] = p^g$, where p is the characteristic of $\mathbf{F}$. Theorem 3.1 shows that for fixed K, the expected run time of our algorithm is heuristically polynomial in $\log r$.

For an abelian variety of dimension g over the field $\mathbf{F}$ of q elements, the group $A(\mathbf{F})$ has roughly q^g elements, and one compares this size to r by setting

$$\rho = \frac{g \log q}{\log r}. \tag{1.1}$$

In cryptographic terms, ρ measures the ratio of a pairing-based system's required bandwidth to its security level, so small ρ-values are desirable. *Supersingular* abelian varieties can achieve ρ-values close to 1, but their embedding degrees are limited to a few values that are too small to be practical [4,8]. Theorem 3.4 discusses the distribution of the (larger) ρ-values we obtain.

In Section 4, we address the issue of the actual construction of abelian varieties corresponding to the Weil numbers found by our algorithm. This is accomplished via the construction in characteristic zero of the abelian varieties having CM by the ring of integers $\mathcal{O}_K$ of K, a hard problem that is far from being algorithmically solved. We discuss the elliptic case $g = 1$, for which reasonable algorithms exist, and the case $g = 2$, for which such algorithms are still in their infancy. For genus $g \geq 3$, we restrict attention to a few families of curves that we can handle at this point. Our final Section 5 provides numerical examples.

2 Weil Numbers Yielding Prescribed Embedding Degrees

Let $\mathbf{F}$ be a field of q elements, A a g-dimensional simple abelian variety over $\mathbf{F}$, and $K = \mathbf{Q}(\pi) \subset \operatorname{End}(A) \otimes \mathbf{Q}$ the number field generated by the Frobenius endomorphism π. Then π is a q-*Weil number* in K: an algebraic integer with the property that all of its embeddings in $\overline{\mathbf{Q}}$ have complex absolute value $\sqrt{q}$.

The q-Weil number π determines the group order of $A(\mathbf{F})$: the $\mathbf{F}$-rational points of A form the kernel of the endomorphism $\pi - 1$, and in the case where $K = \mathbf{Q}(\pi)$ is the full endomorphism algebra $\operatorname{End}(A) \otimes \mathbf{Q}$ we have

$$\#A(\mathbf{F}) = \mathrm{N}_{K/\mathbf{Q}}(\pi - 1).$$

In the case $K = \operatorname{End}(A) \otimes \mathbf{Q}$ we will focus on, K is a *CM-field* of degree $2g$ as in [10, Section 1], i.e., a totally complex quadratic extension of a totally real subfield $K_0 \subset K$.

Proposition 2.1. *Let A, $\mathbf{F}$ and π be as above, and assume $K = \mathbf{Q}(\pi)$ equals $\operatorname{End}_{\mathbf{F}}(A) \otimes \mathbf{Q}$. Let k be a positive integer, Φ_k the k-th cyclotomic polynomial, and $r \nmid qk$ a prime number. If we have*

$$\mathrm{N}_{K/\mathbf{Q}}(\pi - 1) \equiv 0 \pmod{r},$$
$$\Phi_k(\pi\overline{\pi}) \equiv 0 \pmod{r},$$

then A has embedding degree k with respect to r.

Proof. The first condition tells us that r divides $\#A(\mathbf{F})$, the second that the order of $\pi\overline{\pi} = q$ in $(\mathbf{Z}/r\mathbf{Z})^*$, which is the embedding degree of A with respect to r, equals k. □

By Honda-Tate theory [10], all q-Weil numbers arise as Frobenius elements of abelian varieties over $\mathbf{F}$. Thus, we can prove the *existence* of an abelian variety A as in Proposition 2.1 by exhibiting a q-Weil number $\pi \in K$ as in that proposition. The following Lemma states what we need.

Lemma 2.2. *Let π be a q-Weil number. Then there exists a unique isogeny class of simple abelian varieties $A/\mathbf{F}$ with Frobenius π. If $K = \mathbf{Q}(\pi)$ is totally imaginary of degree $2g$ and q is prime, then such A have dimension g, and K is the full endomorphism algebra $\operatorname{End}_{\mathbf{F}}(A) \otimes \mathbf{Q}$. If furthermore q is unramified in K, then A is ordinary.*

Proof. The main theorem of [10] yields existence and uniqueness, and shows that $E = \operatorname{End}_{\mathbf{F}}(A) \otimes \mathbf{Q}$ is a central simple algebra over $K = \mathbf{Q}(\pi)$ satisfying

$$2 \cdot \dim(A) = [E : K]^{\frac{1}{2}}[K : \mathbf{Q}].$$

For K totally imaginary of degree $2g$ and q prime, Waterhouse [12, Theorem 6.1] shows that we have $E = K$ and $\dim(A) = g$. By [12, Prop. 7.1], A is ordinary if and only if $\pi + \overline{\pi}$ is prime to $q = \pi\overline{\pi}$ in $\mathcal{O}_K$. Thus if A is not ordinary, the ideals (π) and $(\overline{\pi})$ have a common divisor $\mathfrak{p} \subset \mathcal{O}_K$ with $\mathfrak{p}^2 \mid q$, so q ramifies in K. □

Example 2.3. Our general construction is motivated by the case where K is a Galois CM-field of degree $2g$, with cyclic Galois group generated by σ. Here σ^g is complex conjugation, so we can construct an element $\pi \in \mathcal{O}_K$ satisfying $\pi\sigma^g(\pi) = \pi\overline{\pi} \in \mathbf{Z}$ by choosing any $\xi \in \mathcal{O}_K$ and letting $\pi = \prod_{i=1}^{g} \sigma^i(\xi)$. For such π, we have $\pi\overline{\pi} = N_{K/\mathbf{Q}}(\xi) \in \mathbf{Z}$. If $N_{K/\mathbf{Q}}(\xi)$ is a prime q, then π is a q-Weil number in K.

Now we wish to impose the conditions of Proposition 2.1 on π. Let r be a rational prime that splits completely in K, and $\mathfrak{r}$ a prime of $\mathcal{O}_K$ over r. For $i = 1, \ldots, 2g$, put $\mathfrak{r}_i = \sigma^{-i}(\mathfrak{r})$; then the factorization of r in $\mathcal{O}_K$ is $r\mathcal{O}_K = \prod_{i=1}^{2g} \mathfrak{r}_i$. If $\alpha_i \in \mathbf{F}_r = \mathcal{O}_K/\mathfrak{r}_i$ is the residue class of ξ modulo $\mathfrak{r}_i$, then $\sigma^i(\xi)$ modulo $\mathfrak{r}$ is also α_i, so the residue class of π modulo $\mathfrak{r}$ is $\prod_{i=1}^{g} \alpha_i$. Furthermore, the residue class of $\pi\overline{\pi}$ modulo $\mathfrak{r}$ is $\prod_{i=1}^{2g} \alpha_i$. If we choose ξ to satisfy

$$\prod_{i=1}^{g} \alpha_i = 1 \in \mathbf{F}_r, \tag{2.4}$$

we find $\pi \equiv 1 \pmod{\mathfrak{r}}$ and thus $N_{K/\mathbf{Q}}(\pi - 1) \equiv 0 \pmod{r}$. By choosing ξ such that in addition

$$\zeta = \prod_{i=1}^{2g} \alpha_i = \prod_{i=g+1}^{2g} \alpha_i \tag{2.5}$$

is a primitive k-th root of unity in $\mathbf{F}_r^*$, we guarantee that $\pi\overline{\pi} = q$ is a primitive k-th root of unity modulo r. Thus we can try to find a Weil number as in Proposition 2.1 by picking residue classes $\alpha_i \in \mathbf{F}_r^*$ for $i = 1, \ldots, 2g$ meeting the two conditions above, computing some 'small' lift $\xi \in \mathcal{O}_K$ with $(\xi \bmod \mathfrak{r}_i) = \alpha_i$, and testing whether $\pi = \prod_{i=1}^{g} \sigma^i(\xi)$ has prime norm. As numbers of moderate size have a high probability of being prime by the prime number theorem, a small number of choices $(\alpha_i)_i$ should suffice. There are $(r-1)^{2g-2}\varphi(k)$ possible choices for $(\alpha_i)_{i=1}^{2g}$, where φ is the Euler totient function, so for $g > 1$ and large r we are very likely to succeed. For $g = 1$, there are only a few choices $(\alpha_1, \alpha_2) = (1, \zeta)$, but one can try various lifts and thus recover what is known as the Cocks-Pinch algorithm [2, Theorem 4.1] for finding pairing-friendly elliptic curves. □

For arbitrary CM-fields K, the appropriate generalization of the map

$$\xi \mapsto \prod_{i=1}^{g} \sigma^i(\xi)$$

in Example 2.3 is provided by the *type norm*. A *CM-type* of a CM-field K of degree $2g$ is a set $\Phi = \{\phi_1, \ldots, \phi_g\}$ of embeddings of K into its normal closure L such that $\Phi \cup \overline{\Phi} = \{\phi_1, \ldots, \phi_g, \overline{\phi_1}, \ldots, \overline{\phi_g}\}$ is the complete set of embeddings of K into L. The *type norm* $N_\Phi : K \to L$ with respect to Φ is the map

$$N_\Phi : x \mapsto \prod_{i=1}^{g} \phi_i(x),$$

which clearly satisfies

$$N_\Phi(x)\overline{N_\Phi(x)} = N_{K/\mathbf{Q}}(x) \in \mathbf{Q}. \tag{2.6}$$

If K is not Galois, the type norm N_Φ does not map K to itself, but to its *reflex field* $\widehat{K}$ with respect to Φ. To end up in K, we can however take the type norm with respect to the *reflex type* Ψ, which we will define now (cf. [9, Section 8]).

Let G be the Galois group of $L/\mathbf{Q}$, and H the subgroup fixing K. Then the $2g$ left cosets of H in G can be viewed as the embeddings of K in L, and this makes the CM-type Φ into a set of g left cosets of H for which we have $G/H = \Phi \cup \overline{\Phi}$. Let S be the union of the left cosets in Φ, and put $\widehat{S} = \{\sigma^{-1} : \sigma \in S\}$. Let $\widehat{H} = \{\gamma \in G : \gamma S = S\}$ be the stabilizer of S in G. Then $\widehat{H}$ defines a subfield $\widehat{K}$ of L, and as we have $\widehat{H} = \{\gamma \in G : \widehat{S}\gamma = \widehat{S}\}$ we can interpret $\widehat{S}$ as a union of left cosets of $\widehat{H}$ inside G. These cosets define a set of embeddings Ψ of $\widehat{K}$ into L. We call $\widehat{K}$ the *reflex field* of (K, Φ) and we call Ψ the *reflex type*.

Lemma 2.7. *The field $\widehat{K}$ is a CM-field. It is generated over $\mathbf{Q}$ by the sums $\sum_{\phi \in \Phi} \phi(x)$ for $x \in K$, and Ψ is a CM-type of $\widehat{K}$. The type norm N_Φ maps K to $\widehat{K}$.*

Proof. The first two statements are proved in [9, Chapter II, Proposition 28] (though the definition of $\widehat{H}$ differs from ours, because Shimura lets G act from the right). For the last statement, notice that for $\gamma \in \widehat{H}$, we have $\gamma S = S$, so $\gamma \prod_{\phi \in \Phi} \phi(x) = \prod_{\phi \in \Phi} \phi(x)$.   □

A CM-type Φ of K is *induced* from a CM-subfield $K' \subset K$ if it is of the form $\Phi = \{\phi : \phi|_{K'} \in \Phi'\}$ for some CM-type Φ' of K'. In other words, Φ is induced from K' if and only if S as above is a union of left cosets of $\mathrm{Gal}(L/K')$. We call Φ *primitive* if it is not induced from a strict subfield of K; primitive CM-types correspond to simple abelian varieties [9]. Notice that the reflex type Ψ is primitive by definition of $\widehat{K}$, and that (K, Φ) is induced from the reflex of its reflex. In particular, if Φ is primitive, then the reflex of its reflex is (K, Φ) itself. For K Galois and Φ primitive we have $\widehat{K} = K$, and the reflex type of Φ is $\Psi = \{\phi^{-1} : \phi \in \Phi\}$.

For CM-fields K of degree 2 or 4 with primitive CM-types, the reflex field $\widehat{K}$ has the same degree as K. This fails to be so for $g \geq 3$.

Lemma 2.8. *If K has degree $2g$, then the degree of $\widehat{K}$ divides $2^g g!$.*

Proof. We have $K = K_0(\sqrt{\eta})$, with K_0 totally real and $\eta \in K$ totally negative. The normal closure L of K is obtained by adjoining to the normal closure $\widetilde{K_0}$ of K_0, which has degree dividing $g!$, the square roots of the g conjugates of η. Thus L is of degree dividing $2^g g!$, and $\widehat{K}$ is a subfield of L.   □

For a 'generic' CM field K the degree of L is exactly $2^g g!$, and $\widehat{K}$ is a field of degree 2^g generated by $\sum_\sigma \sqrt{\sigma(\eta)}$, with σ ranging over $\mathrm{Gal}(K_0/\mathbf{Q})$.

From (2.6) and Lemma 2.7, we find that for every $\xi \in \mathcal{O}_{\widehat{K}}$, the element $\pi = N_\Psi(\xi)$ is an element of $\mathcal{O}_K$ that satisfies $\pi\overline{\pi} \in \mathbf{Z}$. To make π satisfy the conditions of Proposition 2.1, we need to impose conditions modulo r on ξ in $\widehat{K}$. Suppose r splits completely in K, and therefore in its normal closure L and in the reflex field $\widehat{K}$ with respect to Φ. Pick a prime $\mathfrak{R}$ over r in L, and write $\mathfrak{r}_\psi = \psi^{-1}(\mathfrak{R}) \cap \mathcal{O}_{\widehat{K}}$ for $\psi \in \Psi$. Then the factorization of r in $\mathcal{O}_{\widehat{K}}$ is

$$r\mathcal{O}_{\widehat{K}} = \prod_{\psi \in \Psi} \mathfrak{r}_\psi \overline{\mathfrak{r}_\psi}. \tag{2.9}$$

Theorem 2.10. *Let (K, Φ) be a CM-type and $(\widehat{K}, \Psi)$ its reflex. Let $r \equiv 1$ (mod k) be a prime that splits completely in K, and write its factorization in $\mathcal{O}_{\widehat{K}}$ as in (2.9). Given $\xi \in \mathcal{O}_{\widehat{K}}$, write $(\xi \bmod \mathfrak{r}_\psi) = \alpha_\psi \in \mathbf{F}_r$ and $(\xi \bmod \overline{\mathfrak{r}_\psi}) = \beta_\psi \in \mathbf{F}_r$ for $\psi \in \Psi$. If we have*

$$\prod_{\psi \in \Psi} \alpha_\psi = 1 \qquad \text{and} \qquad \prod_{\psi \in \Psi} \beta_\psi = \zeta \qquad (2.11)$$

for some primitive k-th root of unity $\zeta \in \mathbf{F}_r^*$, then $\pi = N_\Psi(\xi) \in \mathcal{O}_K$ satisfies $\pi\overline{\pi} \in \mathbf{Z}$ and

$$N_{K/\mathbf{Q}}(\pi - 1) \equiv 0 \pmod{r},$$
$$\Phi_k(\pi\overline{\pi}) \equiv 0 \pmod{r}.$$

Proof. This is a straightforward generalization of the argument in Example 2.3. The conditions (2.11) generalize (2.4) and (2.5), and imply in the present context that $\pi - 1 \in \mathcal{O}_K$ and $\Phi_k(\pi\overline{\pi}) \in \mathbf{Z}$ are in the prime $\mathfrak{R} \subset \mathcal{O}_L$ over r that underlies the factorization (2.9). ☐

If the element π in Theorem 2.10 generates K and $N_{K/\mathbf{Q}}(\pi)$ is a prime q that is unramified in K, then by Lemma 2.2 π is a q-Weil number corresponding to an ordinary abelian variety A over $\mathbf{F} = \mathbf{F}_q$ with endomorphism algebra K and Frobenius element π. By Proposition 2.1, A has embedding degree k with respect to r. This leads to the following algorithm.

Algorithm 2.12

Input: a CM-field K of degree $2g \geq 4$, a primitive CM-type Φ of K, a positive integer k, and a prime $r \equiv 1$ (mod k) that splits completely in K.

Output: a prime q and a q-Weil number $\pi \in K$ corresponding to an ordinary, simple abelian variety $A/\mathbf{F}$ with embedding degree k with respect to r.

1. Compute a Galois closure L of K and the reflex $(\widehat{K}, \Psi)$ of (K, Φ). Set $\widehat{g} \leftarrow \frac{1}{2} \deg \widehat{K}$ and write $\Psi = \{\psi_1, \psi_2, \ldots, \psi_{\widehat{g}}\}$.
2. Fix a prime $\mathfrak{R} \mid r$ of $\mathcal{O}_L$, and compute the factorization of r in $\mathcal{O}_{\widehat{K}}$ as in (2.9).
3. Compute a primitive k-th root of unity $\zeta \in \mathbf{F}_r^*$.
4. Choose random $\alpha_1, \ldots, \alpha_{\widehat{g}-1}, \beta_1, \ldots, \beta_{\widehat{g}-1} \in \mathbf{F}_r^*$.
5. Set $\alpha_{\widehat{g}} \leftarrow \prod_{i=1}^{\widehat{g}-1} \alpha_i^{-1} \in \mathbf{F}_r^*$ and $\beta_{\widehat{g}} \leftarrow \zeta \prod_{i=1}^{\widehat{g}-1} \beta_i^{-1} \in \mathbf{F}_r^*$.
6. Compute $\xi \in \mathcal{O}_{\widehat{K}}$ such that $(\xi \bmod \mathfrak{r}_{\psi_i}) = \alpha_i$ and $(\xi \bmod \overline{\mathfrak{r}_{\psi_i}}) = \beta_i$ for $i = 1, 2, \ldots, \widehat{g}$.
7. Set $q \leftarrow N_{\widehat{K}/\mathbf{Q}}(\xi)$. If q is not prime, go to Step (4).
8. Set $\pi \leftarrow N_\Psi(\xi)$. If q is not unramified in K, or π does not generate K, go to Step (4).
9. Return q and π.

Remark 2.13. We require $g \geq 2$ in Algorithm 2.12, as the case $g = 1$ is already covered by Example 2.3, and requires a slight adaptation.

The condition that r be prime is for simplicity of presentation only; the algorithm easily extends to square-free values of r that are given as products of splitting primes. Such r are required, for example, by the cryptosystem of [1].

3   Performance of the Algorithm

Theorem 3.1. *If the field K is fixed, then the heuristic expected run time of Algorithm 2.12 is polynomial in $\log r$.*

Proof. The algorithm consists of a precomputation for the field K in Steps (1)–(3), followed by a loop in Steps (4)–(7) that is performed until an element ξ is found that has prime norm $N_{\widehat{K}/\mathbf{Q}}(\xi) = q$, and we also find in Step (8) that q is unramified in K and the type norm $\pi = N_\Psi(\xi)$ generates K.

The primality condition in Step (7) is the 'true' condition that becomes harder to achieve with increasing r, whereas the conditions in Step (8), which are necessary to guarantee correctness of the output, are so extremely likely to be fulfilled (especially in cryptographic applications where K is small and r is large) that they will hardly ever fail in practice and only influence the run time by a constant factor.

As ξ is computed in Step (6) as the lift to $\mathcal{O}_{\widehat{K}}$ of an element $\overline{\xi} \in \mathcal{O}_{\widehat{K}}/r\mathcal{O}_{\widehat{K}} \cong (\mathbf{F}_r)^{2\widehat{g}}$, its norm can be bounded by a constant multiple of $r^{2\widehat{g}}$. Heuristically, $q = N_{\widehat{K}/\mathbf{Q}}(\xi)$ behaves as a random number, so by the prime number theorem it will be prime with probability at least $(2\widehat{g}\log r)^{-1}$, and we expect that we need to repeat the loop in Steps (4)–(7) about $2\widehat{g}\log r$ times before finding ξ of prime norm q. As each of the steps is polynomial in $\log r$, so is the expected run time up to Step (7), and we are done if we show that the conditions in Step (8) are met with some positive probability if K is fixed and r is sufficiently large.

For q being unramified in K, one simply notes that only finitely many primes ramify in the field K (which is fixed) and that q tends to infinity with r, since r divides $N_{K/\mathbf{Q}}(\pi - 1) \leq (\sqrt{q} + 1)^{2g}$.

Finally, we show that π generates K with probability tending to 1 as r tends to infinity. Suppose that for every vector $v \in \{0, 1\}^{\widehat{g}}$ that is not all 0 or 1, we have

$$\textstyle\prod_{i=1}^{\widehat{g}}(\alpha_i/\beta_i)^{v_i} \neq 1. \tag{3.2}$$

This set of $2^{\widehat{g}} - 2$ (dependent) conditions on the $2\widehat{g} - 2$ independent random variables α_i, β_i for $1 \leq i < \widehat{g}$ is satisfied with probability at least $1 - (2^{\widehat{g}} - 2)/(r - 1)$. For any automorphism ϕ of L, the set $\phi \circ \Psi$ is a CM-type of $\widehat{K}$ and there is a $v \in \{0, 1\}^{\widehat{g}}$ such that $v_i = 0$ if $\phi \circ \Psi$ contains ψ_i and $v_i = 1$ otherwise. Then α_i is $(\psi_i(\xi) \bmod \mathfrak{R})$, while β_i is $(\overline{\psi_i(\xi)} \bmod \mathfrak{R})$, so $(\pi/\phi(\pi) \bmod \mathfrak{R})$ is $\prod_{i=1}^{\widehat{g}}(\alpha_i/\beta_i)^{v_i}$. By (3.2), if this expression is 1 then $v = 0$ or $v = 1$, so $\phi \circ \Psi = \Psi$ or $\phi \circ \Psi = \overline{\Psi}$, which by definition of the reflex is equivalent to ϕ or $\overline{\phi}$ being trivial on K, i.e., to ϕ being trivial on the maximal real subfield K_0. Thus if (3.2) holds, then $\phi(\pi) = \pi$ implies that ϕ is trivial on K_0, hence $K_0 \subset \mathbf{Q}(\pi)$. Since $\pi \in K$ is not real (otherwise, $q = \pi^2$ ramifies in K), this implies that $K = \mathbf{Q}(\pi)$. □

In order to maximize the likelihood of finding prime norms, one should minimize the norm of the lift ξ computed in the Chinese Remainder Step (6). This involves minimizing a norm function of degree $2\widehat{g}$ in $2\widehat{g}$ integral variables, which is already infeasible for $\widehat{g} = 2$.

In practice, for given r, one lifts a standard basis of $\mathcal{O}_{\widehat{K}}/r\mathcal{O}_{\widehat{K}} \cong (\mathbf{F}_r)^{2\widehat{g}}$ to $\mathcal{O}_{\widehat{K}}$. Multiplying those lifts by integer representatives for the elements α_i and β_i of $\mathbf{F}_r$, one quickly obtains lifts ξ. We also choose, independently of r, a $\mathbf{Z}$-basis of $\mathcal{O}_{\widehat{K}}$ consisting of elements that are 'small' with respect to all absolute values of $\widehat{K}$. We translate ξ by multiples of r to lie in rF, where F is the fundamental parallelotope in $\widehat{K} \otimes \mathbf{R}$ consisting of those elements that have coordinates in $(-\frac{1}{2}, \frac{1}{2}]$ with respect to our chosen basis.

If we denote the maximum on $F \cap \widehat{K}$ of all complex absolute values of $\widehat{K}$ by $M_{\widehat{K}}$, we have $q = N_{\widehat{K}/\mathbf{Q}}(\xi) \leq (rM_{\widehat{K}})^{2\widehat{g}}$. For the ρ-value (1.1) we find

$$\rho \leq 2g\widehat{g}(1 + \log M_{\widehat{K}}/\log r), \tag{3.3}$$

which is approximately $2g\widehat{g}$ if r gets large with respect to $M_{\widehat{K}}$. We would like ρ to be small, but this is not what one obtains by lifting random admissible choices of $\overline{\xi}$.

Theorem 3.4. *If the field K is fixed and r is large, we expect that (1) the output q of Algorithm 2.12 yields $\rho \approx 2g\widehat{g}$, and (2) an optimal choice of $\xi \in \mathcal{O}_{\widehat{K}}$ satisfying the conditions of Theorem 2.10 yields $\rho \approx 2g$.*

Open Problem 3.5. *Find an efficient algorithm to compute an element $\xi \in \mathcal{O}_{\widehat{K}}$ satisfying the conditions of Theorem 2.10 for which $\rho \approx 2g$.*

We will prove Theorem 3.4 via a series of lemmas. Let $H_{r,k}$ be the subset of the parallelotope $rF \subset \widehat{K} \otimes \mathbf{R}$ consisting of those $\xi \in rF \cap \mathcal{O}_{\widehat{K}}$ that satisfy the two congruence conditions (2.11) for a given embedding degree k. Heuristically, we will treat the elements of $H_{r,k}$ as random elements of rF with respect to the distributions of complex absolute values and norm functions. We will also use the fact that, as $\widehat{K}$ is totally complex of degree $2\widehat{g}$, the $\mathbf{R}$-algebra $\widehat{K} \otimes \mathbf{R}$ is naturally isomorphic to $\mathbf{C}^{\widehat{g}}$. We assume throughout that $g \geq 2$.

Lemma 3.6. *Fix the field K. Under our heuristic assumption, there exists a constant $c_1 > 0$ such that for all $\varepsilon > 0$, the probability that a random $\xi \in H_{r,k}$ satisfies $q < r^{2(\widehat{g}-\varepsilon)}$ is less than $c_1 r^{-\varepsilon}$.*

Proof. The probability that a random ξ lies in the set $V = \{z \in \mathbf{C}^{\widehat{g}} : \prod|z_i|^2 \leq r^{2(\widehat{g}-\varepsilon)}\} \cap rF$ is the quotient of the volume of V by the volume $2^{-\widehat{g}}\sqrt{|\Delta_{\widehat{K}}|}r^{2\widehat{g}}$ of rF, where $\Delta_{\widehat{K}}$ is the discriminant of $\widehat{K}$. Now V is contained inside $W = \{z \in \mathbf{C}^{\widehat{g}} : \prod|z_i|^2 \leq r^{2(\widehat{g}-\varepsilon)}, |z_i| \leq rM_{\widehat{K}}\}$, which has volume

$$(2\pi)^{\widehat{g}} \int_{\substack{x \in [0, rM_{\widehat{K}}]^{\widehat{g}} \\ \prod|x_i|^2 \leq r^{2(\widehat{g}-\varepsilon)}}} \prod |x_i| dx \quad < \quad (2\pi)^{\widehat{g}} \int_{x \in [0, rM_{\widehat{K}}]^{\widehat{g}}} r^{\widehat{g}-\varepsilon} dx \quad = \quad (2\pi M_{\widehat{K}})^{\widehat{g}} r^{2\widehat{g}-\varepsilon},$$

so a random ξ lies in V with probability less than $(4\pi M_{\widehat{K}})^{\widehat{g}}|\Delta_{\widehat{K}}|^{-1/2}r^{-\varepsilon}$. $\qquad \square$

Lemma 3.7. *There exists a number $Q_{\widehat{K}}$, depending only on $\widehat{K}$, such that for any positive real number $X < rQ_{\widehat{K}}$, the expected number of $\xi \in H_{r,k}$ with all absolute values below X is*

$$\frac{\varphi(k)(2\pi)^{\widehat{g}}}{|\Delta_{\widehat{K}}|} \frac{X^{2\widehat{g}}}{r^2}.$$

Proof. Let $Q_{\widehat{K}} > 0$ be a lower bound on $\widehat{K} \setminus F$ for the maximum of all complex absolute values, so the box $V_X \subset \widehat{K} \otimes \mathbf{R}$ consisting of those elements that have all absolute values below X lies completely inside $(X/Q_{\widehat{K}})F \subset rF$. The volume of V_X in $\widehat{K} \otimes \mathbf{R}$ is $(\pi X^2)^{\widehat{g}}$, while rF has volume $2^{-\widehat{g}}\sqrt{|\Delta_{\widehat{K}}|} r^{2\widehat{g}}$. The expected number of $\xi \in H_{r,k}$ satisfying $|\xi| < X$ for all absolute values is $\# H_{r,k} = r^{2\widehat{g}-2}\varphi(k)$ times the quotient of these volumes. $\square$

Lemma 3.8. *Fix the field K. Under our heuristic assumption, there exists a constant c_2 such that for all positive $\varepsilon < 2\widehat{g} - 2$, if r is sufficiently large, then we expect the number of $\xi \in H_{r,k}$ satisfying $N_{\widehat{K}/\mathbf{Q}}(\xi) < r^{2+\varepsilon}$ to be at least $c_2 r^\varepsilon$.*

Proof. Any ξ as in Lemma 3.7 satisfies $N_{\widehat{K}/\mathbf{Q}}(\xi) < X^{2\widehat{g}}$, so we apply the lemma to $X = r^{(1/\widehat{g}+\varepsilon/2\widehat{g})}$, which is less than $rQ_{\widehat{K}}$ for large enough r and $\epsilon < 2\widehat{g}-2$. $\square$

Lemma 3.9. *Fix the field K. Under our heuristic assumption, for all $\varepsilon > 0$, if r is large enough, we expect there to be no $\xi \in H_{r,k}$ satisfying $N_{\widehat{K}/\mathbf{Q}}(\xi) < r^{2-\varepsilon}$.*

Proof. Let $\widehat{\mathcal{O}}$ be the ring of integers of the maximal real subfield of $\widehat{K}$. Let U be the subgroup of norm one elements of $\widehat{\mathcal{O}}^*$. We embed U into $\mathbf{R}^{\widehat{g}}$ by mapping $u \in U$ to the vector $l(u)$ of logarithms of absolute values of u. The image is a complete lattice in the $(\widehat{g} - 1)$-dimensional space of vectors with coordinate sum 0. Fix a fundamental parallelotope F' for this lattice. Let ξ_0 be the element of $H_{r,k}$ of smallest norm. Since the conditions (2.11), as well as the norm of ξ_0, are invariant under multiplication by elements of U, we may assume without loss of generality that $l(\xi_0)$ is inside $F' + \mathbf{C}(1, \ldots, 1)$. Then every difference of two entries of $l(\xi_0)$ is bounded, and hence every quotient of absolute values of ξ_0 is bounded from below by a positive constant c_3 depending only on K. In particular, if m is the maximum of all absolute values of ξ_0, then $N_{\widehat{K}/\mathbf{Q}}(\xi) > (c_3 m)^{2\widehat{g}}$. Now suppose ξ_0 has norm below $r^{2-\varepsilon}$. Then all absolute values of ξ_0 are below $X = r^{(1/\widehat{g}-\varepsilon/2\widehat{g})}/c_3$, and $X < rQ_{\widehat{K}}$ for r sufficiently large. Now Lemma 3.7 implies that the expected number of $\xi \in H_{r,k}$ with all absolute values below X is a constant times $r^{-\varepsilon}$, so for any sufficiently large r we expect there to be no such ξ, a contradiction. $\square$

Proof (of Theorem 3.4). The upper bound $\rho \lesssim 2g\widehat{g}$ follows from (3.3). Lemma 3.6 shows that for any $\varepsilon > 0$, the probability that ρ is smaller than $2g\widehat{g} - \varepsilon$ tends to zero as r tends to infinity, thus proving the lower bound $\rho \gtrsim 2g\widehat{g}$. Lemma 3.8 shows that for any $\varepsilon > 0$, if r is sufficiently large then we expect there to exist a ξ with ρ-value at most $2g + \varepsilon$, thus proving the bound $\rho \lesssim 2g$. Lemma 3.9 shows that we expect $\rho > 2g - \varepsilon$ for the optimal ξ, which proves the bound $\rho \gtrsim 2g$. $\square$

For very small values of r we are able to do a brute-force search for the smallest q by testing all possible values of $\alpha_1, \ldots, \alpha_{\widehat{g}-1}, \beta_1, \ldots, \beta_{\widehat{g}-1}$ in Step 4 of Algorithm 2.12. We performed two such searches, one in dimension 2 and one in dimension 3. The experimental results support our heuristic evidence that $\rho \approx 2g$ is possible with a smart choice in the algorithm, and that $\rho \approx 2g\widehat{g}$ is achieved with a randomized algorithm.

Example 3.10. Take $K = \mathbf{Q}(\zeta_5)$, and let $\varPhi = \{\phi_1, \phi_2\}$ be the CM-type of K defined by $\phi_n(\zeta_5) = e^{2\pi i n/5}$. We ran Algorithm 2.12 with $r = 1021$ and $k = 2$, and tested all possible values of α_1, β_1. The total number of primes q found was 125578, and the corresponding ρ-values were distributed as follows:

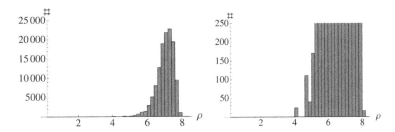

The smallest q found was 2023621, giving a ρ-value of 4.19. The curve over $\mathbf{F} = F_q$ for which the Jacobian has this ρ-value is $y^2 = x^5 + 18$, and the number of points on its Jacobian is 4092747290896.

Example 3.11. Take $K = \mathbf{Q}(\zeta_7)$, and let $\varPhi = \{\phi_1, \phi_2, \phi_3\}$ be the CM-type of K defined by $\phi_i(\zeta_7) = e^{2\pi i/7}$. We ran Algorithm 2.12 with $r = 29$ and $k = 4$, and tested all possible values of $\alpha_1, \alpha_2, \beta_1, \beta_2$. The total number of primes q found was 162643, and the corresponding ρ-values were distributed as follows:

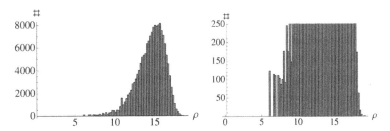

The smallest q found was 911, giving a ρ-value of 6.07. The curve over $\mathbf{F} = F_q$ for which the Jacobian has this ρ-value is $y^2 = x^7 + 34$, and the number of points on its Jacobian is 778417333.

Example 3.12. Take $K = \mathbf{Q}(\zeta_5)$, and let $\varPhi = \{\phi_1, \phi_2\}$ be the CM-type of K defined by $\phi_i(\zeta_5) = e^{2\pi i/5}$. We ran Algorithm 2.12 with $r = 2^{160} + 685$ and $k = 10$, and tested 2^{20} random values of α_1, β_1. The total number of primes q found was 7108. Of these primes, 6509 (91.6%) produced ρ-values between 7.9 and 8.0, while 592 (8.3%) had ρ-values between 7.8 and 7.9. The smallest q found had 623 binary digits, giving a ρ-value of 7.78.

4   Constructing Abelian Varieties with Given Weil Numbers

Our Algorithm 2.12 yields q-Weil numbers $\pi \in K$ that correspond, in the sense of Honda and Tate [10], to isogeny classes of ordinary, simple abelian varieties over prime fields that have a point of order r and embedding degree k with respect to r. It does not give a method to explicitly construct an abelian variety A with Frobenius $\pi \in K$. In this section we focus on the problem of explicitly constructing such varieties using complex multiplication techniques.

The key point of the complex multiplication construction is the fact that every ordinary, simple abelian variety over $\mathbf{F} = \mathbf{F}_q$ with Frobenius $\pi \in K$ arises as the reduction at a prime over q of some abelian variety A_0 in characteristic zero that has CM by the ring of integers of K. Thus if we have fixed our K as in Algorithm 2.12, we can solve the construction problem for all ordinary Weil numbers coming out of the algorithm by compiling the finite list of $\overline{\mathbf{Q}}$-isogeny classes of abelian varieties in characteristic zero having CM by $\mathcal{O}_K$. There will be one $\overline{\mathbf{Q}}$-isogeny class for each equivalence class of primitive CM-types of K, where Φ and Φ' are said to be equivalent if we have $\Phi = \Phi' \circ \sigma$ for an automorphism σ of K. As we can choose our favorite field K of degree $2g$ to produce abelian varieties of dimension g, we can pick fields K for which such lists already occur in the literature.

From representatives of our list of isogeny classes of abelian varieties in characteristic zero having CM by $\mathcal{O}_K$, we obtain a list $\mathcal{A}$ of abelian varieties over $\mathbf{F}$ with CM by $\mathcal{O}_K$ by reducing at some fixed prime $\mathfrak{q}$ over q. Changing the choice of the prime $\mathfrak{q}$ amounts to taking the reduction at $\mathfrak{q}$ of a conjugate abelian variety which also has CM by $\mathcal{O}_K$ and hence is $\overline{\mathbf{F}}$-isogenous to one already in the list.

For every abelian variety $A \in \mathcal{A}$, we compute the set of its twists, i.e., all the varieties up to $\mathbf{F}$-isomorphism that become isomorphic to A over $\overline{\mathbf{F}}$. There is at least one twist B of an element $A \in \mathcal{A}$ satisfying $\#B(\mathbf{F}) = \mathrm{N}_{K/\mathbf{Q}}(\pi - 1)$, and this B has a point of order r and the desired embedding degree.

Note that while efficient point-counting algorithms do not exist for varieties of dimension $g > 1$, we can determine probabilistically whether an abelian variety has a given order by choosing a random point, multiplying by the expected order, and seeing if the result is the identity.

The complexity of the construction problem rapidly increases with the genus $g = [K : \mathbf{Q}]/2$, and it is fair to say that we only have satisfactory general methods at our disposal in very small genus.

In genus one, we are dealing with elliptic curves. The j-invariants of elliptic curves over $\mathbf{C}$ with CM by $\mathcal{O}_K$ are the roots of the *Hilbert class polynomial* of K, which lies in $\mathbf{Z}[X]$. The degree of this polynomial is the class number h_K of K, and it can be computed in time $\widetilde{O}(|\Delta_K|)$.

For genus 2, we have to construct abelian surfaces. Any principally polarized simple abelian surface over $\overline{\mathbf{F}}$ is the Jacobian of a genus 2 curve, and all genus 2 curves are hyperelliptic. There is a theory of class polynomials analogous to that for elliptic curves, as well as several algorithms to compute these polynomials, which lie in $\mathbf{Q}[X]$. The genus 2 algorithms are not as well-developed as those for elliptic curves; at present they can handle only very small quartic CM-fields, and there exists no rigorous run time estimate. From the roots in $\mathbf{F}$ of these polynomials, we can compute the genus 2 curves using Mestre's algorithm.

Any three-dimensional principally polarized simple abelian variety over $\overline{\mathbf{F}}$ is the Jacobian of a genus 3 curve. There are two known families of genus 3 curves over $\mathbf{C}$ whose Jacobians have CM by an order of dimension 6. The first family, due to Weng [14], gives hyperelliptic curves whose Jacobians have CM by a degree-6 field containing $\mathbf{Q}(i)$. The second family, due to Koike and Weng [5], gives Picard curves (curves of the form $y^3 = f(x)$ with $\deg f = 4$) whose Jacobians have CM by a degree-6 field containing $\mathbf{Q}(\zeta_3)$.

Explicit CM-theory is mostly undeveloped for dimension ≥ 3. Moreover, most principally polarized abelian varieties of dimension ≥ 4 are not Jacobians, as the moduli space of Jacobians has dimension $3g - 3$, while the moduli space of abelian varieties has dimension $g(g + 1)/2$. For implementation purposes we prefer Jacobians or even hyperelliptic Jacobians, as these are the only abelian varieties for which group operations can be computed efficiently.

In cases where we cannot compute every abelian variety in characteristic zero with CM by $\mathcal{O}_K$, we use a single such variety A and run Algorithm 2.12 for each different CM-type of K until it yields a prime q for which the reduction of A mod q is in the correct isogeny class. An example for $K = \mathbf{Q}(\zeta_{2p})$ with p prime is given by the Jacobian of $y^2 = x^p + a$, which has dimension $g = (p - 1)/2$.

5   Numerical Examples

We implemented Algorithm 2.12 in MAGMA and used it to compute examples of hyperelliptic curves of genus 2 and 3 over fields of cryptographic size for which the Jacobians are pairing-friendly. The subgroup size r is chosen so that the discrete logarithm problem in $A[r]$ is expected to take roughly 2^{80} steps. The embedding degree k is chosen so that $r^{k/g} \approx 1024$; this would be the ideal embedding degree for the 80-bit security level if we could construct varieties with $\#A(\mathbf{F}) \approx r$. Space constraints prevent us from giving the group orders for each Jacobian, but we note that a set of all possible q-Weil numbers in K, and hence all possible group orders, can be computed from the factorization of q in K.

Example 5.1. Let $\eta = \sqrt{-2 + \sqrt{2}}$ and let K be the degree-4 Galois CM field $\mathbf{Q}(\eta)$. Let $\Phi = \{\phi_1, \phi_2\}$ be the CM type of K such that $\mathrm{Im}(\phi_i(\eta)) > 0$. We ran Algorithm 2.12 with CM type (K, Φ), $r = 2^{160} - 1679$, and $k = 13$. The algorithm output the following field size:

$q =$ 31346057808293157913762344531005275715544680219641338497449500238872300350617165 \
     40892530853973205578151445285706963588204818794198739264123849002104890399459807 \
     46313273247715465151766675702167   (640 bits)

There is a single $\overline{\mathbf{F}}_q$-isomorphism class of curves over $\mathbf{F}_q$ whose Jacobians have CM by $\mathcal{O}_K$ and it has been computed in [11]; the desired twist turns out to be $C : y^2 = -x^5 + 3x^4 + 2x^3 - 6x^2 - 3x + 1$. The ρ-value of $\mathrm{Jac}(C)$ is 7.99.

Example 5.2. Let $\eta = \sqrt{-30 + 2\sqrt{5}}$ and let K be the degree-4 non-Galois CM field $\mathbf{Q}(\eta)$. The reflex field $\widehat{K}$ is $\mathbf{Q}(\omega)$ where $\omega = \sqrt{-15 + 2\sqrt{55}}$. Let Ψ be the CM type of K such that $\mathrm{Im}(\phi_i(\eta)) > 0$. We ran Algorithm 2.12 with the CM type (K, Φ), subgroup size $r = 2^{160} - 1445$, and embedding degree $k = 13$. The algorithm output the following field size:

$q = $ 11091654887169512971365407040293599579976378158973405181635081379157078302130927 \
51652003623786192531077127388944453303584091334492452752693094089192986541533819 \
35518866167783400231181308345981461 (645 bits)

The class polynomials for K can be found in the preprint version of [13]. We used the roots of the class polynomials mod q to construct curves over $\mathbf{F}_q$ with CM by $\mathcal{O}_K$. As K is non-Galois with class number 4, there are 8 isomorphism classes of curves in 2 isogeny classes. We found a curve C in the correct isogeny class with equation $y^2 = x^5 + a_3 x^3 + a_2 x^2 + a_1 x + a_0$, with

$a_3 = $ 37909827361040902434390338072754918705969566622865244598340785379492062293493023 \
07887220632471591953460261515915189503199574055791975955834407879578484212700263 \
2600401437108457032108586548189769
$a_2 = $ 18960350992731066141619447121681062843951822341216980089632110294900985267348927 \
56700435114431697785479098782721806327279074708206429263751983109351250831853735 \
1901282000421070182572671506056432
$a_1 = $ 69337488142924022910219499907432470174331183248226721112535199929650663260487281 \
50177351432967251207037416196614255668796808046612641767922273749125366541534440 \
5882465731376523304907041006464504
$a_0 = $ 31678142561939596895646021753607012342277658384169880961095701825776704126204818 \
48230687778916790603969757571449880417861689471274167016388608712966941178120424 \
3813332617272038494020178561119564

The ρ-value of $\mathrm{Jac}(C)$ is 8.06.

Example 5.3. Let K be the degree-6 Galois CM field $\mathbf{Q}(\zeta_7)$, and let $\Phi = \{\phi_1, \phi_2, \phi_3\}$ be the CM type of K such that $\phi_n(\zeta_7) = e^{2\pi i n/7}$. We used the CM type (K, Φ) to construct a curve C whose Jacobian has embedding degree 17 with respect to $r = 2^{180} - 7427$. Since K has class number 1 and one equivalence class of primitive CM types, there is a unique isomorphism class of curves in characteristic zero whose Jacobians are simple and have CM by K; these curves are given by $y^2 = x^7 + a$. Algorithm 2.12 output the following field size:

$q = $ 15755841381197715359178780201436879305777694686713746395506787614025008121759749 \
72634937716254216816917600718698808129260457040637146802812702044068612772692590 \
77188966205156107806823000096120874915612017184924206843204621759232946263357637 \
19251697987740263891168971441085531481109276328740299111531260484082698571214310 \
33499 (1077 bits)

The equation of the curve C is $y^2 = x^7 + 10$. The ρ-value of $\mathrm{Jac}(C)$ is 17.95.

We conclude with an example of an 8-dimensional abelian variety found using our algorithms. We started with a single CM abelian variety A in characteristic zero and applied our algorithm to different CM-types until we found a prime q for which the reduction has the given embedding degree.

Example 5.4. Let $K = \mathbf{Q}(\zeta_{17})$. We set $r = 1021$ and $k = 10$ and ran Algorithm 2.12 repeatedly with different CM types for K. Given the output, we tested the Jacobians of twists of $y^2 = x^{17} + 1$ for the specified number of points. We found that the curve $y^2 = x^{17} + 30$ has embedding degree 10 with respect to r over the field $\mathbf{F}$ of order

$$q = 6869603508322434614854908535545208978038819437.$$

The CM type was

$$\Phi = \{\phi_1, \phi_3, \phi_5, \phi_6, \phi_8, \phi_{10}, \phi_{13}, \phi_{15}\},$$

where $\phi_n(\zeta_{17}) = e^{2\pi in/17}$. The ρ-value of $\mathrm{Jac}(C)$ is 121.9.

References

1. Boneh, D., Goh, E.-J., Nissim, K.: Evaluating 2-DNF formulas on ciphertexts. In: Kilian, J. (ed.) TCC 2005. LNCS, vol. 3378, pp. 325–341. Springer, Heidelberg (2005)
2. Freeman, D., Scott, M., Teske, E.: A taxonomy of pairing-friendly elliptic curves. In: Cryptology eprint 2006/371, http://eprint.iacr.org
3. Frey, G., Rück, H.: A remark concerning m-divisibility and the discrete logarithm in the divisor class group of curves. Math. Comp. 62, 865–874 (1994)
4. Galbraith, S.: Supersingular curves in cryptography. In: Boyd, C. (ed.) ASIACRYPT 2001. LNCS, vol. 2248, pp. 495–513. Springer, Heidelberg (2001)
5. Koike, K., Weng, A.: Construction of CM Picard curves. Math. Comp. 74, 499–518 (2004)
6. Menezes, A., Okamoto, T., Vanstone, S.: Reducing elliptic curve logarithms to logarithms in a finite field. IEEE Transactions on Information Theory 39, 1639–1646 (1993)
7. Paterson, K.: Cryptography from pairings. In: Blake, I.F., Seroussi, G., Smart, N.P. (eds.) Advances in Elliptic Curve Cryptography, pp. 215–251. Cambridge University Press, Cambridge (2005)
8. Rubin, K., Silverberg, A.: Supersingular abelian varieties in cryptology. In: Yung, M. (ed.) CRYPTO 2002. LNCS, vol. 2442, pp. 336–353. Springer, Heidelberg (2002)
9. Shimura, G.: Abelian Varieties with Complex Multiplication and Modular Functions. Princeton University Press, Princeton (1998)
10. Tate, J.: Classes d'isogénie des variétés abéliennes sur un corps fini (d'après T. Honda). Séminaire Bourbaki 1968/69, Springer Lect. Notes in Math. 179, exposé 352 pp. 95–110 (1971)
11. van Wamelen, P.: Examples of genus two CM curves defined over the rationals. Math. Comp. 68, 307–320 (1999)
12. Waterhouse, W.C.: Abelian varieties over finite fields. Ann. Sci. École Norm. Sup. 2(4), 521–560 (1969)
13. Weng, A.: Constructing hyperelliptic curves of genus 2 suitable for cryptography. Math. Comp. 72, 435–458 (2003)
14. Weng, A.: Hyperelliptic CM-curves of genus 3. Journal of the Ramanujan Mathematical Society 16(4), 339–372 (2001)

Almost Prime Orders
of CM Elliptic Curves Modulo p

Jorge Jiménez Urroz[*]

CRM, Universite de Montreal

Abstract. Given an elliptic curve over $\mathbb{Q}$ with complex multiplication by $\mathcal{O}_K$, the ring of integers of the quadratic imaginary field K, we analyze the integer $d_E =\gcd\{|E(\mathbb{F}_p)| : p$ splits in $\mathcal{O}_K\}$, where $|E(\mathbb{F}_p)|$ is the size of the group of rational $\mathbb{F}_p$ points, and prove that it can be bigger than the common factor that comes from the torsion of the curve. Then, we prove that $\#\{p \le x, p$ splits in $\mathcal{O}_K : \frac{1}{d_E}|E(\mathbb{F}_p)| = P_2\} \gg x/(\log x)^2$ hence extending the results in [16]. This is the best known result in the direction of the Koblitz conjecture about the primality of $|E(\mathbb{F}_p)|$.

1 Introduction and Statements of Results

There is a rich literature in the study of the structure and size of the group of points over finite fields of complex multiplication elliptic curves that is becoming each day more extensive and diverse. One of the reasons to study these groups comes from Cryptography. Indeed, In general, cryptosystems built over the group of points of a certain elliptic curve guarantee a high level of security, with a lower cost in the size of the keys, whenever the order of the group has a big prime divisor. It is in this way that the problem of finding a finite field $\mathbb{F}_p$, and a curve $E/\mathbb{F}_p$ defined over the field, such that $|E(\mathbb{F}_p)|$ has a prime factor as large as possible, arose. In practice one can make a random selection of this pair of a curve and field. However, the theory that one would need to analyse the utility of this random algorithm is complex and neither clear nor complete. Suppose $E/\mathbb{Q}$ is an elliptic curve defined over the rationals, and let $E(\mathbb{F}_p)$ denote the group of $\mathbb{F}_p$ points of the reduced curve modulo p, a prime of good reduction, (from now on we will restrict always to primes of good reduction). Somehow we have to ensure that, for x sufficiently large, many of the elements of the sequence $\hat{A}(x) = \{|E(\mathbb{F}_p)| : p \le x\}$ have a big prime divisor. One important remark at this point is that, since the reduction modulo p injects the torsion subgroup of the curve $E(\mathbb{Q})_{\text{tors}}$ into $E(\mathbb{F}_p)$ for almost all primes p, whenever this is nontrivial, (for E or any of its isogenus curves), almost all the elements of the sequence $\hat{A}(x)$ will have a small common divisor. In this sense, if d is this common factor, we will be considering the more convenient sequence $\mathcal{A}(x) = \{\frac{1}{d}|E(\mathbb{F}_p)| : p \le x\}$.

---

[*] Partially supported by Secretaría de Estado de Universidades e Investigación del Ministerio de Educación y Ciencia of Spain, DGICYT Grants MTM2006-15038-C02-02 and TSI2006-02731.

A.J. van der Poorten and A. Stein (Eds.): ANTS-VIII 2008, LNCS 5011, pp. 74–87, 2008.

This sequence has being widely studied in the literature. In 1988 Koblitz [19] conjectured that for any elliptic curve over the rationals without rational torsion in its $\mathbb{Q}$-isogeny class, the elements in $\mathcal{A}$ not only have a big prime factor very frequently, but in fact infinitely many of them are themselves prime numbers. Concretely if we denote by $\Pi_E(x)$ the function which counts the number of primes in $\mathcal{A}(x)$, then he claims that there exists a constant $c > 0$, depending on the curve, such that $\Pi_E(x) \sim cx/(\log x)^2$ as $x \to \infty$.

But there are other reasons why one would like to know the factorization of the elements in $\mathcal{A}(x)$. In 1977 Lang and Trotter conjectured that, given an elliptic curve E and a nontorsion point $P \in E(\mathbb{Q})$, the density of primes p for which P generates $E(\mathbb{F}_p)$ exists. In these cases the point P is called a primitive point. In particular they predict that the group of $\mathbb{F}_p$-points of the reduced curve mod p is cyclic for many primes p. Since then there has been an extensive study, either of the conjecture itself, or on the cyclicity of the group of $\mathbb{F}_p$-points. A few examples can be found in [3], [6], [11], [20] or [24].

We could find lower bounds for the size of the prime factors, and ensure cyclicity of the group, both at the same time, if we were able to prove that many elements in $\mathcal{A}(x)$ are squarefree with a very small number of prime divisors. In general we say that an integer n is P_r if it is squarefree with at most r prime factors and if $r = 2$ we say our number is almost prime. Finding P_r numbers among the elements of a certain sequence is at the heart of sieve theory. However, it is important to note that, even though using sieve methods is the most efficient way to attack this kind of problems, it does not provide, at least considered in its classical way, lower bounds for the number of primes in certain sequences due to the parity problem. In fact, when $r = 1$, although the result is known on average, (see [2]), there is not a single example of a curve for which the asymptotics predicted by Koblitz have been proved.

For $r > 1$, now with sieve equipment available, the situation is a little bit more promising. Miri and Murty in [21] proved, assuming the Grand Riemann Hipothesis, GRH, that for curves without complex multiplication $|\{P_{16} \in \mathcal{A}(x)\}| \gg x/(\log x)^2$. In [26], (see also [27]), Steuding and Weng improved the previous result giving $|\{P_6 \in \mathcal{A}(x)\}| \gg x/(\log x)^2$ for non-CM curves. They also proved $|\{P_4 \in \mathcal{A}(x)\}| \gg x/(\log x)^2$ in the CM case, but always under GRH, and recently Cojocaru in [7] proved unconditionally that for CM elliptic curves, with $d = 1$ in $\mathcal{A}(x)$, $|\{P_5 \in \mathcal{A}(x)\}| \gg x/(\log x)^2$. The best known result nowadays is due to Iwaniec and the author of this paper in [16], were they prove $|\{P_2 \in \mathcal{A}(x)\}| \gg x/(\log x)^2$ for the elliptic curve $y^2 = x^3 - x$. The main object of this paper is to complete the program initiated in this last reference, by extending the result to any curve with complex multiplication. Therefore we will consider curves over $\mathbb{Q}$ with complex multiplication by O_K, the ring of integers of the quadratic field K. Note that any elliptic curve over $\mathbb{Q}$ can only have complex multiplication by one of the nine imaginary quadratic fields of class number one, namely those with discriminant $D = -3, -4, -8, -7, -11, -19, -43, -67, -163$. Hence, we can summarize the possible equations of CM elliptic curves as

$$y^2 = x^3 + g^2\alpha x + g^3\beta, \quad y^2 = x^3 + g, \text{ or } y^2 = x^3 + gx, \tag{1}$$

where g is any integer so the equation is nonsingular, and α and β are fixed, given in Table 2 below. The first equation is for the case when $D \neq -3, -4$, and the other two are the cases $D = -4$ and $D = -3$ respectively. Moreover, we know that for any prime p of ordinary reduction, the number of $\mathbb{F}_p$ points is given by

$$|E(\mathbb{F}_p)| = p + 1 - (\pi + \overline{\pi}) = N(\pi - 1), \tag{2}$$

for a certain $\pi \in \mathcal{O}_K$ of norm $N(\pi) = p$. On the other hand, the reduction is supersingular for any inert prime in K, i.e. any prime such that $\left(\frac{D}{p}\right) = -1$. Let N_E be the conductor of the curve and let d_E be the integer defined by $d_E = \gcd\{|E(\mathbb{F}_p)|, p \text{ splits in } \mathcal{O}_K, p \nmid 6N_E\}$. Observe that this integer depends on the torsion of the curves in the isogeny class of E. Then, we can prove the following result.

Theorem 1. *Let $E/\mathbb{Q}$ be an elliptic curve with complex multiplication by $\mathcal{O}_K$, the ring of integers of the imaginary quadratic field K. For $x \geq 5$ we have*

$$\left|\left\{p \leq x \,, p \text{ splits in } \mathcal{O}_K \,, \frac{1}{d_E}|E(\mathbb{F}_p)| = P_2\right\}\right| \gg x/(\log x)^2.$$

Theorem 1 is the natural generalization of Theorem 2 in [16], and the proof goes exactly along the same lines of reasoning. However, when considering any elliptic curve of complex multiplication, several interesting facts appear naturally from this generalization which are covered in Section 2. First of all comes the number d_E. In general, in the complex multiplication case, we have a very precise definition of the appropriate prime π to be chosen in (2), (see [25], [1], [13], [23]). From there we can deduce, (and we will do it below), that the integer d_E can be any of the divisors of 24 except 6 and 24. We will describe each of these cases. Also, by looking at (2) we see that Theorem 1 is clearly related to the Twin Prime Conjecture, in this case, in the domain $\mathcal{O}_K$ and, hence, the theorem can be considered as analogous to Chen's celebrated theorem, now in the corresponding domain $\mathcal{O}_K$. Hence, for the proof of the theorem, we will need to adjust the switching principle of sieve theory to this context. This will be done by using two generalizations of the Bombieri-Vinogradov theorem, first to the field K, and then for P_3 type numbers in $\mathcal{O}_K$. For the former, as in [16], we will appeal to [17] which is suitable to our particular case. The second generalization we have mentioned is the content of Proposition 3 of Section 6 below, (see Proposition 5 in [16]). It might be interesting to remark that, in order to improve the previous results in [21], [26], and [7], apart from the Bombieri-Vinogradov Theorem, which in this context is even more efficient that any version of the Riemann Hypothesis, and the switching principle, one needs to increase the level of distribution in the sequence by discarding the inert primes, (which contribute as squares).

Let us finish this introduction by mentioning the relation of Theorem 1 with the study of primitive points. Let E be an elliptic curve with positive rank, and let $P \in E(\mathbb{Q})$ be a point of infinite order. From the work of Gupta and Murty, [11], it is known that for CM curves and under the Grand Riemann Hypothesis, (GRH), the set of primes such that $\bar{P}$, the reduction of P mod p, generates $E(\mathbb{F}_p)$ has a density over the set of primes and it is also known that this density is positive in certain cases. But nothing is known unconditionally. In Section 7 we include some discussion of the (mild) consequences of Theorem 1 in this direction. Finally, Remark 1, is intended to show the more relevant impact of the idea in Section 7 in the inert case.

2   On the Integer d_E

It is well known that the torsion subgroup of the elliptic curve E injects into the reduction modulo p for all but finitely many primes p. In those cases, $|\hat{E}(\mathbb{F}_p)|$ will always be divisible by the order of the torsion, for any $\hat{E}$ in the isogeny class of E. Moreover the restriction to primes in certain congruence classes, the splitting primes, and hence of those π above them, could cause extra divisibility in (2). This is the role that plays d_E in Theorem 1. We devote this section to present the precise value of $d_E = \gcd\{|E(\mathbb{F}_p)|, p$ splits in $\mathcal{O}_K, p \nmid 6N_E\}$.

Proposition 1. *Let $E/\mathbb{Q}$ be an elliptic curve with complex multiplication by $\mathcal{O}_K$, defined by the equation $y^2 = x^3 + g_4 x + g_6$, and conductor N_E. Then $d_E \mid 24$ and its precise value is given in Table 1.*

Table 1. The integer d_E in terms of the equation. Here g is any integer and we write m to denote an integer such that there is no intersection between any two rows of the table.

D	(g_4, g_6)	d_E
-4	$(-g^4, 0), (4g^4, 0)$	8
-4	$(m^2, 0), (-m^2, 0)$	4
-4	$(m, 0)$	2
-8	$(-30g^2, -56g^3)$	2
-3	$(0, g^6), (0, -27g^6)$	12
-3	$(0, m^3)$	4
-3	$(0, m^2), (0, -27m^2)$	3
-3	$(0, m)$	1
-7	$(-140g^2, -784g^3)$	4
-11	$(-1056g^2, -13552g^3)$	1
-19	$(-608g^2, -5776g^3)$	1
-43	$(-13760g^2, -621264g^3)$	1
-67	$(-117920g^2, -15585808g^3)$	1
-163	$(-34790720g^2, -78984748304g^3)$	1

The following corollary may be of independent interest.

Corollary 1. *Let $E/\mathbb{Q}$ and elliptic curve with complex multiplication by $\mathcal{O}_K$ the ring of integers of a field with discriminant $-D > 7$. Then the torsion subgroup $E_{\text{tors}}(\mathbb{Q})$ is trivial.*

It is interesting to observe that, in the cases where $d_E > 1$, the curves have rational points of torsion whose orders do not always coincide with d_E. In other words, d_E does not come wholly from the torsion of the curve, but some part definitely belongs to the complex multiplication. On the other hand, when considering the integer $\gcd\{|E(\mathbb{F}_p)|, p \nmid 6N_E\}$, i.e, considering every prime of good reduction, it is indeed the order of the torsion subgroup. This can be easily checked with the equation. It might be interesting to compare Proposition 1 with Theorem 2 (bis) of [18]. Whereas that theorem is true only for a set of primes of density 1, here we only need a set of primes of Čebotarev type, $\mathcal{P}$, to ensure that whenever m— divides $|E(\mathbb{F}_p)|$ for any $p \in \mathcal{P}$, then m comes from the torsion of the curve.

Proof (of Proposition 1). We will split the proof into different cases depending on the value of the discrimininant D of the CM field of the curve.

• Case $-D > 11$

It is clear that $|(O_K/\lambda O_K)^*| \geq 3$ for any prime $\lambda \in O_K$. Note that this is true since 2 and 3 are inert primes in any of these fields. Moreover ± 1 are the only units so, if $\pi \equiv \alpha \pmod{\lambda}$ is a splitting prime such that neither α nor $-\alpha$ are 1 modulo λ, then $N_p = N(\pi - 1)$ can not be multiple of l for any choice of π above p, where $N(\lambda) = l$. We know that there exist infinitely many such primes $\mathfrak{p}$ by Čebotarev's theorem.

• Case $-D = 11$

We first prove that 3 is not a common divisor of N_p for every p. By [25] we know that, for any given prime p of ordinary reduction, the number of points over $\mathbb{F}_p$ of the curve $E_{11,g}$ defined by $y^2 = 4x^3 - 264g^2x - 847g^3$ is given by

$$N_p = p + 1 + \left(\frac{-3g}{p}\right)\left(\frac{u}{11}\right)u, \tag{3}$$

where $\pi = (u + v\sqrt{-11})/2$ is any prime above p so, in particular, $4p = u^2 + 11v^2$. If we let α, β to be primes above 3 and 11 respectively, then $\pi \equiv m \pmod{\alpha\beta}$ for some integer $0 \leq m \leq 32$ coprime with 33. In this case $\bar{\pi} \equiv m \pmod{\bar{\alpha}\bar{\beta}}$, and so $mu \equiv p + m^2 \pmod{33}$. Suppose $g = -b^2$ for some integer b. Then, taking $m = 13$, $p \equiv 1 \pmod 3$, $u \equiv 2 \pmod 3$ and $u \equiv 4 \pmod{11}$, and we get $N_p \equiv 1 \pmod 3$. If, on the other hand, $-g$ is not a perfect square, then choosing $m = 1$ we have $p \equiv 1 \pmod 3$ and $u \equiv 2 \pmod 3$ so $N_p \equiv 2(1 - (-g/p)) \pmod 3$. It is now enought to choose π such that $(-a/p) = -1$ to get $N_p \equiv 1 \pmod 3$. Again the Čebotarev density theorem guarantees the existence of infinitely many primes with the required properties in each case. In particular $E_{11,a}$ does not have 3 torsion for any a. One can prove this fact easily by showing that $2P = -P$ does not have rational solutions. Observe that, on the other hand, for any prime $p \equiv 2 \pmod 3$ indeed $3 | N_p$ since, in this case, $u \equiv 0 \pmod 3$. For primes other than 3 the argument is the same as in the previous case.

• **Case** $-D \leq 8$

For the rest of the cases of Proposition 1 the arguments are very similar and rely upon three facts, namely Čebotarev's theorem, the formula N_p as a norm in the corresponding field of CM, and the explicit formula for the number of points N_p in terms of characters. Then, straighforward calculations, similar to those made in the previous cases, give the results shown in the table of the proposition. We omit these calculations since they can be easily performed by the reader. We recall that in these cases the only primes that can divide d_E are $2, 3$, and 7 in the case $D = -7$. The argument to discard higher powers of 2 and 3 is also achieved by a proper selection of primes in O_K in certain aritmetic progression. We will include the explicit formula for the number of points for the convenience of the reader. In any event this formula can be found in either [25], [1] or Chapter 18 of [13] for the case $D = -3, -4$, and any of these references would also make interesting reading. In particular, in [23], an explicit formula for the number of points is given, which is valid for CM curves either defined over a field extension, or with a ring of endomorphisms that is strictly smaller than the maximal order of the field.

In order to state the formula we will use the following convention: a prime $\pi \in O_K$, $\pi = (u+v\sqrt{D})/2$, is primary if $\pi \equiv 1 \pmod{2(1+i)}$ and $K = \mathbb{Q}(\sqrt{-1})$, if $\pi \equiv 2 \pmod 3$ and $K = \mathbb{Q}(\sqrt{-3})$, or if $\mathrm{Re}(\pi) > 0$ in all other cases. Then, for the elliptic curve $E := y^2 = x^3 + g_4 x + g_6$ and $a_p = p+1-N_p$ we have the data in Table 2.

Table 2. Formula for the number of points over $\mathbb{F}_p$ in terms of the equation; here $(\cdot)_m$ is the m-th residue symbol, $\chi_{\pi,8}(g) = -(g/p)(-1)^k(-1/U)u$ for $U = u/2$, and $k = [p/8]$, and $\chi_{\pi,d}(g) = \varepsilon(\varepsilon g/p)(u/d)u$ with $\varepsilon = (-1)^{(d^2-1)/2}$ for the rest

D	(g_4, g_6)	a_p
$D = -3$	$(0, g)$	$-\left(\frac{4g}{\pi}\right)_6 \pi - \left(\frac{4g}{\pi}\right)_6 \bar{\pi}$
$D = -4$	$(-g, 0)$	$\left(\frac{g}{\pi}\right)_4 \pi + \left(\frac{g}{\pi}\right)_4 \bar{\pi}$
$D = -8$	$(-5 \cdot 2g^2/3, -14 \cdot 2^2 g^3/27)$	$\chi_{\pi,8}(g)u$
$D = -7$	$(-5 \cdot 7g^2/16, -7^2 g^3/32)$	$\chi_{\pi,7}(g)u$
$D = -11$	$(-2 \cdot 11g^2/3, -7 \cdot 11^2 g^3/108)$	$\chi_{\pi,11}(g)u$
$D = -19$	$(-2 \cdot 19g^2, -19^2 g^3/4)$	$\chi_{\pi,19}(g)u$
$D = -43$	$(-20 \cdot 43g^2, -21 \cdot 43^2 g^3/4)$	$\chi_{\pi,43}(g)u$
$D = -67$	$(-110 \cdot 67g^2, -217 \cdot 67^2 g^3/4)$	$\chi_{\pi,67}(g)u$
$D = -163$	$(-13340 \cdot 163g^2, -185801 \cdot 163^2 g^3/4)$	$\chi_{\pi,163}(g)u$

3   A Weighted Sum for the Sieve Problem

We start with the notation that will be used later. From now on $E/\mathbb{Q}$, given by the equation $y^2 = x^3 + ag^2x + \beta g^3$ as in Table 2, is a curve of complex multiplication by O_K, the maximal order in the field K. To simplify the computations we will consider $(6, g) = 1$. As usual, for any sequence of rational integers C, and a positive number x, we have $C(x) = \{c \in C : c \leq x\}$, and $|C(x)|$ is the number of elements in the set. Given an integer d, the set $C_d = \{c \in C : d|c\}$ consists of the elements of C which are multiples of d and $S(C, d) = |\{c \in C : (c, d) = 1\}|$ is the number of elements in C coprime with d. Analogously we define $\mathfrak{C}_\delta$ and $S(\mathfrak{C}, \delta)$ for $\mathfrak{C} \subset \mathcal{O}_K$ and $\delta \in \mathcal{O}_K$. We will also make several useful conventions. From now on $\lambda, \lambda_1, \lambda_2, \ldots$, denote primes in $\mathcal{O}_K$ and $l, l_1, l_2, \ldots$ the rational primes below them; similarly p, p_0, p_1, p_2, p_3 will be rational primes that split in $\mathcal{O}_K$, and $\pi, \pi_0, \pi_1, \pi_2, \pi_3$ will denote primary primes above them. On the other hand q will be an inert rational prime inert. Finally p_K will denote the unique rational prime which ramifies in O_K. Let

$$P(z) = \prod_{\substack{p < z \\ p, \text{ split}}} p, \text{ and } Q(z) = \prod_{\substack{q < z \\ q, \text{ inert}}} q,$$

where the products refer to the corresponding domain O_K we are considering in each case. As mentioned in the introduction, the proof of Theorem 1 goes along the line of Theorem 2 in [16], hence we give only a sketch of the proof, which can be completed with the details given there. We first translate the problem in terms of integers in the domain $\mathcal{O}_K$. Let $\delta_E = 2(1+i), (1+i)^2, 1+i, \sqrt{-2}, 2\sqrt{-3}, 2, \sqrt{-3}, (1 + \sqrt{-7})/2$, be an integer in the corresponding O_K with $N(\delta_E) = d_E$ whenever $d_E > 1$, and $\delta_E = 1$ otherwise. Let $\chi_\pi(E)$ be the character as given in Table 2, and let α_0 be such that for any $\pi \equiv \alpha_0 \pmod{\delta_E g}$, $\chi_\pi(E) = \zeta \in O_K^*$ is constant and $((\pi - \zeta)/\delta_E, \delta_E g) = 1$. The existence of this α_0 is guaranteed by the corresponding reciprocity law in each field whenever $(6, g) = 1$. Consider

$$\mathcal{P}(x) = \{\pi \text{ prime }, \pi\bar{\pi} \leq x, \pi \equiv \alpha_0 \pmod{\delta_E g}\}$$

and the sequence, (to be sifted later),

$$\mathcal{A}(x) = \left\{a = N\left(\frac{\pi - \zeta}{\delta_E}\right), \pi \in \mathcal{P}(x)\right\}.$$

Observe that, in this case, $\frac{\pi - \zeta}{\delta_E}$ is indeed an integer which is coprime with gd_E. Also let

$$S(x) = \sum_{P_2 \in \mathcal{A}(x)} 1.$$

Then it is clear that $S(x)$ is a constant times the left hand side of the inequality in Theorem 1 and, therefore, it suffices to prove that

$$S(x) \gg x/(\log x)^2. \tag{4}$$

Consider now the weighted sum given by

$$W(x) = \sum_{\substack{a \in \mathcal{A}(x) \\ (a, P(z)Q(z)p_K)=1}} \left\{ 1 - \sum_{\substack{p_0 | a \\ z < p_0 \leq y}} \frac{1}{2} - \sum_{\substack{a = p_1 p_2 p_3 \\ z < p_3 \leq u < p_2 < p_1}} \frac{1}{2} \right\}$$

$$= \sum_{\substack{a \in \mathcal{A}(x) \\ (a, P(z)Q(z)p_K)=1}} 1 - \frac{1}{2} \sum_{z < p_0 \leq y} \sum_{\substack{ap_0 \in \mathcal{A}(x) \\ (a, P(z)Q(z))=1}} 1 - \frac{1}{2} \sum_{\substack{z < p_3 \leq u < p_2 < p_1 \\ p_3 p_2 p_1 \in \mathcal{A}(x)}} 1$$

$$= W_1(x) - \frac{1}{2}W_2(x) - \frac{1}{2}W_3(x), \tag{5}$$

where $z = x^{1/8}$ and $y = x^{1/3}$. As in [16], any term with positive weight in $W(x)$ is either P_2 or divisible by some nontrivial square, and the contribution from non-squarefree elements is negligible. So, in order to prove the theorem, we need the estimation

$$W(x) \gg x/(\log x)^2.$$

We will estimate $W_1(x)$, $W_2(x)$, $W_3(x)$ separately.

4   Lower Bound for $W_1(x)$

$W_1(x)$ is the classical sieve sum and, hence, to estimate it we need to control $|\mathcal{A}_d(x)|$. Using Lemma 3 of [16], we can reduce the study of $|\mathcal{A}_d(x)|$ to that of $S(\mathfrak{A}(x), \kappa)$ for $\kappa | d$, where

$$\mathfrak{A}(x) = \left\{ \frac{\pi - \zeta}{\delta_E} : \pi \in \mathcal{P}(x) \right\}.$$

We now introduce a slight modification of the generalization of the Bombieri-Vinogradov theorem to imaginary quadratic fields, given by Johnson in the corollary on page 203 of [17]. In particular, for a general ideal $\mathfrak{a} \in O_K$, and integer $\alpha \in O_K$, we write

$$\Pi(x; \mathfrak{a}, \alpha) = \sum_{\substack{\pi \equiv \alpha \,(\mathrm{mod}\ \mathfrak{a}) \\ \pi \in \mathcal{P}(x)}} 1,$$

and $\Pi'(x; \mathfrak{a}, \alpha)$ will be the analogous sum but restricted to primary primes.

Proposition 2. *Let $g \in \mathbb{Z}$ and let α_0 be an integer in O_K. Then we have*

$$\sum_{\substack{N(\mathfrak{a}) \leq Q \\ (\mathfrak{a}, g) = 1}} \max_{(\alpha, \mathfrak{a}g) = 1} \left| \Pi(x; \mathfrak{a}g, \alpha) - \frac{|O_K^*|}{\Phi(\mathfrak{a})} \Pi'(x; g, \alpha_0) \right| \ll \frac{x}{(\log x)^A} \tag{6}$$

where $Q = \sqrt{x}/(\log x)^B$, and $\Phi(\mathfrak{a}) = |(O_K/\mathfrak{a})^*|$. Here A is any positive number and B and the implied constant depend only on A.

Proof. The proof follows from the corollary on page 203 of [17] and the triangle inequality.

Following the same reasoning as in Section 4 of [16], we get

$$|\mathcal{A}_d(x)| = \Pi(x; \delta_E g, \alpha_0) h(d) + r_d(x) = \frac{|O_K^*|}{\varphi(\delta_E)} \Pi'(x; g, \alpha_0) h(d) + r_d(x), \tag{7}$$

where $h(\cdot)$ is a multiplicative function such that $h(l) = 0$ for any prime $l | g$ by our selection of α_0, $h(p) = 2/(p-1) + O(1/p^2)$ for splitting primes and for all other primes q we have $h(q) = 1/(q^2 - 1)$. Moreover

$$\sum_{d \leq \sqrt{x}/(\log x)^B} |r_d(x)| \ll \frac{x}{(\log x)^A}. \tag{8}$$

Given that precisely half of the primes split in O_K, we deduce that the density function $h(\cdot)$ satisfies the linear sieve assumption

$$\left(\frac{\log z}{\log w}\right)\left(1 - \frac{L_1}{\log w}\right) \leq \prod_{w \leq p < z} (1 - h(p))^{-1} \leq \left(\frac{\log z}{\log w}\right)\left(1 + \frac{L_2}{\log w}\right), \quad (9)$$

for some constants L_1, L_2. Thus, by (8) and (9), we can apply the linear sieve to $\mathcal{A}(x)$ with level of distribution $D(x) = \sqrt{x}/(\log x)^B$ to deduce, by the Jurkat-Richert Theorem, (see end of Section 4 [16]), that the inequality

$$W_1(x) \geq \left(\tfrac{1}{2}e^\gamma \log 3 - \varepsilon\right) \frac{|O_K^*|}{\varphi(\delta_E)} \Pi'(x; g, \alpha_0) V(z), \quad (10)$$

is valid for any $\varepsilon > 0$ and for x sufficiently large in terms of ε.

5   Upper Bound for W_2

Now, instead of $\mathcal{A}(x)$, the sets to consider in the sieve process are

$$\mathcal{A}_{p_0}(x) = \{a \in \mathcal{A}(x) \, : \, p_0 | a\},$$

for each prime p_0 in the interval $(z, y]$. In this case the number of elements in $\mathcal{A}_{p_0}$ divisible by d is precisely

$$|\mathcal{A}_{dp_0}(x)| = \frac{|O_K^*|}{\varphi(\delta_E)} \Pi'(x; g, \alpha_0) h(dp_0) + r_{dp_0}(x)$$

for $h(\cdot)$ and $r(\cdot)$ as in (7). Now the level of distribution is $D(x)/p_0$ and again by Jurkat Richert and (8) we get

$$\sum_{\substack{a \in \mathcal{A}_{p_0}(x) \\ (a, P(z)Q(z)p_K)=1}} 1 \leq \frac{|O_K^*|}{\varphi(\delta_E)} \Pi'(x; g, \alpha_0) V(z) g(p_0) \{F(s_{p_0}) + o(1)\}. \quad (11)$$

where $s_{p_0} = \log(D(x)/p_0)/\log z$, and $F(s) = 2e^\gamma s^{-1}$ for any $1 \leq s \leq 3$. Summing over all primes, and using partial sumation we obtain, (see Section 4 of [16]),

$$W_2 \leq \left(\tfrac{1}{2}e^\gamma \log 6 + \varepsilon\right) \frac{|O_K^*|}{\varphi(\delta_E)} \Pi'(x; g, \alpha_0) V(z), \quad (12)$$

for any $\varepsilon > 0$, and x sufficiently large depending on ε.

6   Upper Bound for $W_3(x)$

Finally we have to control $W_3(x)$ which counts the number of elements a in $\mathcal{A}(x)$ such that $a = p_1 p_2 p_3$ for splitting primes in a certain range. Consider the set $\mathfrak{P}_\varepsilon(x)$ given by tuples (π_1, π_2, π_3) of primary primes such that $z \leq N(\pi_3) < y \leq N(\pi_2) \leq N(\pi_1)$, and $\pi_1 \pi_2 \pi_3 \equiv \bar{\varepsilon}(\alpha_0 - \zeta)/\delta_E \pmod{g}$, and let

$$\mathcal{B}(x) = \{N(\zeta + \omega) : \omega \in \Omega(x)\},$$

where

$$\Omega(x) = \{\omega = \delta_E \epsilon \pi_1 \pi_2 \pi_3 : \epsilon \in \mathbb{O}_K^*, N(\omega) \leq x, (\pi_1, \pi_2, \pi_3) \in \mathfrak{P}_\varepsilon(x)\}.$$

Then,

$$W_3(x) \leq \sum_{\substack{b \in \mathcal{B}(x) \\ (b, P(\sqrt{x})Q(\sqrt{x}))=1}} 1 + O(\sqrt{x}),$$

and we may now apply sieve theory to the sequence $\mathcal{B}(x)$, in this case with a new sieve parameter $z_0 = \sqrt{x}$. Again we need to estimate $|\mathcal{B}_d(x)|$, the number of elements in $\mathcal{B}(x)$ divisible by $d | P(\sqrt{x})Q(\sqrt{x})$. If $(d, \delta_E g) > 1$, then the set $\mathcal{B}_d(x)$ is trivially empty. For any other d we proceed as before and note that finding an upper bound for $W_3(x)$ boils down to estimating

$$|\mathfrak{B}_\mathfrak{d}(x)| = \sum_{\substack{\omega \in \Omega(x) \\ \omega \equiv -\zeta \pmod{\mathfrak{d}}}} 1. \tag{13}$$

For this purpose we need an analogous Bombieri-Vinogradov Theorem for the numbers in the set $\Omega(x)$. If we let

$$E(x; \alpha, \mathfrak{a}) = \sum_{\substack{\omega \in \Omega(x) \\ \omega \equiv \alpha \pmod{\mathfrak{a}}}} 1 - \frac{1}{\Phi(\mathfrak{a})} \sum_{\substack{\omega \in \Omega(x) \\ (\omega, \mathfrak{a})=1}} 1.$$

then we can prove the following proposition.

Proposition 3. *Let the notation be as above, and $x > 0$. We have*

$$\sum_{N(\mathfrak{a}) \leq Q} \max_{(\alpha, \mathfrak{a})=1} |E(x; \alpha, \mathfrak{a})| \ll \frac{x}{(\log x)^A}, \tag{14}$$

with $Q = \sqrt{x}/(\log x)^B$. Here A is any positive number and B and the implied constant depend only on A.

The proof is exactly as Proposition 5 in [16], though in this case we consider, more generally, characters over $(O_K/\mathfrak{a})^*$. It might be interesting to observe that, since π is in a fixed congruence class modulo $\delta_E g$, we are considering triples π_1, π_2, π_3 such that $\pi_1 \equiv \bar{\varepsilon}(\alpha_0 - \zeta)/(\delta_E \pi_3 \pi_2) \pmod{g}$, (note that it follows immediately that any number $\zeta + \omega \equiv \zeta \pmod{\delta_E}$), and, as there is no restriction in π_2, π_3, this does not affect the Siegel-Walfisz type theorem for π_3, (Inequality (20) on p. 11 in [16]).

Given the above proposition we can write

$$|\mathcal{B}_d(x)| = |\Omega(x)|h(d) + \mathfrak{r}_d(x),$$

where $h(\cdot)$ is the same multiplicative function appearing in $\mathcal{A}_d(x)$, and so, by Jurkat Richert, we get

$$W_3(x) \leq \prod_{l < \sqrt{x}} (1 - h(l)) |\Omega(x)| \{F(1) + o(1)\} < \frac{e^{\gamma}}{2} V(z) |\Omega(x)| \{1 + o(1)\},$$

where we have used $F(s) = 2e^{\gamma} s^{-1}$ as in (11), and (9). To complete the proof we just have to compare $|\Omega(x)|$ with $\frac{|O_K^*|}{\varphi(\delta_E)} \Pi'(x; g, \alpha_0) V(z)$, appearing in (10) and (12). By definition we have

$$|\Omega(x)| \leq |O_K^*| \sum_{z \leq |\pi_3|^2 < y} \sum_{< |\pi_2|^2 < \sqrt{x}/|\pi_3|^2} \Pi'(x/(d_E|\pi_3 \pi_2|^2; g, \xi)$$

where $\xi \equiv \bar{\varepsilon}(\alpha_0 - \zeta)/(\delta_E \pi_3 \pi_2) \pmod{d_E g}$. Asymptotically, as x goes to infinity, the above is the same as

$$\frac{|O_K^*|}{d_E} \Pi'(x; g, \alpha_0) \sum_{z \leq |\pi_3|^2 < y} \sum_{< |\pi_2|^2 < \sqrt{x}/|\pi_3|} \frac{\log x}{|\pi_3 \pi_2|^2 \log(x/(|\pi_3 \pi_2|^2)},$$

A new application of partial summation, together with a change of variables, as in the deduction of (12), gives

$$|\Omega(x)| \leq \left(\frac{|O_K^*|}{d_E} \int_{\frac{1}{8}}^{\frac{1}{3}} \int_{\frac{1}{3}}^{\frac{1-v}{2}} \frac{1}{1 - u - v} \frac{du\,dv}{uv} + \varepsilon \right) \Pi'(x; g, \alpha_0).$$

It remains to combine the previous results, and note that $d_E \geq \varphi(\delta_E)$ to get

$$W_3(x) \leq \left(\frac{ce^{\gamma}}{2\varphi(\delta_E)} |O_K^*| + \varepsilon \right) \Pi'(x; g, \alpha_0) V(z), \tag{15}$$

for some $c < 0.36308373$. Hence Theorem 1 follows on using (10), (12), and (15) in (5).

7   On Primitive Points

Let $E/\mathbb{Q}$ be an elliptic curve with CM by O_K, with positive rank and equation given by (1). For simplicity we restrict ourselves to $D \neq -3, -4, -7, -8$. Let $p \leq x$ be prime, $P \in E(\mathbb{Q})$ of infinite order, and $\bar{P}$ the reduction of P mod p. As mentioned in the introduction, it was conjectured by Lang and Trotter that $\bar{P}$ generates the full group $E(\mathbb{F}_p)$ for a positive density of primes, and this is not known unconditionally in any case. However, in [11] the authors, among other very important results, included an approach to the problem in the following direction, (see Lemma 14 and 17 in that reference).

Theorem 2. *(Gupta-Murty) Let $E/\mathbb{Q}$ be a CM curve with positive rank and let $P \in E(\mathbb{Q})$ be a point of infinite order. Then,*

$$\#\{q \leq x, q \text{ inert } : \ | < \bar{P} > | < x^{1/3 - \varepsilon}\} \ll x^{1 - 3\varepsilon} \quad \text{and}$$
$$\#\{p \leq x, p \text{ splits } : \ | < \bar{P} > | < x^{1/2 - \varepsilon}\} \ll x^{1 - 2\varepsilon}$$

for any $\varepsilon > 0$.

In particular for almost all primes the point P generates a group of order at least $x^{1/3-\varepsilon}$. Here, it might be worthwhile to include the following remark.

Remark 1. Let $E/\mathbb{Q}$ be a CM curve with positive rank and let $P \in E(\mathbb{Q})$ be a point of infinite order. Then,

$$\#\{q \le x, q \text{ inert } : |<\bar{P}>| > x^{0.449}\} \gg x/(\log x)^2.$$

To be precise, this remark does not belong properly to the theory of elliptic curves, but to the classical twin prime conjecture. Indeed, we eed only to consider the sequence $\mathcal{A}_q(x) = \{(q+1)/2 : q \le x, \text{ inert in } O_K\}$ and, the result is the consequence of the best estimates in the constant C such that $\mathcal{A}_q(x) \ge Cx/(\log x)^2$. One can find this type of bounds in [4], and it is also possible to get an even better result with the subsequent paper [29]. Although the bounds in these references hold for the sequence $p + 2$, the arguments can be translated in a straightforward manner to our sequence $\mathcal{A}_q(x)$. We can also apply the same reasoning, this time to primes splitting in K, using Theorem 1. Indeed, we have proved in Theorem 1 that the number of P_2 in the sequence $\mathcal{A}(x)$ of Section 3 is bigger than

$$C\frac{|O_K^*|}{\varphi(\delta_E)}\Pi'(x; g, \alpha_0)V(z),\tag{16}$$

for some constant C. On the other hand, the number of elements counted in $S(\mathcal{A}, P(z)Q(z)p_K)$, with some prime factor between $x^{1/3}$ and x^β is exactly the sum $W_2(x)$ but now with parameters, $\frac{1}{3}, \beta$ and so, it is bounded by the constant

$$\frac{e^\gamma}{2} \int_{x^{1/3}}^{x^\beta} \frac{\log x}{\log(x/t^2)t\log t}dt.$$

One has to choose β appropriately to make this quantity smaller than C. Consider now

$$\mathcal{A}_\beta(x) = \{a \in \mathcal{A}(x), a = P_2, (l, a) = 1 \text{ for } l < z \text{ or } x^{1/3} < l < x^\beta\},$$

then, we can conclude that $|\mathcal{A}_\beta(x)| \gg x/(\log x)^2$. When reducing the curve modulo the primes p counted in $\mathcal{A}_\beta(x)$, and of size about x, $E(\mathbb{F}_p)$ must have one of its two prime factors bigger than x^β, since both cannot be smaller than $x^{1/3}$. On the other hand, by Lemma 14 of [11], the point $\bar{P}$ has to have order bigger than $x^{1/3}$ and, hence, bigger than x^β since it has to be a divisor of a which gives us the corresponding improvement. Although the parameter β that is obtained in this way is much worse than the $1/2 - \varepsilon$ that we deduce from Theorem 2, it is worthwhile to note that, while the theorem ensures the existence of a subgroup of $E(\mathbb{F}_p)$ of big order, the one generated by $\bar{P}$, the nature of the sieving procedure to obtain the elements in $\mathcal{A}_\beta(x)$ guarantee that, in those cases, every subgroup of $E(\mathbb{F}_p)$ has to be big, at least of size $x^{1/8}$. In order to prove the remark, one proceeds in the same way but now with the sequence $(q+1)/2$ and, instead, using the depper sieve techniques as developed in [4], [28] and [29] to get a much better result for the analogous constant C in (16).

Acknowledgments

I would like to thank I. Shparlinski for the suggestion of considering the application included in Section 7, for reading a previous version of the manuscript, and for his a lot of advice subsequently. I would also like to thank C. David and J. González for answering many questions, and for the various conversations that make this job more enjoyable, and A. Srinivasan for her help after a careful reading of a previous version of the manuscript. This work was completed during a stay at CRM in Montreal, Canada. I appreciate the warm hospitality I received during my stay at the center. Also I would like to thank the anonymous referee for helpful comments and suggestions.

References

1. Atkin, A.O.L., Morain, F.: Elliptic curves and primality proving. Math. Comp. 61(203), 29–68 (1993)
2. Balog, A., Cojocaru, A., David, C.: Average twin prime conjecture for elliptic curves, http://arXiv.org/abs/0709.1461
3. Borosh, I., Moreno, C.J., Porta, H.: Elliptic curves over finite fields. II, Math. Comput. 29, 951–964 (1975)
4. Cai, Y.C.: On Chen's theorem (II). J. Number Theory (2007) (Available online)
5. Chen, J.R.: On the representation of a larger even integer as the sum of a prime and the product of at most two primes. Sci. Sinica 16, 157–176 (1973)
6. Cojocaru, A.C.: Questions about the reductions modulo primes of an elliptic curve. In: Number Theory, CRM Proc. Lecture Notes, vol. 36, pp. 61–79. Amer. Math. Soc., Providence, RI (2004)
7. Cojocaru, A.C.: Reductions of an elliptic curve with almost prime orders. Acta Arith. 119, 265–289 (2005)
8. Cojocaru, A.C.: Cyclicity of Elliptic Curves Modulo p. PhD thesis, Queen's University (2002)
9. Greaves, G.: Sieves in Number Theory. Springer, Berlin (2001)
10. Friedlander, J., Iwaniec, H.: The Sieve (preprint)
11. Gupta, R., Murty, M.R.: Primitive points on elliptic curves. Compositio Math. 58, 13–44 (1986)
12. Gupta, R., Murty, M.R.: Cyclicity and generation of points modulo p on elliptic curves. Invent. Math. 101, 225–235 (1990)
13. Ireland, K., Rosen, M.: A Classical Introduction to Modern Number Theory. In: GTM, vol. 84, Springer, Heidelberg (1982)
14. Iwaniec, H.: Sieve Methods (notes for a graduate course in Rutgers University) (1996)
15. Iwaniec, H., Kowalski, E.: Analytic Number Theory, vol. 53. Colloquim Publications, AMS (2004)
16. Iwaniec, H., Jiménez Urroz, J.: Orders of CM elliptic curves modulo p with at most two primes, https://upcommons.upc.edu/e-prints/bitstream/2117/1169/1/p2incmellip906.pdf
17. Johnson, D.: Mean values of Hecke L-functions. J. reine angew. Math. 305, 195–205 (1979)
18. Katz, N.: Galois properties of torsion points on abelian varieties. Invent. Math. 62, 481–502 (1981)

19. Koblitz, N.: Primality of the number of points on an elliptic curve over a finite field. Pacifc J. Math. 131, 157–165 (1988)
20. Lang, S., Trotter, H.: Frobenius distributions in GL2-extensions. Lecture Notes in Math., vol. 504. Springer, Berlin, New York (1976)
21. Miri, S.A., Murty, V.K.: An application of sieve methods to elliptic curves. In: Pandu Rangan, C., Ding, C. (eds.) INDOCRYPT 2001. LNCS, vol. 2247, pp. 91–98. Springer, Heidelberg (2001)
22. Murty, M.R.: Artin's Conjecture and Non-Abelian Sieves. PhD Thesis, MIT (1980)
23. Rubin, K., Silverberg, A.: Point counting on reductions of CM elliptic curves, http://arxiv.org/abs/0706.3711v1
24. Serre, J.-P.: Résumé des cours de 1977–1978, Ann. Collège France. Collège de France, Paris, p. 6770ff (1978)
25. Stark, H.M.: Counting points on CM elliptic curves. Rocky Mountain J. Math. 26(3), 1115–1138 (1996)
26. Steuding, J., Weng, A.: On the number of prime divisors of the order of elliptic curves modulo p. Acta Arith. 117(4), 341–352 (2005)
27. Steuding, J., Weng, A.: Erratum: On the number of prime divisors of the order of elliptic curves modulo p. Acta Arith. 119(4), 407–408 (2005)
28. Wu, J.: Chen's double sieve, Goldbach's conjecture and the twin prime problem. Acta Arith. 114, 215–273 (2004)
29. Wu, J.: Chen's double sieve. Goldbach's conjecture and the twin prime problem 2, http://arXiv.org/abs/0709.3764

Efficiently Computable Distortion Maps for Supersingular Curves

Katsuyuki Takashima

Information Technology R&D Center, Mitsubishi Electric Corporation,
5-1-1, Ofuna, Kamakura, Kanagawa 247-8501, Japan
Takashima.Katsuyuki@aj.MitsubishiElectric.co.jp

Abstract. *Efficiently computable* distortion maps are useful in cryptography. Galbraith-Pujolàs-Ritzenthaler-Smith [6] considered them for supersingular curves of genus 2. They showed that there exists a distortion map in a specific set of efficiently computable endomorphisms for every pair of nontrivial divisors *under some unproven assumptions* for two types of curves. In this paper, we prove that this result holds using a different method *without these assumptions* for both curves with $r > 5$ and $r > 19$ respectively, where r is the prime order of the divisors. In other words, we solve an open problem in [6]. Moreover, we successfully generalize this result to the case $C : Y^2 = X^{2g+1} + 1$ over $\mathbb{F}_p$ for *any g* s.t. $2g + 1$ is prime. In addition, we provide explicit bases of $\mathrm{Jac}_C[r]$ with a new property that seems interesting from the cryptographic viewpoint.

1 Introduction

Let C be a nonsingular projective curve over a finite field $\mathbb{F}$, and let e be a nondegenerate bilinear pairing on its Jacobian $\mathrm{Jac}_C[r]$ for a prime r s.t. $r \mid \sharp\mathrm{Jac}_C(\mathbb{F})$. A distortion map [13] for two nontrivial D and D' in $\mathrm{Jac}_C[r]$, is an endomorphism ϕ on Jac_C s.t. $e(D, \phi(D')) \neq 1$. We say that a curve C is supersingular when Jac_C is supersingular. Galbraith *et al.* [6] showed the existence of a distortion map for supersingular curves (See Theorem 1). In cryptography, an *efficiently computable* distortion map is important, however, its existence has not yet been established for the higher genus curves ([5,6], see [7] also). We will solve an open problem given in [6] on the topic.

An elliptic curve $E : Y^2 = X^3 + 1$ over $\mathbb{F}_p$ where p is prime and $p \equiv 2 \bmod 3$, provides a good starting point for understanding the problem. Let D^* be a nontrivial point in $E(\mathbb{F}_p)[r]$ where a prime $r > 3$, and let an automorphism ρ on E be $(x, y) \mapsto (\zeta x, y)$ using a third root of unity ζ in $\mathbb{F}_{p^2}$. Because $\rho(D^*) \notin E(\mathbb{F}_p)$, the set $\{D^*, \rho(D^*)\}$ is a basis of $E[r] \cong (\mathbb{Z}/r\mathbb{Z})^2$. Then $e(D^*, \rho(D^*)) \neq 1$ since $\dim_{\mathbb{F}_r} E[r] = 2$. Thus, ρ is a distortion map for D^* and D^*. The elliptic curve is the first in a sequence of supersingular curves $C : Y^2 = X^w + 1$ over $\mathbb{F}_p$ where $w := 2g + 1$ is prime and $p \bmod w$ is a generator of $\mathbb{F}_w^*$. Their Jacobians also have a similar action ρ of a w-th root of unity ζ in $\mathbb{F}_{p^{2g}}$. In fact, we will show that an analogous result for ρ holds for the higher genus curves (Corollary 2). However, the argument is not as simple as the genus 1 case.

A.J. van der Poorten and A. Stein (Eds.): ANTS-VIII 2008, LNCS 5011, pp. 88–101, 2008.
© Springer-Verlag Berlin Heidelberg 2008

Let π be the p-th power Frobenius endomorphism on Jac_C. Then $\Delta = \{\pi^i \rho^j \mid 0 \leq i, j \leq 2g - 1\}$ provides a set of natural candidates for a distortion map for a pair of nontrivial divisors. Galbraith *et al.* [6] considered the genus 2 case — $C : Y^2 = X^5 + 1$ over $\mathbb{F}_p$ where $p \equiv 2, 3 \bmod 5$ — in detail. They showed that there exists a distortion map in Δ for every pair of nontrivial divisors *under some unproven assumption*. For the details of the assumption, see Section 3.1.

We prove that the generalization of their result holds for the above curve of *general genus $g \geq 1$ without the unproven assumption* with $r > w$ (Theorem 4). We take a different approach from that of Galbraith *et al.* to solve the problem. First, we construct explicit eigenvectors of π described by Gauss sums. Using arithmetic properties of the Gauss sums, we then show the above result. As a corollary, we prove that the above assumption given in [6] holds (Corollary 1).

Galbraith *et al.* [6] also treated a supersingular curve over a finite field of characteristic 2, $C : Y^2 + Y = X^5 + X^3 + b$ where $b \in \mathbb{F}_2$. They showed the existence of a distortion map in Δ for every pair of divisors under a similar unproven assumption, where Δ is a set of natural candidates for the distortion map as given above. We also prove the existence result *without the assumption* when $r > 19$ (Theorem 10, See Corollary 4 also).

While obtaining the above results, we will obtain explicit bases, which we call *efficiently constructible semi-symplectic bases* of the Jacobians for the Weil pairing. These bases are the first explicit constructions with explicitly known relations among the discrete logarithms of all Weil pairing values as far as we know. For the above curve $C : Y^2 = X^w + 1$, a key step to obtaining the explicit basis is to describe such relations in terms of Jacobi sums (Theorem 7). These bases seem useful for some applications using a rich torsion structure ([4,3]).

Section 2 reviews notation and facts related to circulant matrices. Section 3 defines notions regarding distortion maps from computational and constructive viewpoints, and summarizes the previous results of Galbraith *et al.* Section 4 proves the above results for the curves $C : Y^2 = X^w + 1$ where w is an odd prime and $r > w$. Section 5 shows the results for the curves over $\mathbb{F}_2$ in [6].

2   Circulant and Related Matrices

We fix notation and summarize facts on circulant matrices. See [2] for details. Set

$$
V = \begin{pmatrix} 1 & 1 & \cdots & 1 \\ v_0 & v_1 & \cdots & v_{n-1} \\ \vdots & \vdots & & \vdots \\ v_0^{n-1} & v_1^{n-1} & \cdots & v_{n-1}^{n-1} \end{pmatrix} \quad \text{and} \quad \Gamma = \begin{pmatrix} t_0 & t_1 & \cdots & t_{n-1} \\ t_{n-1} & t_0 & \cdots & t_{n-2} \\ \vdots & \vdots & \ddots & \vdots \\ t_1 & t_2 & \cdots & t_0 \end{pmatrix}. \tag{1}
$$

We recall that the $n \times n$ matrix $V = V(v_0, v_1, \ldots, v_{n-1})$ is a Vandermonde matrix. The matrix $\Gamma = \mathrm{circ}(t_0, t_1, \ldots, t_{n-1})$ in (1), with i-th row the $(i-1)$-th cyclic shift of its first row, is a *circulant* matrix.

The (i, j)-entry of Γ is $t_{j-i \bmod n}$, and is in a finite field $\mathbb{F}$ in this paper. If $n \neq 0$ in $\mathbb{F}$, the eigenvectors of the circulant are given by

$$\mathcal{Z}_i := n^{-1/2} \left(1, \mathfrak{z}^i, \ldots, \mathfrak{z}^{(n-1)i}\right)^{\mathrm{T}}, \quad i = 0, \ldots, n-1, \tag{2}$$

where $\mathfrak{z}$ is a primitive n-th root of unity in $\overline{\mathbb{F}}$, $n^{-1/2} \in \overline{\mathbb{F}}$, and the superscript T denotes transposition. Then the corresponding eigenvalues are given by

$$\eta_i := \sum_{\kappa=0}^{n-1} t_\kappa \mathfrak{z}^{i\kappa}, \quad i = 0, \ldots, n-1, \tag{3}$$

respectively. Let a diagonal matrix Ψ be $\mathrm{diag}(\eta_0, \ldots, \eta_{n-1})$, and let a matrix V be $V(1, \mathfrak{z}, \ldots, \mathfrak{z}^{n-1}) = (\mathcal{Z}_0, \cdots, \mathcal{Z}_{n-1})$. In particular, $\det(V) \neq 0$ when $n \neq 0$ in $\mathbb{F}$. Then $\Gamma V = V\Psi$. In other words, $\Gamma = V\Psi V^{-1}$ and $\Psi = V^{-1}\Gamma V$.

3   Efficiently Computable Distortion Maps

In this paper, let C be a nonsingular projective curve over a finite field $\mathbb{F}$ of genus g and let r be an odd prime number s.t. $r \mid \sharp\mathrm{Jac}_C(\mathbb{F})$. In addition, let $e = e_r$ be a bilinear nondegenerate pairing on $\mathrm{Jac}_C[r]$ whose values are in the multiplicative group $\boldsymbol{\mu}_r$ of order r in some extension of $\mathbb{F}$. For readability, we hereafter use the simple notation e, not e_r. We denote the zero in $\mathrm{Jac}_C[r]$ by $\mathcal{O}$.

3.1   Results of Galbraith *et al.* [6]

According to [6], we define the notion of a distortion map as follows.

Definition 1 ([6]). *For a nondegenerate pairing e on $\mathrm{Jac}_C[r]$ and two points D and D' in $\mathrm{Jac}_C[r]$, an endomorphism ϕ on Jac_C is a distortion map for e, D, and D' if $e(D, \phi(D')) \neq 1$.*

The next Theorem 1 is Theorem 2.1 given in [6], and it assures the existence of a distortion map on a supersingular Jacobian variety of a curve C ([6] proved it for a supersingular abelian variety in general).

Theorem 1 ([6]). *Let Jac_C be supersingular, and let e be a nondegenerate pairing on $\mathrm{Jac}_C[r]$. For every pair of nontrivial D and D' in $\mathrm{Jac}_C[r]$, there exists a distortion map ϕ on Jac_C, i.e., $e(D, \phi(D')) \neq 1$.*

Galbraith *et al.* [6] showed that the endomorphism ring $\mathrm{End}(\mathrm{Jac}_C)$ of a supersingular Jacobian variety has $\mathbb{Z}$-rank $(2g)^2$. Therefore, to clarify our presentation, we define a new notion of a complete set of distortion maps here.

Definition 2. *Let Δ be a subset of $\mathrm{End}(\mathrm{Jac}_C)$ s.t. $\sharp\Delta \leq (2g)^2$. The set Δ is a complete set of efficiently computable distortion maps on $\mathrm{Jac}_C[r]$ if $\mathrm{Jac}_C[r] = \langle \delta(D) \mid \delta \in \Delta \rangle$ spaned as an $\mathbb{F}_r$-vector space for every nontrivial divisor $D \in \mathrm{Jac}_C[r]$, and if all $\delta \in \Delta$ are efficiently computable (or polynomial-time computable, formally).*

Remark 1. If Δ is a complete set of efficiently computable distortion maps, then, for every nondegenerate pairing e, we can efficiently (or in polynomial time) check which $\delta \in \Delta$ is a distortion map for a given pair of divisors of order r. This gives an efficient algorithm for constructing a distortion map given in Theorem 1.

One of the goals in [6] was to find a complete set of efficiently computable distortion maps for the following curves.

For a supersingular curve $C : Y^2 = X^5 + 1$ over $\mathbb{F}_p$ where $p \equiv 2, 3 \bmod 5$, $\mathbb{Q}$-coefficient endomorphism ring $\mathrm{End}^0(\mathrm{Jac}_C) := \mathrm{End}(\mathrm{Jac}_C) \otimes_{\mathbb{Z}} \mathbb{Q}$ is $\mathbb{Q}[\rho, \pi]$ (See [6]). Here, π is the p-th power Frobenius endomorphism, and ρ is the action of a fifth root of unity $\zeta = \zeta_5$, i.e., $\rho : (x, y) \mapsto (\zeta x, y)$ on Jac_C. We notice that $\mathrm{End}(\mathrm{Jac}_C)$ is not necessarily equal to $\mathbb{Z}[\rho, \pi]$. Therefore, Galbraith *et al.* [6] made Assumption 1 for the completeness of $\Delta = \{\pi^i \rho^j \mid 0 \leq i, j \leq 3\}$.

A distortion map ϕ in Theorem 1 is given by $\phi = \sum_{0 \leq i,j \leq 3} \lambda_{i,j} \pi^i \rho^j$ where $\lambda_{i,j} \in \mathbb{Q}$. Let m be the least common multiple of denominators of $\lambda_{i,j}$ ($0 \leq i, j \leq 3$). Then $m\phi \in \mathbb{Z}[\rho, \pi]$. In [6], the following Assumption 1 was made for m, and under Assumption 1, they showed the following Theorem 2.

Assumption 1 ([6]). *The above ϕ may be chosen s.t. $\gcd(m, r) = 1$.*

Theorem 2 ([6]). *Under Assumption 1 (for a nondegenerate pairing), Δ is a complete set of efficiently computable distortion maps on $\mathrm{Jac}_C[r]$.*

We prove that the above theorem holds *without Assumption 1* when $r > 5$ in Theorem 4 in Section 4 (See Corollary 1 also). We notice that $r > 5$ in typical cryptographic applications.

They also discussed efficiently computable distortion maps for another type of curves. For m s.t. $m \equiv \pm 1 \bmod 6$, let q be 2^m. A curve $C : Y^2 + Y = X^5 + X^3 + b$ over $\mathbb{F}_q$ where $b = 0$ or 1 is a supersingular curve of genus 2. Endomorphisms $\sigma_\tau, \sigma_\theta$, and σ_ξ (given in Section 5.1) are efficiently computable on Jac_C. They then proved an analogous result to Theorem 2 under a similar assumption for the curve C and r, that is, the completeness of $\Delta = \{\pi^i, \pi^j \sigma_\tau, \pi^\kappa \sigma_\theta, \pi^l \sigma_\xi \mid 0 \leq i, j, \kappa, l \leq 3\}$ where π is the q-th power Frobenius. We also show the completeness of Δ *without that assumption* when $r > 19$ in Theorem 10 in Section 5.

3.2 The Notion of an Efficiently Constructible Semi-symplectic Basis

In addition to proving the completeness of Δ as above, we will obtain interesting bases of $\mathrm{Jac}_C[r]$. We call them *efficiently constructible semi-symplectic bases.*

Definition 3. *A basis $\{D_0, \ldots, D_{2g-1}\}$ of the $\mathbb{F}_r$-vector space $\mathrm{Jac}_C[r]$ is an efficiently constructible semi-symplectic basis for a nondegenerate skew-symmetric pairing e if $e(D_i, D_j) = 1$ when $i \neq 2g-1-j$, and $e(D_i, D_j) = u$ when $i = 2g-1-j$ and $i = 0, \ldots, g-1$ for some $u \neq 1 \in \mu_r$, and there is an efficient algorithm that outputs the basis taking the parameters of the curve C, r, and e as input.*

Let $\mathcal{B}$ be an efficiently constructible semi-symplectic basis. We can calculate the discrete logarithm of $e(D, D')$ to the base u when we know the coefficients of D and D' expressed in terms of the basis $\mathcal{B}$. The bases in Sections 4.4 and 5.2 are the first explicit constructions with this property as far as we know. In the last section of [4], Galbraith *et al.* suggested a possibility of a new application of pairing with the "rich torsion structure." Our explicit constructions will provide a basic tool for such an application.

3.3   Invariance of the Weil pairing

The main results in this paper (in Sections 4 and 5) are based on the following Fact 1. It shows the invariance of the Weil pairing under the diagonal action of an automorphism. For a proof of Fact 1, refer to [11] p.186 and [10] p.132.

Fact 1. *Let e be the Weil pairing. Then $e(D, D') = e(\phi(D), \phi(D'))$ for all D and $D' \in \mathrm{Jac}_C[r]$, and all automorphisms ϕ on Jac_C.*

4   Curves with Actions of Roots of Unity

Let g be a positive integer s.t. $w := 2g + 1$ is prime, and let p be a odd prime s.t. $p \bmod w$ is a generator of $\mathbb{F}_w^*$. We consider a curve

$$C : Y^2 = X^w + 1$$

over $\mathbb{F}_p$. Then C is a supersingular curve of genus g. Since $\sharp \mathrm{Jac}_C(\mathbb{F}_p) = p^g + 1$, set a prime r s.t. $r \mid p^g + 1$. Therefore, the embedding degree k for r is $2g$, and k is also the full embedding degree (cf. [12]). In other words, $\mathrm{Jac}_C[r] \subset \mathrm{Jac}_C(\mathbb{F}_{p^k})$.

In this section, we show that the natural generalization of Theorem 2 holds without any unproven assumption when $\gcd(r, 2gw) = 1$ (Theorem 4). Moreover, we obtain an efficiently constructible semi-symplectic basis of $\mathrm{Jac}_C[r]$ in Section 4.4. Certainly, as given in [6], the results in this section can be generalized to the twists $Y^2 = X^w + A$ of C where $A \neq 0$. Hereafter, we consider the case that $\gcd(r, 2gw) = 1$. This holds when $r > w = 2g + 1$, which is always satisfied in typical cryptographic applications.

See Chapter 5 in [9] for the facts about Gauss and Jacobi sums used in Sections 4.1 and 4.4, for example.

4.1   Bases $\mathcal{B}$ and $\widetilde{\mathcal{B}}$ of the Vector Space $\mathrm{Jac}_C[r]$ over $\mathbb{F}_r$

First, we choose a nontrivial divisor D^* in $\mathrm{Jac}_C(\mathbb{F}_p)[r]$. Then let D_i be $\pi^i \rho(D^*) = \rho^{a^i}(D^*)$ where $i = 0, \ldots, 2g - 1$. Here, π is the p-th power Frobenius map, $a = p \bmod w$, and ρ is the action of a w-th root of unity $\zeta = \zeta_w$, i.e., $\rho : (x, y) \mapsto (\zeta x, y)$. Let $\mathcal{B}$ be $\{D_i \mid 0 \leq i \leq 2g - 1\}$. Next, we define divisors $\widetilde{D}_j := \sum_{i=0}^{2g-1} (p^j)^i D_i$ $(j = 0, \ldots, 2g - 1)$ using a Vandermonde matrix $V = V(1, p, p^2, \ldots, p^{2g-1})$. Let $\widetilde{\mathcal{B}}$ be $\{\widetilde{D}_i \mid 0 \leq i \leq 2g - 1\}$.

Theorem 3. *Suppose that* $\gcd(r, 2gw) = 1$. *Then both* $\mathcal{B}$ *and* $\widetilde{\mathcal{B}}$ *are bases of the* $\mathbb{F}_r$-*vector space* $\mathrm{Jac}_C[r]$. *Moreover,* $\widetilde{D}_i$ ($\in \widetilde{\mathcal{B}}$) *is an eigenvector of* π *with the eigenvalue* p^{-i} *where* $i = 0, \ldots, 2g - 1$.

Proof. Because $\pi(D_i) = D_{i+1}$, we see $\pi(\widetilde{D}_j) = p^{-j}\widetilde{D}_j$. Then we will prove $\widetilde{D}_j \neq \mathcal{O}$ for $\widetilde{D}_j$ to be an eigenvector of π. Let χ be a nontrivial additive character (4), and let ψ be a multiplicative character (5) of $\mathbb{F}_w$ of order $2g$ since p is of order $k = 2g$ in $\mathbb{F}_r^*$.

$$\chi : \mathbb{F}_w \ni v \mapsto \rho^v \in (\mathbb{F}_r[\rho])^* \subset \mathrm{End}(\mathrm{Jac}_C) \otimes_{\mathbb{Z}} \mathbb{F}_r. \tag{4}$$

$$\psi : \mathbb{F}_w^* = \langle a \rangle \ni a \mapsto p \in \langle p \rangle \subset \mathbb{F}_r^* , \tag{5}$$

and $\psi(0) := 0$. The values of χ are in the group $(\mathbb{F}_r[\rho])^*$ of units in the commutative subring generated by ρ in $\mathrm{End}(\mathrm{Jac}_C) \otimes_{\mathbb{Z}} \mathbb{F}_r$ where $\rho^w = 1$. Since $D_i = \rho^{a^i} D^*$, then $\widetilde{D}_j = \sum_{i=0}^{2g-1} (p^j)^i \rho^{a^i} D^*$. The operator $\sum_{i=0}^{2g-1} (p^j)^i \rho^{a^i}$ is a Gauss sum $G(\psi^j, \chi)$ of a multiplicative character ψ^j and an additive character χ. That is, $\widetilde{D}_j = G(\psi^j, \chi)D^*$. Since $G(\psi^{-j}, \chi)G(\psi^j, \chi) = \psi^{-j}(-1)w = (-1)^j w$ in $\mathbb{F}_r[\rho]$ and $r \neq w$, therefore $G(\psi^{-j}, \chi)\widetilde{D}_j = (-1)^j w D^* \neq \mathcal{O}$. Thus, $\widetilde{D}_j$ is an eigenvector of π with eigenvalue p^{-j}. Since the order of p in $\mathbb{F}_r^*$ is $k = 2g$, the eigenvalues p^{-j} are different from each other. Therefore, $\widetilde{\mathcal{B}}$ is a basis of $\mathrm{Jac}_C[r]$. Because $2g \neq 0 \bmod r$, then $\det(V) \neq 0$ (See Section 2). Hence, $\mathcal{B}$ is also a basis. □

4.2 Completeness of Δ

Lemma 1 gives basic relations of π and ρ, and that lemma is a slight generalization of Lemma 4.2 in [6].

Lemma 1. *Let* π *and* ρ *be as in Section 4.1. Then* $\pi^\ell \rho^j = \rho^{a^\ell j} \pi^\ell$ *for all* $\ell, j \in \mathbb{Z}$.

Proof. From the definition of $a = p \bmod w$, $\pi^\ell \rho = \rho^{a^\ell} \pi^\ell$ holds for $\ell \geq 0$ (and $j = 1$). Then by induction for $j(> 1)$, when $\ell \geq 0$, $\pi^\ell \rho^j = \pi^\ell \rho^{j-1} \rho = \rho^{a^\ell(j-1)} \pi^\ell \rho = \rho^{a^\ell(j-1)} \rho^{a^\ell} \pi^\ell = \rho^{a^\ell j} \pi^\ell$. Since $\pi^\ell \rho^j = \rho^{a^\ell j} \pi^\ell$ if and only if $\pi^\ell \rho^{-j} = \rho^{-a^\ell j} \pi^\ell$, then for negative j and positive ℓ, Lemma holds. For negative ℓ and any $j \in \mathbb{Z}$, using $\ell' = -\ell > 0$, we must show that $\rho^j \pi^{\ell'} = \pi^{\ell'} \rho^{a^{-\ell'} j}$. Let j' be $a^{-\ell'} j \in \mathbb{Z}/w\mathbb{Z}$. Then the equality is $\rho^{a^{\ell'} j'} \pi^{\ell'} = \pi^{\ell'} \rho^{j'}$ where $\ell' > 0$. This has been proved already. □

Theorem 4. *Suppose that* $\gcd(r, 2gw) = 1$. *Then* $\Delta = \{\pi^i \rho^j \mid 0 \leq i, j \leq 2g-1\}$ *is a complete set of efficiently computable distortion maps on* $\mathrm{Jac}_C[r]$.

Corollary 1. *We can choose* ϕ *with* $m = 1$ *in Assumption 1 when* $r > 5$.

Proof of Theorem 4. We show that $\mathrm{Jac}_C[r] = \langle \delta(D) \mid \delta \in \Delta \rangle$ as an $\mathbb{F}_r$-vector space for a nontrivial $D \in \mathrm{Jac}_C[r]$. Expressing D as a linear combination of $\widetilde{D}_j$, $D = \sum_j c_j \widetilde{D}_j$ where some $c_j \neq 0$ because $D \neq \mathcal{O}$. We then define generalizations of the trace map $\mathrm{Tr} = \mathrm{Tr}_{\mathbb{F}_{p^{2g}}/\mathbb{F}_p} = \sum_i \pi^i$. Let Pr_j be $\sum_i p^{ij} \pi^i$ for $j = 0, \ldots, 2g-1$

(then $\mathrm{Tr} = \mathrm{Pr}_0$, and see also Remark after Lemma 3.2 in [5]). By a simple calculation, $\mathrm{Pr}_j(\widetilde{D}_\kappa)$ is $\mathcal{O}$ if $j \neq \kappa$, and is $2g\widetilde{D}_j$ if $j = \kappa$. Let T_j be $G(\psi^{-j}, \chi)\mathrm{Pr}_j$ using the operator $G(\psi^{-j}, \chi)$ in the proof of Theorem 3. Here, by definition, T_j is in the noncommutative ring $\mathbb{F}_r[\pi, \rho]$. Then $T_j(\widetilde{D}_j) = G(\psi^{-j}, \chi)\mathrm{Pr}_j(\widetilde{D}_j) = (-1)^j 2gwD^* \neq \mathcal{O}$, and $T_j(D) = (-1)^j 2gwc_j D^* = c'_j D^* \neq \mathcal{O}$ because $c_j \neq 0$. Thus, $\mathrm{Jac}_C[r] = \langle \rho^i(T_j(D)) \mid i = 0, \ldots, 2g-1 \rangle$ by Theorem 3. The endomorphism $\rho^i T_j$ is an $\mathbb{F}_r$-linear combination of elements in Δ by Lemma 1. Then, $\mathrm{Jac}_C[r] = \langle \delta(D) \mid \delta \in \Delta \rangle$. $\square$

4.3 Representation of the Weil Pairing Matrix by a Circulant Matrix

To obtain an efficiently constructible semi-symplectic basis of $\mathrm{Jac}_C[r]$, first, we show that the matrix of logarithms of the Weil pairing values of D_i's where $i = 0, \ldots, 2g-1$ is essentially reduced to a circulant matrix.

Let $u_{i,j}$ be $e(D_i, D_j)$ for (i, j) s.t. $0 \leq i, j \leq 2g-1$. Then we consider the matrix $W := (\log_u(u_{i,j}))_{i,j}$ where u is $e(D^*, D_0)$ and $\log_u f$ is an integer s s.t. $f = u^s$. If $f \neq u^s$ for any integer s, then $\log_u f$ is undefined. However, we prove that $\log_u(u_{i,j})$ in $\mathbb{F}_r$ can be defined in Theorem 5. We call W the Weil pairing matrix of $\mathcal{B}$ to the base u where $\mathcal{B} = \{D_i\}$.

For $\kappa = 1, \ldots, 2g-1$, let h_κ be $a^\kappa - 1 \bmod w$ ($\neq 0$), and let $\ell_\kappa \in \mathbb{Z}/2g\mathbb{Z}$ be $\log_a(h_\kappa)$, which is well-defined because $h_\kappa \neq 0$ and a is a generator in $\mathbb{F}_w^*$. Since $p \in \mathbb{F}_r$ is of order $2g$ and $\ell_\kappa \in \mathbb{Z}/2g\mathbb{Z}$, the power $t_\kappa := p^{\ell_\kappa} \in \mathbb{F}_r$ is well-defined for $\kappa = 1, \ldots, 2g-1$. In addition, let t_0 be 0. In terms of the multiplicative character ψ given by (5), $t_\kappa = \psi(a^\kappa - 1)$ for $\kappa = 0, \ldots, 2g-1$.

Theorem 5. *The (i,j)-entry of W, i.e., $\log_u(u_{i,j})$, can be defined, and equal to $p^i t_\kappa$ where $\kappa = j - i \bmod 2g$. In other words, $W = \Omega\Gamma$ where $\Omega = \mathrm{diag}(1, p, \ldots, p^{2g-1})$ and $\Gamma = \mathrm{circ}(t_0, t_1, \ldots, t_{2g-1})$ given by (1) by using the above t_κ for $\kappa = 0, \ldots, 2g-1$.*

Proof. Using Fact 1, the Galois invariance of the Weil pairing, and Lemma 1, we show that for $0 \leq i, j \leq 2g-1$,

$$e(D_i, D_j) = e(\rho^{a^i}(D^*), \rho^{a^j}(D^*)) = e(\rho^{a^i}(D^*), \rho^{a^i}\rho^{a^j - a^i}(D^*))$$
$$= e(D^*, \rho^{a^j - a^i}(D^*)) = e(D^*, \rho^{a^i(a^{j-i} - 1)}(D^*)) = e(D^*, \rho^{a^i(a^{j-i}-1)}\pi^i(D^*))$$
$$= e(\pi^i(D^*), \pi^i\rho^{a^{j-i}-1}(D^*)) = e(D^*, \rho^{a^{j-i}-1}(D^*))^{p^i}.$$

Then for $i \neq j$, the above formula is $u_{i,j} = e(D_i, D_j) = e(D^*, \rho^{h_\kappa}(D^*))^{p^i}$, and for $i = j$, it is $e(D_i, D_i) = 1$. Thus, $\log_u(u_{i,j}) = 0 = p^i t_\kappa$ when $\kappa = 0$. When $\kappa \neq 0$, Lemma 1 gives $\pi^{\ell_\kappa}\rho = \rho^{h_\kappa}\pi^{\ell_\kappa}$ because $\ell_\kappa = \log_a(h_\kappa)$. Then $u^{t_\kappa} = e(D^*, \rho^{h_\kappa}(D^*))$, that is, $t_\kappa = \log_u(e(D^*, \rho^{h_\kappa}(D^*)))$. Therefore, $\log_u(u_{i,j}) = p^i t_\kappa$ when $\kappa \neq 0$. Thus, each row of $W = (\log_u(u_{i,j}))_{i,j}$ is the multiplication of that of $\Gamma = \mathrm{circ}(t_0, t_1, \ldots, t_{2g-1})$ and p^i, respectively. Then $W = \Omega\Gamma$ by using $\Omega = \mathrm{diag}(1, p, \ldots, p^{2g-1})$. $\square$

Corollary 2. *Suppose that* $\gcd(r, 2gw) = 1$. *Then the base* $u = e(D^*, D_0) \neq 1$ *and the Weil pairing matrix* W *of* $\mathcal{B} = \{D_i\}$ *to the base* u *is regular.*

Proof. Assume that $e(D^*, D_0) = 1$. Then all $e(D_i, D_j) = 1$ $(0 \leq i, j \leq 2g - 1)$ by Theorem 5. However, since $\mathcal{B}$ is a basis of $\mathrm{Jac}_C[r]$ by Theorem 3, this contradicts the nondegeneracy of the Weil pairing. Thus we proved that $e(D^*, D_0) \neq 1$. Moreover, the nondegeneracy of the pairing also shows the regularity of W.   □

4.4   An Efficiently Constructible Semi-symplectic Basis

Theorem 6 gives the eigenvalues of Γ as Jacobi sums. We use Jacobi sums $J(\psi, \psi^i) = \sum_{\kappa=1}^{2g-1} \psi(1 - a^\kappa)\psi^i(a^\kappa) \in \mathbb{F}_r$ for $i \neq 0$ where the multiplicative character ψ is given by (5) and a $(= p \bmod w)$ is a generator of $\mathbb{F}_w^*$.

Theorem 6. *Suppose that* $\gcd(r, 2gw) = 1$. *Then the eigenvalues of* Γ *are* $\eta_0 = 1$ *and* $\eta_i = -J(\psi, \psi^i)$ *where* $i = 1, \ldots, 2g - 1$.

Proof. We can diagonalize the circulant matrix Γ using the eigenvectors (2) because r does not divide $2g$. The eigenvalues are given by (3). In addition, since p is a primitive $(2g)$-th root of unity in $\mathbb{F}_r^*$, we can use p as $\mathfrak{z}$ in (3). Since $t_\kappa = \psi(a^\kappa - 1)$, we then obtain $\eta_i = \sum_{\kappa=0}^{2g-1} t_\kappa p^{i\kappa} = \sum_{\kappa=1}^{2g-1} \psi(a^\kappa - 1)\psi^i(a^\kappa) = \psi(-1)\sum_{\kappa=1}^{2g-1} \psi(1 - a^\kappa)\psi^i(a^\kappa)$ where $i = 0, \ldots, 2g - 1$. Thus, $\eta_0 = 1$ and $\eta_i = -J(\psi, \psi^i)$ where $i \neq 0$ since $\psi(-1) = -1$.   □

All the eigenvalues are nonzero by Corollary 2 because $\Gamma = \Omega^{-1}W$. Let Ψ be the diagonal matrix $\mathrm{diag}(\eta_0, \ldots, \eta_{2g-1}) = V^{-1}\Gamma V$ (See Section 2).

Theorem 7. *Suppose that* $\gcd(r, 2gw) = 1$. *Then the Weil pairing* $e(\widetilde{D}_i, \widetilde{D}_j) = u^{2g\eta_j} \neq 1$ *if* $i = 2g-1-j$, *and* $e(\widetilde{D}_i, \widetilde{D}_j) = 1$ *if* $i \neq 2g-1-j$ *where* $0 \leq i, j \leq 2g-1$.

Proof. Since $V^{\mathrm{T}} = V$, $W = \Omega\Gamma$ (Theorem 5), $\Gamma = V\Psi V^{-1}$, and W is the Weil pairing matrix of $\mathcal{B}$, the pairing matrix $\widetilde{W}$ of $\widetilde{\mathcal{B}}$ to the base u is

$$\widetilde{W} = VWV^{\mathrm{T}} = VWV = V\Omega\Gamma V = V\Omega V\Psi V^{-1}V = V\Omega V\Psi.$$

The diagonal matrix Ω is equal to $V^{-1}\Pi V$ where Π is the fundamental permutation matrix $\mathrm{circ}(0, 1, 0, \ldots, 0)$. Therefore, $\widetilde{W} = \Pi V^2\Psi = K\Psi$ where $K := \Pi V^2$ is a counterdiagonal matrix of size $2g$ as given below. Hence, since $\Psi = \mathrm{diag}(\eta_0, \ldots, \eta_{2g-1})$, the Weil pairing matrix $\widetilde{W}$ of $\widetilde{\mathcal{B}}$ is a counterdiagonal matrix as follows, where $\eta_i (\neq 0)$ are explicitly given by Theorem 6.

$$K = 2g \cdot \begin{pmatrix} 0 & \cdots & 1 \\ \vdots & \cdot^{\cdot^{\cdot}} & \vdots \\ 1 & \cdots & 0 \end{pmatrix}, \quad \widetilde{W} = 2g \cdot \begin{pmatrix} 0 & \cdots & \eta_{2g-1} \\ \vdots & \cdot^{\cdot^{\cdot}} & \vdots \\ \eta_0 & \cdots & 0 \end{pmatrix}.$$

Since $u \neq 1$ from Corollary 2, we obtain the theorem.   □

If we normalize $\widetilde{D}_i$ to $(2g\eta_{2g-1-i})^{-1}\widetilde{D}_i$ where $i = 0, \ldots, g - 1$, the counterdiagonal entries in $\widetilde{W}$ become ± 1. In other words, each counterdiagonal pairing value is u or u^{-1}. Thus, we obtained an efficiently constructible semi-symplectic basis for the Weil pairing since we can calculate η_i exactly (Theorem 6).

5   Curves of Artin-Schreier Type

In this section, we investigate curves of Artin-Schreier type in [6], whose embedding degree $k = 12$. Let m be an integer s.t. $m \equiv \pm 1 \bmod 6$ and let q be 2^m. Throughout this section, the ground field is $\mathbb{F}_q$. We consider a nonsingular curve C over $\mathbb{F}_q$,

$$C : Y^2 + Y = X^5 + X^3 + b \qquad (b = 0 \text{ or } 1).$$

5.1   Action of an Extra-Special 2-Group

We define polynomials over $\mathbb{F}_2$ as follows:

$$E^+(z) = \beta_6(z)\beta_3^+(z)\beta_1(z) = (z^6 + z^5 + z^3 + z^2 + 1)(z^3 + z + 1)(z + 1),$$
$$E^-(z) = \beta_3^-(z)\beta_2(z) = (z^3 + z^2 + 1)(z^2 + z + 1),$$
$$E(z) = zE^-(z)E^+(z) = z^{16} + z^8 + z^2 + z.$$

These polynomials E, E^-, and E^+ are E_2, E_2^-, and E_2^+ in [8], respectively. For a root ω of the equation $E(z) = 0$, we define an automorphism σ_ω on C as follows:

$$\sigma_\omega : (x, y) \mapsto (x + \omega, y + s_2 x^2 + s_1 x + s_0) \qquad (6)$$

where $s_2 = \omega^8 + \omega^4 + \omega$, $s_1 = \omega^4 + \omega^2$, and s_0 is one of the roots of $s^2 + s = \omega^5 + \omega^3$. Here we note that s_0 and $s_0 + 1$ are two roots of the quadratic equation. Then we can define σ_ω up to ± 1 multiplication. We can verify the fundamental relations

$$\sigma_\omega \sigma_{\omega'} = \pm \sigma_{\omega'} \sigma_\omega = \pm \sigma_{\omega + \omega'}. \qquad (7)$$

In particular, $\sigma_\omega^2 = \pm 1$. Hence, $\mathbb{G} = \langle \pm \sigma_\omega \mid E(\omega) = 0 \rangle$ ($\subset \mathrm{Aut}_C$) is of order 32 $= 2^5$. In fact, $\mathbb{G}$ is an extra-special 2-group that is the central product of the dihedral group of order 8 and the quaternion group of order 8 with identified center (See [8]). We notice that the roots of $E^+(z) = 0$ (and $E^-(z) = 0$) define the automorphisms of order 4 (and 2), respectively [8].

Let $\tau \in \mathbb{F}_{2^6}$ be a root of $\beta_6(z) = 0$. We then set $\xi := \tau^4 + \tau^2$ and $\theta := \tau^4 + \tau^2 + \tau$. Then $\beta_2(\theta) = 0, \beta_3^-(\xi) = 0$, and $\xi = \theta + \tau$. Therefore, $\sigma_\tau^2 = -1, \sigma_\theta^2 = -1, \sigma_\theta^2 = 1$, and $\sigma_\xi^2 = 1$. In addition, we fix the (± 1)-ambiguity of σ_ξ such that $\sigma_\xi = \sigma_\theta \sigma_\tau$. From direct calculations using (6), we obtain the following commutator relations.

$$\begin{aligned}
\sigma_\tau \sigma_\theta &= -\sigma_\theta \sigma_\tau, & \sigma_\tau \sigma_\xi &= -\sigma_\xi \sigma_\tau, & \sigma_\theta \sigma_\xi &= -\sigma_\xi \sigma_\theta, \\
\sigma_1 \sigma_\tau &= -\sigma_\tau \sigma_1, & \sigma_1 \sigma_\theta &= -\sigma_\theta \sigma_1, & \sigma_1 \sigma_\xi &= \sigma_\xi \sigma_1.
\end{aligned} \qquad (8)$$

Here we note that the above relations are satisfied regardless of the (± 1)-ambiguity of σ_ω used above.

Galbraith *et al.* [6] showed that $\mathrm{End}^0(\mathrm{Jac}_C) = \mathbb{Q}[\pi, \sigma_\tau, \sigma_\theta] = \mathbb{Q}(\pi) \oplus \sigma_\tau \mathbb{Q}(\pi) \oplus \sigma_\theta \mathbb{Q}(\pi) \oplus \sigma_\xi \mathbb{Q}(\pi)$. Let Δ and Δ^* be $\{\pi^i, \pi^j \sigma_\theta, \pi^\kappa \sigma_\tau, \pi^l \sigma_\xi \mid 0 \le i, j, \kappa, l \le 3\}$ and $\{\pi^i, \sigma_\theta \pi^j, \sigma_\tau \pi^\kappa, \sigma_\xi \pi^l \mid 0 \le i, j, \kappa, l \le 3\}$, respectively. A distortion map ϕ in Theorem 1 is a $\mathbb{Q}$-linear combination of elements of Δ because $\mathrm{End}^0(\mathrm{Jac}_C) = \mathbb{Q}[\pi, \sigma_\tau, \sigma_\theta]$. They state the following result similar to Theorem 2. Let r be an odd prime s.t. $r \mid \sharp \mathrm{Jac}_C(\mathbb{F}_q)$.

Theorem 8 ([6]). *Assume that the denominators of coefficients of ϕ are coprime to r. Then Δ is a complete set of efficiently computable distortion maps on $\mathrm{Jac}_C[r]$.*

We show that Theorem 8 holds without any unproven assumption when $r > 19$ in Theorem 10 in Section 5.4 (See also Corollary 4).

5.2   An Efficiently Constructible Semi-symplectic Basis

Let polynomials $P_m^{\pm}(T)$ be $T^4 \pm hT^3 + qT^2 \pm qhT + q^2$ where $h = 2^{(m+1)/2}$, respectively. Then the characteristic polynomial of the q-th power Frobenius endomorphism π is $P_m^+(T)$ or $P_m^-(T)$. Therefore, $\sharp\mathrm{Jac}_C(\mathbb{F}_q) = P_m^+(1)$ or $P_m^-(1)$. For a prime r s.t. $r \mid \sharp\mathrm{Jac}_C(\mathbb{F}_q)$, the embedding degree k (the smallest positive integer k s.t. $q^k = 1 \bmod r$) is 12^1. As in Section 4.3, we choose a nontrivial D_1 in $\mathrm{Jac}_C(\mathbb{F}_q)[r]$ at first. Then $D_2 := \sigma_\theta D_1, D_3 := \sigma_\tau D_1, D_4 := \sigma_\xi D_1$. We set $\mathcal{B} := \{D_i \mid i = 1, \ldots, 4\}$, and an $\mathbb{F}_r$-vector subspace $\mathcal{V} := \langle \mathcal{B} \rangle \subset \mathrm{Jac}_C[r]$.

The automorphism σ_1 is defined over $\mathbb{F}_q$, then for D_1, it acts as some scalar multiplication. In addition, since $\sigma_1^2 = -1$ and q^3 is a primitive fourth root of unity, so $\sigma_1 D_1 = \pm q^3 D_1$. We then fix (± 1)-ambiguity of σ_1 s.t. $\sigma_1 D_1 = q^3 D_1$.

In addition to the fundamental relations (7) and the commutator relations (8), we use the following commutator relations with the Frobenius endomorphism π ([6]),

$$\pi\sigma_\omega = \pm\sigma_{\omega^{2m}}\pi, \qquad \pi^3\sigma_\tau\pi^{-3} = \pm\sigma_{\tau^{2^3}} = \pm\sigma_{\tau+1} = \pm\sigma_\tau\sigma_1,$$

$$\pi\sigma_\theta\pi^{-1} = \pm\sigma_{\theta^2} = \pm\sigma_{\theta+1} = \pm\sigma_\theta\sigma_1, \qquad \pi^3\sigma_\xi\pi^{-3} = \sigma_\xi.$$

The last equality is from $\xi^8 = \xi$ and $s_0 = \xi$ or $\xi + 1$ in (6) of σ_ξ.

Lemma 2. *The divisors $D_1 \in \mathrm{Jac}_C(\mathbb{F}_q)[r]$, $D_2 \in \mathrm{Jac}_C(\mathbb{F}_{q^4})[r]$, $D_3 \in \mathrm{Jac}_C(\mathbb{F}_{q^{12}})[r]$, and $D_4 \in \mathrm{Jac}_C(\mathbb{F}_{q^3})[r]$. In addition, $\dim_{\mathbb{F}_r} \mathcal{V} \geq 3$ where $\mathcal{V} = \langle \mathcal{B} \rangle$.*

Proof. By definition, $D_1 \in \mathrm{Jac}_C(\mathbb{F}_q)[r]$. Since $\pi\sigma_\theta\pi^{-1} = \pm\sigma_\theta\sigma_1$, $\pi^2\sigma_\theta\pi^{-2} = -\sigma_\theta$. Then $D_2 \in \mathrm{Jac}_C(\mathbb{F}_{q^4})[r]$ and not defined over a smaller field. Since $\pi^3\sigma_\xi = \sigma_\xi\pi^3$ and $\xi \notin \mathbb{F}_q$, $D_4 \in \mathrm{Jac}_C(\mathbb{F}_{q^3})[r]$ and it is also not defined over a smaller field. From $\pi^3\sigma_\tau\pi^{-3} = \pm\sigma_\tau\sigma_1$, $\pi^6\sigma_\tau\pi^{-6} = -\sigma_\tau$. Thus $D_3 \in \mathrm{Jac}_C(\mathbb{F}_{q^{12}})[r]$. Indeed, since D_1, D_2, and D_4 are linearly independent over $\mathbb{F}_r$, $\dim_{\mathbb{F}_r} \mathcal{V} \geq 3$. $\square$

Theorem 9. *The discrete logarithms of $e(D_i, D_j)$ to the base $e(D_1, D_3)$ are tabulated as*

$$\begin{pmatrix} 0 & 0 & 1 & 0 \\ 0 & 0 & 0 & 1 \\ -1 & 0 & 0 & 0 \\ 0 & -1 & 0 & 0 \end{pmatrix}. \tag{9}$$

---

[1] This is mentioned in [6]. In fact, they have shown that k divides 12 in [6]. For completeness, we show that k is 12 for any prime r s.t. $r \mid \sharp\mathrm{Jac}_C(\mathbb{F}_q)$ in Proposition 1 in Appendix.

Proof. Since $\pi^2 D_2 = -D_2$ and e is Galois invariant, $e(D_1, D_2)^{q^2} = e(D_1, D_2)^{-1}$. From Proposition 1 in Appendix, $q^2 + 1 \not\equiv 0 \bmod r$. Hence, $e(D_1, D_2) = 1$. Similarly, since $\pi^3 D_4 = D_4$, we see that $e(D_1, D_4)^{q^3} = e(D_1, D_4)$ and $e(D_1, D_4) = 1$ as well. Using Fact 1, we can verify that $e(D_i, D_j) = 1$ except for $(i, j) = (1, 3), (3, 1), (2, 4), (4, 2)$. For example, $e(D_3, D_2) = e(\sigma_\tau D_1, \sigma_\theta D_1) = e(\sigma_\tau D_1, \sigma_\tau \sigma_\xi D_1) = e(D_1, \sigma_\xi D_1) = e(D_1, D_4) = 1$. Finally, $e(D_2, D_4) = e(D_1, D_3)$ since $e(D_2, D_4) = e(\sigma_\theta D_1, \sigma_\xi D_1) = e(\sigma_\theta D_1, \sigma_\theta \sigma_\tau D_1) = e(D_1, \sigma_\tau D_1) = e(D_1, D_3)$. □

Corollary 3. *The base $e(D_1, D_3)$ in Theorem 9 is not equal to 1: $e(D_1, D_3) \neq 1$. Consequently, $\mathcal{B}$ is an efficiently constructible semi-symplectic basis of $\mathrm{Jac}_C[r]$ for the Weil pairing.*

Proof. Assume that $e(D_1, D_3) = 1$. From Theorem 9, then $e(D, D') = 1$ for all D and $D' \in V$. This contradicts the nondegeneracy of the Weil pairing because $\dim_{\mathbb{F}_r} V \geq 3$. We then conclude that $e(D_1, D_3) \neq 1$, and $\mathcal{B}$ is an efficiently constructible semi-symplectic basis of $\mathrm{Jac}_C[r]$ for the Weil pairing. □

From Corollary 3, we know that the full embedding degree is also 12 (cf. [3,6,12]).

5.3   Frobenius Action on the Basis $\mathcal{B}$

We determine the action of π on the basis $\mathcal{B}$ for the completeness of Δ and Δ^*.

Lemma 3. *The Frobenius π acts on $\mathcal{B} = \{D_i\}$ as follows: $\pi D_1 = D_1$, $\pi D_2 = \lambda D_2$, $\pi D_3 = \lambda(\mu D_3 + d D_2)$, and $\pi D_4 = \mu D_4 + d D_1$ where $\lambda = q^3$ or $\lambda = -q^3 = q^9$, $\mu = q^4$ or $\mu = q^8$, and some $d \in \mathbb{F}_r$.*

Proof. The first formula is trivial. Since $\pi \sigma_\theta \pi^{-1} = \pm \sigma_\theta \sigma_1$, $\pi \sigma_\theta D_1 = \pm \sigma_\theta \sigma_1 D_1$. In other words, $\pi D_2 = \pm q^3 D_2 = \lambda D_2$ for $\lambda = \pm q^3$.

From Lemma 2, $\pi D_4 \in \mathrm{Jac}_C(\mathbb{F}_{p^3})[r] = \langle D_1, D_4 \rangle$. Because $\pi^3 D_4 = D_4$, we know that $\pi D_4 = \mu D_4 + d D_1$ for $\mu = q^4$ or $\mu = q^8$, and some $d \in \mathbb{F}_r$.

Since $\sigma_\tau = \sigma_\theta \sigma_\xi$, $\pi D_3 = \pi \sigma_\theta \sigma_\xi D_1 = (\pi \sigma_\theta \pi^{-1})(\pi \sigma_\xi) D_1 = \pm \sigma_\theta \sigma_1 (\mu D_4 + d D_1) = \pm \sigma_\theta \sigma_1 (\mu \sigma_\xi D_1 + d D_1) = \pm \sigma_\theta \sigma_1 (\mu \sigma_\xi + d) D_1$. By using $\sigma_\xi \sigma_1 = \sigma_1 \sigma_\xi$ in (8), this becomes $\sigma_\theta (\mu \sigma_\xi + d) \cdot (\pm \sigma_1 D_1)$. Here, this $\pm$ sign and that in the definition in λ above are equal. Therefore, it is $\sigma_\theta (\mu \sigma_\xi + d) \cdot (\lambda D_1) = \lambda(\mu \sigma_\tau + d \sigma_\theta) D_1 = \lambda(\mu D_3 + d D_2)$ because $\sigma_\tau = \sigma_\theta \sigma_\xi$. □

Lemma 4. *For d and μ in Lemma 3, $d^2 = -\mu$. In particluar, $d = q^5$ or $d = -q^5$ when $\mu = q^4$, $d = q$ or $d = -q$ when $\mu = q^8$.*

Proof. From Lemma 3, $\pi^2 D_4 = \mu^2 D_4 + d(\mu + 1) D_1 = \mu^2(D_4 - d D_1)$.

When $m = 1 \bmod 6$, $\pi^i D_4 = \pm \sigma_{\xi^{2i}} D_1$. Then $\pi D_4 = \pm \sigma_{\xi^2} D_1$ and $\pi^2 D_4 = \pm \sigma_{\xi^4} D_1 = \pm \sigma_\xi \sigma_{\xi^2} \sigma_1 D_1 = \pm q^3 \sigma_\xi \sigma_{\xi^2} D_1 = \pm q^3 \sigma_\xi \pi D_4$. When $m = -1 \bmod 6$, $\pi^i D_4 = \pm \sigma_{\xi^{2-i}} D_1$. Similarly, $\pi D_4 = \pm \sigma_{\xi^4} D_1$ and $\pi^2 D_4 = \pm \sigma_{\xi^2} D_1 = \pm \sigma_\xi \sigma_{\xi^4} \sigma_1 D_1 = \pm q^3 \sigma_\xi \pi D_4$. In both cases, $\pi^2 D_4 = \pm q^3 \sigma_\xi \pi D_4 = \pm q^3 \sigma_\xi (\mu D_4 + d D_1) = \pm q^3 (\mu D_1 + d D_4)$ because $\sigma_\xi^2 = 1$.

Therefore, because of the linear independence of D_1 and D_4, $-d\mu^2 = \pm q^3 \mu$, $\mu^2 = \pm q^3 d$. Then $d^2 = -\mu$, and $d = \pm q^5$ when $\mu = q^4$, $d = \pm q$ when $\mu = q^8$. □

5.4 Completeness of Δ and Δ^*

Let $\nu \in \mathbb{F}_r$ be $\frac{d}{\lambda-1}$ and let $\widetilde{D}_4, \widetilde{D}_3$ be $D_4 + \nu D_1, D_3 + \nu D_2$ respectively. Then $\pi \widetilde{D}_4 = \lambda \widetilde{D}_4$ and $\pi \widetilde{D}_3 = \lambda \mu \widetilde{D}_3$. In addition, let $\widetilde{D}_1$ and $\widetilde{D}_2$ be D_1 and D_2, respectively. Consequently, $\widetilde{D}_1, \widetilde{D}_2, \widetilde{D}_3, \widetilde{D}_4$ are eigenvectors of π with eigenvalue $1, \mu, \lambda\mu, \lambda$ respectively. We set $\widetilde{\mathcal{B}} := \{\widetilde{D}_i \mid i = 1, \ldots, 4\}$. The Weil pairing matrix of the basis $\widetilde{\mathcal{B}}$ is also given by (9). Based on the following Lemmas 5 and 6, we show the completeness of Δ^* and Δ when $r > 19$ (Theorem 10).

Lemma 5. *If $r > 19$, then ν is neither 0 nor ± 1.*

Proof. That $\nu \neq 0$ is trivial from Lemma 3. We notice that all equalities in the following are in $\mathbb{F}_r$. Since $\lambda^2 = -1$ and $d^2 = -\mu$ by Lemma 4, $\nu^2 = \left(\frac{d}{\lambda-1}\right)^2 = \frac{\mu}{2\lambda} = \frac{q}{2}, \frac{-q}{2}, \frac{1}{2q}$, or $\frac{-1}{2q}$. Assume that $\nu^2 = 1$, then $q = \pm 2$ or $2q = \pm 1$. We use $2q = h^2$. If $q = 2$, then $h = \pm 2$ and $P_m^{\pm}(1) = 1 \pm 2 + 2 \pm 4 + 4 = 13, 1$. If $q = -2$, then $h = \pm 2q^3 = \pm 16$ and $P_m^{\pm}(1) = 1 \pm 16 - 2 \mp 32 + 4 = -13, 19$. If $2q = 1$, then $h = \pm 1$ and $4P_m^{\pm}(1) = 4 \pm 4h + 4q \pm 4qh + 4q^2 = 4 \pm 4 + 2 \mp 2 + 1 = 9, 5$. If $2q = -1$, then $h = \pm q^3$ and $8h = \pm 1$. Thus, $16P_m^{\pm}(1) = 16 \pm 16h + 16q \pm 16qh + 16q^2 = 16 \pm 2 - 8 \mp 1 + 4 = 13, 11$. This contradicts that $r > 19$. Hence, $\nu^2 \neq 1$. $\qquad\square$

Lemma 6. *Let $D \neq \mathcal{O}$ and D' be in $\mathrm{Jac}_C[r]$, and let D' be expressed as $\sum_{i=1}^{4} c'_i D_i$. If the Weil pairing $e(D, \sigma D') = 1$ for all $\sigma \in \{1, \sigma_\theta, \sigma_\tau, \sigma_\xi\}$, then*

$$\alpha(c'_1, c'_2, c'_3, c'_4) := (c'_1)^2 - (c'_2)^2 + (c'_3)^2 - (c'_4)^2 = 0. \tag{10}$$

Proof. Using the relations (8) and $\sigma_\tau^2 = -1$, etc., we know that

$$
\begin{aligned}
\sigma_\theta(D_1) &= D_2, & \sigma_\theta(D_2) &= D_1, & \sigma_\theta(D_3) &= D_4, & \sigma_\theta(D_4) &= D_3, \\
\sigma_\tau(D_1) &= D_3, & \sigma_\tau(D_2) &= -D_4, & \sigma_\tau(D_3) &= -D_1, & \sigma_\tau(D_4) &= D_2, \\
\sigma_\xi(D_1) &= D_4, & \sigma_\xi(D_2) &= -D_3, & \sigma_\xi(D_3) &= -D_2, & \sigma_\xi(D_4) &= D_1.
\end{aligned}
\tag{11}
$$

Let D be $\sum c_i D_i$. Then $e(D, D') = 1$ implies that $c_1 c'_3 - c_3 c'_1 + c_2 c'_4 - c_4 c'_2 = 0$ from (9) and Corollary 3. Using (11), we obtain similar relations from $e(D, \sigma D') = 1$ for all σ's. That is,

$$
\begin{pmatrix} c_1 & c_2 & c_3 & c_4 \end{pmatrix}
\begin{pmatrix}
c'_3 & c'_4 & -c'_1 & -c'_2 \\
c'_4 & c'_3 & c'_2 & c'_1 \\
-c'_1 & -c'_2 & -c'_3 & -c'_4 \\
-c'_2 & -c'_1 & c'_4 & c'_3
\end{pmatrix}
= \begin{pmatrix} 0 & 0 & 0 & 0 \end{pmatrix}. \tag{12}
$$

Because $D \neq \mathcal{O}$, the determinant of the matrix in the LHS of (12) is 0. It is $-\alpha(c'_1, c'_2, c'_3, c'_4)^2$ where α is defined in (10). Therefore, $\alpha(c'_1, c'_2, c'_3, c'_4) = 0$. $\qquad\square$

Theorem 10. *Both $\Delta^* = \{\pi^i, \sigma_\theta \pi^j, \sigma_\tau \pi^\kappa, \sigma_\xi \pi^l \mid 0 \leq i, j, \kappa, l \leq 3\}$ and $\Delta = \{\pi^i, \pi^j \sigma_\theta, \pi^\kappa \sigma_\tau, \pi^l \sigma_\xi \mid 0 \leq i, j, \kappa, l \leq 3\}$ are complete sets of efficiently computable distortion maps on $\mathrm{Jac}_C[r]$ with $r > 19$.*

Corollary 4. *We can choose ϕ in Theorem 8 whose denominator is 1 when $r > 19$.*

Proof of Theorem 10. We show that $\mathrm{Jac}_C[r] = \langle \delta(D') \mid \delta \in \Delta^* \rangle$ as an $\mathbb{F}_r$-vector space for every nontrivial $D' \in \mathrm{Jac}_C[r]$. First, we express D' in terms of the basis $\widetilde{\mathcal{B}}$, i.e., $D' = \sum \widetilde{c}_i \widetilde{D}_i$ where some $\widetilde{c}_i \neq 0$. Using relations of D_j and $\widetilde{D}_j$,

$$\begin{aligned}
\pi^i(D') &= \widetilde{c}_1 \widetilde{D}_1 + \mu^i \widetilde{c}_2 \widetilde{D}_2 + (\lambda\mu)^i \widetilde{c}_3 \widetilde{D}_3 + \lambda^i \widetilde{c}_4 \widetilde{D}_4 \\
&= (\widetilde{c}_1 + \lambda^i \nu \widetilde{c}_4) D_1 + (\mu^i \widetilde{c}_2 + (\lambda\mu)^i \widetilde{c}_3) D_2 + (\lambda\mu)^i \widetilde{c}_3 D_3 + \lambda^i \widetilde{c}_4 D_4.
\end{aligned}$$

To prove that $\mathrm{Jac}_C[r] = \langle \delta(D') \mid \delta \in \Delta^* \rangle$ by reductio ad absurdum, first, we assume that $e(D, \sigma\pi^i D') = 1$ for the Weil pairing e, all $\sigma \in \{1, \sigma_\theta, \sigma_\tau, \sigma_\xi\}$, and some $D \neq \mathcal{O}$. We then apply Lemma 6 to D and $\pi^i(D')$. After some calculation, we obtain

$$\widetilde{c}_1^2 - \mu^{2i}\widetilde{c}_2^2 - \mu^{2i}\lambda^{2i}(\nu^2 - 1)\widetilde{c}_3^2 + \lambda^{2i}(\nu^2 - 1)\widetilde{c}_4^2 + 2\lambda^i \nu \widetilde{c}_1 \widetilde{c}_4 - 2\mu^{2i}\lambda^i \nu \widetilde{c}_2 \widetilde{c}_3 = 0 \quad (13)$$

for all integers i. For $i = 0, \ldots, 5$, we consider (13) with $\widetilde{c}_1^2, \ldots, \widetilde{c}_4^2, \widetilde{c}_1 \widetilde{c}_4, \widetilde{c}_2 \widetilde{c}_3$ as 6 indeterminates. Since $\nu \neq 0, \pm 1$ from Lemma 5, the coefficient matrix of (13) is the product of the regular diagonal matrix $\mathrm{diag}(1, -1, -(\nu^2 - 1), \nu^2 - 1, 2\nu, -2\nu)$ and the Vandermonde matrix $V = V(1, \mu^2, \mu^2\lambda^2, \lambda^2, \lambda, \lambda\mu^2)$. Since $\mu^2 \in \{q^4, q^8\}, \lambda \in \{q^3, q^9\}, \lambda^2 = q^6, \lambda^2\mu^2 \in \{q^2, q^{10}\}$, and $\lambda\mu^2 \in \{q, q^5, q^7, q^{11}\}$, the determinant of V is not zero, all $\widetilde{c}_i$ are zero, and $D' = \mathcal{O}$. It contradicts. Hence, $e(D, \delta D') \neq 1$ for some $\delta \in \Delta^*$. This concludes the completeness of Δ^*.

For the completeness of Δ, first, we see that $e(\delta D, D') = e(\sigma\pi^i D, D') = e(D, \pi^j \sigma D')^{\pm q^j}$ for $\delta = \sigma\pi^i \in \Delta^*$ where $j = 12 - i$. Hence, the completeness of Δ^* implies that of Δ. □

6   Conclusion

We have proved that a specific set of efficiently computable endomorphisms definitely gives a distortion map for every pair of nontrivial divisors on the curves in [6]. In addition, we treated the general version of the curve here. Moreover, we obtained efficiently constructible semi-symplectic bases for these curves using cyclotomy (Gauss sum, Jacobi sum, etc.) and group-theoretic consideration. The bases will provide a basic tool for a possible new cryptographic application of pairing on a higher dimensional vector space suggested in [4,3].

Acknowledgments. I would like to thank Tatsuaki Okamoto, Takakazu Satoh, Toyohiro Tsurumaru, Shigenori Uchiyama, and the anonymous reviewers of ANTS VIII for their helpful comments. All computer calculations in support of this work were performed using Magma [1].

References

1. Cannon, J.J., Bosma, W. (eds.): Handbook of Magma Functions, 2.13nd edn., pages 4350 (2006)
2. Davis, P.J.: Circulant Matrices, 2nd edn. Chelsea publishing (1994)

3. Freeman, D.: Constructing pairing-friendly genus 2 curves with ordinary Jacobians. In: Takagi, T., Okamoto, T., Okamoto, E., Okamoto, T. (eds.) Pairing 2007. LNCS, vol. 4575, pp. 152–176. Springer, Heidelberg (2007)
4. Galbraith, S.D., Hess, F., Vercauteren, F.: Hyperelliptic pairings. In: Takagi, T., Okamoto, T., Okamoto, E., Okamoto, T. (eds.) Pairing 2007. LNCS, vol. 4575, pp. 108–131. Springer, Heidelberg (2007)
5. Galbraith, S.D., Pujolàs, J.: Distortion maps for genus two curves. In: Proceedings of a Workshop on Mathematical Problems and Techniques in Cryptology, CRM Barcelona, pp. 46–58 (2005)
6. Galbraith, S.D., Pujolàs, J., Ritzenthaler, C., Smith, B.: Distortion maps for genus two curves (2006), http://arXiv.org/abs/math/0611471
7. Galbraith, S.D., Rotger, V.: Easy decision Diffie-Hellman groups. LMS J. Comput. Math. 7, 201–218 (2004)
8. van der Geer, G., van der Vlugt, M.: Reed-Muller codes and supersingular curves I. Compositio Math. 84, 333–367 (1992)
9. Lidl, R., Niederreiter, H.: Finite Fields, 2nd edn. Cambridge University Press, Cambridge (1997)
10. Milne, J.S.: Abelian varieties. In: Cornell, G., Silverman, J.H. (eds.) Arithmetic Geometry, Springer, Heidelberg (1986)
11. Mumford, D.: Abelian Varieties. Oxford University Press, Oxford (1974)
12. Stichtenoth, H., Xing, C.: On the structure of the divisor class group of a class of curves over finite fields. Arch. Math. 65, 141–150 (1995)
13. Verheul, E.: Evidence that XTR is more secure than supersingular elliptic curve cryptosystems (full version of the proceeding in Eurocrypt 2001). J. Crypt. 17, 277–296 (2004)

Appendix

Proposition 1. *The embedding degree k is 12 for every prime r s.t. $r \mid \sharp \mathrm{Jac}_C(\mathbb{F}_q)$ for the curve in Section 5.*

Proof. In [6], they show that k divides 12. Hence, we must show that r does not divide $\Phi_i(q)$ for any divisor i of 12 s.t. $i \neq 12$ where Φ_i is the i-th cyclotomic polynomial. In the following discussion, all equalities mean that in $\mathbb{F}_r$.

If $\Phi_1(q) = 0$ (i.e., $q = 1$), then $P_m^{\pm}(1) = 3 \pm 2h = 0$. $h^2 = 2$ since $2q = h^2$. Thus $3h \pm 4 = 0$. This leads to $1 = 0$, a contradiction. If $\Phi_2(q) = 0$ (i.e., $q = -1$), then $P_m^{\pm}(1) = 1 = 0$. Another contradiction. If $\Phi_3(q) = 0$, then $P_m^{\pm}(1) = \pm h(q+1) = 0$. $q+1 = \Phi_2(q) = 0$ since h is a power of 2. Contradiction.

If $\Phi_4(q) = q^2 + 1 = 0$, then $P_m^{\pm}(1) = \pm qh + q \pm h = 0$. Then using $2q = h^2$, we obtain $\pm h^2 + h \pm 2 = 0$. We solve the simultaneous equations $h^4 + 4 = 4(q^2 + 1) = 0$ and $\pm h^2 + h \pm 2 = 0$. In the case of the plus sign, the remainder of division of the 2 polynomials is $h + 2(= 0)$, and this leads to $r = 2$. That contradicts $q^2 + 1 = 0$ mod r. In the case of the minus sign, the above remainder is $-3h + 6(= 0)$. It leads to $h = 2$ and $r = 2$ (contradiction as above) or $r = 3$. If $r = 3$, then $q^2 + 1 = 2 = 0$ since q is a power of 2. Again, a contradiction.

If $\Phi_6(q) = q^2 - q + 1 = 0$, then $\pm h^2 + 2h \pm 2 = 0$ since $P_m^{\pm}(1) = 0$. We solve the simultaneous equations $h^4 - 2h^2 + 4 = 0$ and $\pm h^2 + 2h \pm 2 = 0$. Both cases of the $\pm$ sign lead to contradictions as above. We have completed the proof. $\square$

On Prime-Order Elliptic Curves with Embedding Degrees $k = 3$, 4, and 6

Koray Karabina and Edlyn Teske

Dept. of Combinatorics and Optimization
University of Waterloo
Waterloo, Ontario, Canada N2L 3G1
{kkarabina,eteske}@uwaterloo.ca

Abstract. We further analyze the solutions to the Diophantine equations from which prime-order elliptic curves of embedding degrees $k = 3, 4$ or 6 (MNT curves) may be obtained. We give an explicit algorithm to generate such curves. We derive a heuristic lower bound for the number $E(z)$ of MNT curves with $k = 6$ and discriminant $D \leq z$, and compare this lower bound with experimental data.

Keywords: Elliptic curves, pairing-based cryptosystems, embedding degree, MNT curves.

1 Introduction

For an elliptic curve E defined over a finite field $\mathbb{F}_q$, let $\#E(\mathbb{F}_q) = n = hr$ be the number of $\mathbb{F}_q$-rational points on E, where r is the largest prime divisor of n, and $\gcd(r, q) = 1$. The set of all points of order r in $E(\bar{\mathbb{F}}_q)$ forms a subgroup of $E(\mathbb{F}_q)$ denoted by $E[r]$. For such an integer r, a bilinear map can be defined from a pair of r-torsion points of E to the group μ_r of rth roots of unity in $\bar{\mathbb{F}}_q$ by

$$e_r : E[r] \times E[r] \mapsto \mu_r.$$

In fact, the multiplicative group μ_r in the above mapping lies in the extension field $\mathbb{F}_{q^k}$ where k is the least positive integer satisfying $k \geq 2$ and $q^k \equiv 1$ (mod r). The above mapping is called the *Weil pairing*, and the integer k is called the *embedding degree* of E.

Pairings such as the Weil pairing (other proposed pairings include the Tate pairing, the Eta pairing [2], or the Ate pairing [7]) are used in many cryptographic applications such as identity based encryption [4], one-round 3-party key agreement protocols [8], and short signature schemes [21]. The computation of pairings requires arithmetic in the finite field $\mathbb{F}_{q^k}$. Therefore, k should be small for the efficiency of the application. On the other hand, the discrete logarithm problem (DLP) in the order-r subgroup of $E(\mathbb{F}_q)$ can be reduced to the DLP in $\mathbb{F}_{q^k}$ [13]. Therefore, k must also be sufficiently large so that the DLP in $\mathbb{F}_{q^k}$ is computationally hard enough for the desired security. In particular, it is reasonable to ask for parameters q, r and k so that the DLP in $E(\mathbb{F}_q)$, and the

A.J. van der Poorten and A. Stein (Eds.): ANTS-VIII 2008, LNCS 5011, pp. 102–117, 2008.

DLP in $\mathbb{F}_{q^k}$ have approximately the same difficulty. Given the best algorithms known and today's computer technology to attack discrete logarithms in elliptic curve groups and in finite field groups, the 80-bit security level can be satisfied by choosing $r \approx 2^{160}$, and $q^k \approx 2^{1024}$. If $E/\mathbb{F}_q$ is of prime order, then $r \approx q$, and thus the 80-bit security level can be achieved if $q \approx 2^{170}$ and $k = 6$.

Now, Miyaji, Nakabayashi, and Takano [14] gave a characterization of prime-order elliptic curves with embedding degree $k = 3, 4$ and 6, in terms of necessary and sufficient conditions on the pair (q, t) where $t = q + 1 - \#E(\mathbb{F}_q)$, the *trace* of E over $\mathbb{F}_q$. Such elliptic curves, if ordinary (i.e., when $\gcd(q, t) = 1$), are nowadays commonly called *MNT curves*.

The only known method to construct MNT curves is to compute suitable integers q and t such that there exists an ordinary elliptic curve $E/\mathbb{F}_q$ of prime order and embedding degree k, and to then use the Complex Multiplication method (or CM method) [1] to find the equation of the curve E over $\mathbb{F}_q$. In fact, all methods known so far to construct ordinary elliptic curves of any order and small embedding degree use the CM method; see [5] for a comprehensive survey. A central equation in this context is the *CM equation*

$$4q - t^2 = DY^2 \tag{1}$$

where D is a positive integer and $Y \in \mathbb{Z}$. If D is square-free, we call D the *Complex Multiplication discriminant* (or *CM discriminant*, or briefly discriminant) of E. Given current algorithms and computing power, the CM method is practical if $D < 10^{10}$ (see [5] for a discussion of this bound).

From (1) Miyaji, Nakabayashi, and Takano [14] derived Pell-type equations, which we subsequently call *MNT equations* (see Section 2). For a fixed embedding degree $k \in \{3, 4, 6\}$ and CM discriminant D, solving the corresponding MNT equation leads to candidate parameters (q, t) for prime-order elliptic curves $E/\mathbb{F}_q$ of trace $t = q + 1 - \#E(\mathbb{F}_q)$, embedding degree k and discriminant D. As, by nature of generalized Pell equations, the solutions of an MNT equation (if sorted by bitsize and enumerated) grow exponentially, MNT curves are very rare. In fact, Luca and Shparlinski [11] gave a heuristic argument that for any upper bound z, there exists only a finite number of MNT curves with discriminant $D \leq z$, regardless of the field size. On the other hand, specific sample curves of cryptographic interest have been found, such as MNT curves of 160-bit, 192-bit, or 256-bit prime order ([17,20]).

Contribution of This Paper. First, we further analyze the solutions of the MNT equations and establish that the MNT curves of embedding degree 6 are given through the solutions in one of the two (if any) solution classes of the MNT equation (Section 3). Based on this analysis we give a complete algorithm (in the appendix) to calculate such solutions that lead to potentially prime-order elliptic curves; we could not find such an explicit algorithm anywhere in the literature. We also point out a one-to-one correspondence between MNT curves of embedding degree 4 and MNT curves of embedding degree 6 (Proposition 1).

Second, building on the work by Luca and Shparlinski [11] who gave a heuristic upper bound on the expected number $E(z)$ of MNT curves with embedding degree 6 and bounded discriminant $D \leq z$, we provide a heuristic lower bound for $E(z)$ (Section 4.2). Specifically, we show that for large enough z we have $E(z) \geq 0.49 \frac{\sqrt{z}}{(\log z)^2}$, which nicely complements the Luca-Shparlinski result that $E(z) \ll z/(\log z)^2$ and corrects the guess [11, p. 559] that $E(z) \leq z^{o(1)}$. Here and throughout, $\log z$ denotes the natural logarithm of z.

Finally, we give numerical data on $E(z)$ over finite fields of bounded characteristic, and compare those data with our new lower bound (Section 4.3). At least for this experimentally verifyable range, our lower bound, once corrected by a constant factor, seems to quite well capture the number of MNT curves of discriminant $D \leq z$.

2   MNT Curves and Their Pell Equations

The Miyaji-Nakabayashi-Takano characterization [14] of MNT curves is summarized in the following theorem.

Theorem 1. *Let $E/\mathbb{F}_q$ be an ordinary elliptic curve defined over a finite field $\mathbb{F}_q$. Let $n = \#E(\mathbb{F}_q)$ be a prime and k the embedding degree of E.*

1. *Suppose $q > 64$. Then $k = 3$ if and only if $q = 12l^2 - 1$ and $t = -1 \pm 6l$ for some $l \in \mathbb{Z}$.*
2. *Suppose $q > 36$. Then $k = 4$ if and only if $q = l^2 + l + 1$ and $t = -l, l + 1$ for some $l \in \mathbb{Z}$.*
3. *Suppose $q > 64$. Then $k = 6$ if and only if $q = 4l^2 + 1$ and $t = 1 \pm 2l$ for some $l \in \mathbb{Z}$.*

Note that for each elliptic curve characterized by Theorem 1 we have exactly two representations. For example ($k = 4$), if $t = -l$ and $q = l^2 + l + 1$ for some integer l, we can also write $l' = -l - 1$ and $t = l' + 1$ and $q = l'^2 + l' + 1$. (See also Proposition 3.)

The characterization from Theorem 1 implies a one-to-one correspondence between MNT curves with embedding degree $k = 4$ and MNT curves with embedding degree $k = 6$.

Proposition 1. *Let $n > 64$ and $q > 64$ be primes. Then n and q represent an elliptic curve $E_6/\mathbb{F}_q$ with embedding degree $k = 6$ and $\#E_6(\mathbb{F}_q) = n$ if and only if n and q represent an elliptic curve $E_4/\mathbb{F}_n$ with embedding degree $k = 4$ and $\#E_4(F_n) = q$.*

Proof. Let $n > 64$ and $q > 64$ represent an elliptic curve $E_6/\mathbb{F}_q$ with $k = 6$ and $\#E_6(\mathbb{F}_q) = n = q + 1 - t$. By Hasse's theorem we have $t^2 \leq 4q$. Now,

$$t^2 \leq 4q \Leftrightarrow t^2 \leq 4(t - 1 + n)$$
$$\Leftrightarrow (t - 2)^2 \leq 4n. \tag{2}$$

Let $n' = q$, $q' = n$, and $t' = q' + 1 - n'$. Then $t' = 2 - t$, and by (2), t' satisfies the Hasse bound with $q' = n$. So let E_4 be an elliptic curve over $\mathbb{F}_{q'}$ with n'

points. Now, by Theorem 1(3) $q = 4l^2 + 1$ for some integer l. If $t = 1 - 2l$, then $q' = q + 1 - t = (2l)^2 + 2l + 1$ and $t' = 2l + 1$, and thus by (2) of Theorem 1, $E_4/\mathbb{F}_{q'}$ has embedding degree $k' = 4$. Replacing l by $-l$ in the last sentence settles the other case, $t = 1 + 2l$.

To prove the converse, let n, q be primes greater than 64 representing an elliptic curve $E_4/\mathbb{F}_q$ with embedding degree $k = 4$ and n points, and let $t = q + 1 - n$. Then by Theorem 1(2) $t = l + 1$ or $t = -l$ for some $l \in \mathbb{Z}$. Since both n, q are odd primes, t must be odd. Thus, l is even if $t = l + 1$, and l is odd if $t = -l$. In the first case, $l = 2m$ and $t = 1 + 2m$ for some integer m, while in the second case, we can write $l = 2(-m) - 1$ and $t = 1 + 2m$ for some $m \in \mathbb{Z}$. We now proceed just as in the first part (starting after (2)).

Now, let us parametrize MNT curves by $(q(l), t(l))$ where $q(l)$ and $t(l)$ are as in Theorem 1. Then, after some elementary manipulation of the corresponding CM equations $4q(l) - t(l)^2 = DY^2$, one can obtain generalized Pell equations which we call the *MNT equations*. In particular:

1. The MNT equation for $k = 3$ is $X^2 - 3DY^2 = 24$, where $t(l) = 6l - 1$ and $X = 6l + 3$, or $t(l) = -6l - 1$ and $X = 6l - 3$.
2. The MNT equation for $k = 4$ is $X^2 - 3DY^2 = -8$, where $t(l) = -l$ and $X = 3l + 2$, or $t(l) = l + 1$ and $X = 3l + 1$.
3. The MNT equation for $k = 6$ is $X^2 - 3DY^2 = -8$. where $t(l) = 2l + 1$ and $X = 6l - 1$, or $t(l) = -2l + 1$ and $X = 6l + 1$.

The *MNT method* then consists of the following: Fix k. Choose $D < 10^{10}$. Solve the MNT equation to (hopefully) find pairs (q, t) such that q is a prime power and of the desired bitlength, and $q + 1 - t$ is prime. Finally, use the CM method to construct the actual curve.

3   Solving the MNT Equations

For solving the MNT equations, we need some facts from the theory of Pell equations and continued fractions. We refer to Mollin's book [15] for more details.

Let $m \in \mathbb{Z}$, $D \in \mathbb{N}$ and D not a perfect square. Then a generalized Pell equation can be given as follows

$$X^2 - DY^2 = m. \tag{3}$$

If $x \in \mathbb{Z}, y \in \mathbb{Z}$ and $x^2 - Dy^2 = m$ then we use both (x, y) and $x + y\sqrt{D}$ to refer to a solution of (3), since $x + y\sqrt{D}$ is an element in the quadratic field $\mathbb{Q}(\sqrt{D})$ with norm $x^2 - Dy^2 = m$. Let $\alpha = x + y\sqrt{D}$ be a solution to (3). If $\gcd(x, y) = 1$ then α is called a *primitive solution*. Two primitive solutions $\alpha_1 = x_1 + y_1\sqrt{D}$ and $\alpha_2 = x_2 + y_2\sqrt{D}$ belong to the same *class* of solutions if there is a solution $\beta = u + v\sqrt{D}$ of $X^2 - DY^2 = 1$ such that $\alpha_1 = \beta\alpha_2$. Now, if $\alpha = x + y\sqrt{D}$ then let α' denote the *conjugate* of α, that is, $\alpha' = x - y\sqrt{D}$. If a primitive solution and its conjugate are in the same class then the class is called *ambiguous*. If $\alpha = x + y\sqrt{D}$ is a solution of (3) for which y is the least positive value in its class then α is called the *fundamental solution* in its class. Note that if the class is not ambiguous then the fundamental solution is determined uniquely. If the class is ambiguous then adding the condition $x \geq 0$ defines the fundamental solution uniquely. Finally, if $\alpha = x + y\sqrt{D}$ is a solution of (3) for which y is the least positive value and x is nonnegative in its class then α is called the *minimal solution* in its class, and it is determined uniquely. If (x, y) is a minimal solution to $X^2 - DY^2 = m$, and (u, v) is a minimal solution to $U^2 - DV^2 = 1$ then all primitive solutions (x_j, y_j) in the class of (x, y) are generated as follows:

$$x_j + y_j\sqrt{D} = \pm(x + y\sqrt{D})(u + v\sqrt{D})^j, \text{where } j \in \mathbb{Z}. \tag{4}$$

Now we show that some Pell-type equations cannot have elements from an ambiguous class as solutions. We will use this result in Section 3.1.

Lemma 1. *Let $m \in \mathbb{Z}$, $m \equiv 0 \pmod 4$, and let D be an odd positive integer, not a perfect square. Then, the set of solutions to $X^2 - DY^2 = m$ does not contain any ambiguous class.*

Proof. Suppose that there is an ambiguous class of solutions. Then there exists a primitive solution $\alpha = x + y\sqrt{D}$ such that α and $\alpha' = x - y\sqrt{D}$ are in the same class. Then $(x^2 + y^2 D)/m$ is an integer ([15, Proposition 6.2.1]), and thus also $2y^2 D/m \in \mathbb{Z}$. But this cannot be true as y is odd, and so is D, while $4|m$.

If $\alpha = (x, y)$ is any solution in a given solution class of $X^2 - DY^2 = m$ then it is known ([16], Theorem 4.2) that there exists an integer P_0 which satisfies $-|m|/2 < P_0 \leq |m|/2$ and

$$P_0 + \sqrt{D} = (x + y\sqrt{D})(s + t\sqrt{D}) \tag{5}$$

for some unique element $s + t\sqrt{D}$. In this case $\alpha = (x, y)$ is said to *belong* to the element P_0.

Remark 1. If α belongs to P_0 and the class containing α is not ambigious, then $\alpha' = (x, -y)$ belongs to $-P_0$. This can be seen by conjugating (5) and then multiplying it by -1, which gives $-P_0 + \sqrt{D} = (x - y\sqrt{D})(-s + t\sqrt{D})$.

3.1   Embedding Degree $k = 6$

In this section we analyze the MNT equation for the case $k = 6$: $X^2 - 3DY^2 = -8$. We let $D' = 3D$ and for future reference rewrite the equation as

$$X^2 - D'Y^2 = -8. \tag{6}$$

We will show that for finding all computable MNT curves with $k = 6$ the following applies:

1. D' should be fixed such that $0 < D' < 3 \cdot 10^{10}$ and $D'/3$ is squarefree. – This is required for the CM method.
2. $D' \equiv 9 \pmod{24}$ and -2 is a square modulo D' (Proposition 2).

3. If there is a solution to $X^2 - D'Y^2 = -8$ then it is enough to find, if it exists, only one minimal solution, say (x_0, y_0) (Theorem 2, Proposition 3).
4. Let (u, v) be a minimal solution to $U^2 - D'V^2 = 1$ and $(x_j, y_j) = \pm(x_0, y_0)(u, v)^j$ the set of all solutions in the same class as (x, y). Then it is enough to consider only one of the solutions (x_j, y_j) and $-(x_j, y_j)$ (Proposition 3).

Proposition 2. *Assume $E/\mathbb{F}_q$ ($q > 64$) is an MNT curve with embedding degree $k = 6$ and CM discriminant D that is constructible with the MNT method. Let $D' = 3D$. Then (6) must have only primitive solutions. Further, $D' \equiv 9$ (mod 24), and -2 must be a square modulo D'.*

Proof. If there exists $E/\mathbb{F}_q$ with $k = 6$ then by Theorem 1(3) there exists some integer l satisfying $4q - t^2 = 12l^2 \pm 4l + 3$. As the CM equation (1) needs to hold, this implies $4l(3l \pm 1) + 3 = DY^2$, and so $DY^2 \equiv 3 \pmod 8$. Hence, $D \equiv 3 \pmod 8$, and $D' \equiv 9 \pmod{24}$. Now, let (x, y) be a solution of (6) with $\gcd(x, y) = d > 1$ and let $x = dx'$, $y = dy'$. Since $d^2(x'^2 - D'y'^2) = -8$ and D' is odd, we must have $d = 2$. Then $x'^2 - D'y'^2 = -2$ and thus $x'^2 - y'^2 \equiv 6 \pmod 8$. But this congruence has no integer solutions, and so any solution of (6) must be primitive. Finally, reducing (6) modulo D' proves that -2 must be a square modulo D'. $\square$

By Proposition 2, the MNT curves with $k = 6$ can only be obtained through the primitive solutions of the equation

$$X^2 - D'Y^2 = -8, \quad \text{where } D' \equiv 9 \pmod{24}. \tag{7}$$

Remark 2. If (x, y) is a primitive solution to (7), then x and y must both be odd. (This is directly implied by the facts that $\gcd(x, y) = 1$ and D' is odd.)

Remark 3. For any solution (x, y) of (6) with x odd we must have $x \equiv \pm 1 \pmod 6$. (Reducing (6) modulo 3 yields $x^2 \equiv 1 \pmod 3$.)

Theorem 2. *Equation (7) either does not have any solution or it has exactly two classes of solutions. In particular, if α is a solution of (7) then α and its conjugate α' represent the two solution classes.*

Proof. If (7) does not have any solution then we are done. Therefore, we shall assume that α is a solution belonging to some class, say P_0. Then, by Lemma 1 and Remark 1, α' is a solution belonging to $-P_0$. If these are the only two solution classes then we are done. So assume that there are more than two solution classes. Now, by the choice of P_0 we have $P_0^2 - D' \equiv 0 \pmod 8$, and $-4 < P_0 \le 4$. Thus, since $D' \equiv 1 \pmod 8$, the only possible values for P_0 which represent the different classes of solutions are $P_0 = \pm 1, \pm 3$. So let $\alpha, \alpha', \beta, \beta'$ correspond to the P_0 values $1, -1, 3, -3$, respectively.

Since α is a solution belonging to class $P_0 = 1$ we can write for some integers s_1, t_1 that

$$1 + \sqrt{D'} = \alpha(s_1 + t_1\sqrt{D'}), \tag{8}$$

and thus by conjugation (see Remark 1)

$$1 - \sqrt{D'} = \alpha'(s_1 - t_1\sqrt{D'}). \tag{9}$$

Now, let $D' \equiv 1 \pmod 8$ and let $\alpha = x + y\sqrt{D'}$. Consider the quadratic field $\mathbb{Q}(\sqrt{D'})$, and its ring of integers R. The prime ideal generated by 2 factors in R as

$$2R = \langle 2, \frac{1 + \sqrt{D'}}{2}\rangle\langle 2, \frac{1 - \sqrt{D'}}{2}\rangle \tag{10}$$

([12, Theorem 25]). Note that $\alpha/2$ and $\alpha'/2$ are both algebraic integers in $Q(\sqrt{D'})$ since, by Remark 2, x and y have the same parity. Also the principal ideals generated by $\alpha/2$ and $\alpha'/2$ are prime ideals since both have norm 2 in $\mathbb{Q}(\sqrt{D'})$. Therefore, (8) and (9) give the inclusion $\langle 2, \frac{1+\sqrt{D'}}{2}\rangle \subseteq \langle \frac{\alpha}{2}\rangle$ and $\langle 2, \frac{1-\sqrt{D'}}{2}\rangle \subseteq \langle \frac{\alpha'}{2}\rangle$, respectively. In fact, we even have equality in both inclusions since all four ideals are nonzero prime ideals, that is, $\langle \frac{\alpha}{2}\rangle = \langle 2, \frac{1+\sqrt{D'}}{2}\rangle$ and $\langle \frac{\alpha'}{2}\rangle = \langle 2, \frac{1-\sqrt{D'}}{2}\rangle$.

Applying a similar reasoning to β and β' yields

$$\langle \frac{\beta'}{2}\rangle = \langle 2, \frac{1 + \sqrt{D'}}{2}\rangle = \langle \frac{\alpha}{2}\rangle \tag{11}$$

and

$$\langle \frac{\beta}{2}\rangle = \langle 2, \frac{1 - \sqrt{D'}}{2}\rangle = \langle \frac{\alpha'}{2}\rangle. \tag{12}$$

It follows from (11) that

$$1 + \sqrt{D'} = \beta'(\frac{s_3 + t_3\sqrt{D'}}{2}) \tag{13}$$

for some integers s_3, t_3 of the same parity. In fact, s_3 and t_3 must be odd since α and β' belong to different solution classes. Similarly, it follows from (12) that

$$3 + \sqrt{D'} = \alpha'(\frac{s_4 + t_4\sqrt{D'}}{2}) \tag{14}$$

for some odd integers s_4 and t_4. Now write $D' = 8n + 1$ for some integer n. If n is odd, then we multiply (13) with its conjugate to obtain $s_3^2 - t_3^2 D' = 4n$. So $s_3^2 - t_3^2 \equiv 4 \pmod 8$, which does not have any solution for odd values of (s_3, t_3). If n is even, then multiplying (14) with its conjugate gives $s_4^2 - t_4^2 D' = 4(n - 1)$, that is, $s_4^2 - t_4^2 \equiv 4 \pmod 8$ which does not have any solution for odd values of (s_4, t_4). Consequently, the assumption that there are more than two solution classes was wrong. This completes the proof.

Proposition 3. *Assume (7) has a solution, and let S and S' denote the two solution classes. Let $\mathcal{E}$ and $\mathcal{E}'$ denote the sets of elliptic curves of embedding degree 6 that correspond to the solutions in S and S', respectively, using the* *correspondence from Section 2: if $(x, y) \in S$ (or S') and $x \equiv 1$ (mod 6), let $l = (x - 1)/6$ and E_x be the elliptic curve over $\mathbb{F}_q$ with trace t where $q = 4l^2 + 1$ and $t = 1 + 2l$, while if $(x, y) \in S$ (or S') and $x \equiv -1$ (mod 6), let $l = (x+1)/6$ and E_x be the elliptic curve over $\mathbb{F}_q$ with trace t where $q = 4l^2 + 1$ and $t = 1 - 2l$. Then $\mathcal{E} = \mathcal{E}'$.*

Proof. Let $E/\mathbb{F}_q \in \mathcal{E}$ with trace t, and $\#E(\mathbb{F}_q) = n$. Then there exists a pair $(x, y) \in S$ such that $x \equiv \pm 1$ (mod 6). Suppose first that $x \equiv 1$ (mod 6), and $l = (x - 1)/6$. Then $q = 4l^2 + 1$, $t = 1 - 2l$ and $n = 4l^2 + 2l + 1$. Now let $(x', y') = (-x, y)$. Since the set of solutions to (7) does not contain any ambiguous class (Lemma 1), we have $(x', y') \in S'$. Further, $x' \equiv -1$ (mod 6). Now let $l' = (x' + 1)/6$, and $q' = 4l'^2 + 1$, $t' = 1 + 2l'$, $n' = 4l'^2 + 2l' + 1$. Let $E'_x \in \mathcal{E}'$ be the corresponding elliptic curve over $\mathbb{F}_q$ with trace t' and n' points. Since $l' = -l$ and thus $q' = q$, $t' = t$ and $n' = n$, we have (up to isogenies) $E_{x'} = E$. The analogous reasoning applies for the case $x \equiv -q$ (mod 6). Thus, $\mathcal{E} \subset \mathcal{E}'$. The converse follows with the same argument.

Summing up, we showed that MNT curves with $k = 6$ are completely characterized through certain primitive solutions of the corresponding MNT equation, $X^2 - 3DY^2 = -8$. Moreover, we showed that this MNT equation either has no primitive solutions or has exactly two solution classes. In the latter case, we proved that the two solution classes lead to the same set of elliptic curves and so it is enough to consider only one of the two solution classes. Also, we gave some necessary conditions on D for the existence of solutions to the MNT equation.

3.2 Embedding Degree $k = 4$

The case of MNT curves with embedding degree $k = 4$ is completely covered by combining the above analysis for the $k = 6$ case with the explicit one-to-one correspondence of Proposition 1 between the MNT curves with embedding degree $k = 6$ and those with $k = 4$.

3.3 Embedding Degree $k = 3$

The analysis of this case is similar to the case $k = 6$. First, we let $D' = 3D$ and rewrite the CM equation for $k = 3$ as

$$X^2 - D'Y^2 = 24.$$

Below, we summarize the results from our analysis [9].

1. D' should be fixed such that $0 < D' < 3 \cdot 10^{10}$ and $D'/3$ is squarefree.
2. $D' \equiv 57$ (mod 72) and 6 is a square modulo D'.
3. If there is a solution to $X^2 - D'Y^2 = 24$ then it is enough to find, if it exists, only one minimal solution, say (x_0, y_0).
4. Let (u, v) be a minimal solution to $U^2 - D'V^2 = 1$. Let $(x_j, y_j) = \pm (x_0, y_0)(u, v)^j$ be the set of all solutions in the same class as (x, y). It is enough to consider only one of the solutions (x_j, y_j) and $-(x_j, y_j)$.

4   Frequency of MNT Curves

In this section we give estimates for the number of (isogeny classes of) MNT curves of bounded CM discriminant. In our discussion, we focus on the case $k = 6$. Following Luca and Shparlinski [11], we define $E(z)$ to be the expected total number of all isogeny classes of MNT curves (over all finite fields) with embedding degree 6 and CM discriminant $D \leq z$. Luca and Shparlinski [11] gave heuristic upper bounds on $E(z)$ which we recall in Section 4.1, while in Section 4.2 we will give a (new) heuristic lower bound.

4.1   The Luca-Shparlinski Upper Bounds

Recall from Sections 2 and 3.1 that in order to find MNT curve parameters with $k = 6$ (for a particular D), one needs to first find a minimal solution (x, y) of (7) as well as the minimal solution, say (u, v), of $U^2 - 3DV^2 = 1$. Then the solutions (x_j, y_j) $(j \in \mathbb{Z})$ in the same class as (x, y) would lead to an integer $l_j = (x_j \pm 1)/6$ (see Remarks 2 and 3). Finally, one checks if $q_j := 4l_j^2 + 1$ and $n_j := q_j \mp 2l_j$ (cf. Theorem 1(3)) satisfy the primality conditions.

Luca and Shparlinski [11] define, for a fixed discriminant D, $N(D)$ as the expected total number of $j \in \mathbb{Z}$ for which q_j is a prime power and n_j is a prime. Then

$$E(z) = \sum_{\substack{D \leq z \\ D \text{ squarefree}}} N(D) .$$

Under the assumption that the primality properties of q_j and n_j are ruled by the prime number theorem (meaning that q_j and n_j are prime with probabilities $1/\log q_j$ and $1/\log n_j$, respectively), Luca and Shparlinski show that $N(D) \ll 1/(\log D)^2$. They conclude that $E(z) \ll z/(\log z)^2$. Further, Luca and Shparlinski suggest a stronger upper bound for $E(z)$ which relies on the conjecture (see [10, p.185]) that there exists a set $\mathcal{D}$ of nonsquare positive integers that has asymptotic density 1 and such that $\lim_{D \in \mathcal{D}} \log \log(u + v\sqrt{3D})/\log \sqrt{D} = 1$. Using this conjecture, Luca and Shparlinski argue that $N(D) \leq 1/(D^{1+o(1)})$ for $D \in \mathcal{D}$, and suggest that $E(z) \leq z^{o(1)}$. We will see below (Theorem 3) that this does not hold.

4.2   A Lower Bound

In this section we give a lower bound for $E(z)$. For this we are going to restrict ourselves to solutions of the MNT equation $X^2 - 3DY^2 = -8$ with $Y = 1$.

Theorem 3. *Assume that the primality properties of $4l^2 + 1$ and $4l^2 \pm 2l + 1$, where $l \in \mathbb{N}$ and such that $(6l \pm 1)^2 = 3D - 8$ for some odd squarefree integer D, are captured by the prime number theorem. Then there exists an integer z_0 such that*

$$E(z) \geq 0.49 \frac{\sqrt{z}}{(\log z)^2} \tag{15}$$

for every $z \geq z_0$.

Proof. Let $\mathcal{F}(z)$ denote the set of odd and squarefree integers $D \in [3, z]$ such that $3D - 8$ is a perfect square, and let $F(z) = \#\mathcal{F}(z)$. For $D \in \mathcal{F}(z)$, let $x_D (> 0)$ such that $x_D^2 = 3D - 8$, and let $l_D \in \mathbb{N}$ such that $x_D = 6l_D + 1$ or $x_D = 6l_D - 1$. Denote $q_D = 4l_D^2 + 1$, and $n_D = 4l_D^2 + 2l_D + 1$ if $x_D = 6l_D + 1$ or $n_D = 4l_D^2 - 2l_D + 1$ otherwise.

An easy calculation shows that if $D \leq z$, then $q_D \leq z/2$ and $n_D \leq 3z/4$. As we assume that the primality properties of both q_D and n_D are captured by the prime number theorem, and since for $z > 17$, the number $\pi(z)$ of primes $\leq z$ satisfies $\pi(z) > z/\log z$, we have

$$\text{Prob}(q_D \text{ and } n_D \text{ prime} \mid q_D = 4l^2 + 1, n_D = 4l^2 \pm 2l + 1, \text{ where}$$
$$l \geq 1 \text{ and } (6l \pm 1)^2 = 3D - 8 \text{ for some squarefree } D \leq z)$$
$$> \frac{1}{\log(z/2)} \cdot \frac{1}{\log(3z/4)} > \frac{1}{(\log z)^2}.$$

Now, by Section 2, the number $G(z)$ of pairs (q_D, n_D) $(D \in \mathcal{F}(z))$ where both q_D and n_D are prime constitutes a lower bound for $E(z)$. Thus,

$$E(z) \geq G(z) \geq F(z) \cdot \frac{1}{(\log z)^2}. \tag{16}$$

To find a lower bound for $F(z)$, first note that $3D - 8$ is a perfect square and D is odd and squarefree, if and only if $D = 12l^2 \pm 4l + 3$ is squarefree (by putting $3D - 8 = (6l \pm 1)^2$). Let $f_+(l) = 12l^2 + 4l + 3$, and $\mathcal{F}_+(z) = \{D \in [5, z] : D = f_+(l) \text{ squarefree}\}$. As $f_+(l)$ is irreducible over $\mathbb{Z}[l]$, there are $\sim c_{f_+} L$ positive integers $l \leq L$ such that $f_+(l)$ is squarefree, where c_{f_+} is a positive constant ([18, Theorem A], [6, Theorem 1]). Now, $5 \leq D = f_+(l) \leq z$ if and only if $1 \leq l \leq \sqrt{\frac{z}{12} - \frac{2}{9}} - \frac{1}{6} =: L_+$. Thus, for each $\varepsilon > 0$ there exists an integer Z_+ such that $(c_{f_+} - \varepsilon)L_+ < \#\mathcal{F}_+(z) < (c_{f_+} + \varepsilon)L_+$ for all $z \geq Z_+$. Doing the analogous with $f_-(l) := 12l^2 - 4l + 3$, and $\mathcal{F}_-(z) := \{D \in [5, z] : D = f_-(l) \text{ squarefree}\}$ and $L_- := \sqrt{\frac{z}{12} - \frac{2}{9}} + \frac{1}{6}$ we find that there exists a positive constant c_{f_-} such that for each $\varepsilon > 0$ there exists an integer Z_- such that $(c_{f_-} - \varepsilon)L_- < \#\mathcal{F}_-(z) < (c_{f_-} + \varepsilon)L_-$ for all $z \geq Z_-$. Thus, since $\mathcal{F}(z) = \mathcal{F}_+(z) \cup \mathcal{F}_-(z) \cup \{3\}$ (disjoint), we obtain

$$F(z) > (c_{f_+} + c_{f_-} - 2\varepsilon)\sqrt{z/12} \tag{17}$$

for all $z \geq z_0 := \max\{Z_+, Z_-\}$. Now, $c_{f_+} = \prod_{p \text{ prime}} (1 - w_{f_+}(p)/p^2)$ where $w_{f_+}(p)$ denotes the number of integers $a \in [1, p^2]$ for which $f_+(a) \equiv 0 \pmod{p^2}$ ([18,6]), and the same holds for c_{f_-} with f_+ replaced by f_- throughout. It can be readily seen that $w_{f_+}(3) = w_{f_-}(3) = 1$ and $w_{f_+}(p), w_{f_-}(p) \in \{0, 2\}$ otherwise. Further, the polynomial $ax^2 + bx + c$ has two solutions modulo p^2 if and only if a is invertible modulo p^2 and $b^2 - 4ac$ is a square modulo p^2. Thus, $f_+(l) \equiv 0 \pmod{p^2}$ $(p > 3)$ has two solutions modulo p^2 if and only if -128 is a quadratic residue modulo p^2. This is the case if and only if $\left(\frac{-2}{p}\right) = 1$, which holds if and only if $p \equiv 1 \pmod 8$ or $p \equiv 3 \pmod 8$. The same reasoning applies to $f_-(l)$. Consequently,

$$c_{f_+} = c_{f_-} = \tfrac{8}{9} \cdot \prod_{p \text{ prime, } p \equiv 1,3 \ (\text{mod } 8)} \left(1 - 2/p^2\right).$$

Now,

$$\prod_{p \text{ prime, } p \leq 10000, \ p \equiv 1,3 \ (\text{mod } 8)} \left(1 - 2/p^2\right) > 0.858146 \,,$$

while the tail can be bounded below by

$$\prod_{p \text{ prime}, p > 10000} \left(1 - 2/p^2\right) \geq \prod_{s > 10000} \left(1 - 4/s^2\right) = \frac{9999 \cdot 10000}{10001 \cdot 10002} > 0.9996.$$

Hence, $c_{f_\pm} > 0.858146 \cdot 0.9996 > 0.8578$. Combined with (17), using $\varepsilon = 0.0008$, this yields $F(z) > 0.857\sqrt{z/3}$ for all $z \geq z_0$. Used along with (16), this completes the proof.

Remark 4. The above lower bound on $E(z)$ can be increased by a constant factor if also solutions to the MNT equation $X^2 - 3DY^2 = -8$ with $Y > 1$ are considered. In fact, for each odd Y such that $X^2 \equiv 3Y^2 - 8 \pmod{6Y^2}$ is solvable, a lower bound for the number $F_Y(z)$ of odd and squarefree integers $D \in [3, z]$ such that $3Y^2 D - 8$ is a perfect square, can be derived in exactly the same way as for $Y = 1$. The corresponding polynomials $f_{Y,\pm}(l)$ are given as $f_{Y,\pm}(l) = 12Y^2 l^2 \pm 4sl + (s^2 + 8)/(3Y^2)$, where $s^2 \equiv 3Y^2 - 8 \pmod{6Y^2}$. They all have (polynomial) discriminant -128, and thus the corresponding c_f-values will differ only by those factors that involve primes $p|Y$. In particular, including the cases $Y = 3, 9$ will raise our lower bound by a factor of $(1 + 1/3 + 1/9)$.

4.3   Experimental Results on $E(z)$

Using the computational algebra system MAGMA [3] we implemented an algorithm to calculate, for given bitsize N and upper discriminant bound z, all (isogeny classes of) MNT elliptic curves of embedding degree 6 and discriminant $D \leq z$ over a finite field q where $q - 1$ is an N-bit prime.

As discussed in Section 3, only those squarefree D such that for $D' = 3D$ we have $D' \equiv 9 \pmod{24}$ and $\left(\frac{-2}{D'}\right) = 1$ need to be considered.

For any such $D \leq z$, our algorithm (Algorithm 3 of the appendix) first calls a Pell equation solver to compute minimal solutions (x, y) and (u, v) to (6) and to the equation $u^2 - 3Dv^2 = 1$, respectively. This Pell equation solver is Algorithm 1 (of the appendix) if $3D > 64$ and Algorithm 2 (of the appendix) if $3D < 64$; both algorithms are taken from Robertson [19]. The minimal solutions (x, y) and (u, v) are used to compute, one by one, all primitive solutions to (6). For each such primitive solution, it is checked if it yields values for q and n such that q is a prime power and of the desired bitsize, and n is prime.

Using Algorithm 3, we first conducted a series of experiments to check the quality of our lower bound on $E(z)$ (Theorem 3).

Let $E_B(z)$ denote the number of (isogeny classes of) MNT elliptic curves with embedding degree $k = 6$ and CM discriminant $D \leq z$ over finite fields $\mathbb{F}_q$ with $q < 2^B$. Then $E_B(z) \leq E(z)$ for all B, and $E(z) = \lim_{B \to \infty} E_B(z)$.

We computed $E_B(z)$ for selected values of B, by running Algorithm 3 with input N, for all $1 \leq N \leq B$. Table 4.3 shows the ratios of $E_B(z)$ and the lower bound (15) for $z = 2^i$, $z \leq 2^{25}$ and $B = 160, 300, 500, 700, 1000$.

Table 1. Ratios $R(B, z)$ of $E_B(z)$ and the lower bound (15) for $E(z)$. Here $E_B(z)$ denotes the number of MNT curves with $k = 6$ and $D \leq z$ over $\mathbb{F}_q$ with $q < 2^B$.

				$R(B, z) = E_B(z)/(0.49\frac{\sqrt{z}}{(\log z)^2})$, where $z = 2^i$.				
i	$B = 25$	$B = 50$	$B = 100$	$B = 160$	$B = 300$	$B = 500$	$B = 700$	$B = 1000$
10	30.64	30.64	30.64	33.70	33.70	33.70	33.70	33.70
11	31.45	34.08	34.08	36.70	36.70	36.70	36.70	36.70
12	26.47	28.68	28.68	30.88	30.88	30.88	30.88	30.88
13	23.80	25.63	25.63	27.46	27.46	27.46	27.46	27.46
14	24.02	27.02	27.02	30.02	30.02	30.02	30.02	30.02
15	23.15	26.81	26.81	30.46	30.46	30.46	30.46	30.46
16	21.57	25.49	26.47	29.41	29.41	29.41	29.41	29.41
17	20.35	24.26	26.61	29.74	29.74	29.74	29.74	29.74
18	19.23	23.57	25.43	27.92	27.92	27.92	27.92	27.92
19	18.57	23.46	25.42	27.86	28.35	28.35	28.35	28.35
20	16.85	21.83	24.51	26.81	27.19	27.19	27.57	27.57
21	15.22	21.20	23.58	25.67	26.87	27.47	28.06	28.06
22	14.83	22.01	26.64	28.73	29.66	30.12	30.81	30.81
23	14.32	22.74	27.40	29.72	30.62	30.98	32.05	32.41
24	13.65	24.12	28.54	30.88	32.12	32.67	33.64	34.05
25	13.11	24.54	29.30	31.52	32.79	33.32	34.17	34.48

Let $R(B, z) = E_B(z)/(0.49\frac{\sqrt{z}}{(\log z)^2})$. As we would expect, $R(B, z)$ is increasing for fixed z as B increases. For the smallest values of B, we also see that $R(B, z)$ is essentially decreasing (for fixed B) as z increases. In fact, we expect that $\lim_{z \to \infty} R(B, z) = 0$ for any fixed value of B, as if $X^2 - DY^2 = -8$, then the resulting field size $q(\leq 2^B)$ is of the order of magnitude of $\sqrt{D}$, which implies that $E_B(z)$ remains constant for large enough z. On the other hand, for larger fixed values of B and in particular along the down-ward diagonal, $R(B, z)$ seems somewhat more stable (around 30, although there is an increase towards the very end). It is tempting to conclude from this that the lower bound (15) for $E(z)$ has indeed the right order of magnitude, and possibly is just off by a factor of around 30. So, let us try to estimate the number of (isogeny classes of) *computable* MNT elliptic curves of embedding degree 6. That is, put $z_0 = 10^{10} (\approx 2^{33})$, and let's boldly assume that $E(z) = 30 \cdot (0.49\frac{\sqrt{z}}{(\log z)^2})$. Then $E(z_0) \approx 30 \cdot 92.4 = 2772$. For comparison, we found that $E_{2^{25}}(2^{10}) = 10$, $E_{2^{1000}}(2^{10}) = 11$, $E_{2^{25}}(2^{24}) = 124$ and $E_{2^{1000}}(2^{25}) = 326$.

As prime-order elliptic curves over fields of bitsize $155 - 170$ approximately match the security level of SKIPJACK (i.e., the 80-bit symmetric key security level), we found it of interest to calculate the number of (isogeny classes of) MNT elliptic curves over $155 - 170$-bit fields. But the smallest discriminant for which we found an MNT curve in the desired bit range has 21 bits, with the next two such MNT curves appearing for 24-bit discriminants. These data certainly do not allow for a meaningful extrapolation to $z = 10^{10}$.

5   Conclusion

Our analysis in this paper brought us closer to the true nature of the function $E(z)$, the number of prime-order elliptic curves over finite fields with embedding degree $k = 6$ (MNT curves) and discriminant $D \leq z$. However, it would be nice to be able to estimate the number of MNT curves of bounded discrimant *and* given bit-size. Our experimental data for the cryptographically interesting range are too limited to encourage any predictions.

Acknowledgements. The authors thank Florian Luca and Igor Shparlinski for their feedback on an earlier version of this paper, which helped us to improve the statement and proof of Theorem 3.

References

1. Atkin, A.O.L., Morain, F.: Elliptic curves and primality proving. Math. Comp. 61(203), 29–68 (1993)
2. Barreto, P.S.L.M., Galbraith, S., O'hEigeartaigh, C., Scott, M.: Efficient pairing computation on supersingular abelian varieties. Designs, Codes and Cryptography 42, 239–271 (2007)
3. Computational Algebra Group: The Magma computational algebra system for algebra, number theory and geometry. School of Mathematics and Statistics, University of Sydney, http://magma.maths.usyd.edu.au/magma
4. Franklin, M., Boneh, D.: Identity based encryption from the Weil pairing. In: Franklin, M. (ed.) CRYPTO 2004. LNCS, vol. 3152, pp. 41–55. Springer, Heidelberg (2004)
5. Freeman, D., Scott, M., Teske, E.: A taxonomy of pairing-friendly elliptic curves. Cryptology ePrint Archive Report 2006/372 (2006), http://eprint.iacr.org/2006/372/
6. Granville, A.: ABC allows us to count squarefrees. International Mathematical Research Notices 19, 991–1009 (1998)
7. Hess, F., Smart, N., Vercauteren, F.: The Eta pairing revisited. IEEE Transactions on Information Theory 52, 4595–4602 (2006)
8. Joux, A.: A one round protocol for tripartite Diffie-Hellman. In: Bosma, W. (ed.) ANTS 2000. LNCS, vol. 1838, pp. 383–394. Springer, Heidelberg (2000)
9. Karabina, K.: On prime-order elliptic curves with embedding degrees 3,4 and 6. Master's thesis, University of Waterloo (2006), http://uwspace.uwaterloo.ca/handle/10012/2671

10. Lenstra Jr., H.W.: Solving the Pell equation. Notices Amer. Math. Soc. 49, 182–192 (2002)
11. Luca, F., Shparlinski, I.E.: Elliptic curves with low embedding degree. Journal of Cryptology 19, 553–562 (2006)
12. Marcus, D.A.: Number fields. Springer, New York (1977)
13. Menezes, A., Okamoto, T., Vanstone, S.: Reducing elliptic curve logarithms to logarithms in a finite field. IEEE Transactions on Information Theory 39, 1639–1646 (1993)
14. Miyaji, A., Nakabayashi, M., Takano, S.: New explicit conditions of elliptic curve traces for FR-reduction. IEICE Trans. Fundamentals E84-A, 1234–1243 (2001)
15. Mollin, R.A.: Fundamental number theory with applications. CRC Press, Boca Raton (1998)
16. Mollin, R.A.: Simple continued fraction solutions for Diophantine equations. Expositiones Mathematicae 19, 55–73 (2001)
17. Page, D., Smart, N.P., Vercauteren, F.: A comparison of MNT curves and supersingular curves. Applicable Algebra in Engineering, Communication and Computing 17, 379–392 (2006)
18. Ricci, G.: Ricerche aritmetiche sui polinomi. Rend. Circ. Mat. Palermo. 57, 433–475 (1933)
19. Robertson, J.P.: Solving the generalized Pell equation $x^2 - dy^2 = n$ (2004), http://hometown.aol.com/jpr2718/
20. Scott, M., Barreto, P.S.L.M.: Generating more MNT elliptic curves. Designs, Codes and Cryptography 38, 209–217 (2006)
21. Shacham, H., Boneh, D., Lynn, B.: Short signatures from the Weil pairing. In: Boyd, C. (ed.) ASIACRYPT 2001. LNCS, vol. 2248, pp. 514–532. Springer, Heidelberg (2001)

Appendix: Algorithms

We present two Pell equation solver algorithms: Algorithms 1 and 2; and one algorithm for finding suitable MNT curve parameters for embedding degree $k = 6$: Algorithm 3. Our reference for the first two algorithms is Robertson's paper [19]. Algorithm 3 uses these two algorithms and the facts developed in this paper.

Algorithm 1. Pell Equation Solver

Input: $D \in \mathbb{Z}$, $m \in \mathbb{Z}\backslash\{0\} : D > m^2$, D is not a perfect square

Output: all minimal positive solutions $(x, y) : x^2 - Dy^2 = m$

1: $B_{-1} \leftarrow 0$, $G_{-1} \leftarrow 1$
2: $P_0 \leftarrow 0$, $Q_0 \leftarrow 1$, $a_0 \leftarrow \lfloor \sqrt{D} \rfloor$, $B_0 \leftarrow 1$, $G_0 \leftarrow a_0$
3: $i \leftarrow 0$
4: **repeat**
5:     $i \leftarrow i + 1$
6:     $P_i \leftarrow a_{i-1}Q_{i-1} - P_{i-1}$
7:     $Q_i \leftarrow (D - P_i^2)/Q_{i-1}$
8:     $a_i \leftarrow \lfloor (P_i + \sqrt{D})/Q_i \rfloor$
9:     $B_i \leftarrow a_iB_{i-1} + B_{i-2}$
10:     $G_i \leftarrow a_iG_{i-1} + G_{i-2}$
11: **until** $Q_i = 1$ and $i \equiv 0 \pmod{2}$
12: $s \leftarrow 0$
13: **for** $0 \le j \le i - 1$ **do**
14:     **if** $G_j^2 - DB_j^2 = m/f^2$ for some $f > 0$ **then**
15:         Output: (fG_j, fB_j)
16:         $s \leftarrow 1$
17:     **end if**
18: **end for**
19: **if** $s == 0$ **then**
20:     Output: No solutions exist
21: **end if**

Algorithm 2. Pell Equation Solver 2

Input: $D \in \mathbb{Z}$, $m \in \mathbb{Z}\backslash\{0\} : D \le m^2$, D is not a perfect square

Output: all fundamental solutions $(x, y) : x^2 - Dy^2 = m$

1: Find a minimal solution (u, v) to $U^2 - DV^2 = 1$ using Algorithm 1 with inputs D, 1.
2: **if** $m > 0$ **then**
3:     $L_1 \leftarrow 0$, $L_2 \leftarrow \sqrt{m(u - 1)/(2D)}$
4: **else**
5:     $L_1 \leftarrow \sqrt{(-m)/D}$, $L_2 \leftarrow \sqrt{(-m)(v + 1)/(2D)}$
6: **end if**
7: **for** $L_1 \le y \le L_2$ **do**
8:     **if** $m + Dy^2$ is a square **then**
9:         $x \leftarrow \sqrt{m + Dy^2}$
10:         **if** (x, y) and $(-x, y)$ are not in the same class **then**
11:             Output: $(x, y), (-x, y)$
12:         **else**
13:             Output: (x, y)
14:         **end if**
15:     **end if**
16: **end for**

Algorithm 3. Elliptic curve parameters, embedding degree $k = 6$

Input: N, z

Output: EC parameters (q, n, D) where $q-1$ is an N-bit prime, $q^6 \equiv 1 \pmod{n}$ but $q^i \not\equiv 1 \pmod{n}$ for $1 \leq i \leq 5$, and $D \leq z$ (where $4q - t^2 = DY^2$)

1: **for** $0 < D' \leq 3z$, $D'/3$ squarefree, $D' \equiv 9 \pmod{24}$, -2 is a square modulo D' **do**

2:    **if** $D' > 64$ **then**

3:        find a minimal solution, (x_0, y_0), to $X^2 - D'Y^2 = -8$ by using Algorithm 1 with input D', -8.

4:    **else**

5:        find a minimal solution, (x_0, y_0), to $X^2 - D'Y^2 = -8$ by using Algorithm 2 with input D', -8.

6:    **end if**

7:    find a minimal solution, (u, v), to $U^2 - D'V^2 = 1$ by using Algorithm 1 with input D', 1.

8:    $x \leftarrow x_0$, $y \leftarrow y_0$

9:    **if** $x \equiv \pm 1 \pmod{6}$ **then**

10:        **while** $|x| \leq 2^{\lceil N/2 \rceil}$ **do**

11:            $l \leftarrow (x \mp 1)/6$

12:            **if** $(N - 3)/2 \leq \log_2 l < (N - 2)/2$ **then**

13:                $q \leftarrow 4l^2 + 1$, $n \leftarrow 4l^2 \mp 2l + 1$

14:                **if** q and n are primes **then**

15:                    Output $(q, n, D'/3)$

16:                **end if**

17:            **end if**

18:            $\tilde{x} \leftarrow x$

19:            $x \leftarrow xu + yvD'$

20:            $y \leftarrow \tilde{x}v + uy$

21:        **end while**

22:    **end if**

23:    $x \leftarrow x_0 u - y_0 v D'$, $y \leftarrow uy_0 - x_0 v$

24:    **if** $x \equiv \pm 1 \pmod{6}$ **then**

25:        **while** $|x| \leq 2^{\lceil N/2 \rceil}$ **do**

26:            $l \leftarrow (x \mp 1)/6$

27:            **if** $(N - 3)/2 \leq \log_2 l < (N - 2)/2$ **then**

28:                $q \leftarrow 4l^2 + 1$, $n \leftarrow 4l^2 \mp 2l + 1$

29:                **if** q and n are primes **then**

30:                    Output $(q, n, D'/3)$

31:                **end if**

32:            **end if**

33:            $\tilde{x} \leftarrow x$

34:            $x \leftarrow xu - yvD'$

35:            $y \leftarrow uy - \tilde{x}v$

36:        **end while**

37:    **end if**

38: **end for**

Computing in Component Groups
of Elliptic Curves

J.E. Cremona

Mathematics Institute, University of Warwick, Coventry, CV4 7AL, UK
J.E.Cremona@warwick.ac.uk

Abstract. Let K be a p-adic local field and E an elliptic curve defined over K. The component group of E is the group $E(K)/E^0(K)$, where $E^0(K)$ denotes the subgroup of points of good reduction; this is known to be finite, cyclic if E has multiplicative reduction, and of order at most 4 if E has additive reduction. We show how to compute explicitly an isomorphism $E(K)/E^0(K) \cong \mathbb{Z}/N\mathbb{Z}$ or $E(K)/E^0(K) \cong \mathbb{Z}/2\mathbb{Z} \times \mathbb{Z}/2\mathbb{Z}$.

1 Introduction

Let K be a p-adic local field (that is, a finite extension of $\mathbb{Q}_p$ for some prime p), with ring of integers R, uniformizer π, residue field $k = R/(\pi)$ and valuation function v. Let E be an elliptic curve defined over K. The component group of E is the finite abelian group $\Phi = E(K)/E^0(K)$, where $E^0(K)$ denotes the subgroup of points of good reduction.

When E has split multiplicative reduction, we have $\Phi \cong \mathbb{Z}/N\mathbb{Z}$, where $N = v(\Delta)$ and Δ is the discriminant of a minimal model for E. In all other cases, Φ has order at most 4, so is isomorphic to $\mathbb{Z}/n\mathbb{Z}$ with $n \in \{1, 2, 3, 4\}$ or to $\mathbb{Z}/2\mathbb{Z} \times \mathbb{Z}/2\mathbb{Z}$. The order of Φ is called the Tamagawa number of E/K, usually denoted c or c_p.

In this note we will show how to make the isomorphism $\kappa\colon E(K)/E^0(K) \to A$ explicit, where A is the one of the above standard abelian groups.

The most interesting case is that of split multiplicative reduction. Here the map κ is almost determined by a formula for the (local) height in [3]. Specifically, if the minimal Weierstrass equation for E has coefficients a_1, a_2, a_3, a_4, a_6 as usual, for a point $P = (x, y) \in E(K) \setminus E^0(K)$ we have $\kappa(P) = \pm n \pmod{N}$, where $n = \min\{v(2y + a_1 x + a_3), N/2\}$, and $0 < n \le N/2$. In computing heights, of course, one need not distinguish between P and $-P$, but for our purposes this is essential. We show how to determine the appropriate sign in a consistent way to give an isomorphism $\kappa\colon E(K)/E^0(K) \cong \mathbb{Z}/N\mathbb{Z}$. Note that for an individual point this is not a well-defined question since negation gives an automorphism of $\mathbb{Z}/N\mathbb{Z}$; but when comparing the values of κ at two or more points it is important. We first establish the formula for Tate curves, and then see how to apply it to a general minimal Weierstrass model.

We also make some remarks about the other reduction types, which are much simpler to deal with, and also the real case.

A.J. van der Poorten and A. Stein (Eds.): ANTS-VIII 2008, LNCS 5011, pp. 118–124, 2008.

One application for this, which was our motivation, occurs in the determination of the full Mordell-Weil group $E(K)$, where E is an elliptic curve defined over a number field K. Given a subgroup B of $E(K)$ of full rank, generated by r independent points P_i for $1 \leq i \leq r$, one method for extending this to a $\mathbb{Z}$-basis for $E(K)$ (modulo torsion) requires determining the index in B of $B \cap \bigcap_{p \leq \infty} E^0(\mathbb{Q}_p)$. [For $p = \infty$, we denote as usual $\mathbb{R} = \mathbb{Q}_\infty$, and then $E^0(\mathbb{Q}_p)$ denotes the connected component of the identity in $E(\mathbb{R})$.] The component group maps κ for each prime p may be used for this, and are accordingly implemented in our program mwrank [1].

We use standard notation for Weierstrass equations of elliptic curves throughout.

2   The Split Multiplicative Case

We refer to [4, Chapter V] for the theory of the Tate parametrization of elliptic curves with split multiplicative reduction.

2.1   The Case of Tate Curves

For each $q \in K^*$ with $|q| < 1$ we define the Tate curve E_q by its Weierstrass equation

$$Y^2 + XY = X^3 + a_4 X + a_6,$$

where $a_4 = a_4(q)$ and $a_6 = a_6(q)$ are given by explicit power series in q. We have $v(\Delta) = v(a_6) = N$, where $N = v(q) > 0$, and $v(a_4) \geq N$. Also, $v(c_4) = v(c_6) = 0$.

Reducing modulo π^N, the equation becomes $Y(Y + X) \equiv X^3$; the linear factors Y, $Y + X$ give the distinct tangents at the node $(0,0)$ on the reduced curve over k.

Theorem 1. *The map* $\kappa \colon E(K) \to \mathbb{Z}/N\mathbb{Z}$ *given by*

$$\kappa(P) = \begin{cases} 0 & \text{if } P \in E^0(K) \\ +n & \text{if } P = (x,y) \notin E^0(K) \text{ and } n = v(x+y) < v(y) \\ -n & \text{if } P = (x,y) \notin E^0(K) \text{ and } n = v(y) < v(x+y) \\ N/2 & \text{if } P = (x,y) \notin E^0(K) \text{ and } v(y) = v(x+y) \end{cases}$$

induces an isomorphism $E(K)/E^0(K) \cong \mathbb{Z}/N\mathbb{Z}$. *The integer* n *here always satisfies* $0 < n < N/2$. *The last case only occurs when* N *is even, and then* $v(y) \geq N/2$.

Remark. This is compatible with the result from [3] quoted in the introduction, which here says that $\kappa(P) = \pm n$, where $n = \min\{v(2y+x), N/2\}$. What we have done is decompose $2y + x$ as $y + (y + x)$, where the summands come from the tangent lines at the singular point, and consider the valuations of each summand separately.

Proof. Recall that the Tate parametrization gives an isomorphism $\varphi \colon K^*/q^{\mathbb{Z}}R^* \cong E(K)/E^0(K)$, and that κ is determined by $\kappa(\varphi(u)) = v(u) \pmod{N}$ for $u \in K^*$.

Let $P = \varphi(u) = (x, y)$. Then $x = X(u, q)$ and $y = Y(u, q)$, where $X(u, q)$ and $Y(u, q)$ are power series given in [4, §V.3, Theorem 3.1(c)]:

$$x = \frac{u}{(1-u)^2} + \sum_{m \geq 1} \left(\frac{q^m u}{(1 - q^m u)^2} + \frac{q^m/u}{(q^m/u - 1)^2} - \frac{2mq^m}{1 - q^m} \right);$$

$$y = \frac{u^2}{(1-u)^3} + \sum_{m \geq 1} \left(\frac{(q^m u)^2}{(1 - q^m u)^3} + \frac{q^m/u}{(q^m/u - 1)^3} + \frac{mq^m}{1 - q^m} \right).$$

First suppose that $v(u) = n$ with $0 < n < N/2$. The first series shows that $v(x) = n$, since the term outside the sum has valuation n, while all those in the sum have strictly greater valuation. Regarding y, the term outside the sum has valuation $2n$ and all those in the sum have strictly greater valuation, except possibly the term $\frac{q^m/u}{(q^m/u-1)^3}$ for $m = 1$, which has valuation $N - n > n$. Considering the three cases $N - n > 2n$, $N - n = 2n$, $n < N - n < 2n$, we find that

$$v(y) = 2n \qquad \qquad \text{if } 0 < n < N/3;$$
$$v(y) \geq 2n \qquad \qquad \text{if } n = N/3;$$
$$n < v(y) = N - n < 2n \qquad \text{if } N/3 < n < N/2.$$

It follows that $\kappa(P) = n$ with $n = v(y + x) = v(x) < v(y)$ as required. (We have $P \in V_n$ in the notation of [4, p.434].)

Next suppose that $v(u) = -n$ with $0 < n < N/2$. Now $v(u^{-1}) = n$ and $\varphi(u^{-1}) = -P = (x, -y - x)$, so by the first case we have $\kappa(P) = -\kappa(-P) = -n$, where $n = v(y) = v(x) < v(x + y)$ as required. (We have $P \in U_n$ in the notation of [4, p.434].)

Finally suppose that N is even and $v(u) = N/2$. Now we have $v(y) = N/2$, while both $v(x) \geq N/2$ and $v(x + y) \geq N/2$, so $N/2 = v(y) \leq v(x + y)$. (We have $P \in W$ in the notation of [4, p.434].)

2.2   The General Case

Let E with split multiplicative reduction be given by the minimal Weierstrass equation $F(X, Y) = 0$, where

$$F(X, Y) = Y^2 + a_1 XY + a_3 Y - (X^3 + a_2 X^2 + a_4 X + a_6).$$

Thus $a_i \in R$, $v(\Delta) = N > 0$ and $v(c_4) = 0$. Define

$$x_0 = c_4^{-1}(18b_6 - b_2 b_4);$$
$$y_0 = c_4^{-1}(a_1^3 a_4 - 2a_1^2 a_2 a_3 + 4a_1 a_2 a_4 + 3a_1 a_3^2 - 36a_1 a_6 - 8a_2^2 a_3 + 24a_3 a_4)$$
$$= -\frac{1}{2}(a_1 x_0 + a_3).$$

Our result is as follows.

Theorem 2. *Let α_1, α_2 be the roots of $T^2 + a_1 T - (a_2 + 3x_0)$; these lie in R and are distinct. For $P = (x, y) \in E(K) \setminus E^0(K)$, set*

$$n_i = v((y - y_0) - \alpha_i(x - x_0))$$

for $i = 1, 2$. Then $\kappa(P) \in \mathbb{Z}/N\mathbb{Z}$ is given by

$$\kappa(P) = \begin{cases} +n & \text{if } n = n_2 < n_1; \\ -n & \text{if } n = n_1 < n_2; \\ N/2 & \text{if } n_1 = n_2, \end{cases}$$

where in the first two cases we have $0 < n < N/2$, and the last case can only occur when N is even.

Remarks. Note that in order to determine $\kappa(P)$ we need to compute the quantities x_0, y_0, α_i only modulo π^N (or even $\pi^{\lceil N/2 \rceil}$), and that these depend only on E, not on P. Also, if we interchange the order of the roots α_i the only effect is to replace $\kappa(P)$ by $-\kappa(P)$ consistently, which is harmless since negation is an automorphism of $\mathbb{Z}/N\mathbb{Z}$. Finally note that

$$[(y - y_0) - \alpha_i(x - x_0)] + [(y - y_0) - \alpha_2(x - x_0)] = 2y + a_1 x - (2y_0 + a_1 x_0)$$
$$= 2y + a_1 x + a_3,$$

so this result is compatible with the formula from [3] quoted in the introduction.

Proof. With x_0, y_0 as given we may check that $F(x_0, y_0) \equiv F_X(x_0, y_0) \equiv F_Y(x_0, y_0) \equiv 0 \pmod{\pi^N}$. (Here the subscripts denote derivatives.) In other words, (x_0, y_0) reduces to a singular point, not just modulo π but modulo π^N. As in the first step of Tate's algorithm (where normally one only requires x_0 and y_0 modulo π), we shift the origin by setting $X = X' + x_0$ and $Y = Y' + y_0$. This results in a new Weierstrass equation with coefficients a'_i satisfying $a'_1 = a_1$, $a'_2 = a_2 + 3x_0$, $b'_2 = b_2 + 12x_0 \in R^*$, and

$$a'_3 \equiv a'_4 \equiv a'_6 \equiv b'_4 \equiv b'_6 \equiv b'_8 \equiv 0 \pmod{\pi^N}.$$

Since we have split multiplicative reduction, the quadratic $T^2 + a'_1 T - a'_2$, whose discriminant is b'_2, splits modulo π and hence by Hensel's Lemma splits over K. The roots α_1, α_2 lie in R, and $\alpha_1 - \alpha_2 \in R^*$ since $(\alpha_1 - \alpha_2)^2 = b'_2$.

Now set $\beta_i = (\alpha_1 - \alpha_2)^{-1}(a'_4 - \alpha_i a'_3)$ for $i = 1, 2$. Then $\beta_i \equiv 0 \pmod{\pi^N}$ and we may check that

$$F = (Y' - \alpha_1 X' - \beta_1)(Y' - \alpha_2 X' + \beta_2) - (X'^3 + b'_8/b'_2)$$
$$\equiv (Y' - \alpha_1 X')(Y' - \alpha_2 X') - X'^3$$
$$\equiv Y''(Y'' + a''_1 X') - X'^3 \pmod{\pi^N},$$

where we have set $Y' = Y'' + \alpha_1 X' + \beta_1$ and $a''_1 = \alpha_1 - \alpha_2$. (Here we have used: $\beta_1 - \beta_2 = -a'_3$, $\alpha_1\beta_2 - \alpha_2\beta_1 = a'_4$, and $b'_2(a'_6 - \beta_1\beta_2) = b'_8$.) After a further scaling by the unit a''_1, this has the form of a Tate curve.

Applying the result of the previous section, we see that $\kappa(P)$ is given in terms of the valuations of y'' and $y'' + a_1''x''$. Now

$$y'' \equiv y' - \alpha_1 x' \equiv (y - y_0) - \alpha_1(x - x_0) \pmod{\pi^N}$$

and

$$y'' + a_1''x'' \equiv y' - \alpha_2 x' \equiv (y - y_0) - \alpha_2(x - x_0) \pmod{\pi^N},$$

which implies the result as stated.

2.3   Example

Let E be the elliptic curve defined over $\mathbb{Q}$ denoted 8025j1 in the tables [2], whose Weierstrass equation is

$$Y^2 + Y = X^3 + X^2 + 2242417292X + 12640098293119.$$

Take $P = (335021/4, 224570633/8)$, a generator of the Mordell-Weil group $E(\mathbb{Q})$ which is isomorphic to $\mathbb{Z}$.

We consider E over $K = \mathbb{Q}_3$, where it has split multiplicative reduction of type I_{31}. Thus $N = 31$. We compute $x_0 = 556930682563112$ and $y_0 = 308836698141973$ modulo 3^{31}, and $\alpha_1 \equiv -\alpha_2 \equiv 256142918648120$. Now for the point P, we find

$$(y - y_0) - \alpha_1(x - x_0) \equiv 446797736663247 \pmod{3^{31}},$$
$$(y - y_0) - \alpha_2(x - x_0) \equiv 325294064834346 \pmod{3^{31}},$$

with valuations $n_1 = 12$ and $n_2 = 6$, so $\kappa(P) = +6 \pmod{31}$.

To test our implementation of the computation of κ, we computed $\kappa(iP)$ independently for $1 \le i \le 30$, checking that $\kappa(iP) \equiv 6i \pmod{31}$. The results are given in the following table:

i	1	2	3	4	5	6	7	8	9	10	11	12	13	14	15
e_1	12	19	13	7	1	10	20	14	8	2	8	20	15	9	3
e_2	6	12	18	14	2	5	11	17	16	4	4	10	16	18	6
$\kappa(iP)$	6	12	-13	-7	-1	5	11	-14	-8	-2	4	10	-15	-9	-3

i	30	29	28	27	26	25	24	23	22	21	20	19	18	17	16
e_1	6	12	18	14	2	5	11	17	16	4	4	10	16	18	6
e_2	12	19	13	7	1	10	20	14	8	2	8	20	15	9	3
$\kappa(iP)$	-6	-12	13	7	1	-5	-11	14	8	2	-4	-10	15	9	3

3   Other Reduction Types

For completeness we will now discuss the other reduction types, as well as $K = \mathbb{R}$.

3.1   Types Where Φ Is Trivial

When the reduction type is I_1 (good reduction), II or II*, the component group Φ is trivial, i.e. $c = 1$. This is also the case for non-split multiplicative reduction of type I_m when m is odd, and in the "non-split" cases for types IV, IV*, and I_0^*. Here there is nothing to be done.

3.2   Types Where $\Phi \cong \mathbb{Z}/2\mathbb{Z}$

When the reduction type is non-split multiplicative of type I_m when m is even, III or III*, and some cases of type I_m, we have $\Phi \cong \mathbb{Z}/2\mathbb{Z}$. Here all we need do is define $\kappa(P) = 0$ if P has good reduction and 1 otherwise.

3.3   Types Where $\Phi \cong \mathbb{Z}/3\mathbb{Z}$

When the reduction type is IV or IV* we may have $\Phi \cong \mathbb{Z}/3\mathbb{Z}$ in the "split" case. Our task is to see how to distinguish the two nontrivial components or cosets of $E^0(K)$ in $E(K)$.

First consider Type IV. After translating the model so that the singular point is $(0,0)$ (mod π), as in the first step of Tate's algorithm, the quadratic $h(T) = T^2 + \pi^{-1}a_3T - \pi^{-2}a_6$ has distinct roots in the residue field k (since if the roots only lie in a quadratic extension of k then $c = 1$ and Φ is trivial: the "non-split" case). Let α_1, α_2 be the roots of $h(T)$. Then any point $P = (x, y)$ of bad reduction has $y \equiv \alpha_i\pi$ (mod π^2) for $i \in \{1, 2\}$, as may be seen by reducing the Weierstrass equation modulo π^2. These two cases distinguish the two components, and we may define $\kappa(P) = i$ (mod 3).

We may translate this condition to apply to the original coordinates of the point: if the singular point is (x_0, y_0) (mod π) then for $P = (x, y) \in E(K) \setminus E^0(K)$ the value of $y - y_0$ lies in one of two distinct residue classes modulo π^2, which we may label arbitrarily and use to distinguish the nonzero values of κ. However, this is hardly worthwhile in practice: instead we may simply define $\kappa(P_1) = 1$ (mod 3) for the first point P_1 of bad reduction we encounter, and then for subsequent such points P we have $\kappa(P) = \pm 1$ according as $P - P_1$ does or does not have good reduction.

This latter strategy is certainly to be preferred for the case IV*, where (referring to Tate's algorithm) a second change of variables may be required. Otherwise we would need to determine y_0 (mod π^2) and use the value of $y - y_0$ (mod π^3) to distinguish the cases.

3.4   Types Where $\Phi \cong \mathbb{Z}/4\mathbb{Z}$

This can only occur with Type I_m^* when m is odd. Since this route in Tate's algorithm is the most subtle, rather than analyze the situation in more detail we can proceed as follows.

Set $\kappa(P) = 0$ if P has good reduction; otherwise set $\kappa(P) = 2$ if $2P$ has good reduction; otherwise $\kappa(P) = \pm 1$. A simple strategy, similar to that used for the $\mathbb{Z}/3\mathbb{Z}$ case, may be used to distinguish the latter in practice.

3.5   Types Where $\Phi \cong \mathbb{Z}/2\mathbb{Z} \times \mathbb{Z}/2\mathbb{Z}$

This can occur with Type I_m^* when m is even (including $m = 0$). Noting that the automorphism group of Φ includes all permutations of its nontrivial elements, we may proceed as follows:

Set $\kappa(P) = (0,0)$ if P has good reduction; otherwise set $\kappa(P_1) = (1,0)$ for the first point P_1 of bad reduction and $\kappa(P_2) = (0,1)$ for the first point P_2 such that neither P_2 nor $P_1 + P_2$ has good reduction. Now we can determine $\kappa(P)$ for all P simply by testing P, $P + P_1$ and $P + P_2$ for good reduction.

In case Type I_0^*, the nonzero values of $\kappa(P)$ may also be distinguished by the residue of $x - x_0 \pmod{\pi^2}$, where as usual $(x_0, y_0) \pmod{\pi}$ is the singular point on the reduction; but we have not attempted to extend this to a criterion for $m > 0$.

3.6   The Real Case

For completeness we finish by mentioning the case $K = \mathbb{R}$, where the component group is trivial if $\Delta < 0$ and has order 2 when $\Delta > 0$. In the latter case we may test whether a given point $P = (x,y)$ lies in $E^0(\mathbb{R})$ by checking whether $g'(x) > 0$ and $g''(x) > 0$, where $g(X) = 4X^3 + b_2 X^2 + 2b_4 X + b_6$; note that this may be done using exact arithmetic when E is defined over $\mathbb{Q}$ and $P \in E(\mathbb{Q})$, and so does not rely on approximating the real 2-torsion points.

References

1. Cremona, J.E.: mwrank and related programs for elliptic curves over **Q** (1990–2008),
   http://www.warwick.ac.uk/staff/J.E.Cremona/mwrank/index.html
2. Cremona, J.E.: Tables of elliptic curves (1990–2008),
   http://www.warwick.ac.uk/staff/J.E.Cremona/ftp/data/INDEX.html
3. Silverman, J.H.: Computing heights on elliptic curves. Math. Comp. 51, 339–358 (1988)
4. Silverman, J.H.: Advanced Topics in the Arithmetic of Elliptic Curves. In: Graduate Texts in Mathematics, vol. 151, Springer, New York (1994)

Some Improvements to 4-Descent
on an Elliptic Curve

Tom Fisher

University of Cambridge, DPMMS, Centre for Mathematical Sciences,
Wilberforce Road, Cambridge CB3 0WB, UK
T.A.Fisher@dpmms.cam.ac.uk
http://www.dpmms.cam.ac.uk/~taf1000

Abstract. The theory of 4-descent on elliptic curves has been developed
in the PhD theses of Siksek [18], Womack [21] and Stamminger [20].
Prompted by our use of 4-descent in the search for generators of large
height on elliptic curves of rank at least 2, we explain how to cut down
the number of class group and unit group calculations required, by using
the group law on the 4-Selmer group.

1 Introduction

Let E be an elliptic curve over a number field K. A 2-descent (see e.g. [3], [5],
[19]) furnishes us with a list of quartics $g(X) \in K[X]$ representing the everywhere
locally soluble 2-coverings of E, and hence the elements of the 2-Selmer group
$S^{(2)}(E/K)$. If we are unable to resolve the existence of K-rational points on the
curves $Y^2 = g(X)$, then it may be necessary to perform a 4-descent. Cassels [4]
has constructed a pairing on $S^{(2)}(E/K)$ whose kernel is the image of $[2]_*$ in the
exact sequence

$$E[2](K) \longrightarrow S^{(2)}(E/K) \xrightarrow{\iota_*} S^{(4)}(E/K) \xrightarrow{[2]_*} S^{(2)}(E/K) \ . \qquad (1)$$

We have checked [12] that this pairing agrees with the usual Cassels-Tate pairing
on $\text{III}(E/K)[2]$. An improved method for computing the pairing has recently
been found by Steve Donnelly [8].

Computing this pairing is sufficient to determine the structure of $S^{(4)}(E/K)$
as an abelian group, but if our aim is to find generators of $E(K)$ of large height,
then we also need to find equations for the 4-coverings parametrised by this
group. For this we use the theory of 4-descent, as developed in [14], [21] and [20].
Each quartic $g(X)$ has an associated flex algebra[1] $F = K[X]/(g(X))$, which is
usually a degree 4 field extension of K. The existing methods of 4-descent (as
implemented in Magma [2] by Tom Womack, and improved by Mark Watkins)
require us to compute the class group and units for the flex field of every quartic
in the image of $[2]_*$. In this article we explain how to cut down the number of class

---

[1] We keep the terminology of [7, Paper 1]. Were we to use a term specific to 2-descent
then "ramification algebra" would seem more appropriate.

A.J. van der Poorten and A. Stein (Eds.): ANTS-VIII 2008, LNCS 5011, pp. 125–138, 2008.

group and unit group calculations, by using the group law on $S^{(4)}(E/K)$. This is a non-trivial task since by properties of the obstruction map [7], [15], we expect to have to solve an explicit form of the local-to-global principle for the Brauer group $\mathrm{Br}(K)$. We also give a test for equivalence of 4-coverings (generalising the tests for 2-coverings and 3-coverings given in [5], [6] and [9]).

Even when the calculation of class groups and unit groups does finish, the output may be unmanageably large. We get round this by using a method described in §2, to find good representatives for elements of $K^\times/(K^\times)^n$. This technique is not specific to descent calculations on elliptic curves.

2   Selmer Groups of Number Fields

Let K be a number field of degree $[K : \mathbb{Q}] = d$ and let S be a finite set of primes of K. The n-Selmer group

$$K(S, n) = \{x(K^\times)^n \in K^\times/(K^\times)^n : \mathrm{ord}_{\mathfrak{p}}(x) \equiv 0 \pmod{n} \text{ for all } \mathfrak{p} \notin S\}$$

plays an important role in the construction of number fields via Kummer theory, and in the theory of descent on elliptic curves.

The height of an algebraic integer x in K is $H(x) = \prod_{i=1}^{d} \max(|\sigma_i(x)|, 1)$ where $\sigma_1, \ldots, \sigma_d$ are the distinct embeddings of K into $\mathbb{C}$. We write r_1 (resp. r_2) for the number of real (resp. complex) places, and Δ_K for the absolute discriminant. The Minkowski bound is

$$m_K = \left(\frac{4}{\pi}\right)^{r_2} \frac{d!}{d^d} \sqrt{|\Delta_K|} \ .$$

Theorem 2.1. *Let $n \geq 1$ be an integer. Let $\alpha \in K^\times$ with $(\alpha) = \mathfrak{b}\mathfrak{c}^n$ and $\mathfrak{b}$ an integral ideal. Then there exists $\beta \in \mathfrak{b}$ with $\alpha\beta^{-1} \in (K^\times)^n$ and*

$$H(\beta) \leq \max(m_K^n N\mathfrak{b}, \exp(nd)) \ .$$

The proof uses two lemmas.

Lemma 2.2. *If $a_1, \ldots, a_d$ are positive real numbers with $\sum_{i=1}^{d} a_i \leq dc^{1/d}$ then*

$$\prod_{i=1}^{d} \max(a_i, 1) \leq \max(c, \exp(d)) \ .$$

Proof. We may assume that $a_i \geq 1$ for $1 \leq i \leq r$ and $a_i < 1$ for $r + 1 \leq i \leq d$. By the inequality of the arithmetic and geometric means we obtain

$$\prod_{i=1}^{d} \max(a_i, 1) = \prod_{i=1}^{r} a_i \leq f(r/d)$$

where $f(x) = x^{-dx}c^x$. If $\log(c) \geq d$ then $f'(x) \geq 0$ for all $0 < x \leq 1$. Thus $f(r/d) \leq f(1) = c$. On the other hand if $\log(c) \leq d$ we obtain

$$\log f(x) \leq dx(1 - \log x) \leq d \ . \qquad \square$$

We extend the embeddings $\sigma_i : K \to \mathbb{C}$ to maps defined on $K \otimes_{\mathbb{Q}} \mathbb{R}$.

Lemma 2.3. *Let Λ be a lattice in $K \otimes_{\mathbb{Q}} \mathbb{R}$ of covolume V. Then there exists non-zero $\xi \in \Lambda$ with*

$$\sum_{i=1}^{d} |\sigma_i(\xi)| \leq \left(\left(\frac{4}{\pi} \right)^{r_2} d! \, V \right)^{1/d} .$$

Proof. This is a standard application of Minkowski's convex body theorem. $\square$

The usual application of Lemma 2.3 is to show that every fractional ideal $\mathfrak{b}$ in K contains an element β with $|N_{K/\mathbb{Q}}(\beta)| \leq m_K N\mathfrak{b}$.

Proof of Theorem 2.1. Let $|\cdot|$ be the map on $K \otimes_{\mathbb{Q}} \mathbb{R} \cong \mathbb{R}^{r_1} \oplus \mathbb{C}^{r_2}$ given componentwise by $x \mapsto |x|$. We apply Lemma 2.3 to the lattice $\Lambda = |\alpha|^{1/n} \mathfrak{c}^{-1}$ and let $\beta = \frac{\alpha}{|\alpha|} \xi^n$. The covolume of Λ is

$$|N_{K/\mathbb{Q}}(\alpha)|^{1/n} (N\mathfrak{c})^{-1} \sqrt{|\Delta_K|} = (N\mathfrak{b})^{1/n} \sqrt{|\Delta_K|} .$$

Thus β satisfies

$$\sum_{i=1}^{d} |\sigma_i(\beta)|^{1/n} \leq d \left(m_K (N\mathfrak{b})^{1/n} \right)^{1/d} .$$

Since $\beta \in \mathfrak{b}$ is an algebraic integer, we deduce by Lemma 2.2 that

$$H(\beta)^{1/n} \leq \max(m_K (N\mathfrak{b})^{1/n}, \exp(d))$$

as required. $\square$

Theorem 2.1 shows that every element of $K(S, n)$ is represented by an element of K of height at most

$$\max \left(m_K^n \left(\prod_{\mathfrak{p} \in S} N\mathfrak{p} \right)^{n-1}, \exp(nd) \right) . \tag{2}$$

Since there are only finitely many elements of K of height less than a given bound, this gives a new proof that $K(S, n)$ is finite. More importantly for us, replacing Minkowski's convex body theorem by the LLL algorithm, we obtain an algorithm for computing small representatives of Selmer group elements from large ones. This is particularly useful when using Magma's function pSelmerGroup (so $n = p$ a prime here) which returns a list of "small" elements of $K^{\times}$, and a list of exponents to which they must be multiplied to give generators for $K(S, p)$. In many examples of interest to us, multiplying out directly in $K^{\times}$ gives elements of unfeasibly large height. Using our algorithm (after every few multiplications) eliminates this problem. Moreover, the process can be arranged so that the only factorisations required are of the original list of "small" elements.

In principle one could also compute $K(S, n)$ by searching up to the bound (2), but of course this would be absurdly slow in practice.

3   Background on Quadric Intersections

Let $\mathcal{QI}(K)$ be the space of "quadric intersections" i.e. pairs of homogeneous polynomials of degree 2 in $K[x_1, x_2, x_3, x_4]$. Given $(A, B) \in \mathcal{QI}(K)$ we identify A and B with their matrices of second partial derivatives, and compute

$$g(X) = \det(AX + B) = aX^4 + bX^3 + cX^2 + dX + e \ .$$

The invariants of the quartic $g(X)$ are $I = 12ae - 3bd + c^2$ and $J = 72ace - 27ad^2 - 27b^2e + 9bcd - 2c^3$, and the invariants of (A, B) are $c_4 = I$ and $c_6 = \frac{1}{2}J$. It is well known (see [1]) that if $\Delta = (c_4^3 - c_6^2)/1728$ is non-zero then the curves $C_2 = \{Y^2 = g(X)\}$ and $C_4 = \{A = B = 0\} \subset \mathbb{P}^3$ are smooth curves of genus one with Jacobian

$$E : \quad y^2 = x^3 - 27c_4 x - 54c_6 \ . \tag{3}$$

Moreover C_4 is a 2-covering of C_2 (see [1], [14]) the composite $C_4 \to C_2 \xrightarrow{X} \mathbb{P}^1$ being given by $-T_1/T_2$ where T_1 and T_2 are the quadrics determined by

$$\mathrm{adj}((\mathrm{adj}A)X + (\mathrm{adj}B)) = a^2 AX^3 + aT_1 X^2 + eT_2 X + e^2 B.$$

Following [6], we say that quartics $g_1, g_2 \in K[X]$ are K-equivalent if their homogenisations satisfy $g_1 = \mu^2 g_2 \circ M$ for some $\mu \in K^\times$ and $M \in \mathrm{GL}_2(K)$. Quadric intersections $(A, B), (A', B') \in \mathcal{QI}(K)$ are K-equivalent if

$$(A', B') = (m_{11} A \circ N + m_{12} B \circ N, m_{21} A \circ N + m_{22} B \circ N)$$

for some $(M, N) \in \mathcal{G}_4(K) := \mathrm{GL}_2(K) \times \mathrm{GL}_4(K)$. It is routine to check that the quartics associated to equivalent quadric intersections are themselves equivalent.

In the course of a 4-descent, a 2-covering C_4 of C_2 is computed as follows. Let C_2 have equation $Y^2 = g(X)$ and flex algebra $F = K[\theta] = K[X]/(g(X))$. Suppose we are given $\xi \in F^\times$ with $N_{F/K}(\xi) \equiv a \mod (K^\times)^2$ where a is the leading coefficient of g. (The existence of such a ξ is clearly necessary for the existence of K-rational points on C_2.) We consider the equation

$$X - \theta = \xi(x_1 + x_2\theta + x_3\theta^2 + x_4\theta^3)^2 \ .$$

A quadric intersection, defining a 2-covering C_4 of C_2, is obtained by expanding in powers of θ and taking the coefficients of θ^2 and θ^3. The answer only depends (up to K-equivalence) on the class of ξ in $F^\times/K^\times(F^\times)^2$. Using the method of §2 to find a good representative for this class, can significantly decrease the time subsequently taken to find a good choice of co-ordinates on $\mathbb{P}^3$, that is, to minimise and reduce the quadric intersection (using the algorithms in [21]).

4   Galois Cohomology

We keep the notation and conventions of [7, Paper I]. Let $\pi : C \to E$ be the 2-covering corresponding to $\xi \in H^1(K, E[2])$. The flex algebra of ξ is

$F = \text{Map}_K(\varPhi, \overline{K})$ where $\varPhi$ is the fibre of π above 0_E. We note that C is a torsor under E, and $\varPhi$ is a torsor under $E[2]$. Let $\langle \xi \rangle$ be the subgroup of $H^1(K, E[2])$ generated by ξ, and let $\cup$ be the map $H^1(K, E[2]) \times H^1(K, E[2]) \to \text{Br}(K)[2]$ induced by cup product and the Weil pairing $e_2 : E[2] \times E[2] \to \mu_2$. The following theorem is a variant of a standard result (see for example [17], [20]).

Theorem 4.1. *There is a canonical isomorphism*

$$\ker\left(\frac{H^1(K, E[2])}{\langle \xi \rangle} \xrightarrow{\cup \xi} \text{Br}(K) \right) \cong \ker\left(F^\times/K^\times(F^\times)^2 \xrightarrow{N_{F/K}} K^\times/(K^\times)^2 \right) .$$

Proof. Let $\overline{F} = F \otimes_K \overline{K}$. We may identify $\overline{F} = \text{Map}(\varPhi, \overline{K})$ and $\mu_2(\overline{F}) = \text{Map}(\varPhi, \mu_2)$. These are identifications as Galois modules, the action of Galois being given by $\sigma(f) = (P \mapsto \sigma(f(\sigma^{-1}P)))$. An easy generalisation of Hilbert's theorem 90 shows that $H^1(K, \overline{F}^\times) = 0$ and hence $H^1(K, \mu_2(\overline{F})) = F^\times/(F^\times)^2$. We define $N : \text{Map}(\varPhi, \mu_2) \to \mu_2$ by $N(f) = \prod_{P \in \varPhi} f(P)$. The constant maps give an inclusion $\mu_2 \to \text{Map}(\varPhi, \mu_2)$ with quotient X (say). We thus have short exact sequences of Galois modules

$$0 \longrightarrow \mu_2 \longrightarrow \text{Map}(\varPhi, \mu_2) \xrightarrow{q} X \longrightarrow 0$$

and

$$0 \longrightarrow E[2] \xrightarrow{w} X \xrightarrow{N} \mu_2 \longrightarrow 0$$

where $w(T)$ is the class of $P \mapsto e_2(P - P_0, T)$, for any fixed choice of $P_0 \in \varPhi$. Taking the long exact sequences of Galois cohomology we obtain a diagram

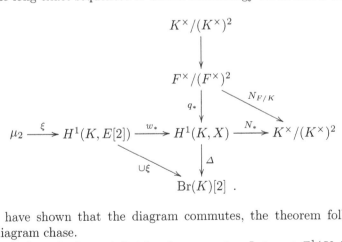

Once we have shown that the diagram commutes, the theorem follows by a routine diagram chase.

We check that the lower left triangle commutes. Let $\eta \in Z^1(K, E[2])$ be a cocycle. Then $w_*(\eta)_\sigma$ is the map $P \mapsto e_2(P - P_0, \eta_\sigma)$. Applying the connecting map Δ gives $a \in Z^2(K, \mu_2)$ with

$$\begin{aligned}
a_{\sigma\tau} &= e_2(P - \sigma(P_0), \sigma(\eta_\tau)) \, e_2(P - P_0, \eta_\sigma) \, e_2(P - P_0, \eta_{\sigma\tau})^{-1} \\
&= e_2(P_0 - \sigma(P_0), \sigma(\eta_\tau)) \, e_2(P - P_0, \sigma(\eta_\tau) + \eta_\sigma - \eta_{\sigma\tau}) \\
&= e_2(\xi_\sigma, \sigma(\eta_\tau)) .
\end{aligned}$$

This is the cup product of ξ and η. The commutativity of the upper right triangle is clear. □

The case $\xi = 0$ of Theorem 4.1 is well-known. In this case F is the étale algebra $K \times L$ of $E[2]$ where $E : Y^2 = f(X)$ and $L = K[X]/(f(X))$.

Corollary 4.2. *There is a canonical isomorphism*

$$H^1(K, E[2]) \cong \ker \left(L^\times / (L^\times)^2 \overset{N_{L/K}}{\longrightarrow} K^\times / (K^\times)^2 \right) .$$

The following theorem, due to Steve Donnelly, gives an explicit description of the isomorphism of Theorem 4.1 (in one direction). We make the identification of Corollary 4.2 so that now ξ is represented by some $\alpha \in L^\times$. Let LF be the tensor product $L \otimes_K F$ and let $L[\sqrt{\alpha}]$ be the algebra $L[X]/(X^2 - \alpha)$. By the formulae in [5, §3] there is a natural inclusion $L[\sqrt{\alpha}] \subset LF$. (If $\mathrm{Gal}(F/K) \cong S_4$ then L is the resolvent cubic field, LF is the usual composite of fields, and we are quoting that α is a square in LF.) Let τ be the non-trivial automorphism of $L[\sqrt{\alpha}]$ that fixes L.

Theorem 4.3. *Let* $\delta \in F^\times$ *with* $N_{F/K}(\delta) = k^2$ *for some* $k \in K$. *Suppose we are given* $\nu \in L[\sqrt{\alpha}]^\times$ *with* $N_{LF/L[\sqrt{\alpha}]}(\delta)/k = \tau(\nu)/\nu$. *Then*

$$\beta := N_{LF/L[\sqrt{\alpha}]}(\delta)\nu^2 = k N_{L[\sqrt{\alpha}]/L}(\nu) \in L^\times \tag{4}$$

represents an element of $\ker \left(L^\times / (L^\times)^2 \overset{N_{L/K}}{\longrightarrow} K^\times / (K^\times)^2 \right)$ *mapping to* δ *under the isomorphisms of Theorem 4.1 and Corollary 4.2.*

Proof. We identify

$$LF = L \otimes_K F = \mathrm{Map}_K((E[2] \setminus \{0\}) \times \Phi, \overline{K}) .$$

Then $N_{LF/L[\sqrt{\alpha}]}(\delta)$ is the map $(T, P) \mapsto \delta(P)\delta(T + P)$. So fixing a base point $P_0 \in \Phi$ we can rewrite the first equality in (4) as

$$\beta(P - P_0) = \delta(P)\delta(P_0)\nu(P - P_0, P)^2 \tag{5}$$

for all $P \in \Phi$ with $P \neq P_0$.

The image of β in $H^1(K, X)$ is represented by a cocycle (ψ_σ) where

$$\psi_\sigma(P) = \begin{cases} \dfrac{\sigma\sqrt{\beta}}{\sqrt{\beta}}(P - P_0) & \text{if } P \neq P_0 \\ 1 & \text{if } P = P_0 . \end{cases}$$

It follows by (5) that

$$\psi_\sigma(P) = \frac{\sigma\sqrt{\delta}}{\sqrt{\delta}}(P)\frac{\sigma\sqrt{\delta}}{\sqrt{\delta}}(P_0)$$

for all $P \in \Phi$. (The case $P = P_0$ is just $1 = (\pm 1)^2$.) By the definition of X we may ignore the term involving P_0, and so (ψ_σ) also represents the image of δ in $H^1(K, X)$. □

Remark 4.4. If $\varepsilon = N_{LF/L[\sqrt{\alpha}]}(\delta)/k$ then $N_{L[\sqrt{\alpha}]/L}(\varepsilon) = 1$. So by Hilbert's theorem 90 there exists $\nu \in L[\sqrt{\alpha}]^\times$ with $\varepsilon = \tau(\nu)/\nu$. The construction of Theorem 4.3 therefore gives a well-defined map

$$\ker\left(F^\times/K^\times(F^\times)^2 \xrightarrow{N_{F/K}} K^\times/(K^\times)^2\right) \to L^\times/\{1,\alpha\}(L^\times)^2 .$$

The ambiguity up to multiplication by α is predicted by Theorem 4.1, and in this construction comes from the arbitrary choice of sign for k.

5   Testing Equivalence of 4-Coverings

Let $g(X) \in K[X]$ be a (non-singular) quartic with flex algebra $F = K[\theta] = K[X]/(g(X))$. We put $\mathcal{QI}(K)^{\det=g} = \{(A,B) \in \mathcal{QI}(K) : \det(AX+B) = g(X)\}$. If $(A,B) \in \mathcal{QI}(K)^{\det=g}$ then keeping the notation of §3 we define

$$\mathcal{Q} = \theta^{-1}eA + T_1 + \theta T_2 + \theta^2 aB \qquad (6)$$

with suitable modifications if $ae = 0$. (For example if $e = 0$ then the "$\theta = 0$ component" of $\mathcal{Q}$ is $-dA + T_1$.) Then $\mathcal{Q}$ is a rank 1 quadratic form, $i.e.$ $\mathcal{Q} = \xi\ell^2$ for some $\xi \in F^\times$ and $\ell \in F[x_1, x_2, x_3, x_4]$ a linear form. This defines a map

$$\lambda \; : \; \mathcal{QI}(K)^{\det=g} \longrightarrow F^\times/(F^\times)^2; \quad (A,B) \mapsto \xi$$

inverse to the construction of §3.

Lemma 5.1. *Quadric intersections in $\mathcal{QI}(K)^{\det=g}$ define isomorphic coverings of $C_2 = \{Y^2 = g(X)\}$ if and only if they are related by a transformation $(\mu I_2, N) \in \mathcal{G}_4(K)$ with $\mu^2 \det(N) = 1$.*

Proof. If $\pi : C_4 \to C_2$ is the 2-covering defined by $(A,B) \in \mathcal{QI}(K)^{\det=g}$ and $P_0 \in C_2$ is a ramification point of $C_2 \to \mathbb{P}^1$ then the divisor $\pi^*(P_0)$ is a hyperplane section of C_4 (in fact cut out by the linear form ℓ). So if a pair of quadric intersections determine isomorphic 2-coverings of C_2, then they must be K-equivalent. Moreover, the equivalence $(M,N) \in \mathcal{G}_4(K)$ is of the form described since, by definition of a 2-covering, the induced self-equivalence of g must be trivial as an automorphism of C_2.   □

If $(A_0, B_0) \in \mathcal{QI}(K)^{\det=g}$ defines $C_4 \subset \mathbb{P}^3$ then the 2-coverings of C_2 are parametrised as twists of $C_4 \to C_2$ by $H^1(K, E[2])$. This defines a map

$$\phi_0 \; : \; \frac{\mathcal{QI}(K)^{\det=g}}{\{(\mu I_2, N) \in \mathcal{G}_4(K) : \mu^2 \det N = 1\}} \longrightarrow H^1(K, E[2]) .$$

We find that quotienting out by the transformations with $\mu^2 \det N = -1$ corresponds to quotienting out by $\langle \xi_2 \rangle$ where $\xi_2 \in H^1(K, E[2])$ is the class of g.

Theorem 5.2. *The following diagram is commutative.*

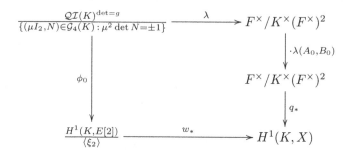

Proof. This is a variant of [20, Theorem 6.1.4]. Let $\mathcal{Q}_0 = \xi_0 \ell_0^2$ and $\mathcal{Q}_1 = \xi_1 \ell_1^2$ be the rank 1 quadratic forms determined by (A_0, B_0) and (A, B). If $(\mu I_2, N) \in \mathcal{G}_4(\overline{K})$ relates (A, B) and (A_0, B_0) then by properties of the Weil pairing

$$w_*(\phi_0(A, B)) = \left(\sigma \mapsto \frac{\ell_0 \circ \sigma(N) N^{-1}}{\ell_0} = \frac{\sigma(\ell_0 \circ N)}{\ell_0 \circ N} \right) .$$

Since $\mathcal{Q}_1 = \mu \, \mathcal{Q}_0 \circ N$, this works out as $q_*(\xi_0 \xi_1)$. □

The maps ϕ_0 and w_* of Theorem 5.2 are injective. It follows that λ is injective. So to test whether a pair of quadric intersections $(A_1, B_1), (A_2, B_2) \in \mathcal{QI}(K)$ are equivalent we proceed as follows. We have implemented this test in the case $K = \mathbb{Q}$ and contributed it to Magma [2].

Step 1. Let $g_i(X) = \det(A_i X + B_i)$ for $i = 1, 2$. We test whether g_1 and g_2 are equivalent, using one of the tests in [5], [6]. We are now reduced to the case $g_1 = g_2$. (If there is more than one equivalence between g_1 and g_2 then we must repeat the remaining steps for each of these.)

Step 2. Compute $\xi_i = \lambda(A_i, B_i)$ for $i = 1, 2$ by evaluating the quadratic form (6) at points in $\mathbb{P}^3(K)$. It helps with Step 3 if we use several points in $\mathbb{P}^3(K)$ to give several representatives for the class of ξ_i in $F^\times/(F^\times)^2$. (Spurious prime factors can then be removed from consideration by computing gcd's.)

Step 3. Let S be a finite set of primes of K, including all primes that ramify in F. We enlarge S so that $\xi_1, \xi_2 \in F(S', 2)$ where S' is the set of primes of F above S.

Step 4. The quadric intersections are equivalent if and only if $\xi_1 \xi_2^{-1}$ is in the image of the natural map $K(S, 2) \to F(S', 2)$. We cut down the subgroup of $K(S, 2)$ to be considered by reducing modulo some random primes, and then loop over all possibilities.

In the case that (A_1, B_1) and (A_2, B_2) are equivalent, we can reduce to the case $\mathcal{Q}_1 = \xi_1 \ell_1^2$ and $\mathcal{Q}_2 = \xi_2 \ell_2^2$ with $\xi_1 \xi_2^{-1} \in K$. Then solving $\ell_1 \circ N = \ell_2$ for $N \in \mathrm{Mat}_4(K)$, gives the change of co-ordinates relating the two quadric intersections. This transformation is also returned by our Magma function.

6   Adding 2-Selmer and 4-Selmer Elements

In §8 we describe a general method for adding 4-Selmer group elements. This involves solving an explicit form of the local-to-global principle for $\mathrm{Br}(K)$. But in the special case where we add 2-Selmer and 4-Selmer elements, no such problem need be solved. This is essentially because (by a theorem of Zarhin [22] relating the cup product in Theorem 4.1 to the obstruction map in [7, Paper I], [15]) we have already solved all the conics we need when doing the original 2-descent. To make this explicit we have found the following partial description of the isomorphism of Theorem 4.1.

Let $g(X) \in K[X]$ be a quartic with invariants I and J. Let $L = K[\varphi]$ where φ is a root of $f(X) = X^3 - 3IX + J$. We assume that the discriminant $\Delta_0 = 27(4I^3 - J^2)$ is non-zero. Formulae in [5], [6] allow us to represent g by $\alpha = a_0 + a_1\varphi \in L^\times$ with $a_0, a_1 \in K$ and $N_{L/K}(\alpha) \in (K^\times)^2$. We assume $\alpha \notin (L^\times)^2$. As in §4 we put $F = K[X]/(g(X))$ and $LF = L \otimes_K F$.

Theorem 6.1. *If $\beta, \gamma \in L^\times$ are linear in φ with $N_{L/K}(\beta), N_{L/K}(\gamma) \in (K^\times)^2$ and $\alpha\beta\gamma \in (L^\times)^2$, then the isomorphisms of Theorem 4.1 and Corollary 4.2 map each of β and γ to the class of*

$$\delta := \mathrm{Tr}_{LF/F}\left(\frac{\sqrt{\alpha}}{f'(\varphi)}\right) \mathrm{Tr}_{LF/F}\left(\frac{\sqrt{\beta\gamma}}{f'(\varphi)}\right) \in F^\times .$$

Proof. Let $\varphi_1 = \varphi, \varphi_2, \varphi_3$ be the K-conjugates of φ, and likewise for α, β, γ, m where $\alpha\beta\gamma = m^2$. Using that α, β, γ are linear in φ we compute

$$N_{LF/L[\sqrt{\alpha}]}(\delta) = \frac{(\sqrt{\alpha_2} - \sqrt{\alpha_3})^2(\sqrt{\beta_2\gamma_3} - \sqrt{\beta_3\gamma_2})^2}{\Delta_0 \, (\varphi_2 - \varphi_3)^2} .$$

The hypotheses of Theorem 4.3 are therefore satisfied with

$$k = \frac{(\alpha_2 - \alpha_3)(\beta_2\gamma_3 - \beta_3\gamma_2)}{\Delta_0 \, (\varphi_2 - \varphi_3)^2} = \frac{a_1(b_1c_0 - b_0c_1)}{\Delta_0} \in K^\times$$

and (swapping β and γ if necessary to avoid dividing by zero)

$$\nu^{-1} = \frac{(\sqrt{\alpha_2} - \sqrt{\alpha_3})(\sqrt{\beta_2\gamma_3} - \sqrt{\beta_3\gamma_2})}{\sqrt{\alpha_2\beta_2\gamma_3} + \sqrt{\alpha_3\beta_3\gamma_2}} = 1 - \frac{m_2\beta_3 + m_3\beta_2}{m_2\gamma_3 + m_3\gamma_2}\sqrt{\frac{\gamma_2\gamma_3}{\beta_2\beta_3}} .$$

We are done since

$$(\sqrt{\alpha_2\beta_2\gamma_3} + \sqrt{\alpha_3\beta_3\gamma_2})^2 = \frac{(m_2\gamma_3 + m_3\gamma_2)^2}{\gamma_2\gamma_3} \equiv \gamma \mod (L^\times)^2 . \qquad \square$$

We give an example in the case $K = \mathbb{Q}$. The quartics

$$\begin{aligned} g_1(X) &= -675X^4 - 7970X^3 - 18923X^2 + 27176X - 7848 \\ g_2(X) &= -5483X^4 + 10470X^3 + 8869X^2 - 13240X - 8768 \\ g_3(X) &= -3728X^4 - 8536X^3 + 9037X^2 + 15940X - 13000 \end{aligned}$$

have invariants $I = 1071426889$ and $J = 70141299507574$. Moreover they sum to zero in $S^{(2)}(E/\mathbb{Q})$ where $E : -3Y^2 = f(X) = X^3 - 3IX + J$. Let $L = \mathbb{Q}(\varphi) = \mathbb{Q}[X]/(f(X))$ and $F_1 = \mathbb{Q}(\theta) = \mathbb{Q}[X]/(g_1(X))$. We use the existing FourDescent routine in Magma to compute 2-coverings D_i of $C_i = \{Y^2 = g_i(X)\}$ for $i = 2, 3$ and then add these using the method of §8 to give a 2-covering D_1 of $C_1 = \{Y^2 = g_1(X)\}$. By a formula in [5] the quartics g_1, g_2, g_3 are represented by

$$
\begin{aligned}
\alpha &= -900\varphi + 29459500 \\
\beta &= (-21932\varphi + 717892516)/3 \\
\gamma &= (-14912\varphi + 488109376)/3
\end{aligned}
$$

in $L^\times/(L^\times)^2$. Theorem 6.1 and the map λ in §5 convert C_2 and D_1 to

$$
\delta = 265659750\theta^3 + 3276444415\theta^2 + 917786936\theta - 582546987
$$

and $\xi_1 = 47250\theta^3 + 591650\theta^2 + 1684960\theta - 106600$ in $F_1^\times/\mathbb{Q}^\times(F_1^\times)^2$. We then multiply δ and ξ_1 in $F_1^\times$ and recover a new 2-covering D_1' of C_1 by the method of §3. By Theorem 5.2 this new 4-covering of E represents the sum of $\iota_*(C_2)$ and D_1 in $S^{(4)}(E/\mathbb{Q})$ where ι_* is the map in (1). Notice that at no stage of the computation of D_1 and D_1' did we need to find the class group and units of F_1, although it is only for much larger examples that this saving becomes worthwhile.

7   Computing the Action of the Jacobian

In this section we generalise the formulae of [9, §7] from 3-coverings to 4-coverings. The main new ingredient is a certain generalisation of the Hessian, introduced in [10]. This is an $\mathrm{SL}_2(K) \times \mathrm{SL}_4(K)$-equivariant polynomial map $H : \mathcal{QI}(K) \to \mathcal{QI}(K)$. In the notation of §3 it is given by

$$
H : (A, B) \mapsto (6T_2 - cA - 3bB, \; 6T_1 - cB - 3dA) . \tag{7}
$$

The analogue of the Hesse pencil of plane cubics, is the "Hesse family" of quadric intersections

$$
U(a, b) = (a(x_1^2 + x_3^2) - 2bx_2x_4, \; a(x_2^2 + x_4^2) - 2bx_1x_3)
$$

with invariants

$$
\begin{aligned}
c_4(a, b) &= 2^8(a^8 + 14a^4b^4 + b^8) \\
c_6(a, b) &= -2^{12}(a^{12} - 33a^8b^4 - 33a^4b^8 + b^{12}) \\
\Delta(a, b) &= 2^{20}a^4b^4(a^4 - b^4)^4
\end{aligned}
$$

and Hessian $U(a', b')$ where $a' = -2^4a(a^4 - 5b^4)$ and $b' = 2^4b(5a^4 - b^4)$.

If $U \in \mathcal{QI}(K)$ is a non-singular quadric intersection with Jacobian E, then the pencil of quadric intersections spanned by U and its Hessian is a twist of the Hesse family. So there are exactly six singular fibres, and each singular fibre is a "square" (really a quadrilateral spanning $\mathbb{P}^3$). Each square is uniquely the intersection of a pair of rank 2 quadrics and the union of these quadrics is the set of fixed planes for the action of M_T on $\mathbb{P}^3$ for some $T \in E[4] \setminus E[2]$. So there is a Galois equivariant bijection between the syzygetic squares and the cyclic subgroups of $E[4]$ of order 4. (Our terminology generalises that in [13, §II.7].)

Lemma 7.1. *Let U be a non-singular quadric intersection with invariants c_4, c_6 and Hessian H. Let $T = (x_T, y_T)$ be a point of order 4 on the Jacobian (3). Then the syzygetic square corresponding to $\pm T$ is defined by $\mathcal{S} = \frac{1}{3}x_T U + H$, and this quadric intersection satisfies $H(\mathcal{S}) = \nu_T^2 \mathcal{S}$ where*

$$\nu_T = (x_T^4 - 54c_4 x_T^2 - 216c_6 x_T - 243c_4^2)/(18y_T) \ .$$

Proof. We may assume that U belongs to the Hesse family and that $T = (2^4 3(a^4 - 5b^4), 2^7 3^3 i(a^4 - b^4)b^2)$. The lemma follows by direct calculation. $\square$

Let $C \subset \mathbb{P}^3$ be a genus one normal curve of degree 4, defined over K, and with Jacobian E. Let L/K be any field extension. Given $T \in E(L)$ a point of order 4, we aim to construct $M_T \in \mathrm{GL}_4(L)$ describing the action of T on C. We start with a quadric intersection U defining C. Then we compute the syzygetic square $\mathcal{S} = \frac{1}{3}x_T U + H$ as described in Lemma 7.1. Making a change of co-ordinates (defined over K) we may assume

- The point $(1 : 0 : 0 : 0)$ does not lie on either of the rank 2 quadrics whose intersection is the syzygetic square.
- The line $\{x_3 = x_4 = 0\}$ does not meet either diagonal of the square.

Let A and B be the rank 2 quadrics in the pencil spanned by $\mathcal{S}$, scaled so that the coefficient of x_1^2 is 1 in each case. These quadrics are defined over a field L' with $[L' : L] \leq 2$, and are easily found by factoring the determinant of a generic quadric in the pencil. We factor A and B over $\overline{K}$ as

$$A = (x_1 + \alpha_1 x_2 + \beta_1 x_3 + \gamma_1 x_4)(x_1 + \alpha_3 x_2 + \beta_3 x_3 + \gamma_3 x_4)$$
$$B = (x_1 + \alpha_2 x_2 + \beta_2 x_3 + \gamma_2 x_4)(x_1 + \alpha_4 x_2 + \beta_4 x_3 + \gamma_4 x_4) \ .$$

Then we put

$$P = \begin{pmatrix} 1 & \alpha_1 & \beta_1 & \gamma_1 \\ 1 & \alpha_2 & \beta_2 & \gamma_2 \\ 1 & \alpha_3 & \beta_3 & \gamma_3 \\ 1 & \alpha_4 & \beta_4 & \gamma_4 \end{pmatrix}$$

and $\xi = \alpha_1 - i\alpha_2 - \alpha_3 + i\alpha_4$ where $i = \sqrt{-1}$.

Theorem 7.2. *If $\xi \neq 0$ then the matrix*

$$M_T = \xi P^{-1} \begin{pmatrix} 1 & & & \\ & i & & \\ & & -1 & \\ & & & -i \end{pmatrix} P$$

belongs to $\mathrm{GL}_4(L)$ and describes the action of T (or $-T$) on C.

Proof. The image of this matrix in PGL_4 has order 4, and acts on $\mathbb{P}^3$ with fixed planes defined by the linear factors of A and B. So the second statement is clear. Theorem 7.3 shows that M_T has entries in L. (It may also be checked directly that each entry is fixed by $\mathrm{Gal}(L'(i)/L)$.) $\square$

Any polynomial in the α_i, β_i, γ_i invariant under the action of $C_2 \times C_2$ that swaps the subscripts $1 \leftrightarrow 3$ and $2 \leftrightarrow 4$ may be rewritten as a polynomial in the coefficients of A and B. We write $A = \sum_{i \leq j} a_{ij} x_i x_j$ and $B = \sum_{i \leq j} b_{ij} x_i x_j$. Then by computer algebra we find an expression for $\kappa = (\alpha_1 - \alpha_3)(\alpha_2 - \alpha_4) \det(P)$ as a polynomial in the a_{ij} and b_{ij}, and likewise for the entries of

$$M_1 = (\alpha_2 - \alpha_4)\mathrm{adj}(P)\mathrm{Diag}(1,0,-1,0)P$$

and

$$M_2 = (\alpha_1 - \alpha_3)\mathrm{adj}(P)\mathrm{Diag}(0,1,0,-1)P \ .$$

Let $\mathcal{S} = (\lambda_1 A + \mu_1 B, \lambda_2 A + \mu_2 B)$ with $\lambda_i, \mu_i \in L'$. Then $\kappa \in L$, whereas if A and B are not defined over L then $\mathrm{Gal}(L'/L)$ interchanges $\lambda_1 \leftrightarrow \lambda_2$, $\mu_1 \leftrightarrow \mu_2$ and $M_1 \leftrightarrow M_2$.

Theorem 7.3. *The matrix M_T of Theorem 7.2 is given by*

$$M_T = \frac{a_{12}^2 - 4a_{22}}{\kappa} M_1 + \frac{b_{12}^2 - 4b_{22}}{\kappa} M_2 \pm \frac{\lambda_1 \mu_2 - \lambda_2 \mu_1}{\nu_T}(M_1 - M_2)$$

where $\nu_T = (x_T^4 - 54c_4 x_T^2 - 216c_6 x_T - 243c_4^2)/(18y_T)$.

Proof. By our choice of co-ordinates we have $\alpha_1 \neq \alpha_3$ and $\alpha_2 \neq \alpha_4$. So $\kappa \in L$ is non-zero. We compute

$$\begin{aligned}
\kappa M_T &= \xi(\alpha_1 - \alpha_3)(\alpha_2 - \alpha_4)\mathrm{adj}(P)\mathrm{Diag}(1,i,-1,-i)P \\
&= \xi(\alpha_1 - \alpha_3)M_1 + i\xi(\alpha_2 - \alpha_4)M_2 \\
&= (\alpha_1 - \alpha_3)^2 M_1 + (\alpha_2 - \alpha_4)^2 M_2 - i\kappa(\det P)^{-1}(M_1 - M_2) \ .
\end{aligned}$$

Since $H(x_1 x_3, x_2 x_4) = (-x_1 x_3, -x_2 x_4)$ we have

$$H(\mathcal{S}) = -(\lambda_1 \mu_2 - \lambda_2 \mu_1)^2 \det(P)^2 \mathcal{S} \ .$$

By Lemma 7.1 we deduce $\nu_T = \pm i(\lambda_1 \mu_2 - \lambda_2 \mu_1)\det(P)$, and substituting this into the above expression for κM_T completes the proof of the theorem. $\square$

By our choice of co-ordinates it is impossible that both $\xi = \alpha_1 - i\alpha_2 - \alpha_3 + i\alpha_4$ and $\xi' = \alpha_1 + i\alpha_2 - \alpha_3 - i\alpha_4$ vanish. So if our formula for M_T gives the zero matrix, we can instead use the formula for M_{-T} and take the inverse.

8   Adding 4-Selmer Group Elements

Finally we outline how the theory in [7] can be used to add elements of $S^{(4)}(E/K)$. (Of course, the method in §6 should be used in preference whenever it applies.) Let

$C \subset \mathbb{P}^3$ be a 4-covering of E. We embed E in $\mathbb{P}^3$ via $(x, y) \mapsto (1 : x : y : x^2)$. In [11, §6.2] we gave a practical algorithm for computing $B \in \mathrm{GL}_4(\overline{K})$ describing a change of co-ordinates on $\mathbb{P}^3$ taking C to E.

Now let R be the étale algebra of $E[4]$. Applying the formulae of §7 over each constituent field of R, we compute $M, M' \in \mathrm{GL}_4(R)$ describing the actions of $E[4]$ on $E \subset \mathbb{P}^3$ and $C \subset \mathbb{P}^3$ respectively. We scale these matrices by using the method of §2 to find good representatives for their determinants in $R^\times/(R^\times)^4$. These matrices now determine $\gamma \in \overline{R}^\times = \mathrm{Map}(E[4], \overline{K}^\times)$ by the rule

$$BM'_T B^{-1} = \gamma(T)M_T$$

for all $T \in E[4]$. It is shown in [7, Paper I] that we may identify $H^1(K, E[4])$ with a certain subquotient of $(R \otimes R)^\times$. Our 4-covering corresponds to $\rho \in (R \otimes R)^\times$ given by the rule

$$\rho(S, T) = \frac{\gamma(S)\gamma(T)}{\gamma(S+T)} \tag{8}$$

for all $S, T \in E[4]$. So if 4-coverings C_1 and C_2 determine $\gamma_1, \gamma_2 \in \overline{R}^\times$, then their sum (by the group law of $H^1(K, E[4])$) corresponds to the product $\gamma_1\gamma_2$.

It remains to explain how, if C is everywhere locally soluble, we can recover equations for $C \subset \mathbb{P}^3$ from $\gamma \in \overline{R}^\times$. Let $\varepsilon \in (R \otimes R)^\times$ be the element determined by $\varepsilon(S, T)I_4 = M_S M_T M_{S+T}^{-1}$ for all $S, T \in E[4]$, and let ρ be given by (8). We view $R \otimes R$ as an R-algebra via the comultiplication $R \to R \otimes R$ and write $\mathrm{Tr} : R \otimes R \to R$ for the corresponding trace map. In [7, Paper I] we defined the obstruction algebra $A_\rho = (R, +, *_{\varepsilon\rho})$ to be the K-vector space R equipped with a new multiplication $z_1 *_{\varepsilon\rho} z_2 = \mathrm{Tr}(\varepsilon\rho.(z_1 \otimes z_2))$.

In our situation, we already have a trivialisation of A_ρ over $\overline{K}$, namely the isomorphism of $\overline{K}$-algebras $A_\rho \otimes_K \overline{K} \cong \mathrm{Mat}_4(\overline{K})$ given by

$$z \mapsto \sum_{T \in E[4]} z(T)\gamma(T)M_T .$$

So picking a basis $r_1, \ldots, r_{16}$ for R gives matrices $M_1, \ldots, M_{16} \in \mathrm{Mat}_4(\overline{K})$. We then compute structure constants $c_{ijk} \in K$ for the obstruction algebra A_ρ by the rule $M_i M_j = \sum_{k=1}^{16} c_{ijk} M_k$.

Our only implementation so far is in the case $K = \mathbb{Q}$. In practice we fix an embedding $\overline{\mathbb{Q}} \subset \mathbb{C}$, and so γ is represented by a 16-tuple of complex numbers (to some precision). In [7, Paper III] we will explain how to choose a basis for R so that the structure constants c_{ijk} are (reasonably small) integers. This makes it easy to recognise them from their floating point approximations.

Since C is everywhere locally soluble, it is guaranteed by class field theory that there is an isomorphism of K-algebras $A_\rho \cong \mathrm{Mat}_4(K)$. We must find such an isomorphism explicitly, and for this we use the method of Pílniková [16], who reduces the problem to that of solving conics over (at most quadratic) extensions of K. Finally any one of the three methods in [7, Paper I, §5] may be used to recover equations for C. In practice we use the Hesse pencil method, which by virtue of the Hessian (7) has a natural generalisation from 3-descent to 4-descent.

Acknowledgements

I would like to thank John Cremona, Michael Stoll and Denis Simon for many useful discussions in connection with this work, and Steve Donnelly for providing me with the construction described in Theorem 4.3 and Remark 4.4. All computer calculations in support of this work were performed using MAGMA [2].

References

1. An, S.Y., Kim, S.Y., Marshall, D.C., Marshall, S.H., McCallum, W.G., Perlis, A.R.: Jacobians of genus one curves. J. Number Theory 90(2), 304–315 (2001)
2. Bosma, W., Cannon, J., Playoust, C.: The Magma algebra system I: The user language. J. Symbolic Comput. 24, 235–265 (1997), http://magma.maths.usyd.edu.au/magma/
3. Cassels, J.W.S.: Lectures on Elliptic Curves. LMS Student Texts, vol. 24. CUP Cambridge (1991)
4. Cassels, J.W.S.: Second descents for elliptic curves. J. reine angew. Math. 494, 101–127 (1998)
5. Cremona, J.E.: Classical invariants and 2-descent on elliptic curves. J. Symbolic Comput. 31, 71–87 (2001)
6. Cremona, J.E., Fisher, T.A.: On the equivalence of binary quartics (submitted)
7. Cremona, J.E., Fisher, T.A., O'Neil, C., Simon, D., Stoll, M.: Explicit n-descent on elliptic curves, I Algebra. J. reine angew. Math. 615, 121–155 (2008), II Geometry, to appear in J. reine angew. Math.; III Algorithms (in preparation)
8. Donnelly, S.: Computing the Cassels-Tate pairing (in preparation)
9. Fisher, T.A.: Testing equivalence of ternary cubics. In: Hess, F., Pauli, S., Pohst, M. (eds.) ANTS 2006. LNCS, vol. 4076, pp. 333–345. Springer, Heidelberg (2006)
10. Fisher, T.A.: The Hessian of a genus one curve (preprint)
11. Fisher, T.A.: Finding rational points on elliptic curves using 6-descent and 12-descent (submitted)
12. Fisher, T.A., Schaefer, E.F., Stoll, M.: The yoga of the Cassels-Tate pairing (submitted)
13. Hilbert, D.: Theory of Algebraic Invariants. CUP, Cambridge (1993)
14. Merriman, J.R., Siksek, S., Smart, N.P.: Explicit 4-descents on an elliptic curve. Acta Arith. 77(4), 385–404 (1996)
15. O'Neil, C.: The period-index obstruction for elliptic curves. J. Number Theory 95(2), 329–339 (2002)
16. Pílniková, J.: Trivializing a central simple algebra of degree 4 over the rational numbers. J. Symbolic Comput. 42(6), 579–586 (2007)
17. Poonen, B., Schaefer, E.F.: Explicit descent for Jacobians of cyclic covers of the projective line. J. reine angew. Math. 488, 141–188 (1997)
18. Siksek, S.: Descent on Curves of Genus 1. PhD thesis, University of Exeter (1995)
19. Simon, D.: Computing the rank of elliptic curves over number fields. LMS J. Comput. Math. 5, 7–17 (2002)
20. Stamminger, S.: Explicit 8-Descent on Elliptic Curves. PhD thesis, International University Bremen (2005)
21. Womack, T.: Explicit Descent on Elliptic Curves. PhD thesis, University of Nottingham (2003)
22. Zarhin, Y.G.: Noncommutative cohomology and Mumford groups. Math. Notes 15, 241–244 (1974)

Computing a Lower Bound for the Canonical Height on Elliptic Curves over Totally Real Number Fields

Thotsaphon Thongjunthug

Mathematics Institute, University of Warwick, Coventry CV4 7AL, UK
T.Thongjunthug@warwick.ac.uk

Abstract. Computing a lower bound for the canonical height is a crucial step in determining a Mordell–Weil basis of an elliptic curve. This paper presents a new algorithm for computing such lower bound, which can be applied to any elliptic curves over totally real number fields. The algorithm is illustrated via some examples.

1 Introduction

Computing a lower bound for the canonical height is a crucial step in determining a set of generators in Mordell–Weil basis (See [7] for full detail). To be precise, the task of explicit computation of Mordell–Weil basis for $E(K)$, where K is a number field, consists of:

1. A 2-descent (or possibly higher m-descent) is used to determine $P_1, \ldots, P_s$, a basis for $E(K)/2E(K)$ (or $E(K)/mE(K)$ respectively).
2. A lower bound $\lambda > 0$ for the canonical height $\hat{h}(P)$ is determined. This together with the geometry of numbers yields an upper bound on the index n of the subgroup of $E(K)$ spanned by $P_1, \ldots, P_s$.
3. A sieving procedure is used to deduce a Mordell–Weil basis for $E(K)$.

In Step 2, we certainly wish to have the index n as small as possible. In particular, $P_1, \ldots, P_s$ will certainly be a Mordell–Weil basis of $E(K)$ if $n < 2$. It then turns out that, in order to have a *smaller* index, we need to have a *larger* value of the lower bound. This can be seen easily from the following theorem.

Theorem 1. *Let E be an elliptic curve over K. Suppose that $E(K)$ contains no points P of infinite order with $\hat{h}(P) \leq \lambda$ for some $\lambda > 0$. Suppose that $P_1, \ldots, P_s$ generate a sublattice of $E(K)/E_{\mathrm{tors}}(K)$ of full rank $s \geq 1$. Then the index n of the span of $P_1, \ldots, P_s$ in such sublattice satisfies*

$$n \leq R(P_1, \ldots, P_s)^{1/2} (\gamma_s / \lambda)^{s/2} \ ,$$

where $R(P_1, \ldots, P_s) = \det(\langle P_i, P_j \rangle)_{1 \leq i,j \leq s}$ and

$$\langle P_i, P_j \rangle = \frac{1}{2} (\hat{h}(P_i + P_j) - \hat{h}(P_i) - \hat{h}(P_j)) \ .$$

A.J. van der Poorten and A. Stein (Eds.): ANTS-VIII 2008, LNCS 5011, pp. 139–152, 2008.

Moreover,

$$\gamma_1^1 = 1, \quad \gamma_2^2 = 4/3, \quad \gamma_3^3 = 2, \quad \gamma_4^4 = 4,$$
$$\gamma_5^5 = 8, \quad \gamma_6^6 = 64/3, \quad \gamma_7^7 = 64, \quad \gamma_8^8 = 2^8,$$

and $\gamma_s = (4/\pi)\Gamma(s/2+1)^{2/s}$ *for* $s \geq 9$.

Proof. See [7, Theorem 3.1]. □

In the past, a number of explicit lower bounds for the canonical height on $E(K)$ have been proposed, including [6, Theorem 0.3]. Although this lower bound has some good properties and is model-independent, it is rather not suitable to computation. For $K = \mathbb{Q}$, there is recently a better lower bound given by Cremona and Siksek [5]. This paper is therefore a generalisation of their work. In particular, we will focus on the case when K is a *totally real* number field.

This work is part of my forthcoming PhD thesis. I wish to thank my supervisor Dr Samir Siksek for all his useful suggestions during the preparation of this paper. I am also indebted to the Development and Promotion of Science and Technology Talent Project (DPST), Ministry of Education of Thailand, for their sponsorship and financial support for my postgraduate study.

1.1 Points of Good Reduction

Suppose K is a totally real number field of degree $r = [K : \mathbb{Q}]$. Let E be an elliptic curve defined over K with discriminant Δ. We define the map

$$\phi : E(K) \rightarrow \prod_{v \in S} E^{(v)}(K_v) ,$$

with $S = \{\infty_1, \ldots, \infty_r\} \cup \{\mathfrak{p} : \mathfrak{p} \mid \Delta\}$, in such a way that P is mapped into its corresponding point on each real embedding $E^1, \ldots, E^r$ (according as the archimedean places $\infty_1, \ldots, \infty_r$ on K) and its corresponding point on each $E^{(v)}$, a *minimal model* of E at a non-archimedean place v. It is well-known that if K has class number greater than 1, E may not have a *globally minimal* model, i.e. $E^{(v)}$ may differ for different v.

Instead of working directly on $E(K)$, the method we use is to determine a lower bound μ for the canonical height of non-torsion points on the subgroup

$$E_{\mathrm{gr}}(K) = \phi^{-1}\left(\prod_{v \in S} E_0^{(v)}(K_v)\right) ,$$

where $E_0^{(v)}(K_v)$ is the connected component of the identity for archimedean v, and the set of points of good reduction for non-archimedean v. In other words, $E_{\mathrm{gr}}(K)$ is the set of points of good reduction on every $E^{(v)}(K_v)$.

Once μ is determined, we can easily deduce the lower bound for the canonical height on the whole $E(K)$: let c be the least common multiple of the Tamagawa indices $c_v = [E^{(v)}(K_v) : E_0^{(v)}(K_v)]$ (including at $v = \infty_1, \ldots, \infty_r$). This is well-defined since $c_v = 1$ for almost all places v. Then the lower bound for the canonical height of all non-torsion points in $E(K)$ is given by $\lambda = \mu/c^2$.

Remark 1. Let v be a non-archimedean place. Suppose E is given by a Weierstrass equation with all coefficients in $\mathcal{O}_v = \{x \in K : \mathrm{ord}_v(x) \geq 0\}$. Let Δ and c_4 be the constants as defined in Section 2. Then E is minimal at v if either $\mathrm{ord}_v(\Delta) < 12$, or $\mathrm{ord}_v(c_4) < 4$.

2 Heights

Throughout this paper, we first define the usual constants of an elliptic curve

$$E: \quad y^2 + a_1 xy + a_3 y = x^3 + a_2 x^2 + a_4 x + a_6 \ ,$$

with $a_1, a_2, a_3, a_4, a_6 \in \mathcal{O}_K$, in the following way (See [8, p.46]):

$$
\begin{aligned}
b_2 &= a_1^2 + 4a_2, \quad b_4 = 2a_4 + a_1 a_3, \\
b_6 &= a_3^2 + 4a_6, \quad b_8 = a_1^2 a_6 + 4a_2 a_6 - a_1 a_3 a_4 + a_2 a_3^2 - a_4^2, \\
c_4 &= b_2^2 - 24 b_4, \quad c_6 = -b_2^3 + 36 b_2 b_4 - 216 b_6, \\
\Delta &= -b_2^2 b_8 - 8 b_4^3 - 27 b_6^2 + 9 b_2 b_4 b_6 \ .
\end{aligned}
$$

Also let

$$f(P) = 4x(P)^3 + b_2 x(P)^2 + 2b_4 x(P) + b_6, \quad g(P) = x(P)^4 - b_4 x(P)^2 - 2b_6 x(P) - b_8,$$

so that $x(2P) = g(P)/f(P)$.

In this paper, we use the definition of local and canonical heights as in [4], which is analogous to the one in Cremona's book [3]. This has the same normalisation as the one implemented in MAGMA package, so that both heights can be compared directly. Note that normalisation of heights varies in literature. In particular, our normalisation is twice the one used in Silverman's paper [9].

Denote M_K the set of all places of K. For $P \in E(K)$, define the *naive height* of P by

$$H_K(P) = \prod_{v \in M_K} \max\{1, |x(P)|_v\}^{n_v} \ ,$$

where $n_v = [K_v : \mathbb{Q}_v]$. Observe that

$$H_K(2P) = \prod_{v \in M_K} \max\{|f(P)|_v, |g(P)|_v\}^{n_v} \ .$$

The archimedean places $\infty_1, \infty_2, \dots, \infty_r$ correspond to the real embeddings $\sigma_1, \sigma_2, \dots, \sigma_r : K \to \mathbb{R}$, while all non-archimedean places are simply all prime ideals $\mathfrak{p}$ in $\mathcal{O}_K$. For $x \in K$ and $v \in M_K$, the *absolute value* of x at v is given by

$$
|x|_v = \begin{cases} |\sigma_j(x)| & \text{if } v = \infty_j \ , \\ \mathcal{N}(\mathfrak{p})^{-\mathrm{ord}_{\mathfrak{p}}(x)/n_{\mathfrak{p}}} & \text{if } v = \mathfrak{p}, \text{ a prime ideal } , \end{cases}
$$

where $\mathcal{N}(\mathfrak{p})$ is the norm of $\mathfrak{p}$. It is verified that this definition satisfies all axioms of valuation theory and the product formula $\prod_{v \in M_K} |x|_v^{n_v} = 1$. From now on, we shall denote $|x|_{\infty_j}$ by $|x|_j$.

The *logarithmic height* of P is then defined by

$$h(P) = \frac{1}{r} \log H_K(P) \ .$$

With these definitions, it can be deduced that

$$h(2P) - 4h(P) = \frac{1}{r} \sum_{v \in M_K} n_v \log \Phi_v(P) \ ,$$

where

$$\Phi_v(P) = \begin{cases} \dfrac{\max\{|f(P)|_v, |g(P)|_v\}}{\max\{1, |x(P)|_v\}^4} & \text{if } P \neq O \ , \\ 1 & \text{if } P = O \ . \end{cases}$$

Using the definition of *canonical height*:

$$\hat{h}(P) = \lim_{n \to \infty} \frac{h(2^n P)}{4^n} \ ,$$

and the telescoping sum trick, we have

$$\hat{h}(P) = h(P) + \left[\frac{h(2P)}{4} - h(P)\right] + \left[\frac{h(2^2 P)}{4^2} - \frac{h(2P)}{4}\right] + \ldots = \frac{1}{r} \sum_{v \in M_K} n_v \lambda_v(P) \ ,$$

where

$$\lambda_v(P) = \log \max\{1, |x(P)|_v\} + \sum_{i=0}^{\infty} \frac{\log \Phi_v(2^i P)}{4^{i+1}} \ . \tag{1}$$

Such function $\lambda_v : E(K_v) \to \mathbb{R}$ is called the *local height* at v. This allows us to obtain $\hat{h}(P)$ by combining the contribution of λ_v on each local model $E(K_v)$.

2.1 The Non-archimedean Local Heights

We shall first consider the properties of λ_v when v is non-archimedean (i.e. $v = \mathfrak{p}$).

For $P \in E(K)$, let $P^{(\mathfrak{p})}$ be its corresponding point (via the map ϕ) on the minimal model $E^{(\mathfrak{p})}$. Let $\lambda_\mathfrak{p}$ be the local height associated to E, and $\lambda_\mathfrak{p}^{(\mathfrak{p})}$ be the local height associated to $E^{(\mathfrak{p})}$. Assume that E is integral and $E^{(\mathfrak{p})}$ has all coefficients in $\mathcal{O}_\mathfrak{p}$, we denote Δ and $\Delta^{(\mathfrak{p})}$ the discriminants of E and $E^{(\mathfrak{p})}$ respectively. These values are related by $\Delta = \left(u^{(\mathfrak{p})}\right)^{12} \Delta^{(\mathfrak{p})}$, for some $u^{(\mathfrak{p})} \in \mathcal{O}_\mathfrak{p}$.

The following lemma illustrates the relation between $\lambda_\mathfrak{p}$ and $\lambda_\mathfrak{p}^{(\mathfrak{p})}$.

Lemma 1

$$\lambda_\mathfrak{p}(P) = \lambda_\mathfrak{p}^{(\mathfrak{p})}(P^{(\mathfrak{p})}) + \frac{1}{6} \log |\Delta/\Delta^{(\mathfrak{p})}|_\mathfrak{p} \ .$$

Proof. See [4, Lemma 4].   □

Now for $P \in E_{gr}(K)$, it follows that $P^{(\mathfrak{p})} \in E_0^{(\mathfrak{p})}(K_\mathfrak{p})$ at every prime ideal $\mathfrak{p}$. In this case, we can easily compute $\lambda_\mathfrak{p}^{(\mathfrak{p})}(P^{(\mathfrak{p})})$ with the following lemma.

Lemma 2. *Let $\mathfrak{p}$ be a prime ideal and $P^{(\mathfrak{p})} \in E_0^{(\mathfrak{p})}(K_\mathfrak{p}) \setminus \{O\}$ (i.e. P is a point of good reduction). Then*

$$\lambda_\mathfrak{p}^{(\mathfrak{p})}(P^{(\mathfrak{p})}) = \log \max\{1, |x(P^{(\mathfrak{p})})|_\mathfrak{p}\} \ .$$

Proof. This is a standard result. See, for example, in [9, Section 5]. □

Note that we may write the principal ideal $\langle x(P^{(\mathfrak{p})})\rangle = AB^{-1}$, where A, B are coprime integral ideals. We call B the *denominator ideal* of $x(P^{(\mathfrak{p})})$, denoted by $\operatorname{denom}(x(P^{(\mathfrak{p})}))$.

The next result is immediate from above lemmas and the definition of $\hat{h}(P)$.

Lemma 3. *Suppose $P \in E_{gr}(K) \setminus \{O\}$. Then*

$$\hat{h}(P) = \frac{1}{r}\left(\sum_{j=1}^r \lambda_{\infty_j}(P) + L(P) - \frac{1}{6}\log\mathcal{N}\left(\prod_\mathfrak{p} \mathfrak{p}^{\operatorname{ord}_\mathfrak{p}(\Delta/\Delta^{(\mathfrak{p})})} \right) \right) ,$$

where

$$L(P) = \log\mathcal{N}\left(\prod_{\mathfrak{p}|\operatorname{denom}(x(P^{(\mathfrak{p})}))} \mathfrak{p}^{-\operatorname{ord}_\mathfrak{p}(x(P^{(\mathfrak{p})}))} \right) .$$

Proof. From the definition of $\hat{h}(P)$, we have

$$\hat{h}(P) = \frac{1}{r} \sum_{v \in M_K} n_v \lambda_v(P) = \frac{1}{r}\left(\sum_{j=1}^r \lambda_{\infty_j}(P) + \sum_\mathfrak{p} n_\mathfrak{p} \lambda_\mathfrak{p}(P) \right) , \qquad (2)$$

where (2) follows after we note that

$$n_{\infty_j} = [K_{\infty_j} : \mathbb{Q}_{\infty_j}] = [\mathbb{R} : \mathbb{R}] = 1, \quad \text{for } j = 1, \ldots, r \ .$$

From Lemma 1, we have

$$\sum_\mathfrak{p} n_\mathfrak{p} \lambda_\mathfrak{p}(P) = \sum_\mathfrak{p} n_\mathfrak{p} \lambda_\mathfrak{p}^{(\mathfrak{p})}(P^{(\mathfrak{p})}) + \frac{1}{6}\sum_\mathfrak{p} n_\mathfrak{p} \log|\Delta/\Delta^{(\mathfrak{p})}|_\mathfrak{p}$$

$$= \sum_\mathfrak{p} n_\mathfrak{p} \log\max\{1, |x(P^{(\mathfrak{p})})|_\mathfrak{p}\} + \frac{1}{6}\sum_\mathfrak{p} n_\mathfrak{p} \log|\Delta/\Delta^{(\mathfrak{p})}|_\mathfrak{p} \ . \quad (3)$$

The last equality follows from Lemma 2, since by assumption $P \in E_{gr}(K)$ (so that $P^{(\mathfrak{p})} \in E_0^{(\mathfrak{p})}(K_\mathfrak{p})$ for all $\mathfrak{p}$). Now recall that

$$|x(P^{(\mathfrak{p})})|_\mathfrak{p} = \mathcal{N}(\mathfrak{p})^{-\operatorname{ord}_\mathfrak{p}(x(P^{(\mathfrak{p})}))/n_\mathfrak{p}} \ .$$

Then for every $\mathfrak{p}$ such that $|x(P^{(\mathfrak{p})})|_\mathfrak{p} \leq 1$, the term $\log\{1, |x(P^{(\mathfrak{p})})|_\mathfrak{p}\}$ will vanish. Thus all $\mathfrak{p}$ that yield a non-zero value to the first sum in (3) are ones such that $|x(P^{(\mathfrak{p})})|_\mathfrak{p} > 1$, i.e. those which divide the denominator ideal of $x(P^{(\mathfrak{p})})$. By definition of absolute value and this fact, the first sum in (3) becomes

$$\sum_{\mathfrak{p}} n_\mathfrak{p} \log\max\{1, |x(P^{(\mathfrak{p})})|_\mathfrak{p}\} = \log\mathcal{N}\left(\prod_{\mathfrak{p}|\mathrm{denom}(x(P^{(\mathfrak{p})}))} \mathfrak{p}^{-\mathrm{ord}_\mathfrak{p}(x(P^{(\mathfrak{p})}))}\right) = L(P).$$

Similarly, the second sum in (3) becomes

$$\frac{1}{6}\sum_{\mathfrak{p}} n_\mathfrak{p} \log|\Delta/\Delta^{(\mathfrak{p})}|_\mathfrak{p} = -\frac{1}{6}\log\mathcal{N}\left(\prod_{\mathfrak{p}} \mathfrak{p}^{\mathrm{ord}_\mathfrak{p}(\Delta/\Delta^{(\mathfrak{p})})}\right).$$

Combining these two equalities with (2) yields the result.   □

2.2   The Archimedean Local Height Difference

We now consider the archimedean local heights λ_v, i.e. when $v = \infty_1, \ldots, \infty_r$. For $j = 1, \ldots, r$, define

$$\alpha_j^{-3} = \inf_{P \in E_0^j(\mathbb{R})} \Phi_{\infty_j}(P).$$

The exponent -3 is introduced to simplify expressions appearing later. These $\alpha_1, \ldots, \alpha_r$ can be easily computed by method given in [7] with some adjustment.

The following lemma follows directly from the definition of local height.

Lemma 4. *If $P \in E_0^j(\mathbb{R}) \setminus \{O\}$, then*

$$\log\max\{1, |x(P)|_j\} - \lambda_{\infty_j}(P) \leq \log\alpha_j.$$

Proof. Rearrange (1) and use the fact that

$$\sum_{i=0}^{\infty} \frac{\log\Phi_{\infty_j}(2^i P)}{4^{i+1}} \geq \sum_{i=0}^{\infty} \frac{\log(\alpha_j^{-3})}{4^{i+1}} = -\log\alpha_j.$$ □

3   Multiplication by n

In this section, we will derive a lower estimate for the contribution that multiplication by n makes towards $\hat{h}(nP)$. This will be useful later in the next section.

Let $k_\mathfrak{p}$ be the residue class field of $\mathfrak{p}$, and $e_\mathfrak{p}$ be the exponent of the group $E_{\mathrm{ns}}^{(\mathfrak{p})}(k_\mathfrak{p}) \cong E_0^{(\mathfrak{p})}(K_\mathfrak{p})/E_1^{(\mathfrak{p})}(K_\mathfrak{p})$. Define

$$D_E(n) = \sum_{\substack{\mathfrak{p}\text{ prime}\\ e_\mathfrak{p}|n}} 2(1 + \mathrm{ord}_{c(\mathfrak{p})}(n/e_\mathfrak{p}))\log\mathcal{N}(\mathfrak{p}),$$

where $c(\mathfrak{p})$ is the characteristic of $k_\mathfrak{p}$. Note that k_p is a finite field, so $c(\mathfrak{p})$ is always a prime number. In particular, $\mathcal{N}(\mathfrak{p}) = |k_\mathfrak{p}| \leq c(\mathfrak{p})^r$.

Proposition 1. *If $e_{\mathfrak{p}} \mid n$, then $\mathcal{N}(\mathfrak{p}) \leq (n+1)^{\max\{2,r\}}$. Hence $D_E(n)$ is finite. Moreover, if P is a non-torsion point in $E_{\mathrm{gr}}(K)$ and $n \geq 1$, then*

$$\hat{h}(nP) \geq \frac{1}{r}\left(\sum_{j=1}^{r} \lambda_{\infty_j}(nP) + D_E(n) - \frac{1}{6}\log\mathcal{N}\left(\prod_{\mathfrak{p}} \mathfrak{p}^{\mathrm{ord}_{\mathfrak{p}}(\Delta/\Delta^{(\mathfrak{p})})}\right)\right) .$$

Proof. Suppose $e_{\mathfrak{p}} \mid n$. If $E^{(\mathfrak{p})}$ has bad reduction at $\mathfrak{p}$, then $e_{\mathfrak{p}}$ is $c(\mathfrak{p})$, $\mathcal{N}(\mathfrak{p}) - 1$, or $\mathcal{N}(\mathfrak{p}) + 1$ depending on whether $E^{(\mathfrak{p})}$ has additive, non-split multiplicative, or split multiplicative reduction at $\mathfrak{p}$. In either case, this implies

$$n \geq e_{\mathfrak{p}} \geq \mathcal{N}(\mathfrak{p})^{1/r} - 1 ,$$

and thus $\mathcal{N}(\mathfrak{p}) \leq (n+1)^r$. Now for $\mathfrak{p}$ at which $E^{(\mathfrak{p})}$ has good reduction, we have

$$E_{\mathrm{ns}}^{(\mathfrak{p})}(k_{\mathfrak{p}}) = E^{(\mathfrak{p})}(k_{\mathfrak{p}}) \cong \mathbb{Z}/d_1\mathbb{Z} \oplus \mathbb{Z}/d_2\mathbb{Z} ,$$

where $d_1 \mid d_2$ and $d_2 = e_{\mathfrak{p}}$. Hence by Hasse's theorem,

$$(\sqrt{\mathcal{N}(\mathfrak{p})} - 1)^2 \leq |E_{\mathrm{ns}}^{(\mathfrak{p})}(k_{\mathfrak{p}})| = d_1 d_2 \leq e_{\mathfrak{p}}^2 \leq n^2 .$$

Thus $\mathcal{N}(\mathfrak{p}) \leq (n+1)^2$. Putting this together yields $\mathcal{N}(\mathfrak{p}) \leq (n+1)^{\max\{2,r\}}$.

The second part follows directly from Lemma 3 once we can show that $L(nP) \geq D_E(n)$. To show this, first note that $P \in E_{\mathrm{gr}}(K)$ implies $P^{(\mathfrak{p})} \in E_0^{(\mathfrak{p})}(K_{\mathfrak{p}})$ for every $\mathfrak{p}$. Define $E_n^{(\mathfrak{p})}(K_{\mathfrak{p}}) = \{P \in E_0^{(\mathfrak{p})}(K_{\mathfrak{p}}) : \mathrm{ord}_{\mathfrak{p}}(x(P)) \leq -2n\}$. Then it is known (see [2, Lemma 7.3.28]) that for all $n \geq 1$,

$$E_n^{(\mathfrak{p})}(K_{\mathfrak{p}})/E_{n+1}^{(\mathfrak{p})}(K_{\mathfrak{p}}) \cong k_{\mathfrak{p}}^+ \cong (\mathbb{Z}/c(\mathfrak{p})\mathbb{Z})^t ,$$

for some $t \in \mathbb{Z}^+$. Let $e(\mathfrak{p}) = \mathrm{ord}_{c(\mathfrak{p})}(n/e_{\mathfrak{p}})$. Then $nP^{(\mathfrak{p})} \in E_{e(\mathfrak{p})+1}^{(\mathfrak{p})}(K_{\mathfrak{p}})$, i.e.

$$\mathrm{ord}_{\mathfrak{p}}(\mathrm{denom}(x(nP^{(\mathfrak{p})}))) \geq 2(e(\mathfrak{p}) + 1) .$$

This implies that $e_{\mathfrak{p}} \mid n$ is equivalent to $\mathfrak{p} \mid \mathrm{denom}(x(nP^{(\mathfrak{p})}))$. Hence

$$\prod_{\mathfrak{p}\mid\mathrm{denom}(x(nP^{(\mathfrak{p})}))} \mathcal{N}(\mathfrak{p})^{-\mathrm{ord}_{\mathfrak{p}}(x(nP^{(\mathfrak{p})}))} \geq \prod_{\substack{\mathfrak{p}\text{ prime}\\ e_{\mathfrak{p}}\mid n}} \mathcal{N}(\mathfrak{p})^{2(e(\mathfrak{p})+1)} .$$

Taking logarithm both sides proves our claim. $\square$

4   A Bound for Multiples of Points of Good Reduction

We now wish to show whether a given $\mu > 0$ satisfies $\hat{h}(P) > \mu$ for all non-torsion $P \in E_{\mathrm{gr}}(K)$. Suppose there exists a non-torsion $P \in E_{\mathrm{gr}}(K)$ with $\hat{h}(P) \leq \mu$. Then for each $E^j(\mathbb{R})$ we will obtain a sequence of inequalities satisfied by the x-coordinates of the multiples nP, for $n = 1, \ldots, k$. With suitable μ and k, the system of inequalities on some $E^j(\mathbb{R})$ may have no solution, which implies $h(P) > \mu$. In this section we will show how to derive such inequalities.

Let α_j and D_E be defined as before. For $\mu > 0$ and $n \in \mathbb{Z}^+$, define

$$B_n(\mu) = \exp\left(rn^2\mu - D_E(n) + \sum_{j=1}^r \log\alpha_j + \frac{1}{6}\log\mathcal{N}\left(\prod_{\mathfrak{p}} \mathfrak{p}^{\mathrm{ord}_\mathfrak{p}(\Delta/\Delta^{(\mathfrak{p})})}\right)\right).$$

Proposition 2. *If $B_n(\mu) < 1$ then $\hat{h}(P) > \mu$ for all non-torsion points on $E_{\mathrm{gr}}(K)$. On the other hand, if $B_n(\mu) \geq 1$ then for all non-torsion points $P \in E_{\mathrm{gr}}(K)$ with $\hat{h}(P) \leq \mu$, we have*

$$|x(nP)|_j \leq B_n(\mu) \ ,$$

for all $j = 1, \ldots, r$.

Proof. Suppose there exists a non-torsion point $P \in E_{\mathrm{gr}}(K)$ with $\hat{h}(P) \leq \mu$. From Lemma 4, we have

$$\log\max\{1, |x(nP)|_j\} - \lambda_{\infty_j}(nP) \leq \log\alpha_j \ ,$$

for all $j = 1, \ldots, r$. This implies that

$$\sum_{j=1}^r \log\max\{1, |x(nP)|_j\} \leq \sum_{j=1}^r \lambda_{\infty_j}(nP) + \sum_{j=1}^r \log\alpha_j \ . \tag{4}$$

By Proposition 1 and our assumption that $\hat{h}(P) \leq \mu$, we have

$$\sum_{j=1}^r \lambda_{\infty_j}(nP) \leq r\hat{h}(nP) - D_E(n) + \frac{1}{6}\log\mathcal{N}\left(\prod_{\mathfrak{p}} \mathfrak{p}^{\mathrm{ord}_\mathfrak{p}(\Delta/\Delta^{(\mathfrak{p})})}\right)$$

$$\leq rn^2\mu - D_E(n) + \frac{1}{6}\log\mathcal{N}\left(\prod_{\mathfrak{p}} \mathfrak{p}^{\mathrm{ord}_\mathfrak{p}(\Delta/\Delta^{(\mathfrak{p})})}\right) \ . \tag{5}$$

Combining (4) and (5) and taking exponential, we obtain

$$\prod_{j=1}^r \max\{1, |x(nP)|_j\} \leq B_n(\mu) \ .$$

Clearly the left-hand side of this inequality is at least 1. Thus, if $B_n(\mu) < 1$ we simply obtain a contradiction, i.e. $\hat{h}(P) > \mu$ for every non-torsion $P \in E_{\mathrm{gr}}(K)$.

On the other hand, by considering all different cases of $|x(nP)|_j$, it is easy to see that every case implies that $|x(nP)|_j \leq B_n(\mu)$ for all $j = 1, \ldots, r$. □

Corollary 1. *Let $\mathfrak{q}$ be a prime ideal such that*

$$\mathcal{N}(\mathfrak{q}) > \prod_{j=1}^r \alpha_j^{1/2} \cdot \mathcal{N}\left(\prod_{\mathfrak{p}} \mathfrak{p}^{\mathrm{ord}_\mathfrak{p}(\Delta/\Delta^{(\mathfrak{p})})}\right)^{1/12} \ , \tag{6}$$

and set $n = e_{\mathfrak{q}}$ and

$$\mu_0 = \frac{1}{rn^2}\left(D_E(n) - \sum_{j=1}^{r}\log\alpha_j - \frac{1}{6}\log\mathcal{N}\left(\prod_{\mathfrak{p}}\mathfrak{p}^{\mathrm{ord}_{\mathfrak{p}}(\Delta/\Delta^{(\mathfrak{p})})}\right)\right) .$$

Then $\mu_0 > 0$, and in particular, $\hat{h}(P) \geq \mu_0$ for all non-torsion point $P \in E_{\mathrm{gr}}(K)$.

Proof. Suppose $\mathfrak{q}$ is a prime ideal satisfying (6). By definition of $D_E(n)$, we have

$$D_E(n) \geq 2\log\mathcal{N}(\mathfrak{q}) > \sum_{j=1}^{r}\log\alpha_j + \frac{1}{6}\log\mathcal{N}\left(\prod_{\mathfrak{p}}\mathfrak{p}^{\mathrm{ord}_{\mathfrak{p}}(\Delta/\Delta^{(\mathfrak{p})})}\right) ,$$

which implies that $\mu_0 > 0$. Then for any $\mu < \mu_0$, we have

$$rn^2\mu - D_E(n) + \sum_{j=1}^{r}\log\alpha_j + \frac{1}{6}\log\mathcal{N}\left(\prod_{\mathfrak{p}}\mathfrak{p}^{\mathrm{ord}_{\mathfrak{p}}(\Delta/\Delta^{(\mathfrak{p})})}\right)$$

$$< rn^2\mu_0 - D_E(n) + \sum_{j=1}^{r}\log\alpha_j + \frac{1}{6}\log\mathcal{N}\left(\prod_{\mathfrak{p}}\mathfrak{p}^{\mathrm{ord}_{\mathfrak{p}}(\Delta/\Delta^{(\mathfrak{p})})}\right) = 0 ,$$

and thus $B_n(\mu) < 1$. Hence $\hat{h}(P) > \mu$ for all non-torsion point $P \in E_{\mathrm{gr}}(K)$ by Proposition 2. Since this is true for all $\mu < \mu_0$, then $\hat{h}(P) \geq \mu_0$ as required. □

It is possible to derive a lower bound for any points on $E_{\mathrm{gr}}(K)$ by Corollary 1 alone. However, our practical experience shows that the bound derived from this corollary itself is not as good as the bound obtained by collecting more information on $x(nP)$. This claim will be illustrated later in our examples.

5   Solving Inequalities Involving the Multiples of Points

From Proposition 2, we know that every non-torsion point $P \in E_{\mathrm{gr}}(K)$ with $\hat{h}(P) \leq \mu$ must satisfy $|x(nP)|_j \leq B_n(\mu)$ for all $j = 1, \ldots, r$. This means that we need to consider r elliptic curves over $\mathbb{R}$, say

$$E^j : \quad y^2 + \sigma_j(a_1)xy + \sigma_j(a_3)y = x^3 + \sigma_j(a_2)x^2 + \sigma_j(a_4)x + \sigma_j(a_6) ,$$

for $j = 1, \ldots, r$. In other words, we need to consider $\sigma_j(nP)$ over $E_0^j(\mathbb{R})$. To prove that $\hat{h}(P) > \mu$ for all non-torsion $P \in E_{\mathrm{gr}}(K)$, we shall derive a contradiction from these inequalities using an application of *elliptic logarithm*.

5.1   Elliptic Logarithm

An elliptic logarithm is an isomorphism $\varphi : E_0(\mathbb{R}) \to \mathbb{R}/\mathbb{Z} \cong [0, 1)$. This can be rapidly computed by method of arithmetic-geometric means. In our program, we use the algorithm in Cohen's book [1, Algorithm 7.4.8] for this computation.

We wish to apply elliptic logarithm to solving our inequalities on these r real embeddings. For $j = 1, \ldots, r$, let

$$\text{On } E^j: \quad f_j(x) = 4x^3 + \sigma_j(b_2)x^2 + 2\sigma_j(b_4)x + \sigma_j(b_6) \ .$$

Note that we can rewrite the Weierstrass equation of E^j as

$$f_j(x) = (2y + \sigma_j(a_1)x + \sigma_j(a_3))^2 \ .$$

Denote β_j the largest real root of f_j. On each E^j, we define the corresponding elliptic logarithm φ_j as follows: let

$$\Omega_j = 2 \int_{\beta_j}^{\infty} \frac{dx}{\sqrt{f_j(x)}} \ .$$

Then for a point $P = (\xi, \eta) \in E_0^j(\mathbb{R})$ with $2\eta + \sigma_j(a_1)\xi + \sigma_j(a_3) \geq 0$, we let

$$\varphi_j(P) = \frac{1}{\Omega_j} \int_{\xi}^{\infty} \frac{dx}{\sqrt{f_j(x)}} \ ,$$

otherwise, let $\varphi_j(P) = 1 - \varphi_j(-P)$.

Suppose that ξ is a real number satisfying $\xi \geq \beta_j$. Then there exists η such that $2\eta + \sigma_j(a_1)\xi + \sigma_j(a_3) \geq 0$ and $(\xi, \eta) \in E_0^j(\mathbb{R})$. Define

$$\psi_j(\xi) = \varphi_j((\xi, \eta)) \in [1/2, 1) \ .$$

In words, $\psi_j(\xi)$ is the elliptic logarithm of the "higher" of the two points on $E_0^j(\mathbb{R})$ with x-coordinate ξ.

For real ξ_1, ξ_2 with $\xi_1 < \xi_2$, we define the subset $\mathcal{S}^j \subset [0, 1)$ as follows:

$$\mathcal{S}^j(\xi_1, \xi_2) = \begin{cases} \emptyset & \text{if } \xi_2 < \beta_j \ , \\ [1 - \psi_j(\xi_2), \psi_j(\xi_2)] & \text{if } \xi_1 < \beta_j \leq \xi_2 \ , \\ [1 - \psi_j(\xi_2), 1 - \psi_j(\xi_1)] \cup [\psi_j(\xi_1), \psi_j(\xi_2)] & \text{if } \xi_1 \geq \beta_j \ . \end{cases}$$

The following lemma is clear.

Lemma 5. *Suppose $\xi_1 < \xi_2$ are real numbers. Then $P \in E_0^j(\mathbb{R})$ satisfies $\xi_1 \leq x(P) \leq \xi_2$ if and only if $\varphi_j(P) \in \mathcal{S}^j(\xi_1, \xi_2)$.*

If $\bigcup[a_i, b_i]$ is a disjoint union of intervals and $t \in \mathbb{R}$, we define

$$t + \bigcup[a_i, b_i] = \bigcup[a_i + t, b_i + t], \quad t\bigcup[a_i, b_i] = \bigcup[ta_i, tb_i] \quad (\text{for } t > 0) \ .$$

Proposition 3. *Suppose $\xi_1 < \xi_2$ are real numbers, and $n > 0$ is an integer. Let*

$$\mathcal{S}_n^j(\xi_1, \xi_2) = \bigcup_{t=0}^{n-1} \left(\frac{t}{n} + \frac{1}{n}\mathcal{S}^j(\xi_1, \xi_2) \right) \ .$$

Then $P \in E_0^j(\mathbb{R})$ satisfies $\xi_1 \leq x(nP) \leq \xi_2$ if and only $\varphi_j(P) \in \mathcal{S}_n^j(\xi_1, \xi_2)$.

Proof. By Lemma 5, $P \in E_0^j(\mathbb{R})$ satisfies $\xi_1 \leq x(P) \leq \xi_2$ if and only if $\varphi_j(P) \in \mathcal{S}^j(\xi_1, \xi_2)$. Denote the multiplication-by-n map on $\mathbb{R}/\mathbb{Z}$ by ν_n. If $\delta \in [0, 1)$, then

$$\nu_n^{-1}(\delta) = \left\{ \frac{t}{n} + \frac{\delta}{n} : t = 0, 1, 2, \ldots, n-1 \right\} .$$

But since φ_j is an isomorphism, we have $\varphi_j(nP) = n\varphi_j(P)$ (mod 1). Hence

$$\varphi_j(nP) \in \mathcal{S}^j(\xi_1, \xi_2) \Longleftrightarrow \varphi_j(P) \in \nu_n^{-1}(\mathcal{S}^j(\xi_1, \xi_2)) = \mathcal{S}_n^j(\xi_1, \xi_2) . \qquad \square$$

6   The Algorithm

Combining all results we have so far, we obtain our main theorem.

Theorem 2. *Given $\mu > 0$. If $B_n(\mu) < 1$ for some $n \in \mathbb{Z}^+$, then $\hat{h}(P) > \mu$ for every non-torsion point $P \in E_{\mathrm{gr}}(K)$. Otherwise, if $B_n(\mu) \geq 1$ for $n = 1, \ldots, k$, then every non-torsion point $P \in E_{\mathrm{gr}}(K)$ such that $\hat{h}(P) \leq \mu$ satisfies*

$$\varphi_j(\sigma_j P) \in \bigcap_{n=1}^{k} \mathcal{S}_n^j(-B_n(\mu), B_n(\mu)) ,$$

for all $j = 1, \ldots, r$. In particular, if one of above r intersections is empty, then $\hat{h}(P) > \mu$ for all non-torsion $P \in E_{\mathrm{gr}}(K)$.

To use the algorithm, first we give an initial lower bound μ and the number of steps k. In practice, we find that the initial choice of $\mu = 1$ and $k = 5$ is useful.

We start by computing $B_n(\mu)$ for $n = 1, \ldots, k$. If $B_n(\mu) < 1$ for some n, then we deduce that $\hat{h}(P) > \mu$ for every non-torsion $P \in E_{\mathrm{gr}}(K)$. Otherwise, we compute $\bigcap_{n=1}^{k} \mathcal{S}_n^j(-B_n(\mu), B_n(\mu))$ for $j = 1, \ldots, r$. If the intersection is empty for some j, then again $\hat{h}(P) > \mu$ for every non-torsion $P \in E_{\mathrm{gr}}(K)$. However, if none of r intersections is empty, we fail to show that μ is a lower bound.

We can refine μ further until a sufficient accuracy is achieved: if μ is shown to be a lower bound, we increase μ by some factor, say, 1.1. Otherwise, we decrease μ and increase k, say, by multiplying μ by 0.9 and increasing k by 1. Then we repeat the above with new μ (and possibly new k).

Finally, we return the last value of μ which is known to be a lower bound.

7   Remark

Unlike [6], our lower bound is not model-independent. For example, the values α_j defined in Section 2.2 depend on b_2, b_4, b_6, and b_8. Thus we may obtain different values of lower bound if we work with different models of E. At this point, we are however not to decide which model of E maximises the lower bound. Moreover, our formulae can be simplified if E is a globally minimal model. Note that this may not be the case if E is defined over a field K of class number at least 2.

8   Examples

We have implemented our algorithm in MAGMA to illustrate some examples.

Example 1. Consider the elliptic curve E over $K = \mathbb{Q}(\sqrt{2})$ given by

$$E: \quad y^2 = f(x) = x^3 + x + (1 + 2\sqrt{2}) .$$

The discriminant Δ of E is $-3952 - 1728\sqrt{2}$. Moreover, $\langle \Delta \rangle = \mathfrak{p}_1^8 \mathfrak{p}_2^2 \mathfrak{p}_3$, where

$$\mathfrak{p}_1 = \langle \sqrt{2} \rangle, \quad \mathfrak{p}_2 = \langle 7, 3 + \sqrt{2} \rangle, \quad \mathfrak{p}_3 = \langle 769, 636 + \sqrt{2} \rangle .$$

Hence by Remark 1, E is minimal at every prime ideal, and thus it is globally minimal. Our program shows that for any non-torsion point $P \in E_{\mathrm{gr}}(K)$,

$$\hat{h}(P) > 0.2415 .$$

This is obtained after a number of refinements as shown in Table 1.

Table 1. Illustration of algorithm for Example 1

Initial μ	Initial k	Is any $B_n(\mu) < 1$?	Is any intersection empty?	Is μ a lower bound?	Next μ	Next k
1.0000	5	No	No	Fail	0.5000	6
0.5000	6	No	No	Fail	0.2500	7
0.2500	7	No	No	Fail	0.1250	8
0.1250	8	Yes	Skipped	Yes	0.1875	8
0.1875	8	No	Yes	Yes	0.2187	8
0.2187	8	No	Yes	Yes	0.2343	8
0.2343	8	No	Yes	Yes	0.2421	8
0.2421	8	No	No	Fail	0.2382	9
0.2382	9	No	Yes	Yes	0.2402	9
0.2402	9	No	Yes	Yes	0.2412	9
0.2412	9	No	Yes	Yes	0.2416	9
0.2416	9	No	No	Fail	0.2414	10
0.2414	10	No	Yes	Yes	0.2415	10
0.2415	10	No	No	Fail	0.2415	11
0.2415	11	No	Yes	Yes		

On the other hand, the lower bound for $E_{\mathrm{gr}}(K)$ derived from Corollary 1 is not as good as this one. In this example, we have

$$\alpha_1 = 1.096562, \quad \alpha_2 = 1.001830 ,$$

which gives $\alpha_1 \alpha_2 = 1.098569$ We now choose a prime ideal $\mathfrak{p}$ whose norm is greater than $\sqrt{\alpha_1 \alpha_2}$, and set $n = e_{\mathfrak{p}}$. To minimise n, we choose $\mathfrak{p} = \langle \sqrt{2} \rangle$ to get $n = e_{\mathfrak{p}} = 2$. Then we have $D_E(2) = 1.386294$ and finally

$$\mu_0 = (1.386294 - \log(1.098569))/8 = 0.1615 .$$

The Tamagawa indices at $\mathfrak{p}_1, \mathfrak{p}_2, \mathfrak{p}_3$ are 4, 2, and 1 respectively. Moreover, since $\sigma_1(f)$ and $\sigma_2(f)$ both have one real root, we have $c_{\infty_1} = c_{\infty_2} = 1$. Hence $c = 4$, and thus for any non-torsion point $P \in E(K)$,

$$\hat{h}(P) > 0.2415/16 = 0.0150 \ .$$

It can be checked that the torsion subgroup of $E(K)$ is trivial, and the point $P = (1, 1 + \sqrt{2}) \in E(K)$. Using MAGMA, we know that $\hat{h}(P) = 0.5033$, and the rank of $E(K)$ is at most 1. Hence $E(K)$ has rank 1. By Theorem 1, we obtain

$$n = [E(K) : \langle P \rangle] \leq \sqrt{0.5033/0.0150} = 5.7739 \ .$$

Example 2. Consider the elliptic curve E over $K = \mathbb{Q}(\sqrt{7})$ defined by

$$E: \quad y^2 + (3 + 3\sqrt{7})xy + y = f(x) = x^3 + (26 + 4\sqrt{7})x^2 + x \ .$$

The discriminant Δ of E is $-937513 - 299394\sqrt{7}$. Moreover, $\langle \Delta \rangle = \mathfrak{p}_1 \mathfrak{p}_2 \mathfrak{p}_3$, where

$$\mathfrak{p}_1 = \langle 4219, 1083 + \sqrt{7} \rangle, \quad \mathfrak{p}_2 = \langle 4657, 3544 + \sqrt{7} \rangle, \quad \mathfrak{p}_3 = \langle 12799, 5358 + \sqrt{7} \rangle \ .$$

Hence by Remark 1, E is minimal at every prime ideal $\mathfrak{p}$, so it is a globally minimal model. Our program shows that for any non-torsion point $P \in E_{\mathrm{gr}}(K)$,

$$\hat{h}(P) > 0.1415 \ .$$

The Tamagawa indices at $\mathfrak{p}_1, \mathfrak{p}_2, \mathfrak{p}_3$ are all 1. Also $c_{\infty_1} = c_{\infty_2} = 2$ since both $\sigma_1(f)$ and $\sigma_2(f)$ have 3 real roots. Hence $c = 2$. Then for any non-torsion $P \in E(K)$, we have
$$\hat{h}(P) > 0.1415/4 = 0.0353 \ .$$

In this example, the torsion subgroup of $E(K)$ is trivial. Let $P_1 = (0, 0)$ and $P_2 = (1, \sqrt{7})$. It can be verified that both points are on $E(K)$, and

$$\hat{h}(P_1) = 0.8051, \quad \hat{h}(P_2) = 1.4957 \ .$$

Hence by computing the height pairing matrix, we have

$$R(P_1, P_2) = \det \begin{pmatrix} \langle P_1, P_1 \rangle & \langle P_1, P_2 \rangle \\ \langle P_2, P_1 \rangle & \langle P_2, P_2 \rangle \end{pmatrix} = \begin{vmatrix} 0.8051 & -0.1941 \\ -0.1941 & 1.4957 \end{vmatrix} = 1.1665 \neq 0 \ .$$

Therefore P_1 and P_2 are independent. From MAGMA, we know that the rank of $E(K)$ is at most 2. Hence $E(K)$ has rank 2. By Theorem 1, we finally obtain

$$n = [E(K) : \langle P_1, P_2 \rangle] \leq \frac{(\sqrt{1.1665})(2/\sqrt{3})}{0.0353} = 35.2450 \ .$$

Example 3. Let E be the elliptic curve over $K = \mathbb{Q}(\sqrt{10})$ given by

$$E: \quad y^2 = f(x) = x^3 + 125 \ .$$

Note that K has class number 2. By decomposing the discriminant Δ of E, it can be seen that $\langle \Delta \rangle = \langle -2^4 3^3 5^6 \rangle = \mathfrak{p}_1^{12} \mathfrak{p}_2^3 \mathfrak{p}_3^3 \mathfrak{p}_4^8$, where

$$\mathfrak{p}_1 = \langle 5, \sqrt{10} \rangle, \quad \mathfrak{p}_2 = \langle 3, 4 + \sqrt{10} \rangle, \quad \mathfrak{p}_3 = \langle 3, 2 + \sqrt{10} \rangle, \quad \mathfrak{p}_4 = \langle 2, \sqrt{10} \rangle .$$

By calculating the constant c_4 of E, we have $c_4 = 0$ and so $\mathrm{ord}_\mathfrak{p}(c_4) = \infty \not< 4$. Hence by Remark 1, E is minimal everywhere except at $\mathfrak{p}_1$. By substituting

$$x = (\sqrt{10})^2 x', \quad y = (\sqrt{10})^3 y',$$

we have a new elliptic curve $E' : y'^2 = x'^3 + 1/8$. Now E' is minimal at $\mathfrak{p}_1$ and elsewhere, except at all prime ideals dividing 2. Thus we let $E^{(\mathfrak{p}_1)} = E'$ and $E^{(\mathfrak{p})} = E$ for any $\mathfrak{p} \neq \mathfrak{p}_1$ in our computation. Our program shows that

$$\hat{h}(P) > 0.2859 ,$$

for every non-torsion $P \in E_{\mathrm{gr}}(K)$.

The Tamagawa indices at $\mathfrak{p}_1, \mathfrak{p}_2, \mathfrak{p}_3, \mathfrak{p}_4$ are 1, 2, 2, and 1 respectively. Moreover, $\sigma_1(f)$ and $\sigma_2(f)$ both have only one real root, so $c_{\infty_1} = c_{\infty_2} = 1$. Thus $c = 2$, and hence for any non-torsion point $P \in E(K)$, we have

$$\hat{h}(P) > 0.2859/(2^2) = 0.0714 .$$

It can be checked that the point $P = (5, 5\sqrt{10}) \in E(K)$. From MAGMA, we know that $\hat{h}(P) = 0.6532$, and the rank of $E(K)$ is at most 1. Hence $E(K)$ must have rank 1. Finally by Theorem 1, we have

$$n = [E(K) : \langle P \rangle] \leq \sqrt{0.6532/0.0714} = 3.0229 .$$

References

1. Cohen, H.: A Course in Computational Algebraic Number Theory, Graduate Texts in Mathematics, vol. 138. Springer, Heidelberg (1993)
2. Cohen, H.: Number Theory. vol. 1: tools and Diophantine equations. Graduate Texts in Mathematics, vol. 239. Springer, Heidelberg (2007)
3. Cremona, J.E.: Algorithms for modular elliptic curves, 2nd edn. Cambridge University Press, Cambridge (1997)
4. Cremona, J.E., Prickett, M., Siksek, S.: Height difference bounds for elliptic curves over number fields. J. Number Theory 116, 42–68 (2006)
5. Cremona, J., Siksek, S.: Computing a lower bound for the canonical height on elliptic curves over Q. In: Hess, F., Pauli, S., Pohst, M. (eds.) ANTS 2006. LNCS, vol. 4076, pp. 275–286. Springer, Heidelberg (2006)
6. Hindry, M., Silverman, J.H.: The canonical height and integral points on elliptic curves. Invent. Math. 93, 419–450 (1988)
7. Siksek, S.: Infinite descent on elliptic curves. Rocky Mountain J. Math. 25, 1501–1538 (1995)
8. Silverman, J.H.: The arithmetic of elliptic curves. Graduate Texts in Mathematics, vol. 106. Springer, Heidelberg (1986)
9. Silverman, J.H.: Computing heights on elliptic curves. Math. Comp. 51, 339–358 (1988)

Faster Multiplication in $GF(2)[x]$

Richard P. Brent[1], Pierrick Gaudry[2],
Emmanuel Thomé[3], and Paul Zimmermann[3]

[1] Australian National University, Canberra, Australia
[2] LORIA/CNRS, Vandœuvre-lès-Nancy, France
[3] INRIA Nancy - Grand Est, Villers-lès-Nancy, France

Abstract. In this paper, we discuss an implementation of various algorithms for multiplying polynomials in $GF(2)[x]$: variants of the window methods, Karatsuba's, Toom-Cook's, Schönhage's and Cantor's algorithms. For most of them, we propose improvements that lead to practical speedups.

Introduction

The arithmetic of polynomials over a finite field plays a central role in algorithmic number theory. In particular, the multiplication of polynomials over $GF(2)$ has received much attention in the literature, both in hardware and in software. It is indeed a key operation for cryptographic applications [22], for polynomial factorisation or irreducibility tests [8,3]. Some applications are less known, for example in integer factorisation, where multiplication in $GF(2)[x]$ can speed up Berlekamp-Massey's algorithm inside the (block) Wiedemann algorithm [20,1].

We focus here on the classical dense representation — called "binary polynomial" — where a polynomial of degree $n-1$ is represented by the bit-sequence of its n coefficients. We also focus on software implementations, using classical instructions provided by modern processors, for example in the C language.

Several authors already made significant contributions to this subject. Apart from the classical $O(n^2)$ algorithm, and Karatsuba's algorithm which readily extends to $GF(2)[x]$, Schönhage in 1977 and Cantor in 1989 proposed algorithms of complexity $O(n \log n \log \log n)$ and $O(n(\log n)^{1.5849\cdots})$ respectively [18,4]. In [16], Montgomery invented Karatsuba-like formulæ splitting the inputs into more than two parts; the key feature of those formulæ is that they involve no division, thus work over any field. More recently, Bodrato [2] proposed good schemes for Toom-Cook 3, 4, and 5, which are useful cases of the Toom-Cook class of algorithms [7,21]. A detailed bibliography on multiplication and factorisation in $GF(2)[x]$ can be found in [9].

Discussions on implementation issues are found in some textbooks such as [6,12]. On the software side, von zur Gathen and Gerhard [9] designed a software tool called BiPolAr, and managed to factor polynomials of degree up to $1\,000\,000$, but BiPolAr no longer seems to exist. The reference implementation for the last decade is the NTL library designed by Victor Shoup [19].

A.J. van der Poorten and A. Stein (Eds.): ANTS-VIII 2008, LNCS 5011, pp. 153–166, 2008.

The contributions of this paper are the following: (a) the "double-table" algorithm for the word-by-word multiplication and its extension to two words using the SSE-2 instruction set (§1); (b) the "word-aligned" variants of the Toom-Cook algorithm (§2); (c) a new view of Cantor's algorithm, showing in particular that a larger base field can be used, together with a truncated variant avoiding the "staircase effect" (§3.1); (d) a variant of Schönhage's algorithm (§3.2) and a splitting technique to improve it (§3.3); (e) finally a detailed comparison of our implementation with previous literature and current software (§4).

Notation: w denotes the machine word size (usually $w = 32$ or 64), and we consider polynomials in $GF(2)[x]$. A polynomial of degree less than d is represented by a sequence of d bits, which are stored in $\lceil d/w \rceil$ consecutive words.

The code that we developed for this paper, and for the paper [3], is contained in the GF2X package, available under the GNU General Public License from http://wwwmaths.anu.edu.au/~brent/gf2x.html.

1   The Base Case (Small Degree)

We first focus on the "base case", that is, routines that multiply full words (32, 64 or 128 bits). Such routines eventually act as building blocks for algorithms dealing with larger degrees. Since modern processors do not provide suitable hardware primitives, one has to implement them in software.

Note that the treatment of "small degree" in general has also to deal with sizes which are not multiples of the machine word size: what is the best strategy to multiply, e.g., 140-bit polynomials? This case is not handled here.

1.1   Word by Word Multiplication (mul1)

Multiplication of two polynomials $a(x)$ and $b(x)$ of degree at most $w - 1$ can be performed efficiently with a "window" method, similar to base-2^s exponentiation, where the constant s denotes the window size. This algorithm was found in version 5.1a of NTL, which used $s = 2$, and is here generalized to any value of s:

1. Store in a table the multiples of b by all 2^s polynomials of degree $< s$.
2. Scan bits of a, s at a time. The corresponding table data are shifted and accumulated in the result.

Note that Step 1 discards the high coefficients of $b(x)$, which is of course undesired[1], if $b(x)$ has degree $w - 1$. The computation must eventually be "repaired" with additional steps which are performed at the end.

The "repair step" (Step 3) exploits the following observation. Whenever bit $w - j$ of b is set (where $0 < j < s$), then bits at position j' of a, where j' mod $s \geq j$, contribute to a missing bit at position $w + j' - j$ in the product. Therefore only the high result word has to be fixed. Moreover, for each j, $0 < j < s$,

---

[1] The multiples of b are stored in one word, i.e., modulo 2^w; alternatively, one could store them in two words, but that would be much slower.

the fixing can be performed by an exclusive-or involving values easily derived from a: selecting bits at indices j' with $j' \bmod s \geq j$ can be done inductively by successive shifts of a, masked with an appropriate value.

```
mul1(ulong a, ulong b)
```
multiplies polynomials a *and* b. *The result goes in* l *(low part) and* h *(high part).*
```
        ulong u[2^s] = { 0, b, 0, ... };              /* Step 1 (tabulate) */
        for(int i = 2 ; i < 2^s ; i += 2)
            u[i] = u[i >> 1] << 1; u[i + 1] = u[i] ^ b;
        ulong g = u[a & (2^s - 1)], l = g, h = 0;      /* Step 2 (multiply) */
        for(int i = s ; i < w ; i += s)
            g = u[a >> i & (2^s - 1)]; l ^= g << i; h ^= g >> (w - i);
        ulong m = (2^s - 2) × (1 + 2^s + 2^{2s} + 2^{3s} + ···) mod 2^w; /* Step 3 (repair) */
        for(int j = 1 ; j < s ; j++)
            a = (a << 1) & m;
            if (bit w - j of b is set) h ^= a;
        return l, h;
```

Fig. 1. Word-by-word multiplication with repair steps

The pseudo-code in Fig. 1 illustrates the word-by-word multiplication algorithm (in practice s and w will be fixed for a given processor, thus the for-loops will be replaced by sequences of instructions). There are many alternatives for organizing the operations. For example, Step 1 can also be performed with a Gray code walk. In Step 2, the bits of a may be scanned either from right to left, or in reverse order. For an efficient implementation, the `if` statement within Step 3 should be replaced by a masking operation to avoid branching[2]:

$$h \text{ ^= } a \text{ \& } -(((\text{long}) \text{ (b } << \text{ (j-1))}) < 0);$$

A non trivial improvement of the repair steps comes from the observation that Steps 2 and 3 of Fig. 1 operate associatively on the result registers l and h. The two steps can therefore be swapped. Going further, Step 1 and the repair steps are independent. Interleaving of the code lines is therefore possible and has actually been found to yield a small speed improvement. The GF2X package includes an example of such an interleaved code.

The double-table algorithm. In the `mul1` algorithm above, the choice of the window size s is subject to some trade-off. Step 1 should not be expanded unreasonably, since it costs 2^s, both in code size and memory footprint. It is possible, without modifying Step 1, to operate as if the window size were $2s$ instead of s. Within Step 2, replace the computation of the temporary variable g by:

---

[2] In the C language, the expression $(x < 0)$ is translated into the `setb` x86 assembly instruction, or some similar instruction on other architectures, which does not perform any branching.

$$g = \texttt{u[a >> i \& } (2^s - 1)\texttt{]} \; \texttt{\^{}} \; \texttt{u[a >> (i+s) \& } (2^s - 1)\texttt{]} \texttt{ << s}$$

so that the table is used twice to extract $2s$ bits (the index i thus increases by $2s$ at each loop). Step 1 is faster, but Step 2 is noticeably more expensive than if a window size of $2s$ were effectively used.

A more meaningful comparison can be made with window size s: there is no difference in Step 1. A detailed operation count for Step 2, counting loads as well as bitwise operations &, ^, <<, and >> yields 7 operations for every s bits of inputs for the code of Fig. 1, compared to 12 operations for every $2s$ bits of input for the "double-table" variant. A tiny improvement of 2 operations for every $2s$ bits of input is thus obtained. On the other hand, the "double-table" variant has more expensive repair steps. It is therefore reasonable to expect that this variant is worthwhile only when s is small, which is what has been observed experimentally (an example cut-off value being $s = 4$).

1.2   Extending to a `mul2` Algorithm

Modern processors can operate on wider types, for instance 128-bit registers are accessible with the SSE-2 instruction set on the Pentium 4 and Athlon 64 CPUs. However, not all operations are possible on these wide types. In particular, arithmetic shifts by arbitrary values are not supported on the full 128-bit registers with SSE-2. This precludes a direct adaptation of our `mul1` routine to a `mul2` routine (at least with the SSE-2 instruction set). We discuss here how to work around this difficulty in order to provide an efficient `mul2` routine.

To start with, the algorithm above can be extended in a straightforward way so as to perform a $k \times 1$ multiplication (k words by one word). Step 1 is unaffected by this change, since it depends only on the second operand. In particular, a 2×1 multiplication can be obtained in this manner.

Following this, a 2×2 `mul2` multiplication is no more than two 2×1 multiplications, where only the second operand changes. In other words, those two multiplications can be performed in a "single-instruction, multiple-data" (SIMD) manner, which corresponds well to the spirit of the instruction set extensions introducing wider types. In practice, a 128-bit wide data type is regarded as a vector containing two 64-bit machine words. Two 2×1 multiplications are performed in parallel using an exact translation of the code in Fig. 1. The choice of splitting the wide register into two parts is fortunate in that all the required instructions are supported by the SSE-2 instruction set.

1.3   Larger Base Case

To multiply two binary polynomials of n words for small n, it makes sense to write some special code for each value of n, as in the NTL library, which contains hard-coded Karatsuba routines for $2 \leq n \leq 8$ [19,22]. We wrote such hard-coded routines for $3 \leq n \leq 9$, based on the above `mul1` and `mul2` routines.

2   Medium Degree

For medium degrees, a generic implementation of Karatsuba's or Toom-Cook's algorithm has to be used. By "generic" we mean that the number n of words of the input polynomials is an argument of the corresponding routine. This section shows how to use Toom-Cook without any extension field, then discusses the word-aligned variant, and concludes with the unbalanced variant.

2.1   Toom-Cook without Extension Field

A common misbelief is that Toom-Cook's algorithm cannot be used to multiply binary polynomials, because Toom-Cook 3 (TC3) requires 5 evaluation points, and we have only 3, with both elements of GF(2) and ∞. In fact, any power of the transcendental variable x can be used as evaluation point. For example TC3 can use $0, 1, \infty, x, x^{-1}$. This was discovered by Michel Quercia and the last author a few years ago, and implemented in the `irred-ntl` patch for NTL [23]. This idea was then generalized by Bodrato [2] to any polynomial in x; in particular Bodrato shows it is preferable to choose $0, 1, \infty, x, 1 + x$ for TC3.

A small drawback of using polynomials in x as evaluation points is that the degrees of the recursive calls increase slightly. For example, with points $0, 1, \infty, x, x^{-1}$ to multiply two polynomials of degree less than $3n$ by TC3, the evaluations at x and x^{-1} might have up to $n + 2$ non-zero coefficients. In any case, this will increase the size of the recursive calls by at most one word.

For Toom-Cook 3-way, we use Bodrato's code; and for Toom-Cook 4-way, we use a code originally written by Marco Bodrato, which we helped to debug[3].

2.2   Word-Aligned Variants

In the classical Toom-Cook setting over the integers, one usually chooses $0, 1, 2, 1/2, \infty$ for TC3. The word-aligned variant uses $0, 1, 2^w, 2^{-w}, \infty$, where w is the word-size in bits. This idea was used by Michel Quercia in his `numerix` library[4], and was independently rediscovered by David Harvey [13]. The advantage is that no shifts have to be performed in the evaluation and interpolation phases, at the expense of a few extra words in the recursive calls.

The same idea can be used for binary polynomials, simply replacing 2 by x. Our implementation **TC3W** uses $0, 1, x^w, x^{-w}, \infty$ as evaluation points (Fig. 2). Here again, there is a slight increase in size compared to using x and x^{-1}: polynomials of $3n$ words will yield two recursive calls of $n + 2$ words for x^w and x^{-w}, instead of $n + 1$ words for x and x^{-1}. The interpolation phase requires two exact divisions by $x^w + 1$, which can be performed very efficiently.

2.3   Unbalanced Variants

When using the Toom-Cook idea to multiply polynomials $a(x)$ and $b(x)$, it is not necessary to assume that $\deg a = \deg b$. We only need to evaluate $a(x)$ and

---

[3] http://bodrato.it/toom-cook/binary/
[4] http://pauillac.inria.fr/~quercia/cdrom/bibs/, version 0.21a, March 2005.

TC3W(a, b)

Multiplies polynomials $A = a_2 X^2 + a_1 X + a_0$ *and* $B = b_2 X^2 + b_1 X + b_0$ *in* $\mathrm{GF}(2)[x]$
    Let $W = x^w$ (assume X is a power of W for efficiency).
    $c_0 \leftarrow a_1 W + a_2 W^2$, $c_4 \leftarrow b_1 W + b_2 W^2$, $c_5 \leftarrow a_0 + a_1 + a_2$, $c_2 \leftarrow b_0 + b_1 + b_2$
    $c_1 \leftarrow c_2 \times c_5$, $c_5 \leftarrow c_5 + c_0$, $c_2 \leftarrow c_2 + c_4$, $c_0 \leftarrow c_0 + a_0$
    $c_4 \leftarrow c_4 + b_0$, $c_3 \leftarrow c_2 \times c_5$, $c_2 \leftarrow c_0 \times c_4$, $c_0 \leftarrow a_0 \times b_0$
    $c_4 \leftarrow a_2 \times b_2$, $c_3 \leftarrow c_3 + c_2$, $c_2 \leftarrow c_2 + c_0$, $c_2 \leftarrow c_2/W + c_3$
    $c_2 \leftarrow (c_2 + (1 + W^3)c_4)/(1 + W)$, $c_1 \leftarrow c_1 + c_0$, $c_3 \leftarrow c_3 + c_1$
    $c_3 \leftarrow c_3/(W^2 + W)$, $c_1 \leftarrow c_1 + c_2 + c_4$, $c_2 \leftarrow c_2 + c_3$
Return $c_4 X^4 + c_3 X^3 + c_2 X^2 + c_1 X + c_0$.

Fig. 2. Word-aligned Toom-Cook 3-way variant (all divisions are exact)

$b(x)$ at $\deg a + \deg b + 1$ points in order to be able to reconstruct the product $a(x)b(x)$. This is pointed out by Bodrato [2], who gives (amongst others) the case $\deg a = 3$, $\deg b = 1$. This case is of particular interest because in sub-quadratic polynomial GCD algorithms, of interest for fast polynomial factorisation [3, 17], it often happens that we need to multiply polynomials a and b where the size of a is about twice the size of b.

We have implemented a word-aligned version **TC3U** of this case, using the same evaluation points 0, 1, x^w, x^{-w}, ∞ as for **TC3W**, and following the algorithm given in [2, p. 125]. If a has size $4n$ words and b has size $2n$ words, then one call to **TC3U** reduces the multiplication $a \times b$ to 5 multiplications of polynomials of size $n + O(1)$. In contrast, two applications of Karatsuba's algorithm would require 6 such multiplications, so for large n we expect a speedup of about 17% over the use of Karatsuba's algorithm.

3 Large Degrees

In this section we discuss two efficient algorithms for large degrees, due to Cantor and Schönhage [4, 18]. A third approach would be to use segmentation, also known as Kronecker-Schönhage's trick, but it is not competitive in our context.

3.1 Cantor's Algorithm

Overview of the Algorithm. Cantor's algorithm provides an efficient method to compute with polynomials over finite fields of the form $F_k = \mathrm{GF}(2^{2^k})$. Cantor proposes to perform a polynomial multiplication in $F_k[x]$ using an evaluation/interpolation strategy. The set of evaluation points is carefully chosen to form an additive subgroup of F_k. The reason for the good complexity of Cantor's algorithm is that polynomials whose roots form an additive subgroup are sparse: only the monomials whose degree is a power of 2 can occur. Therefore it is possible to build a subproduct tree, where each internal node corresponds to a translate of an additive subgroup of F_k, and the cost of going up and down the tree will be almost linear due to sparsity.

We refer to [4, 8] for a detailed description of the algorithm, but we give a description of the subproduct tree, since this is useful for explaining our improvements. Let us define a sequence of polynomials $s_i(x)$ over $\mathrm{GF}(2)$ as follows:

$$s_0(x) = x, \quad \text{and for all } i > 0, \; s_{i+1}(x) = s_i(x)^2 + s_i(x).$$

The s_i are linearized polynomials of degree 2^i, and for all i, $s_i(x) \mid s_{i+1}(x)$. Furthermore, one can show that for all k, $s_{2^k}(x)$ is equal to $x^{2^{2^k}} + x$, whose roots are exactly the elements of F_k. Therefore, for $0 \le i \le 2^k$, the set of roots of s_i is a subvector-space W_i of F_k of dimension i. For multiplying two polynomials whose product has a degree less than 2^i, it is enough to evaluate/interpolate at the elements of W_i, that is to work modulo $s_i(x)$. Therefore the root node of the subproduct tree is $s_i(x)$. Its child nodes are $s_{i-1}(x)$ and $s_{i-1}(x) + 1$ whose product gives $s_i(x)$. More generally, a node $s_j(x) + \alpha$ is the product of $s_{j-1}(x) + \alpha'$ and $s_{j-1}(x) + \alpha' + 1$, where α' verifies $\alpha'^2 + \alpha' = \alpha$. For instance, the following diagram shows the subproduct tree for $s_3(x)$, where $\langle 1 = \beta_1, \beta_2, \beta_3 \rangle$ are elements of F_k that form a basis of W_3. Hence the leaves correspond exactly to the elements of W_3. In this example, we have to assume $k \ge 2$, so that $\beta_i \in F_k$.

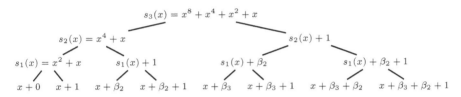

Let c_j be the number of non-zero coefficients of $s_j(x)$. The cost of evaluating a polynomial at all the points of W_i is then $O(2^i \sum_{j=1}^{i} c_j)$ operations in F_k. The interpolation step has identical complexity. The numbers c_j are linked to the numbers of odd binomial coefficients, and one can show that $C_i = \sum_{j=1}^{i} c_j$ is $O(i^{\log_2(3)}) = O(i^{1.5849\dots})$. Putting this together, one gets a complexity of $O(n(\log n)^{1.5849\dots})$ operations in F_k for multiplying polynomials of degree $n < 2^{2^k}$ with coefficients in F_k.

In order to multiply arbitrary degree polynomials over $\mathrm{GF}(2)$, it is possible to clump the input polynomials into polynomials over an appropriate F_k, so that the previous algorithm can be applied. Let $a(x)$ and $b(x)$ be polynomials over $\mathrm{GF}(2)$ whose product has degree less than n. Let k be an integer such that $2^{k-1}2^{2^k} \ge n$. Then one can build a polynomial $A(x) = \sum A_i x^i$ over F_k, where A_i is obtained by taking the i-th block of 2^{k-1} coefficients in $a(x)$. Similarly, one constructs a polynomial $B(x)$ from the bits of $b(x)$. Then the product $a(x)b(x)$ in $\mathrm{GF}(2)[x]$ can be read from the product $A(x)B(x)$ in $F_k[x]$, since the result coefficients do not wrap around (in F_k). This strategy produces a general multiplication algorithm for polynomials in $\mathrm{GF}(2)[x]$ with a bit-complexity of $O(n(\log n)^{1.5849\dots})$.

Using a Larger Base Field. When multiplying binary polynomials, a natural choice for the finite field F_k is to take k as small as possible. For instance, in [8], the cases $k = 4$ and $k = 5$ are considered. The case $k = 4$ is limited to computing a product of 2^{19} bits, and the case $k = 5$ is limited to 2^{36} bits, that is 8 GB (not a big concern for the moment). The authors of [8] remarked experimentally that their $k = 5$ implementation was almost as fast as their $k = 4$ implementation for inputs such that both methods were available.

This behaviour can be explained by analyzing the different costs involved when using F_k or F_{k+1} for doing the same operation. Let M_i (resp. A_i) denote the number of field multiplications (resp. additions) in one multipoint evaluation phase of Cantor's algorithm when 2^i points are used. Then M_i and A_i verify

$$M_i = (i - 1)2^{i-1}, \quad \text{and} \quad A_i = 2^{i-1}C_{i-1}.$$

Using F_{k+1} allows chunks that are twice as large as when using F_k, so that the degrees of the polynomials considered when working with F_{k+1} are twice as small as those involved when working with F_k. Therefore one has to compare $M_i m_k$ with $M_{i-1}m_{k+1}$ and $A_i a_k$ with $A_{i-1}a_{k+1}$, where m_k (resp. a_k) is the cost of a multiplication (resp. an addition) in F_k.

Since A_i is superlinear and a_k is linear (in 2^i resp. in 2^k), if we consider only additions, there is a clear gain in using F_{k+1} instead of F_k. As for multiplications, an asymptotical analysis, based on a recursive use of Cantor's algorithm, leads to choosing the smallest possible value of k. However, as long as 2^k does not exceed the machine word size, the cost m_k should grow roughly linearly with 2^k. In practice, since we are using the 128-bit multimedia instruction sets, up to $k = 7$, the growth of m_k is more than balanced by the decay of M_i.

In the following table, we give some data for computing a product of $N = 16\,384$ bits and a product of $N = 524\,288$ bits. For each choice of k, we give the cost m_k (in Intel Core2 CPU cycles) of a multiplication in F_k, with the mpF$_q$ library [11]. Then we give A_i and M_i for the corresponding value of i required to perform the product.

k	2^k	m_k (in cycles)	$N = 16\,384$			$N = 524\,288$		
			i	M_i	A_i	i	M_i	A_i
4	16	32	11	10\,240	26\,624	16	491\,520	2\,129\,920
5	32	40	10	4\,608	11\,776	15	229\,376	819\,200
6	64	77	9	2\,048	5\,120	14	106\,496	352\,256
7	128	157	8	896	2\,432	13	49\,152	147\,456

The Truncated Cantor Transform. In its plain version, Cantor's algorithm has a big granularity: the curve of its running time is a staircase, with a big jump at inputs whose sizes are powers of 2. In [8], a solution is proposed (based on some unpublished work by Reischert): for each integer $\ell \geq 1$ one can get a variant of Cantor's algorithm that evaluates the inputs modulo $x^\ell - \alpha$, for all α in a set W_i. The transformations are similar to the ones in Cantor's algorithm, and the pointwise multiplications are handled with Karatsuba's algorithm. For a given ℓ, the curve of the running time is again a staircase, but the jumps are at different positions for each ℓ. Therefore, for a given size, it is better to choose an ℓ, such that we are close to (and less than) a jump.

We have designed another approach to smooth the running time curve. This is an adaptation of van der Hoeven's truncated Fourier transform [14]. Van der Hoeven describes his technique at the butterfly level. Instead, we take the general idea, and restate it using polynomial language.

Let n be the degree of the polynomial over F_k that we want to compute. Assuming n is not a power of 2, let i be such that $2^{i-1} < n < 2^i$. The idea of the truncated transform is to evaluate the two input polynomials at just the required number of points of W_i: as in [14], we choose to evaluate at the n points that correspond to the n left-most leaves in the subproduct tree. Let us consider the polynomial $P_n(x)$ of degree n whose roots are exactly those n points. Clearly $P_n(x)$ divides $s_i(x)$. Furthermore, due to the fact that we consider the left-most n leaves, $P_n(x)$ can be written as a product of at most i polynomials of the form $s_j(x) + \alpha$, following the binary expansion of the integer n: $P_n = q_{i-1}q_{i-2}\cdots q_0$, where q_j is either 1 or a polynomial $s_j(x) + \alpha$ of degree 2^j, for some α in F_k.

The multi-evaluation step is easily adapted to take advantage of the fact that only n points are wanted: when going down the tree, if the subtree under the right child of some node contains only leaves of index $\geq n$, then the computation modulo the corresponding subtree is skipped. The next step of Cantor's algorithm is the pointwise multiplication of the two input polynomials evaluated at points of W_i. Again this is trivially adapted, since we have just to restrict it to the first n points of evaluation. Then comes the interpolation step. This is the tricky part, just like the inverse truncated Fourier transform in van der Hoeven's algorithm. We do it in two steps:

1. Assuming that all the values at the $2^i - n$ ignored points are 0, do the same interpolation computation as in Cantor's algorithm. Denote the result by f.
2. Correct the resulting polynomial by reducing f modulo P_n.

In step 1, a polynomial f with 2^i coefficients is computed. By construction, this f is congruent to the polynomial we seek modulo P_n and congruent to 0 modulo s_i/P_n. Therefore, in step 2, the polynomial f of degree $2^i - 1$ (or less) is reduced modulo P_n, in order to get the output of degree $n - 1$ (or less).

Step 1 is easy: as in the multi-evaluation step, we skip the computations that involve zeros. Step 2 is more complicated: we can not really compute P_n and reduce f modulo P_n in a naive way, since P_n is (a priori) a dense polynomial over F_k. But using the decomposition of P_n as a product of the sparse polynomials q_j, we can compute the remainder within the appropriate complexity.

3.2   Schönhage's Algorithm

Fig. 3 describes our implementation of Schönhage's algorithm [18] for the multiplication of binary polynomials. It slightly differs from the original algorithm, which was designed to be applied recursively; in our experiments — up to degree 30 million — we found out that TC4 was more efficient for the recursive calls. More precisely, Schönhage's original algorithm reduces a product modulo $x^{2N} + x^N + 1$ to $2K$ products modulo $x^{2L} + x^L + 1$, where K is a power of 3,

---

FFTMul(a, b, N, K)   *Assumes $K = 3^k$, and K divides N.*
Multiplies polynomials a *and* b *modulo* $x^N + 1$, *with a transform of length K.*
  1. Let $N = KM$, and write $a \sum_{i=0}^{K-1} a_i x^{iM}$, where $\deg a_i < M$ (idem for b)
  2. Let L be the smallest multiple of $K/3$ larger or equal to M
  3. Consider a_i, b_i in $R := \mathrm{GF}(2)[x]/(x^{2L} + x^L + 1)$, and let $\omega = x^{L/3^{k-1}} \in R$
  4. Compute $\hat{a}_i = \sum_{j=0}^{K-1} \omega^{ij} a_j$ in R for $0 \le i < K$ (idem for b)
  5. Compute $\hat{c}_i = \hat{a}_i \hat{b}_i$ in R for $0 \le i < K$
  6. Compute $c_\ell = \sum_{i=0}^{K-1} \omega^{-\ell i} \hat{c}_i$ in R for $0 \le \ell < K$
  7. Return $c = \sum_{\ell=0}^{K-1} c_\ell x^{\ell M}$.

Fig. 3. Our variant of Schönhage's algorithm

$L \ge N/K$, and N, L are multiples of K. If one replaces N and K respectively by $3N$ and $3K$ in Fig. 3, our variant reduces one product modulo $x^{3N} + 1$ to $3K$ products modulo $x^{2L} + x^L + 1$, with the same constraints on K and L.

A few practical remarks about this algorithm and its implementation: the forward and backward transforms (steps 4 and 6) use a fast algorithm with $O(K \log K)$ arithmetic operations in R. In the backward transform (step 6), we use the fact that $\omega^K = 1$ in R, thus $\omega^{-\ell i} = \omega^{-\ell i \bmod K}$. It is crucial to have an efficient arithmetic in R, i.e., modulo $x^{2L} + x^L + 1$. The required operations are multiplication by x^j with $0 \le j < 3L$ in steps 4 and 6, and plain multiplication in step 5. A major difference from Schönhage-Strassen's algorithm (SSA) for integer multiplication is that here K is a power of *three* instead of a power of two. In SSA, the analog of R is the ring of integers modulo $2^L + 1$, with L divisible by $K = 2^k$. As a consequence, in SSA one usually takes L to be a multiple of the numbers of bits per word — usually 32 or 64 —, which simplifies the arithmetic modulo $2^L + 1$ [10]. However assuming L is a multiple of 32 or 64 here, in addition to being a multiple of $K/3 = 3^{k-1}$, would lead to huge values of L, hence an inefficient implementation. Therefore the arithmetic modulo $x^{2L} + x^L + 1$ may not impose any constraint on L, which makes it tricky to implement.

Following [10], we can define the *efficiency* of the transform by the ratio $M/L \le 1$. The highest this ratio is, the more efficient the algorithm is. As an example, to multiply polynomials of degree less than $r = 6\,972\,593$, one can take $N = 13\,948\,686 = 2126K$ with $K = 6\,561 = 3^8$. The value of N is only 0.025% larger than the maximal product degree $2r - 2$, which is close to optimal. The corresponding value of L is $2\,187$, which gives an efficiency M/L of about 97%. One thus has to compute $K = 6561$ products modulo $x^{4374} + x^{2187} + 1$, corresponding to polynomials of 69 words on a 64-bit processor.

3.3 The Splitting Approach to FFT Multiplication

Due to the constraints on the possible values of N in Algorithm **FFTMul**, the running time (as a function of the degree of the product ab) follows a "staircase". Thus, it is often worthwhile to split a multiplication into two smaller multiplications and then reconstruct the product.

FFTReconstruct(c′, c″, N, N′, N″)
Reconstructs the product c of length N from wrapped products c′ of length N′ and c″ of length N″, assuming N′ > N″ > N/2. The result overwrites c′.
1. $\delta := N' - N''$
2. For $i := N - N' - 1$ downto 0 do
   $\{c'_{i+N'} := c'_{i+\delta} \oplus c''_{i+\delta}; \quad c'_i := c'_i \oplus c'_{i+N'}\}$
3. Return $c := c'$

Fig. 4. Reconstructing the product with the splitting approach

More precisely, choose $N' > N'' > \deg(c)/2$, where $c = ab$ is the desired product, and N', N'' are chosen as small as possible subject to the constraints of Algorithm **FFTMul**. Calling Algorithm **FFTMul** twice, with arguments $N = N'$ and $N = N''$, we obtain $c' = c \bmod (x^{N'}+1)$ and $c'' = c \bmod (x^{N''}+1)$. Now it is easy to reconstruct the desired product c from its "wrapped" versions c' and c''. Bit-serial pseudocode is given in Fig. 4.

It is possible to implement the reconstruction loop efficiently using full-word operations provided $N' - N'' \geq w$. Thus, the reconstruction cost is negligible in comparison to the cost of **FFTMul** calls.

4   Experimental Results

The experiments reported here were made on a 2.66Ghz Intel Core 2 processor, using gcc 4.1.2. A first tuning program compares all Toom-Cook variants from

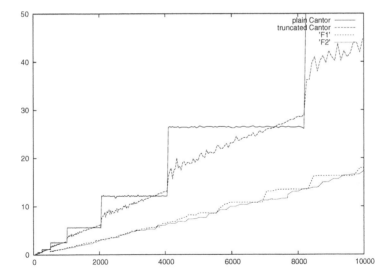

Fig. 5. Comparison of the running times of the plain Cantor algorithm, its truncated variant, our variant of Schönhage's algorithm (F1), and its splitting approach (F2). The horizontal axis represents 64-bit words, the vertical axis represents milliseconds.

Karatsuba to TC4, and determines the best one for each size. The following table gives for each algorithm the range in words where it is used, and the percentage of word sizes where it is used in this range.

Algorithm	Karatsuba	TC3	TC3W	TC4
Word range	10-65 (50%)	21-1749 (5%)	18-1760 (45%)	166-2000 (59%)

A second tuning program compared the best Karatsuba or Toom-Cook algorithm with both FFT variants (classical and splitting approach): the FFT is first used for 2461 words, and TC4 is last used for 3295 words.

In Fig. 5 the running times are given for our plain implementation of Cantor's algorithm over F_7, and its truncated variant. We see that the overhead induced by handling P_n as a product implies that the truncated version should not be used for sizes that are close to (and less than) a power of 2. We remark that

Table 1. Comparison of the multiplication routines for small degrees with existing software packages (average cycle counts on an Intel Core2 CPU)

	$N = 64$	128	192	256	320	384	448	512
NTL 5.4.1	99	368	703	1 130	1 787	2 182	3 070	3 517
LIDIA 2.2.0	117	317	787	988	1 926	2 416	2 849	3 019
ZEN 3.0	158	480	1 005	1 703	2 629	3 677	4 960	6 433
this paper	54	132	364	410	806	850	1 242	1 287

Table 2. Comparison in cycles with the literature and software packages for the multiplication of N-bit polynomials over GF(2): the timings of [16,8,17] were multiplied by the given clock frequency. Kn means n-term Karatsuba-like formula. In [8] we took the best timings from Table 7.1, and the degrees in [17] are slightly smaller. $F1(K)$ is the algorithm of Fig. 3 with parameter $K = 3^k$; $F2(K)$ is the splitting variant described in Section 3.3 with two calls to $F1(K)$.

reference	[16]	[8]	[17]	NTL 5.4.1	LIDIA 2.2.0	this paper
processor	Pentium 4	UltraSparc1	IBM RS6k	Core 2	Core 2	Core 2
$N = 1\,536$	1.1e5 [K3]			1.1e4	2.5e4	1.0e4 [TC3]
4 096	4.9e5 [K4]			5.3e4	9.4e4	3.9e4 [K2]
8 000			1.3e6	1.6e5	2.8e5	1.1e5 [TC3W]
10 240	2.2e6 [K5]			2.6e5	5.8e5	1.9e5 [TC3W]
16 384		5.7e6	3.4e6	4.8e5	8.6e5	3.3e5 [TC3W]
24 576	8.3e6 [K6]			9.3e5	2.1e6	5.9e5 [TC3W]
32 768		1.9e7	8.7e6	1.4e6	2.6e6	9.3e5 [TC4]
57 344	3.3e7 [K7]			3.8e6	7.3e6	2.4e6 [TC4]
65 536		4.7e7	1.7e7	4.3e6	7.8e6	2.6e6 [TC4]
131 072		1.0e8	4.1e7	1.3e7	2.3e7	7.2e6 [TC4]
262 144		2.3e8	9.0e7	4.0e7	6.9e7	1.9e7 [F2(243)]
524 288		5.2e8		1.2e8	2.1e8	3.7e7 [F1(729)]
1 048 576		1.1e9		3.8e8	6.1e8	7.4e7[F2(729)]

this overhead is more visible for small sizes than for large sizes. This figure also compares our variant of Schönhage's algorithm (Fig. 3) with the splitting approach: the latter is faster in most cases, and both are faster than Cantor's algorithm by a factor of about two. It appears from Fig. 5 that a truncated variant of Schönhage's algorithm would not save much time, if any, over the splitting approach.

Tables 1 and 2 compare our timings with existing software or published material. Table 1 compares the basic multiplication routines involving a fixed small number of words. Table 2 compares the results obtained with previous ones published in the literature. Since previous authors used 32-bit computers, and we use a 64-bit computer, the cycle counts corresponding to references [16, 8, 17] should be divided by 2 to account for this difference. Nevertheless this would not affect the comparison.

5   Conclusion

This paper presents the current state-of-the-art for multiplication in GF(2)[x]. We have implemented and compared different algorithms from the literature, and invented some new variants.

The new algorithms were already used successfully to find two new primitive trinomials of record degree 24 036 583 (the previous record was 6 972 593), see [3].

Concerning the comparison between the algorithms of Schönhage and Cantor, our conclusion differs from the following excerpt from [8]: *The timings of Reischert (1995) indicate that in his implementation, it [Schönhage's algorithm] beats Cantor's method for degrees above 500,000, and for degrees around 40,000,000, Schönhage's algorithm is faster than Cantor's by a factor of $\approx \frac{3}{2}$*. Indeed, Fig. 5 shows that Schönhage's algorithm is consistently faster by a factor of about 2, already for a few thousand bits. However, a major difference is that, in Schönhage's algorithm, the pointwise products are quite expensive, whereas they are inexpensive in Cantor's algorithm. For example, still on a 2.66Ghz Intel Core 2, to multiply two polynomials with a result of 2^{20} bits, Schönhage's algorithm with $K = 729$ takes 28ms, including 18ms for the pointwise products modulo $x^{5832} + x^{2916} + 1$; Cantor's algorithm takes 57ms, including only 2.3ms for the pointwise products. In a context where a given Fourier transform is used many times, for example in the block Wiedemann algorithm used in the "linear algebra" phase of the Number Field Sieve integer factorisation algorithm, Cantor's algorithm may be competitive.

Acknowledgements

The word-aligned variant of TC3 for GF(2)[x] was discussed with Marco Bodrato. The authors thank Joachim von zur Gathen and the anonymous referees for their useful remarks. The work of the first author was supported by the Australian Research Council.

References

1. Aoki, K., Franke, J., Kleinjung, T., Lenstra, A., Osvik, D.A.: A kilobit special number field sieve factorization. In: Kurosawa, K. (ed.) ASIACRYPT 2007. LNCS, vol. 4833, pp. 1–12. Springer, Heidelberg (2007)
2. Bodrato, M.: Towards optimal Toom-Cook multiplication for univariate and multivariate polynomials in characteristic 2 and 0. In: Carlet, C., Sunar, B. (eds.) WAIFI 2007. LNCS, vol. 4547, pp. 116–133. Springer, Heidelberg (2007)
3. Brent, R.P., Zimmermann, P.: A multi-level blocking distinct degree factorization algorithm. Research Report 6331, INRIA (2007)
4. Cantor, D.G.: On arithmetical algorithms over finite fields. J. Combinatorial Theory, Series A 50, 285–300 (1989)
5. Chabaud, F., Lercier, R.: ZEN, a toolbox for fast computation in finite extensions over finite rings, http://sourceforge.net/projects/zenfact
6. Cohen, H., Frey, G., Avanzi, R., Doche, C., Lange, T., Nguyen, K., Vercauteren, F.: Handbook of Elliptic and Hyperelliptic Curve Cryptography. In: Discrete Mathematics and its Applications, Chapman & Hall/CRC (2005)
7. Cook, S.A.: On the Minimum Computation Time of Functions. PhD thesis, Harvard University (1966)
8. von zur Gathen, J., Gerhard, J.: Arithmetic and factorization of polynomials over F_2. In: Proceedings of ISSAC 1996, Zürich, Switzerland, pp. 1–9 (1996)
9. von zur Gathen, J., Gerhard, J.: Polynomial factorization over F_2. Math. Comp. 71(240), 1677–1698 (2002)
10. Gaudry, P., Kruppa, A., Zimmermann, P.: A GMP-based implementation of Schönhage-Strassen's large integer multiplication algorithm. In: Proceedings of ISSAC 2007, Waterloo, Ontario, Canada, pp. 167–174 (2007)
11. Gaudry, P., Thomé, E.: The mpFq library and implementing curve-based key exchanges. In: Proceedings of SPEED, pp. 49–64 (2007)
12. Hankerson, D., Menezes, A.J., Vanstone, S.: Guide to Elliptic Curve Cryptography. Springer Professional Computing. Springer, Heidelberg (2004)
13. Harvey, D.: Avoiding expensive scalar divisions in the Toom-3 multiplication algorithm, 10 pages (Manuscript) (August 2007)
14. van der Hoeven, J.: The truncated Fourier transform and applications. In: Gutierrez, J. (ed.) Proceedings of ISSAC 2004, Santander, 2004, pp. 290–296 (2004)
15. THE LiDIA GROUP. LiDIA, A C++ Library For Computational Number Theory, Version 2.2.0 (2006)
16. Montgomery, P.L.: Five, six, and seven-term Karatsuba-like formulae. IEEE Trans. Comput. 54(3), 362–369 (2005)
17. Roelse, P.: Factoring high-degree polynomials over F_2 with Niederreiter's algorithm on the IBM SP2. Math. Comp. 68(226), 869–880 (1999)
18. Schönhage, A.: Schnelle Multiplikation von Polynomen über Körpern der Charakteristik 2. Acta Inf. 7, 395–398 (1977)
19. Shoup, V.: NTL: A library for doing number theory, Version 5.4.1 (2007), http://www.shoup.net/ntl/
20. Thomé, E.: Subquadratic computation of vector generating polynomials and improvement of the block Wiedemann algorithm. J. Symb. Comp. 33, 757–775 (2002)
21. Toom, A.L.: The complexity of a scheme of functional elements realizing the multiplication of integers. Soviet Mathematics 3, 714–716 (1963)
22. Weimerskirch, A., Stebila, D., Shantz, S.C.: Generic GF(2) arithmetic in software and its application to ECC. In: Safavi-Naini, R., Seberry, J. (eds.) ACISP 2003. LNCS, vol. 2727, pp. 79–92. Springer, Heidelberg (2003)
23. Zimmermann, P.: Irred-ntl patch, http://www.loria.fr/~zimmerma/irred/

Predicting the Sieving Effort for the Number Field Sieve

Willemien Ekkelkamp[1,2]

[1] CWI, P.O. Box 94079, 1090 GB Amsterdam, The Netherlands
[2] Leiden University, P.O. Box 9512, 2300 RA Leiden, The Netherlands
W.H.Ekkelkamp@cwi.nl

Abstract. We present a new method for predicting the sieving effort for the number field sieve (NFS) in practice. This method takes relations from a short sieving test as input and simulates relations according to this test. After removing singletons, we decide how many relations we need for the factorization according to the simulation and this gives a good estimate for the real sieving. Experiments show that our estimate is within 2 % of the real data.

1 Introduction

One of the most popular methods for factoring large numbers is the number field sieve [4], as this is the fastest algorithm known so far. In order to estimate the most time-consuming step of this method, namely the sieving step in which the so-called relations are generated, one looks at actual sieving times for numbers of comparable size. If these are not available, one could try to extrapolate actual sieving times for smaller numbers, using the formula for the running time $L(N)$ of this method, where N is the number to be factored. We have

$$L(N) = \exp(((64/9)^{1/3} + o(1))(\log N)^{1/3}(\log\log N)^{2/3}), \text{ as } N \to \infty ,$$

where the logarithms are natural. These estimates can be 10–30 % off.

In this paper we present a method for predicting the number of relations needed for factoring a given number in practice within 2 % of the actual number of relations needed. With 'in practice' we mean: on a given computer, for a given implementation, and for a given choice of the parameters in the NFS. This allows us to predict the actually required sieving time within 2 %. Our method is based on a short sieving test and a very cheap simulation of the relations needed for the factorization. By applying this method for various choices of the parameters of the number field sieve, it is possible to find an optimal choice of the parameters, e.g., in terms of minimal sieving time or in terms of minimizing the size of the resulting matrix. Before going into details we give a short overview of the NFS in order to show where our method fits in.

The NFS consists of the following four steps. First we select two irreducible polynomials $f_1(x)$ and $f_2(x)$, $f_1, f_2 \in \mathbb{Z}[x]$, and an integer $m < N$, such that

$$f_1(m) \equiv f_2(m) \equiv 0 \ (\mathrm{mod}\ N) .$$

A.J. van der Poorten and A. Stein (Eds.): ANTS-VIII 2008, LNCS 5011, pp. 167–179, 2008.
© Springer-Verlag Berlin Heidelberg 2008

Polynomials with 'small' integer coefficients are preferred, because the values of these polynomials are smaller on average and smoother (i.e. having smaller prime factors on average) than the values of polynomials with large integer coefficients. Usually $f_1(x)$ is a linear polynomial and $f_2(x)$ a higher degree polynomial, referred to as rational side and algebraic side, respectively. If N is of a special form (e.g., $c^n \pm 1$) then we can use this to get a polynomial $f_2(x)$ with very small coefficients. In that case we talk about the special number field sieve (SNFS), else we talk about the general number field sieve (GNFS). By α_1 and α_2 we denote roots of $f_1(x)$ and $f_2(x)$, respectively.

The second step is the relation collection. We choose a factorbase FB of primes below the bound F and a large primes bound L; for ease of exposition we take the same bounds on both the rational side and the algebraic side. Then we search for pairs (a, b) such that $\gcd(a, b) = 1$, and such that both $F_1(a, b) = b^{\deg(f_1)} f_1(a/b)$ and $F_2(a, b) = b^{\deg(f_2)} f_2(a/b)$ have all their prime factors below F and at most two prime factors between F and L, the so-called large primes. These pairs (a, b) are referred to below as relations (a_i, b_i).

There are many possibilities for the relation collection, the fastest of which are based on sieving. Two sieving methods in particular are widely used, namely line sieving and lattice sieving. For line sieving we select a rectangular sieve area of points (a, b) and the sieving is done per horizontal line. For lattice sieving we select an interval of so-called special primes and for each special prime we only sieve those pairs (a, b) for which this special prime divides $b^{\deg(f_2)} f_2(a/b)$; for each special prime these pairs form a lattice in the sieving area. In case of SNFS the special prime is chosen on the rational side.

The third step consists of linear algebra to construct a set S of indices i such that the two products $\prod_{i \in S}(a_i - b_i \alpha_1)$ and $\prod_{i \in S}(a_i - b_i \alpha_2)$ are both squares of products of prime ideals. This product comes from the fact that $b^{\deg(f_1)} f_1(a/b)$ is the norm of the algebraic number $a - b\alpha_1$, multiplied with the leading coefficient of $f_1(x)$. The principal ideal $(a - b\alpha_1)$ factors into the product of prime ideals in the number field $\mathbb{Q}(\alpha_1)$. The situation is similar for f_2.

The last step is the square root step. We determine algebraic numbers $\alpha'_1 \in \mathbb{Q}(\alpha_1)$ and $\alpha'_2 \in \mathbb{Q}(\alpha_2)$ such that $(\alpha'_1)^2 = \prod_{i \in S}(a_i - b_i \alpha_1)$ and $(\alpha'_2)^2 = \prod_{i \in S}(a_i - b_i \alpha_2)$. Then we use the homomorphisms $\phi_{\alpha_1} : \mathbb{Q}(\alpha_1) \to \mathbb{Z}/N\mathbb{Z}$ and $\phi_{\alpha_2} : \mathbb{Q}(\alpha_2) \to \mathbb{Z}/N\mathbb{Z}$ with $\phi_{\alpha_1}(\alpha_1) = \phi_{\alpha_2}(\alpha_2) = m$ to get $\phi_{\alpha_1}(\alpha'_1)^2 = \phi_{\alpha_1}((\alpha'_1)^2) = \phi_{\alpha_1}(\prod_{i \in S}(a_i - b_i \alpha_1)) \equiv \prod_{i \in S}((a_i - b_i m) \equiv \phi_{\alpha_2}(\alpha'_2)^2 (\mathrm{mod}\ N)$. Now compute $\gcd(\phi_{\alpha_1}(\alpha'_1) - \phi_{\alpha_2}(\alpha'_2), N)$ to obtain a factor of N. If this gives the trivial factorization, continue with the next set of indices, otherwise we have found a nontrivial factorization of N. For more details of the NFS, see e.g., [3], [4], or [5].

Our method works as follows. After choosing polynomials, bounds F and L, and a sieve area, we perform a sieve test for a relatively short period of time. For a 120-digit N one could sieve for ten minutes or so, but for larger numbers one may spend considerably more time on the sieve test. Based on the relations in this sieve test we simulate as many relations as are necessary for factoring the number. The simulation uses a random number generator and functions that describe the underlying distribution of the large primes, and this can be done fast.

During the simulation of the relations, we regularly remove the singletons from all the relations simulated so far. As soon as the number of relations left after singleton removal exceeds the number of primes in the relations we stop and it turns out that the total number of relations simulated so far gives us a good estimate of the actual number of relations that we need to factor our number.

The number of useful relations after singleton removal grows in a hard-to-predict fashion as a function of the number of relations found. This growth behaviour differs from number to number, which makes it hard to predict the overall sieving time: for instance, even estimates based on factoring times of numbers of comparable size can easily be 10 % off. Our method, however, which is purely based on the individual behavior of the relations found for the number to be factored, allows us to predict how the number of useful relations will behave as a function of the number of relations found, thereby giving us a tool to accurately predict the overall sieving time.

The simulations in this paper were carried out on a Intel® Core™2 Duo with 2 GB of memory. The line sieving data sets were generated with the NFS software package of CWI. The lattice sieving data sets were given by Bruce Dodson and Thorsten Kleinjung.

In Section 2 we describe how we simulate the relations. Section 3 is about the singleton removal and about how to decide when we have enough relations to factor the given number. In Section 4 we compare results of the simulation with real factorizations and Section 5 contains the conclusions and our intentions for future work.

2   Simulating Relations

Before we start with the simulation, we run a short sieving test. In order to get a representative selection of the actual relations, we ensure that the points we are sieving in this test are spread over the entire sieving area. The parameters for the sieving are set in such a way that we have at most two large primes both on the rational side and on the algebraic side. In the case of lattice sieving we have one additional special prime on one of the sides. In this section we describe the process of simulating relations both for line sieving and for lattice sieving. Note that we only simulate the large primes; for the primes in the factorbase we use a correction as will be explained in Section 3.

The first step after the sieving test consists of splitting the relations according to the number of large primes. The set of relations with i large primes on the rational side and j large primes on the algebraic side is denoted by $r_i a_j$ for $i, j \in \{0, 1, 2\}$. This leads to nine different sets and the mutual ratios of their cardinalities determine the ratios by which we will simulate the relations. In the case of lattice sieving we split the relations in the same way, ignoring the special prime.

Next we take a closer look at the relations in each set and specify a model that fits the distribution of the large primes in these sets as closely as we can accomplish. To clarify this, we explain for each set how to simulate the relations in that set, for the case of line sieving.

$\underline{r_0 a_0}$: We count the number of relations in this set.

$\underline{r_1 a_0}$: We started with sorting all the large primes and put them in an array. Our first experiments with simulating the large primes (and removing singletons) concentrated on the large primes at hand. We tried linear interpolation between two consecutive large primes, Lagrange polynomials, and splines, but all these local approaches did not give a satisfying result; the result after singleton removal was too far from the real data. We then tried a more global approach, looking at all the large primes and see if we could find a distribution for them. We found that an exponential distribution simulates best the distribution of these large primes over the interval $[F, L]$ (cf. [2], Ch. 6) and the result after singleton removal was satisfactory. The inverse of this distribution function is given by

$$G(x) = F - a \log \left(1 - x \left(1 - e^{\frac{F-L}{a}} \right) \right), 0 \leq x \leq 1 , \tag{1}$$

where a is the average of the large primes in the set $r_1 a_0$. Note that $G(0) = F$ and $G(1) = L$. In order to generate primes according to the actual distribution of the large primes, we generate a random number between 0 and 1, substitute this number in $G(x)$, round the number $G(x)$ to the nearest prime, and repeat this for each prime that we want to generate.

To avoid expensive prime tests, we work with the index of the primes p, defined as $i_p = \pi(p)$, rather than with the prime itself. This index can be found by using a look-up table or the approximation $i_p \approx \frac{p}{\log p} + \frac{p}{\log^2 p} + \frac{2p}{\log^3 p}$ [6]. Experiments showed that this third order approximation gives almost the same results as looking up indices in a table. It is especially more efficient to use this approximation when L is large. For working with indices, we have to adjust (1); we write i_F for the index of the first prime above F, and i_L for the index of the prime just below L, and a' for the average of the indices of the large primes in the set $r_1 a_0$. Then the formula becomes

$$G(x) = i_F - a' \log \left(1 - x \left(1 - e^{\frac{i_F - i_L}{a'}} \right) \right) . \tag{2}$$

To illustrate that the distribution of the large primes is approximated well by (2) we have generated the following graph (Fig. 1), which consists of two sorted sets. One set consists of the indices of the primes of the original sieving data and the other set consists of the indices simulated with help of (2). The line of the simulated data is the one which lies below the other line (of the original data) around position 7000.

The necessary number of relations in the set $r_1 a_0$ depends on how many relations we have to generate in total.

$\underline{r_0 a_1}$: We would like to use the same idea as we used for $r_1 a_0$, but now we have to deal with algebraic primes. This means that not all primes can occur, and that each prime that does occur can have up to d different roots, where d is the degree of the polynomial $f_2(x)$. This yields pairs of a prime and a root which we denote by $(prime, root)$. Luckily, (heuristically) the amount of pairs $(prime, root)$ with $F < prime < L$ is about equal to the amount of primes between F and L. This

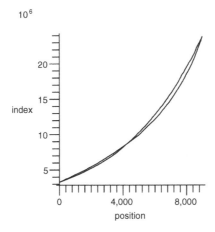

10^6

Fig. 1. Comparing original and simulated data

implies that we do not have to simulate pairs with a certain subset of indices, as we may assume that all indices can occur in the simulation. We found that an exponential distribution fits here as well, so here we use the same approach as we did for $r_1 a_0$.

$r_1 a_1$: We know now how to simulate $r_1 a_0$ and $r_0 a_1$, and we assume that the value of the index on the rational side is independent of the value of the index on the algebraic side. We combine both approaches: using (2), generate a random number and compute the corresponding rational index, generate a new random number (do not use the first random number as input for the random number generator) and compute the algebraic index.

$r_2 a_0$: Here we have to deal with two large primes on the rational side, denoted by q_1 and q_2 with $q_1 > q_2$. We started with sorting the list with q_1 and (to our surprise) we found that a linear distribution fits these data well. So the distribution function of the index i_{q_1} of q_1 is given by

$$H_1(x) = i_F + x(i_L - i_F) \; ,$$

where x is a number between 0 and 1.

We continued with q_2 and sorted them. Here, an exponential distribution fits the data, but now we have to take into account that $q_2 < q_1$. Remember that we need an average value for the exponential distribution, but we cannot use all q_2-values. Instead of using one average value, we make a list of averages a_{q_2} of the sorted q_2-indices, where $a_{q_2}[j]$ contains the average of the first j q_2-indices.

Now we describe how to simulate elements of $r_2 a_0$. We begin with a random number between 0 and 1 and compute $H_1(x)$, which gives us an index i_{q_1} of q_1. We look up this index in the sorted list of q_2-indices and the corresponding position j tells us which average we should use for computing the index i_{q_2} of q_2.

We generate a new random number between 0 and 1 and substitute it for x in the following formula $H_2(x)$, which is an adjusted form of $G(x)$:

$$H_2(x) = i_F - a_{q_2}[j] \log \left(1 - x \left(1 - e^{\frac{i_F - i_{q_1}}{a_{q_2}[j]}}\right)\right) .$$

This gives us an index i_{q_2} of q_2 that is smaller than the index we generated for q_1.

Our observation of a linear distribution of the largest prime and an exponential distribution of the second prime may not be as one would expect theoretically, but this might very well be a consequence of sieving in practice. For example, products of size approximately L^2 factor most of the time as one prime below L and one prime above L and are discarded. Thus most sievers do not spend much time on factors of this size. It may turn out to be the case that a siever with different implementation choices gives rise to different distributions, which needs to be investigated further.

To illustrate the distribution of the products of the two large primes for the dataset of $13,220+$ (cf. Section 4) found by our implementation of the siever, we added for each relation in $r_2 a_0$ the indices of the two large primes and split the interval $[2i_F, 2i_L]$ in ten equal subintervals (labeled $s = 1, \ldots, 10$). For each subinterval we counted the number of relations for which the sum of the two indices of the two large primes lies in this subinterval: see Table 1.

Table 1. Distribution of the sum of the indices $(13,220+)$

s	1	2	3	4	5	6	7	8	9	10
# relations	120780	161735	148757	133845	121967	78725	39253	20710	8107	0

The zero in the last column is due to one of the bounds in the siever, which was set at $F^{0.1}L^{1.9}$ instead of L^2.

$\underline{r_0 a_2}$: We know how to deal with $r_2 a_0$ and we apply the same approach to $r_0 a_2$, as we can make the same transition as we made from $r_1 a_0$ to $r_0 a_1$.

Sorting the list with q_1 showed that we could indeed use a linear distribution and the sorted list with q_2 showed that an exponential distribution fitted here. Now we simulate elements of $r_2 a_0$ in the same way as elements of $r_0 a_2$.

$\underline{r_1 a_2}$: As with $r_1 a_1$, we assume that the rational side and the algebraic side are independent. Here we combine the approaches of $r_1 a_0$ and $r_0 a_2$ to get the elements of $r_1 a_2$.

$\underline{r_2 a_1}$: Combine the approaches of $r_2 a_0$ and $r_0 a_1$ to get the elements of $r_2 a_1$.

$\underline{r_2 a_2}$: As in the previous two sets, we combine two approaches, this time $r_2 a_0$ and $r_0 a_2$.

Summarizing, our simulation model consists of four assumptions:

- the rational side and the algebraic side are independent,
- the rational side and the algebraic side are equivalent,

- a model for one large prime (described in $r_1 a_0$),
- a model for two large primes (described in $r_2 a_0$).

In case of lattice sieving, we simulate the relations in the same way and add a special prime to all the relations in the following way. We compute the average number of relations per pair (*special prime, root*) in the sieving test. Then we divide the number of relations we want to simulate by this average and this gives the total number of special primes in our simulation. Then we select an appropriate interval from which the special primes are chosen. Divide this interval in a (small) number of sections: per section select randomly the special primes and add each of these special primes to a relation. By dividing in sections (and simulating the same amount of relations per section) we make sure that the entire interval of special primes is covered, but by choosing randomly in each section, we get enough variation in the amount of relations per special prime. If the interval of the special primes is very large, it might become necessary to decrease the number of relations per section. In our example this was not the case, but a well-chosen sieve test will give this information.

It is possible to use different factorbase bounds for the rational primes and the algebraic primes, bound the product of the two large primes on the same side, etc. All these details in the sieving influence the relations, but once the general model is known, it is relatively easy to adjust it to match the details.

3   The Stop Criterion

We now know how to simulate relations, but how many should we simulate?

In order to factor the number N we have to find dependencies in a matrix, which is determined by the relations, as mentioned in the introduction in the third step of the NFS. Every column is identified with a prime $\leq L$ (rational and algebraic primes). Suppose each row represents a relation. If a prime occurs an odd number of times in that relation, we put a one in the column of that prime and a zero otherwise. After representing all relations in this matrix, we remove those relations with a 1 that is the only 1 in the entire column, the so-called singletons. This may generate new singletons, so this singleton removal step is repeated until all primes occur at least twice. In practice, this is done before actually building a matrix.

For our stop criterion it is enough to know when we have enough relations, i.e. *when the number of relations after singleton removal exceeds the number of different primes that occur in the remaining relations.*

After the singleton removal, we count how many relations are left and how many different large primes occur in these relations. We define the percentage oversquareness O_r after singleton removal (s.r.) as

$$O_r := \frac{n_r}{n_l + n_F - n_f} \times 100 \ ,$$

where n_r is the number of relations after singleton removal, n_l is the number of different large primes after singleton removal, n_F is the number of primes in the factorbase, approximated by $\pi(F_{\text{rat}}) + \pi(F_{\text{alg}})$, and n_f is the number of free relations from factorbase elements. We have ([3], Ch. 3):

$$n_f = \frac{1}{g}\pi(\min(F_{\text{rat}}, F_{\text{alg}})) \ ,$$

where g is the order of the Galois group of $f_1(x)f_2(x)$. If $O_r \geq 100\,\%$ we may expect to find a dependency in the matrix, and we may stop with simulating relations. To make practically sure to find a dependency, we may stop at $102\,\%$. Even a larger percentage is allowed if one would like to have more choice in the relations that can form a dependency and subsequently form a smaller matrix in the linear algebra step.

One final point concerns lattice sieving. It is well known that lattice sieving produces lots of duplicates, especially when it involves many special primes. We treat our relations as if there are no duplicates, but that implies that in the case of lattice sieving we have to add a certain number of relations to the relations that we should collect in the sieving stage. This number can be computed as in [1]. The basic idea in [1] is to run a small sieve test and find out which relations have more than one prime in the special primes interval. If such a relation would be found by more than one lattice in the sieving area (remember that each special prime gives rise to a lattice in the sieving area), than this gives a duplicate relation.

4   Experiments

We have applied our method to several real data sets (coming from factored numbers) and show that this gives good results. We have carried out two types of experiments.

First we assumed that the complete data set is given and we wanted to know if the simulation gave the same oversquareness when simulating the same number of relations as is contained in the original data set. For the simulation we used $0.1\,\%$ of the original data.

Secondly we assumed that only a small percentage ($0.1\,\%$) of the original data is known. Based on this data we simulated relations until $O_r \geq 100\,\%$. Then we compared this with the oversquareness of the same number of original relations.

This $0.1\,\%$ is somewhat arbitrary. We came to it in the following way: we started a simulation based on $100\,\%$ real data and lowered this percentage in the next experiment until results after singleton removal were too far from the real data. We went down as far as $0.01\,\%$, but this percentage did not always give good results, unless we would have been satisfied with an estimate within $5\,\%$ of the real data (although some experiments with $0.01\,\%$ of the real data were even as good as the ones based on $0.1\,\%$ of the real data).

4.1   Line Sieving

Some relevant parameters for all the real data sets in this section are given in Table 2, where M stands for million. Numbers are written in the format $a, b+$ or

Table 2. Sieving parameters (line sieving)

number	# dec. digits	F	L	g	$n_F - n_f$
13,220+	117	30M	400M	120	3700941
26,142+	124	30M	250M	120	3700941
19,183−	131	30M	250M	18	3613192
66,129+	136	35M	300M	18	4175312
80,123−	150	55M	450M	18	6383294

$a, b-$, meaning $a^b + 1$ or $a^b - 1$. In the case of GNFS, some prime factors were already known and for the remaining factors it was more efficient to use GNFS instead of SNFS.

The experiments for the first two GNFS data sets $13,220+$ and $26,142+$ are in Table 3. Here, O stands for the original data and S for the simulated data. Table 3 shows that the numbers were oversieved, but the simulated data show about the same oversquareness. In Table 4, we computed the relative difference $(S-O)/O \times 100\,\%$ of the entries in the S- and O-column of Table 3. We see that our predictions of the number of relations after s.r., the number of large primes after s.r., and the oversquareness are close to the real data to about $1\,\%$.

Table 3. Experiments line sieving

GNFS	13,220+ O	13,220+ S	26,142+ O	26,142+ S
# relations before s.r.	35 496 483	35 496 483	23 580 294	23 580 294
# relations after s.r.	21 320 864	21 394 640	15 150 790	15 253 825
# large primes after s.r.	13 781 518	13 950 420	9 448 082	9 397 751
oversquareness (%)	121.96	121.21	115.22	116.45

Table 4. Relative differences of Table 3 results

GNFS	13,220+	26,142+
relations after s.r. (%)	0.35	0.68
large primes after s.r. (%)	1.22	−0.53
oversquareness (%)	−0.61	1.07

We give the following timings for these experiments: simulation of the relations, singleton removal, and real sieving time (Table 5). For the actual sieving we used multiple machines and added the sieving times of each machine. As we used $0.1\,\%$ data, we have to keep in mind that we need to add $0.1\,\%$ of the sieving time to a complete experiment, which consists of generating a small data set, simulate a big data set, and remove singletons. When we change parameters in the NFS we have to generate a new data set.

Roughly speaking, we can say that one prediction of the total sieving time (for a given choice of the NFS parameters) with our method costs less than one CPU hour, whereas the actual sieving costs several hundreds of CPU hours.

Table 5. Timings

GNFS	13,220+	26,142+
simulation (sec.)	224	156
singleton removal (sec.)	927	573
actual sieving (hrs.)	316	709

Now for our second type of experiments, we assume that we only have a small sieve test of the number to be factored. When are we in the neighbourhood of 100 % oversquareness according to our simulation and will the real data agree with our simulation? We started to simulate 5M, 10M, ... relations (with increment 5M) and for these numbers we computed the oversquareness O_r; when O_r approached the 100 % bound we decreased the increment to 1M. Table 6 gives the number of relations for which O_r is closest to 100 % and the next O_r (for 1M more relations), both for the simulated data and the original data. This may of course be refined.

Table 6. Around 100 % oversquareness (GNFS)

# rel. before s.r.	O_r S (%)	O_r O (%)	rel. diff. (%)
28M (13,220+)	99.66	99.87	−0.21
29M (13,220+)	103.15	103.29	−0.14
20M (26,142+)	100.57	99.24	1.34
21M (26,142+)	105.38	104.03	1.30

Table 7. Experiments line sieving

SNFS	# rel. before s.r.	# rel. after s.r.	# l.p. after s.r.	oversquareness (%)
19,183− O	21 259 569	11 887 312	7 849 531	103.70
19,183− S	21 259 569	12 156 537	7 936 726	105.25
66,129+ O	26 226 688	15 377 495	10 036 942	108.20
66,129+ S	26 226 688	15 656 253	10 123 695	109.49
80,123− O	36 552 655	20 288 292	12 810 641	105.70
80,123− S	36 552 655	20 648 909	12 973 952	106.67

For SNFS the higher degree polynomial has small coefficients. Tables 7–10 show the same kind of data as Tables 3–6, but now for SNFS. We start in Table 7 with the complete data set and simulate the same number of relations. Table 8 gives the relative differences of the results of the experiments in Table 7. The timings are given in Table 9.

In Table 10 we simulate the number of relations that leads to an oversquareness around 100 %. We compare this number with the real data and give the differences in oversquareness.

Table 8. Relative differences of Table 7 results

SNFS	19,183−	66,129+	80,123−
relations after s.r. (%)	2.26	1.81	1.78
large primes after s.r. (%)	1.11	0.86	1.27
oversquareness (%)	1.49	1.19	0.92

Table 9. Timings

SNFS	19,183−	66,129+	80,123−
simulation (sec.)	128	166	223
singleton removal (sec.)	487	603	771
sieving (hrs.)	154	197	200

Table 10. Around 100 % oversquareness (SNFS)

# rel. before s.r.	O_r S (%)	O_r O (%)	rel. diff. (%)
20M (19,183−)	99.22	97.71	1.55
21M (19,183−)	104.06	102.51	1.51
23M (66,129+)	96.44	95.35	1.14
24M (66,129+)	100.72	99.60	1.12
34M (80,123−)	99.93	98.66	1.29
35M (80,123−)	102.82	101.50	1.30

All these data sets were generated with the NFS software package of CWI, and the models for describing the underlying distributions were the same for SNFS and GNFS, as described in Section 2.

4.2 Lattice Sieving

For lattice sieving we used a data set from Bruce Dodson (7,333−, SNFS). Besides the factorbase bound and the large primes bound, we have two intervals for the special primes. These are given in Table 11.

Table 11. Sieving parameters (lattice sieving)

	7,333−
# dec. digits	177
F	16 777 215
L	250 000 000
special primes	[16 777 333, 29 120 617]
	[60 000 013, 73 747 441]
g	6
$n_F - n_f$	1 976 740

Table 12. Oversquareness 7,333−

# rel. before s.r.	O_r S (%)	O_r O (%)	rel. diff. (%)
17M (7,333−)	98.34	97.45	0.91
18M (7,333−)	103.96	103.08	0.85
25 112 543 (7,333−)	135.39	136.64	−0.91

As we are now dealing with lattice sieving, we have an extra (special) prime to simulate, in the way described in Section 2. Fortunately, the distribution of the other large primes did not change. The results of our experiments are given in Table 12, based on 0.023 % original data. The last line in this table is the total number of relations without duplicates. In total 26 024 921 relations were sieved.

Apart from receiving a lattice sieving data set from Bruce Dodson, we also received lattice sieving data sets from Thorsten Kleinjung. Unfortunately the model described in this paper for the large primes does not yield satisfactory results for the latter data sets.

5   Conclusions and Future Work

Our experiments show that our simulation of the relations works well. Based on a small fraction of the sieving data, we obtain a good model of the distribution of the large primes in the relations. Combined with singleton removal, our estimation of the oversquareness is within 2 % of the real data. Thus we cheaply obtain a good estimate of the number of necessary relations for factoring a given number on a given computer, and hence of the actual computing time. Therefore, this method is a useful tool for optimizing parameters in the number field sieve, and we actually are using it in our practical factorization work.

Future work will include finding the correct model for the lattice sieve data sets of Kleinjung and check to which extent this model depends on the implementation of the siever. A second objective is to find a theoretical explanation for the occurrence of the various distributions (linear, exponential, ...) of the large primes. Another objective will be to find the optimal oversquareness for minimizing the resulting matrix. Once these issues are properly understood we intend to develop a tool to determine bounds F and L that optimize the overall effort for relation collection and matrix processing with respect to the available resources.

Acknowledgements

The author thanks Arjen Lenstra for suggesting the idea to predict the sieving effort by simulating relations on the basis of a short sieving test. She thanks Marie-Colette van Lieshout for suggesting several statistical models including the model which is used in Section 2, $r_1 a_0$, and Dag Arne Osvik for providing the singleton removal code for relations written in a special format.

The author thanks Arjen Lenstra, Herman te Riele, and Rob Tijdeman for reading the paper and giving constructive criticism and comments, Bruce Dodson and Thorsten Kleinjung for sharing data sets, and the anonymous referees for carefully reading the paper and suggesting clarifications.

Part of this research was carried out while the author was visiting École Polytechnique Fédérale de Lausanne in August 2006. She thanks Arjen Lenstra and EPFL for the hospitality during this visit.

References

1. Aoki, K., Franke, J., Kleinjung, T., Lenstra, A., Osvik, D.A.: A kilobit special number field sieve factorization. In: Kurosawa, K. (ed.) ASIACRYPT 2007. LNCS, vol. 4833, pp. 1–12. Springer, Heidelberg (2007)
2. Breiman, L.: Statistics: With a View Toward Applications. Houghton Mifflin Company, Boston (1973)
3. Elkenbracht-Huizing, M.: The Number Field Sieve. PhD thesis, University of Leiden (1997)
4. Lenstra, A.K., Lenstra Jr., H.W. (eds.): The Development of the Number Field Sieve. Lecture Notes in Math., vol. 1554. Springer, Berlin (1993)
5. Montgomery, P.L.: A survey of modern integer factorization algorithms. CWI Quarterly 7/4, 337–366 (1994)
6. Panaitopel, L.: A formula for $\pi(x)$ applied to a result of Koninck-Ivić. Nieuw Arch. Wiskunde 5/1, 55–56 (2000)

Improved Stage 2 to P ± 1 Factoring Algorithms

Peter L. Montgomery[1] and Alexander Kruppa[2]

[1] Microsoft Research, One Microsoft Way, Redmond, WA 98052 USA
pmontgom@cwi.nl
[2] LORIA, Campus Scientifique, BP 239, 54506 Vandœuvre-lès-Nancy Cedex, France
kruppaal@loria.fr

Abstract. Some implementations of stage 2 of the P–1 method of factorization use convolutions. We describe a space-efficient implementation, allowing convolution lengths around 2^{23} and stage 2 limit around 10^{16} while attempting to factor 230-digit numbers on modern PC's. We describe arithmetic algorithms on reciprocal polynomials. We present adjustments for the P+1 algorithm. We list some new findings.

Keywords: Integer factorization, convolution, discrete Fourier transform, number theoretic transform, P–1, P+1, multipoint polynomial evaluation, reciprocal polynomials.

1 Introduction

John Pollard introduced the P–1 algorithm for factoring an odd composite integer N in 1974 [11, §4]. It hopes that some prime factor p of N has smooth $p-1$. It picks $b_0 \not\equiv \pm 1 \pmod{N}$ and coprime to N and outputs $b_1 = b_0^e \bmod N$ for some positive exponent e. This exponent might be divisible by all prime powers below a bound B_1. Stage 1 succeeds if $(p-1) \mid e$, in which case $b_1 \equiv 1 \pmod{p}$ by Fermat's little theorem. The algorithm recovers p by computing $\gcd(b_1 - 1, N)$ (except in rare cases when this GCD is composite). When this GCD is 1, we hope that $p - 1 = qn$ where n divides e and q is not too large. Then

$$b_1^q \equiv (b_0^e)^q = b_0^{eq} = (b_0^{nq})^{e/n} = \left(b_0^{p-1}\right)^{e/n} \equiv 1^{e/n} = 1 \pmod{p}, \qquad (1)$$

so p divides $\gcd(b_1^q - 1, N)$. Stage 2 of P–1 tries to find p when $q > 1$ but q is not too large. The search bound for q is called B_2.

Pollard [11] tests each prime q in $[B_1, B_2]$ individually. If q_1 and q_2 are successive primes, then look up $b_1^{q_2-q_1} \bmod N$ in a small table. Given $b_1^{q_1} \bmod N$, form $b_1^{q_2} \bmod N$ and test $\gcd(b_1^{q_2} - 1, N)$. He observes that one can combine GCD tests: if $p \mid \gcd(x, N)$ or $p \mid \gcd(y, N)$, then $p \mid \gcd(xy \bmod N, N)$. His stage 2 cost is two modular multiplications per q, one GCD with N at the end, and a few multiplications to build the table.

Montgomery [7] uses two sets S_1 and S_2, such that each prime q in $[B_1, B_2]$ divides a nonzero difference $s_1 - s_2$ where $s_1 \in S_1$ and $s_2 \in S_2$. He forms $b_1^{s_1} - b_1^{s_2}$ using two table look-ups, saving one modular multiplication per q. Sometimes one

A.J. van der Poorten and A. Stein (Eds.): ANTS-VIII 2008, LNCS 5011, pp. 180–195, 2008.
© Springer-Verlag Berlin Heidelberg 2008

$s_1 - s_2$ works for multiple q. Montgomery adapts his scheme to Hugh Williams's P+1 method and Hendrik Lenstra's elliptic curve method (ECM).

These changes lower the constant of proportionality, but stage 2 still uses $O(\pi(B_2) - \pi(B_1))$ (number of primes between B_1 and B_2) operations modulo N.

The end of [11] suggests an FFT continuation to P−1. Silverman [8, p. 844] implements it, using a circular convolution to evaluate a polynomial along a geometric progression. It costs $O(\sqrt{B_2} \log B_2)$ operations to build and multiply two polynomials of degree $O(\sqrt{B_2})$, compared to $O(B_2/\log B_2)$ primes below B_2, so [8] beats [7] when B_2 is large.

Montgomery's dissertation [9] describes an FFT continuation to ECM. He takes the GCD of two polynomials. Zimmermann [13] implements another FFT continuation to ECM, based on evaluating a polynomial at arbitrary points. These cost an extra factor of $\log B_2$ when the points are not a geometric progression. Zimmermann adapts his implementation to $P \pm 1$ methods.

Like [8], we evaluate a polynomial along geometric progressions. We exploit patterns in its roots to generate its coefficients quickly. Those patterns are not present in ECM, so these techniques do not apply there. We aim for low memory overhead, saving it for convolution inputs and outputs (which are elements of $\mathbb{Z}/N\mathbb{Z}$). Using memory efficiently lets us raise the convolution length ℓ. Many intermediate results are reciprocal polynomials, which need about half the storage and can be multiplied using weighted convolutions.

Doubling ℓ costs slightly over twice as much time per convolution, but each longer convolution extends the search for q (and effective B_2) fourfold. Silverman's 1989 implementation used 42 megabytes and allowed 250-digit inputs. It repeatedly evaluated a polynomial of degree 15360 at $8 \cdot 17408$ points in geometric progression, using $\ell = 32768$. This enabled him to achieve $B_2 \approx 10^{10}$.

Today's (2008) PC memories are 100 times as large as that used in [8]. With this extra memory, we achieve $\ell = 2^{23}$, a growth factor of 256. With the same number of convolutions (individually longer lengths but running on faster hardware) our B_2 advances by a factor of $256^2 \approx 6.6e4$. Supercomputers with huge shared memories do spectacularly.

Section 12 gives some new results, including a record 60-digit P+1 factor.

2   P+1 Algorithm

Hugh Williams [12] introduced a P+1 factoring algorithm. It finds a prime factor p of N when $p + 1$ (rather than $p - 1$) is smooth. It is modeled after P−1.

One variant of the P+1 algorithm chooses $P_0 \in \mathbb{Z}/N\mathbb{Z}$ and lets the indeterminate α_0 be a zero of the quadratic $\alpha_0^2 - P_0\alpha_0 + 1$. We hope this quadratic is irreducible modulo p. If so, its second root in $\mathbb{F}_{p^2}$ will be α_0^p. The product of its roots is the constant term 1. Hence $\alpha_0^{p+1} \equiv 1 \pmod{p}$ when we choose well.

Stage 1 of the P+1 algorithm computes $P_1 = \alpha_1 + \alpha_1^{-1}$ where $\alpha_1 \equiv \alpha_0^e$ $\pmod{N}$ for some exponent e. If $\gcd(P_1 - 2, N) > 1$, then the algorithm succeeds. Stage 2 of P+1 hopes that $\alpha_1^q \equiv 1 \pmod{p}$ for some prime q, not too large, and some prime p dividing N.

Most techniques herein adapt to P+1, but some computations take place in an extension ring, raising memory usage if we use the same convolution sizes.

2.1 Chebyshev Polynomials

Although the theory behind P+1 mentions α_0 and $\alpha_1 = \alpha_0^e$, an implementation manipulates primarily values of $\alpha_0^n + \alpha_0^{-n}$ and $\alpha_1^n + \alpha_1^{-n}$ for various integers n rather than the corresponding values (in an extension ring) of α_0^n and α_1^n.

For integer n, the Chebyshev polynomials V_n and U_n are determined by $V_n(X + X^{-1}) = X^n + X^{-n}$ and $(X - X^{-1})U_n(X + X^{-1}) = X^n - X^{-n}$. The use of these polynomials shortens many formulas, such as

$$P_1 \equiv \alpha_1 + \alpha_1^{-1} \equiv \alpha_0^e + \alpha_0^{-e} = V_e(\alpha_0 + \alpha_0^{-1}) = V_e(P_0) \pmod{N}.$$

These polynomials have integer coefficients, so $P_1 \equiv V_e(P_0) \pmod{N}$ is in the base ring $\mathbb{Z}/N\mathbb{Z}$ even when α_0 and α_1 are not.

The Chebyshev polynomials satisfy many identities, including

$$
\begin{aligned}
V_{mn}(X) &= V_m(V_n(X)), \\
U_{m+n}(X) &= U_m(X)\,V_n(X) - U_{m-n}(X), \\
U_{m+n}(X) &= V_m(X)\,U_n(X) + U_{m-n}(X), \\
V_{m+n}(X) &= V_m(X)\,V_n(X) - V_{m-n}(X), \\
V_{m+n}(X) &= (X^2 - 4)\,U_m(X)\,U_n(X) + V_{m-n}(X).
\end{aligned}
\tag{2}
$$

(3)

3 Overview of Stage 2 Algorithm

Our algorithm performs multipoint evaluation of polynomials by convolutions. Its inputs are the output of stage 1 (b_1 for P–1 or P_1 for P+1), and the desired stage 2 interval $[B_1, B_2]$.

The algorithm chooses a highly composite odd integer P. It checks for q in arithmetic progressions with common difference $2P$. There are $\phi(P)$ such progressions to check when $\gcd(q, 2P) = 1$.

We need an even convolution length $\ell_{\max}$ (determined primarily by memory constraints) and a factorization $\phi(P) = s_1 s_2$ where s_1 is even and $0 < s_1 < \ell_{\max}$. Sections 5, 9.1 and 11 have sample values.

Our polynomial evaluations will need approximately

$$s_2 \left\lceil \frac{B_2 - B_1}{2P(\ell_{\max} - s_1)} \right\rceil \approx \frac{\phi(P)}{2P} \frac{B_2 - B_1}{s_1(\ell_{\max} - s_1)} \tag{4}$$

convolutions of length $\ell_{\max}$. We prefer a small $\phi(P)/P$ to keep (4) low. We also prefer s_1 near $\ell_{\max}/2$, say $0.3 \le s_1/\ell_{\max} \le 0.7$.

Using a factorization of $(\mathbb{Z}/P\mathbb{Z})^*$ as described in §5, it constructs two sets S_1 and S_2 of integers such that

(a) $|S_1| = s_1$ and $|S_2| = s_2$.
(b) S_1 is symmetric around 0. If $k \in S_1$, then $-k \in S_1$.

(c) If $k \in \mathbb{Z}$ and $\gcd(k, P) = 1$, then there exist unique $k_1 \in S_1$ and $k_2 \in S_2$ such that $k \equiv k_1 + k_2 \pmod{P}$.

Once S_1 and S_2 are chosen, it computes the coefficients of

$$f(X) = X^{-s_1/2} \prod_{k_1 \in S_1} (X - b_1^{2k_1}) \bmod N \tag{5}$$

by the method in §7. Since S_1 is symmetric around zero, this $f(X)$ is symmetric in X and $1/X$.

For each $k_2 \in S_2$ it evaluates (the numerators of) all

$$f(b_1^{2k_2 + (2m+1)P}) \bmod N \tag{6}$$

for $\ell_{\max} - s_1$ consecutive values of m as described in §8, and checks the product of these outputs for a nontrivial GCD with N. This checks $s_1(\ell_{\max} - s_1)$ (not necessarily prime) candidates, hoping to find q.

For the P+1 method, replace (5) by $f(X) = X^{-s_1/2} \prod_{k_1 \in S_1} (X - \alpha_1^{2k_1}) \bmod N$. Similarly, replace b_1 by α_1 in (6). The polynomial f is still over $\mathbb{Z}/N\mathbb{Z}$ since each product $(X - \alpha_1^{2k_1})(X - \alpha_1^{-2k_1}) = X^2 - V_{2k_1}(P_1) + 1 \in (\mathbb{Z}/N\mathbb{Z})[X]$ but the multipoint evaluation works in an extension ring. See §8.1.

4 Justification

Let p be an unknown prime factor of N. As in (1), assume $b_1^q \equiv 1 \pmod{p}$ where q is not too large, and $\gcd(q, 2P) = 1$.

The selection of S_1 and S_2 ensures there exist $k_1 \in S_1$ and $k_2 \in S_2$ such that $(q - P)/2 \equiv k_1 + k_2 \pmod{P}$. That is,

$$q = P + 2k_1 + 2k_2 + 2mP = 2k_1 + 2k_2 + (2m + 1)P \tag{7}$$

for some integer m. We can bound m knowing bounds on q, k_1, k_2, as detailed in §5. Both $b_1^{\mp 2k_1}$ are roots of $f \pmod{p}$. Hence

$$f(b_1^{2k_2 + (2m+1)P}) = f(b_1^{q - 2k_1}) \equiv f(b_1^{-2k_1}) \equiv 0 \pmod{p}. \tag{8}$$

For the P+1 method, if $\alpha_1^q \equiv 1 \pmod{p}$, then (8) evaluates f at $X = \alpha_1^{2k_2 + (2m+1)P} = \alpha_1^{q - 2k_1}$. The factor $X - \alpha_1^{-2k_1}$ of $f(X)$ evaluates to $r^{-2k_1}(\alpha_1^q - 1)$, which is zero modulo p even in the extension ring.

5 Selection of S_1 and S_2

Let "+" of two sets denote the set of sums. By the Chinese Remainder Theorem,

$$(\mathbb{Z}/(mn)\mathbb{Z})^* = n(\mathbb{Z}/m\mathbb{Z})^* + m(\mathbb{Z}/n\mathbb{Z})^* \text{ if } \gcd(m, n) = 1. \tag{9}$$

This is independent of the representatives: if $S \equiv (\mathbb{Z}/m\mathbb{Z})^*$ (mod m) and $T \equiv (\mathbb{Z}/n\mathbb{Z})^*$ (mod n), then $nS + mT \equiv (\mathbb{Z}/(mn)\mathbb{Z})^*$ (mod mn). For prime powers, $(\mathbb{Z}/p^k\mathbb{Z})^* = (\mathbb{Z}/p\mathbb{Z})^* + \sum_{i=1}^{k-1} p^i(\mathbb{Z}/p\mathbb{Z})$.

We choose S_1 and S_2 so that $S_1 + S_2 \equiv (\mathbb{Z}/P\mathbb{Z})^*$ (mod P) which ensures that all values coprime to P, in particular all primes, in the stage 2 interval are covered. One way uses a factorization $mn = P$ and (9). Other choices are available by factoring individual $(\mathbb{Z}/p\mathbb{Z})^*$, $p \mid P$, into smaller sets of sums.

Let $R_n = \{2i - n - 1 : 1 \leq i \leq n\}$ be the arithmetic progression centered at 0 of length n and common difference 2. For odd primes p, a set of representatives of $(\mathbb{Z}/p\mathbb{Z})^*$ is R_{p-1}. Its cardinality is composite for $p \neq 3$ and the set can be factored into arithmetic progressions of prime length by $R_{mn} = R_m + mR_n$. If $p \equiv 3$ (mod 4), alternatively $\frac{p+1}{4}R_2 + \frac{1}{2}R_{(p-1)/2}$ can be chosen as a set of representatives with smaller absolute values.

When evaluating (6) for all $m_1 \leq m < m_2$ and $k_2 \in S_2$, the highest exponent coprime to P that is *not* covered at the low end of the stage 2 range will be $2\max(S_1 + S_2) + (2m_1 - 1)P$. Similarly, the smallest value at the high end of the stage 2 range not covered is $2\min(S_1 + S_2) + (2m_2 + 1)P$. Hence, for a given choice of P, S_1, S_2, m_1 and m_2, all primes in $[(2m_1 - 1)P + 2\max(S_1 + S_2) + 1, (2m_2 + 1)P + 2\min(S_1 + S_2) - 1]$ are covered.

Choose parameters that minimize $s_2 \cdot \ell_{\max}$ so that $[B_1, B_2]$ is covered, $\ell_{\max}$ is permissible by available memory, and, given several choices, $(2m_2 + 1)P + 2\min(S_1 + S_2)$ is maximal.

For example, to cover the interval $[1000, 500000]$ with $\ell_{\max} = 512$, we might choose $P = 1155$, $s_1 = 240$, $s_2 = 2$, $m_1 = -1$, $m_2 = 271$. With $S_1 = 231(\{-1, 1\} + \{-2, 2\}) + 165(\{-2, 2\} + \{-1, 0, 1\}) + 105(\{-3, 3\} + \{-2, -1, 0, 1, 2\})$ and $S_2 = 385\{-1, 1\}$, we have $\max(S_1 + S_2) = -\min(S_1 + S_2) = 2098$ and thus cover all primes in $[-3 \cdot 1155 + 4196 + 1, \ 541 \cdot 1155 - 4196 - 1] = [732, 620658]$.

6   Circular Convolutions and Polynomial Multiplication

Let R be a ring and ℓ a positive integer. All rings herein are assumed commutative with 1. A *circular convolution* of length ℓ over R multiplies two polynomials $f_1(X)$ and $f_2(X)$ of degree at most $\ell - 1$ in $R[X]$, returning $f_1(X)f_2(X)$ mod $X^\ell - 1$. When $\deg(f_1) + \deg(f_2) < \ell$, this gives an exact product.

If R has a primitive ℓ-th root ω of unity, and if ℓ is not a zero divisor in R, then one convolution algorithm uses the discrete Fourier transform (DFT) [1, chapter 7]. Fix ω. A *forward DFT* evaluates all $f_1(\omega^i)$ for $0 \leq i < \ell$. Another forward DFT evaluates all ℓ values of $f_2(\omega^i)$. Multiply these pointwise. Then an *inverse DFT* interpolates to find a polynomial $f_3 \in R[X]$ of degree at most $\ell - 1$ with $f_3(\omega^i) = f_1(\omega^i)f_2(\omega^i)$ for all i. Return f_3.

If ℓ is a power of 2 and we use a fast Fourier transform (FFT) algorithm for the forward and inverse DFTs, then the convolution takes $O(\ell \log \ell)$ operations in a suitable ring, compared to $O(\ell^2)$ ring operations for the naïve algorithm.

6.1   Convolutions over $\mathbb{Z}/N\mathbb{Z}$

The DFT cannot be used directly when $R = \mathbb{Z}/N\mathbb{Z}$, since we don't know a suitable ω. As in [13, p. 534], we consider two ways to do the convolutions.

Montgomery [8, §4] suggests a number theoretic transform (NTT). He treats the input polynomial coefficients as integers in $[0, N - 1]$ and multiplies the polynomials over $\mathbb{Z}$. The product polynomial, reduced modulo $X^\ell - 1$, has coefficients in $[0, \ell(N - 1)^2]$. Select distinct NTT primes p_j that each fit into one machine word such that $\prod_j p_j > \ell(N - 1)^2$. Require each $p_j \equiv 1 \pmod{\ell}$, so a primitive ℓ-th root of unity exists. Do the convolution modulo each p_j and use the Chinese Remainder Theorem (CRT) to determine the product over $\mathbb{Z}$ modulo $X^\ell - 1$. Reduce this product modulo N. Montgomery's dissertation [9, chapter 8] describes these computations in detail.

The convolution codes need interfaces to (1) zero a DFT buffer (2) insert an entry modulo N in a DFT buffer, i.e. reduce it modulo the NTT primes, (3) perform a forward, in-place, DFT on a buffer, (4) multiply two DFT buffers pointwise, overwriting an input, and perform an in-place inverse DFT on the product, and (5) extract a product coefficient modulo N via a CRT computation and reduction modulo N.

The Kronecker-Schönhage convolution algorithm uses fast integer multiplication. See §11. Nussbaumer [10] gives other convolution algorithms.

6.2   Reciprocal Laurent Polynomials and Weighted NTT

Define a *reciprocal Laurent polynomial* (RLP) in X to be an expression $a_0 + \sum_{j=1}^{d} a_j \cdot (X^j + X^{-j}) = a_0 + \sum_{j=1}^{d} a_j V_j(X + X^{-1})$ for scalars a_j in a ring. It is *monic* if $a_d = 1$. It is said to have degree $2d$ if $a_d \neq 0$. The degree is always even. A monic RLP of degree $2d$ fits in d coefficients (excluding the leading 1).

While manipulating RLPs of degree at most $2d$, the *standard basis* is $\{1\} \cup \{X^j + X^{-j} : 1 \leq j \leq d\} = \{1\} \cup \{V_j(Y) : 1 \leq j \leq d\}$ where $Y = X + X^{-1}$.

Let $Q(X) = q_0 + \sum_{j=1}^{d_q} q_j(X^j + X^{-j})$ be an RLP of degree at most $2d_q$ and likewise $R(X)$ an RLP of degree at most $2d_r$. To obtain the product RLP $S(X) = Q(x)R(x) = s_0 + \sum_{j=1}^{d_s} s_j(X^j + X^{-j})$ of degree at most $2d_s = 2(d_q + d_r)$, choose a convolution length $\ell > d_s$ and perform a weighted convolution product [4] by computing $\tilde{S}(wX) = Q(wX)R(wX) \bmod (X^\ell - 1)$ for a suitable w. Suppose $\tilde{S}(wX) = \sum_{i=0}^{\ell-1} \tilde{s}_i X^i = S(wX) \bmod (X^\ell - 1)$. If $0 \leq i \leq d_s$, then the coefficient of X^i or X^{-i} in $Q(X)R(X)$ is s_i. The coefficient of X^i in $Q(wX)R(wX)$ is $s_i w^i$, whereas its coefficient of X^{-i} is s_i/w^i. When $0 \leq i < \ell$, the coefficient $\tilde{s}_i$ of X^i in $\tilde{S}(wX)$ has a contribution $s_i w^i$ from X^i in $Q(wX)R(wX)$ (if $i \leq d_s$) as well as $s_{\ell-i}/w^{\ell-i}$ from $X^{i-\ell}$ (if $d_s < \ell - i$). This translates to $\tilde{s}_i = s_i w^i$ when $0 \leq i < \ell - d_s$, which we can solve for s_i. When instead $\ell - d_s \leq i \leq d_s$, we find $\tilde{s}_i = s_i w^i + s_{\ell-i}/w^{\ell-i}$. Replacing i by $\ell - i$ gives the system $\begin{pmatrix} 1 & w^{-\ell} \\ w^{-\ell} & 1 \end{pmatrix}\begin{pmatrix} s_i \\ s_{\ell-i} \end{pmatrix} = \begin{pmatrix} \tilde{s}_i/w^i \\ \tilde{s}_{\ell-i}/w^{\ell-i} \end{pmatrix}$. There is a unique solution when $w \neq 0$ and the matrix is invertible.

This leads to the algorithm in Figure 1. It flows like the interface in §6.1.

Input: RLPs $Q(X) = q_0 + \sum_{j=1}^{d_q} q_j(X^j + X^{-j})$ of degree at most $2d_q$,
and $R(X) = r_0 + \sum_{j=1}^{d_r} r_j(X^j + X^{-j})$ of degree at most $2d_r$,
both in standard basis. A convolution length $\ell > d_q + d_r$
Output: RLP $S(X) = s_0 + \sum_{j=1}^{d_s} s_j(X^j + X^{-j}) = Q(X)R(X)$ of degree at most
$2d_s = 2d_q + 2d_r$ in standard basis. Output may overlap input
Auxiliary storage: NTT arrays M and M', each with ℓ elements per p_j.
A squaring may use the same array for M and M'

Zero M and M'
For each NTT prime p_j
    Choose w_j with $w_j^\ell \not\equiv 0, \pm 1 \pmod{p_j}$
    Set $M_{j,0} := q_0 \bmod p_j$ and $M'_{j,0} := r_0 \bmod p_j$
For $1 \le i \le d_q$ (any order)
    For each p_j
        Set $M_{j,i} := w_j^i q_i \bmod p_j$ and $M_{j,\ell-i} := w_j^{-i} q_i \bmod p_j$
Do similarly with R and M'
For each p_j
    Perform forward NTTs of length ℓ modulo p_j on $M_{j,*}$ and $M'_{j,*}$.
    Multiply elementwise $M_{j,*} := M_{j,*} * M'_{j,*}$ and perform inverse NTT on $M_{j,*}$
    For $1 \le i \le \ell - d_s - 1$ set $M_{j,i} := w_j^{-i} M_{j,i} \pmod{p_j}$
    For $\ell - d_s \le i \le \lfloor \ell/2 \rfloor$
        Set $\begin{pmatrix} M_{j,i} \\ M_{j,\ell-i} \end{pmatrix} := \begin{pmatrix} 1 & -w^{-\ell} \\ -w^{-\ell} & 1 \end{pmatrix} \begin{pmatrix} w^{-i} M_{j,i}/(1 - w^{-2\ell}) \\ w^{i-\ell} M_{j,\ell-i}/(1 - w^{-2\ell}) \end{pmatrix} \bmod p_j$
For $0 \le i \le d_s$ perform CRT on $M_{*,i}$ residues to obtain s_i, store in output

---

Fig. 1. NTT-Based Multiplication Algorithm for Reciprocal Laurent Polynomials

Our code chooses the NTT primes $p_j \equiv 1 \pmod{3\ell}$. We require $3 \nmid \ell$. Our w_j is a primitive cube root of unity. Multiplications by 1 are omitted. When $3 \nmid i$, we use $w_j^i q_i + w_j^{-i} q_i \equiv -q_i \pmod{p_j}$ to save a multiply.

Substituting $X = e^{i\theta}$ where $i^2 = -1$ gives

$$Q(e^{i\theta})R(e^{i\theta}) = \left(q_0 + 2\sum_{j=1}^{d_q} \cos j\theta \right) \left(r_0 + 2\sum_{j=1}^{d_r} \cos j\theta \right).$$

These cosine series can be multiplied using discrete cosine transforms, in approximately the same auxiliary space needed by the weighted convolutions. We did not implement that approach.

6.3 Multiplying General Polynomials by RLPs

In section 8 we will construct an RLP $h(X)$ which will later be multiplied by various $g(X)$. The length-ℓ DFT of $h(X)$ evaluates $h(\omega^i)$ for $0 \le i < \ell$. However since $h(X)$ is reciprocal, $h(\omega^i) = h(\omega^{\ell-i})$ and the DFT has only $\ell/2 + 1$ distinct coefficients. In signal processing, the DFT of a signal extended symmetrically around the center of each endpoint is called a Discrete Cosine Transform of type I. Using a DCT–I algorithm [2], we could compute the coefficients $h(\omega^i)$ for $0 \le i \le \ell/2$ with a length $\ell/2 + 1$ transform. We have not implemented this.

Instead we compute the full DFT of the RLP (using $X^\ell = 1$ to avoid negative exponents). To conserve memory, we store only the $\ell/2+1$ distinct DFT output coefficients for later use.

7 Computing Coefficients of f

Assume the P+1 algorithm. The monic RLP $f(X)$ in (5), with roots α_1^{2k} where $k \in S_1$, can be constructed using the decomposition of S_1. The coefficients of f will always be in the base ring since $P_1 \in \mathbb{Z}/N\mathbb{Z}$.

For the P–1 algorithm, set $\alpha_1 = b_1$ and $P_1 = b_1 + b_1^{-1}$. The rest of the construction of f for P–1 is identical to that for P+1.

Assume S_1 and S_2 are built as in §5, say $S_1 = T_1 + T_2 + \cdots + T_m$ where each T_j has an arithmetic progression of prime length, centered at zero. At least one of these has even cardinality since $s_1 = |S_1| = \prod_j |T_j|$ is even. Renumber the T_j so $|T_1| = 2$ and $|T_2| \geq |T_3| \geq \cdots \geq |T_m|$.

If $T_1 = \{-k_1, k_1\}$, then initialize $F_1(X) = X + X^{-1} - \alpha_1^{2k_1} - \alpha_1^{-2k_1} = X + X^{-1} - V_{2k_1}(P_1)$, a monic RLP in X of degree 2.

Suppose $1 \leq j < m$. Given the coefficients of the monic RLP $F_j(X)$ with roots $\alpha_1^{2k_1}$ for $k_1 \in T_1 + \cdots + T_j$, we want to construct

$$F_{j+1}(X) = \prod_{k_2 \in T_{j+1}} F_j(\alpha_1^{2k_2} X). \tag{10}$$

The set T_{j+1} is assumed to be an arithmetic progression of prime length $t = |T_{j+1}|$ centered at zero with common difference k, say $T_{j+1} = \{(-1-t)k/2 + ik : 1 \leq i \leq t\}$. If t is even, k is even to ensure integer elements. On the right of (10), group pairs $\pm k_2$ when $k_2 \neq 0$. We need the coefficients of

$$F_{j+1}(X) = \begin{cases} F_j(\alpha_1^{-k} X)\, F_j(\alpha_1^k X), & \text{if } t = 2; \\ F_j(X) \prod_{i=1}^{(t-1)/2} \big(F_j(\alpha_1^{2ki} X)\, F_j(\alpha_1^{-2ki} X)\big), & \text{if } t \text{ is odd.} \end{cases}$$

Let $d = \deg(F_j)$, an even number. The monic input F_j has $d/2$ coefficients in $\mathbb{Z}/N\mathbb{Z}$ (plus the leading 1). The output F_{j+1} will have $td/2 = \deg(F_{j+1})/2$ such coefficients.

Products such as $F_j(\alpha_1^{2ki} X)\, F_j(\alpha_1^{-2ki} X)$ can be formed by the method in §7.1, using d coefficients to store each product. The interface can pass $\alpha_1^{2ki} + \alpha_1^{-2ki} = V_{2ki}(P_1) \in \mathbb{Z}/N\mathbb{Z}$ as a parameter instead of $\alpha_1^{\pm 2ki}$.

For odd t, the algorithm in §7.1 forms $(t-1)/2$ such monic products each with d output coefficients. We still need to multiply by the input F_j. Overall we store $(d/2) + \frac{t-1}{2}d = td/2$ coefficients. Later these $(t+1)/2$ monic RLPs can be multiplied in pairs, with products overwriting the inputs, until F_{j+1} (with $td/2$ coefficients plus the leading 1) is ready.

All polynomial products needed for (10), including those in §7.1, have output degree at most $t \deg(F_j) = \deg(F_{j+1})$, which divides the final $\deg(F_m) = s_1$. The polynomial coefficients are saved in the (MZNZ) buffer of §9. The (MDFT) buffer allows convolution length $\ell_{\max}/2$, which is adequate when an RLP product has degree up to $2(\ell_{\max}/2)-1 \geq s_1$. A smaller length might be better for a particular product.

7.1 Scaling by a Power and Its Inverse

Let $F(X)$ be a monic RLP of even degree d, say $F(X) = c_0 + \sum_{i=1}^{d/2} c_i(X^i + X^{-i})$, where each $c_i \in \mathbb{Z}/N\mathbb{Z}$ and $c_{d/2} = 1$. Given $Q \in \mathbb{Z}/N\mathbb{Z}$, where $Q = \gamma + \gamma^{-1}$ for some unknown γ, we want the d coefficients (excluding the leading 1) of $F(\gamma X)\, F(\gamma^{-1}X)$ mod N in place of the $d/2$ such coefficients of F. We are allowed a few scalar temporaries and any storage internal to the polynomial multiplier.

Denote $Y = X + X^{-1}$. Rewrite, while pretending to know γ,

$$F(\gamma X) = c_0 + \sum_{i=1}^{d/2} c_i(\gamma^i X^i + \gamma^{-i}X^{-i})$$

$$= c_0 + \sum_{i=1}^{d/2} \frac{c_i}{2}\left((\gamma^i + \gamma^{-i})(X^i + X^{-i}) + (\gamma^i - \gamma^{-i})(X^i - X^{-i})\right)$$

$$= c_0 + \sum_{i=1}^{d/2} \frac{c_i}{2}\left(V_i(Q)V_i(Y) + (\gamma - \gamma^{-1})U_i(Q)(X - X^{-1})U_i(Y)\right).$$

Replace γ by γ^{-1} and multiply to get

$$F(\gamma X)\, F(\gamma^{-1}X) = G^2 - (\gamma - \gamma^{-1})^2(X - X^{-1})^2\, H^2$$
$$= G^2 - (Q^2 - 4)(X - X^{-1})^2\, H^2, \tag{11}$$

where

$$G = c_0 + \sum_{i=1}^{d/2} c_i \frac{V_i(Q)}{2} V_i(Y), \qquad H = \sum_{i=1}^{d/2} c_i \frac{U_i(Q)}{2} U_i(Y).$$

This G is a (not necessarily monic) RLP of degree at most d in the standard basis, with coefficients in $\mathbb{Z}/N\mathbb{Z}$. This H is another RLP, of degree at most $d-2$, but using the basis $\{U_i(Y) : 1 \leq i \leq d/2\}$. Starting with the coefficient of $U_{d/2}(Y)$, we can repeatedly use $U_{j+1}(Y) = V_j(Y)U_1(Y) + U_{j-1}(Y) = V_j(Y) + U_{j-1}(Y)$ for $j > 0$, along with $U_1(Y) = 1$ and $U_0(Y) = 0$, to convert H to standard basis. This conversion costs $O(d)$ additions in $\mathbb{Z}/N\mathbb{Z}$.

Use (3) and (2) to evaluate $V_i(Q)/2$ and $U_i(Q)/2$ for consecutive i as you evaluate the $d/2 + 1$ coefficients of G and the $d/2$ coefficients of H. Using the memory model in §9, and the algorithm in Figure 1, write the NTT images of the standard-basis coefficients of G and H to different parts of (MDFT). Later retrieve the $d-1$ coefficients of H^2 and the $d+1$ coefficients of G^2 as you finish the (11) computation. Discard the leading 1.

8 Multipoint Polynomial Evaluation

We have constructed $f = F_m$ in (5). The monic RLP $f(X)$ has degree s_1, say $f(X) = f_0 + \sum_{j=1}^{s_1/2} f_j \cdot (X^j + X^{-j}) = \sum_{j=-s_1/2}^{s_1/2} f_j X^j$ where $f_j = f_{-j} \in \mathbb{Z}/N\mathbb{Z}$.

Assuming the P–1 method (otherwise see §8.1), compute $r = b_1^P \in \mathbb{Z}/N\mathbb{Z}$. Set $\ell = \ell_{\max}$ and $M = \ell - 1 - s_1/2$.

Equation (6) needs $\gcd(f(X), N)$ where $X = b_1^{2k_2+(2m+1)P}$, for several consecutive m, say $m_1 \le m < m_2$. By setting $x_0 = b_1^{2k_2+(2m_1+1)P}$, the arguments to f become $x_0 b_1^{2mP} = x_0 r^{2m}$ for $0 \le m < m_2 - m_1$. The points of evaluation form a geometric progression with ratio r^2. We can evaluate these for $0 \le m < \ell - 1 - s_1$ with one convolution of length ℓ and $O(\ell)$ setup cost [1, exercise 8.27].

To be precise, set $h_j = r^{-j^2} f_j$ for $-s_1/2 \le j \le s_1/2$. Then $h_j = h_{-j}$. Set $h(X) = \sum_{j=-s_1/2}^{s_1/2} h_j X^j$, an RLP. The construction of h does not reference x_0 — we reuse h as x_0 varies.

Let $g_i = x_0^{M-i} r^{(M-i)^2}$ for $0 \le i \le \ell - 1$ and $g(X) = \sum_{i=0}^{\ell-1} g_i X^i$.

All nonzero coefficients in $g(X)h(X)$ have exponents from $0 - s_1/2$ to $(\ell-1) + s_1/2$. Suppose $0 \le m \le \ell - 1 - s_1$. Then $M - m - \ell = -1 - s_1/2 - m < -s_1/2$ whereas $M - m + \ell = (\ell - 1 + s_1/2) + (\ell - s_1 - m) > \ell - 1 + s_1/2$. The coefficient of X^{M-m} in $g(X)h(X)$, reduced modulo $X^\ell - 1$, is

$$\sum_{\substack{0 \le i \le \ell-1 \\ -s_1/2 \le j \le s_1/2 \\ i+j \equiv M-m \,(\mathrm{mod}\ \ell)}} g_i h_j \quad = \sum_{\substack{0 \le i \le \ell-1 \\ -s_1/2 \le j \le s_1/2 \\ i+j = M-m}} g_i h_j = \sum_{j=-s_1/2}^{s_1/2} g_{M-m-j} h_j$$

$$= \sum_{j=-s_1/2}^{s_1/2} x_0^{m+j} r^{(m+j)^2} r^{-j^2} f_j = \sum_{j=-s_1/2}^{s_1/2} x_0^m r^{m^2} \left(x_0 r^{2m}\right)^j f_j = x_0^m r^{m^2} f(x_0 r^{2m}).$$

Since we want only $\gcd(f(x_0\, r^{2m}), N)$, the $x_0^m\, r^{m^2}$ factors are harmless.

We can compute successive $g_{\ell-i}$ with two ring multiplications each since the ratios $g_{\ell-1-i}/g_{\ell-i} = x_0\, r^{2i-s_1-1}$ form a geometric progression.

8.1 Adaptation for P+1 Algorithm

If we replace b_1 with α_1, then r becomes α_1^P, which satisfies $r + r^{-1} = V_P(P_1)$. The above algebra evaluates f at powers of α_1. However α_1, r, h_j, x_0, and g_i lie in an extension ring.

Arithmetic in the extension ring can use a basis $\{1, \sqrt{\Delta}\}$ where $\Delta = P_1^2 - 4$. The element α_1 maps to $(P_1 + \sqrt{\Delta})/2$. A product $(c_0 + c_1\sqrt{\Delta})(d_0 + d_1\sqrt{\Delta})$ where $c_0, c_1, d_0, d_1 \in \mathbb{Z}/N\mathbb{Z}$ can be done using four base-ring multiplications: $c_0 d_0$, $c_1 d_1$, $(c_0 + c_1)(d_0 + d_1)$, $c_1 d_1 \Delta$, plus five base-ring additions.

We define linear transformations E_1, E_2 on $(\mathbb{Z}/N\mathbb{Z})[\sqrt{\Delta}]$ so that $E_1(c_0 + c_1\sqrt{\Delta}) = c_0$ and $E_2(c_0 + c_1\sqrt{\Delta}) = c_1$ for all $c_0, c_1 \in \mathbb{Z}/N\mathbb{Z}$. Extend E_1 and E_2 to polynomials by applying them to each coefficient.

To compute r^{n^2} for successive n, we use recurrences. We observe

$$r^{n^2} = r^{(n-1)^2+2} \cdot V_{2n-3}(r + r^{-1}) - r^{(n-2)^2+2},$$

$$r^{n^2+2} = r^{(n-1)^2+2} \cdot V_{2n-1}(r + r^{-1}) - r^{(n-2)^2} .$$

After initializing the variables $\mathbf{r1}[i] := r^{i^2}$, $\mathbf{r2}[i] := r^{i^2+2}$, $\mathbf{v}[i] := V_{2i+1}(r + r^{-1})$ for two consecutive i, we can compute $\mathbf{r1}[i] = r^{i^2}$ for larger i in sequence by

$$\mathbf{r1}[i] := \mathbf{r2}[i-1] \cdot \mathbf{v}[i-2] - \mathbf{r2}[i-2],$$

$$\mathbf{r2}[i] := \mathbf{r2}[i-1] \cdot \mathbf{v}[i-1] - \mathbf{r1}[i-2], \tag{12}$$

$$\mathbf{v}[i] := \mathbf{v}[i-1] \cdot V_2(r + 1/r) - \mathbf{v}[i-2] .$$

Since we won't use $\mathbf{v}[i-2]$ and $\mathbf{r2}[i-2]$ again, we can overwrite them with $\mathbf{v}[i]$ and $\mathbf{r2}[i]$. For the computation of r^{-n^2} where r has norm 1, we can use r^{-1} as input, by taking the conjugate.

All $\mathbf{v}[i]$ are in the base ring but $\mathbf{r1}[i]$ and $\mathbf{r2}[i]$ are in the extension ring. Each application of (12) takes five base-ring multiplications (compared to two multiplications per r^{n^2} in the P–1 algorithm).

We can compute successive $g_i = x_0^{M-i} r^{(M-i)^2}$ similarly. One solution to (12) is $\mathbf{r1}[i] = g_i$, $\mathbf{r2}[i] = r^2 g_i$, $\mathbf{v}[i] = x_0 r^{2M-2i-1} + x_0^{-1} r^{1+2i-2M}$. Again each $\mathbf{v}[i]$ is in the base ring, so (12) needs only five base-ring multiplications.

If we try to follow this approach for the multipoint evaluation, we need twice as much space for an element of $(\mathbb{Z}/N\mathbb{Z})[\sqrt{\Delta}]$ as one of $\mathbb{Z}/N\mathbb{Z}$. We also need a convolution routine for the extension ring.

If p divides the coefficient of X^{M-m} in $g(X)h(X)$, then p divides both coordinates thereof. The coefficients of $g(X)h(X)$ occasionally lie in the base ring, making $E_2(g(X)h(X))$ a poor choice for the gcd with N. Instead we compute

$$E_1(g(X)h(X)) = E_1(g(X))E_1(h(X)) + \Delta E_2(g(X))E_2(h(X)) . \tag{13}$$

The RLPs $E_1(h(X))$ and $E_2(\Delta h(X))$ can be computed once and for each the $\ell_{\max}/2 + 1$ distinct coefficients of its length-$\ell_{\max}$ DFT saved in (MHDFT). To compute $E_2(\Delta h(X))$, multiply $E_2(\mathbf{r1}[i])$ and $E_2(\mathbf{r2}[i])$ by Δ after initializing for two consecutive i. Then apply (12).

Later, as each g_i is computed we insert the NTT image of $E_2(g_i)$ into (MDFT) while saving $E_1(g_i)$ in (MZNZ) for later use. After forming $E_2(g(X))E_1(h(X))$, retrieve and save coefficients of X^{M-m} for $0 \le m \le \ell - 1 - s_1$. Store these in (MZNZ) while moving the entire saved $E_1(g_i)$ into the (now available) (MDFT) buffer. Form the $E_1(g(X))E_2(\Delta h(X))$ product and the sum in (13).

9   Memory Allocation Model

We aim to fit our major data into the following:

(MZNZ)
:   An array with $s_1/2$ elements of $\mathbb{Z}/N\mathbb{Z}$, for convolution inputs and outputs. This is used during polynomial construction. This is not needed during P–1 evaluation. During P+1 evaluation, it grows to $\ell_{\max}$ elements of $\mathbb{Z}/N\mathbb{Z}$.

(MDFT)
:   An NTT array holding $\ell_{\max}$ values modulo each prime p_j, for use during DWTs.

Section 7.1 does two overlapping squarings, whereas §7 multiplies two arbitrary RLPs. Each product degree is at most $\deg(f) = s_1$. The algorithm in Figure 1 needs $\ell \geq s_1/2$ and might use convolution length $\ell = \ell_{\max}/2$, assuming $\ell_{\max}$ is even. Two arrays of this length fit in (MDFT).

After f has been constructed, (MDFT) is used for NTT transforms with length up to $\ell_{\max}$.

(MHDFT)

Section 8 scales the coefficients of f by powers of r to build h. Then it builds and stores a length-ℓ DFT of h, where $\ell = \ell_{\max}$. This transform output normally needs ℓ elements per p_j for P$-$1 and 2ℓ elements per p_j for P+1. The symmetry of h lets us cut these needs almost in half, to $\ell/2+1$ elements for P$-$1 and $\ell + 2$ elements for P+1.

During the construction of F_{j+1} from F_j, if we need to multiply pairs of monic RLPs occupying adjacent locations within (MZNZ) (without the leading 1's), we use (MDFT) and the algorithm in Figure 1. The outputs overwrite the inputs within (MZNZ).

During polynomial evaluation for P$-$1, we need only (MHDFT) and (MDFT). Send the NTT image of each g_i coefficient to (MDFT) as g_i is computed. When (MDFT) fills (with $\ell_{\max}$ entries), do a length-$\ell_{\max}$ forward DFT on (MDFT), pointwise multiply by the saved DFT output from h in (MHDFT), and do an inverse DFT in (MDFT). Retrieve each needed polynomial coefficient, compute their product, and take a GCD with N.

9.1   Potentially Large B_2

Nowadays (2008) a typical PC memory is 4 gigabytes. The median size of composite cofactors N in the Cunningham project **http://homes.cerias.purdue. edu/~ssw/cun/index.html** is about 230 decimal digits, which fits in twelve 64-bit words (called *quadwords*). Table 1 estimates the memory requirements during stage 2, when factoring a 230-digit number, for both polynomial construction and polynomial evaluation phases, assuming convolutions use the NTT approach in §6.1. The product of our NTT prime moduli must be at least $\ell_{\max}(N - 1)^2$.

Table 1. Estimated memory usage (quadwords) while factoring 230-digit number

Array name	Construct f. Both $P \pm 1$	Build h.		Evaluate f.	
(MZNZ)	$12(s_1/2)$	$12(s_1/2)$		0	(P$-$1)
				$12\ell_{\max}$	(P+1)
(MDFT)	$25\ell_{\max}$	$25\ell_{\max}$		$25\ell_{\max}$	
(MHDFT)	0	$25(\ell_{\max}/2 + 1)$	(P$-$1)	$25(\ell_{\max}/2 + 1)$	(P$-$1)
		$25(\ell_{\max} + 2)$	(P+1)	$25(\ell_{\max} + 2)$	(P+1)
Totals, if $s_1 = \ell_{\max}/2$	$28\ell_{\max} + O(1)$	$40.5\ell_{\max} + O(1)$	(P$-$1)	$37.5\ell_{\max} + O(1)$	(P$-$1)
		$53\ell_{\max} + O(1)$	(P+1)	$62\ell_{\max} + O(1)$	(P+1)

If $N^2 \ell_{max}$ is below $0.99 \cdot (2^{63})^{25} \approx 10^{474}$, then it will suffice to have 25 NTT primes, each 63 or 64 bits.

The P–1 polynomial construction phase uses an estimated $40.5\ell_{max}$ quadwords, vs. $37.5\ell_{max}$ quadwords during polynomial evaluation. We can reduce the overall maximum to $37.5\ell_{max}$ by taking the (full) DFT transform of h in (MDFT), and releasing the (MZNZ) storage before allocating (MHDFT).

Four gigabytes is 537 million quadwords. A possible value is $\ell_{max} = 2^{23}$, which needs 315 million quadwords. When transform length $3 \cdot 2^k$ is supported, we could use $\ell_{max} = 3 \cdot 2^{22}$, which needs 472 million quadwords.

We might use $P = 3 \cdot 5 \cdot 7 \cdot 11 \cdot 13 \cdot 17 \cdot 19 \cdot 23 = 111546435$, for which $\phi(P) = 36495360 = 2^{13} \cdot 3^4 \cdot 5 \cdot 11$. We choose $s_2 \mid \phi(P)$ so that s_2 is close to $\phi(P)/(\ell_{max}/2) \approx 8.7$, i.e. $s_2 = 9$ and $s_1 = 4055040$, giving $s_1/\ell_{max} \approx 0.48$.

We can do 9 convolutions, one for each $k_2 \in S_2$. We will be able to find $p \mid N$ if $b_1^q \equiv 1 \pmod{p}$ where q satisfies (7) with $m < \ell_{max} - s_1 = 4333568$. As described in §5, the effective value of B_2 will be about $9.66 \cdot 10^{14}$.

Larger systems can search further in little more time.

10   Opportunities for Parallelization

Modern PC's are multi-core, typically with 2–4 CPUs (cores) and a shared memory. When running on such systems, it is desirable to utilize multiple cores.

While building $h(X)$ and $g(X)$ in §8, each core can process a contiguous block of subscripts. Use the explicit formulas to compute r^{-j^2} or g_i for the first two elements of a block, and the recurrences elsewhere.

If convolutions use NTT's and the number of processors divides the number of primes, then allocate the primes evenly across the processors. The (MDFT) and (MHDFT) buffers in §9 can have separate subbuffers for each prime. On NUMA architectures, the memory for each subbuffer should be allocated locally to the processor that will process it. Accesses to remote memory occur only when converting the h_j and g_i to residues modulo small primes, and when reconstructing the coefficients of $g(x)h(x)$ with the CRT.

11   Our Implementation

Our implementation is based on GMP-ECM, an implementation of P–1, P+1, and the Elliptic Curve Method for integer factorization. It uses the GMP library [5] for arbitrary precision arithmetic. The code for stage 1 of P–1 and P+1 is unchanged; the code for the new stage 2 has been written from scratch and will replace the previous implementation [13] which used product trees of cost $O\left(n(\log n)^2\right)$ modular multiplications for building polynomials of degree n and a variant of Montgomery's POLYEVAL [9] algorithm for multipoint evaluation which has cost $O\left(n(\log n)^2\right)$ modular multiplications and $O(n \log n)$ memory. The practical limit for B_2 was about $10^{14} - 10^{15}$.

GMP-ECM includes modular arithmetic routines, using e.g. Montgomery's REDC [6], or fast reduction modulo a number of the form $2^n \pm 1$. It also includes routines for polynomial arithmetic, in particular convolution products. One algorithm available for this purpose is a small prime NTT/CRT, using the "Explicit CRT" [3] variant which speed reduction modulo N after the CRT step but requires 2 or 3 additional small primes. Its current implementation allows only power-of-two transform lengths. Another is Kronecker-Schönhage's segmentation method [13], which is faster than the NTT if the modulus is large and the convolution length is comparatively small, and it works for any convolution length. Its main disadvantage is significantly higher memory use, reducing the possible convolution length.

On a 2.4 GHz Opteron with 8 GB memory, P–1 stage 2 on a 230-digit composite cofactor of $12^{254} + 1$ with $B_2 = 1.2 \cdot 10^{15}$, using the NTT with 27 primes for the convolution, can use $P = 64579515$, $\ell_{\max} = 2^{24}$, $s_1 = 7434240$, $s_2 = 3$ and takes 1738 seconds while P+1 stage 2 takes 3356 seconds. Using multi-threading to use both cpus on the same machine, P–1 stage 2 with the same parameters takes 1753 seconds cpu and 941 seconds elapsed time while P+1 takes 3390 seconds cpu and 2323 seconds elapsed time. For comparison, the previous implementation of P–1 stage 2 in GMP-ECM [13] needs to use a polynomial $F(X)$ of degree 1013760 and 80 blocks for $B_2 = 10^{15}$ and takes 34080 seconds on one cpu of the same machine.

On a 2.6 GHz Opteron with 8 cores and 32 GB of memory, a multi-threaded P–1 stage 2 on the same input number with the same parameters takes 1661 seconds cpu and 269 seconds elapsed time, while P+1 takes 3409 seconds cpu and 642 seconds elapsed time. With $B_2 = 1.34 \cdot 10^{16}$, $P = 198843645$, $\ell_{\max} = 2^{26}$, $s_1 = 33177600$, $s_2 = 2$, P–1 stage 2 takes 5483 seconds cpu and 922 elapsed time while P+1 takes 10089 seconds cpu and 2192 seconds elapsed time.

12   Some Results

We ran at least one of $P \pm 1$ on over 1500 composite cofactors, including

(a) Richard Brent's tables with $b^n \pm 1$ factorizations for $13 \leq b \leq 99$;
(b) Fibonacci and Lucas numbers F_n and L_n with $n < 2000$, or $n < 10000$ and cofactor size $< 10^{300}$;
(c) Cunningham cofactors of $12^n \pm 1$ with $n < 300$;
(d) Cunningham cofactors c300 and larger.

The B_1 and B_2 values varied, with 10^{11} and 10^{16} being typical. Table 2 has new large prime factors p and the largest factors of the corresponding $p \pm 1$.

The 52-digit factor of $47^{146} + 1$ and the 60-digit factor of L_{2366} each set a new record for the P+1 factoring algorithm upon their discovery. The previous record was a 48-digit factor of L_{1849}, found by the second author in March 2003.

The 53-digit factor of $24^{142} + 1$ has $q = 12750725834505143$, a 17-digit prime. To our knowledge, this is the largest prime in the group order associated with any factor found by the P–1, P+1 or Elliptic Curve methods of factorization.

The largest q reported in Table 2 of [8] is $q = 6496749983$ (10 digits), for a 19-digit factor p of $2^{895} + 1$. That table includes a 34-digit factor of the Fibonacci number F_{575}, which was the P–1 record in 1989.

Table 2. Large $P \pm 1$ factors found

Input	Factor p found	Size
Method	Largest factors of $p \pm 1$	
$73^{109} - 1$	76227040047863715568322367158695720006439518152299	c191
P–1	$12491 \cdot 37987 \cdot 156059 \cdot 2244509 \cdot 462832247372839$	p50
$68^{118} + 1$	750668634803774062109771018320047658050507374932508$9^*$	c151
P–1	$22807 \cdot 480587 \cdot 14334767 \cdot 89294369 \cdot 4649376803 \cdot 5380282339$	p52
$24^{142} + 1$	204890474274505790519896836864533701541268201046245 37	c183
P–1	$4959947 \cdot 7216081 \cdot 16915319 \cdot 17286223 \cdot 12750725834505143$	p53
$47^{146} + 1$	79864788660358229882201629788746313352749574950084 01	c235
P+1	$20540953 \cdot 56417663 \cdot 1231471331 \cdot 1632221953 \cdot 843497917739$	p52
L_{2366}	725516237739635905037132916171116034279215026146021770250523	c290
P+1	$932677 \cdot 62754121 \cdot 19882583417 \cdot 751245344783 \cdot 483576618980159$	p60

* = Found during stage 1

The largest P–1 factor reported in [13, pp. 538–539] is a 58-digit factor of $2^{2098} + 1$ with $q = 9909876848747$ (13 digits). Site **http://www.loria.fr/~zimmerma/records/Pminus1.html** has other records, including a 66-digit factor of $960^{119} - 1$ found by P–1 for which $q = 2110402817$ (only ten digits).

The first author ran stage 1 with $B_1 = 10^{11}$ for the p53 of $24^{142} + 1$ in Table 2. It took 44 hours on a 2200 MHz AMD Athlon processor in 32-bit mode at CWI.

Table 3. Timing for stage 2 of $24^{142} + 1$ factorization

Operation	Minutes (per CPU)	Parameters
Compute f	22	$P = 198843645$
Compute h	2	$\ell_{\max} = 2^{26}$
Compute DCT–I(h)	8	$s_1 = 33177600$
Compute all g_i	6 (twice)	$s_2 = 1$
Compute $g * h$	17 (twice)	$m_1 = 246$
Test for non-trivial GCD	2 (twice)	
Total	$32 + 2 \cdot 25 = 82$	

Stage 2 was run by the second author on an 8-core, 32 Gb node of the Grid5000 network. Table 3 shows where the time went. The overall stage 2 time is $8 \cdot 82 = 656$ minutes, about 25% of the stage 1 CPU time.

Acknowledgements

We thank Paul Zimmermann for his advice and guidance; and thank the reviewers for their comments. We are grateful to the Centrum voor Wiskunde en Informatica (CWI, Amsterdam) and to INRIA for providing huge amounts of computer time for this work.

Experiments presented in this paper were carried out using the Grid'5000 experimental testbed, an initiative from the French Ministry of Research through the ACI GRID incentive action, INRIA, CNRS and RENATER and other contributing partners (see `https://www.grid5000.fr`).

References

1. Aho, A.V., Hopcroft, J.E., Ullman, J.D.: The Design and Analysis of Computer Algorithms. Addison-Wesley, Reading (1974)
2. Baszenski, G., Tasche, M.: Fast polynomial multiplication and convolutions related to the discrete cosine transform. Linear Algebra and its Applications 252, 1–25 (1997)
3. Bernstein, D.J., Sorenson, J.P.: Modular exponentiation via the explicit Chinese remainder theorem. Math. Comp. 76, 443–454 (2007)
4. Crandall, R., Fagin, B.: Discrete weighted transforms and large-integer arithmetic. Math. Comp. 62, 305–324 (1994)
5. Granlund, T.: GNU MP: The GNU Multiple Precision Arithmetic Library, `http://gmplib.org/`
6. Montgomery, P.L.: Modular multiplication without trial division. Math. Comp. 44, 519–521 (1985)
7. Montgomery, P.L.: Speeding the Pollard and elliptic curve methods of factorization. Math. Comp. 48, 243–264 (1987)
8. Montgomery, P.L., Silverman, R.D.: An FFT extension to the $P - 1$ factoring algorithm. Math. Comp. 54, 839–854 (1990)
9. Montgomery, P.L.: An FFT Extension to the Elliptic Curve Method of Factorization. UCLA dissertation (1992), `ftp://ftp.cwi.nl/pub/pmontgom`
10. Nussbaumer, H.J.: Fast Fourier Transform and convolution algorithms, 2nd edn. Springer, Heidelberg (1982)
11. Pollard, J.M.: Theorems on factorization and primality testing. Proc. Cambridge Philosophical Society 76, 521–528 (1974)
12. Williams, H.C.: A $p + 1$ method of factoring. Math. Comp. 39, 225–234 (1982)
13. Zimmermann, P., Dodson, B.: 20 years of ECM. In: Hess, F., Pauli, S., Pohst, M. (eds.) ANTS 2006. LNCS, vol. 4076, pp. 525–542. Springer, Heidelberg (2006)

Shimura Curve Computations Via K3 Surfaces of Néron–Severi Rank at Least 19

Noam D. Elkies*

Department of Mathematics, Harvard University, Cambridge, MA 02138
elkies@math.harvard.edu

1 Introduction

In [E1] we introduced several computational challenges concerning Shimura curves, and some techniques to partly address them. The challenges are: obtain explicit equations for Shimura curves and natural maps between them; determine a Schwarzian equation on each curve (a.k.a. Picard–Fuchs equation, a linear second-order differential equation with a basis of solutions whose ratio inverts the quotient map from the upper half-plane to the curve); and locate CM (complex multiplication) points on the curves. We identified some curves, maps, and Schwarzian equations using the maps' ramification behavior; located some CM points as images of fixed points of involutions; and conjecturally computed others by numerically solving the Schwarzian equations.

But these approaches are limited in several ways: we must start with a Shimura curve with very few elliptic points (not many more than the minimum of three); maps of high degree are hard to recover from their ramification behavior, limiting the range of provable CM coordinates; and these methods give no access to the abelian varieties with quaternionic multiplication (QM) parametrized by Shimura curves. Other approaches somewhat extend the range where our challenges can be met. Detailed theoretical knowledge of the arithmetic of Shimura curves makes it possible to identify some such curves of genus at most 2 far beyond the range of [E1] (see e.g. [Rob, GR]), though not their Schwarzian equations or CM points. Roberts [Rob] showed in principle how to find CM coordinates using product formulas analogous to those of [GZ] for differences between CM j-invariants, but such formulas have yet to be used to verify and extend the tables of [E1]. Errthum [Er] recently used Borcherds products to verify all the conjectural rational coordinates for CM points tabulated in [E1] for the curves associated to the quaternion algebras over $\mathbf{Q}$ ramified at $\{2,3\}$ and $\{2,5\}$; it is not yet clear how readily this technique might extend to more complicated Shimura curves. The p-adic numerical techniques of [E3] give access to further maps and CM points. Finally, in the $\{2,3\}$ and $\{2,5\}$ cases Hashimoto and Murabayashi had already parametrized the relevant QM abelian surfaces in 1995 [HM], but apparently such computations have not been pushed further since then.

* Supported in part by NSF grant DMS-0501029.

In this paper we introduce a new approach, which exploits the fact that some Shimura curves also parametrize K3 surfaces of Néron–Severi rank at least 19. "Singular" K3 surfaces, those whose Néron–Severi rank attains the characteristic-zero maximum of 20, then correspond to CM points on the curve. We first encountered such parametrizations while searching for elliptic K3 surfaces of maximal Mordell–Weil rank over $\mathbf{Q}(t)$ (see [E4]), for which we used the K3 surface corresponding to a rational non-CM point on the Shimura curve $X(6, 79)/\langle w_{6 \cdot 79} \rangle$ of genus 2. The feasibility of this computation suggested that such parametrizations might be used systematically in Shimura curve computations.

This approach is limited to Shimura curves associated to quaternion algebras over $\mathbf{Q}$. Within that important special case, though, we can compute curves and CM points that were previously far beyond reach. The periods of the K3 surfaces should also allow the computation of Schwarzian equations as in [LY], though we have not attempted this yet. We do, however, find the corresponding QM surfaces using Kumar's recent formulas [Ku] that make explicit Dolgachev's correspondence [Do] between Jacobians of genus-2 curves and certain K3 surfaces of rank at least 17. The parametrizations do get harder as the level of the Shimura curve grows, but it is still much easier to parametrize the K3 surfaces than to work directly with the QM abelian varieties — apparently because the level, reflected in the discriminant of the Néron–Severi group, is spread over 19 Néron–Severi generators rather than the handful of generators of the endomorphism ring.[1] In this paper we illustrate this with several examples of such computations for the curves $X(N, 1)$ and their quotients. As the example of $X(6, 79)/\langle w_{6 \cdot 79} \rangle$ shows, the technique also applies to Shimura curves not covered by $X(N, 1)$, but already for $X(N, 1)$ there is so much new data that we can only offer a small sample here: the full set of results can be made available online but is much too large for conventional publication. Since we shall not work with $X(N, M)$ for $M > 1$, we abbreviate the usual notation $X(N, 1)$ to $X(N)$ here.

The rest of this paper is organized as follows. In the next section, we review the necessary background, drawn mostly from [Vi, Rot2, BHPV], concerning Shimura curves, the abelian and K3 surfaces that they parametrize, and the structure of elliptic K3 surfaces in characteristic zero; then give A. Kumar's explicit formulas for Dolgachev's correspondence, which we use to recover Clebsch–Igusa coordinates for QM Jacobians from our K3 parametrizations; and finally describe some of our techniques for computing such parametrizations. In the remaining sections we illustrate these techniques in the four cases $N = 6$, $N = 14$, $N = 57$, and $N = 206$. For $N = 6$ we find explicit elliptic models for our family of K3 surfaces S parametrized by $X(6)/\langle w_6 \rangle$, locate a few CM points to find the double cover $X(6) \rightarrow X(6)/\langle w_6 \rangle$, transform S to find an elliptic model with essential lattice $N_{\mathrm{ess}} \supset E_7 \oplus E_8$ to which we can apply Kumar's formulas, and verify that our results are consistent with previous computations of CM points [E1] and Clebsch–Igusa coordinates [HM]. For $N = 14$ we exhibit S and verify the

---

[1] It would be interesting to quantify the computational complexity of such computations in terms of the level and the CM discriminant; we have not attempted such an analysis.

location of a CM point that we computed numerically in [E1] but could not prove using the techniques of [E1, E3]. For $N = 57$, the first case for which $X(N)/\langle w_N \rangle$ has positive genus, we exhibit the K3 surfaces parametrized by this curve, and locate all its rational CM points. For $N = 206$, the last case for which $X(N)/\langle w_N \rangle$ has genus zero, we exhibit the corresponding family of K3 surfaces and the hyperelliptic curves $X(206)$ and $X(206)/\langle w_2 \rangle$, $X(206)/\langle w_{103} \rangle$ covering the rational curves $X(206)/\langle w_{206} \rangle$ and $X(206)/\langle w_2, w_{103} \rangle$.

2 Definitions and Techniques

Quaternion Algebras over Q, Shimura Curves, and QM Abelian Surfaces. Fix a squarefree integer $N > 0$ with an even number of prime factors. There is then a unique indefinite quaternion algebra $\mathsf{A}/\mathbf{Q}$ whose finite ramified primes are precisely the factors of N. Let $\mathcal{O}$ be a maximal order in A. Since A is indefinite, all maximal orders are conjugate in A, and conjugate orders will be equivalent for our purposes. Let $\mathcal{O}_1^*$ be the group of units of reduced norm 1 in $\mathcal{O}$; let Γ be the arithmetic subgroup $\mathcal{O}_1^*/\{\pm 1\}$ of $\mathsf{A}^*/\mathbf{Q}^*$; and let Γ^* be the normalizer of Γ in the positive-norm subgroup of $\mathsf{A}^*/\mathbf{Q}^*$. If $N = 1$ then $\Gamma^* = \Gamma$; otherwise Γ^*/Γ is an abelian group of exponent 2, and for each factor $d|N$ there is a unique element $w_d \in \Gamma^*/\Gamma$ whose lifts to A^* have reduced norms in $d \cdot \mathbf{Q}^{*2}$.

Because A is indefinite, $\mathsf{A} \otimes_{\mathbf{Q}} \mathbf{R}$ is isomorphic with the matrix algebra $M_2(\mathbf{R})$, so the positive-norm subgroup of $\mathsf{A}^*/\mathbf{Q}^*$ is contained in $\mathrm{PSL}_2(\mathbf{R})$ and acts on the upper half-plane $\mathcal{H}$. The quotient $\mathcal{H}/\Gamma$ is then a complex model of the Shimura curve associated to Γ, usually called $X(N, 1)$. In [E1] we called this curve $\mathcal{X}(1)$ in analogy with the classical modular curve $X(1)$ (see below), since N was fixed and we studied Shimura curves that we called $\mathcal{X}_0(p)$, $\mathcal{X}_1(p)$, etc., associated with various congruence subgroups of $\mathsf{A}^*/\mathbf{Q}^*$. In this paper we restrict attention to $\mathcal{H}/\Gamma$ and its quotients by subgroups of Γ^*/Γ; thus we return to the usual notation, but simplify it to $X(N)$ because we do not need $X(N, M)$ for $M > 1$. If $N = 1$ then $\mathsf{A} \cong M_2(\mathbf{Q})$, and we may take $\mathcal{O} = M_2(\mathbf{Z})$, when $\Gamma = \Gamma^* = \mathrm{PSL}_2(\mathbf{Z})$ and $\mathcal{H}$ must be extended by its rational cusps before we can identify $\mathcal{H}/\Gamma$ with $X(1)$. Here we study curves $X(N)$ and their quotients only for $N > 1$, and these curves have no cusps.

The Shimura curve $X(N)$ associated to a quaternion algebra over $\mathbf{Q}$ has a reasonably simple moduli description. Fix a positive anti-involution ϱ of A of the form $\varrho(\beta) = \mu^{-1}\bar{\beta}\mu$ for some $\mu \in \mathcal{O}$ with $\mu^2 + N = 0$. Then $X(N)$ parametrizes pairs (A, ι) where A is a principally polarized abelian surface and ι is an embedding of $\mathcal{O}$ into the ring $\mathrm{End}(A)$ of endomorphisms of A, such that the Rosati involution is given by ϱ. See [Rot2, §2 and Prop. 4.1]. This gives $X(N)$ the structure of an algebraic curve over $\mathbf{Q}$.

An abelian surface with an action of a (not necessarily maximal) order in a quaternion algebra is said to have "quaternionic multiplication" (QM). A *complex multiplication* (CM) point of $X(N)$ is a point, necessarily defined over $\bar{\mathbf{Q}}$, for which A has complex multiplication, i.e. is isogenous with the square of a

CM elliptic curve. We shall use the QM abelian surfaces A to find models for the Shimura curves $X(N)$ and locate some of their CM points.

When $N = 1$, an abelian surface together with an action of $\mathcal{O} \cong M_2(\mathbf{Z})$ is just the square of an elliptic curve, so we recover the classical modular curve $X(1)$. We henceforth fix $N > 1$. Then the group Γ^*/Γ, acting on $X(N)$ by involutions that we also call w_d, is nontrivial. These involutions are again defined over $\mathbf{Q}$, taking (A, ι) to (A_d, ι_d) for some A_d isogenous with A. Specifically, A_d is the quotient of A by the subgroup of the d-torsion group $A[d]$ annihilated by the two-sided ideal of $\mathcal{O}$ consisting of elements whose norm is divisible by d, and the principal polarization on A_d is $1/d$ times the pull-back of the principal polarization on A. In particular A_d is CM if and only if A is. Hence the notion of a CM point makes sense on the quotient of $X(N)$ by Γ^*/Γ or by any subgroup of Γ^*/Γ. If a CM point of discriminant $-D$ on $X(N)/(\Gamma^*/\Gamma)$ is rational then the class group of $\mathbf{Q}(\sqrt{-D})$ must be generated by the classes of primes lying over factors $p|D$ that also divide N. Thus the class group has exponent 1 or 2 and bounded size; in particular, only finitely many D can arise. In each of the cases $N = 6, 14, 57$, and 206 that we treat in this paper, N has two prime factors, so the class number is at most 4 and we can cite Arno [Ar] to prove that a list of discriminants of rational CM points is complete. When N has 4 or 6 prime factors we can use Watkins' solution of the class number problem up to 100 [Wa].

We have $A_N \cong A$ as principally polarized abelian surfaces, but for $N > 1$ the embeddings ι, ι_N are not equivalent for generic QM surfaces A. When we pass from A to its Kummer surface we shall lose the distinction between ι and ι_N, and so will at first obtain only the quotient curve $X(N)/\langle w_N \rangle$. We shall determine its double cover $X(N)$ by locating the branch points, which are the CM points on $X(N)/\langle w_N \rangle$ for which A is isomorphic to the product of two elliptic curves with CM by the quadratic imaginary order of discriminant $-N$ or $-4N$; the arithmetic behavior of other CM points will then pin down the cover, including the right quadratic twist over $\mathbf{Q}$.

An abelian surface with QM by $\mathcal{O}$ has at least one principal polarization, and the number of principal polarizations of a generic surface with QM by $\mathcal{O}$ was computed in [Rot1, Theorem 1.4 and §6] in terms of the class number of $\mathbf{Q}(\sqrt{-N})$. Each of these yields a map from $X(N)/\langle w_N \rangle$ to $\mathcal{A}_2$, the moduli three-fold of principally polarized abelian surfaces. This map is either generically $1:1$ or generically $2:1$, and in the $2:1$ case it factors through an involution $w_d = w_{d'}$ on $X(N)/\langle w_N \rangle$ where $d, d' > 1$ are integers such that $N = dd'$ and

$$A \cong \left(\frac{-N, d}{\mathbf{Q}} \right) \left[= \left(\frac{d, d'}{\mathbf{Q}} \right) \right]. \tag{1}$$

(See the last paragraph of [Rot2, §4], which also notes that a $2:1$ map occurs for $N = 6$ and $N = 10$, each of which has a unique choice of polarization. In the other cases $N = 14, 57, 206$ that we study in this paper, only $1:1$ maps arise, because the criterion (1) is not satisfied.) We aim to determine at least one of the maps $X(N)/\langle w_N \rangle \to \mathcal{A}_2$ in terms of the Clebsch–Igusa coordinates on $\mathcal{A}_2$, and thus to find the moduli of the generic abelian surface with endomorphisms by $\mathcal{O}$.[2]

K3 Surfaces, Elliptic K3 Surfaces, and the Dolgachev–Kumar Correspondence.

Let F be a field of characteristic zero. Recall that a *K3 surface* over F is a smooth, complete, simply connected algebraic surface S/F with trivial canonical class. The *Néron–Severi group* $\mathrm{NS}(S) = \mathrm{NS}_{\overline{F}}(S)$ is the group of divisors on S defined over the algebraic closure $\overline{F}$, modulo algebraic equivalence. For a K3 surface this is a free abelian group whose rank, the *Picard number* $\rho = \rho(S)$, is in $\{1, 2, 3, \ldots, 20\}$. The intersection pairing gives $\mathrm{NS}(S)$ the structure of an integral lattice; by the index theorem for surfaces, this lattice has signature $(1, \rho - 1)$, and for a K3 surface the lattice is *even*: $v \cdot v \equiv 0 \bmod 2$ for all $v \in \mathrm{NS}(S)$. Over $\mathbf{C}$, the cycle class map embeds $\mathrm{NS}(S)$ into the "K3 lattice" $H^2(S, \mathbf{Z}) \cong \mathrm{II}_{3,19} \cong U^3 \oplus E_8\langle -1\rangle^2$, where $U = \mathrm{II}_{1,1}$ is the "hyperbolic plane" (the indefinite rank-2 lattice with Gram matrix $\left(\begin{smallmatrix} 0 & 1 \\ 1 & 0 \end{smallmatrix}\right)$), and $E_8\langle -1\rangle$ is the E_8 root lattice made negative-definite by multiplying the inner product by -1. The Torelli theorem of Piateckii-Shapiro and Šafarevič [PSS] describes the moduli of K3 surfaces, at least over $\mathbf{C}$: the embedding of $\mathrm{NS}(S)$ into $\mathrm{II}_{3,19}$ is primitive, that is, realizes $\mathrm{NS}(S)$ as the intersection of $\mathrm{II}_{3,19}$ with a $\mathbf{Q}$-vector subspace of $\mathrm{II}_{3,19} \otimes \mathbf{Q}$; for every such lattice L of signature $(1, \rho - 1)$, there is a nonempty (coarse) moduli space of pairs (S, ι), where $\iota : L \to \mathrm{NS}(S)$ is a primitive embedding consistent with the intersection pairing; and each component of the moduli space has dimension $20 - \rho$. Moreover, for $\rho = 20, 19, 18, 17$ these moduli spaces repeat some more familiar ones: isogenous pairs of CM elliptic curves for $\rho = 20$, elliptic and Shimura modular curves for $\rho = 19$, moduli of abelian surfaces with real multiplication or isogenous to products of two elliptic curves for $\rho = 18$, and moduli of abelian surfaces for certain cases of $\rho = 17$. Note the consequence that an algebraic family of K3 surfaces in characteristic zero with $\rho \geq 19$ whose members are not all $\overline{F}$-isomorphic must have $\rho = 19$ generically, else there would be a positive-dimensional family of K3 surfaces with $\rho \geq 20$.

An *elliptic* K3 surface S/F is a K3 surface together with a rational map $t : S \to \mathbf{P}^1$, defined over F, whose generic fiber is an elliptic curve. The classes of the zero-section s_0 and fiber f in $\mathrm{NS}(S)$ then satisfy $s_0 \cdot s_0 = -2$, $s_0 \cdot f = 1$, and $f \cdot f = 0$, and thus generate a copy of U in $\mathrm{NS}(S)$ defined over F. Conversely, *any* copy of U in $\mathrm{NS}(S)$ defined over F yields a model of S as an elliptic surface: one of the standard isotropic generators or its negative is effective, and has 2 independent sections, whose ratio gives the desired map to $\mathbf{P}^1$. We often use this construction to transform one elliptic model of S to another that would be harder to compute directly. (Warning: in general one might have to subtract some base locus from the effective generator to recover the fiber class f.)

Since $\mathrm{disc}(U) = -1$ is invertible, we have $\mathrm{NS}(S) = \langle s_0, f\rangle \oplus \langle s_0, f\rangle^{\perp}$, with the orthogonal complement $\langle s_0, f\rangle^{\perp}$ having signature $(0, \rho - 2)$; we thus write

---

[2] Alas we cannot say simply "find the generic abelian surface with endomorphisms by $\mathcal{O}$", even up to quadratic twist, because there are abelian surfaces with rational moduli but no model over $\mathbf{Q}$.

$\langle s_0, f \rangle^\perp = N_{\text{ess}}\langle -1 \rangle$ for some positive-definite even lattice N_{ess}, the "essential lattice" of the elliptic K3 surface. A vector $v \in N_{\text{ess}}$ of norm 2, corresponding to $v \in \langle s_0, f \rangle^\perp$ with $v \cdot v = -2$, is called a "root" of N_{ess}; let $R \subseteq N_{\text{ess}}$ be the sublattice generated by the roots. This root sublattice decomposes uniquely as a direct sum of simple root lattices A_n ($n \geq 1$), D_n ($n \geq 4$), or E_n ($6 \leq n \leq 8$). These simple factors biject with reducible fibers, each factor being the sublattice of N_{ess} generated by the components of its reducible fiber that do not meet s_0. The graph whose vertices are these components, and whose edges are their intersections, is then the A_n, D_n, or E_n root diagram; if the identity component and its intersection(s) are included in the graph then the extended root diagram $\tilde{A}_n$, $\tilde{D}_n$, or $\tilde{E}_n$ results. The quotient group N_{ess}/R is isomorphic with the Mordell–Weil group of the surface over $\overline{F}(t)$; the isomorphism takes a point P to the projection of the corresponding section s_P to $\langle s_0, f \rangle^\perp$, and the quadratic form on the Mordell–Weil group induced from the pairing on N_{ess} is the canonical height. Thus the Mordell–Weil regulator is $\tau^2 \operatorname{disc}(N_{\text{ess}}) / \operatorname{disc}(R) = \tau^2 |\operatorname{disc}(\text{NS}(S))| / \operatorname{disc}(R)$, where τ is the size of the torsion subgroup of the Mordell–Weil group.

An elliptic K3 surface has Weierstrass equation $Y^2 = X^3 + A(t)X + B(t)$ for polynomials A, B of degrees at most 8, 12 with no common factor of multiplicity at least 4 and 6 respectively, and such that either $\deg(A) > 4$ or $\deg(B) > 6$ (i.e., such that the condition on common factors holds also at $t = \infty$ when A, B are considered as bivariate homogeneous polynomials of degrees 8, 12). The reducible fibers then occur at multiple roots of the discriminant $\Delta = -16(4A^3 + 27B^2)$ where B does not vanish to order exactly 1 (and at $t = \infty$ if $\deg \Delta \leq 22$ and $\deg B \neq 11$). To obtain a smooth model for S we may start from the surface $Y^2 = X^3 + A(t)X + B(t)$ in the $\mathbf{P}^2$ bundle $\mathbf{P}(O(0) \oplus O(2) \oplus O(3))$ over $\mathbf{P}^1$ with coordinates $(1 : X : Y)$, and resolve the reducible fibers, as exhibited in Tate's algorithm [Ta], which also gives the corresponding Kodaira types and simple root lattices. This information can then be used to calculate the canonical height on the Mordell–Weil group, as in [Si].

The *Kummer surface* $\text{Km}(A)$ of an abelian surface A is obtained by blowing up the $16 = 2^4$ double points of $A/\{\pm 1\}$, and is a K3 surface with Picard number $\rho(\text{Km}(A)) = \rho(A) + 16 \geq 17$. In general $\text{NS}(\text{Km}(A))$ need not consist of divisors defined over F, even when $\text{NS}(A)$ does, because each 2-torsion point of A yields a double point of $A/\{\pm 1\}$ whose blow-up contributes to $\text{NS}(\text{Km}(A))$, and typically $\text{Gal}(\overline{F}/F)$ acts nontrivially on $A[2]$. But when A is principally polarized Dolgachev [Do] constructs another K3 surface S_A/F, related with $\text{Km}(A)$ by degree-2 maps defined over $\overline{F}$, together with a rank-17 sublattice of $\text{NS}(S_A)$ that is isomorphic with $U \oplus E_7 \oplus E_8$ and consists of divisor classes defined over F. It is these surfaces that we parametrize to get at the Shimura curves $X(N)$.

If A has QM then $\rho(A) \geq 3$, with equality for non-CM surfaces, so $\rho(S_A) = \rho(\text{Km}(A)) \geq 19$. When A has endomorphisms by $\mathcal{O}$, we obtain a sublattice $L_N \subseteq \text{NS}(S_A)$ of signature $(1, 18)$ and discriminant $2N$. This even lattice L_N is characterized by its signature and discriminant together with the following condition: for each odd $p|N$ the dual lattice L_N^* contains a vector of norm c/p for some $c \in \mathbf{Z}$ such that $\chi_p(c) = -\chi_p(-2N/p)$, where χ_p is the Legendre symbol

$(\cdot/p)$; equivalently, N^*_{ess} contains a vector of norm c/p with $\chi_p(c) = -\chi_p(+2N/p)$. There is a corresponding local condition at 2, but it holds automatically once the conditions at all odd $p|N$ are satisfied; likewise when N is odd it is enough to check all but one $p|N$. The Shimura curve $X(N)/\langle w_N \rangle$ parametrizes pairs (S, ι) where S is a K3 surface with $\rho(S) \geq 19$ and ι is an embedding $L_N \hookrightarrow NS(S)$. If $\rho(S) = 20$ then (S, ι) corresponds to a CM point on $X(N)/\langle w_N \rangle$ whose discriminant equals disc(NS(S)). The CM points of discriminant $-N$ or $-4N$ are the branch points of the double cover $X(N)$ of $X(N)/\langle w_N \rangle$. The arithmetic of other CM points then determines the cover; for instance, if $X(N)/\langle w_N \rangle$ is rational, we know $X(N)$ up to quadratic twist, and then a rational CM point of discriminant $D \neq -N, -4N$ lifts to a pair conjugate over $\mathbf{Q}(\sqrt{-D})$.

The correspondence between A and S_A was made explicit by Kumar [Ku, Theorem 5.2]. Let A be the Jacobian of a genus-2 curve C, and let I_2, I_4, I_6, I_{10} be the Clebsch–Igusa invariants of C. (If a principally polarized abelian surface A is not a Jacobian then it is the product of two elliptic curves, and thus cannot have QM unless it is a CM surface.) We give an elliptic model of S_A with $N_{ess} = R = E_7 \oplus E_8$, using a coordinate t on $\mathbf{P}^1$ that puts the E_7 and E_8 fibers at $t = 0$ and $t = \infty$. Any such surface has the formula

$$Y^2 = X^3 + (at^4 + a't^3)X + (b''t^7 + bt^6 + b't^5) \qquad (2)$$

for some a, a', b, b', b'' with $a', b'' \neq 0$. (There are five parameters, but the moduli space has dimension only $5 - 2 = 3$ as expected, because multiplying t by a nonzero scalar yields an isomorphic surface, and multiplying a, a' by λ^2 and b, b' by λ^3 for some $\lambda \neq 0$ yields a quadratic twist with the same moduli.) Kumar shows that setting

$$(a, a', b, b', b'') = \left(-I_4/12, -1, (I_2I_4 - 3I_6)/108, I_2/24, I_{10}/4\right) \qquad (3)$$

in (2) yields the surface $S_{J(C)}$. Starting from any surface (2) we may scale (t, X, Y) to $(-a't, a'^2X, a'^3Y)$ and divide through by a'^6 to obtain an equation of the same form with $a' = -1$; doing this and solving (3) for the Clebsch–Igusa invariants I_i, we find

$$(I_2, I_4, I_6, I_{10}) = (-24b'/a', -12a, 96ab'/a' - 36b, -4a'b''). \qquad (4)$$

If A has QM by $\mathcal{O}$, but is not CM, then the elliptic surface (2) has a Mordell–Weil group of rank 2 and regulator N, with each choice of polarization of A corresponding to a different Mordell–Weil lattice. The polarizations for which the map $X(N)/\langle w_N \rangle \to \mathcal{A}_2$ factors through some w_d are those for which the lattice has an involution other than -1. When this happens, two points on $X(N)/\langle w_N \rangle$ related by w_d yield the same surface (2) but a different choice of Mordell–Weil generators. For example, when $N = 6$ and $N = 10$ these lattices have Gram matrices $\frac{1}{2}\begin{pmatrix} 5 & 1 \\ 1 & 5 \end{pmatrix}$ and $\frac{1}{2}\begin{pmatrix} 8 & 0 \\ 0 & 5 \end{pmatrix}$ respectively.

Some Computational Tricks. Often we need elliptic surfaces with an A_n fiber for moderately large n, that is, for which $4A^3 + 27B^2$ vanishes to moderately large order $n + 1$ at some $t = t_0$ at which neither A nor B vanishes. Thus we have approximately $(A, B) = (-3a^2, 2a^3)$ near $t = t_0$. Usually one lets a be a polynomial that locally approximates $(-A/3)^{1/2}$ at $t = t_0$, and writes

$$(A, B) = (-3(a^2 + 2b), 2(a^3 + 3ab) + c) \tag{5}$$

for some b, c of valuations $v(b) = \nu$, $v(c) = 2\nu$ at t_0. Then $v(\Delta) \geq 2\nu$ always, and $v(\Delta) \geq 3\nu$ if and only if $v(3b^2 - ac) \geq 3\nu$; also if $\mu < \nu$ then $v(\Delta) = 2\nu + \mu$ if and only if $v(3b^2 - ac) = 2\nu + \mu$. See [Ha]; this was also the starting point of our analysis in [E2]. For our purposes it is more convenient to allow extended Weierstrass form and write the surface as

$$Y^2 = X^3 + a(t)X^2 + 2b(t)X + c(t) \tag{6}$$

with polynomials a, b, c of degrees at most $4, 8, 12$ such that $(v(b), v(c)) = (\nu, 2\nu)$. Translating X by $-a/3$ shows that this is equivalent to (5), with a, b divided by 3 (so $\mu = v(b^2 - ac)$ in (6)). But (6) tends to produce simpler formulas, both for the surface itself and for the components of the fiber, which are rational if and only if a is a square. For instance, the Shioda–Hall surface with an A_{18} fiber [Sh, Ha] can be written simply as

$$Y^2 = X^3 + (t^4 + 3t^3 + 6t^2 + 7t + 4)X^2 - 2(t^3 + 2t^2 + 3t + 2)X + (t^2 + t + 1)$$

with the A_{18} fiber at infinity, and this is the quadratic twist that makes all of NS(S) defined over $\mathbf{Q}$. The same applies to D_n, when $A' := A/t^2$ and $B' := B/t^3$ are polynomials such that $4A'^3 + 27B'^2$ has valuation $n - 4$. See for instance (19) below. When we want singular fibers at several t values we use an extended Weierstrass form (6) for which $(v(b), v(c)) = (\nu, 2\nu)$ holds (possibly with different ν) at each of these t.

Having parametrized our elliptic surface S with $L_N \hookrightarrow \mathrm{NS}(S)$, we seek specializations of rank 20 to locate CM points. In all but finitely many cases S has an extra Mordell–Weil generator. In the exceptional cases, either some of the reducible fibers merge, or one of those fibers becomes more singular, or there is an extra A_1 fiber. Such CM points are easy to locate, though some mergers require renormalization to obtain a smooth model and find the CM discriminant D, as we shall see. When there is an extra Mordell–Weil generator, its height is at least $|D|/2N$, but usually not much larger. (Equality holds if and only if the extra generator is orthogonal to the generic Mordell–Weil lattice; in particular this happens if S has generic Mordell–Weil rank zero.) The larger the height of the extra generator, the harder it typically is to find the surface. This has the curious consequence that while the difficulty of parametrizing S increases with N, the CM points actually become easier to find. In some cases we cannot solve for the coefficients directly. We thus adapt the methods of [E3], exhaustively searching for a solution modulo a small prime p and then lifting it to a p-adic solution to enough accuracy to recognize the underlying rational numbers. We choose the smallest p such that $\chi_p(-D) = +1$, so that reduction mod p does not raise the Picard number, and we can save a factor of p in the exhaustive search by first counting points mod p on each candidate S to identify the one with the correct CM.

For large N we use the following variation of the p-adic lifting method to find the Shimura curve $X(N)/\langle w_N \rangle$ and the surfaces S parametrized by it. First choose some indefinite primitive sublattice $L' \subset L_N$ and parametrize all S with $\mathrm{NS}(S) \supseteq L'$. Search in that family modulo a small prime p to find a surface S_0 with the desired L_N. Let f_1, f_2 be simple rational functions on the (S, L') moduli space. We hope that the degrees, call them d_i, of the restriction of f_i to $X(N)/\langle w_N \rangle$ are positive but small; that f_1 is locally $1:1$ on the point of $X(N)/\langle w_N \rangle$ parametrizing S_0; and that the map $(f_1, f_2) : X(N)/\langle w_N \rangle \to \mathbf{A}^2$ is generically $1:1$ to its image in the affine plane. For various small lifts $\tilde{f}_1$ of $f_1(S_0)$ to $\mathbf{Q}$, lift S_0 to a surface $S/\mathbf{Q}_p$ with $f_1(S) = \tilde{f}_1$, compute $f_2(S)$ to high p-adic precision, and use lattice reduction to recognize $f_2(S)$ as the solution of a polynomial equation $F(f_2) = 0$ of degree (at most) d_1. Discard the few cases where the degree is not maximal, and solve simultaneous linear equations to guess the coefficients of F as polynomials of degree at most d_2 in $\tilde{f}_1$. At this point we have a birational model $F(f_1, f_2) = 0$ for $X(N)/\langle w_N \rangle$. Then recover a smooth model of the curve (using Magma if necessary), recognize the remaining coefficients of S as rational functions by solving a few more linear equations, and verify that the surface has the desired embedding $L_N \hookrightarrow \mathrm{NS}(S)$.

3 $N = 6$: The First Shimura Curve

The K3 Surfaces. We take $N_{\mathrm{ess}} = R = A_2 \oplus D_7 \oplus E_8$, which has discriminant $3 \cdot 4 \cdot 1 = 12 = 2N$, and the correct behavior at 3 because A_2^* contains vectors of norm $2/3$ with $\chi_3(2) = -\chi_3(2 \cdot 6/3)[= -1]$. We choose the rational coordinate t on $\mathbf{P}^1$ such that the A_2, D_7, and E_8 fibers are at $t = 1$, 0, and ∞ respectively. If we relax the condition at $t = 1$ by asking only that the discriminant vanish to order at least 2 rather than 3 then the general such surface can be written as

$$Y^2 = X^3 + (a_0 + a_1 t)t X^2 + 2a_0 b t^3 (t-1) X + a_0 b^2 t^5 (t-1)^2 \qquad (7)$$

for some a_0, a_1, b, with $a_1 b \neq 0$ lest the surface be too singular at $t = 0$. The discriminant is then $t^9 (t-1)^2 \Delta_1(t)$ with Δ_1 a cubic polynomial such that $\Delta(1) = -64 a_0 a_1 (a_0 + a_1)^2 b^2$. Thus $\Delta_1(1) = 0$ if and only if $a_1 = 0$ or $a_0 + a_1 = 0$. In the latter case the surface has additive reduction at $t = 1$. Hence we must have $a_1 = 0$. The non-identity components of the resulting A_2 fiber at $t = 1$ then have $X = O(t - 1)$; we calculate that $X = x_1(t-1) + O((t-1)^2)$ makes $Y^2 = (x_1 + b)^2 a_0 (t-1)^2 + O(t-1)^3$. Therefore these components are rational if and only if a_0 is a square. We can then replace (X, Y, b) by $(a_0 X, a_0^{3/2} Y, a_0 b)$ in (7) to obtain the formula

$$Y^2 = X^3 + t X^2 + 2b t^3 (t-1) X + b^2 t^5 (t-1)^2 \qquad (8)$$

for the general elliptic K3 surface with $N_{\mathrm{ess}} = R = A_2 D_7 E_8$ and rational A_2 components. The two components of the D_7 fiber farthest from the identity component then have $X = bt^2 + O(t^3)$, so $Y^2 = b^3 t^6 + O(t^7)$; thus these components are both rational as well if and only b is a square, say $b = r^2$. Then b and r are rational coordinates on the Shimura curves $X(6)/\langle w_2, w_3 \rangle$ and $X(6)/\langle w_6 \rangle$ respectively, with the involution $w_2 = w_3$ on $X(6)/\langle w_6 \rangle$ taking r to $-r$.

The elliptic surface (8) has discriminant $\Delta = 16b^3t^9(t-1)^3(27b(t^2-t)-4)$. Thus the formula (8) fails at $b = 0$, and also of course at $b = \infty$. Near each of these two points we change variables to obtain a formula that extends smoothly to $b = 0$ or $b = \infty$ as well. These formulas require extracting respectively a fourth and third root of β, presumably because $b = 0$ and $b = \infty$ are elliptic points of the Shimura curve. For small b, we take $b = \beta^4$ and replace (t, X, Y) by $(t/\beta^2, X/\beta^2, Y/\beta^3)$ to obtain

$$Y^2 = X^3 + tX^2 + 2t^3(t-\beta^2)X + t^5(t-\beta^2)^2, \qquad (9)$$

with the A_2 fiber at $t = \beta^2$ rather than $t = 1$. When $\beta = 0$, this fiber merges with the D_7 fiber at $t = 0$ to form a D_{10} fiber, but we still have a K3 surface, namely $Y^2 = X^3 + tX^2 + 2t^4X + t^7$, with $L = R = D_{10} \oplus E_8$. This is the CM point of discriminant -4. For large b, we write $b = 1/\beta^3$ and replace (X, Y) by $(X/\beta^2, Y/\beta^3)$ to obtain

$$Y^2 = X^3 + \beta^2 tX^2 + 2\beta t^3(t-1)X + t^5(t-1)^2; \qquad (10)$$

then taking $\beta \to 0$ yields the surface $Y^2 = X^3 + t^5(t-1)^2$ with $N_{\text{ess}} = R_0 = A_2 \oplus E_8 \oplus E_8$: the $t = 0$ fiber changes from D_7 to E_8, and the $t = 1$ fiber becomes additive but still contributes A_2 to R (Kodaira type IV rather than I_3). This is the CM point of discriminant -3.

Two More CM Points. The factor $27b(t^2-t)-4$ of Δ is a quadratic polynomial in t of discriminant $27b(27b+16)$. Hence at $b = -16/27$ we have $N_{\text{ess}} = R = A_1 \oplus A_2 \oplus D_7 \oplus E_8$, and we have located the CM point of discriminant -24. Three points fix a rational coordinate on $\mathbf{P}^1$, so we can compare with the coordinate used in [E1, Table 1], which puts the CM points of discriminant -3, -4, and -24 at ∞, 1, and 0 respectively; thus that coordinate is $1 + 27b/16$. This also confirms that $X(6)$ is obtained by extracting a square root of $-(27r^2+16)$.

We next locate a CM point of discriminant -19 by finding b for which the surface (8) has a section s_P of canonical height $19/12$. This is the smallest possible canonical height for a surface with $R = A_2 \oplus D_7 \oplus E_8$, because the naïve height is at least 4 and the height corrections at the A_2 and D_7 fibers can reduce it by at most $2/3$ and $7/4$ respectively, reaching $4 - 2/3 - 7/4 = 19/12$. Let $(X(t), Y(t))$ be the coordinates of a point P of height $19/12$. Then $X(t)$ and $Y(t)$ are polynomials of degree at most 4 and 6 respectively (else s_P intersects s_0 and the naïve height exceeds 4), and X vanishes at $t = 1$ (so s_P passes through a non-identity component of the A_2 fiber) and has the form $bt^2 + O(t^3)$ at $t = 0$ (so s_P meets one of the components of the D_7 fiber farthest from the identity component). That is, $X = b(t^2 - t^3)(1 + t_1 t)$ for some t_1. Substituting this into (8) and dividing by the known square factor $(t^4 - t^3)^2$ yields b^3 times

$$-t_1^3 t^4 + (t_1^3 - 3t_1^2)t^3 + 3(t_1^2 - t_1)t^2 + ((3t_1 - 1) + b^{-1}t_1^2)t + 1, \qquad (11)$$

so we seek b, t_1 such that the quartic (11) is a square. We expand its square root in a Taylor expansion about $t = 0$ and set the t^3 and t^4 coefficients equal to zero. This gives a pair of polynomial equations in b and t_1, which we solve by taking a resultant with respect to t_1. Eliminating a spurious multiple solution at $b = 0$, we finally obtain $(b, t_1) = (81/64, -9)$, and confirm that this makes (11) a square, namely $(27t^2 - 18t - 1)^2$. Therefore $81/64$ is the b-coordinate of a CM point of discriminant -19. Then $1 + 27b/16 = 3211/2^{10}$, same as the value obtained in [E1].

Clebsch–Igusa Coordinates. The next diagram shows the graph whose vertices are the zero-section (circled) and components of reducible fibers of an elliptic K3 surface S with $N_{\mathrm{ess}} = A_2 \oplus D_7 \oplus E_8$, and whose edges are intersections between pairs of these rational curves on the surface. Eight of the vertices form an extended root diagram of type $\tilde{E}_7$, and are marked with their multiplicities in a reducible fiber of type E_7 of an alternative elliptic model for S. We may take either of the unmarked vertices of the $\tilde{D}_7$ subgraph as the zero-section. Then the essential lattice of the new model includes an E_8 root diagram as well as the forced E_7. We can thus apply Kumar's formulas to this model once we compute its coefficients.

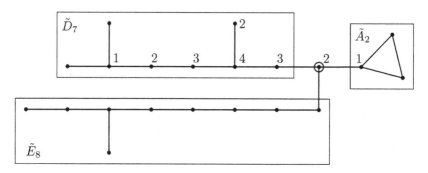

Fig. 1. An $\tilde{E}_7$ divisor supported on the zero-section and fiber components of an $A_2 D_7 E_8$ surface

The sections of the $\tilde{E}_7$ divisor are generated by 1 and $u := X/(t^4 - t^3) + b/t$. Thus $u : S \rightarrow \mathbf{P}^1$ gives the new elliptic fibration. Taking $X = (t^3 - t^2)(tu - b)$ in (8) and dividing by $(t^4 - t^3)^2$ yields $Y_1^2 = Q(t)$ for some quartic Q. Using standard formulas for the Jacobian of such a curve, and bringing the resulting surface into Weierstrass form, we obtain a formula (2) with (a, a', b, b', b'') replaced by $(-3b, 1, -2b^2, -(b + 1), -b^3)$. As expected this surface has Mordell–Weil rank 2 with generators of height $5/2$, namely

$$\left(r^6 t^4 + 2(r^4 + r^3)t^3 + (r^2 + 1)t^2, \; r^9 t^6 + 3(r^7 + r^6)t^5 + 3(r^5 + r^4 + r^3)t^4 + (r^3 + 1)t^3\right)$$

and the image of this section under $r \leftrightarrow -r$ (recall that $b = r^2$). The formula (4) yields the Clebsch–Igusa coordinates

$$(I_2, I_4, I_6, I_{10}) = ((24b + 1), 36b, 72b(5b + 4), 4b^3). \tag{12}$$

4 $N = 14$: The CM Point of Discriminant -67

The K3 Surfaces. Here we take $N_{\text{ess}} = R = A_3 \oplus A_6 \oplus E_8$, which has discriminant $4 \cdot 7 \cdot 1 = 28 = 2N$, and the correct behavior at 7 because A_6^* contains vectors of norm $6/7$ with $\chi_7(6) = -\chi_7(2 \cdot 14/7)[= -1]$. We put the A_3, A_6, and E_8 fibers at $t = 1$, 0, and ∞ respectively. We then seek an extended Weierstrass form (6) with a, b, c of degrees $2, 4, 7$ such that $t^3 - t^2 | b$, $(t^3 - t^2)^2 | c$, and $(t^3 - t^2)^6 | b^2 - ac$. This gives at least A_3, A_5, E_8. It is then easy to impose the extra condition $t^7 | \Delta$, and we obtain $a = \lambda((s+1)t^2 + (3s^2 + 2s)t + s^3)$, $b = \lambda^2(s+1)((4s+2)t + 2s^2)(t^3 - t^2)$, $c = \lambda^3(s+1)^2(t+s)(t^3 - t^2)^2$ for some s, λ. The twist λ must be chosen so that $a(0)$ and $a(1)$ are both squares; this is possible if and only if $s^2 + s$ is a square, so $s = r^2/(2r+1)$ for some r. Thus r and s are rational coordinates on $X(14)/\langle w_{14} \rangle$ and $X(14)/\langle w_2, w_7 \rangle$ respectively, with the involution $w_2 = w_7$ on $X(14)/\langle w_{14} \rangle$ taking r to $-r/(2r+1)$. The formula in terms of r is cleaner if we let the A_3 fiber move from $t = 1$; putting it at $t = 2r + 1$ yields

$$a = ((r+1)^2)t^2 + (3r^4 + 4r^3 + 2r^2)t + r^6),$$
$$b = 2(r+1)^2((2r^2 + 2r + 1)t + r^4)(t - (2r+1))t^2, \qquad (13)$$
$$c = (r+1)^4(t - (2r+1))^2(t + r^2)t^4.$$

Easy CM Points. At $r = 0$, the A_6 fiber becomes E_7, so we have a CM point with $D = -8$; at $r = -1/2$, the A_3 and A_6 fibers merge to A_{10}, giving a CM point with $D = -11$. These have $s = 0$, $s = \infty$ respectively. There is an extra A_1 fiber when $11s^2 + 3s + 8 = 0$; the roots of this irreducible quadratic give the CM points with $D = -56$ (and their lifts to $X(14)/\langle w_{14} \rangle$ are the branch points of the double cover $X(14)$). In [E1] we gave a rational coordinate t on $X(14)/\langle w_2, w_7 \rangle$ for which the CM points of discriminants -8, -11, and -56 had $t = 0$, $t = -1$, and $16t^2 + 13t + 8 = 0$ respectively. Therefore that t is our $-s/(s+1)$.

A Harder CM Point. At the CM point of discriminant -67 our surface has a section of height $67/28 = 4 - (3/4) - (6/7)$. Thus $Y^2 = X^3 + aX^2 + bX + c$ has a solution in polynomials X, Y of degrees $4, 6$ with $X(0) = X(2h + 1) = 0$ and Y having valuation exactly 1 at $t = 0$ and $t = 2h + 1$. An exhaustive search mod 17 quickly finds an example, whose lift to $\mathbf{Q}_{17}$ then yields $r = -35/44$ with

$$X = \frac{3^4}{5^2 \, 2^{25}} \, t \, (22t + 13) \, (527076t^2 + 760364t + 275625).$$

Thus $s = -1225/1144$, and $-s/(s+1)$ confirms the entry $-1225/81$ in the $|D| = 67$ row of [E1, Table 5].

5 $N = 57$: The First Curve $X(N)/w_N$ of Positive Genus

The K3 Surfaces. We cannot have $N_{\text{ess}} = R$ here because there is no root lattice of rank 17 and discriminant $6 \cdot 19$. Instead we take for R the rank-16 lattice

$A_5 \oplus A_{11}$ of discriminant $6 \cdot 12 = 72$, and require an infinite cyclic Mordell–Weil group N_{ess}/R with a generator corresponding to a section that meets the A_5 and A_{11} fibers in non-identity components farthest from the A_5 identity and nearest the A_{11} identity respectively, and does not meet the zero-section (i.e., for which X is a polynomial of degree at most 4 in t). Such a point has canonical height

$$4 - \frac{3 \cdot 3}{6} - \frac{1 \cdot 11}{12} = \frac{19}{12} = \frac{2N}{\mathrm{disc}\, R}. \tag{14}$$

Thus $\mathrm{disc}(N_{\mathrm{ess}})$ has the desired discriminant $2N$. We may check the local conditions by noting that A_5^* contains a vector of norm $4/3$ that remains in N_{ess}^*, and $\chi_3(4) = -\chi_3(2 \cdot 57/3)[= +1]$. We put the A_5 fiber at $t = 0$ and the A_{11} fiber at $t = \infty$. We eventually obtain the following parametrization in terms of a coordinate r on the rational curve $X(57)/\langle w_3, w_{19} \rangle$: let

$$
\begin{aligned}
p(r) &= 4(r-1)(r^2-2) + 1, \\
d &= (r^2-1)^2(9t + (2r-1)p(r)), \\
c &= 9t^2 - (2r-1)(8r^2 + 4r - 22)t + (2r-1)^2 p(r), \\
b &= (t - (r^2 - 2r))c + d, \\
a &= (t - (r^2 - 2r))^2 c + 2(t - (r^2 - 2r))d + (r^2 - 1)^4((4r+4)t + p(r));
\end{aligned}
\tag{15}
$$

Then the surface is

$$Y^2 = X^3 + aX^2 + 8(r-1)^4(r+1)^5 bt^2 X + 16(r-1)^8(r+1)^{10} ct^4, \tag{16}$$

with a section of height $19/12$ at

$$X = -\frac{4(r-1)^4(r+1)^5(2r-1)t^2}{(r^2 - r + 1)^2} + \frac{4(r-2)(r+1)^4 t^3}{r^2 - r + 1}. \tag{17}$$

The components of the A_{11} fiber are rational because the leading coefficient of a is 9, a square; the constant coefficient is $(r^2 - r + 1)^4 p(r)$, so $X(57)/\langle w_{57} \rangle$ is obtained by extracting a square root of $p(r)$. This gives the elliptic curve with coefficients $[a_1, a_2, a_3, a_4, a_6] = [0, -1, 1, -2, 2]$, whose conductor is 57 as expected (see e.g. Cremona's tables [Cr] where this curve appears as 57-A1(E)).

This curve has rank 1, with generator $P = (2, 1)$. The point at infinity is the CM point of discriminant -19; this may be seen by substituting $1/s$ for r and $(t/s^3, X/s^{12}, Y/s^{18})$ for (t, X, Y), then letting $s \to 0$ to obtain the surface

$$Y^2 = X^3 + (9t^4 - 16t^3 + 4t)X^2 + (72t^5 - 128t^4)X + (144t^6 - 256t^5) \tag{18}$$

with a D_6 fiber at $t = 0$ rather than an A_5. Then we still have a section $(X, Y) = (4t^3 - 8t^2, (3t-5)(t^4 - t^3))$ of height $19/12$, but there is a 2-torsion point $(X, Y) = (-4t, 0)$ so $\mathrm{disc}(N_{\mathrm{ess}}) = -\mathrm{disc}(\mathrm{NS}(S))$ drops to $4 \cdot 12 \cdot (19/12)/2^2 = 19$. The remaining rational CM points on $X(57)/\langle w_{57} \rangle$ come in six pairs $\pm nP$:

n	1	2	3	4	5	8
r	2	1	-1	0	$5/4$	$13/9$
$-D$	7	4	16	28	43	163

The last three of these have extra sections $X = -4t$, $X = 0$, and

$$X = -28 \cdot 11^3 \left((t^2/3^6) + (415454t/3^{18})\right)$$

respectively. At $r = 2$, the A_{11} fiber becomes an A_{12} and our generic Mordell–Weil generator becomes divisible by 3; the new generator $(-972t, 26244t^2)$ has height $4 - (5/6) - (40/13) = 7/78$, so disc $N_{\mathrm{ess}} = 6 \cdot 13 \cdot (7/78) = 7$. At $r = 1$, the A_5 and A_{11} fibers together with the section all merge to form a D_{18} fiber: let $r = 1 + s$ and change (t, X) to $(st - 1, -8s^3X)$, divide by $(-2s)^9$, and let $s \to 0$ to obtain the second Shioda–Hall surface

$$X^3 + (t^3 + 8t)X^2 - (32t^2 + 128)X + 256t \tag{19}$$

with a D_{18} fiber at $t = \infty$ [Sh, Ha]. At $t = -1$, the reducible fibers again merge, this time forming an A_{17} while the Mordell–Weil generator's height drops to $4 - (4 \cdot 14/18) = 8/9$, whence $\mathrm{disc}(N_{\mathrm{ess}}) = 16$.

We find four more rational CM values of r that do not lift to rational points on X(57)/$\langle w_{57} \rangle$, namely $r = 5$, $1/2$, $17/16$, $-7/4$, for discriminants -123, -24, -267, and $-627 = -11 \cdot 57$ respectively. The first of these again has an A_{12} fiber, this time with the section of height $4 - (9/6) - (12/13) = 41/26$; the second has a rational section at $X = 0$; in the remaining two cases we find the extra section by p-adic search:

$$X = -\frac{11^3 3^2}{2^{21} 91^2} t^2 (7840t^2 - 2037t + 3267) \tag{20}$$

for $r = 17/16$, and

$$X = \frac{3^5 11^4 t^2 q(t)}{2^{12}(81920t^3 + 9216t^2 + 23868t + 39339)^2} \tag{21}$$

for $r = -7/4$, where $q(t)$ is the quintic

$$419430400t^5 + 2846883840t^4 + 17148174336t^3$$
$$+ 78784560576t^2 + 175272616341t - 12882888.$$

Using [Ar] we can show that there are no further rational CM values.

6 $N = 206$: The Last Curve X(N)/w_N of Genus Zero

Summary of Results. Again we take N_{ess} of rank 16 and an infinite cyclic Mordell–Weil group, here $R = A_2 \oplus A_4 \oplus A_{10}$ with a Mordell–Weil generator of height $412/165 = 6 - (1 \cdot 2/3) - (2 \cdot 3)/5 - (2 \cdot 9)/11$. With the reducible fibers placed at $1, 0, \infty$ as usual, the choice of R means $\Delta = t^5(t - 1)^3\Delta_1$ with Δ_1 of degree $24 - (3 + 5 + 11) = 5$ and $\Delta_1(0), \Delta_1(1) \neq 0$; the generator must then have $X(t) = X_1(t)/(t - t_0)^2$ for some sextic X_1 and some $t_0 \neq 0, 1$, with the corresponding section passing through a non-identity component of the A_2 fiber and components at distance 2 from the identity of A_4 and A_{10}. We eventually succeed in parametrizing such surfaces, finding a rational coordinate on the modular curve $X(206)/\langle w_{206} \rangle$. These elliptic models do not readily exhibit the involution $w_2 = w_{103}$ on this curve, so we recover this involution from the fact that it must permute the branch points of the double cover $X(206)$ of $X(206)/\langle w_{206} \rangle$. We locate these branch points as simple zeros of the discriminant of Δ_1. As expected, there are 20 (this is the class number of $\mathbf{Q}(\sqrt{-206})$), forming a single Galois orbit. We find a unique involution of the projective line $X(206)/\langle w_{206} \rangle$ that permutes these zeros. This involution has two fixed points, so we switch to a rational coordinate r on $X(206)/\langle w_{206} \rangle$ that makes the involution $r \leftrightarrow -r$. Then $r_0 := r^2$ is a rational coordinate on $X(206)/\langle w_2, w_{103} \rangle$, and the 20 branch points are the roots of $P_{10}(r^2)$ where P_{10} is the degree-10 polynomial

$$P_{10}(r_0) = 8r_0^{10} - 13r_0^9 - 42r_0^8 - 331r_0^7 - 220r_0^6 + 733r_0^5 \qquad (22)$$
$$+ 6646r_0^4 + 19883r_0^3 + 28840r_0^2 + 18224r_0 + 4096.$$

As a further check on the computation, P_{10} has dihedral Galois group, discriminant $-2^{138}103^7$, and field discriminant $-2^{12}103^5$, while $P_{10}(r^2)$ has discriminant $2^{311}103^{14}$ and field discriminant $2^{27}103^{10}$. We find that $r = 0, \pm 1, \pm 2, \infty$ give CM points of discriminants $D = -4, -19, -163, -8$ respectively; evaluating $P_{10}(r^2)$ at any of these points gives $-D$ times a square, showing that the Shimura curve $X(206)$ has the equation $s^2 = -P_{10}(r^2)$ over $\mathbf{Q}$. The curves $X(206)/\langle w_2 \rangle$, $X(206)/\langle w_{103} \rangle$ are then the double covers $s_0^2 = -P_{10}(r_0)$, $s_0'^2 = -r_0 P_{10}(r_0)$ of the r_0-line $X(206)/\langle w_2, w_{103} \rangle$ (in that order, because w_{103} cannot fix a CM point of discriminant -4 or -8).

Acknowledgements

I thank Benedict H. Gross, Joseph Harris, John Voight, Abhinav Kumar, and Matthias Schütt for enlightening discussion and correspondence, and for several references concerning Shimura curves and K3 surfaces. I thank M. Schütt, Jeechul Woo, and the referees for carefully reading an earlier version of the paper and suggesting many corrections and improvements. The symbolic and numerical computations reported here were carried out using the packages GP, MAXIMA, and Magma.

References

[Ar]   Arno, S.: The Imaginary Quadratic Fields of Class Number 4. Acta Arith. 40, 321–334 (1992)

[BHPV]   Barth, W.P., Hulek, K., Peters, C.A.M., van de Ven, A.: Compact Complex Surfaces, 2nd edn. Springer, Berlin (2004)

[Cr]   Cremona, J.E.: Algorithms for Modular Elliptic Curves, Cambridge University Press, Cambridge (1992); 2nd edn. (1997), http://www.warwick.ac.uk/staff/J.E.Cremona/book/fulltext/index.html

[Do]   Dolgachev, I., Galluzzi, F., Lombardo, G.: Correspondences between K3 surfaces. Michigan Math. J. 52(2), 267–277 (2004)

[E1]   Elkies, N.D.: Shimura curve computations. In: Buhler, J.P. (ed.) ANTS 1998. LNCS, vol. 1423, pp. 1–47. Springer, Heidelberg (1998), http://arXiv.org/abs/math/0005160

[E2]   Elkies, N.D.: Rational points near curves and small nonzero $|x^3 - y^2|$ via lattice reduction. In: Bosma, W. (ed.) ANTS 2000. LNCS, vol. 1838, pp. 33–63. Springer, Heidelberg (2000), http://arXiv.org/abs/math/0005139

[E3]   Elkies, N.D.: Shimura curves for level-3 subgroups of the $(2, 3, 7)$ triangle group, and some other examples. In: Hess, F., Pauli, S., Pohst, M. (eds.) ANTS 2006. LNCS, vol. 4076, pp. 302–316. Springer, Heidelberg (2006), http://arXiv.org/abs/math/0409020

[E4]   Elkies, N.D.: Three lectures on elliptic surfaces and curves of high rank. Oberwolfach lecture notes (2007), http://arXiv.org/abs/0709.2908

[Er]   Errthum, E.: Singular Moduli of Shimura Curves. PhD thesis, Univ. of Maryland (2007), http://arxiv.org/abs/0711.4316

[GR]   González, J., Rotger, V.: Equations of Shimura curves of genus two. International Math. Research Notices 14, 661–674 (2004)

[GZ]   Gross, B.H., Zagier, D.: On singular moduli. J. für die reine und angew. Math. 335, 191–220 (1985)

[Ha]   Hall, M.: The Diophantine equation $x^3 - y^2 = k$. In: Atkin, A., Birch, B. (eds.) Computers in Number Theory, pp. 173–198. Academic Press, London (1971)

[HM]   Hashimoto, K.-i., Murabayashi, N.: Shimura curves as intersections of Humbert surfaces and defining equations of QM-curves of genus two. Tohoku Math. Journal (2) 47(2), 271–296 (1995)

[Ku]   Kumar, A.: K3 Surfaces of High Rank. PhD thesis, Harvard (2006)

[LY]   Lian, B.H., Yau, S.-T.: Mirror maps, modular relations and hypergeometric series I (preprint, 1995), http://arXiv.org/abs/hep-th/9507151

[PSS]   Piateckii-Shapiro, I., Šafarevič, I.R.: A Torelli theorem for algebraic surfaces of type K3 [Russian]. Izv. Akad. Nauk SSSR, Ser. Mat. 35, 530–572 (1971)

[Rob]   Roberts, D.P.: Shimura Curves Analogous to $X_0(N)$. PhD thesis, Harvard (1989)

[Rot1] Rotger, V.: Shimura curves embedded in Igusa's threefold. In: Cremona, J., Lario, J.-C., Quer, J., Ribet, K. (eds.) Modular curves and abelian varieties. Progress in Math., vol. 224, pp. 263–276. Birkhäuser, Basel (2004), http://arXiv.org/abs/math/0312435

[Rot2] Rotger, V.: Modular Shimura varieties and forgetful maps. Trans. Amer. Math. Soc. 356, 1535–1550 (2004), http://arXiv.org/abs/math/0303163

[Sh]   Shioda, T.: The elliptic K3 surfaces with a maximal singular fibre. C. R. Acad. Sci. Paris Ser. I 337, 461–466 (2003)

[Si]   Silverman, J.H.: Computing Heights on Elliptic Curves. Math. Comp. 51(183), 339–358 (1988)

[Ta]   Tate, J.: Algorithm for determining the type of a singular fiber in an elliptic pencil. In: Birch, B.J., Kuyk, W. (eds.) Modular Functions of One Variable IV (Antwerp, 1972). Lect. Notes in Math., vol. 476, pp. 33–52. Springer, Berlin (1975)

[Vi]   Vignéras, M.-F.: Arithmétique des Algèbres de Quaternions. Lect. Notes in Math., vol. 800. Springer, Berlin (1980)

[Wa]   Watkins, M.: Class numbers of imaginary quadratic fields. Math. Comp. 73, 907–938 (2003)

K3 Surfaces of Picard Rank One
and Degree Two

Andreas-Stephan Elsenhans and Jörg Jahnel

Universität Göttingen, Mathematisches Institut, Bunsenstraße 3–5,
D-37073 Göttingen, Germany[*]
elsenhan@uni-math.gwdg.de, jahnel@uni-math.gwdg.de

Abstract. We construct explicit examples of K3 surfaces over $\mathbb{Q}$ which are of degree 2 and geometric Picard rank 1. We construct, particularly, examples of the form $w^2 = \det M$ where M is a (3×3)-matrix of ternary quadratic forms.

1 Introduction

A K3 surface is a simply connected, projective algebraic surface with trivial canonical class. If $S \subset \mathbf{P}^n$ is a K3 surface then its degree is automatically even. For every even number $d > 0$, there exists a K3 surface $S \subset \mathbf{P}^n$ of degree d.

Examples 1. A K3 surface of degree two is a double cover of $\mathbf{P}^2$, ramified in a smooth sextic. K3 surfaces of degree four are smooth quartics in $\mathbf{P}^3$. A K3 surface of degree six is a smooth complete intersection of a quadric and a cubic in $\mathbf{P}^4$. And, finally, K3 surfaces of degree eight are smooth complete intersections of three quadrics in $\mathbf{P}^5$.

The Picard group of a K3 surface is isomorphic to $\mathbb{Z}^n$ where n may range from 1 to 20. It is generally known that a generic K3 surface over $\mathbb{C}$ is of Picard rank one. This does, however, not yet imply that there exists a K3 surface over $\mathbb{Q}$ the geometric Picard rank of which is equal to one. The point is, genericity means that there are countably many exceptional subvarieties in moduli space.

It seems that the first explicit examples of K3 surfaces of geometric Picard rank one have been constructed as late as in 2005 [vL]. All these examples are of degree four.

The goal of this article is to provide explicit examples of K3 surfaces over $\mathbb{Q}$ which are of geometric Picard rank one and degree two.
For that, let first $\mathscr{S}$ be a K3 surface over a finite field $\mathbb{F}_q$. Then, we have the first Chern class homomorphism

$$c_1 \colon \operatorname{Pic}(\mathscr{S}_{\overline{\mathbb{F}}_q}) \longrightarrow H^2_{\text{ét}}(\mathscr{S}_{\overline{\mathbb{F}}_q}, \overline{\mathbb{Q}}_l(1))$$

---

[*] The computer part of this work was executed on the Sun Fire V20z Servers of the Gauß Laboratory for Scientific Computing at the Göttingen Mathematisches Institut. Both authors are grateful to Prof. Y. Tschinkel for the permission to use these machines as well as to the system administrators for their support.

into l-adic cohomology at our disposal. There is a natural operation of the Frobenius on $H^2_{\text{ét}}(\mathscr{S}_{\overline{\mathbb{F}}_q}, \overline{\mathbb{Q}}_l(1))$. All eigenvalues are of absolute value 1. The Frobenius operation on the Picard group is compatible with the operation on cohomology.

Every divisor is defined over a finite extension of the ground field. Consequently, on the subspace $\text{Pic}(\mathscr{S}_{\overline{\mathbb{F}}_q}) \otimes_{\mathbb{Z}} \overline{\mathbb{Q}}_l \hookrightarrow H^2_{\text{ét}}(\mathscr{S}_{\overline{\mathbb{F}}_q}, \overline{\mathbb{Q}}_l(1))$, all eigenvalues are roots of unity. Those correspond to eigenvalues of the Frobenius operation on $H^2_{\text{ét}}(\mathscr{S}_{\overline{\mathbb{F}}_q}, \overline{\mathbb{Q}}_l)$ which are of the form $q\zeta$ for ζ a root of unity.

We may therefore estimate the rank of the Picard group $\text{Pic}(\mathscr{S}_{\overline{\mathbb{F}}_q})$ from above by counting how many eigenvalues are of this particular form. It is conjectured that this estimate is always sharp but we avoid having to make use of this.

Estimates from below may be obtained by explicitly constructing divisors. Under certain circumstances, it is possible, in that way, to determine $\text{rk Pic}(\mathscr{S}_{\overline{\mathbb{F}}_q})$, exactly.

Our general strategy is to use reduction modulo p. We apply the inequality

$$\text{rk Pic}(S_{\overline{\mathbb{Q}}}) \leq \text{rk Pic}(S_{\overline{\mathbb{F}}_p})$$

which is true for every smooth variety S over $\mathbb{Q}$ and every prime p of good reduction [Fu, Example 20.3.6, 19.3.1.iii) and iv))]. Having constructed an example with $\text{rk Pic}(S_{\overline{\mathbb{F}}_3}) = \text{rk Pic}(S_{\overline{\mathbb{F}}_5}) = 2$, we use the same technique as in [vL] to deduce $\text{rk Pic}(S_{\overline{\mathbb{Q}}}) = 1$.

Remark 2. Let S be a K3 surface over $\mathbb{Q}$ of degree two and geometric Picard rank one. Then, S cannot be isomorphic, not even over $\overline{\mathbb{Q}}$, to a K3 surface $S' \subset \mathbf{P}^3$ of degree 4.

Indeed, $\text{Pic}(S_{\overline{\mathbb{Q}}}) = \mathbb{Z} \cdot \langle \mathscr{L} \rangle$ and $\deg S = 2$ mean that the intersection form on $\text{Pic}(S_{\overline{\mathbb{Q}}})$ is given by $\langle \mathscr{L}^{\otimes n}, \mathscr{L}^{\otimes m} \rangle := 2nm$. The self-intersection numbers of divisors on $S_{\overline{\mathbb{Q}}}$ are of the form $2n^2$ which is always different from 4.

2   Lower Bounds for the Picard Rank

In order to estimate the rank of the Picard group from below, we need to explicitly construct divisors. Calculating discriminants, it is possible to show that the corresponding divisor classes are linearly independent.

Notation 3. Let k be an algebraically closed field of characteristic $\neq 2$. In the projective plane $\mathbf{P}^2_k$, let a smooth curve B of degree 6 be given by $f_6(x, y, z) = 0$. Then, $w^2 = f_6(x, y, z)$ defines a K3 surface $\mathscr{S}$ in a weighted projective space. We have a double cover $\pi \colon \mathscr{S} \to \mathbf{P}^2$ ramified at $\pi^{-1}(B)$.

Construction 4. i) One possible construction with respect to our aims is to start with a branch curve "$f_6 = 0$" which allows a tritangent line G. The pull-back of G to the K3 surface $\mathscr{S}$ is a divisor splitting into two irreducible components. The corresponding divisor classes are linearly independent.

ii) A second possibility is to use a conic which is tangent to the branch sextic in six points.

Both constructions yield a lower bound of 2 for the rank of the Picard group.

Tritangent. Assume, the line G is a tritangent to the sextic given by $f_6 = 0$. This means, the restriction of f_6 to $G \cong \mathbf{P}^1$ is a section of $\mathscr{O}(6)$, the divisor of which is divisible by 2 in $\mathrm{Div}(G)$. As G is of genus 0, this implies $f_6|_G$ is the square of a section $f \in \Gamma(G, \mathscr{O}(3))$. The form f_6 may, therefore, be written as $f_6 = \widetilde{f}^2 + l q_5$ for l a linear form defining G, $\widetilde{f}$ a cubic form lifting f, and a quintic form q_5.

Consequently, the restriction of π to $\pi^{-1}(G)$ is given by an equation of the form $w^2 = f^2(s, t)$. Hence, we have $\pi^*(G) = D_1 + D_2$ where D_1 and D_2 are the two irreducible divisors given by $w = \pm f(s, t)$. Both curves are isomorphic to G. In particular, they are projective lines.

The adjunction formula shows $-2 = D_1(D_1 + K) = D_1^2$. Analogously, one sees $D_2^2 = -2$. Finally, we have $G^2 = 1$. It follows that $(D_1 + D_2)^2 = 2$ which yields $D_1 D_2 = 3$. Thus, for the discriminant, we find

$$\mathrm{Disc}\langle D_1, D_2 \rangle = \begin{vmatrix} -2 & 3 \\ 3 & -2 \end{vmatrix} = -5 \neq 0$$

guaranteeing $\mathrm{rk}\,\mathrm{Pic}(\mathscr{S}) \geq 2$.

Remark 5. We note explicitly that this argument works without modification if two or all three points of tangency coincide.

Conic Tangent in Six Points. If C is a conic tangent to the branch curve "$f_6 = 0$" in six points then, for the same reasons as above, we have $\pi^*(C) = C_1 + C_2$ where C_1 and C_2 are irreducible divisors. Again, C_1 and C_2 are isomorphic to C and, therefore, of genus 0. This shows $C_1^2 = C_1^2 = -2$.

We have another divisor at our disposal, the pull-back $D := \pi^*(G)$ of a line in $\mathbf{P}_k^2$. $G^2 = 1$ implies that $D^2 = 2$. Further, we have $GC = 2$ which implies $D(C_1 + C_2) = 4$ and $DC_1 = 2$. For the discriminant, we obtain

$$\mathrm{Disc}\langle C_1, D \rangle = \begin{vmatrix} -2 & 2 \\ 2 & 2 \end{vmatrix} = -8 \neq 0.$$

Consequently, $\mathrm{rk}\,\mathrm{Pic}(\mathscr{S}) \geq 2$ in this case, too.

Remark 6. There is no further refinement of $\langle C_1, D \rangle$ to a lattice in $\mathrm{Pic}(\mathscr{S})$ of discriminant (-2). Indeed, the self-intersection number of a curve on a K3 surface is always even. Hence, the discriminant of an arbitrary rank two lattice in $\mathrm{Pic}(\mathscr{S})$ is of the shape $\left|\begin{smallmatrix} 2a & c \\ c & 2b \end{smallmatrix}\right| = 4ab - c^2$ for $a, b \in \mathbb{Z}$. The quadratic form on the right hand side does not represent integers which are 1 or 2 modulo 4.

The discriminant of the lattice spanned by C_1 and C_2 turns out to be $\mathrm{Disc}\langle C_1, C_2 \rangle = \left|\begin{smallmatrix} -2 & 6 \\ 6 & -2 \end{smallmatrix}\right| = -32 \neq 0$ which would be completely sufficient for our purposes.

Remark 7. Further tritangents or further conics which are tangent in six points lead to even larger Picard groups.

Detection of Tritangents. The property of a line of being a tritangent may easily be written down as an algebraic condition. Therefore, tritangents may be searched for, in practice, by investigating a Gröbner base.

More precisely, a general line in $\mathbf{P}^2$ can be described by a parametrization

$$g_{a,b} \colon t \mapsto [1 : t : (a + bt)].$$

$g_{a,b}$ is a (possibly degenerate) tritangent of the sextic given by $f_6 = 0$ if and only if $f_6 \circ g_{a,b}$ is a perfect square in $\overline{\mathbb{F}}_q[t]$. This means,

$$f_6(g_{a,b}(t)) = (c_0 + c_1 t + c_2 t^2 + c_3 t^3)^2$$

is an equation which encodes the tritangent property of $g_{a,b}$. Comparing coefficients, this yields a system of seven equations in c_0, c_1, c_2, and c_3 which is solvable if and only if $g_{a,b}$ is a tritangent.

The latter may be understood as well as a system of equations in a, b, c_0, c_1, c_2, and c_3 encoding the existence of a tritangent of the form above. Corresponding to this system of equations, there is an ideal $I \subseteq \mathbb{F}_q[a, b, c_0, c_1, c_2, c_3]$ given explicitly by seven generators.

The remaining one-dimensional family of lines may be treated analogously using the parametrizations $g_a \colon t \mapsto [1 : a : t]$ and $g \colon t \mapsto [0 : 1 : t]$. Similarly, this leads to ideals $I' \subseteq \mathbb{F}_q[a, c_0, c_1, c_2, c_3]$ and $I'' \subseteq \mathbb{F}_q[c_0, c_1, c_2, c_3]$.

Thus, there is a simple method to find out whether the sextic given by $f_6 = 0$ has a tritangent or not.

Algorithm 8 (Given a sextic form f_6 over $\mathbb{F}_q$, this algorithm decides whether the curve given by $f_6 = 0$ has a tritangent).

i) Compute a Gröbner base for the ideal $I \subseteq \mathbb{F}_q[a, b, c_0, c_1, c_2, c_3]$, described above.

ii) Compute a Gröbner base for the ideal $I' \subseteq \mathbb{F}_q[a, c_0, c_1, c_2, c_3]$.

iii) Compute a Gröbner base for the ideal $I'' \subseteq \mathbb{F}_q[c_0, c_1, c_2, c_3]$.

iv) If it turns out that actually all three ideals are equal to the unit ideal then output that the curve given has no tritangent. Otherwise, output that a tritangent was detected.

Remark 9. There are a few obvious refinements.

i) For example, given the Gröbner bases, it is easy to calculate the lengths of the quotient rings $\mathbb{F}_q[a, b, c_0, c_1, c_2, c_3]/I$, $\mathbb{F}_q[a, c_0, c_1, c_2, c_3]/I'$, and $\mathbb{F}_q[c_0, c_1, c_2, c_3]/I''$. Each of them is twice the number of the corresponding tritangents.

ii) Usually, from the Gröbner bases, the tritangents may be read off directly.

Remark 10. We ran Algorithm 8 using `Magma`. The time required to compute a Gröbner base as needed over a finite field is usually a few seconds.

Remark 11. The existence of a tritangent is a codimension one condition. Over small ground fields, one occasionally finds tritangents on randomly chosen examples.

Searching for Conics Tangent in Six Points. A non-degenerate conic in $\mathbf{P}^2$ allows a parametrization of the form

$$c: t \mapsto [(c_0 + c_1 t + c_2 t^2) : (d_0 + d_1 t + d_2 t^2) : (e_0 + e_1 t + e_2 t^2)].$$

With the sextic given by $f_6 = 0$, all intersection multiplicities are even if and only if $f_6 \circ c$ is a perfect square in $\overline{\mathbb{F}}_q[t]$. This may easily be checked by factoring $f_6 \circ c$.

Algorithm 12 (Given a sextic form f_6 over $\mathbb{F}_q$, this algorithm decides whether the curve given by $f_6 = 0$ allows a conic defined over $\mathbb{F}_q$ which is tangent in six points).

i) In a precomputation, generate a list of parametrizations, one for each of the $q^2(q^3 - 1)$ non-degenerate conics defined over $\mathbb{F}_q$.

ii) Run through the list. For each parametrization, factorize the univariate polynomial $f_6 \circ c$ into irreducible factors. If it turns out to be a perfect square then output that a conic which is tangent in six points has been found.

Remarks 13. a) For very small q, this algorithm is extremely efficient. We need it only for $q = 3$ and 5.

b) A general method, analogous to the one for tritangents, to find conics defined over $\overline{\mathbb{F}}_q$ does not succeed. The required Gröbner base computation becomes too large.

3   An Upper Bound for the Geometric Picard Rank

In this section, we consider a K3 surface $\mathscr{S}$ over a finite field $\mathbb{F}_p$. A method to understand the operation of the Frobenius ϕ on the l-adic cohomology $H^2_{\text{ét}}(\mathscr{S}_{\overline{\mathbb{F}}_p}, \overline{\mathbb{Q}}_l) \cong \overline{\mathbb{Q}}_l^{22}$ works as follows.

The Lefschetz Trace Formula. Count the points on $\mathscr{S}$ over $\mathbb{F}_{p^d}$ and apply the Lefschetz trace formula [Mi] to compute the trace of the Frobenius $\phi_{\mathbb{F}_{p^d}} = \phi^d$. In our situation, this yields

$$\text{Tr}(\phi^d) = \#\mathscr{S}(\mathbb{F}_{p^d}) - p^{2d} - 1.$$

We have $\text{Tr}(\phi^d) = \lambda_1^d + \cdots + \lambda_{22}^d =: \sigma_d(\lambda_1, \ldots, \lambda_{22})$ when we denote the eigenvalues of ϕ by $\lambda_1, \ldots, \lambda_{22}$. Newton's identity [Ze]

$$s_k(\lambda_1, \ldots, \lambda_{22}) = \frac{1}{k} \sum_{r=0}^{k-1} (-1)^{k+r+1} \sigma_{k-r}(\lambda_1, \ldots, \lambda_{22}) s_r(\lambda_1, \ldots, \lambda_{22})$$

shows that, doing this for $d = 1, \ldots, k$, one obtains enough information to determine the coefficient $(-1)^k s_k$ of t^{22-k} of the characteristic polynomial f_p of ϕ.

Remark 14. Observe that we also have the functional equation

$$(*) \qquad\qquad p^{22} f_p(t) = \pm t^{22} f_p(p^2/t)$$

at our disposal. It may be used to convert the coefficient of t^i into the one of t^{22-i}.

Algorithms for Counting Points. The number $\#\mathscr{S}(\mathbb{F}_q)$ of points may be determined as the sum

$$\sum_{[x:y:z]\in\mathbf{P}^2(\mathbb{F}_q)} \left[1 + \chi\big(f_6(x,y,z)\big)\right].$$

Here, χ is the quadratic character of $\mathbb{F}_q^*$. The sum is well-defined since $f_6(x,y,z)$ is uniquely determined up to a sixth-power residue. To count the points naively, one would need $q^2 + q + 1$ evaluations of f_6 and χ.

Here, an obvious possibility for optimization arises. We may use symmetry: If f_6 is defined over $\mathbb{F}_p$ then the summands for $[x : y : z]$ and $\phi([x : y : z])$ are equal.

Algorithm 15 (Point counting).

i) Precompute a list which contains exactly one representative for each Galois orbit of $\mathbb{F}_q$. Equip each member y with an additional marker s_y indicating the size of its orbit.

ii) Let $[0 : y : z]$ run through all $\mathbb{F}_q$-rational points on the projective line and add up the values of $[1 + \chi\big(f_6(0,y,z)\big)]$ to a sum Z.

iii) In an iterated loop, let y run through the precomputed list and z through the whole of $\mathbb{F}_q$. Add up Z and all values of $s_y \cdot [1 + \chi\big(f_6(1,y,z)\big)]$.

Remark 16. Over $\mathbb{F}_{p^d}$, we save a factor of about d as, on the affine chart "$x \neq 0$", we put in for y only values from a fundamental domain of the Frobenius.

A second possibility for optimization is to use decoupling: Suppose, f_6 is decoupled, i.e., it contains only monomials of the form $x^i y^{6-i}$ or $x^i z^{6-i}$. Then, on the affine chart "$x \neq 0$", the form f_6 may be written as $f_6(1,y,z) = g(y) + h(z)$. If f_6 is defined over $\mathbb{F}_p$ then we still may use symmetry. The ranges of g and h are invariant under the operation of Frobenius. There is an algorithm as follows.

Algorithm 17 (Point counting – decoupled situation).

i) For the function g, generate a list A of its values. For each $u \in A$, store the number $n_A(u)$ indicating how many times it is adopted by g.

ii) For the function h, generate a list B of its values. For each $v \in B$, store the number $n_B(v)$ indicating how many times it is adopted by h.

iii) Modify the table for g. For each orbit $F = \{u_1, \ldots, u_e\}$ of the Frobenius, delete all elements except one, say u_1. Multiply $n_A(u_1)$ by $\#F$.

iv) Tabulate the quadratic character χ.

v) Let $[0 : y : z]$ run through all $\mathbb{F}_q$-rational points on the projective line and add up the values of $[1 + \chi\big(f_6(0,y,z)\big)]$ to a sum Z.

vi) Use the table for χ and the tables built up in steps i) through iii) to compute the sum

$$\sum_{u\in A}\sum_{v\in B} \chi(u+v)\cdot n_A(u)\cdot n_B(v).$$

vii) Add $q^2 + Z$ to the number obtained.

Remarks 18. i) The tables for g and h may be built up in $O(q \log q)$ steps.

ii) Statistically, after steps i) and ii) the sizes of A and B are approximately $(1 - 1/e) \cdot q = (1 - 1/e) \cdot p^d$. Step iii) reduces the size of A almost to $(1 - 1/e) \cdot p^d / d$. After all the preparations, we therefore expect about $(1 - 1/e)^2 \cdot q^2 / d$ additions to be executed in step vi).

The advantage of a decoupled situation is, therefore, not only that evaluations of the polynomial f_6 in $\mathbb{F}_{p^d}$ get replaced by additions. Furthermore, the expected number of additions is only about 40% of the number of evaluations of f_6 required by Algorithm 15.

Remark 19. We implemented the point counting algorithms in C. The optimization realized in Algorithm 15 allows to determine the number of $\mathbb{F}_{3^{10}}$-rational points on $\mathscr{S}$ within half an hour on an AMD Opteron processor.

In a decoupled situation, the number of $\mathbb{F}_{5^9}$-rational points may be counted within two hours by Algorithm 17. In a few cases, we determined the numbers of points over $\mathbb{F}_{5^{10}}$. This took around two days. Using Algorithm 15, the same counts would have taken around one day or 25 days, respectively.

This shows, using the methods above, we may effectively compute the traces of $\phi_{\mathbb{F}_{p^d}} = \phi^d$ for $d = 1, \ldots, 9, (10)$.

Remark 20. In Algorithm 17, the sum calculated in step vi) is nothing but $\sum_{w \in \mathbb{F}_q} \chi(w) \cdot (n_A * n_B)(w)$. It might be on option to compute the convolution $n_A * n_B$ using FFT. We expect that, concerning running times, this might lead to a certain gain. On the other hand, such an algorithm would require a lot more space than Algorithm 17.

This possible use of FFT could be of interest from a theoretical point of view. It is well-known that, in most applications, FFT is used on large cyclic groups. Here, however, the group is $(\mathbb{F}_{p^d}, +) \cong (\mathbb{Z}/p\mathbb{Z})^d$ for p very small.

An Upper Bound for $\mathrm{rk}\,\mathrm{Pic}(\mathscr{S}_{\overline{\mathbb{F}}_p})$ Having Counted till $d = 10$

We know that f_p, the characteristic polynomial of the Frobenius, has a zero at p since the pull-back of a line in $\mathbf{P}^2$ is a divisor defined over $\mathbb{F}_p$. Suppose, we determined $\mathrm{Tr}(\phi^d)$ for $d = 1, \ldots, 10$. Then, we may use the following algorithm.

Algorithm 21 (Upper bound for $\mathrm{rk}\,\mathrm{Pic}(\mathscr{S}_{\overline{\mathbb{F}}_p})$).

i) First, assume the minus sign in the functional equation $(*)$. Then, f_p automatically has coefficient 0 at t^{11}. Therefore, the numbers of points counted suffice in this case to determine f_p, completely.

ii) Then, assume that, on the other hand, the plus sign is present in $(*)$. In this case, the data collected immediately allow to compute all coefficients of f_p, except that at t^{11}. Use the known zero at p to determine that final coefficient.

iii) Use the numerical test, provided by Algorithm 23 below, to decide which sign is actually present.

iv) Factor $f_p(pt)$ into irreducible polynomials. Check which of the factors are cyclotomic polynomials, add their degrees, and output that sum as an upper bound for $\mathrm{rk}\,\mathrm{Pic}(\mathscr{S}_{\overline{\mathbb{F}}_p})$. If step iii) had failed then work with both candidates for f_p and output the maximum.

Verifying $\operatorname{rk}\operatorname{Pic}(\mathscr{S}_{\overline{\mathbb{F}}_p}) = 2$ Having Counted Only till $d = 9$

Assume, $\mathscr{S}$ is a K3 surface over $\mathbb{F}_p$ given by Construction 4.i) or ii). We, therefore, know that the rank of the Picard group is at least equal to 2. We assume that the divisor constructed by pull-back splits already over $\mathbb{F}_p$. This ensures p is a double zero of f_p.

Suppose, we determined $\operatorname{Tr}(\phi^d)$ for $d = 1, \ldots, 9$. Then, there is the following algorithm.

Algorithm 22 (Verifying $\operatorname{rk}\operatorname{Pic}(\mathscr{S}_{\overline{\mathbb{F}}_p}) = 2$).

i) First, assume the minus sign in the functional equation $(*)$. This forces another zero of f_p at $(-p)$. The data collected are then sufficient to determine f_p, completely. Algorithm 23 below may indicate a contradiction. Otherwise, output `FAIL` and terminate prematurely. (In this case, we could still find an upper bound for $\operatorname{rk}\operatorname{Pic}(\mathscr{S}_{\overline{\mathbb{F}}_p})$ which is, however, at least equal to 4.)

ii) As we have the plus sign in $(*)$, the data immediately suffice to compute all coefficients of f_p, with the exception of those at t^{10}, t^{11}, and t^{12}. The functional equation yields a linear relation for the three remaining coefficients of f_p. From the known double zero at p, one computes another linear condition.

iii) Let n run through all natural numbers such that $\varphi(n) \leq 20$. (The largest such n is 66.)

Assume, in addition, that there is another zero of the form $p\zeta_n$. This yields further linear relations. Inspecting this system of linear equations, one either achieves a contradiction or determines all three remaining coefficients. In the latter case, Algorithm 23 may indicate a contradiction. Otherwise, output `FAIL` and terminate prematurely.

iv) Output that $\operatorname{rk}\operatorname{Pic}(\mathscr{S}_{\overline{\mathbb{F}}_p}) = 2$.

Algorithm 23 (A numerical test – Given a polynomial f, this test may prove that f is not the characteristic polynomial of the Frobenius).

i) Given $f \in \mathbb{Z}[t]$ of degree 22, calculate all its zeroes as complex floating point numbers.

ii) If at least one of them is of an absolute value clearly different from p then output that f can not be the characteristic polynomial of the Frobenius for any K3 surface over $\mathbb{F}_p$. Otherwise, output `FAIL`.

Remark 24. Consequently, the equality $\operatorname{rk}\operatorname{Pic}(\mathscr{S}_{\overline{\mathbb{F}}_p}) = 2$ may be effectively provable having determined $\operatorname{Tr}(\phi^d)$ for $d = 1, \ldots, 9$, only. This is of importance since point counting over $\mathbb{F}_{5^{10}}$ is not that fast, even in a decoupled situation.

Possible Values of the Upper Bound. This approach will always yield an even number for the upper bound of the geometric Picard rank. Indeed, the bound we use is

$$\operatorname{rk}\operatorname{Pic}(\mathscr{S}_{\overline{\mathbb{F}}_p}) \leq \dim(H^2_{\text{ét}}(\mathscr{S}_{\overline{\mathbb{F}}_p}, \overline{\mathbb{Q}_l})) - \#\{ \text{ zeroes of } f_p \text{ not of the form } \zeta_n p \} .$$

The relevant zeroes come in pairs of complex conjugate numbers. Hence, for a K3 surface the bound is always even.

Remark 25. There is a famous conjecture due to John Tate [Ta] which implies that the canonical injection $c_1 \colon \operatorname{Pic}(\mathscr{S}_{\overline{\mathbb{F}}_p}) \to H^2_{\text{ét}}(\mathscr{S}_{\overline{\mathbb{F}}_p}, \overline{\mathbb{Q}}_l(1))$ maps actually onto the sum of all eigenspaces for the eigenvalues which are roots of unity. Together with the conjecture of J.-P. Serre claiming that the Frobenius operation on étale cohomology is always semisimple, this would imply that the bound above is actually sharp.

It is a somewhat surprising consequence of the Tate conjecture that the Picard rank of a K3 surface over $\overline{\mathbb{F}}_p$ is always even. For us, this is bad news. The obvious strategy to prove $\operatorname{rk}\operatorname{Pic}(S_{\overline{\mathbb{Q}}}) = 1$ for a K3 surface S over $\mathbb{Q}$ would be to verify $\operatorname{rk}\operatorname{Pic}(S_{\overline{\mathbb{F}}_p}) = 1$ for a suitable place p of good reduction. The Tate conjecture, however, indicates that there is no hope for such an approach.

4   Proving rk Pic($S_{\overline{\mathbb{Q}}}$) = 1

Using the methods described above, on one hand, we can construct even upper bounds for the Picard rank. On the other hand, we can generate lower bounds by explicitly stating divisors. In an optimal situation, this may establish an equality $\operatorname{rk}\operatorname{Pic}(S_{\overline{\mathbb{F}}_p}) = 2$.

How is it possible that way to reach Picard rank 1 for a surface S defined over $\mathbb{Q}$? For this, a technique due to R. van Luijk [vL, Remark 2] is helpful.

Lemma 26. *Assume that we are given a K3 surface $\mathscr{S}^{(3)}$ over $\mathbb{F}_3$ and a K3 surface $\mathscr{S}^{(5)}$ over $\mathbb{F}_5$ which are both of geometric Picard rank 2. Suppose further that the discriminants of the intersection forms on $\operatorname{Pic}(\mathscr{S}^{(3)}_{\overline{\mathbb{F}}_3})$ and $\operatorname{Pic}(\mathscr{S}^{(5)}_{\overline{\mathbb{F}}_5})$ are essentially different, i.e., their quotient is not a perfect square in $\mathbb{Q}$.*

Then, every K3 surface S over $\mathbb{Q}$ such that its reduction at 3 is isomorphic to $\mathscr{S}^{(3)}$ and its reduction at 5 is isomorphic to $\mathscr{S}^{(5)}$ is of geometric Picard rank one.

Proof. The reduction maps $\iota_p \colon \operatorname{Pic}(S_{\overline{\mathbb{Q}}}) \to \operatorname{Pic}(S_{\overline{\mathbb{F}}_p}) = \operatorname{Pic}(\mathscr{S}^{(p)}_{\overline{\mathbb{F}}_p})$ are injective [Fu, Example 20.3.6]. Observe here, $\operatorname{Pic}(S_{\overline{\mathbb{Q}}})$ is equal to the group of divisors on $S_{\overline{\mathbb{Q}}}$ modulo numerical equivalence.

This immediately leads to the bound $\operatorname{rk}\operatorname{Pic}(S_{\overline{\mathbb{Q}}}) \leq 2$. Assume, by contradiction, that equality holds. Then, the reductions of $\operatorname{Pic}(S_{\overline{\mathbb{Q}}})$ are sublattices of maximal rank in both, $\operatorname{Pic}(S_{\overline{\mathbb{F}}_3}) = \operatorname{Pic}(\mathscr{S}^{(3)}_{\overline{\mathbb{F}}_3})$ and $\operatorname{Pic}(S_{\overline{\mathbb{F}}_5}) = \operatorname{Pic}(\mathscr{S}^{(5)}_{\overline{\mathbb{F}}_5})$.

The intersection product is compatible with reduction. Therefore, the quotients $\operatorname{Disc}\operatorname{Pic}(S_{\overline{\mathbb{Q}}})/\operatorname{Disc}\operatorname{Pic}(\mathscr{S}^{(3)}_{\overline{\mathbb{F}}_3})$ and $\operatorname{Disc}\operatorname{Pic}(S_{\overline{\mathbb{Q}}})/\operatorname{Disc}\operatorname{Pic}(\mathscr{S}^{(5)}_{\overline{\mathbb{F}}_5})$ are perfect squares. This is a contradiction to the assumption.   □

Remark 27. Suppose that $\mathscr{S}^{(3)}$ and $\mathscr{S}^{(5)}$ are K3 surfaces of degree two given by explicit branch sextics in $\mathbf{P}^2$. Then, using the Chinese Remainder Theorem, they can easily be combined to a K3 surface S over $\mathbb{Q}$.

Assume $\operatorname{rk}\operatorname{Pic}(\mathscr{S}^{(3)}_{\overline{\mathbb{F}}_3}) = 2$ and $\operatorname{rk}\operatorname{Pic}(\mathscr{S}^{(5)}_{\overline{\mathbb{F}}_5}) = 2$. If one of the two branch sextics allows a conic tangent in six points and the other a tritangent then the discriminants of the intersection forms on $\operatorname{Pic}(\mathscr{S}^{(3)}_{\overline{\mathbb{F}}_3})$ and $\operatorname{Pic}(\mathscr{S}^{(5)}_{\overline{\mathbb{F}}_5})$ are essentially different as shown in section 2.

5 An Example

Examples 28. We consider two particular K3 surfaces.

i) By $\mathscr{X}^0$, we denote the surface over $\mathbb{F}_3$ given by the equation

$$w^2 = (y^3 - x^2 y)^2 \\ + (x^2 + y^2 + z^2)(2x^3 y + x^3 z + 2x^2 yz + x^2 z^2 + 2xy^3 + 2y^4 + z^4).$$

ii) Further, let $\mathscr{Y}^0$ be the K3 surface over $\mathbb{F}_5$ given by

$$w^2 = x^5 y + x^4 y^2 + 2x^3 y^3 + x^2 y^4 + xy^5 + 4y^6 + 2x^5 z + 2x^4 z^2 + 4x^3 z^3 + 2xz^5 + 4z^6.$$

Theorem 29. *Let S be any K3 surface over $\mathbb{Q}$ such that its reduction modulo 3 is isomorphic to $\mathscr{X}^0$ and its reduction modulo 5 is isomorphic to $\mathscr{Y}^0$. Then, $\mathrm{rk}\,\mathrm{Pic}(S_{\overline{\mathbb{Q}}}) = 1$.*

Proof. We follow the strategy described in Remark 27. For the branch locus of $\mathscr{X}^0$, the conic given by $x^2 + y^2 + z^2 = 0$ is tangent in six points. The branch locus of Y_0 has a tritangent given by $z - 2y = 0$. It meets the branch locus at $[1:0:0]$, $[1:3:1]$, and $[0:1:2]$.

It remains necessary to show that $\mathrm{rk}\,\mathrm{Pic}(\mathscr{X}^0_{\overline{\mathbb{F}_3}}) \leq 2$ and $\mathrm{rk}\,\mathrm{Pic}(\mathscr{Y}^0_{\overline{\mathbb{F}_5}}) \leq 2$. To verify the first assertion, we ran Algorithm 21 together with Algorithm 15 for counting the points. For the second assertion, we applied Algorithm 22 and Algorithm 17. Note that, for $\mathscr{Y}^0$, the sextic form on the right hand side is decoupled. □

Corollary 30. *Let S be the K3 surface given by*

$$w^2 = 11x^5 y + 7x^5 z + x^4 y^2 + 5x^4 yz + 7x^4 z^2 + 7x^3 y^3 + 10x^3 y^2 z + 5x^3 yz^2 + 4x^3 z^3 \\ + 6x^2 y^4 + 5x^2 y^3 z + 10x^2 y^2 z^2 + 5x^2 yz^3 + 5x^2 z^4 + 11xy^5 + 5xy^3 z^2 + 12xz^5 \\ + 9y^6 + 5y^4 z^2 + 10y^2 z^4 + 4z^6.$$

i) *Then, $\mathrm{rk}\,\mathrm{Pic}(S_{\overline{\mathbb{Q}}}) = 1$.*
ii) *Further, $S(\mathbb{Q}) \neq \emptyset$. $[2\,;\,0:0:1]$ and $[3\,;\,0:1:0]$ are examples of $\mathbb{Q}$-rational points on S.*

Remark 31. a) For the K3 surface $\mathscr{X}^0$, the assumption of the negative sign leads to zeroes the absolute values of which range (without scaling) from 2.598 to 3.464. Thus, the sign in the functional equation is positive. For the decomposition of the characteristic polynomial f_p of the Frobenius, we find (after scaling to zeroes of absolute value 1)

$$(t-1)^2(3t^{20} + 2t^{19} + 2t^{18} + 2t^{17} + t^{16} - 2t^{13} - 2t^{12} - t^{11} - 2t^{10} \\ - t^9 - 2t^8 - 2t^7 + t^4 + 2t^3 + 2t^2 + 2t + 3)/3$$

with an irreducible polynomial of degree 20.

b) For the K3 surface $\mathscr{Y}^0$, the assumption of the negative sign leads to zeroes the absolute values of which range (without scaling) from 3.908 to 6.398. The sign in the functional equation is therefore positive. For the decomposition of the scaled characteristic polynomial of the Frobenius, we find

$$(t-1)^2(5t^{20} - 5t^{19} - 5t^{18} + 10t^{17} - 2t^{16} - 3t^{15} + 4t^{14} - 2t^{13} - 2t^{12} + t^{11}$$
$$+ 3t^{10} + t^9 - 2t^8 - 2t^7 + 4t^6 - 3t^5 - 2t^4 + 10t^3 - 5t^2 - 5t + 5)/5 \,.$$

c) For $\mathscr{X}^0$ and $\mathscr{Y}^0$, the sextics appearing on the right hand side are smooth. This was checked by a Gröbner base computation. The numbers of points and the traces of the Frobenius we determined are reproduced in table 1.

6   An Example in Determinantal Form

Lemma 32. *Let M be a matrix of the particular shape*

$$M := \begin{pmatrix} l^2 & q & 0 \\ c & a & b \\ d & 0 & a \end{pmatrix}.$$

Here, l is supposed to be an arbitrary linear form. a, b, c, d, and q are arbitrary quadratic forms, q being non-degenerate and not a multiple of a.

Then, $q(x, y, z) = 0$ defines a smooth conic meeting the sextic given by $\det(M(x, y, z)) = 0$ only with even multiplicities.

Proof. This may be seen by observing the congruence

$$\det(M) \equiv l^2 a^2 \pmod{q}\,. \qquad \square$$

Examples 33. i) Let $\mathscr{X}$ be the K3 surface over $\mathbb{F}_3$ given by $w^2 = f_6(x, y, z)$ for

$$f_6(x, y, z) = \det \begin{pmatrix} l^2 & q & 0 \\ c & a & b \\ d & 0 & a \end{pmatrix} = \begin{cases} x^6 + 2x^5y + 2x^5z + 2x^4y^2 + x^4yz + x^4z^2 + x^3y^2z \\ + 2x^3yz^2 + 2x^3z^3 + x^2y^4 + x^2y^3z + 2x^2yz^3 + xy^5 + xy^4z \\ + xy^3z^2 + xyz^4 + xz^5 + 2y^6 + 2y^5z + 2y^4z^2 + y^3z^3 + yz^5\,. \end{cases}$$

Here, we put

$$q = x^2 + y^2 + z^2, \qquad l = 2x + y + z,$$
$$a = x^2 + xy + 2z^2, \qquad b = xy + y^2 + yz + 2z^2,$$
$$c = xy + 2xz + z^2, \qquad d = 2xy + 2xz + 2y^2 + 2z^2.$$

Then, the conic given by $q = 0$ meets the ramification locus such that all intersection multiplicities are even.

ii) Let $\mathscr{Y}$ be the K3 surface over $\mathbb{F}_5$ given by $w^2 = f_6(x, y, z)$ for

$$f_6(x, y, z) = \det \begin{pmatrix} 0 & 2x^2 + 2xy + 4y^2 & 4x^2 + 2xz \\ 4x^2 + 2xz + 4z^2 & 0 & x^2 + 2xy + 4y^2 \\ 2x^2 + xy + 4y^2 & x^2 + 2z^2 & 0 \end{pmatrix}$$

$$= 4x^5y + x^4y^2 + 2x^3y^3 + 2x^2y^4 + 4y^6 + x^5z + 2x^4z^2 + xz^5.$$

There appears a degenerate tritangent G given by $x = 0$. It meets the branch sextic at $[0 : 0 : 1]$ with intersection multiplicity 6. The divisor $\pi^*(G)$ splits already over $\mathbb{F}_5$.

Remark 34. Over $\mathbb{F}_5$, we intended to construct examples of K3 surfaces of the form $w^2 = \det(M(x, y, z))$ where $M(x, y, z)$ is a (3×3)-matrix the entries of which are quadratic forms.

In order to be able to execute investigations over $\mathbb{F}_5$ in a reasonable amount of time, we needed a decoupled right hand side. This means, $f_6 := \det(M(x, y, z))$ must not contain monomials containing both y and z. In determinantal form, this may easily be achieved by choosing M of the particular structure

$$M(x, y, z) := \begin{pmatrix} 0 & q_1(x, y) & r_1(x, z) \\ r_2(x, z) & 0 & q_2(x, y) \\ q_3(x, y) & r_3(x, z) & 0 \end{pmatrix}.$$

Then, the determinant has the form $\det M = q_1 q_2 q_3 + r_1 r_2 r_3$.

Note that, in r_1, the monomial z^2 is missing. This causes that, in f_6, the coefficient of z^6 is equal to zero. Therefore, the line given by $x = 0$ meets the sextic "$\det M(x, y, z) = 0$" in only one point.

Theorem 35. *Let S be any K3 surface over $\mathbb{Q}$ such that its reduction modulo 3 is isomorphic to $\mathscr{X}$ and its reduction modulo 5 is isomorphic to $\mathscr{Y}$. Then, $\operatorname{rk} \operatorname{Pic}(S_{\overline{\mathbb{Q}}}) = 1$.*

Proof. It remains necessary to show that $\operatorname{rk} \operatorname{Pic}(\mathscr{X}_{\overline{\mathbb{F}}_3}) \leq 2$ and $\operatorname{rk} \operatorname{Pic}(\mathscr{Y}_{\overline{\mathbb{F}}_5}) \leq 2$. To verify the first assertion, we ran Algorithm 21 together with Algorithm 15 for counting the points. For the second assertion, we applied Algorithm 22 and Algorithm 17. Note that, for $\mathscr{Y}$, the sextic form on the right hand side is decoupled. $\qquad \square$

Corollary 36. *Let S be the K3 surface given by*

$$w^2 = \det \begin{pmatrix} 10x^2 + 10xy + 10xz + 10y^2 + 5yz + 10z^2 & 7x^2 + 12xy + 4y^2 + 70z^2 & 9x^2 + 12xz \\ 9x^2 + 10xy + 2xz + 4z^2 & 10x^2 + 10xy + 5z^2 & 6x^2 + 7xy + 4y^2 + 10yz + 5z^2 \\ 12x^2 + 11xy + 5xz + 14y^2 + 5z^2 & 6x^2 + 12z^2 & 10x^2 + 10xy + 5z^2 \end{pmatrix}$$

$$= -80x^6 + 194x^5y - 424x^5z + 941x^4y^2 - 125x^4yz - 863x^4z^2$$
$$+ 3222x^3y^3 + 520x^3y^2z - 1735x^3yz^2 + 1040x^3z^3$$
$$+ 3292x^2y^4 + 1180x^2y^3z + 8370x^2y^2z^2 + 8510x^2yz^3 + 210x^2z^4$$
$$+ 1240xy^5 + 2200xy^4z + 10900xy^3z^2 + 7320xy^2z^3 + 2170xyz^4 + 976xz^5$$
$$+ 224y^6 + 560y^5z + 3800y^4z^2 + 8560y^3z^3 + 4890y^2z^4 + 2125yz^5.$$

i) *Then,* $\operatorname{rk}\operatorname{Pic}(S_{\overline{\mathbb{Q}}}) = 1$.

ii) *Further,* $S(\mathbb{Q}) \neq \emptyset$. *For example,* $[0 \; ; \; 0 : 0 : 1] \in S(\mathbb{Q})$.

Remark 37. a) For $\mathscr{X}$, the assumption of the negative sign leads to zeroes the absolute values of which range (without scaling) from 2.609 to 3.450. Thus, we have the positive sign in the functional equation. The decomposition of the characteristic polynomial (after scaling to zeroes of absolute value 1) is

$$(t-1)^2(3t^{20} + t^{18} - 2t^{17} - t^{15} + t^{13} - t^{12} + 3t^{11} + 3t^9 - t^8 + t^7 - t^5 - 2t^3 + t^2 + 3)/3$$

with an irreducible degree 20 polynomial. Therefore, the geometric Picard rank is equal to 2.

b) For $\mathscr{Y}$, the assumption of the negative sign leads to zeroes the absolute values of which range (without scaling) from 4.350 to 5.748. The sign in the functional equation is therefore positive. The decomposition of the scaled characteristic polynomial is

$$(t-1)^2(5t^{20} + 5t^{19} - 2t^{18} - 2t^{17} + 2t^{16} - 2t^{15} - 3t^{14} - 2t^{12} + 3t^{10}$$
$$- 2t^8 - 3t^6 - 2t^5 + 2t^4 - 2t^3 - 2t^2 + 5t + 5)/5.$$

Consequently, the geometric Picard rank is equal to 2.

c) We list the numbers of points and the traces of the Frobenius we determined in table 1.

Details on the Experiments. i) Choosing l, a, b, c, d, and q randomly, we had generated a sample of 30 examples over $\mathbb{F}_3$. For each of them, by inspecting the ideal of the singular locus, we had checked that the branch sextic is smooth. Further, they had passed the tests described in section 2 to exclude the existence of a tritangent or a second conic tangent in six points.

For exactly five of the 30 examples, we found an upper bound of two for the geometric Picard rank. Example 33. i) reproduces one of them. The running time was around 30 minutes per example.

ii) We had randomly generated a series of 30 examples over $\mathbb{F}_5$ in which the branch locus is smooth and does neither allow a conic tangent in six points nor further tritangents.

Table 1. Numbers of points and traces of the Frobenius

	$\mathscr{X}^0$		$\mathscr{Y}^0$		$\mathscr{X}$		$\mathscr{Y}$	
d	$\#\mathscr{X}^0(\mathbb{F}_{3^d})$	$\operatorname{Tr}(\phi^d)$	$\#\mathscr{Y}^0(\mathbb{F}_{5^d})$	$\operatorname{Tr}(\phi^d)$	$\#\mathscr{X}(\mathbb{F}_{3^d})$	$\operatorname{Tr}(\phi^d)$	$\#\mathscr{Y}(\mathbb{F}_{5^d})$	$\operatorname{Tr}(\phi^d)$
1	14	4	41	15	16	6	31	5
2	92	10	751	125	94	12	721	95
3	758	28	15 626	0	838	108	15 751	125
4	6 752	190	392 251	1 625	6 742	180	391 701	1 075
5	59 834	784	9 759 376	-6 250	59 671	621	9 781 251	15 625
6	532 820	1 378	244 134 376	-6 250	533 818	2 376	244 155 751	15 125
7	4 796 120	13 150	6 103 312 501	-203 125	4 781 674	-1 296	6 103 878 126	362 500
8	43 068 728	22 006	152 589 156 251	1 265 625	43 081 390	34 668	152 589 507 501	1 616 875
9	387 421 463	973	3 814 704 296 876	7 031 250	387 322 075	-98 415	3 814 693 734 376	-3 531 250
10	3 487 077 812	293 410	95 367 474 609 376	42 968 750	3 486 694 249	-90 153	95 367 469 575 001	37 934 375

For each of them, we determined the numbers of points over the fields $\mathbb{F}_{5^d}$ for $d \leq 9$. The method described in section 3 above showed $\mathrm{rk}\,\mathrm{Pic}(S_{\overline{\mathbb{F}}_5}) = 2$ for two of the examples. For these, we further determined the numbers of points over $\mathbb{F}_{5^{10}}$. Example 33.ii) is one of the two.

The code was running for two hours per example which were almost completely needed for point counting. The time required to identify and factorize the characteristic polynomials of the Frobenii was negligible. The point counting over $\mathbb{F}_{5^{10}}$ took around two days of CPU time per example.

iii) The numbers of points counted and the traces of the Frobenius computed in the examples are listed in the table below.

References

[Be] Beauville, A.: Surfaces algébriques complexes. In: Astérisque 54, Société Mathématique de France, Paris (1978)

[EJ] Elsenhans, A.-S., Jahnel, J.: The Asymptotics of Points of Bounded Height on Diagonal Cubic and Quartic Threefolds. In: Hess, F., Pauli, S., Pohst, M. (eds.) ANTS 2006. LNCS, vol. 4076, pp. 317–332. Springer, Heidelberg (2006)

[Fu] Fulton, W.: Intersection theory. Springer, Berlin (1984)

[Li] Lieberman, D.I.: Numerical and homological equivalence of algebraic cycles on Hodge manifolds. Amer. J. Math. 90, 366–374 (1968)

[vL] van Luijk, R.: K3 surfaces with Picard number one and infinitely many rational points. Algebra & Number Theory 1, 1–15 (2007)

[Mi] Milne, J.S.: Étale Cohomology. Princeton University Press, Princeton (1980)

[Pe] Persson, U.: Double sextics and singular K3 surfaces. In: Algebraic Geometry, Sitges (Barcelona) 1983. Lecture Notes in Math., vol. 1124, pp. 262–328. Springer, Berlin (1985)

[Ta] Tate, J.: Conjectures on algebraic cycles in l-adic cohomology. In: Motives, Proc. Sympos. Pure Math., vol. 55-1, pp. 71–83. Amer. Math. Soc., Providence (1994)

[Ze] Zeilberger, D.: A combinatorial proof of Newtons's identities. Discrete Math. 49, 319 (1984)

Number Fields Ramified at One Prime

John W. Jones[1] and David P. Roberts[2]

[1] Dept. of Mathematics and Statistics, Arizona State Univ., Tempe, AZ 85287
jj@asu.edu
[2] Div. of Science and Mathematics, Univ. of Minnesota–Morris, Morris, MN 56267
roberts@morris.umn.edu

Abstract. For G a finite group and p a prime, a G-p field is a Galois number field K with $\mathrm{Gal}(K/\mathbf{Q}) \cong G$ and $\mathrm{disc}(K) = \pm p^a$ for some a. We study the existence of G-p fields for fixed G and varying p.

For G a finite group and p a prime, we define a G-p field to be a Galois number field $K \subset \mathbf{C}$ satisfying $\mathrm{Gal}(K/\mathbf{Q}) \cong G$ and $\mathrm{disc}(K) = \pm p^a$ for some a. Let $\mathcal{K}_{G,p}$ denote the finite, and often empty, set of G-p fields.

The sets $\mathcal{K}_{G,p}$ have been studied mainly from the point of view of fixing p and varying G; see [Har94], for example. We take the opposite point of view, as we fix G and let p vary. Given a finite group G, we let $\mathcal{P}_G$ be the sequence of primes where each prime p is listed $|\mathcal{K}_{G,p}|$ times. We determine, for various groups G, the first few primes in $\mathcal{P}_G$ and their corresponding fields. Only the primes p dividing $|G|$ can be wildly ramified in a G-p field, and so the sequences $\mathcal{P}_G$ which are infinite are dominated by tamely ramified fields.

In Sections 1, 2, and 3, we consider the cases when G is solvable with length 1, 2, and ≥ 3 respectively, using mainly class field theory. Section 4 deals with the much more difficult case of non-solvable groups, with results obtained by complete computer searches for certain polynomials in degrees 5, 6, and 7.

In Section 5, we consider a remarkable $\mathrm{PGL}_2(7)$-53 field given by an octic polynomial from the literature. We show that the generalized Riemann hypothesis implies that in fact $\mathcal{P}_{\mathrm{PGL}_2(7)}$ begins with 53. Sections 6 and 7 construct fields for the first primes in $\mathcal{P}_G$ for more groups G by considering extensions of fields previously found. Finally in Section 8, we conjecture that $\mathcal{P}_G$ always has a density, and this density is positive if and only if G^{ab} is cyclic.

As a matter of notation, we present G-p fields as splitting fields of polynomials $f(x) \in \mathbf{Z}[x]$, with $f(x)$ chosen to have minimal degree. When $\mathcal{K}_{G,p}$ has exactly one element, we denote this element by $K_{G,p}$. To avoid a proliferation of subscripts, we impose the convention that m represents a cyclic group of order m. Finally, for odd primes p let $\hat{p} = (-1)^{(p-1)/4}p$, so that $K_{2,p}$ is $\mathbf{Q}(\sqrt{\hat{p}})$.

One reason that number fields ramified only at one prime are interesting is that general considerations simplify in this context. For example, the formalism of quadratic lifting as in Section 7 becomes near-trivial. A more specific reason is that algebraic automorphic forms ramified at no primes give rise to number fields ramified at one prime via associated p-adic Galois representations. For example, the fields $K_{S_3,23}$, $K_{S_3,31}$, $K_{\tilde{S}_4,59}$ and $K_{\mathrm{SL}_2^{\pm}(11),11}$ here all arise in this way in the

A.J. van der Poorten and A. Stein (Eds.): ANTS-VIII 2008, LNCS 5011, pp. 226–239, 2008.
© Springer-Verlag Berlin Heidelberg 2008

context of classical modular forms of level one [SD73]. We expect that some of the other fields presented in this paper will likewise arise in similar studies of automorphic forms on larger groups.

Most of the computations carried out for this paper made use of pari/gp [PAR06], in both library and command line modes.

1   Abelian Groups

For n a positive integer, set $\zeta_n = e^{2\pi i/n}$, a primitive n^{th} root of unity. The field $\mathbf{Q}(\zeta_n)$ is abelian over $\mathbf{Q}$, with Galois group $(\mathbf{Z}/n)^\times$, where $g \in (\mathbf{Z}/n)^\times$ sends ζ to ζ^g. The Kronecker-Weber theorem says that any finite abelian extension F of $\mathbf{Q}$ is contained in some $\mathbf{Q}(\zeta_n)$ with n divisible by exactly the set of primes ramifying in $F/\mathbf{Q}$. These classical facts let one quickly determine $\mathcal{K}_{G,p}$ for abelian G, and we record the results for future reference.

Proposition 1.1. *Let p be a prime and G a finite abelian group of order $d = p^a m$, with $\gcd(p, m) = 1$.*

1. *If p is odd, there exists a G-p field if and only if G is cyclic and $m \mid p - 1$. In this case, $|\mathcal{K}_{G,p}| = 1$, and $K_{G,p}/\mathbf{Q}$ is tamely ramified if and only if $a = 0$.*
2. *There exists a G-2 field if and only if for some $j \geq 1$, $G \cong 2^j$ or $G \cong 2^j \times 2$. One has $|\mathcal{K}_{2^j \times 2,2}| = 1$, $|\mathcal{K}_{2,2}| = 3$, and, for $j \geq 2$, $|\mathcal{K}_{2^j,2}| = 2$. All fields in $\mathcal{K}_{G,2}$ are wildly ramified.*

For odd p, a defining polynomial for $K_{d,p}$ is given by the minimal polynomial of the trace $\mathrm{Tr}_{\mathbf{Q}(\zeta_{p^{a+1}})/K_{d,p}}(\zeta_{p^{a+1}})$. Explicitly,

$$f_{d,p}(x) = \prod_{u=1}^{d} (x - \sum_{j=1}^{(p-1)/m} e^{2\pi i g^{u+dj}/p^{a+1}}),$$

where g is a generator for the cyclic group $(\mathbf{Z}/p^{a+1})^\times$. For example,

$$f_{7,7}(x) = x^7 - 21x^5 - 21x^4 + 91x^3 + 112x^2 - 84x - 97,$$
$$f_{7,29}(x) = x^7 + x^6 - 12x^5 - 7x^4 + 28x^3 + 14x^2 - 9x + 1.$$

Irreducibility of $f_{d,p}(x)$ follows from the stronger fact that $f_{d,p}(x + (p-1)/m)$ is p-Eisenstein.

2   Length Two Solvable Groups

The next case beyond abelian groups is solvable groups G of length two. This case also essentially reduces to a classical chapter in the theory of cyclotomic fields. Let K be a G-p field.

The case $p = 2$ needs to be treated separately; it doesn't yield any fields for the G considered explicitly below. We restrict to the case of odd p, in which case G^{ab} is necessarily cyclic and there is a unique field $K_{G^{\mathrm{ab}},p} = K^{G'} \subseteq K$. The cyclicity of G^{ab} forces G to be a semidirect product $G' {:} G^{\mathrm{ab}}$. The following is very similar to statement discovered independently in [Hoe07].

Proposition 2.1. *Let K be a G-p field with $p \neq 2$, so that $G = G' {:} G^{\mathrm{ab}}$ as above. Then if p does not divide $|G'|$, the extension $K/K_{G^{\mathrm{ab}},p}$ is unramified.*

The proof considers the compositum $K\mathbf{Q}(\zeta_{p^k})$ where $K_{G^{\mathrm{ab}},p} \subseteq \mathbf{Q}(\zeta_{p^k})$ and then shows and uses that $\mathbf{Q}(\zeta_{p^k})$ has no tame totally ramified extensions.

Proposition 2.1 says that when p is odd and $|G'|$ is coprime to p the set $\mathcal{K}_{G,p}$ is indexed by G^{ab}-stable quotients of $\mathrm{Cl}(K_{G^{\mathrm{ab}},p})$ which are G^{ab}-equivariantly isomorphic to G'. Defining polynomials for fields in $\mathcal{K}_{G,p}$ can then often be computed using explicit class field theory functions in gp.

The simplest case is this setting is dihedral groups $D_\ell = \ell{:}2$ with ℓ and p different odd primes. The group 2 must act on $\mathrm{Cl}(K_{2,p})$ by negation. If the quotient by multiples of ℓ, $\mathrm{Cl}(K_{2,p})/\ell$, is isomorphic to ℓ^r, then $\mathcal{K}_{D_\ell,p}$ has the structure of an $(r-1)$-dimensional projective space over $\mathbf{F}_\ell$ and thus $|\mathcal{K}_{D_\ell,p}| = (\ell^r - 1)/(\ell - 1)$. The general case is similar, but group-theoretically more complicated. In particular, one has to keep careful track of how G^{ab} acts.

In the table below, we present some cases where G is a length two solvable group with G^{ab} acting faithfully and indecomposably on G'. In this setting, $|G^{\mathrm{ab}}|$ and $|G'|$ are forced to be coprime. We list the first few primes p for which there is a tame G-p extension. If there happens to be a wildly ramified G-p extension as well, we record the prime in the column p_w. A prime listed as $p^{(j)}$ signifies that there are j different G-p fields.

G	G^{ab}	p_w	Tame Primes
S_3	2	3	$23, 31, 59, 83, 107, 139, 199, 211, 229, 239, 257, 283, 307, 331, 367$
D_5	2		$47, 79, 103, 127, 131, 179, 227, 239, 347, 401, 439, 443, 479, 523$
D_7	2	7	$71, 151, 223, 251, 431, 463, 467, 487, 503, 577, 587, 743, 811, 827$
D_{11}	2	11	$167, 271, 659, 839, 967, 1283, 1297, 1303, 1307, 1459, 1531, 1583$
D_{13}	2		$191, 263, 607, 631, 727, 1019, 1439, 1451, 1499, 1667, 1907, 2131$
A_4	3		$163, 277, 349, 397, 547, 607, 709, 853, 937, 1009, 1399, 1699, 1777$
$7{:}3$	3		$313, 877, 1129, 1567, 1831, 1987, 2437, 2557, 3217, 3571, 4219$
F_5	4	$5^{(2)}$	$101, 157, 173, 181, 197, 349, 373, 421, 457, 461, 613, 641, 653^{(2)}$
$3^2{:}4$	4		$149, 293, 661, 733, 1373, 1381, 1613, 1621, 1733, 1973, 2861, 3109$
F_7	6	7	$211, 463, 487, 619, 877, 907, 991, 1069, 1171, 1231, 1303, 1381$

Sample defining polynomials are

$$f_{7:3,313}(x) = x^7 - x^6 - 15x^5 + 20x^4 + 33x^3 - 22x^2 - 32x - 8,$$
$$f_{3^2:4,149}(x) = x^6 - 5x^4 - 9x^3 - 31x^2 - 52x - 17.$$

Note that direct application of class field theory would give defining polynomials of degree 21 and 12 respectively.

Suppose G is such that the action of G^{ab} on G' is faithful and decomposes G' as $N_1 \times N_2$. Suppose G^{ab} acts on N_i through a faithful action of its quotient $Q_i \cong G^{\mathrm{ab}}/H_i$. Put $G_i = N_i{:}Q_i$. Then $\mathcal{K}_{G,p}$ can be constructed directly from $\mathcal{K}_{G_1,p}$ and $\mathcal{K}_{G_2,p}$ by taking composita. The simplest case is when $|N_1|$ and $|N_2|$

are coprime. Then $\mathcal{K}_{G,p}$ consists of the composita $K_1 K_2$ as K_i runs over $\mathcal{K}_{G_i,p}$. In particular, $|\mathcal{K}_{G,p}| = |\mathcal{K}_{G_1,p}| \cdot |\mathcal{K}_{G_2,p}|$. Similarly, if G^{ab} acts on G' through a faithful action of its quotient Q, then $\mathcal{K}_{G,p}$ is empty if $\mathcal{K}_{G^{\mathrm{ab}},p}$ is empty, and otherwise consists of $K_{G^{\mathrm{ab}},p} K$ with K running over $\mathcal{K}_{G':Q,p}$. First primes for some groups of these composed types are

G	$3^2{:}2$	$2^2{:}6$	$3{:}4$	$(3 \times 2^2){:}6$	$3{:}8$	$7{:}4$	$3^3{:}2$	$3^4{:}2$
	$\frac{1}{2}S_3^2$	$A_4 2$	$\frac{1}{2}4S_3$	$A_4 S_3$	$\frac{1}{2}S_3 8$	$\frac{1}{2}D_7 4$	$\frac{1}{4}S_3^3$	$\frac{1}{8}S_3^4$
p	3299	163	229	547	257	577	3,321,607	1,876,623,871

A sample defining polynomial is

$$f_{(3\times 2^2):6,547}(x) = f_{A_4,547}(x) f_{S_3,547}(x) = (x^4 - 21x^2 - 3x + 100)(x^3 - x^2 - 3x - 4).$$

On the table, the first description of G gives $G'{:}G^{\mathrm{ab}}$ and the second emphasizes the compositum structure. The case of $3^r{:}2 = \frac{1}{2^{r-1}} S_3^r$ has been studied intensively in the literature. With gp, it is easy to determine that $p = 3,321,607$ is minimal for $3^3{:}2$. The smallest p for $3^4{:}2$ comes from [Bel04].

3   General Solvable Groups: The Case $G = S_4$

For a general solvable group G, one can in principle proceed inductively via the quotients of G using p-ray class groups. For F a number field, let $\mathrm{Cl}_p(F)$ be the p-ray class group of F. This group is infinite, as e.g. $\mathrm{Cl}_p(\mathbf{Q}) = \tilde{\mathbf{Z}}_p^\times$. However, for any positive integer m, the quotient $\mathrm{Cl}_p(F)/m$ is finite. Let $\tilde{F}$ be a maximal abelian extension ramified only at p with Galois group killed by m. Then by class field theory $\mathrm{Gal}(\tilde{F}/F) = \mathrm{Cl}_p(F)/m$.

To carry out the induction efficiently, one works with typically non-Galois number fields F of degree as low as possible. As an example that we will return to in Section 8, take G to be the length three solvable group S_4. To compute $\mathcal{K}_{S_4,p}$, we start from a list of cubic polynomials $f_i(x)$ with splitting fields running over $\mathcal{K}_{S_3,p}$. We compute $\mathrm{Cl}_2(\mathbf{Q}[x]/f_i(x))/2 \cong 2^r$. For each of the $2^r - 1$ order two quotients of this group, we find a corresponding sextic polynomial $g_{i,j}(x)$. For fixed i, there will be one polynomial with Galois group $6T_2 \cong S_3$. The remaining sextic polynomial are grouped in pairs, according to whether they have the same splitting field in $\mathcal{K}_{S_4,p}$. One member of each pair has Galois group $6T_7 \cong S_4$ and the other has Galois group $6T_8 \cong S_4$. The first primes in $\mathcal{P}_{S_4}$ are

$\lambda_p \backslash s$	0	1	2
4	$2713, 2777^{(2)}, 2857, 3137$	$59, 107, 139, 283^{(2)}, 307$	$229^{(2)}, 733, 1373, 1901$
211	$2777, 7537, 8069, 10273$	$283, 331, 491, 563, 643$	$229, 257, 761, 1129^{(2)}$

Here primes are sorted according to the quartic ramification partitions λ_p and $\lambda_\infty = 2^s 1^{4-2s}$, as explained in the next section. For a given cubic resolvent, let m_4 and m_{211} be the number of corresponding S_4 fields with the indicated λ_p. From the underlying group theory, the possibilities for (m_4, m_{211}) are $(0, 2^j - 1)$

and $(2^j, 2^j - 1)$ for any $j \geq 0$. There are thirteen primes ≤ 307 on the S_3 list, and the table illustrates the possibilities $(0,0)$, $(1,0)$, $(2,1)$, and $(0,1)$ by $\{23, 31, 83, 199, 211, 239\}$, $\{59, 107, 139, 307\}$, $\{229, 283\}$, and $\{257\}$ respectively.

4   Non-solvable Groups

In [JR99, JR03] we describe how one can computationally determine all primitive extensions of $\mathbf{Q}$ of a given degree n which are unramified outside a given finite set of primes by means of a targeted Hunter search. Here we employ this method to find the first several G-p fields with $G = A_5$, S_5, A_6, S_6, $GL_2(3)$, and S_7. All together, the results presented in this section represent several months of CPU time. In each case, the first step is to quickly verify that there are are no wildly ramified fields.

For a tamely ramified prime p, the ramification possibilities are indexed by partitions of n, with $\lambda = 11 \cdots 11$ indicating unramified and $\lambda = n$ totally ramified. If the partition has $|\lambda|$ parts, then the degree n field $\mathbf{Q}[x]/f(x)$ has discriminant $\hat{p}^{n-|\lambda|}$. For fixed n and varying λ, the search for all such fields has run time roughly proportional to $p^{-|\lambda|(n-2)/4}$. As usual, we say that λ is even or odd according to the parity of its number of even parts, i.e. according to the parity of $n - |\lambda|$.

For $G = A_n$ we can naturally restrict attention to even λ. For $G = S_n$ we can restrict attention to odd λ, since the Galois fields sought contain the ramified quadratic $\mathbf{Q}(\sqrt{\hat{p}})$. Similarly for the septic group $GL_3(2)$, we need only search $\lambda = 7$, 421, 331, and 22111. Finally, the fields $\mathbf{Q}[x]/f(x)$ sought have local root numbers ϵ_∞ and ϵ_p with $\epsilon_\infty \epsilon_p = 1$; see e.g. [JR06]. One has $\epsilon_\infty = (-i)^s$ with s the number of complex places. If all parts of λ_p are odd then $\epsilon_p = 1$. Thus whenever all parts of λ_p are odd, the fields we seek are totally real; this fact reduces search times by a substantial factor in each degree.

We now describe our results in degrees 5, 6, and 7 in turn. For the purposes of the next section, the last three columns of the tables give p-ray class group information in terms of elementary divisors. For a field F, we let $\mathrm{Cl}_p^t(F)$ be the tame part of its p-ray class group. Thus $\mathrm{Cl}_p^t(\mathbf{Q})$ is cyclic of order $p-1$; in general, $\mathrm{Cl}_p^t(F)$ is finite because of the tameness condition. To focus on the information which turns out to be interesting, we define $\mathrm{cl}_p(F)$ to be the product of the 2- and 3-primary parts of $\mathrm{Cl}_p^t(F)$ and abbreviate $\mathrm{cl}_p(\mathbf{Q}) = \mathrm{cl}_p$. Further degree-specific information is given below.

The four fields in our A_5 table with $\lambda = 221$ are all in Table 1 of [BK94], which lists all non-real A_5 fields with discriminant $\leq 2083^2$. The paper [DM06], for the purposes of constructing even Galois representations of prime conductor, focuses on totally real fields with $\lambda_p = 5$, 311, and 221 under the respective assumptions that $p \equiv 1$ modulo 5, 3, and 4 respectively. It finds the first primes in these three cases to be 1951, 10267, and 13613. Thus [DM06] skips over our fields with primes 1039 and 4253 because of its congruence conditions.

Theorem 4.1. *There are exactly five A_5-p fields with $p \le 1553$ and five S_5-p fields with $p \le 317$ as listed below. Moreover, the minimal prime for an A_5-p field with $\lambda = 311$ is $p = 4253$.*

p	λ	s	$f_{A_5,p}(x)$	$\mathrm{cl}_p(F_5)$	$\mathrm{cl}_p(F_6)$	cl_p
653	221	2	$x^5 + 3x^3 - 6x^2 + 2x - 1$	$4\cdot3$	$8\cdot2$	4
1039	5	0	$x^5 - 2x^4 - 414x^3 + 4945x^2 - 16574x + 5191$	$2\cdot2\cdot3\cdot3$	$8\cdot2\cdot3$	$2\cdot3$
1061	221	2	$x^5 - x^4 - 4x^3 + 15x^2 + 32x + 16$	$4\cdot3$	$8\cdot2$	4
1381	221	2	$x^5 - 2x^4 + 8x^3 - 18x^2 - x - 36$	$4\cdot3\cdot3$	$8\cdot2\cdot3$	$4\cdot3$
1553	221	2	$x^5 - x^3 - 6x^2 + 16x - 1$	$16\cdot9$	$8\cdot4$	4
			$\vdots$			
4253	311	0	$x^5 - 2x^4 - 10x^3 + 23x^2 - 6x - 4$	$2\cdot2\cdot4\cdot3$	$8\cdot4$	4

p	λ	s	$f_{S_5,p}(x)$	$\mathrm{cl}_p(F_5)$	$\mathrm{cl}_p(F_6)$	cl_p
101	32	2	$x^5 - x^4 - 6x^3 + x^2 + 18x - 4$	$4\cdot2$	$4\cdot4$	4
151	32	1	$x^5 - 2x^4 - x^3 + 7x^2 - 13x + 7$	$2\cdot3$	$8\cdot3$	$2\cdot3$
269	41	2	$x^5 - x^4 - 15x^3 - 11x^2 + 11x - 10$	4	$4\cdot4$	4
281	32	2	$x^5 - 2x^4 + 17x^3 - 25x^2 + 38x - 13$	8	$16\cdot2$	8
317	41	2	$x^5 - 2x^4 - 14x^3 + 28x^2 + 75x - 175$	4	$4\cdot4\cdot2$	4

A Galois A_5 or S_5 field can be presented by either an irreducible quintic or sextic polynomial, with corresponding fields $F_5 = \mathbf{Q}[x]/f_5(x)$ and $F_6 = \mathbf{Q}[x]/f_6(x)$. One can pass back and forth between F_5 and F_6 through sextic twinning, as explained with examples in e.g. [JR99]. In our cases, the maps $\mathrm{Cl}_p^t(F_n) \to \mathrm{Cl}_p^t(\mathbf{Q})$ are isomorphisms on ℓ primary parts for $\ell \ne 2, 3$.

Similarly, a Galois A_6 and S_6 field corresponds to a pair of non-isomorphic sextic fields interchanged by sextic twinning. Below we give a defining polynomial for one of these fields F_6 but not its twin F_6^t. Exactly as in the quintic case, the parts of $\mathrm{Cl}_p^t(F)$ and $\mathrm{Cl}_p^t(F^t)$ not coming from $\mathrm{Cl}_p(\mathbf{Q})$ are entirely 2- and 3-primary.

A sextic A_6 or S_6 field and its twin will have the same ramification partition λ_p with the exception of the interchanges $6 \leftrightarrow 321$, $33 \leftrightarrow 3111$, $222 \leftrightarrow 21111$. The interchanges help in conducting targeted Hunter searches since one needs only search the second partition which is easier in each case.

Theorem 4.2. *There are exactly two Galois A_6-p fields with $p \le 1677$ and seven Galois S_6-p fields with $p \le 1423$ as listed below. Moreover, the minimal prime for an A_6-p field with $\lambda = 2211$ is $p = 3929$.*

p	λ	s	$f_{A_6,p}(x)$	$\mathrm{cl}_p(F_6)$	$\mathrm{cl}_p(F_6^t)$	cl_p
1579	42	2	$x^6 - x^5 + 41x^4 - 349x^3 + 12x^2 + 3099x + 2851$	$2\cdot3\cdot3$	$2\cdot2\cdot3\cdot3$	$2\cdot3$
1667	42	2	$x^6 - 2x^5 - 39x^4 + 60x^3 + 380x^2 + 1267x + 100$	$2\cdot3$	$2\cdot2\cdot3$	2
			$\vdots$			
3929	2211	2	$x^6 - x^5 - 3x^4 + 9x^3 - 8x^2 + 2x - 1$	$8\cdot8\cdot3$	$8\cdot2\cdot3$	8

p	λ	s	$f_{S_6,p}(x)$	$\text{cl}_p(F_6)$	$\text{cl}_p(F_6^t)$	cl_p
197	6	2	$x^6 + 788x - 197$	$4{\cdot}2$	$4{\cdot}2$	4
593	321	2	$x^6 - 2x^3 - x^2 + 58x - 88$	$16{\cdot}2$	$16{\cdot}2$	16
929	321	2	$x^6 - 3x^5 - x^4 + 4x^3 + 56x - 32$	$32{\cdot}2$	$32{\cdot}2$	32
977	6	2	$x^6 - 977x^3 + 7816x^2 - 20517x + 17586$	16	16	16
1109	321	2	$x^6 - 10x^3 - 61x^2 - 41x - 218$	$4{\cdot}4$	4	4
1301	411	2	$x^6 - 2x^5 + 5x^4 - 36x^3 - 24x^2 + 32x - 57$	4	$4{\cdot}2$	4
1409	321	2	$x^6 - x^5 - 7x^4 - 30x^3 - 41x^2 - 177x + 191$	128	$256{\cdot}2{\cdot}2$	128

In our septic cases we give the entire tame class groups because for the second S_7 field the prime 5 also behaves non-trivially.

Theorem 4.3. *There is exactly one* $\mathrm{GL}_3(2)$-p *field with* $p \le 227$ *and exactly two* S_7-p *fields with* $p \le 191$.

p	λ	s	$f_{\mathrm{GL}_3(2),p}(x)$	$\mathrm{Cl}_p^t(F_7)$	$\mathrm{Cl}_p^t(F_7^t)$	$\mathrm{Cl}_p^t(\mathbf{Q})$
227	421	3	$x^7 + 2x^5 - 4x^4 - 5x^3 - 4x^2 - 3x + 10$	$2{\cdot}2{\cdot}2{\cdot}113$	$2{\cdot}2{\cdot}113$	$2{\cdot}113$

p	λ	s	$f_{S_7,p}(x)$	$\mathrm{Cl}_p^t(F_7)$		Cl_p^t
163	52	3	$x^7 - 2x^6 - 19x^4 + 65x^3 + 39x^2 + 3x + 1$	$2{\cdot}81$		$2{\cdot}81$
191	3211	3	$x^7 - 2x^6 + x^5 - x^4 + 3x^3 - 8x^2 + 7x - 2$	$2{\cdot}5{\cdot}5{\cdot}19$		$2{\cdot}5{\cdot}19$

5 $\mathrm{PGL}_2(7)$

The Klüners-Malle website [KM01] contains the polynomial

$$f_0(x) = x^8 - x^7 + 3x^6 - 3x^5 + 2x^4 - 2x^3 + 5x^2 + 5x + 1$$

defining a $\mathrm{PGL}_2(7)$-53 field K_0 with octic ramification partition 611. In comparison with our previous results on first elements of $\mathcal{P}_G$ for nonsolvable G, the prime 53 is remarkably small. In fact,

Proposition 5.1. *Assuming the generalized Riemann hypothesis, K_0 is the only Galois* $\mathrm{PGL}_2(7)$-p *field with* $p \le 53$.

Proof. Let $f(x) \in \mathbf{Z}[x]$ be an octic polynomial defining a $\mathrm{PGL}_2(7)$-p field with $p \le 53$. We will use Odlyzko's GRH bounds [Odl76] to prove that $K = K_0$. To start, since K has degree 336, its root discriminant is at least 24.838.

We first consider the case where $p \notin \{2, 3, 7\}$ so that ramification is tame. Let λ_p be the octic ramification partition of K, and let e be the least common multiple of its parts. As λ_p must be odd and match a cycle type in $\mathrm{PGL}_2(7)$, the possibilities are $\lambda_p = 22211, 611$, or 8. The root discriminant of K is then $p^{(e-1)/e}$ where $e = 2$, 6, or 8. Thus $p \ge 24.838^{e/(e-1)}$ which works out to $p > 616.926$,

$p > 47.221$, and $p > 39.302$ in the three cases. Thus either $e = 6$ and $p = 53$ or $e = 8$ and $p \in \{41, 43, 47, 53\}$.

Suppose for the next two paragraphs that $e = 8$. Then the p-adic field $\mathbf{Q}_p[x]/f(x)$ is a totally ramified octic extension of $\mathbf{Q}_p$ whose associated Galois group D_p is a subgroup of $\mathrm{PGL}_2(7)$. But, a totally ramified octic extension of $\mathbf{Q}_p$ has $D_p = 8T_1$, $8T_8$, $8T_7$, or $8T_6$ depending on whether $p \equiv 1$, 3, 5, or 7 respectively modulo 8. Since $8T_8$ and $8T_7$ are not isomorphic to subgroups of $\mathrm{PGL}_2(7)$, one must have $p \equiv \pm 1 \pmod 8$ when $e = 8$. Thus $p \in \{41, 47\}$.

If p were 41, then the compositum $K_{4,41}K$ would have degree $(336 \cdot 4)/2 = 672$ and root discriminant $41^{7/8} \approx 25.8$. But a degree 672 field has root discriminant ≥ 27.328, a contradiction proving $p \neq 41$. Similarly, if p were 47, then the compositum $K_{D_5,47}K$ would have degree $(336 \cdot 10)/2 = 1680$ and root discriminant $47^{7/8} \approx 29.05$. But a degree 1680 field has root discriminant ≥ 29.992 by [Odl76], a contradiction proving $p \neq 47$. Thus, in fact, $e \neq 8$.

Now suppose $p \in \{2, 3, 7\}$. For $p = 2$ and 3, there are a number of possibilities for the decomposition group D_p. However the maximum possible root discriminant for K would be $2^4 = 16$ and $3^{13/6} \approx 10.81$ respectively, each of which is less than 24.838. For $p = 7$, the field K is not totally real because it would contain $\mathbf{Q}(\sqrt{-7})$. So Khare's theorem [Kha06] applies, showing that there would exist a modular form of level 1 over $\overline{\mathbf{F}}_7$ associated to K. But by [Ser75], representations associated to such modular forms are reducible.

Finally, suppose K were a $\mathrm{PGL}_2(7)$-53 extension different from K_0. Then the compositum K_0K would have degree $336^2/2 = 56448$ and so root discriminant at least 36.613 by Odlyzko's bounds. However also K_0K has root discriminant $53^{5/6} \approx 27.35$, a contradiction proving that in fact $K = K_0$. □

6   Groups of the Form $2^r.G$ and $3.G$ for Non-solvable G

In this section, we start from the groups G of Section 4. We use the fields there and the corresponding class group information to construct $\tilde{G}$-p fields with $\tilde{G}$ of the form $2^r.G$ and $3.G$.

Proposition 6.1. *The polynomials displayed in this section define $\tilde{G}$-p fields, with p as small as possible for the given $\tilde{G}$.*

In each case except for $\tilde{G} = 3.A_6$, there is only one Galois field corresponding to the minimal prime; for $3.A_6$ there are three, differing by cubic twists. The fields in Section 4 often give rise to the next few primes in these $\mathcal{P}_{\tilde{G}}$ as well.

For the case $\tilde{G} = 2^r.G$, we considered all the fields $F = \mathbf{Q}[x]/f(x)$ for each $f(x)$ appearing in Theorems 4.1, 4.2, and 4.3. For each such field F, we computed the quadratic extension corresponding to each order two character of $\mathrm{cl}_p(F)$. Among the defining polynomials found were

$$f^{34}_{2^4.A_5, 1039}(x) = x^{10} - 149x^8 - 15640x^6 - 50311x^4 - 36993x^2 - 1369,$$
$$f^{37}_{2^4.S_5, 101}(x) = x^{10} + 2px^6 - 32px^4 + p^2x^2 - p^2,$$

$$f_{2^5.S_6,197}^{285}(x) = x^{12} + 4px^8 - 4px^6 + 3p^2x^2 + 4p^2,$$

$$f_{2^3.\,\mathrm{GL}_3(2),227}^{33}(x) = x^{14} + 33x^{12} + px^{10} + 3px^8 - 52px^6 - 62px^4 + p^2x^2 - p^2.$$

Here and below, subscripts indicate $\tilde{G}$ and p. Superscripts give the T-number of $\tilde{G}$ to remove ambiguities. Also we express coefficients by factoring out as many p's as possible. This makes the p-Newton polygon visible, and thus sometimes gives information about p-adic ramification. For example, $f_{2^4.S_5,101}^{37}(x)$ factors over $\mathbf{Q}_{101}$ as a totally ramified sextic times a totally ramified quartic; thus the discriminant of the given decic field is 101^8.

Necessarily, to ensure minimality in the sextic and septic cases, we also worked with the twin fields F^t, likewise using $\mathrm{cl}_p(F^t)$. Defining polynomials appearing here were

$$f_{2^5.A_6,1667}^{277}(x) = x^{12} + 341x^{10} - 303x^8 + 10158x^6 - 2998x^4 + 216x^2 + 1,$$

$$f_{2^6.A_6,1579}^{286}(x) = x^{12} - 109x^{10} + 1100x^8 + 2649x^6 - 567637x^4 + 661px^2 - 4356p,$$

$$f_{2^5.S_6,197}^{287}(x) = x^{12} + 9x^{10} - 75x^8 - 9x^6 + 3px^4 - 2px^2 + p.$$

The two degree 2^5 extensions of $K_{S_6,197}$ are disjoint. The group $14T_{33}$ is the non-split extension of $\mathrm{GL}_3(2)$ by 2^3, to be distinguished from the semidirect product $14T_{34} \cong 8T_{48}$.

The case $\tilde{G} = 3.G$ is attractive because one can quickly understand the 3-ranks of all the class groups printed in Theorems 4.1, 4.2, and 4.3. First, if $p \equiv 1$ (3), the extension $K_{3,p}$ contributes 1 to the 3-rank in all three columns. Second, in the A_5 cases, the abelian extension $F_{15} \cong K^V$ over $F_5 \cong K^{A_4}$ contributes an extra 1 to the the 3-rank of $\mathrm{cl}_p(F_5)$. This accounts for the full 3-rank except in the three A_6 cases. The extra 3's printed in the columns $\mathrm{cl}_p(F_6)$ and $\mathrm{cl}_p(F_6^t)$ are all accounted for by fields with Galois group the exceptional cover $3.A_6$. Specializing the lifting results of [Rob], an A_6-p field with defining polynomial $f(x)$ embeds in a $3.A_6$-p field if and only if $\mathbf{Q}_p[x]/f(x)$ is not the product of two non-isomorphic cubic fields. This is the case for all three of our A_6 fields, and a defining polynomial for the smallest prime is

$$f_{3.A_6,1579,a}(x) = x^{18} - 6x^{17} - 23x^{16} + 211x^{15} - 283x^{14} - 115x^{13} - 2146x^{12} +$$
$$6909x^{11} - 3119x^{10} + 9687x^9 - 35475x^8 - 3061x^7 + 47135x^6 + 14267x^5$$
$$- 13368x^4 - 19592x^3 - 10421x^2 - 4728x - 297.$$

When $p \equiv 1$ (3), each non-obstructed field in $\mathcal{K}_{A_6,p}$ gives three fields in $\mathcal{K}_{3.A_6,p}$, differing by cubic twists. When $p \equiv 2$ (3), each non-obstructed field in $\mathcal{K}_{A_6,p}$ gives rise to just one field in $\mathcal{K}_{3.A_6,p}$.

There is a similar but more complicated theory of lifting from S_6 fields to $3.S_6$ fields [Rob]. The first step in our setting is to look at the 3-ranks of $\mathrm{cl}_p(F_6 \otimes \mathbf{Q}(\sqrt{p}))$ and $\mathrm{cl}_p(F_6^t \otimes \mathbf{Q}(\sqrt{p}))$. A necessary condition for the existence of a $3.S_6$ field is that both of these 3-ranks are at least 1. This occurs first for $p = 593$. Indeed there is a unique lift, with defining polynomial

$$f_{3.S_6,593}(x) = x^{18} - 4x^{17} - 15x^{16} + 131x^{15} + 50x^{14} - 2686x^{13} + 1430x^{12} + 32366x^{11}$$
$$- 37880x^{10} - 282470x^9 + 672468x^8 + 2272632x^7 - 6021114x^6 - 15149054x^5$$
$$+ 18548349x^4 + 59752280x^3 + 15265273x^2 - 89821887x - 96674958.$$

7   Groups of the Form 2.G

Let K be a G-p field. Let $\tilde{G}$ be a non-split double cover of G. The quadratic embedding problem in our context asks whether K embeds in a $\tilde{G}$-p field $\tilde{K}$. This section is similar in nature to the last; however, degrees here are forced to be larger, and relevant class groups often cannot be computed. We replace the class group considerations with a more theoretical treatment.

Let $c \in G$ be complex conjugation, and let $\{c_1, c_2\}$ be its preimage in $\tilde{G}$. An obviously necessary condition for the existence of $\tilde{K}$ is that c_1 and c_2 both have order ≤ 2; the other possibility is that they both have order 4. For general K, there can be local obstructions not only at ∞, but also at any prime ramifying in K with even ramification index. However, in general, by the known structure of the 2-torsion in the Brauer group of $\mathbf{Q}$, the set of obstructed places is even, and there are no further global obstructions. In our one-prime context, p is obstructed exactly when ∞ is obstructed. Thus the above necessary condition is also sufficient for the existence of $\tilde{K}$.

When the embedding problem is known to be solvable, we compute defining polynomials for the quadratic overfields as follows. As usual, for K, we have the flexibility of considering defining polynomials corresponding to any subgroup H of G such that the intersection of H with its conjugates is trivial. To be able to pass to the desired overfields, we need to choose H so that the induced double cover $\tilde{H} \to H$ is split. Following e.g. Section 5.4.4 of [Coh00], we carefully choose a degree $[G : H]$ defining polynomial $f(x)$ so that the splitting field for $f(x^2)$ solves the embedding problem. If $\sqrt{\tilde{p}} \in K$, then $\mathcal{K}_{\tilde{G},p}$ consists of one field, the splitting field of $f(x^2)$. If $\sqrt{\tilde{p}} \notin K$, then $\mathcal{K}_{\tilde{G},p}$ consists of two fields, the splitting fields of $f(x^2)$ and $f(\hat{p}x^2)$. If one field is real and the other is imaginary, we distinguish the two by the subscripts r and i respectively; if both have the same type, we use instead a and b. A particularly simple case is when the ramification index at p is odd, i.e. when all parts of the ramification partition for $f(x)$ are odd. Then K automatically embeds into two different $\tilde{K}$.

A general construction lets one "cancel obstructions" by working in a two element group as follows. Consider two different embedding problems of our type, $(K_1, G_1, \tilde{G}_1)$ and $(K_2, G_2, \tilde{G}_2)$, with z_i the order two element in the kernel of $\tilde{G}_i \to G_i$. Let $F = K_1 \cap K_2$ with $Q = \mathrm{Gal}(F/\mathbf{Q})$. Then the Galois group of the compositum is a fiber product: $\mathrm{Gal}(K_1 K_2/\mathbf{Q}) = G_1 \times_Q G_2$. One has a product embedding problem of our type $(K_1 K_2, G_1 \times_Q G_2, \tilde{G}_1 *_Q \tilde{G}_2)$, where $\tilde{G}_1 *_Q \tilde{G}_2$ is the central product $(\tilde{G}_1 \times_Q \tilde{G}_2)/\langle(z_1, z_2)\rangle$. The product embedding problem is obstructed if and only if exactly one of the factor embedding problems is obstructed. We will use this construction with $(K_1, G_1, \tilde{G}_1)$ an obstructed embedding problem with non-abelian G_1, and $(K_2, G_2, \tilde{G}_2)$ the obstructed embedding problem $(K_{2^r,p}, 2^r, 2^{r+1})$ with $r = \mathrm{ord}_2(p-1)$.

In the rest of this section, we will combine the general theory just reviewed with earlier results of this paper, in particular proving the following proposition.

Proposition 7.1. *The polynomials displayed in the remaining portion of this section define $\tilde{G}$-p fields, with p as small as possible for the given $\tilde{G}$.*

In our discussion, we will also identify first primes for some other groups, without producing defining polynomials.

In general, for $n \geq 4$, the group A_n has a unique non-split double cover $\tilde{A}_n$. This double cover extends to two distinct double covers of S_n. These two double covers are distinguished by the cycle types $2^s 1^{n-s}$ of the splitting involutions: $s \equiv 0, 1$ (4) for $\tilde{S}_n$ and $s \equiv 0, 3$ (4) for $\hat{S}_n$. In the case of $n = 6$, the two double covers are interchanged by sextic twinning, reflecting the involution in conjugacy classes $222 \leftrightarrow 21111$.

In the A_4 case, the only possible ramification partition at p is 31, which yields the odd ramification index 3. Thus any A_4-p field is totally real and embeds into two $\tilde{A}_4$ fields. An S_4 field embeds in an $\tilde{S}_4$ field if and only if $s \leq 1$ and in a $\hat{S}_4$ field if and only if $s = 0$. So the first primes in these cases are 59 and 2713, by the table in Section 3. Since all elements of order 2 in S_4 lift to elements of order 4 in $\hat{S}_4$, the largest H we can take is 3. Defining polynomials are

$$f_{\tilde{A}_4,163,i}(x) = x^8 + 9x^6 + 23x^4 + 14x^2 + 1,$$

$$f_{\tilde{S}_4,59}(x) = x^8 + 7x^6 + 58x^4 - 52x^2 - 283,$$

$$f_{\hat{S}_4,2713}(x) = x^{16} + 1773179px^{14} + 748029721760px^{12} + 158386491521428px^{10}$$
$$+ 464227394803676px^8 + 170883278708p^2x^6 + 23421860p^3x^4 + 739p^4x^2 + p^4.$$

An alternative point of view on the first two fields just displayed comes from $\tilde{A}_4 \cong SL_2(3)$ and $\tilde{S}_4 \cong GL_2(3)$.

The first A_5 field in Theorem 4.1 yields the first two $\tilde{A}_5 * \tilde{4}$ fields, twists of one another at $p = 653$; the minimal degree is 48, beyond the reach of our computations. The second A_5 field and the first two S_5 fields in Theorem 4.1 yield

$$f_{\tilde{A}_5,1039,r}(x) = x^{24} - 1378x^{22} + 530449x^{20} - 61379655x^{18} + 1188832770x^{16}$$
$$- 9638857366x^{14} + 38717668417x^{12} - 76991153229x^{10} + 64169595698x^8$$
$$- 10073672645x^6 + 435756634x^4 - 150625x^2 + 1,$$

$$f_{\tilde{S}_5*2\tilde{4},101}(x) = x^{24} + p(2x^{22} + 5183x^{20} + 5018386x^{18} + 1719346983x^{16}$$
$$+ 31145667541x^{14} + 191170958302x^{12} + 470365101611x^{10}$$
$$+ 19509244311x^8 - 98676327x^6 - 10345828x^4 - 139569x^2 + 121)$$

$$f_{\tilde{S}_5,151}(x) = x^{40} - 33x^{38} - 398x^{36} + 5788x^{34} + 180619x^{32} - 1960647x^{30} - 10306409x^{28}$$
$$+ 85964700x^{26} + 499284483x^{24} - 3672894736x^{22} + 3925357724x^{20}$$
$$+ 1667363482x^{18} + 5017492392x^{16} + 2279641280x^{14} + 1575477871x^{12}$$
$$+ 714220278x^{10} - 48630589x^8 - 48329892x^6 - 11843px^4 + 155px^2 - p.$$

Here again, there is an alternative viewpoint: $\tilde{A}_5 \cong SL_2(5)$, and $\tilde{S}_5*2\tilde{4} \cong GL_2(5)$.

From Theorem 4.2, the first p for $\tilde{A}_6 * \tilde{2}$, $\tilde{A}_6 * \tilde{4}$, and $\tilde{S}_6 \cong \hat{S}_6$ respectively are 1579, 3929, and 197. The minimal degree is 80 in each case, and corresponds to an action on $\mathbf{F}_9^2 - \{(0,0)\}$ via $\tilde{A}_6 = SL_2(9)$.

From Theorem 4.3, the first two primes for $\hat{S}_7$ are 163 and 191. Here $H = 7{:}6$. and so the minimal degree is 120. The other lift $\tilde{S}_7$ has $H = 7{:}3$, and so requires the even larger degree 240; it also requires larger primes as both 163 and 191 are obstructed. The first prime for $\mathrm{SL}_2(7) * \tilde{2}$ is 227, with defining polynomial

$$f_{\mathrm{SL}_2(7)*\tilde{2},227}(x) = x^{32} + 351px^{30} + 9952243px^{28} + 144266253px^{26}$$
$$+ 45335657253px^{24} - 1671679993p^2x^{22} + 2492032310p^2x^{20} + 873353354p^2x^{18}$$
$$+ 37974755524p^2x^{16} + 104438863p^3x^{14} + 243444277p^3x^{12} - 91558170p^3x^{10}$$
$$+ 19220043p^3x^8 + 15382p^4x^6 + 2530p^4x^4 + 64p^4x^2 + p^4.$$

This polynomial was calculated in two quadratic steps, starting from an octic polynomial.

For q a prime power congruent to 3 modulo 4, a non-split quadratic lift of $\mathrm{PGL}_2(q)$ is the group $\mathrm{SL}_2^{\pm}(q)$ of matrices of determinant plus or minus one. From Proposition 5.1, under GRH the group $\mathrm{SL}_2^{\pm}(7) *_2 \tilde{4}$ first appears for $p = 53$. The minimal degree is 64. Finally, consider the $\mathrm{PGL}_2(11)$-11 field corresponding to 11-torsion points on the first elliptic curve $X_0(11)$. This field is perhaps the most classical example in the subject of number fields ramified at one prime; a defining dodecic equation can be obtained by substituting $J = -64/297$ into Equation 325a of a 1888 paper of Kiepert [Kie88]. We find that a remarkably simple equation for the $\mathrm{SL}_2^{\pm}(11)$ quadratic overfield is

$$f_{\mathrm{SL}_2^{\pm}(11),11}(x) = x^{24} + 90p^2x^{12} - 640p^2x^8 + 2280p^2x^6 - 512p^2x^4 + 2432px^2 - p^3.$$

8 A Density Conjecture

If G^{ab} is non-cyclic, then $\mathcal{K}_{G,p}$ can only be non-empty for $p = 2$. We close with a conjecture which addresses the behavior of $|\mathcal{K}_{G,p}|$ in the non-trivial case that G^{ab} is cyclic. Our conjecture is inspired by a conjecture of Malle [Mal02] which deals with fields of general discriminant, not just prime power absolute discriminant.

Conjecture 1. *Let G be a finite group with $|G| > 1$ and G^{ab} cyclic. Then the ratio $\sum_{p \le x} |\mathcal{K}_{G,p}| / \sum_{p \le x} 1$ tends to a positive limit δ_G as $x \to \infty$.*

The conjecture is certainly true if G is the cyclic group m. In fact, $\mathcal{P}_m^{\mathrm{tame}}$ is the set of primes congruent to 1 modulo m ,and so $\delta_m = 1/\phi(m)$.

Bhargava [Bha07] has a heuristic which, when transposed from general fields to fields with prime power absolute discriminant, gives a formula for δ_{S_n}. Assume $n \ge 3$ and, for the moment, $n \ne 6$ so that S_n has no non-trivial outer automorphisms. Then any Galois extension K of $\mathbf{Q}$ with Galois group S_n has a well-defined involutory partition $\lambda_\infty = 2^s 1^{n-2s}$ corresponding to complex conjugation. Similarly, if K is not wildly ramified at p then it has a well-defined partition λ_p corresponding to the tame p-adic ramification. If K is ramified at p only, then λ_p must be an odd partition. The density that Bhargava's transposed heuristic gives for S_n-p fields with the indicated invariants is

$$\delta_{S_n,s,\lambda_p} = \frac{1}{(n-2s)!s!2^{s+1}}. \tag{8.1}$$

Here $1/((n-2s)!s!2^s)$ is the fraction of elements in S_n with cycle type $2^s 1^{n-2s}$. The extra factor of 2 in the denominator of (8.1) can be thought of as coming from the global root number condition $\epsilon_\infty \epsilon_p = 1$. Note that the right side of (8.1) is independent of λ_p.

Summing over the possible s and then multiplying by the number of possible λ_p gives the conjectured value for δ_{S_n}. For $n = 6$, all these considerations would go through without change if we were working with isomorphism classes of sextic fields. However, we have placed the focus on Galois fields, and there is one Galois field for each twin pair of sextic fields. Accordingly, we need to divide the right side of (8.1) by 2. The final conjectured values in degrees ≤ 7 are then

n	3	4	5	6	7
δ_{S_n}	$0.\overline{3}$	$0.41\overline{6}$	0.325	$0.131\overline{94}$	$0.16\overline{1}$

Bhargava's heuristic can be recast more group-theoretically to give a conjectural formula for δ_G for arbitrary G. For length two solvable groups $G = \ell^r : G^{\mathrm{ab}}$, the δ_G one obtains is the same as that given by the Cohen-Lenstra heuristics applied to the ℓ part of class groups of fields of the form $K_{G^{\mathrm{ab}}, p}$. Thus we expect e.g. $\delta_{A_4} = 1/8$ and hence, by automatic lifting and twisting as described in the previous section, $\delta_{\tilde{A}_4} = 1/4$.

The computations in [tRW03] for S_3-p fields for several billion primes are strongly supportive of Cohen-Lenstra heuristics in this setting, and hence our expectation $\delta_{S_3} = 1/3$. We have carried out similar computations for S_4-p fields for the first 10^6 primes ≥ 5:

	S_3		S_4					
(s, λ)	$(0, 21)$	$(1, 21)$	$(0, 211)$	$(1, 211)$	$(2, 211)$	$(0, 4)$	$(1, 4)$	$(2, 4)$
10^2	.02	.21	0	.03	.02	0	.12	.02
10^3	.050	.193	.002	.056	.031	.013	.077	.034
10^4	.0634	.2080	.0080	.0698	.0399	.0161	.0965	.0462
10^5	.06911	.22714	.01047	.08589	.04567	.01676	.10525	.04837
10^6	.073965	.234667	.013471	.097131	.050874	.018186	.111884	.052834
∞	$.08\overline{3}$	.25	$.0208\overline{3}$	.125	.0625	$.0208\overline{3}$	.125	.0625

Our S_4 data roughly tracks the slowly convergent S_3 data. There are more fields with $\lambda_p = 4$ than with $\lambda_p = 211$, corresponding to the asymmetry noted in Section 3; we expect this discrepancy to go away in the limit. For each λ_p, the dependence on s already agrees well with the expected limiting ratios $1 : 6 : 3$.

For $n = 5$ through 7, our very small initial segments of $\mathcal{P}_G$ all have smaller density than our conjectured value of δ_{S_n}. This is to be expected, given the behavior for $n = 3$ and 4. However, our determination of first primes 59, 101, 197, 163 at least reflects our expectation $\delta_{S_4} > \delta_{S_5} > \delta_{S_6} < \delta_{S_7}$, including the perhaps surprising inequality $\delta_{S_6} < \delta_{S_7}$.

References

[Bel04]   Belabas, K.: On quadratic fields with large 3-rank. Math. Comp. 73(248), 2061–2074 (2004)

[Bha07]   Bhargava, M.: Mass formulae for extensions of local fields, and conjectures on the density of number field discriminants. Int. Math. Res. Notices, rnm052–20 (2007)

[BK94]   Basmaji, J., Kiming, I.: A table of A_5-fields. In: On Artin's Conjecture for Odd 2-Dimensional Representations. Lecture Notes in Math., vol. 1585, pp. 37–46, pp. 122–141. Springer, Berlin (1994)

[Coh00]   Cohen, H.: Advanced Topics in Computational Number Theory. Graduate Texts in Mathematics, vol. 193. Springer, New York (2000)

[DM06]   Doud, D., Moore, M.W.: Even icosahedral Galois representations of prime conductor. J. Number Theory 118(1), 62–70 (2006)

[Har94]   Harbater, D.: Galois groups with prescribed ramification. In: Arithmetic Geometry (Tempe, AZ, 1993), Contemp. Math., vol. 174, pp. 35–60. Amer. Math. Soc., Providence (1994)

[Hoe07]   Hoelscher, J.-L.: Galois extensions ramified at one prime. PhD thesis, University of Pennsylvania (2007)

[JR99]   Jones, J.W., Roberts, D.P.: Sextic number fields with discriminant $(-1)^j 2^a 3^b$. In: Number Theory (Ottawa, ON, 1996), CRM Proc. Lecture Notes, vol. 19, pp. 141–172. Amer. Math. Soc., Providence (1999)

[JR03]   Jones, J.W., Roberts, D.P.: Septic fields with discriminant $\pm 2^a 3^b$. Math. Comp. 72(244), 1975–1985 (2003)

[JR06]   Jones, J.W., Roberts, D.P.: A database of local fields. J. Symbolic Comput. 41(1), 80–97 (2006)

[Kha06]   Khare, C.: Serre's modularity conjecture: the level one case. Duke Math. J. 134(3), 557–589 (2006)

[Kie88]   Kiepert, L.: Ueber die Transformation der elliptischen Functionen bei zusammengesetztem Transformationsgrade. Math. Ann. 32(1), 1–135 (1888)

[KM01]   Klüners, J., Malle, G.: A database for field extensions of the rationals. LMS J. Comput. Math. 4, 182–196 (2001)

[Mal02]   Malle, G.: On the distribution of Galois groups. J. Number Theory 92(2), 315–329 (2002)

[Odl76]   Odlyzko, A.: Table 2: Unconditional bounds for discriminants (1976), http://www.dtc.umn.edu/~odlyzko/unpublished/discr.bound.table2

[PAR06]   The PARI Group, Bordeaux. PARI/GP, Version 2.3.2 (2006)

[Rob]   Roberts, D.P.: 3.G number fields for sextic and septic groups G (in preparation)

[SD73]   Swinnerton-Dyer, H.P.F.: On l-adic representations and congruences for coefficients of modular forms. In: Modular Functions of One Variable, III (Proc. Internat. Summer School, Univ. Antwerp, 1972). Lecture Notes in Math., vol. 350, pp. 1–55. Springer, Berlin (1973)

[Ser75]   Serre, J.-P.: Valeurs propres des opérateurs de Hecke modulo l. In: Journées Arithmétiques de Bordeaux (Conf., Univ. Bordeaux, 1974), Astérisque 24–25, Soc. Math. France, Paris, pp. 109–117 (1975)

[tRW03]   te Riele, H., Williams, H.: New computations concerning the Cohen-Lenstra heuristics. Experiment. Math. 12(1), 99–113 (2003)

An Explicit Construction of Initial Perfect Quadratic Forms over Some Families of Totally Real Number Fields

Alar Leibak

Department of Mathematics at Tallinn University of Technology
alar@staff.ttu.ee

Abstract. In this paper we construct initial perfect quadratic forms over certain families of totally real number fields $\mathbb{K}$. We assume that the number field $\mathbb{K}$ is either the maximal totally real subfield of a cyclotomic field $\mathbb{Q}(\zeta_n)$, where $3 \nmid n$ is the product of distinct odd primes $p_1, \ldots, p_k$, or $\mathbb{K} = \mathbb{Q}(\sqrt{m_1}, \ldots, \sqrt{m_k})$, where $m_1, \ldots, m_k$ are pairwise relatively prime, square-free positive integers with all or all but one congruent to 1 modulo 4. These perfect forms can be used to find all perfect quadratic forms of given rank (up to equivalence and proportion) over the field $\mathbb{K}$ by applying the generalization of Voronoi's algorithm.

1 Introduction

Let $f(x_1, \ldots, x_m) = \sum_{i,j=1}^{m} a_{ij} x_i x_j$ $(a_{ij} = a_{ji})$ be a positive definite quadratic form with $a_{ij} \in \mathbb{R}$. The *minimum*[1] *of f*, denoted by $m(f)$, is defined to be

$$\min\{f(v_1, \ldots, v_m) | (v_1, \ldots, v_m)^T \in \mathbb{Z}^m \setminus \{0\}\}.$$

Write $M(f) = \{(v_1, \ldots, v_m)^T \in \mathbb{Z}^m | f(v_1, \ldots, v_m) = m(f)\}$. This set is called *the set of minimal vectors of f*.

Definition 1. *A positive definite quadratic form* $f(x_1, \ldots, x_m) = \sum_{i,j=1}^{m} a_{ij} x_i x_j$ $(a_{ij} = a_{ji})$ *is called perfect, if the system of linear equations*

$$\sum_{i=1}^{m} a_{ii} v_i^2 + 2 \sum_{i=1}^{m} \sum_{j>i}^{m} a_{ij} v_i v_j = m(f), \quad (v_1, \ldots, v_m)^T \in M(f) \qquad (1)$$

with indeterminates a_{ij} yields the unique solutions f.

Hence, if f is perfect, then $M(f)$ contains at least $\frac{m(m+1)}{2}$ pairs of minimal vectors v and $-v$. Let L be a lattice in $\mathbb{R}^m$ with a Gram matrix $A = (a_{ij})$ i.e. there is a basis $b_1, \ldots, b_m$ of L such that $b_i \cdot b_j = a_{ij}$, where $\cdot$ denotes the usual dot product in $\mathbb{R}^m$. With the lattice L we associate the quadratic form $f(x_1, \ldots, x_m) = (x_1, \ldots, x_m) A (x_1, \ldots, x_m)^T = (x_1 b_1 + \ldots + x_m b_m) \cdot (x_1 b_1 + \ldots + x_m b_m)$, where

---

[1] Sometimes this is called the *arithmetical minimum* of f.

A.J. van der Poorten and A. Stein (Eds.): ANTS-VIII 2008, LNCS 5011, pp. 240–252, 2008.

$x_1, \ldots, x_m \in \mathbb{Z}$. Therefore, the square of the length of $x_1 b_1 + \ldots + x_m b_m$ equals to $f(x_1, \ldots, x_m)$ and there is a one-to-one correspondence between $M(f)$ and the set[2] of vectors of shortest length in $L \setminus \{0\}$, denoted by $M(L)$. The number $\frac{1}{2}|M(L)|$ is called *the kissing number* of L (see [2] for details). We call a lattice *perfect* if the associated quadratic form is perfect. It follows from the work by Voronoï [11], that if L runs over all lattices in $\mathbb{R}^m$, then $\sup |M(L)|$ is attained at a perfect lattice. He proved also that if the lattice L gives the densest lattice packing of spheres in $\mathbb{R}^m$, then L is perfect ([8,11]). These few examples motivate the study of perfect quadratic forms and perfect lattices.

Voronoï presented an algorithm for finding all perfect quadratic forms (up to equivalence[3] and scaling) of m variables [11]. The main idea of the Voronoï's algorithm is to determine *perfect neighbours* of a given perfect form f. For the convenience of the reader, we recall here the method. *Perfect polyhedra*

$$\Pi_f = \left\{ \sum_{v \in M(f)} \rho_v v v^T | \rho_v \geq 0, v \in M(f) \right\}, \qquad f \text{ is perfect}$$

give a partition of the set $\mathrm{Sym}_{m, \geq}(\mathbb{R})$ of all real symmetric positive semi-definite $m \times m$ matrices. Let $\mathrm{Sym}_m(\mathbb{R})$ denote the linear space of all real symmetric $m \times m$ matrices equipped with the non-degenerate bilinear form $\langle A, B \rangle = \mathrm{TR}(AB)$, $A, B \in \mathrm{Sym}_m(\mathbb{R})$, where TR stands for the trace of matrix. As usual, we write $H^\perp$ for the orthogonal complement of $\emptyset \neq H \subseteq \mathrm{Sym}_m(\mathbb{R})$ with respect to the bilinear form $\langle \cdot, \cdot \rangle$. The perfect forms f and g with m variables are called *neighbouring forms* if

$$\dim_\mathbb{R} \mathrm{span}\{v v^T | v \in M(f) \cap M(g)\} = \dim_\mathbb{R} \mathrm{Sym}_m(\mathbb{R}) - 1.$$

Moreover, if f and g are neighbouring forms, then the perfect polyhedra Π_f and Π_g share a common face $\mathcal{F}$ (see [1,8,11]). Starting from a perfect quadratic form f, we determine all faces of Π_f. With each face $\mathcal{F}$ of Π_f we associate a *facet vector* ψ i.e. an element in $\mathcal{F}^\perp$ what is directed towards the interior of Π_f. If f and g are neighbouring forms along the face $\mathcal{F} \subset \Pi_f$ with the facet vector ψ and $m(f) = m(g)$, then there exists $\lambda > 0$ such that $g = f + \lambda \psi$ (see [1,8,11] for details). Starting from the *initial perfect form*

$$f_0(x_1, \ldots, x_m) = \sum_{i=1}^{m} x_i^2 + \sum_{i=1}^{m} \sum_{j>i}^{m} x_i x_j \tag{2}$$

we determine all perfect neighbours of f_0 with the same arithmetical minimum. Then we apply the Voronoï algorithm to the perfect neighbours of f_0 and so on. The number of perfect forms of the given rank m is finite [11, §7], therefore we obtain the complete list of perfect forms after applying the Voronoï algorithm to

---

[2] The set $M(L)$ is called also *the sphere of L*.
[3] The quadratic forms $f(x) = x^T A x$ and $g(x) = x^T B x$ are called *equivalent* if there exists $S \in \mathrm{GL}_m(\mathbb{Z})$, such that $B = S^T A S$.

the finite number of perfect forms only. One should note that both the number of facets of a perfect polyhedron and the number of inequivalent perfect forms grow very rapidly. For example, the perfect polyhedron of the quadratic form E_8 has 25075566937584 facets and there are 10916 perfect forms (up to equivalence and scaling) of 8 variables! (See Table 1 and Theorems 1.1-2 in [10].)

Voronoï theory can be generalized to number fields as well (see [3,6] for more details). In this paper we consider so called *additive generalization* only (see [6]).

Let $\mathbb{K}$ be a totally real number field of degree r and let $\sigma_1, \ldots, \sigma_r$ be its embeddings into $\mathbb{R}$. Write $\mathcal{O}_{\mathbb{K}}$ for the ring of algebraic integers in $\mathbb{K}$. By a Humbert tuple $(f_1, \ldots, f_r)$ of rank[4] m over $\mathbb{K}$ we mean the tuple of positive definite quadratic forms of rank m, that is, $f_1, \ldots, f_r$ are positive definite quadratic forms of m variables.

Definition 2. *The minimum $m(f)$ of a Humbert tuple $f = (f_1, \ldots, f_r)$ is*

$$m(f) = \min \left\{ \sum_{i=1}^{r} f_i(\sigma_i(v_1), \ldots, \sigma_i(v_m)) \, | (v_1, \ldots, v_m)^T \in \mathcal{O}_{\mathbb{K}}^m \setminus \{0\} \right\}.$$

The set of minimal vectors of f, denoted by $M(f)$, is defined to be the set of vectors $(v_1, \ldots, v_m)^T \in \mathcal{O}_{\mathbb{K}}^m$, such that $\sum_{i=1}^{r} f_i(\sigma_i(v_1), \ldots, \sigma_i(v_m)) = m(f)$. Hence, if f is given, then $m(f)$ and $M(f)$ are uniquely determined. Let us consider the inverse problem. Given $m(f)$ and $M(f)$ we have the following system of linear equations

$$\sum_{i=1}^{r} f_i(\sigma_i(v_1), \ldots, \sigma_i(v_m)) = m(f), \quad (v_1, \ldots, v_m)^T \in M(f), \qquad (3)$$

with indeterminates $f_{i,kl} \in \mathbb{R}$ ($1 \leq i \leq r$, $1 \leq k \leq l \leq m$). If the set $M(f)$ is large enough, then the system (3) has unique solution, i.e. there exists only one Humbert tuple with the given minimum $m(f)$ and the set of minimal vectors $M(f)$.

Definition 3. *A Humbert tuple $f = (f_1, \ldots, f_r)$ over $\mathbb{K}$ is called perfect if it is uniquely determined by $m(f)$ and $M(f)$, i.e. the system of linear equations (3) with indeterminates $f_{i,kl}$ yields the unique solution f.*

(There exist other definitions for perfection [3,8]).

If f is a perfect Humbert tuple of rank m over $\mathbb{K}$, then $M(f)$ contains at least $r\frac{m(m+1)}{2}$ pairs of the minimal vectors of the form $\pm v$, $v \in \mathcal{O}_{\mathbb{K}}^m$.

Let $g(x_1, \ldots, x_m) = \sum_{i,j=1}^{m} g_{ij} x_i x_j$ ($g_{ij} = g_{ji}$) be a quadratic form over $\mathbb{K}$, that is $g_{ij} \in \mathbb{K}$. By a $\sigma_k(g)$ we mean the quadratic form $\sum_{i,j=1}^{m} \sigma_k(g_{ij}) x_i x_j$.

Definition 4. *A quadratic form g over $\mathbb{K}$ is called positive definite, if $\sigma_k(g)$ is positive definite for each $1 \leq k \leq r$.*

---

[4] By the rank of a Humbert tuple $(f_1, \ldots, f_r)$ we mean the rank of the Gram matrix of a quadratic form f_i ($1 \leq i \leq r$). It is required that $f_1, \ldots, f_r$ have the same rank.

As $\mathbb{K}$ being totally real, it follows from Corollary 2 in [6] that if $(f_1,\ldots,f_r)$ is a perfect Humbert tuple of rank m over $\mathbb{K}$, then it is proportional to a tuple $(\sigma_1(g),\ldots,\sigma_r(g))$, where g is a positive definite quadratic form of rank m over $\mathbb{K}$. A positive definite quadratic form g over $\mathbb{K}$ is called *perfect*, if the Humbert tuple $(\sigma_1(g),\ldots,\sigma_r(g))$ is perfect. Clearly, if f is a perfect Humber tuple and $\lambda > 0$, then λf is perfect as well. Therefore, we have one-to-one correspondence between positive definite perfect quadratic forms of rank m over $\mathbb{K}$ and perfect Humbert tuples of rank m over $\mathbb{K}$ modulo positive scalars.

Let $g(x_1,\ldots,x_k)$ be a positive definite quadratic form over $\mathbb{K}$. If $x_k \in \mathcal{O}_{\mathbb{K}}$, then we set $x_k = x_{k,1}\omega_1 + \ldots + x_{k,r}\omega_r$, where $x_{k,1},\ldots,x_{k,r} \in \mathbb{Z}$ and $\omega_1,\ldots,\omega_r$ is a $\mathbb{Z}$-basis of $\mathcal{O}_{\mathbb{K}}$. With this notation,

$$F(x_{1,1},\ldots,x_{m,r}) = \sum_{i=1}^{r} \sigma_i(g)(\sigma_i(x_1),\ldots,\sigma_i(x_m))$$

is a positive definite rational quadratic form of mr variables. If g is perfect (over $\mathbb{K}$), then $|M(F)| = |M(g)| \geq r\frac{m(m+1)}{2}$. There are examples of algebraic quadratic forms, such that the corresponding rational quadratic forms are critical i.e. the corresponding lattice packings of spheres are the densest ones (see [6, Examples 2-3]). Further, if the rational quadratic form F is a critical quadratic form, then the corresponding quadratic form g with coefficients in $\mathbb{K}$ is perfect (see [6] for more details). This motivates the study of perfect forms over number fields.

From now on, we restrict ourselves to positive definite quadratic forms over $\mathbb{K}$. With each positive definite quadratic form g of rank m over $\mathbb{K}$ we associate a symmetric $m \times m$ matrix $A = (a_{ij})$ with $a_{ij} \in \mathbb{K}$ for all $1 \leq i,j \leq r$, such that $g(x) = x^T A x$. For simplicity of notations, we continue to write $m(g)$ and $M(g)$ for the minimum of g and the set of minimal vectors of g respectively. The positive definite quadratic forms $f(x) = x^T B x$ and $g(x) = x^T A x$ of rank m are called *equivalent*, if there exists $M \in \mathrm{GL}_m(\mathcal{O}_{\mathbb{K}})$, such that $B = M^T A M$. Let $\mathrm{Sym}_m(\mathbb{K})$ denote the set of all symmetric $m \times m$ matrices with entries in $\mathbb{K}$. The linear space $\mathrm{Sym}_m(\mathbb{K})$ is equipped with the non-degenerate bilinear form

$$\langle A, B \rangle = \mathrm{Tr}_{\mathbb{K}/\mathbb{Q}}\left(\mathrm{TR}(AB)\right), \qquad A, B \in \mathrm{Sym}_m(\mathbb{K}) \tag{4}$$

where TR denotes the trace of a matrix. This definition agrees with the *classical* one i.e. if $A, B \in \mathrm{Sym}_m(\mathbb{R})$. By abuse of notations, we continue to write $H^{\perp}$ for the orthogonal complement of $\emptyset \neq H \subseteq \mathrm{Sym}_m(\mathbb{K})$ with respect to the bilinear form $\langle \cdot, \cdot \rangle$.

Let g be a positive definite quadratic form of rank m over $\mathbb{K}$. The dimension of $\mathbb{Q}$-linear space $\mathrm{span}\{vv^T | v \in M(g)\}$ is called the *perfection rank* of g. An equivalent definition for perfection says that g is perfect, if

$$\dim_{\mathbb{Q}} \mathrm{span}\{vv^T | v \in M(g)\} = \dim_{\mathbb{Q}} \mathrm{Sym}_m(\mathbb{K}).$$

One main problem in the theory of perfect forms is enumerating all perfect forms of the given rank m (up to equivalence and scaling by positive scalar). In the case of positive definite quadratic forms over the reals the enumeration can be done by applying Voronoi's algorithm. Ong generalized the *classical* Voronoi's algorithm (i.e. the Voronoi's algorithm for perfect forms over $\mathbb{R}$) for quadratic forms over real quadratic fields [9]. Her generalization holds for totally real number fields as well and it is almost the same as the classical Voronoï algorithm (the perfect polyhedra are contained in $\mathrm{Sym}_m(\mathbb{K})$ and the bilinear form is defined by (4)). As in the classical case, one needs an initial perfect form of rank m to apply the generalization of Voronoi's algorithm. This leads to the following problem.

Problem 1. How to find an initial perfect form of rank m over $\mathbb{K}$ ($m \geq 1$)?

A general (*and robust*) solution is as follows. Take any positive definite quadratic form φ_0 of rank m over $\mathbb{K}$ and increase its perfection rank by applying the Voronoi's method, that is, if $0 \neq g_0 \in \mathrm{span}\{vv^T | v \in M(\varphi_0)\}^\perp$, then there exists a rational number $\delta_0 > 0$ such that

$$\dim_{\mathbb{Q}} \mathrm{span}\{vv^T | v \in M(\varphi_0 + \delta_0 g_0)\} > \dim_{\mathbb{Q}} \mathrm{span}\{vv^T | v \in M(\varphi_0)\}$$

and $M(\varphi_0) \subset M(\varphi_0 + \delta_0 g_0)$. Let $\varphi_1 = \varphi_0 + \delta g_0$. If φ_1 is not perfect, then we can find $\varphi_2 = \varphi_1 + \delta_1 g_1$, $0 \neq g_1 \in \mathrm{span}\{vv^T | v \in M(\varphi_0)\}^\perp$, such that

$$\dim_{\mathbb{Q}} \mathrm{span}\{vv^T | v \in M(\varphi_2)\} > \dim_{\mathbb{Q}} \mathrm{span}\{vv^T | v \in M(\varphi_1)\}$$

and $M(\varphi_1) \subset M(\varphi_2)$. We continue in this fashion to obtain the sequence of quadratic forms of rank m, say $\varphi_0, \varphi_1, \ldots, \varphi_\ell$, which stops at the perfect form φ_ℓ (see also [8, Theorem 9.1.9]). Recall that the explicit formula for δ_k is

$$\delta_k = \inf \left\{ \frac{\mathrm{Tr}_{\mathbb{K}/\mathbb{Q}} \varphi_k(v) - m(\varphi_k)}{-\mathrm{Tr}_{\mathbb{K}/\mathbb{Q}} g_k(v)} \; | v \in \mathcal{O}_{\mathbb{K}}^m \wedge \mathrm{Tr}_{\mathbb{K}/\mathbb{Q}} g_k(v) < 0 \right\}.$$

In practice, there are more efficient ways for computing δ_k (see [8, Section 7.8]). The main computational disadvantage of this process is that at each step the computation of short vectors of a quadratic form of rank m is required in order to determine the rational number δ_k (see [8, Section 7.8] for more details).

Definition 5 ([5, §7.1]). *A lattice L is of E-type if for any lattice L' we have*

$$M(L \otimes L') \subseteq \{x \otimes y | x \in M(L), y \in M(L')\}.$$

For deeper discussion of lattices of E-type we refer reader to [5].

The Problem 1 can be reduced to the problem of finding a unary perfect form due to the following theorem.

Theorem 1 ([6, Theorem 1]). *Let $\mathbb{K}$ be a totally real algebraic number field and let $\mathcal{O}_{\mathbb{K}}$ denote its ring of integers. Let ax^2 be a perfect unary quadratic form over $\mathcal{O}_{\mathbb{K}}$ with lattice L_a over $\mathbb{Z}$ and let g be a perfect quadratic form over $\mathbb{Z}$ with lattice L. If L_a or L is of then the quadratic form ag is perfect over $\mathbb{K}$.*

As the classical initial perfect form (2) is of E-type (see [5, Theorem 7.1.2]) we have an explicit construction for the initial perfect form over $\mathbb{K}$ of rank m with $m > 1$.

Remark 1. The quadratic forms

$$f_D(x_1,\dots,x_m) = \sum_{i=1}^{m} x_i^2 - \sum_{i=1}^{m-2} x_i x_{i+1} - x_{m-2}x_m \quad (m \geq 4)$$

$$f_E(x_1,\dots,x_m) = \sum_{i=1}^{m} x_i^2 - \sum_{i=1}^{m-2} x_i x_{i+1} - x_{m-3}x_m \quad (m \in \{6,7,8\})$$

being both perfect (see [4, Theorem 5 at p. 404]) and of E-type. If ax^2 is a perfect unary form over $\mathbb{K}$, then the quadratic forms af_D and af_E are perfect over $\mathbb{K}$. Therefore, they can be used as the initial perfect forms for the generalized Voronoi's algorithm as well.

Therefore we have the following problem.

Problem 2. How to find a unary perfect quadratic form over $\mathbb{K}$?

We can solve this as we explained in the *solution* for Problem 1. In this case we work with unary forms only, but we would not avoid the computation of short vectors.

The purpose of this paper is to present a partial solution to this problem that includes neither the Voronoi's method for increasing the perfection rank nor computation of short vectors. Assuming that $\mathbb{K}$ is either $\mathbb{Q}(\sqrt{m_1},\dots,\sqrt{m_k})$, where $m_1,\dots,m_k$ are pairwise relatively prime, square-free positive integers with all or all but one congruent to 1 modulo 4, or the maximal totally real subfield of a cyclotomic field $\mathbb{Q}(\zeta_n)$, where n is the product of distinct odd primes which are at least 5, we present the explicit construction of a unary perfect form over $\mathbb{K}$.

In the case $\mathbb{K} = \mathbb{Q}(\sqrt{m})$ with square-free $m > 1$ the problem of finding a unary perfect form is solved already in [7]. For the convenience of the reader, we recall here the result.

Theorem 2 (Theorem 1 in [7]). *Let $D > 1$ be a square-free integer.*

1. *Suppose that $|k^2 - D|$ attains minimum at integer $k > 0$. If $D \equiv 2 \,(\mathrm{mod}\,4)$ or $D \equiv 3 \,(\mathrm{mod}\,4)$, then the unary form $ax^2 = (a_1 + a_2\sqrt{D})x^2$, with*

$$a_1 = 2kD, \qquad a_2 = k^2 + D - 1,$$

is perfect and $\{1, k - \sqrt{D}\} \subseteq M(ax^2)$.

2. *Let $k > 0$ be the smallest integer, such that $|(2k - 1)^2 - D|$ is minimal. If $D \equiv 1 \,(\mathrm{mod}\,4)$, then the unary form $ax^2 = (a_1 + a_2\frac{1+\sqrt{D}}{2})x^2$, with*

$$a_1 = 1 - k^2 + (1 + D)k - \frac{1 + 3D}{4}, \qquad a_2 = 2k^2 - 2k + \frac{1 + D}{2} - 2$$

is perfect and $\{1, -k + \frac{1+\sqrt{D}}{2}\} \subseteq M(ax^2)$.

We can now formulate our main results.

Theorem 3. *Let* $m_1, \ldots, m_k$ *be pairwise relatively prime, square-free positive integers with all or all but one congruent to 1 modulo 4. The unary quadratic form*

$$(a_1 \cdots a_k) x^2,$$

where $a_i x^2$ *is a unary perfect quadratic form over* $\mathbb{Q}(\sqrt{m_i})$ *for all* $1 \leq i \leq k$, *is perfect over* $\mathbb{Q}(\sqrt{m_1}, \ldots, \sqrt{m_k})$.

Theorem 4. *Let* $n > 1$ *be a square-free odd integer* $n = p_1 \cdots p_k$ *and* $3 \nmid n$. *The unary quadratic form*

$$\left(\prod_{i=1}^{k} (2 - \zeta_{p_i} - \zeta_{p_i}^{-1}) \right) x^2$$

is perfect over $\mathbb{Q}(\zeta_n + \zeta_n^{-1})$, *where* ζ_{p_i} *is a primitive* p_i-*th root of unity and* ζ_n *is a primitive* n-*th root of unity.*[5]

It is worth pointing out that combining Theorem 3 with Theorem 2 (Theorem 4 with Theorem 6) we obtain the explicit construction of the minimal vectors of the initial perfect form given in Theorem 3 (Theorem 4 respectively).

2   Definitions and Notation

Let $f(x_1, \ldots, x_m)$ be a positive definite quadratic form with rational coefficients. It will cause no confusion if we denote the minimum of f by $m(f)$, where

$$m(f) = \min\{f(X) | X \in \mathbb{Z}^m \setminus \{0\}\}.$$

Let L be a lattice in $\mathbb{Q}^m$ with a positive definite Gram matrix A. For the simplicity of notation, we let $M(L)$ stand for the set of minimal vectors of the lattice L, that is the set of minimal vectors of the quadratic form $f(X) = X^T A X$. Write B_L for the bilinear form $L \times L \to \mathbb{Q}$ of the lattice L. If B_L is not given, then we assume that B_L is the usual dot product in $\mathbb{R}^m$.

Example 1. Let $\mathbb{K}$ be a totally real number field with degree r. The ring of algebraic integers $\mathcal{O}_{\mathbb{K}}$ can be considered as the lattice $\mathbb{Z}^r$ equipped with the bilinear form

$$B((x_1, \ldots, x_r), (y_1, \ldots, y_r)) = \operatorname{Tr}_{\mathbb{K}/\mathbb{Q}}(a(x_1\omega_1 + \cdots + x_r\omega_r)(y_1\omega_1 + \cdots + y_r\omega_r)),$$

where $a \in \mathbb{K}$ and $\omega_1, \ldots, \omega_r$ is a $\mathbb{Z}$-basis of $\mathcal{O}_{\mathbb{K}}$. The Gram matrix of this lattice is $A = (a_{ij}) = (\operatorname{Tr}_{\mathbb{K}/\mathbb{Q}}(a\omega_i\omega_j))$ $(1 \leq i, j \leq r)$. If a is totally positive, then both the Gram matrix A and the quadratic form $f(x_1, \ldots, x_r) = B((x_1, \ldots, x_r), (x_1, \ldots, x_r))$ are positive definite.

By the tensor product of the lattices $L \subset \mathbb{Q}^m$ and $L' \subset \mathbb{Q}^n$ we mean the lattice $L \otimes L'$ in $\mathbb{Q}L \otimes \mathbb{Q}L'$ equipped with the bilinear form

$$B_{L \otimes L'}(l_1 \otimes l_1', l_2 \otimes l_2') = B_L(l_1, l_2) \cdot B_{L'}(l_1', l_2') \quad l_1, l_2 \in L, l_1', l_2' \in L'.$$

---

[5] This theorem is a corrected version of Theorem 4 in [6].

Let $f(l) = B_L(l, l)$, where $l \in L$, and let $f'(l') = B_{L'}(l', l')$, where $l' \in L'$. The tensor product of lattices L and L' gives us the tensor product of the quadratic forms f and f'

$$(f \otimes f')(l \otimes l') = B_{L \otimes L'}(l \otimes l', l \otimes l') = B_L(l, l) \cdot B_{L'}(l', l') = f(l) \cdot f'(l').$$

Definition 6. *A positive definite quadratic form* $f(X) = X^T A X$ *is of E-type if* A *is the Gram matrix of a lattice of E-type (see Definition 5).*

3   Theorems

In order to obtain the main results we need the following theorem.

Theorem 5. *Let* $\mathbb{K}_1$ *and* $\mathbb{K}_2$ *be totally real number fields with degree* r_1 *and* r_2 *respectively. Let* $a_i x^2$ *be a perfect quadratic forms over* $\mathbb{K}_i$ *respectively* $(i = 1, 2)$. *If*

1. $\mathbb{K}_1$ *and* $\mathbb{K}_2$ *are linearly disjoint (i.e. if* $\alpha_1, \ldots, \alpha_{r_1}$ *is a basis of* $\mathbb{K}_1$ *over* $\mathbb{Q}$ *and* $\beta_1, \ldots, \beta_{r_2}$ *is a basis of* $\mathbb{K}_2$ *over* $\mathbb{Q}$, *then* $\{\alpha_i \beta_j\}$ *is a basis of* $\mathbb{K}_1 \mathbb{K}_2$ *over* $\mathbb{Q}$);
2. $\gcd(\mathrm{disc}(\mathbb{K}_1), \mathrm{disc}(\mathbb{K}_2)) = 1$;
3. $\{v \cdot w \mid v \in M(a_1 x^2), w \in M(a_2 x^2)\} \subseteq M((a_1 a_2) x^2)$,

then $(a_1 a_2) x^2$ *is a perfect quadratic form over* $\mathbb{K}_1 \mathbb{K}_2$.

Proof. Let $\sigma_1, \ldots, \sigma_{r_1}$ be the embeddings of $\mathbb{K}_1$ into $\mathbb{R}$ and let $\tau_1, \ldots, \tau_{r_2}$ be the embeddings of $\mathbb{K}_2$ into $\mathbb{R}$. Hence, any embedding of $\mathbb{K}_1 \mathbb{K}_2$ into $\mathbb{R}$ can be uniquely written as $\sigma_i \tau_j$. Seeking a contradiction, suppose $(a_1 a_2) x^2$ is not perfect over $\mathbb{K}_1 \mathbb{K}_2$. Write $\mu = m((a_1 a_2) x^2)$. Therefore the system

$$\begin{cases} \sigma_1(a_1)\tau_1(a_2)\sigma_1(v_1^2)\tau_1(w_1^2) + \cdots + \sigma_{r_1}(a_1)\tau_{r_2}(a_2)\sigma_{r_1}(v_1^2)\tau_{r_2}(w_1^2) = \mu, \\ \qquad\qquad\qquad\qquad\qquad\qquad\qquad\qquad\qquad\qquad \vdots \\ \sigma_1(a_1)\tau_1(a_2)\sigma_1(v_{r_1}^2)\tau_1(w_{r_2}^2) + \cdots + \sigma_{r_1}(a_1)\tau_{r_2}(a_2)\sigma_{r_1}(v_{r_1}^2)\tau_{r_2}(w_{r_2}^2) = \mu \end{cases}$$

with $v_1, \ldots, v_{r_1} \in M(a_1 x^2)$ and $w_1, \ldots, w_{r_2} \in M(a_2 x^2)$ does not yield to the unique solution. Hence $\det(\sigma_i(v_k^2)\tau_j(w_l^2)) = 0$. From

$$\det(\sigma_i(v_k^2)\tau_j(w_l^2)) = \det(\sigma_i(v_k^2)) \det(\tau_j(w_l^2)),$$

it follows that $\det(\sigma_i(v_k^2)) \det(\tau_j(w_l^2)) = 0$, which is impossible, since both $a_1 x^2$ and $a_2 x^2$ are perfect. Hence $(a_1 a_2) x^2$ is perfect. $\qquad\square$

3.1   The Case $\mathbb{K} = \mathbb{Q}(\sqrt{m_1}, \ldots, \sqrt{m_k})$

Proof (of Theorem 3). The proof is by induction on k. If $k = 1$, then Theorem 3 coincides with Theorem 2.

Assume that Theorem 3 is true for $\mathbb{Q}(\sqrt{m_1}, \ldots, \sqrt{m_{k-1}})$. Let $a_i x^2$ be a perfect unary form over the quadratic field $\mathbb{Q}(\sqrt{m_i})$. The unary form $(a_1 \cdots a_{k-1}) x^2$ over

$\mathbb{Q}(\sqrt{m_1}, \ldots, \sqrt{m_{k-1}})$ is perfect by the hypothesis. Clearly, $\mathbb{K}_1 = \mathbb{Q}(\sqrt{m_1}, \ldots, \sqrt{m_{k-1}})$ and $\mathbb{K}_2 = \mathbb{Q}(\sqrt{m_k})$ are linearly disjoint and they have mutually prime discriminants. Hence $\mathcal{O}_{\mathbb{K}_1 \mathbb{K}_2} = \mathcal{O}_{\mathbb{K}_1} \mathcal{O}_{\mathbb{K}_2} = \mathcal{O}_{\mathbb{K}_1} \otimes \mathcal{O}_{\mathbb{K}_2}$ and $(a_1 \cdots a_{k-1})x^2 \otimes a_k y^2 = (a_1 \cdots a_k)z^2$. Write $\ell = 2^{k-1}$. Let $\{1, \omega_2, \ldots, \omega_\ell\}$ be a $\mathbb{Z}$-basis of $\mathcal{O}_{\mathbb{K}_1}$ and let $\{1, \omega_k\}$ be a $\mathbb{Z}$-basis of $\mathcal{O}_{\mathbb{K}_2}$. Consider the rational quadratic forms

$$f_1(x_1, \ldots, x_\ell) = \mathrm{Tr}_{\mathbb{K}_1/\mathbb{Q}}((a_1 \cdots a_{k-1})(x_1 + x_2\omega_2 + \cdots + x_\ell\omega_\ell)^2)$$

and $f_2(x_1', x_2') = \mathrm{Tr}_{\mathbb{K}_2/\mathbb{Q}}(a_k(x_1' + x_2'\omega_k)^2)$ with $x_1, \ldots, x_\ell, x_1', x_2' \in \mathbb{Z}$. Write $\mathbb{K} = \mathbb{Q}(\sqrt{m_1}, \ldots, \sqrt{m_k})$. An easy computation shows that

$$\mathrm{Tr}_{\mathbb{K}/\mathbb{Q}}((a_1 \cdots a_k)(x_1 + \cdots + x_\ell\omega_\ell)^2(x_1' + x_2'\omega_k)^2) =$$
$$= (f_1 \otimes f_2)((x_1, \ldots, x_\ell) \otimes (x_1', x_2')) = f_1(x_1, \ldots, x_\ell) \cdot f_2(x_1', x_2')$$

for all $x_1, \ldots, x_\ell, x_1', x_2' \in \mathbb{Z}$. From this we obtain $m((a_1 \cdots a_k)x^2) = m(f_1 \otimes f_2)$. We have $m(f_1 \otimes f_2) = m(f_1) \cdot m(f_2)$, because f_2 is of E-type by [5, Theorem 7.1.1]. Since $m(f_1) = m((a_1 \cdots a_{k-1})x^2)$ and $m(f_2) = m(a_k x^2)$, we conclude that

$$\{v \cdot w | v \in M((a_1 \cdots a_{k-1})x^2), w \in M(a_k x^2)\} \subseteq M((a_1 \cdots a_k)x^2)$$

as required. Theorem 5 now shows that $(a_1 \cdots a_k)x^2$ is perfect over $\mathbb{K}$, which proves the theorem. $\qquad\square$

3.2   The Case $\mathbb{K} = \mathbb{Q}(\zeta_n + \zeta_n^{-1})$

Theorem 6 ([6, Theorem 4]). *Let ζ_p be a primitive p-th root of unity, where $p > 3$ is a prime. The unary quadratic form $(2 - \zeta_p - \zeta_p^{-1})x^2$ is a perfect quadratic form over $\mathbb{Q}(\zeta_p + \zeta_p^{-1})$. Moreover, $\varepsilon \in \mathbb{Z}[\zeta_p + \zeta_p^{-1}]^*$ is a minimal vector of $(2 - \zeta_p - \zeta_p^{-1})x^2$ iff $\sigma(2 - \zeta_p - \zeta_p^{-1}) = (2 - \zeta_p - \zeta_p^{-1})\varepsilon^2$ holds for some $\sigma \in \mathrm{Gal}(\mathbb{Q}(\zeta_p + \zeta_p^{-1})/\mathbb{Q})$.*

Proof. Let ζ be a primitive p-th root of unity ($p > 3$). Write $\mathbb{K} = \mathbb{Q}(\zeta + \zeta^{-1})$. The proof will be divided into three steps.

Step 1. We show there exist $r = \frac{p-1}{2}$ units $\varepsilon_1, \ldots, \varepsilon_r$ in $\mathbb{Z}[\zeta + \zeta^{-1}]$, such that

$$\mathrm{Tr}((2 - \zeta - \zeta^{-1})\varepsilon_i^2) = \mathrm{Tr}((2 - \zeta - \zeta^{-1})\varepsilon_j^2), \quad \text{for each } 1 \leq i, j \leq r .$$

We start with the observation that

$$2 - (\zeta + \zeta^{-1}) = (1 - \zeta) \cdot (1 - \zeta^{-1}) = (1 - \zeta) \cdot \overline{(1 - \zeta)} = \mathrm{Nm}_{\mathbb{Q}(\zeta)/\mathbb{K}}(1 - \zeta).$$

Suppose $\sigma \in \mathrm{Gal}(\mathbb{K}/\mathbb{Q})$. Let us consider the fraction $\frac{\sigma(2 - (\zeta + \zeta^{-1}))}{2 - (\zeta + \zeta^{-1})}$. Since $\sigma, \bar{\ } \in \mathrm{Gal}(\mathbb{Q}(\zeta)/\mathbb{Q})$ and $\mathrm{Gal}(\mathbb{Q}(\zeta)/\mathbb{Q})$ is an Abelian group, we have

$$\frac{\sigma(2 - (\zeta + \zeta^{-1}))}{2 - (\zeta + \zeta^{-1})} = \frac{\sigma(1 - \zeta)}{1 - \zeta} \cdot \frac{\sigma\overline{(1 - \zeta)}}{\overline{1 - \zeta}} = \frac{\sigma(1 - \zeta)}{1 - \zeta} \cdot \overline{\frac{\sigma(1 - \zeta)}{1 - \zeta}} = \frac{1 - \zeta^k}{1 - \zeta} \cdot \overline{\frac{1 - \zeta^k}{1 - \zeta}},$$

where $1 \le r \le k$. But $\dfrac{1 - \zeta^k}{1 - \zeta} \in \mathbb{Z}[\zeta]^*$ since

$$\frac{1 - \zeta^k}{1 - \zeta} = 1 + \zeta + \ldots + \zeta^{k-1} \in \mathbb{Z}[\zeta] \quad \text{and} \quad \mathrm{Nm}_{\mathbb{Q}(\zeta)/\mathbb{Q}}\left(\frac{1 - \zeta^k}{1 - \zeta}\right) = 1.$$

Therefore

$$\frac{\sigma(1 - \zeta)}{1 - \zeta} \cdot \frac{\sigma(\overline{1 - \zeta})}{\overline{1 - \zeta}} = \zeta^b \varepsilon \cdot \overline{\zeta^b \varepsilon} = \zeta^b \overline{\zeta^b} \varepsilon^2 = \varepsilon^2, \quad \varepsilon \in \mathbb{Z}[\zeta + \zeta^{-1}]^*$$

by [12, Proposition 1.5]. Hence

$$\sigma(2 - (\zeta + \zeta^{-1})) = (2 - (\zeta + \zeta^{-1}))\varepsilon^2, \quad \varepsilon \in \mathbb{Z}[\zeta + \zeta^{-1}]^*.$$

Since $|\mathrm{Gal}(\mathbb{K}/\mathbb{Q})| = r$, there exist r units $\varepsilon_1, \ldots, \varepsilon_r$ as required. Moreover we may take $\varepsilon_1 = 1$.

Step 2. We prove that $\mathrm{Tr}_{\mathbb{K}/\mathbb{Q}}\left((2 - \zeta - \zeta^{-1})\beta^2\right) \ge p$ for any $0 \ne \beta \in \mathbb{Z}[\zeta + \zeta^{-1}]$.
Write $P^r = p\mathbb{Z}[\zeta + \zeta^{-1}]$. Clearly $2 - \zeta - \zeta^{-1} \in P$. Since p is the only prime that ramifies in $\mathbb{Z}[\zeta + \zeta^{-1}]$, it follows that

$$\sigma(2 - \zeta - \zeta^{-1}) \in P \quad \text{for each} \quad \sigma \in \mathrm{Gal}(\mathbb{K}/\mathbb{Q}).$$

Therefore $\mathrm{Tr}_{\mathbb{K}/\mathbb{Q}}\left((2 - \zeta - \zeta^{-1})\beta^2\right) \in P \cap \mathbb{Z} = p\mathbb{Z}$ i.e. $p | \mathrm{Tr}_{\mathbb{K}/\mathbb{Q}}\left((2 - \zeta - \zeta^{-1})\beta^2\right)$ as claimed.

Step 3. If we prove that the vectors

$$(1, \ldots, 1)^t, (\sigma_1(\varepsilon_2^2), \ldots, \sigma_r(\varepsilon_2^2))^t, \ldots, (\sigma_1(\varepsilon_r^2), \ldots, \sigma_r(\varepsilon_r^2))^t$$

are linearly independent over $\mathbb{R}$, then the theorem follows. Let $1, \omega_2, \ldots, \omega_r$ be the $\mathbb{Z}$-basis of $\mathbb{Z}[\zeta + \zeta^{-1}]$. As

$$\begin{pmatrix} 1 & 1 & \ldots & 1 \\ \sigma_1(\omega_2) & \sigma_2(\omega_2) & \ldots & \sigma_r(\omega_2) \\ \vdots & \vdots & \vdots & \vdots \\ \sigma_1(\omega_r) & \sigma_2(\omega_r) & \ldots & \sigma_r(\omega_r) \end{pmatrix} \begin{pmatrix} 1 & \sigma_1(\varepsilon_2^2) & \ldots & \sigma_1(\varepsilon_r^2) \\ 1 & \sigma_2(\varepsilon_2^2) & \ldots & \sigma_2(\varepsilon_r^2) \\ \vdots & \vdots & \vdots & \vdots \\ 1 & \sigma_r(\varepsilon_2^2) & \ldots & \sigma_r(\varepsilon_r^2) \end{pmatrix} =$$

$$= \begin{pmatrix} \mathrm{Tr}_{\mathbb{K}/\mathbb{Q}} 1 & \mathrm{Tr}_{\mathbb{K}/\mathbb{Q}} \varepsilon_2^2 & \ldots & \mathrm{Tr}_{\mathbb{K}/\mathbb{Q}} \varepsilon_r^2 \\ \mathrm{Tr}_{\mathbb{K}/\mathbb{Q}} \omega_2 & \mathrm{Tr}_{\mathbb{K}/\mathbb{Q}} \omega_2 \varepsilon_2^2 & \ldots & \mathrm{Tr}_{\mathbb{K}/\mathbb{Q}} \omega_2 \varepsilon_r^2 \\ \vdots & \vdots & \vdots & \vdots \\ \mathrm{Tr}_{\mathbb{K}/\mathbb{Q}} \omega_r & \mathrm{Tr}_{\mathbb{K}/\mathbb{Q}} \omega_r \varepsilon_2^2 & \ldots & \mathrm{Tr}_{\mathbb{K}/\mathbb{Q}} \omega_r \varepsilon_r^2 \end{pmatrix} \in \mathrm{Mat}_{r \times r}(\mathbb{Z})$$

we have that the columns of the matrix

$$\begin{pmatrix} 1 & \sigma_1(\varepsilon_2^2) & \ldots & \sigma_1(\varepsilon_r^2) \\ 1 & \sigma_2(\varepsilon_2^2) & \ldots & \sigma_2(\varepsilon_r^2) \\ \vdots & \vdots & \vdots & \vdots \\ 1 & \sigma_r(\varepsilon_2^2) & \ldots & \sigma_r(\varepsilon_r^2) \end{pmatrix}$$

should be linearly independent. Seeking a contradiction, assume there exists a tuple $(\beta_1, \ldots, \beta_r)^t \in \mathbb{Q}^r$ such that

$$\begin{pmatrix} 1 & \sigma_1(\varepsilon_2^2) & \cdots & \sigma_1(\varepsilon_r^2) \\ 1 & \sigma_2(\varepsilon_2^2) & \cdots & \sigma_2(\varepsilon_r^2) \\ \vdots & \vdots & \vdots & \vdots \\ 1 & \sigma_r(\varepsilon_2^2) & \cdots & \sigma_r(\varepsilon_r^2) \end{pmatrix} \begin{pmatrix} \beta_1 \\ \beta_2 \\ \vdots \\ \beta_r \end{pmatrix} = \begin{pmatrix} 0 \\ 0 \\ \vdots \\ 0 \end{pmatrix}.$$

Therefore we have also

$$\begin{pmatrix} \mathrm{Tr}_{\mathbb{K}/\mathbb{Q}}\, 1 & \mathrm{Tr}_{\mathbb{K}/\mathbb{Q}}\, \varepsilon_2^2 & \cdots & \mathrm{Tr}_{\mathbb{K}/\mathbb{Q}}\, \varepsilon_r^2 \\ \mathrm{Tr}_{\mathbb{K}/\mathbb{Q}}\, w_2 & \mathrm{Tr}_{\mathbb{K}/\mathbb{Q}}\, w_2\varepsilon_2^2 & \cdots & \mathrm{Tr}_{\mathbb{K}/\mathbb{Q}}\, w_2\varepsilon_r^2 \\ \vdots & \vdots & \vdots & \vdots \\ \mathrm{Tr}_{\mathbb{K}/\mathbb{Q}}\, w_r & \mathrm{Tr}_{\mathbb{K}/\mathbb{Q}}\, w_r\varepsilon_2^2 & \cdots & \mathrm{Tr}_{\mathbb{K}/\mathbb{Q}}\, w_r\varepsilon_r^2 \end{pmatrix} \begin{pmatrix} \beta_1 \\ \beta_2 \\ \vdots \\ \beta_r \end{pmatrix} = \begin{pmatrix} 0 \\ 0 \\ \vdots \\ 0 \end{pmatrix}.$$

Write $\varepsilon_1 = 1$ and $a = 2 - \zeta - \zeta^{-1}$. We have

$$0 = \sum_{i=1}^r \beta_i \varepsilon_i^2 = \left(\sum_{i=1}^r \beta_i \varepsilon_i^2 \right) a = \sum_{i=1}^r \beta_i \varepsilon_i^2 a = \sum_{i=1}^r \beta_i \sigma_i(a).$$

After taking the trace, we obtain

$$0 = \sum_{i=1}^r \beta_i \mathrm{Tr}_{\mathbb{K}/\mathbb{Q}}(a) \implies 0 = \sum_{i=1}^r \beta_i.$$

Thus

$$0 = \sum_{i=1}^r \beta_i \sigma_i(a) = \sum_{i=1}^r \beta_i \sigma_i(2 - \zeta - \zeta^{-1}) = -\sum_{i=1}^{p-1} \beta_i \zeta^i, \quad \beta_i = \beta_{p-i}.$$

Since $\{1, \zeta, \ldots, \zeta^{p-1}\}$ is a basis of $\mathbb{Q}(\zeta)$ we conclude that $\beta_1 = \beta_2 = \ldots = \beta_r = 0$, as required. $\square$

Theorem 7. *Let ζ_p denote a primitive p-th root of unity, where $p \geq 5$ is a prime. Write $\mathbb{K} = \mathbb{Q}(\zeta_p + \zeta_p^{-1})$. The rational quadratic form*

$$f(x_1, \ldots, x_r) =$$
$$= \mathrm{Tr}_{\mathbb{K}/\mathbb{Q}}\left((2 - \zeta_p - \zeta_p^{-1})(x_1 + x_2(\zeta_p + \zeta_p^{-1}) + \cdots + x_r(\zeta_p + \zeta_p^{-1})^{r-1})^2 \right) (5)$$

where $r = \frac{1}{2}(p-1)$, is of E-type.

Proof. After expanding (5) we obtain

$$f(x_1, \ldots, x_r) = \sum_{i=1}^r \mathrm{Tr}_{\mathbb{K}/\mathbb{Q}}\left((2 - \zeta_p - \zeta_p^{-1})(\zeta_p + \zeta_p^{-1})^{2i-2} \right) x_i^2$$

$$+ 2\sum_{i=1}^r \sum_{j>i}^r \mathrm{Tr}_{\mathbb{K}/\mathbb{Q}}\left((2 - \zeta_p - \zeta_p^{-1})(\zeta_p + \zeta_p^{-1})^{i+j-2} \right) x_i x_j.$$

Let P denote the ideal in $\mathbb{Z}[\zeta_p + \zeta_p^{-1}]$ such that $p\mathbb{Z}[\zeta_p + \zeta_p^{-1}] = P^r$. Since $2 - \zeta_p - \zeta_p^{-1} \in P$, we see that $(2 - \zeta_p - \zeta_p^{-1})(\zeta_p + \zeta_p^{-1})^{i+j-2} \in P$ for all $1 \le i, j \le r$. This gives $p | \mathrm{Tr}_{\mathbb{K}/\mathbb{Q}} \left((2 - \zeta_p - \zeta_p^{-1})(\zeta_p + \zeta_p^{-1})^{i+j-2} \right)$ for all $1 \le i, j \le r$. We have

$$f(x_1, \ldots, x_r) = p \left(x_1^2 + \sum_{i=2}^{r} g_{ii} x_i^2 + 2 \sum_{i=1}^{r} \sum_{j>i}^{r} g_{ij} x_i x_j \right) \qquad g_{ij} \in \mathbb{Z},$$

because $\mathrm{Tr}_{\mathbb{K}/\mathbb{Q}}(2 - \zeta_p - \zeta_p^{-1}) = p$. Hence $m(\frac{1}{p}f) = 1$. Therefore $\frac{1}{p}f$ is of E-type by Theorem 7.1.2 in [5]. As f is proportional to the quadratic form of E-type we have that f is of E-type, and the proof is complete.   □

Proof (of Theorem 4). The proof is by induction on k. If $k = 1$, then the theorem follows immediately from the Theorem 6. Let $k > 1$ and assume the theorem is true for $k - 1$. Set $m = n/p_k = p_1 \cdots p_{k-1}$. Hence $\mathbb{Q}(\zeta_n + \zeta_n^{-1}) = \mathbb{Q}(\zeta_m + \zeta_m^{-1})\mathbb{Q}(\zeta_{p_k} + \zeta_{p_k}^{-1})$ and $\mathbb{Z}[\zeta_n + \zeta_n^{-1}] = \mathbb{Z}[\zeta_m + \zeta_m^{-1}]\mathbb{Z}[\zeta_{p_k} + \zeta_{p_k}^{-1}] = \mathbb{Z}[\zeta_m + \zeta_m^{-1}] \otimes \mathbb{Z}[\zeta_{p_k} + \zeta_{p_k}^{-1}]$ by the hypothesis. To shorten notations, we write $\mathbb{K}$ instead of $\mathbb{Q}(\zeta_n + \zeta_n^{-1})$, $\mathbb{K}_1$ instead of $\mathbb{Q}(\zeta_m + \zeta_m^{-1})$ and $\mathbb{K}_2$ instead of $\mathbb{Q}(\zeta_{p_k} + \zeta_{p_k}^{-1})$. Put $a = \prod_{i=1}^{k-1}(2 - \zeta_{p_i} - \zeta_{p_i}^{-1})$ and $a_k = 2 - \zeta_{p_k} - \zeta_{p_k}^{-1}$. By assumption, ax^2 is perfect over $\mathbb{Z}[\zeta_m + \zeta_m^{-1}]$. Since $v \otimes w = vw$ in $\mathbb{Z}[\zeta_n + \zeta_n^{-1}]$ for all $v \in \mathbb{Z}[\zeta_m + \zeta_m^{-1}]$ and $w \in \mathbb{Z}[\zeta_{p_k} + \zeta_{p_k}^{-1}]$, we have $ax^2 \otimes a_k y^2 = (aa_k)z^2$. Our next goal is to show that $m((aa_k)x^2) = m(ax^2) \cdot m(a_k x^2)$. Let $r = \phi(p_k)/2$ and $d = \phi(m)/2$. Consider the following rational quadratic forms

$$
\begin{aligned}
f(x_1, \ldots, x_r) &= \\
&= \mathrm{Tr}_{\mathbb{K}_2/\mathbb{Q}} \left((2 - \zeta_{p_k} - \zeta_{p_k}^{-1})(x_1 + x_2(\zeta_{p_k} + \zeta_{p_k}^{-1}) + \cdots + x_r(\zeta_{p_k} + \zeta_{p_k}^{-1})^{r-1})^2 \right), \\
f'(x_1, \ldots, x_d) &= \\
&= \mathrm{Tr}_{\mathbb{K}_1/\mathbb{Q}} \left((2 - \zeta_m - \zeta_m^{-1})(x_1 + x_2(\zeta_m + \zeta_m^{-1}) + \cdots + x_d(\zeta_m + \zeta_m^{-1})^{d-1})^2 \right).
\end{aligned}
$$

By definition, $m(ax^2) = m(f')$ and $m(a_k x^2) = m(f)$. An easy computation shows that

$$\mathrm{Tr}_{\mathbb{K}/\mathbb{Q}} \left((2 - \zeta_n - \zeta_n^{-1})(x_1 + x_2(\zeta_n + \zeta_n^{-1}) + \cdots + x_{rd}(\zeta_n + \zeta_n^{-1})^{rd-1})^2 \right) =$$
$$= f(x_1, \ldots, x_r) \cdot f'(x_1, \ldots, x_d).$$

This gives $m((aa_k)x^2) = m(f \otimes f')$. We conclude from Theorem 7 that f is of E-type, hence $m((aa_k)x^2) = m(f \otimes f') = m(f) \cdot m(f')$, and finally

$$\{v \cdot w | v \in M(ax^2), w \in M(a_k x^2)\} \subseteq M((aa_k)x^2).$$

From Theorem 5 it follows that

$$(aa_k)x^2 = \left(\prod_{i=1}^{k}(2 - \zeta_{p_i} - \zeta_{p_i}^{-1}) \right) x^2$$

is perfect over $\mathbb{Q}(\zeta_n + \zeta_n^{-1})$, which proves the theorem.   □

References

1. Barnes, E.S.: The complete enumeration of extreme senary forms. Phil. Trans. Roy. Soc. London 249, 461–506 (1957)
2. Conway, J.H., Sloane, N.: Sphere packings, lattices and groups, 3rd edn. Grundlehren der mathematischen Wissenschaften, vol. 290. Springer, Heidelberg (1999)
3. Coulangeon, R.: Voronoï theory over algebraic number fields. Monographies de l'Enseignement Mathématique 37, 147–162 (2001)
4. Gruber, P., Lekkerkerker, C.: Geometry of numbers. North-Holland, Amsterdam (1987)
5. Kitaoka, Y.: Arithmetic of quadratic forms. Cambridge University Press, Cambridge (1993)
6. Leibak, A.: On additive generalization of Voronoï's theory for algebraic number fields. Proc. Estonian Acad. Sci. Phys. Math. 54(4), 195–211 (2005)
7. Leibak, A.: The complete enumeration of binary perfect forms over algebraic number field $\mathbb{Q}(\sqrt{6})$. Proc. of Estonian Acad. of Sci. Phys. Math. 54(4), 212–234 (2005)
8. Martinet, J.: Perfect lattices in euclidean spaces. Grundlehren der mathematischen Wissenschaften, vol. 327. Springer, Heidelberg (2003)
9. Ong, H.E.: Perfect quadratic forms over real quadratic number fields. Geometriae Dedicata 20, 51–77 (1986)
10. Sikiric, M.D., Schürmann, A., Vallentin, F.: Classification of eight dimensional perfect forms. Electron. Res. Announc. Amer. Math. Soc. 13, 21–32 (2007)
11. Voronoï, G.: Sur quelques propriétés des formes quadratiques positives parfaites. J. reine angew. Math. 133, 97–178 (1908)
12. Washington, L.C.: Introduction to Cyclotomic Fields, 2nd edn. Graduate Texts in Mathematics, vol. 83. Springer, Berlin (1997)

Functorial Properties of Stark Units in Multiquadratic Extensions

Jonathan W. Sands and Brett A. Tangedal

University of Vermont, Burlington VT, 05401, USA
jwsands@uvm.edu
University of North Carolina at Greensboro, Greensboro NC, 27402, USA
batanged@uncg.edu

Abstract. The goal of this paper is to present computations investigating the "functorial" properties of Stark units, that is, how specific roots of Stark units from certain subfields of a top field are placed with respect to the top field and how these roots relate to the Stark unit in the top field. This type of question is of particular relevance to gaining a better understanding of the somewhat mysterious "abelian condition" in Stark's Conjecture.

1 Introduction

The original development of Stark's Conjecture (throughout this paper, "Stark's Conjecture" always refers strictly to the original rank one abelian conjecture of Stark appearing in [St3] and [Ta1], where the distinguished prime splitting completely is archimedean; higher rank and non-abelian generalizations of Stark's Conjecture have been considered by Rubin, Popescu, and Chinburg, among others) was focused on the match-up between the logarithms of absolute values of the Galois conjugates of a special unit in the top field (these "Stark units" are conjectured to always exist) and the first derivatives of specific partial zeta-functions evaluated at $s = 0$. This match-up is referred to as the "preliminary version" of Stark's Conjecture below. The abelian condition was the last refinement made to Stark's Conjecture and first appeared in [Ta1] after a slightly weaker "central condition" was announced in [St3]. The preliminary version plus the abelian condition gives the final version of Stark's Conjecture. It often happens that the Stark unit is a square in the top field (see [DH] for a partial explanation) and in this case the abelian condition holds trivially if there are only two roots of unity in the top field.

The strongest progress made overall thus far towards proving Stark's Conjecture is in the case where the Galois group of the relative extension of number fields under consideration is an elementary abelian 2-group. The preliminary version of Stark's Conjecture has been proved in this setting in [DST2] but the abelian condition is still unproven in general. In Section 2, we construct and study a special collection of extensions having Galois group isomorphic to $\mathbb{Z}_2 \times \mathbb{Z}_2 \times \mathbb{Z}_2$ over real quadratic base fields for which the abelian condition is

A.J. van der Poorten and A. Stein (Eds.): ANTS-VIII 2008, LNCS 5011, pp. 253–267, 2008.
© Springer-Verlag Berlin Heidelberg 2008

nontrivial and not known to hold by the theorems in [DST2]. We computationally verify the abelian condition to hold in all of these examples and record at the same time the special circumstances required for the abelian condition to hold in terms of where the 4th roots of Stark units from certain subfields are placed inside the top field.

In the remainder of this section we state the preliminary and final versions of Stark's Conjecture mentioned above along with a version intermediate between these two that only applies over totally real base fields. This intermediate version leads to an efficient algorithm for computing Stark units over totally real base fields. Let $\mathbb{Z}$, $\mathbb{Q}$, $\mathbb{R}$, $\mathbb{C}$, $\mathbb{C}^\times$, and $\mathbb{Z}_m$ denote the set of rational integers, rational numbers, real numbers, complex numbers, nonzero complex numbers, and $\mathbb{Z}/m\mathbb{Z}$ for a fixed integer $m \geq 2$, respectively. A finite abelian group A is always given in terms of its invariant factor decomposition $A \cong \mathbb{Z}_{n_1} \times \cdots \times \mathbb{Z}_{n_r}$, where $n_j \geq 2$ for all j and $n_{i+1} | n_i$ for $1 \leq i < r$. The symbol $\overline{\mathbb{Q}}$ denotes a fixed algebraic closure of $\mathbb{Q}$, considered abstractly as opposed to being considered as a subfield of $\mathbb{C}$. Let F be a field with $\mathbb{Q} \subseteq \mathsf{F} \subset \overline{\mathbb{Q}}$ and $[\mathsf{F} : \mathbb{Q}] = m < \infty$. If there are r_1 embeddings of F into $\mathbb{R}$ and r_2 pairs of complex conjugate embeddings of F into $\mathbb{C}$, then $m = r_1 + 2r_2$ and F has signature $[r_1, r_2]$. Let $\mathcal{O}_\mathsf{F}$ denote the integral closure of $\mathbb{Z}$ in F and let $\mathcal{E}(\mathsf{F})$ denote the multiplicative group of units in $\mathcal{O}_\mathsf{F}$. The discriminant of the field F is denoted by d_F. By an "integral ideal of F", we mean a nonzero ideal $\mathfrak{a} \subseteq \mathcal{O}_\mathsf{F}$. If $r_1 \geq 1$, we label the r_1 embeddings of F into $\mathbb{R}$ by

$$e_1 : \mathsf{F} \hookrightarrow \mathbb{R}, \quad \ldots, \quad e_{r_1} : \mathsf{F} \hookrightarrow \mathbb{R}.$$

Let $\mathsf{F}^{(i)}$, for $1 \leq i \leq r_1$, denote the image of F inside $\mathbb{R}$ under the ith embedding. Any particular ordering of these embeddings will do for now and once fixed we have the corresponding archimedean (or "infinite") primes $\mathfrak{p}_\infty^{(1)}, \ldots, \mathfrak{p}_\infty^{(r_1)}$. Given an integral ideal $\mathfrak{m}$ of F and a formal product $\mathfrak{m}_\infty$ of infinite primes taken from the set $\{\mathfrak{p}_\infty^{(1)}, \ldots, \mathfrak{p}_\infty^{(r_1)}\}$, we define a generalized modulus $\widetilde{\mathfrak{m}} = \mathfrak{m}\mathfrak{m}_\infty$ and the corresponding ray class group $H(\widetilde{\mathfrak{m}})$, which is a finite abelian group. We denote the set of all homomorphisms from $H(\widetilde{\mathfrak{m}})$ to $\mathbb{C}^\times$ by $\widehat{H(\widetilde{\mathfrak{m}})}$ (the elements of $\widehat{H(\widetilde{\mathfrak{m}})}$ are the so-called "ray class characters modulo $\widetilde{\mathfrak{m}}$"). Given $\chi \in \widehat{H(\widetilde{\mathfrak{m}})}$, there is an associated abelian L-function $L_{\widetilde{\mathfrak{m}}}(s, \chi)$ defined for $\Re(s) > 1$ by

$$L_{\widetilde{\mathfrak{m}}}(s, \chi) = \sum \frac{\chi(\mathfrak{a})}{N\mathfrak{a}^s},$$

with the sum running over all integral ideals $\mathfrak{a}$ of F with $(\mathfrak{a}, \mathfrak{m}) = (1)$ ($\chi(\mathfrak{a})$ is the image in $\mathbb{C}^\times$ under χ of the class in $H(\widetilde{\mathfrak{m}})$ to which $\mathfrak{a}$ belongs; $N\mathfrak{a}$ is the norm of $\mathfrak{a}$). The conductor $\mathfrak{f}(\chi)$ of χ is a product of an integral ideal $\mathfrak{f}_\chi$ ($\mathfrak{f}_\chi$ divides $\mathfrak{m}$) with a formal product $\mathfrak{f}_{\chi,\infty}$ of infinite primes (each infinite prime appearing in $\mathfrak{f}_{\chi,\infty}$ also appears in $\mathfrak{m}_\infty$). The unique (primitive) character defined modulo $\mathfrak{f}(\chi)$ equivalent to χ will be denoted by χ_{pr}.

The L-function $L_{\widetilde{\mathfrak{m}}}(s, \chi)$ has a meromorphic continuation to the whole complex plane (still denoted by $L_{\widetilde{\mathfrak{m}}}(s, \chi)$) and is analytic at $s = 0$ on this larger domain. The higher rank generalizations of Stark's Conjecture mentioned above involve the first nonzero coefficient in the Taylor series expansion of $L_{\widetilde{\mathfrak{m}}}(s, \chi)$ about $s = 0$ and a general purpose algorithm for computing this first nonzero coefficient was given in [DT] (see also [Co]). If χ_{pr} is nontrivial, then the order of the zero of $L_{\mathfrak{f}(\chi)}(s, \chi_{pr})$ at $s = 0$ is denoted by $r(\chi_{pr})$ and is equal to $r_1 + r_2 - q(\chi)$, where $q(\chi)$ is the number of infinite primes appearing in the formal product $\mathfrak{f}_{\chi,\infty}$ (see Section 2 of [DT]). Clearly, $r_1 - q(\chi) \geq 0$. Since

$$L_{\widetilde{\mathfrak{m}}}(s, \chi) = L_{\mathfrak{f}(\chi)}(s, \chi_{pr}) \prod_{\mathfrak{p}} (1 - \chi_{pr}(\mathfrak{p}) N\mathfrak{p}^{-s}), \tag{1}$$

with $\mathfrak{p}$ running over all prime ideals dividing $\mathfrak{m}$ but not $\mathfrak{f}_\chi$, the order $r(\chi)$ of the zero of $L_{\widetilde{\mathfrak{m}}}(s, \chi)$ at $s = 0$ satisfies $r(\chi) \geq r(\chi_{pr})$. If χ_0 is the trivial character in $\widehat{H(\widetilde{\mathfrak{m}})}$, and there are t distinct prime ideals dividing $\mathfrak{m}$, then $r(\chi_0) = r_1 + r_2 + t - 1$.

Let X denote a subgroup of characters in $\widehat{H(\widetilde{\mathfrak{m}})}$ containing at least one nontrivial character. If $r_2 \geq 2$, then for all nontrivial characters $\chi \in \mathsf{X}$ we have $r(\chi) \geq 2$. The prescription for Stark's first order zero (or "rank one") abelian conjecture is that $r(\chi) \geq 1$ for all $\chi \in \mathsf{X}$ and $r(\chi) = 1$ for at least one such χ. In [St3], Stark described two general situations that follow this prescription that are classified as Type I and Type II below.

(I) F has signature $[m, 0]$ (that is, F is totally real), $\mathfrak{m}_\infty$ is a product of exactly $m - 1$ of the infinite primes of F, and $r(\chi) \geq 1$ for all $\chi \in \mathsf{X}$.
(II) F has signature $[m - 2, 1]$, $\mathfrak{m}_\infty$ is the product of all $m - 2$ real infinite primes of F, and $r(\chi) \geq 1$ for all $\chi \in \mathsf{X}$.

(Note: By far the most interesting situations of either Type I or Type II are those where $\mathfrak{f}_{\chi,\infty} = \mathfrak{m}_\infty$ for at least one nontrivial $\chi \in \mathsf{X}$.) For the remainder of the paper, we let F denote a field, $\widetilde{\mathfrak{m}}$ a generalized modulus, and X a nontrivial subgroup of $\widehat{H(\widetilde{\mathfrak{m}})}$, satisfying the conditions of the Type I or Type II classification. By class field theory, there exists a unique Galois extension field K of F having the following properties:

(i) $\mathsf{F} \subset \mathsf{K} \subset \overline{\mathbb{Q}}$ and $\mathrm{Gal}(\mathsf{K}/\mathsf{F}) \cong \mathsf{X}$.
(ii) A prime ideal $\mathfrak{p} \subset \mathcal{O}_\mathsf{F}$ with $(\mathfrak{p}, \mathfrak{m}) = (1)$ splits completely in K if and only if $\chi(\mathfrak{p}) = 1$ for all $\chi \in \mathsf{X}$ (a prime ideal dividing $\mathfrak{m}$ might split completely if it does not divide the conductor $\mathfrak{f}(\mathsf{K}/\mathsf{F})$ of the extension). This characterization of the primes splitting completely in a Galois extension K of F (outside of a finite number) defines K uniquely by a theorem of Bauer (see [Ja], Cor. 5.5).
(iii) The relative discriminant $d(\mathsf{K}/\mathsf{F})$ of the extension K/F has a simple expression using the conductor-discriminant formula (see [Ha]): $d(\mathsf{K}/\mathsf{F}) = \prod_{\chi \in \mathsf{X}} \mathfrak{f}_\chi$.
(iv) For a Type I situation, the one infinite prime of F missing from the formal product $\mathfrak{m}_\infty$ splits in the extension K/F. For a Type II situation, the unique complex infinite prime of F automatically splits in K/F. An infinite prime $\mathfrak{p}_\infty^{(i)}$ appearing in $\mathfrak{m}_\infty$ (for either Type I or II) ramifies in K/F if and only if it appears in $\mathfrak{f}_{\chi,\infty}$ for at least one $\chi \in \mathsf{X}$.

Class field theory guarantees the existence of K and gives all of the above information about K but gives no explicit means of actually constructing K. Stark's Conjecture offers the exciting prospect of being able to give an explicit construction of any field K corresponding to a Type I or II situation (see [DST1] and [DTvW], respectively).

Partial zeta-functions play a special role with respect to the constructive aspect of Stark's Conjecture. Let $\mathcal{C}$ be a given class in $H(\widetilde{\mathfrak{m}})$. The partial zeta-function $\zeta_{\widetilde{\mathfrak{m}}}(s,\mathcal{C})$ corresponding to $\mathcal{C}$ is defined for $\Re(s) > 1$ by

$$\zeta_{\widetilde{\mathfrak{m}}}(s,\mathcal{C}) = \sum \frac{1}{\mathrm{N}\mathfrak{a}^s},$$

with the sum running over all integral ideals $\mathfrak{a} \in \mathcal{C}$ (for $\mathfrak{a} \in \mathcal{C}$, we have $(\mathfrak{a},\mathfrak{m}) = (1)$ by definition). Let $J = \{\mathcal{C}_1,\ldots,\mathcal{C}_k\}$ denote the subgroup of classes in $H(\widetilde{\mathfrak{m}})$ on which the group X is trivial, that is, for each $\mathcal{C}_i \in J$, we have $\chi(\mathcal{C}_i) = 1$ for all $\chi \in \mathsf{X}$. For each coset $\mathcal{C}J$ in $H(\widetilde{\mathfrak{m}})/J$, we define

$$\zeta_{\widetilde{\mathfrak{m}}}(s,\mathcal{C}J) = \frac{1}{n}\sum_{\chi\in\mathsf{X}} \overline{\chi}(\mathcal{C}J)L_{\widetilde{\mathfrak{m}}}(s,\chi), \tag{2}$$

where $n = |\mathsf{X}|$, the cardinality of the set X. If $\mathcal{C}J = \{\mathcal{C}'_1,\ldots,\mathcal{C}'_k\}$, then $\zeta_{\widetilde{\mathfrak{m}}}(s,\mathcal{C}J) = \sum_{i=1}^{k}\zeta_{\widetilde{\mathfrak{m}}}(s,\mathcal{C}'_i)$, and every prime ideal $\mathfrak{p}$ in any one of the classes in $\mathcal{C}J$ has the same Frobenius automorphism $\sigma_{\mathfrak{p}} \in \mathrm{Gal}(\mathsf{K}/\mathsf{F})$. In this way, the Artin map sets up an isomorphism $Ar : H(\widetilde{\mathfrak{m}})/J \to \mathrm{Gal}(\mathsf{K}/\mathsf{F})$. For each $\sigma \in \mathrm{Gal}(\mathsf{K}/\mathsf{F})$, this isomorphism allows us to define $\zeta_{\widetilde{\mathfrak{m}}}(s,\sigma) := \zeta_{\widetilde{\mathfrak{m}}}(s,\mathcal{C}J)$, where $\sigma = Ar(\mathcal{C}J)$. Each function $\zeta_{\widetilde{\mathfrak{m}}}(s,\sigma)$ has a meromorphic continuation to all of $\mathbb{C}$ and is analytic at $s = 0$ by Eq. (2). Since $r(\chi) \geq 1$ for all $\chi \in \mathsf{X}$, we also see that each $\zeta_{\widetilde{\mathfrak{m}}}(s,\sigma)$ has at least a first order zero at $s = 0$.

A few more preparations are needed before stating Stark's Conjecture. Let K/F be an extension corresponding to a Type I or II situation as above, where it is assumed that the base field F is neither $\mathbb{Q}$ nor a complex quadratic field (Stark's Conjecture has already been proved over these base fields). In order to focus on only the most interesting situations, we make the additional assumption that $r(\chi) = 1$ for at least one nontrivial $\chi \in \mathsf{X}$, which implies that $\mathfrak{f}_{\chi,\infty} = \mathfrak{m}_\infty$ for this character. The following conventions are fixed for Type I and II situations:

(I) Set $\nu = 1$. Renumber the real embeddings of F such that $\mathfrak{p}_\infty^{(1)}$ is the unique infinite prime not appearing in $\mathfrak{m}_\infty$. For each j with $1 \leq j \leq r_1 = m$, choose an embedding $\mathsf{K} \hookrightarrow \mathsf{K}^{(j)} \subset \mathbb{C}$ extending the embedding $\mathsf{F} \hookrightarrow \mathsf{F}^{(j)} \subset \mathbb{R}$. It is important to note that $\mathsf{K} \hookrightarrow \mathsf{K}^{(1)}$ is a real embedding.

(II) Set $\nu = 2$. Let $\mathsf{F} \hookrightarrow \mathsf{F}^{(1)} \subset \mathbb{C}$ be a fixed nonreal embedding of F into $\mathbb{C}$, and let $\mathsf{F} \hookrightarrow \mathsf{F}^{(j)} \subset \mathbb{R}$, for $2 \leq j \leq m-1$, denote the real embeddings of F. Again, choose one embedding of the top field $\mathsf{K} \hookrightarrow \mathsf{K}^{(j)} \subset \mathbb{C}$ extending each embedding $\mathsf{F} \hookrightarrow \mathsf{F}^{(j)}$, $1 \leq j \leq m-1$, of the base field.

Let $\alpha^{(j)}$ denote the image of $\alpha \in \mathsf{K}$ in $\mathbb{C}$ under the jth embedding of K. The symbol $|\ |$ denotes the usual absolute value on $\mathbb{C}$. Let w_K denote the number of roots of unity in K. We may now state simultaneously for either Type I or II situations the

Preliminary Version of Stark's Rank One Abelian Conjecture: *There exists a unit $\varepsilon \in \mathcal{E}(K)$ such that*

1. $|\sigma(\varepsilon)^{(j)}| = 1$ *for all $j \geq 2$ and all $\sigma \in \mathrm{Gal}(K/F)$, and*
2. $\log |\sigma(\varepsilon)^{(1)}|^{\nu} = -w_K \zeta'_{\mathfrak{m}}(0, \sigma)$ *for all $\sigma \in \mathrm{Gal}(K/F)$.*

The choice of ν made above simply ensures that we take the logarithm of the *normalized* valuation with respect to the first specified archimedean prime of K of the Galois conjugates of the "Stark unit" $\varepsilon \in \mathcal{E}(K)$, which is unique up to a root of unity in K, assuming it exists. The abelian condition states in addition that

3. $K(\varepsilon^{1/w_K})$ *is an abelian extension of* F.

As noted earlier, the final version of Stark's Conjecture is all three parts together. There is an equivalent formulation of the abelian condition that is of great importance from the computational point of view. We state it here only for the special case where $w_K = 2$ (for the general version, see [Ta2], pp. 83-4).

Abelian Condition when $w_K = 2$: *If K/F is a relative abelian extension with $G = \mathrm{Gal}(K/F)$ and $w_K = 2$, then $K(\sqrt{\varepsilon})$ is an abelian extension of F iff*

a. *for each $\sigma \in G$, we have $\varepsilon/\sigma(\varepsilon) = \alpha_\sigma^2$ for some $\alpha_\sigma \in K$, and*
b. *$\alpha_\sigma \sigma(\alpha_\rho) = \alpha_\rho \rho(\alpha_\sigma)$ for all $\sigma, \rho \in G$.*

If G is cyclic, say $G = \{\sigma^0, \sigma^1, \ldots, \sigma^{n-1}\}$, and $\alpha_1 \in K$ satisfies $\varepsilon/\sigma(\varepsilon) = \alpha_1^2$, then part b holds automatically, where $\alpha_2 = \alpha_1 \sigma(\alpha_1)$, $\alpha_3 = \alpha_1 \sigma(\alpha_1) \sigma^2(\alpha_1), \ldots$ satisfy $\varepsilon/\sigma^j(\varepsilon) = \alpha_j^2$ for $j = 2, 3, \ldots, n-1$.

For a Type I situation, since all Galois conjugates of ε with respect to the embedding $K \hookrightarrow K^{(1)}$ are real numbers (also $w_K = 2$, since K has a real embedding), it is natural to wonder if all of these conjugates can be positive simultaneously (if so, the absolute values in part 2 of the preliminary version can be dropped). This is indeed the content of the

Intermediate Version of Stark's Rank One Abelian Conjecture (for Type I situations only): *There exists a unit $\varepsilon \in \mathcal{E}(K)$ such that*

1. $|\sigma(\varepsilon)^{(j)}| = 1$ *for all $j \geq 2$ and all $\sigma \in \mathrm{Gal}(K/F)$, and*
2. $\sigma(\varepsilon)^{(1)} = \exp(-2\zeta'_{\mathfrak{m}}(0, \sigma))$ *for all $\sigma \in \mathrm{Gal}(K/F)$.*

We refer to this as an intermediate version not only because it is stronger than the preliminary version but also because an important piece (how important will be seen in Section 2) of the abelian condition is already being hinted at, namely, part a of the abelian condition can only be satisfied if all Galois conjugates of ε with respect to the embedding $K \hookrightarrow K^{(1)}$ are of the same sign. It can be proved that part 1 of the intermediate version is superfluous in that it is a consequence of part 2 in Type I situations (see p. 369 of [St1] for the proof when F is real quadratic). However, we have retained part 1 here for emphasis because of the crucial role it plays in computing Stark units in Type I situations.

Since all of the computations in this paper are carried out over real quadratic base fields, we give a quick orientation on how the intermediate version of Stark's Conjecture is used to compute Stark units in this special setting. To a given real quadratic field F of discriminant d_F we associate a canonically defined polynomial $f[d_F]$ as follows:

$$f[d_F] = \begin{cases} x^2 - d_F/4 & \text{if } d_F \equiv 0 \bmod (4), \\ x^2 - x - (d_F - 1)/4 & \text{if } d_F \equiv 1 \bmod (4). \end{cases}$$

If $\theta \in \overline{\mathbb{Q}}$ is a root of $f[d_F]$, then $F = \mathbb{Q}(\theta)$ and $[1, \theta]$ is an integral basis for $\mathcal{O}_F$. A given $f[d_F]$ always has one positive real root, denoted by $\theta^{(1)}$, and one negative real root, denoted by $\theta^{(2)}$. This convention allows us to fix the two real embeddings of F into $\mathbb{R}$ as follows:

$$e_1 : F \hookrightarrow \mathbb{R} \quad \text{is defined by the map} \quad a + b\theta \mapsto a + b\theta^{(1)}, \ (a, b \in \mathbb{Q}),$$
$$e_2 : F \hookrightarrow \mathbb{R} \quad \text{is defined by the map} \quad a + b\theta \mapsto a + b\theta^{(2)}.$$

The two infinite primes corresponding to the two real embeddings of F are denoted by $\mathfrak{p}_\infty^{(1)}$ and $\mathfrak{p}_\infty^{(2)}$, respectively. If $f(x) = \alpha_n x^n + \cdots + \alpha_1 x + \alpha_0 \in F[x]$, let $f^{(i)}(x) := \alpha_n^{(i)} x^n + \cdots + \alpha_1^{(i)} x + \alpha_0^{(i)} \in \mathbb{R}[x]$ for $i = 1$ and $i = 2$. Given an integral ideal $\mathfrak{m} \subset \mathcal{O}_F$, we compute the ray class group $H(\widetilde{\mathfrak{m}}) = H(\mathfrak{m}\mathfrak{p}_\infty^{(2)})$ (we used PARI/GP [GP] for all of our computations). Given a subgroup X of n ray class characters modulo $\widetilde{\mathfrak{m}}$ such that $\mathfrak{f}_{\chi,\infty} = \mathfrak{p}_\infty^{(2)}$ for at least one nontrivial $\chi \in X$, we compute $\zeta'_{\widetilde{\mathfrak{m}}}(0, \mathcal{C}J)$ using Eq. (2) for each of the n cosets $\mathcal{C}J$ in $H(\widetilde{\mathfrak{m}})/J$. The intermediate version of Stark's Conjecture says that the polynomial

$$f^{(1)}(x) = \prod_{\mathcal{C}J \in H(\widetilde{\mathfrak{m}})/J} (x - \exp(-2\zeta'_{\widetilde{\mathfrak{m}}}(0, \mathcal{C}J))) = x^n - \alpha_{n-1}^{(1)} x^{n-1} + \cdots + \alpha_0^{(1)}$$

has the special property that $\alpha_0, \ldots, \alpha_{n-1} \in \mathcal{O}_F$. For each j with $0 \leq j \leq n-1$, we have a real number approximation β_j to $\alpha_j^{(1)}$ computed to high precision. Part 1 of the conjecture gives us a positive number B such that $|\alpha_j^{(2)}| \leq B$ (for example, $|\alpha_{n-1}^{(2)}| \leq n$). Assuming $\alpha = a + b\theta \in \mathcal{O}_F$ and $\beta, B \in \mathbb{R}$ are such that $|a + b\theta^{(1)} - \beta| < \delta$ and $|a + b\theta^{(2)}| \leq B$ for a small positive $\delta \in \mathbb{R}$, we may find $a, b \in \mathbb{Z}$ as follows. We first note that $|b(\theta^{(1)} - \theta^{(2)}) - \beta| < B + \delta$. Since $\theta^{(1)} - \theta^{(2)} = \sqrt{d_F}$, we have

$$\frac{\beta}{\sqrt{d_F}} - \frac{B + \delta}{\sqrt{d_F}} < b < \frac{\beta}{\sqrt{d_F}} + \frac{B + \delta}{\sqrt{d_F}}.$$

There should be exactly one integer value of b within this range for which the number $b\theta^{(1)} - \beta$ is very close to some integer c. This is our b, and $a = -c$. It is rather remarkable that the range of possible values of b shrinks with increasing values of d_F (assuming B and δ stay the same). We emphasize that *all* of the computations described above take place working *completely* within the field F. Once we obtain the polynomial $f(x) \in \mathcal{O}_F[x]$, we may carry out an independent verification that any given root ρ of $f(x)$ generates a subfield $F(\rho)$ of the unique Galois extension field K of F with K corresponding to the group X by class field theory. In most cases, Stark's Conjecture predicts that $K = F(\rho)$. This prediction is made, for example, when $L'_{\mathfrak{m}}(0, \chi) \neq 0$ for all $\chi \in X$ with $\mathfrak{f}_{\chi,\infty} = \mathfrak{p}_\infty^{(2)}$ (see Theorem 1 on p. 66 of [St2]).

2 The Abelian Condition for Some Multiquadratic Extensions

If L/F is a relative Galois extension of fields with $\mathrm{Gal}(L/F)$ isomorphic to a direct product of m copies of $\mathbb{Z}_2$, we say that L/F is a multiquadratic extension of rank m. The preliminary version of Stark's Conjecture was proved for multiquadratic extensions of arbitrary rank in [DST2]. However, it is possible to construct multiquadratic extensions for which the abelian condition is nontrivial and not known to hold by the theorems in [DST2]. In this section, we construct and study a special collection of such extensions of rank 3 over real quadratic base fields. We note that the final version of Stark's Conjecture has been proved completely for all multiquadratic extensions of rank 1 and 2 (see Section 3 and Theorem 4, respectively, in [DST2]). The final version has also been proved (see Remark 2 on p. 85 of [DST2]) for multiquadratic extensions L/F in which no prime above 2 ramifies. We therefore fix a prime ideal $\mathfrak{p} \subset \mathcal{O}_F$ lying over the prime 2 and construct extensions of F (F denotes a fixed real quadratic field for the remainder of this section) for which we know $\mathfrak{p}$ ramifies. The final version is also known to hold if more than a certain number of finite primes of F ramify in the extension L/F (again, see Remark 2, op. cit.); we therefore minimize the ramification by ensuring that $\mathfrak{p}$ is the *only* finite prime in F that ramifies in L/F. Following the conventions established for real quadratic fields at the end of Section 1, we set $\mathfrak{m}_\infty$ equal to the infinite prime designated there as $\mathfrak{p}_\infty^{(2)}$. The extensions L/F we construct will therefore be such that $\mathfrak{p}_\infty^{(1)}$ splits and $\mathfrak{p}_\infty^{(2)}$ ramifies. In terms of the notation in Section 1 of [DST2], the distinguished prime $\mathfrak{v}$ of F splitting completely in L/F is $\mathfrak{p}_\infty^{(1)}$, $S = \{\mathfrak{p}_\infty^{(1)}, \mathfrak{p}_\infty^{(2)}, \mathfrak{p}\}$, $|S| = 3$, and $S_{fin} = \{\mathfrak{p}\}$.

By Theorem 3 of [DST2], we are interested only in the *maximal* multiquadratic extension L/F unramified outside of $\mathfrak{p}$ and $\mathfrak{p}_\infty^{(2)}$. The maximal such extension is characterized as being the composite of all relative quadratic extensions K of F ramified at most above $\mathfrak{p}$ and $\mathfrak{p}_\infty^{(2)}$. If K/F is a relative quadratic extension of this type, the conductor $\mathfrak{f}(K/F)$ is of the form $\mathfrak{p}^a$ or $\mathfrak{p}^a \mathfrak{p}_\infty^{(2)}$. Our immediate goal is to find an upper bound b such that $a \leq b$. This bound proves that only finitely many such relative quadratic extensions exist. We also conclude that the conductor $\mathfrak{f}(L/F)$ divides $\mathfrak{p}^b \mathfrak{p}_\infty^{(2)}$ since this conductor is the least common multiple of the conductors of the relative quadratic extensions K/F with $F \subset K \subseteq L$.

Lemma 1. *Let* F *be a real quadratic field and* K/F *a relative quadratic extension ramified at most above* $\mathfrak{p}$ *and* $\mathfrak{p}_\infty^{(2)}$. *Set* $b = 3$ *if the prime 2 splits or is inert in* F *and set* $b = 5$ *if 2 ramifies in* F. *If* $\mathfrak{p}^a$ *is the finite part of the conductor* $\mathfrak{f}(K/F)$, *then* $a \leq b$.

Proof. By the conductor-discriminant formula, the finite part of the conductor $\mathfrak{f}(K/F)$ is equal to the relative discriminant $d(K/F)$. If $\mathfrak{p}$ does not ramify in K/F, then $a = 0$. If $\mathfrak{p}$ does ramify in K/F, then $\mathfrak{p}\mathcal{O}_K = \mathfrak{P}^2$. If s is the exact exponent to which $\mathfrak{P}^s$ divides the ideal $(2) \subset \mathcal{O}_K$, then $s = 2$ when 2 is split or inert in F and $s = 4$ when 2 ramifies in F. By Proposition 6.4 on p. 262 of [Na], the relative different $\mathcal{D}(K/F)$ can not be divisible by a higher power of $\mathfrak{P}$ than $\mathfrak{P}^{s+1}$. Observing that the relative norm from K to F of $\mathfrak{P}$ is $\mathfrak{p}$ completes the proof.

The process used for finding interesting examples was to run through the list of real quadratic fields F (ordered by discriminant) and for each prime ideal $\mathfrak{p} \subset \mathcal{O}_F$ lying over the prime 2 to compute the ray class group $H(\widetilde{\mathfrak{m}})$ in each case, where $\widetilde{\mathfrak{m}} = \mathfrak{p}^b\mathfrak{p}_\infty^{(2)}$ and b is given as in Lemma 1. We find 72 such ray class groups in this way having exactly 3 invariant factors with d_F in the range $5 \leq d_F \leq 1365$ (the first such ray class group with 4 invariant factors occurs with $d_F = 1365$). It is worth noting that in all but 2 of these 72 examples, the prime 2 was either inert or ramified in F. In the two exceptional cases ($d_F = 1105$ and 1241) there was only one prime ideal $\mathfrak{p}$ above 2 for which the corresponding ray class group had 3 invariant factors and therefore in all 72 examples the prime $\mathfrak{p}$ is uniquely determined. The 3rd invariant factor $\mathbb{Z}_{n_3}$ is $\mathbb{Z}_2$ in all 72 examples. In each of these examples, there is a uniquely determined set of 8 characters X in the group $\widehat{H(\widetilde{\mathfrak{m}})}$ of order 1 or 2, and clearly X forms a group isomorphic to $\mathbb{Z}_2 \times \mathbb{Z}_2 \times \mathbb{Z}_2$. It is of interest that in all 72 examples the ideal class group $\mathrm{Cl}(F)$ of F is nontrivial and has order a power of 2. We define the S_{fin}-class group $\mathrm{Cl}_F(S_{fin})$ of F to be the quotient group $\mathrm{Cl}(F)/\langle\mathfrak{p}\rangle$, where $\langle\mathfrak{p}\rangle$ is the subgroup of $\mathrm{Cl}(F)$ generated by the ideal class to which $\mathfrak{p}$ belongs. In all 72 examples, $\mathrm{Cl}_F(S_{fin})$ is a nontrivial cyclic group and so the 2-rank $r_F(S)$ of this group is equal to one in each example.

We will limit ourselves to the discriminant range $5 \leq d_F \leq 1364$ in the remainder of this section and to the 72 examples mentioned above where F, $\mathfrak{p}$, $\widetilde{\mathfrak{m}}$, and X are all uniquely determined. There are 26 such examples where $\mathfrak{p}_\infty^{(2)}$ does not appear in the conductor of any character $\chi \in X$, namely, $d_F = 165, 285, 357, 429, 476, 645, 741, 780, 805, 840, 861, 885, 924, 952, 957, 1005, 1020, 1045, 1085, 1148, 1173, 1221, 1245, 1288, 1309, 1320$. In these 26 examples, $r(\chi) \geq 2$ for all $\chi \in X$ and because of this we remove these from further consideration and focus *only* on the remaining 46 examples. Exactly 4 of the 8 characters in X then have $\mathfrak{p}_\infty^{(2)}$ in their conductors and we label these as χ_1, χ_2, χ_3, and χ_4. We label the trivial character as χ_0 and the remaining characters as χ_5, χ_6, and χ_7. A ray class character defined over a real quadratic field can not have its conductor exactly equal to $\mathfrak{p}_\infty^{(2)}$ and so the conductors of χ_1, χ_2, χ_3, and χ_4 are all divisible by $\mathfrak{p}$. Therefore, by Eq. (1), each of these 4 characters satisfies the relation $L_{\widetilde{\mathfrak{m}}}(s, \chi) = L_{\mathfrak{f}(\chi)}(s, \chi_{pr})$ even though an individual such χ defined modulo $\widetilde{\mathfrak{m}}$ might not be primitive. Therefore, $r(\chi_j) = 1$ for $1 \leq j \leq 4$.

We have $r(\chi_0) = 2$ and $r(\chi_j) \geq 2$ for $j = 5$, 6, and 7. For convenience, we set $L_j := L'_{\overline{\mathfrak{m}}}(0, \chi_j) \neq 0$ for $1 \leq j \leq 4$. Corresponding to each of the seven subgroups of characters $X_j = \{\chi_0, \chi_j\}$, $1 \leq j \leq 7$, there is a relative quadratic extension field K_j of F satisfying properties (i)–(iv) in Section 1. Note in particular that both $\mathfrak{p}$ and $\mathfrak{p}_\infty^{(2)}$ ramify in K_1, K_2, K_3, and K_4 (these fields all have signature $[2, 1]$). As mentioned earlier, the final version of Stark's Conjecture has been proved for relative quadratic extensions (see Section 3 of [DST2]) and we let ε_j denote the Stark unit in K_j for $1 \leq j \leq 4$. The fields K_5, K_6, and K_7 all have signature $[4, 0]$ (there are Stark units associated to these fields as well but they are all equal to 1). Corresponding to the group of characters X is an extension field L of F with $\mathrm{Gal}(L/F) \cong \mathbb{Z}_2 \times \mathbb{Z}_2 \times \mathbb{Z}_2$. The seven relative quadratic extensions of F contained in L are precisely those mentioned above. There are also precisely 7 quartic extensions of F contained in L all having Galois group isomorphic to $\mathbb{Z}_2 \times \mathbb{Z}_2$ over F. We label the first 6 of these as K_{12}, K_{13}, K_{14}, K_{23}, K_{24}, and K_{34} since each K_{ij} here is the composite of K_i and K_j, with $1 \leq i < j \leq 4$. We note that both $\mathfrak{p}$ and $\mathfrak{p}_\infty^{(2)}$ ramify in all 6 of these extensions over F and that these fields all have signature $[4, 2]$. Again, the final version of Stark's Conjecture has been proved for these 6 extensions of F by Theorem 4 of [DST2] and we label the 6 corresponding Stark units as ε_{ij}, $1 \leq i < j \leq 4$, with $\varepsilon_{ij} \in K_{ij}$ (the relation between the ε_{ij}'s and the ε_i's will be described below). The 7th quartic extension of F contained in L is denoted by K_{re} since it is the composite of K_5, K_6, and K_7 and therefore totally real of signature $[8, 0]$. The nontrivial automorphism of L/K_{re} is denoted by τ and we will observe the special role it plays in Stark's Conjecture below. The signature of L is $[8, 4]$.

Stark's Conjecture for the extension L/F requires the specification of two embeddings $L \hookrightarrow L^{(1)} \subset \mathbb{R}$ and $L \hookrightarrow L^{(2)} \subset \mathbb{C}$ extending, respectively, $F \hookrightarrow F^{(1)} \subset \mathbb{R}$ and $F \hookrightarrow F^{(2)} \subset \mathbb{C}$. In order to gain a coherent view of how all of the Stark units here fit together, we assume throughout that the 1st embedding of any intermediate field between F and L (into $\mathbb{R}$!) is obtained upon restriction of the embedding $L \hookrightarrow L^{(1)}$ and similarly for the 2nd embeddings of all fields upon restriction of $L \hookrightarrow L^{(2)}$. Note that for every $\alpha \in L$ we have $\tau(\alpha)^{(2)} = \overline{\alpha^{(2)}}$, which is why τ is referred to as "complex conjugation at $\mathfrak{p}_\infty^{(2)}$ in L/F". The Stark units ε_1, ε_2, ε_3, and ε_4 mentioned above may all be expressed in terms of more basic units $\eta_j \in \mathcal{E}(K_j)$ for $1 \leq j \leq 4$. By the main theorem in Section 3 of [DST2], $\varepsilon_j = \eta_j^{M_j}$ for $1 \leq j \leq 4$ (recall that $|S| = 3$), where M_j is a positive integer for each j. We also have $\eta_j^{(1)} > 0$ and $\tau(\eta_j)^{(1)} > 0$ for $1 \leq j \leq 4$ (note that τ restricts to the nontrivial automorphism of K_j over F for $1 \leq j \leq 4$) as well as $|\eta_j^{(2)}| = |\tau(\eta_j)^{(2)}| = 1$ for all j. In Corollary 2 on p. 93 of [DST2] it is proven that $\frac{M_j}{4} \in \frac{1}{2}\mathbb{Z}$ for $1 \leq j \leq 4$ (we have $m = 3$ and $w_L = 2$ in our setup). This proves that $2 \mid M_j$ for $1 \leq j \leq 4$ and therefore ε_j is a square in K_j for $1 \leq j \leq 4$. For $1 \leq j \leq 4$, let $\sqrt{\varepsilon_j}$ denote an element in K_j whose square is ε_j and such that $\sqrt{\varepsilon_j}^{(1)} > 0$ and $\tau(\sqrt{\varepsilon_j})^{(1)} > 0$ (we have $|\sqrt{\varepsilon_j}^{(2)}| = |\tau(\sqrt{\varepsilon_j})^{(2)}| = 1$ as well). Indeed, $\sqrt{\varepsilon_j}^{(1)} = \exp(-L_j/2)$ and $\tau(\sqrt{\varepsilon_j})^{(1)} = \exp(L_j/2)$ for $j = 1, 2, 3$, and 4 by Stark's Conjecture and we may find a quartic polynomial $f_j(x) \in \mathbb{Z}[x]$ satisfied by each $\sqrt{\varepsilon_j}$ using the method described at the end of Section 1 (these four polynomials have an important function that we will come back to at the end of this section). It is not difficult to prove at this point that $\varepsilon_{ij} = \sqrt{\varepsilon_i}\sqrt{\varepsilon_j}$ for $1 \leq i < j \leq 4$. Corollary 2 of [DST2] also implies that $[\mathcal{L} : L] = 2$ since $r_F(S) = 1$ in our examples (see p. 90 in [DST2] for the definition of $\mathcal{L}$). We mention here that Theorem 1 in [DST2] does not apply to the 46 examples we are considering since $|S|$ is equal to $m + 1 - r_F(S)$.

We require one final observation about the η_j's (see the first part of Section 8 in [DST2]); we have either $\sqrt{\eta_j} \in L$ or $\sqrt{-\eta_j} \in L$ for each j. We must have $\sqrt{\eta_j} \in L$ for each j in the range $1 \leq j \leq 4$ since the field $K_j(\sqrt{-\eta_j})$ is totally complex for each j. This means that for each j in the range $1 \leq j \leq 4$, there is an element $\sqrt[4]{\varepsilon_j} \in L$ whose 4th power is equal to ε_j. We can (and do!) choose these 4th roots such that $\sqrt[4]{\varepsilon_j}^{(1)} > 0$ for $j = 1, 2, 3$, and 4. Note that $\sqrt[4]{\varepsilon_j}^{(1)} = \exp(-L_j/4)$ for $1 \leq j \leq 4$.

The product $\prod_{j=1}^{4} \sqrt[4]{\varepsilon_j}$, which we denote simply by ε, lies in $\mathcal{E}(L)$ and we will prove in Proposition 1 below that ε satisfies the preliminary version of Stark's Conjecture for the extension L/F. For example, if σ_0 denotes the trivial automorphism of $\mathrm{Gal}(L/F)$, then $\zeta'_{\mathfrak{m}}(0, \sigma_0) = (L_1 + L_2 + L_3 + L_4)/8$ by Eq. (2), and $\varepsilon^{(1)} = \prod_{j=1}^{4} \exp(-L_j/4) = \exp(-2\zeta'_{\mathfrak{m}}(0, \sigma_0)) > 0$. However, even if we prove that $|\sigma(\varepsilon)^{(1)}| = \exp(-2\zeta'_{\mathfrak{m}}(0, \sigma))$ holds for all $\sigma \in \mathrm{Gal}(L/F)$ we can not simply remove the absolute value sign on the left hand side for a given nontrivial $\sigma \in \mathrm{Gal}(L/F)$ since $\sigma(\sqrt[4]{\varepsilon_j})^{(1)}$ can easily be negative for some j in the range $1 \leq j \leq 4$. This consideration is indeed the starting point for where the methods of [DST2] fall short in proving the abelian condition; it still needs to be proven that only an *even* number of negative values can appear among the numbers $\sigma(\sqrt[4]{\varepsilon_j})^{(1)}$, $1 \leq j \leq 4$, for any given $\sigma \in \mathrm{Gal}(L/F)$.

The following lemma will be used often.

Lemma 2. *We have* $\tau(\sqrt[4]{\varepsilon_j}) = 1/\sqrt[4]{\varepsilon_j}$ *for each j in the range $1 \leq j \leq 4$.*

Proof. Assume j is fixed in the range $1 \leq j \leq 4$. From the discussion above, $|\sqrt[4]{\varepsilon_j}^{(2)}| = 1$, which we rewrite as $(\sqrt[4]{\varepsilon_j})^{(2)} \cdot \overline{(\sqrt[4]{\varepsilon_j})^{(2)}} = 1$. Since $\tau(\sqrt[4]{\varepsilon_j})^{(2)} = \overline{\sqrt[4]{\varepsilon_j}^{(2)}}$, we have $\left[\sqrt[4]{\varepsilon_j} \cdot \tau(\sqrt[4]{\varepsilon_j})\right]^{(2)} = 1$. The proof is complete since $L \hookrightarrow L^{(2)}$ is an embedding.

We denote the nontrivial automorphism of L/K_{ij} for fixed i and j satisfying $1 \leq i < j \leq 4$ by σ_{ij}. The complete list of all elements in $\mathrm{Gal}(L/F)$ is then $\{\sigma_0, \sigma_{12}, \sigma_{13}, \sigma_{14}, \sigma_{23}, \sigma_{24}, \sigma_{34}, \tau\}$. We define $N_j = F(\sqrt[4]{\varepsilon_j})$ for $1 \leq j \leq 4$ and note that N_j is either a quartic or quadratic extension of F contained in L. Since $K_j \subseteq N_j$ for $1 \leq j \leq 4$, we see that $N_j \neq K_{re}$. Variations on the following lemma will be used extensively below.

Lemma 3. *Assume that $N_1 = K_{12}$. Then*

(a) $\sigma_0(\sqrt[4]{\varepsilon_1}) = \sigma_{12}(\sqrt[4]{\varepsilon_1}) = \sqrt[4]{\varepsilon_1}$.
(b) $\tau(\sqrt[4]{\varepsilon_1}) = \tau \circ \sigma_{12}(\sqrt[4]{\varepsilon_1}) = 1/\sqrt[4]{\varepsilon_1}$ *(note that $\tau \circ \sigma_{12} = \sigma_{34}$).*

(c) $\sigma_{13}(\sqrt[4]{\varepsilon_1}) = \sigma_{14}(\sqrt[4]{\varepsilon_1}) = -\sqrt[4]{\varepsilon_1}$.
(d) $\tau \circ \sigma_{13}(\sqrt[4]{\varepsilon_1}) = \tau \circ \sigma_{14}(\sqrt[4]{\varepsilon_1}) = -1/\sqrt[4]{\varepsilon_1}$ ($\tau \circ \sigma_{13} = \sigma_{24}$ and $\tau \circ \sigma_{14} = \sigma_{23}$).

Proof. Part (a) holds by definition. Part (b) holds by Lemma 2 and we have $\tau \circ \sigma_{12} = \sigma_{34}$ since τ and σ_{12} both restrict to the nontrivial automorphism in $\mathrm{Gal}(K_3/F)$ and $\mathrm{Gal}(K_4/F)$. For part (c), we note that $\sigma_{13}, \sigma_{14} \in \mathrm{Gal}(L/K_1)$. For $\sigma \in \mathrm{Gal}(L/K_1)$, we have $\sqrt{\varepsilon_1} = \sigma(\sqrt{\varepsilon_1}) = \sigma((\sqrt[4]{\varepsilon_1})^2) = \left[\sigma(\sqrt[4]{\varepsilon_1})\right]^2$. Therefore, $\sigma(\sqrt[4]{\varepsilon_1}) = \pm\sqrt[4]{\varepsilon_1}$. Part (d) now follows from the previous parts.

Since each N_j, $1 \le j \le 4$, can be one of exactly 4 distinct intermediate fields between F and L, there are 256 possible arrangements of these 4 fields inside L. The following proposition holds independently of how the fields N_1, N_2, N_3, and N_4 are situated within L. This proposition is just a specialization of Theorem 2 from [DST2] to the set of examples presently under discussion. These examples have been constructed in such a way that the other major theorems of [DST2] (namely, Theorems 1, 3, and 4) do not apply. Therefore, the following proposition represents the strongest result provable with respect to these examples using the methods of [DST2]. As we will see below, further *theoretical* progress towards proving the final version of Stark's Conjecture for these examples must begin by being able to prove something about how the fields N_1, N_2, N_3, and N_4 are situated within L. This means a better understanding of the functorial properties of Stark units is required, a direction not addressed in [DST2] nor elsewhere, to the present authors' knowledge.

Proposition 1. *The preliminary version of Stark's rank one abelian conjecture for the extension* L/F *holds with the element* $\varepsilon = \prod_{j=1}^{4} \sqrt[4]{\varepsilon_j} \in \mathcal{E}(L)$.

Proof. For each $\sigma \in \mathrm{Gal}(L/F)$, it suffices to prove that

$$|\sigma(\sqrt[4]{\varepsilon_j})^{(1)}| = \exp\left[-\chi_j(\sigma)L_j/4\right] \tag{3}$$

holds separately for each $j = 1, 2, 3$, or 4 irregardless of where each N_j lies in L. We consider the case $j = 1$ in detail. We already know by Lemma 3 how each $\sigma \in \mathrm{Gal}(L/F)$ acts on $\sqrt[4]{\varepsilon_1}$ when $N_1 = K_{12}$. If $N_1 = K_1$, then σ_0, σ_{12}, σ_{13}, and σ_{14} all fix $\sqrt[4]{\varepsilon_1}$, whereas σ_{23}, σ_{24}, σ_{34}, and τ all send $\sqrt[4]{\varepsilon_1}$ to $1/\sqrt[4]{\varepsilon_1}$. Note that $\chi_1(\sigma_0) = \chi_1(\sigma_{12}) = \chi_1(\sigma_{13}) = \chi_1(\sigma_{14}) = 1$ and $\chi_1(\sigma_{23}) = \chi_1(\sigma_{24}) = \chi_1(\sigma_{34}) = \chi_1(\tau) = -1$, and therefore Eq. (3) holds when $N_1 = K_1$ and $N_1 = K_{12}$. To see that (3) also holds when either $N_1 = K_{13}$ or K_{14} we record the following information which follows from the same type of arguments as in Lemma 3. If $N_1 = K_{13}$, then σ_0 and σ_{13} fix $\sqrt[4]{\varepsilon_1}$, σ_{12} and σ_{14} send $\sqrt[4]{\varepsilon_1}$ to $-\sqrt[4]{\varepsilon_1}$, σ_{24} and τ send $\sqrt[4]{\varepsilon_1}$ to $1/\sqrt[4]{\varepsilon_1}$, and σ_{23} and σ_{34} send $\sqrt[4]{\varepsilon_1}$ to $-1/\sqrt[4]{\varepsilon_1}$. If $N_1 = K_{14}$, then σ_0 and σ_{14} fix $\sqrt[4]{\varepsilon_1}$, σ_{12} and σ_{13} send $\sqrt[4]{\varepsilon_1}$ to $-\sqrt[4]{\varepsilon_1}$, σ_{23} and τ send $\sqrt[4]{\varepsilon_1}$ to $1/\sqrt[4]{\varepsilon_1}$, and σ_{24} and σ_{34} send $\sqrt[4]{\varepsilon_1}$ to $-1/\sqrt[4]{\varepsilon_1}$. This completes the proof.

The 46 interesting examples mentioned earlier in this section fall into 3 general classes:

A. The 4 fields N_1, N_2, N_3, and N_4 are all quartic extensions of F.
B. Exactly one of the N_j's is a quadratic extension of F.
C. Exactly two of the N_j's are quadratic extensions of F.

There are 29 examples in class A ($d_F = 85, 136, 204, 205, 221, 365, 408, 445,$ $485, 492, 493, 629, 680, 748, 776, 876, 901, 904, 949, 965, 984, 1037, 1105, 1157,$ $1164, 1165, 1205, 1261, 1292$). There are 10 examples in class B ($d_F = 264, 328,$ $456, 520, 584, 712, 1032, 1096, 1160, 1241$). There are 7 examples in class C ($d_F = 533, 565, 685, 1068, 1189, 1285, 1356$). Theorem 1 below summarizes the various arrangements of the N_j's that allow for part a of the abelian condition to be satisfied. All other possibilities are eliminated. We find, for example, that part a of the abelian condition can not hold if exactly three of the N_j's are quadratic extensions of F. Apparently, all four of the N_j's could be quadratic (possibility D in Theorem 1), however this possibility was not observed in the examples we computed.

For class A examples, we may renumber the fields if necessary in such a way that $N_1 = K_{12}$. For class B examples, we may assume without loss of generality that $N_4 = K_4$ is the one N_j that is a quadratic extension of F. Similarly, we assume for class C examples that $N_3 = K_3$ and $N_4 = K_4$. The following convenient shorthand notation is adopted: $(N_2, N_3, N_4) = (12, 13, 24)$, for example, means that $N_2 = K_{12}$, $N_3 = K_{13}$, and $N_4 = K_{24}$. For the α's appearing in the abelian condition, we write α_{ij} instead of $\alpha_{\sigma_{ij}}$ for $1 \leq i < j \leq 4$.

Theorem 1. *Let* $\varepsilon = \prod_{j=1}^{4} \sqrt[4]{\varepsilon_j}$. *We have* $\sigma(\varepsilon)^{(1)} > 0$ *for all* $\sigma \in \mathrm{Gal}(L/F)$ *only for the following ordered arrangements of the fields* N_1, N_2, N_3, *and* N_4 :

A. *Class A examples, assuming that* $N_1 = K_{12}$:
   $(N_2, N_3, N_4) = (12, 13, 24), (12, 23, 14), (12, 34, 34), (23, 23, 34), (23, 34, 14),$
   $(24, 13, 34), (24, 34, 24)$.
B. *Class B examples, assuming that* $N_4 = K_4$:
   $(N_1, N_2, N_3) = (12, 23, 13), (12, 24, 23), (13, 12, 23), (13, 23, 34), (14, 12, 13),$
   $(14, 24, 34)$.
C. *Class C examples, assuming that* $N_3 = K_3$ *and* $N_4 = K_4$:
   $(N_1, N_2) = (12, 12), (13, 24), (14, 23)$.
D. $N_j = K_j$ *for* $j = 1, 2, 3,$ *and* 4.

For all of these arrangements, part a of the abelian condition holds with $\alpha_{\sigma_0} = 1$, $\alpha_\tau = \varepsilon$, *and* $\alpha_{ij} = \sqrt[4]{\varepsilon_k} \cdot \sqrt[4]{\varepsilon_l}$ *for* $1 \leq i < j \leq 4$, *where in each case* $\{k, l\} = \{1, 2, 3, 4\} \setminus \{i, j\}$.

Proof. The proof follows a case by case analysis and so we just consider a specific example. For a class C example, all 8 Galois conjugates of both $\sqrt[4]{\varepsilon_3}$ and $\sqrt[4]{\varepsilon_4}$ are positive with respect to the first embedding $L \hookrightarrow L^{(1)}$ since $N_3 = K_3$ and $N_4 = K_4$ by assumption. For a given choice of N_1 and N_2, we just need to check that $\sigma(\sqrt[4]{\varepsilon_1}\sqrt[4]{\varepsilon_2})^{(1)} > 0$ for all $\sigma \in \mathrm{Gal}(L/F)$. For example, if $N_1 = K_{12}$ and $N_2 = K_{23}$, then $\sigma_{12}(\sqrt[4]{\varepsilon_1})^{(1)} = \sqrt[4]{\varepsilon_1}^{(1)} > 0$ and $\sigma_{12}(\sqrt[4]{\varepsilon_2})^{(1)} = -\sqrt[4]{\varepsilon_2}^{(1)} < 0$, which eliminates this arrangement for class C examples. By this same type of analysis, we find that part a of the abelian condition can not hold if exactly three of the N_j's are quadratic extensions of F.

Since $\varepsilon/\sigma(\varepsilon) = \alpha_\sigma^2$ by definition, clearly $\alpha_{\sigma_0} = 1$, and by Lemma 2 we have $\alpha_\tau = \varepsilon$. Assuming that $\sigma(\varepsilon)^{(1)} > 0$ for all $\sigma \in \mathrm{Gal}(L/F)$, we have $\sigma_{ij}(\varepsilon) = \sqrt[4]{\overline{\varepsilon_i}}\sqrt[4]{\overline{\varepsilon_j}}/\sqrt[4]{\overline{\varepsilon_k}}\sqrt[4]{\overline{\varepsilon_l}}$ for $1 \leq i < j \leq 4$, where in each case $\{k,l\} = \{1,2,3,4\}\backslash\{i,j\}$. This completes the proof.

With respect to the set of examples presently under discussion, Theorem 1 demonstrates that if $\varepsilon = \prod_{j=1}^4 \sqrt[4]{\overline{\varepsilon_j}}$ satisfies the intermediate version of Stark's Conjecture, then part a of the abelian condition is satisfied for ε automatically. This type of result is not true in general, namely, if $\varepsilon \in \mathcal{E}(L)$ satisfies the intermediate version of Stark's Conjecture for a Type I extension of fields L/F, then part a of the abelian condition does not necessarily hold for ε. Theorem 2 below demonstrates that even with respect to the set of examples presently under discussion, if $\varepsilon = \prod_{j=1}^4 \sqrt[4]{\overline{\varepsilon_j}}$ satisfies the intermediate version of Stark's Conjecture, then it still might not satisfy part b of the abelian condition. It is interesting to note that the first derivatives of the partial zeta-functions at $s = 0$ uniquely determine the Stark unit $\varepsilon \in \mathcal{E}(L)$ predicted to exist by the intermediate version of Stark's Conjecture for a Type I extension of fields L/F. In other words, the first derivatives of the partial zeta-functions at $s = 0$ give you *all* of the information necessary to compute the corresponding Stark unit. Because of this, it is natural to wonder if parts a and b of the abelian condition can somehow be formulated directly in terms of the underlying properties of the partial zeta-functions (or L-functions).

Assume that $\sigma(\varepsilon)^{(1)} > 0$ for all $\sigma \in \mathrm{Gal}(L/F)$. For a given pair $\{i,j\}$ with $1 \leq i < j \leq 4$, recall that the Stark unit ε_{ij} associated to the extension K_{ij}/F is equal to $\sqrt{\overline{\varepsilon_i}}\sqrt{\overline{\varepsilon_j}}$ and therefore $\varepsilon_{ij} = \alpha_{kl}^2$, where $\{k,l\} = \{1,2,3,4\} \setminus \{i,j\}$. We also verify that the relative norm from L to K_{ij} of ε is equal to ε_{ij}. Therefore, if $\varepsilon = \beta^2$ for some $\beta \in L$, then $\varepsilon_{ij} = N_{L/K_{ij}}(\beta)^2$ and so $\alpha_{kl} \in K_{ij}$. This implies that if there exists an α_{kl} for some k,l satisfying $1 \leq k < l \leq 4$ *not* fixed by σ_{ij}, then ε is not a square in L. In the computations, we find an α_{kl} that is not fixed by *any* nontrivial automorphism $\sigma \in \mathrm{Gal}(L/F)$.

Theorem 2. *Let* $\varepsilon = \prod_{j=1}^4 \sqrt[4]{\overline{\varepsilon_j}}$. *Among the ordered arrangements of the fields* N_1, N_2, N_3, *and* N_4 *listed in Theorem 1, only the following also satisfy part b of the abelian condition* :

A. *Class* A *examples, assuming that* $N_1 = K_{12}$:
   $(N_2, N_3, N_4) = (12, 34, 34),\ (23, 34, 14),\ (24, 13, 34)$.
B. *Class* B *examples, assuming that* $N_4 = K_4$:
   $(N_1, N_2, N_3) = (12, 24, 23),\ (13, 23, 34),\ (14, 12, 13),\ (14, 24, 34)$.
C. *Class* C *examples, assuming that* $N_3 = K_3$ *and* $N_4 = K_4$:
   $(N_1, N_2) = (12, 12)$.
D. $N_j = K_j$ *for* $j = 1, 2, 3$, *and* 4.

The Stark unit ε *is a square in* L *for class* A *examples when* $(N_2, N_3, N_4) = (12, 34, 34)$, *class* B *examples when* $(N_1, N_2, N_3) = (14, 24, 34)$, *and in case* D. *Otherwise,* ε *is not a square in* L.

Proof. Part b of the abelian condition clearly holds if one of the two automorphisms is the trivial automorphism. We now show that for any i, j satisfying $1 \leq i < j \leq 4$, the relation $\alpha_\tau \tau(\alpha_{ij}) = \alpha_{ij} \sigma_{ij}(\alpha_\tau)$ holds for all ordered arrangements listed in Theorem 1. The left hand side is equal to $\varepsilon / \sqrt[4]{\varepsilon_k} \sqrt[4]{\varepsilon_l} = \sqrt[4]{\varepsilon_i} \sqrt[4]{\varepsilon_j}$. Examining the last piece in the proof of Theorem 1, we see that the right hand side is also equal to $\sqrt[4]{\varepsilon_i} \sqrt[4]{\varepsilon_j}$. The ordered arrangements from Theorem 1 that fail to satisfy part b of the abelian condition all fail to satisfy the relation

$$\alpha_{12} \sigma_{12}(\alpha_{13}) = \alpha_{13} \sigma_{13}(\alpha_{12}). \tag{4}$$

For example, for the class B situation with $(N_1, N_2, N_3) = (12, 23, 13)$ and $N_4 = K_4$, the left hand side of (4) is equal to $\sqrt[4]{\varepsilon_3} \sqrt[4]{\varepsilon_4} \cdot (-\sqrt[4]{\varepsilon_2})/\sqrt[4]{\varepsilon_4}$. The right hand side, however, is equal to $\sqrt[4]{\varepsilon_2} \sqrt[4]{\varepsilon_4} \cdot (\sqrt[4]{\varepsilon_3})/\sqrt[4]{\varepsilon_4}$. The ordered arrangements from Theorem 1 for which relation (4) holds satisfy part b of the abelian condition completely since τ, σ_{12}, and σ_{13} generate the Galois group $\mathrm{Gal}(L/F)$ (see p. 83 of [Ta2]).

To see, for example, that ε is not a square in L when $(N_1, N_2, N_3, N_4) = (12, 23, 34, 14)$, we simply verify that α_{23} is not fixed by any nontrivial automorphism $\sigma \in \mathrm{Gal}(L/F)$. This completes the proof. $\qed$

The main theorem of this section is

Theorem 3. *For the unit* $\varepsilon = \prod_{j=1}^4 \sqrt[4]{\varepsilon_j} \in \mathcal{E}(L)$, *the field* $L(\sqrt{\varepsilon})$ *is an abelian extension of* F *for all 46 of the examples mentioned earlier in this section.*

Proof. Since the proof is computational, we say a little more about how the computations were carried out over F. From the four nonzero values $L_j = L'_{\tilde{\mathfrak{m}}}(0, \chi_j)$, $1 \leq j \leq 4$, we compute a polynomial $f(x) \in \mathcal{O}_F[x]$ of degree 8 assuming the intermediate version of Stark's Conjecture as described in Section 1. We then need to verify that any given root ρ of $f(x)$ generates the field L corresponding to X by class field theory over F. Actually, Stark's Conjecture predicts that $L = \mathbb{Q}(\rho)$ (see p. 66 of [St2]) and we use PARI to compute the basic information associated to $\mathbb{Q}(\rho)$. We then verify that the polynomials $f_j(x)$ satisfied by $\sqrt{\varepsilon_j}$ for $1 \leq j \leq 4$ each have a linear factor in $\mathbb{Q}(\rho)[x]$. This not only gives us elements corresponding to the $\sqrt{\varepsilon_j}$'s in the field $\mathbb{Q}(\rho)$ but also proves that $\mathbb{Q}(\rho) = L$ since the octic field extension corresponding to X by class field theory is generated over F by the four elements $\sqrt{\varepsilon_1}$, $\sqrt{\varepsilon_2}$, $\sqrt{\varepsilon_3}$, and $\sqrt{\varepsilon_4}$. We choose the distinguished first embedding of $L = \mathbb{Q}(\rho)$ into $\mathbb{R}$ in such a way that $\rho^{(1)} = \prod_{j=1}^4 \exp(-L_j/4)$. We then compute 4 elements $\beta_1, \beta_2, \beta_3, \beta_4 \in L$ that are positive with respect to the first embedding and whose squares equal $\sqrt{\varepsilon_1}$, $\sqrt{\varepsilon_2}$, $\sqrt{\varepsilon_3}$, and $\sqrt{\varepsilon_4}$, respectively. A verification is then made that the product of the 4 β's in L is indeed equal to ρ. All that remains to finally verify the abelian condition is that the four fields $N_j = F(\beta_j)$, $1 \leq j \leq 4$, are arranged within L as in Theorem 2.

There are 9 class A examples ($d_F = 205, 221, 445, 876, 901, 904, 1164, 1205, 1292$) and 5 class B examples ($d_F = 264, 456, 584, 712, 1032$) such that ε is a square in L. For the other 32 examples, ε is not a square in L and the abelian condition is nontrivial and not known to hold by the theorems in [DST2].

Acknowledgements

We would like to thank an anonymous referee for several comments that allowed us to considerably improve the clarity of our presentation.

References

[Co]   Cohen, H.: Advanced Topics in Computational Number Theory. Springer, New York (2000)

[DH]   Dummit, D.S., Hayes, D.R.: Checking the p-adic Stark Conjecture when p is Archimedean. In: Cohen, H. (ed.) ANTS 1996. LNCS, vol. 1122, pp. 91–97. Springer, Heidelberg (1996)

[DST1]   Dummit, D.S., Sands, J.W., Tangedal, B.A.: Computing Stark units for totally real cubic fields. Math. Comp. 66, 1239–1267 (1997)

[DST2]   Dummit, D.S., Sands, J.W., Tangedal, B.A.: Stark's conjecture in multiquadratic extensions, revisited. J. Théor. Nombres Bordeaux 15, 83–97 (2003)

[DT]   Dummit, D.S., Tangedal, B.A.: Computing the lead term of an abelian L-function. In: Buhler, J.P. (ed.) ANTS 1998. LNCS, vol. 1423, pp. 400–411. Springer, Heidelberg (1998)

[DTvW]   Dummit, D.S., Tangedal, B.A., van Wamelen, P.B.: Stark's conjecture over complex cubic number fields. Math. Comp. 73, 1525–1546 (2004)

[GP]   Batut, C., Belabas, K., Bernardi, D., Cohen, H., Olivier, M.: User's guide to PARI/GP version 2.1.3 (2000)

[Ha]   Hasse, H.: Vorlesungen über Klassenkörpertheorie. Physica-Verlag, Würzburg (1967)

[Ja]   Janusz, G.J.: Algebraic Number Fields. Academic Press, New York (1973)

[Na]   Narkiewicz, W.: Elementary and Analytic Theory of Algebraic Numbers, 3rd edn. Springer, New York (2004)

[St1]   Stark, H.M.: Class fields for real quadratic fields and L-series at 1. In: Fröhlich, A. (ed.) Algebraic Number Fields, pp. 355–375. Academic Press, London (1977)

[St2]   Stark, H.M.: L-functions at $s = 1$. III. Totally real fields and Hilbert's Twelfth Problem. Advances in Math. 22, 64–84 (1976)

[St3]   Stark, H.M.: L-functions at $s = 1$. IV. First derivatives at $s = 0$. Advances in Math. 35, 197–235 (1980)

[Ta1]   Tate, J.: On Stark's conjectures on the behavior of $L(s, \chi)$ at $s = 0$. J. Fac. Sci. Univ. Tokyo Sect. IA Math. 28(3), 963–978 (1981)

[Ta2]   Tate, J.: Les Conjectures de Stark sur les Fonctions L d'Artin en $s = 0$, Notes d'un cours à Orsay rédigées par Dominique Bernardi et Norbert Schappacher, Birkhäuser, Boston (1984)

Enumeration of Totally Real Number Fields
of Bounded Root Discriminant

John Voight

Department of Mathematics and Statistics,
University of Vermont, Burlington, VT 05401
jvoight@gmail.com

Abstract. We enumerate all totally real number fields F with root discriminant $\delta_F \leq 14$. There are 1229 such fields, each with degree $[F : \mathbb{Q}] \leq 9$.

In this article, we consider the following problem.

Problem 1. Given $B \in \mathbb{R}_{>0}$, enumerate the set $NF(B)$ of totally real number fields F with root discriminant $\delta_F \leq B$, up to isomorphism.

To solve Problem 1, for each $n \in \mathbb{Z}_{>0}$ we enumerate the set

$$NF(n, B) = \{F \in NF(B) : [F : \mathbb{Q}] = n\}$$

which is finite (a result originally due to Minkowski). If F is a totally real field of degree $n = [F : \mathbb{Q}]$, then by the Odlyzko bounds [27], we have $\delta_F \geq 4\pi e^{1+\gamma} - O(n^{-2/3})$ where γ is Euler's constant; thus for $B < 4\pi e^{1+\gamma} < 60.840$, we have $NF(n, B) = \emptyset$ for n sufficiently large and so the set $NF(B)$ is finite. Assuming the generalized Riemann hypothesis (GRH), we have the improvement $\delta_F \geq 8\pi e^{\gamma+\pi/2} - O(\log^{-2} n)$ and hence $NF(B)$ is conjecturally finite for all $B < 8\pi e^{\gamma+\pi/2} < 215.333$. On the other hand, for B sufficiently large, the set $NF(B)$ is infinite: Martin [23] has constructed an infinite tower of totally real fields with root discriminant $\delta \approx 913.493$ (a long-standing previous record was held by Martinet [25] with $\delta \approx 1058.56$). The value

$$\liminf_{n\to\infty} \min\{\delta_F : F \in NF(n, B)\}$$

is presently unknown. If B is such that $\#NF(B) = \infty$, then to solve Problem 1 we enumerate the set $NF(B) = \bigcup_n NF(n, B)$ by increasing degree.

Our restriction to the case of totally real fields is not necessary: one may place alternative constraints on the signature of the fields F under consideration (or even analogous p-adic conditions). However, we believe that Problem 1 remains one of particular interest. First of all, it is a natural boundary case: by comparison, Hajir-Maire [14,15] have constructed an unramified tower of totally complex number fields with root discriminant ≈ 82.100, which comes within a factor 2 of the GRH-conditional Odlyzko bound of $8\pi e^{\gamma} \approx 44.763$. Secondly, in studying

A.J. van der Poorten and A. Stein (Eds.): ANTS-VIII 2008, LNCS 5011, pp. 268–281, 2008.

certain problems in arithmetic geometry and number theory—for example, in the enumeration of arithmetic Fuchsian groups [21] and the computational investigation of the Stark conjecture and its generalizations—provably complete and extensive tables of totally real fields are useful, if not outright essential. Indeed, existing strategies for finding towers with small root discriminant as above often start by finding a good candidate base field selected from existing tables.

The main result of this note is the following theorem, which solves Problem 1 for $\delta = 14$.

Theorem 2. *We have* $\#NF(14) = 1229$.

The complete list of these fields is available online [35]; the octic and nonic fields ($n = 8, 9$) are recorded in Tables 4–5 in §4, and there are no dectic fields ($NF(14, 10) = \emptyset$). For a comparison of this theorem with existing results, see §1.2.

The note is organized as follows. In §1, we set up the notation and background. In §2, we describe the computation of primitive fields $F \in NF(14)$; we compare well-known methods and provide some improvements. In §3, we discuss the extension of these ideas to imprimitive fields, and we report timing details on the computation. Finally, in §4 we tabulate the fields F.

The author wishes to thank: Jürgen Klüners, Noam Elkies, Claus Fieker, Kiran Kedlaya, Gunter Malle, and David Dummit for useful discussions; William Stein, Robert Bradshaw, Craig Citro, Yi Qiang, and the rest of the Sage development team for computational support (NSF Grant No. 0555776); and Larry Kost and Helen Read for their technical assistance.

1   Background

1.1   Initial Bounds

Let F denote a totally real field of degree $n = [F : \mathbb{Q}]$ with discriminant d_F and root discriminant $\delta_F = d_F^{1/n}$. By the unconditional Odlyzko bounds [27] (see also Martinet [24]), if $n \geq 11$ then $\delta_F > 14.083$, thus if $F \in NF(14)$ then $n \leq 10$.

The lower bounds for δ_F in the remaining degrees are summarized in Table 1: for each degree $2 \leq n \leq 10$, we list the unconditional Odlyzko bound $B_O = B_O(n)$, the GRH-conditional Odlyzko bound (for comparison only, as computed by Cohen-Diaz y Diaz-Olivier [7]), and the bound $\delta_F \leq \Delta$ that we employ.

Table 1. Degree and Root Discriminant Bounds

n	2	3	4	5	6	7	8	9	10
B_O	> 2.223	3.610	5.067	6.523	7.941	9.301	10.596	11.823	12.985
B_O (GRH)	> 2.227	3.633	5.127	6.644	8.148	9.617	11.042	12.418	13.736
Δ	30	25	20	17	16	15.5	15	14.5	14

1.2 Previous Work

There has been an extensive amount of work done on the problem of enumerating number fields—we refer to [18] for a discussion and bibliography.

1. The KASH and PARI groups [16] have computed tables of number fields of all signatures with degrees ≤ 7: in degrees $6, 7$, they enumerate totally real fields up to discriminants $10^7, 15 \cdot 10^7$, respectively (corresponding to root discriminants $14.67, 14.71$, respectively).
2. Malle [22] has computed all totally real primitive number fields of discriminant $d_F \leq 10^9$ (giving root discriminants $31.6, 19.3, 13.3, 10$ for degrees $6, 7, 8, 9$). This was reported to take several years of CPU-time on a SUN workstation.
3. The database by Klüners-Malle [17] contains polynomials for all transitive groups up to degree 15 (including possible combinations of signature and Galois group); up to degree 7, the fields with minimal (absolute) discriminant with given Galois group and signature have been included.
4. Roblot [30] constructs abelian extensions of totally real fields of degrees 4 to 48 (following Cohen-Diaz y Diaz-Olivier [6]) with small root discriminant.

The first two of these allow us only to determine $NF(10)$ (if we also separately compute the imprimitive fields); the latter two, though very valuable for certain applications, are in a different spirit than our approach. Therefore our theorem substantially extends the complete list of fields in degrees 7–9.

2 Enumeration of Totally Real Fields

2.1 General Methods

The general method for enumerating number fields is well-known (see Cohen [4, §9.3]). We define the Minkowski norm on a number field F by $T_2(\alpha) = \sum_{i=1}^{n} |\alpha_i|^2$ for $\alpha \in F$, where $\alpha_1, \alpha_2, \ldots, \alpha_n$ are the conjugates of α in $\mathbb{C}$. The norm T_2 gives $\mathbb{Z}_F$ the structure of a lattice of rank n. In this lattice, the element 1 is a shortest vector, and an application of the geometry of numbers to the quotient lattice $\mathbb{Z}_F/\mathbb{Z}$ yields the following result.

Lemma 3 (Hunter). *There exists $\alpha \in \mathbb{Z}_F \setminus \mathbb{Z}$ such that $0 \leq \mathrm{Tr}(\alpha) \leq n/2$ and*

$$T_2(\alpha) \leq \frac{\mathrm{Tr}(\alpha)^2}{n} + \gamma_{n-1} \left(\frac{|d_F|}{n} \right)^{1/(n-1)}$$

where γ_{n-1} is the $(n-1)$th Hermite constant.

Remark 4. The values of the Hermite constant are known for $n \leq 8$ (given by the lattices $A_1, A_2, A_3, D_4, D_5, E_6, E_7, E_8$): we have $\gamma_n^n = 1, 4/3, 2, 4, 8, 64/3, 64, 256$ (see Conway and Sloane [9]) for $n = 1, \ldots, 8$; the best known upper bounds for $n = 9, 10$ are given by Cohn and Elkies [8].

Therefore, if we want to enumerate all number fields F of degree n with $|d_F| \leq B$, an application of Lemma 3 yields $\alpha \in \mathbb{Z}_F \setminus \mathbb{Z}$ such that $T_2(\alpha) \leq C$ for some $C \in \mathbb{R}_{>0}$ depending only on n, B. We thus obtain bounds on the power sums

$$|S_k(\alpha)| = \left| \sum_{i=1}^{n} \alpha_i^k \right| \leq T_k(\alpha) = \sum_{i=1}^{n} |\alpha_i|^k \leq nC^{k/2},$$

and hence bounds on the coefficients $a_i \in \mathbb{Z}$ of the characteristic polynomial

$$f(x) = \prod_{i=1}^{n}(x - \alpha_i) = x^n + a_{n-1}x^{n-1} + \cdots + a_0$$

of α by Newton's relations:

$$S_k + \sum_{i=1}^{k-1} a_{n-1}S_{k-i} + ka_{n-k} = 0. \tag{1}$$

This then yields a finite set $NS(n, B)$ of polynomials $f(x) \in \mathbb{Z}[x]$ such that every F is represented as $\mathbb{Q}[x]/(f(x))$ for some $f(x) \in NS(n, B)$, and in principle each $f(x)$ can then be checked individually. We note that it is possible that α as given by Hunter's theorem may only generate a subfield $\mathbb{Q} \subset \mathbb{Q}(\alpha) \subsetneq F$ if F is imprimitive: for a treatment of this case, see §3.

The size of the set $NS(n, B)$ is $O(B^{n(n+2)/4})$ (see Cohen [4, §9.4]), and the exponential factor in n makes this direct method impractical for large n or B. Note, however, that it is sharp for $n = 2$: we have $NF(2, B) \sim (6/\pi^2)B^2$ (as $B \to \infty$), and indeed, in this case one can reduce to simply listing squarefree integers. For other small values of n, better algorithms are known: following Davenport-Heilbronn, Belabas [2] has given an algorithm for cubic fields; Cohen-Diaz y Diaz-Olivier [7] use Kummer theory for quartic fields; and by work of Bhargava [3], in principle one should similarly be able to treat the case of quintic fields. No known method improves on this asymptotic complexity for general n, though some possible progress has been made by Ellenberg-Venkatesh [12].

2.2   Improved Methods for Totally Real Fields

We now restrict to the case that F is totally real. Several methods can then be employed to improve the bounds given above—although we only improve on the implied constant in the size of the set $NS(n, B)$ of examined polynomials, these improvements are essential for practical computations.

Basic Bounds. From Lemma 3, we have $0 \leq a_{n-1} = -\operatorname{Tr}(\alpha) \leq \lfloor n/2 \rfloor$ and

$$a_{n-2} = \frac{1}{2}a_{n-1}^2 - \frac{1}{2}T_2(\alpha) \geq \frac{1}{2}\left(1 - \frac{1}{n}\right)a_{n-1}^2 - \frac{\gamma_{n-1}}{2}\left(\frac{B}{n}\right)^{1/(n-1)}.$$

For an upper bound on a_{n-2}, we apply the following result.

Lemma 5 (Smyth [32]). *If γ is a totally positive algebraic integer, then*

$$\operatorname{Tr}(\gamma) > 1.7719[\mathbb{Q}(\gamma) : \mathbb{Q}]$$

unless γ is a root of one of the following polynomials:

$$x-1, x^2-3x+1, x^3-5x^2+6x-1, x^4-7x^3+13x^2-7x+1, x^4-7x^3+14x^2-8x+1.$$

Remark 6. The best known bound of the above sort is due to Aguirre-Bilbao-Peral [1], who give $\operatorname{Tr}(\gamma) > 1.780022[\mathbb{Q}(\gamma) : \mathbb{Q}]$ with 14 possible explicit exceptions. For our purposes (and for simplicity), the result of Smyth will suffice.

Excluding these finitely many cases, we apply Lemma 5 to the totally positive algebraic integer α^2, using the fact that $T_2(\alpha) = \operatorname{Tr}(\alpha^2)$, to obtain the upper bound $a_{n-2} < a_{n-1}^2/2 - 0.88595n$.

Rolle's Theorem. Now, given values $a_{n-1}, a_{n-2}, \ldots, a_{n-k}$ for the coefficients of $f(x)$ for some $k \geq 2$, we deduce bounds for a_{n-k-1} using Rolle's theorem—this elementary idea can already be found in Takeuchi [33] and Klüners-Malle [18, §3.1]. Let

$$f_i(x) = \frac{f^{(n-i)}(x)}{(n-i)!} = g_i(x) + a_{n-i}$$

for $i = 0, \ldots, n$. Consider first the case $k = 2$. Then

$$g_3(x) = \frac{n(n-1)(n-2)}{6}x^3 + \frac{(n-1)(n-2)}{2}a_{n-1}x^2 + (n-2)a_{n-2}x.$$

Let $\beta_1 < \beta_2$ denote the roots of $f_2(x)$. Then by Rolle's theorem,

$$f_3(\beta_1) = g_3(\beta_1) + a_{n-3} > 0 \quad \text{and} \quad f_3(\beta_2) = g_3(\beta_2) + a_{n-3} < 0$$

hence $-g_3(\beta_1) < a_{n-3} < -g_3(\beta_2)$. In a similar way, if $\beta_1^{(k)} < \cdots < \beta_k^{(k)}$ denote the roots of $f_k(x)$, then we find that

$$-\min_{\substack{1 \leq i \leq k \\ i \not\equiv k \ (2)}} g_{k+1}(\beta_i^{(k)}) < a_{n-k-1} < -\max_{\substack{1 \leq i \leq k \\ i \equiv k \ (2)}} g_{k+1}(\beta_i^{(k)}).$$

Lagrange Multipliers. We can obtain further bounds as follows. We note that if the roots of f are bounded below by $\beta_0^{(k)}$ (resp. bounded above by $\beta_{k+1}^{(k)}$), then

$$f_k(\beta_0^{(k)}) = g_k(\beta_0^{(k)}) + a_{n-k} > 0$$

(with a similar inequality for $\beta_{k+1}^{(k)}$), and these combine with the above to yield

$$-\min_{\substack{0 \leq i \leq k+1 \\ i \not\equiv k \ (2)}} g_{k+1}(\beta_i^{(k)}) < a_{n-k-1} < -\max_{\substack{0 \leq i \leq k+1 \\ i \equiv k \ (2)}} g_{k+1}(\beta_i^{(k)}). \tag{2}$$

We can compute $\beta_0^{(k)}, \beta_{k+1}^{(k)}$ by the method of Lagrange multipliers, which were first introduced in this general context by Pohst [29] (see Remark 7). The values $a_{n-1}, \ldots, a_{n-k} \in \mathbb{Z}$ determine the power sums s_i for $i = 1, \ldots, k$ by Newton's relations (1). Now the set of all $x = (x_i) \in \mathbb{R}^n$ such that $S_i(x) = s_i$ is closed and bounded, and therefore by symmetry the minimum (resp. maximum) value of the function x_n on this set yields the bound $\beta_0^{(k)}$ (resp. $\beta_{k+1}^{(k)}$). By the method of Lagrange multipliers, we find easily that if $x \in \mathbb{R}^n$ yields such an extremum, then there are at most $k-1$ distinct values among $x_1, \ldots, x_{n-1}$, from which we obtain a finite set of possibilities for the extremum x.

For example, in the case $k = 2$, the extrema are obtained from the equations

$$(n-1)x_1 + x_n = s_1 = -a_{n-1} \quad \text{and} \quad (n-1)x_1^2 + x_n^2 = s_2 = a_{n-1}^2 - 2a_{n-2}$$

which yields simply

$$\beta_0^{(2)}, \beta_3^{(2)} = \frac{1}{n}\left(-a_{n-1} \pm (n-1)\sqrt{a_{n-1}^2 - 2\left(1 + \frac{1}{n-1}\right)a_{n-2}}\right).$$

(It is easy to show that this always improves upon the trivial bounds used by Takeuchi [33].) For $k = 3$, for each partition of $n-1$ into 2 parts, one obtains a system of equations which via elimination theory yield a (somewhat lengthy but explicitly given) degree 6 equation for x_n. For $k \geq 4$, we can continue in a similar way but we instead solve the system numerically, e.g., using the method of homotopy continuation as implemented by the package PHCpack developed by Verschelde [34]; in practice, we do not significantly improve on these bounds whenever $k \geq 5$, and even for $k = 5$, if n is small then it often is more expensive to compute the improved bounds than to simply set $\beta_0^{(k)} = \beta_0^{(k-1)}$ and $\beta_{k+1}^{(k)} = \beta_k^{(k-1)}$.

Remark 7. Pohst's original use of Lagrange multipliers, which applies to number fields of arbitrary signature, instead sought the extrema of the power sum S_{k+1} to bound the coefficient a_{n-k-1}. The bounds given by Rolle's theorem for totally real fields are not only easier to compute (especially in higher degree) but in most cases turn out to be strictly stronger. We similarly find that many other bounds typically employed in this situation (e.g., those arising from the positive definiteness of T_2 on $\mathbb{Z}[\alpha]$) are also always weaker.

2.3 Algorithmic Details

Our algorithm to solve Problem 1 then runs as follows. We first apply the basic bounds from §2.2 to specify finitely many values of a_{n-1}, a_{n-2}. For each such pair, we use Rolle's theorem and the method of Lagrange multipliers to bound each of the coefficients inductively. Note that if $k \geq 3$ is odd and $a_{n-1} = a_{n-3} = \cdots = a_{n-(k-2)} = 0$, then replacing α by $-\alpha$ we may assume that $a_{n-k} \geq 0$.

For each polynomial $f \in NS(n, B)$ that emerges from these bounds, we test it to see if it corresponds to a field $F \in NF(n, B)$. We treat each of these latter two tasks in turn.

Calculation of Real Roots. In the computation of the bounds (2), we use Newton's method to iteratively compute approximations to the roots $\beta_i^{(k)}$, using the fact that the roots of a polynomial are interlaced with those of its derivative, i.e. $\beta_{i-1}^{(k-1)} < \beta_i^{(k)} < \beta_i^{(k-1)}$ for $i = 1, \ldots, k$. Note that by Rolle's theorem, we will either find a simple root in this open interval or we will converge to one of the endpoints, say $\beta_i^{(k)} = \beta_i^{(k-1)}$, and then necessarily $\beta_i^{(k)} = \beta_i^{(k-1)}$ as well, which implies that $f_k(x)$ is not squarefree and hence the entire coefficient range may be discarded immediately. It is therefore possible to very quickly compute an approximate root which differs from the actual root $\beta_i^{(k+1)}$ by at most some fixed $\epsilon > 0$. We choose ϵ small enough to give a reasonable approximation but not so small as to waste time in Newton's method (say, $\epsilon = 10^{-4}$). We deal with the possibility of precision loss by bounding the value $g_{k+1}(\beta_i^{(k)})$ in (2) using elementary calculus; we leave the details to the reader.

Testing Polynomials. For each $f \in NS(n, B)$, we test each of the following in turn.

1. We first employ an "easy irreducibility test": We rule out polynomials f divisible by any of the factors: $x, x \pm 1, x \pm 2, x^2 \pm x - 1, x^2 - 2$. In the latter three cases, we first evaluate the polynomial at an approximation to the values $(1 \pm \sqrt{5})/2, \sqrt{2}$, respectively, and then evaluate f at these roots using exact arithmetic. (Some benefit is gained by hard coding this latter evaluation.)

2. We then compute the discriminant $d = \text{disc}(f)$. If $d \leq 0$, then f is not a real separable polynomial, so we discard f.

3. If $F = \mathbb{Q}[\alpha] = \mathbb{Q}[x]/(f(x)) \in NF(n, B)$, then for some $a \in \mathbb{Z}$ we have $B_O(n)^n < d_F = d/a^2 < B^n$ where B_O is the Odlyzko bound (see §1). Therefore using trial division we can quickly determine if there exists such an $a^2 \mid d$; if not, then we discard f.

4. Next, we check if f is irreducible, and discard f otherwise.

5. By the preceding two steps, an a-maximal order containing $\mathbb{Z}[\alpha]$ is in fact the maximal order $\mathbb{Z}_F$ of the field F. If $\text{disc}(\mathbb{Z}_F) = d_F > B$, we discard f.

6. Apply the POLRED algorithm of Cohen-Diaz y Diaz [5]: embed $\mathbb{Z}_F \subset \mathbb{R}^n$ by Minkowski (as in §1.1) and use LLL-reduction [20] to compute a small element $\alpha_{\text{red}} \in \mathbb{Z}_F$ such that $\mathbb{Q}(\alpha) = \mathbb{Q}(\alpha_{\text{red}}) = F$. Add the minimal polynomial $f_{\text{red}}(x)$ of α_{red} to the list $NF(n, B)$ (along with the discriminant d_F), if it does not already appear.

We expect that almost all isomorphic fields will be identified in Step 6 by computing a reduced polynomial. For reasons of efficiency, we wait until the space $NS(n, B)$ has been exhausted to do a final comparison with each pair of polynomials with the same discriminant to see if they are isomorphic. Finally, we add the exceptional fields coming from Lemma 5, if relevant.

Remark 8. Although Step 1 is seemingly trivial, it rules out a surprisingly significant number of polynomials f—indeed, nearly all reducible polynomials are discarded by this step in higher degrees. Indeed, if $T_2(f) = \sum_i \alpha_i^2$ (where α_i are the roots of f) is small compared to $\deg(f) = n$, then f is likely to be reducible and moreover divisible by a polynomial g with $T_2(g)$ also small. It would be interesting to give a precise statement which explains this phenomenon.

2.4   Implementation Details

For the implementation of our algorithm, we use the computer algebra system Sage [31], which utilizes PARI [28] for Steps 4–6 above. Since speed was of the absolute essence, we found that the use of Cython (developed by Stein and Bradshaw) allowed us to develop a carefully optimized and low-level implementation of the bounds coming from Rolle's theorem and Lagrange multiplier method 2. We used the DSage package (due to Qiang) which allowed for the distribution of the compution to many machines; as a result, our computational time comes from a variety of processors (Opteron 1.8GHz, Athlon Dual Core 2.0GHz, and Celeron 2.53GHz), including a cluster of 30 machines at the University of Vermont.

   In low and intermediate degrees, where we expect comparatively many fields, we find that the running time is dominated by the computation of the maximal order (Step 5), followed by the check for irreducibility (Step 4); this explains the ordering of the steps as above. By contrast, in higher degrees, where we expect few fields but must search in an exponentially large space, most of the time is spent in the calculation of real roots and in Step 1. Further timing details can be found in Table 2.

Table 2. Timing data

n	2	3	4	5	6	7	8	9	10
$\Delta(n)$	30	25	20	17	16	15.5	15	14.5	14
f	443	4922	57721	244600	3242209	1.7×10^7	1.2×10^8	9.5×10^8	2.5×10^9
Irred f	418	2523	27234	157613	2710965	1.6×10^7	1.1×10^8	9.0×10^8	2.5×10^9
$f, d_F \leq B$	418	1573	5665	4497	1288	4839	3016	506	0
F	273	630	1273	674	802	301	164	15	0
Total time	0.2s	2.2s	26.8s	1m25s	17m3s	2h59m	1d4.5h	17d21h	193d
Imprim f	0	0	7059	0	62532	0	239404	15658	945866
Imprim F	0	0	702	0	420	0	100	6	0
Time	-	-	4m22s	-	8m38s	-	1h56m	16m53s	11h27m
Total fields	273	630	1578	674	827	301	164	15	0

3   Imprimitive Fields

In this section, we extend the ideas of the previous section to imprimitive fields F, i.e. those fields F containing a nontrivial subfield. Suppose that F is an extension of E with $[F : E] = m$ and $[E : \mathbb{Q}] = d$. Since $\delta_F \geq \delta_E$, if $F \in NF(B)$ then $E \in NF(B)$ as well, and thus we proceed by induction on E. For each such subfield E, we proceed in an analogous fashion. We let

$$f(x) = x^m + a_{m-1}x^{m-1} + \cdots + a_1 x + a_0$$

be the minimal polynomial of an element $\alpha \in \mathbb{Z}_F$ with $F = E(\alpha)$ and $a_i \in \mathbb{Z}_E$.

3.1   Extension of Bounds

Basic Bounds. We begin with a relative version of Hunter's theorem. We denote by E_∞ the set of infinite places of E.

Lemma 9 (Martinet [26]). *There exists $\alpha \in \mathbb{Z}_F \setminus \mathbb{Z}_E$ such that*

$$T_2(\alpha) \leq \frac{1}{m} \sum_{\sigma \in E_\infty} \left|\sigma\left(\mathrm{Tr}_{F/E}\, \alpha\right)\right|^2 + \gamma_{n-d}\left(\frac{|d_F|}{m^d |d_E|}\right)^{1/(n-d)}. \qquad (3)$$

The inequality of Lemma 9 remains true for any element of the set $\mu_E \alpha + \mathbb{Z}_E$, where μ_E denotes the roots of unity in E. This allows us to choose $\mathrm{Tr}_{F/E}\, \alpha = -a_{m-1}$ among any choice of representatives from $\mathbb{Z}_E/m\mathbb{Z}_E$ (up to a root of unity); we choose the value of a_{m-1} which minimizes

$$\sum_{\sigma \in E_\infty} \left|\sigma\left(\mathrm{Tr}_{F/E}\, \alpha\right)\right|^2 = \sum_{\sigma \in E_\infty} \sigma(a_{m-1})^2,$$

which is a positive definite quadratic form on $\mathbb{Z}_E$; such a value can be found easily using the LLL-algorithm.

Now suppose that F is totally real. Then $\sum_{\sigma \in E_\infty} |\sigma(a_{m-1})|^2 = \mathrm{Tr}_{E/\mathbb{Q}}\, a_{m-1}^2$, and we have 2^d or $\lceil m^d/2 \rceil$ possibilities for a_{m-1}, according as $m = 2$ or otherwise. For each value of a_{m-1}, we have $T_2(\alpha) \in \mathbb{Z}$ bounded from above by Lemma 9 and from below by Lemma 5 since $\mathrm{Tr}_{E/\mathbb{Q}}(\alpha^2) = T_2(\alpha) > 1.7719n$. If we denote $\mathrm{Tr}_{F/E}\, \alpha^2 = t_2$, then by Newton's relations, we have $t_2 = a_{m-1}^2 - 2a_{m-2}$, and hence $\mathrm{Tr}_{E/\mathbb{Q}}\, t_2 = T_2(\alpha)$ and $t_2 \equiv a_{m-1}^2 \pmod 2$. In particular, $t_2 \in \mathbb{Z}_E$ is totally positive and has bounded trace, leaving only finitely many possibilities: indeed, if we embed $\mathbb{Z}_E \hookrightarrow \mathbb{R}^d$ by Minkowski, these inequalities define a parallelopiped in the positive orthant.

Lattice Points in Boxes. One option to enumerate the possible values of t_2 is to enumerate all lattice points in a sphere of radius given by (3) using the

Fincke-Pohst algorithm [13]. However, one ends up enumerating far more than what one needs in this fashion, and so we look to do better. The problem we need to solve is the following.

Problem 10. Given a lattice $L \subset \mathbb{R}^d$ of rank d and a convex polytope P of finite volume, enumerate the set $P \cap L$.

Here we must allow the lattice L to be represented numerically; to avoid issues of precision loss, one supposes without loss of generality that $\partial P \cap L = \emptyset$.

There exists a vast literature on the classical problem of the enumeration of integer lattice points in rational convex polytopes (see e.g., De Loera [10]), as well as several implementations [11,19]. (In many cases, these authors are concerned primarily with simply counting the number of lattice points, but their methods equally allow their enumeration.)

In order to take advantage of these methods to solve Problem 10, we compute an LLL-reduced basis $\gamma = \gamma_1, \ldots, \gamma_d$ of L, and we perform the change of variables $\phi : \mathbb{R}^d \to \mathbb{R}^d$ which maps $\gamma_i \mapsto e_i$ where e_i is the ith coordinate vector. The image $\phi(P)$ is again a convex polytope. We then compute a rational polytope Q (i.e. a polytope with integer vertices) containing $\phi(P)$ by rounding the vertices to the nearest integer point as follows. For each pair of vertices $v, w \in P$ such that the line $\ell(v, w)$ containing v and w is not contained in a proper face of P, we round the ith coordinates $\phi(v)_i$ down and $\phi(w)_i$ up if $\phi(v)_i \leq \phi(w)_i$, and otherwise round in the opposite directions. The convex hull Q of these rounded vertices clearly contains $\phi(P \cap L)$, and is therefore amenable to enumeration using the methods above.

We note that in the case where P is a parallelopiped, for each vertex v there is a unique opposite vertex w such that the line $\ell(v, w)$ is not contained in a proper face, so the convex hull Q will also form a parallelopiped.

Coefficient Bounds and Testing Polynomials. The bounds in §2 apply *mutatis mutandis* to the relative situation. For example, given $a_{m-1}, \ldots, a_{m-k}$ for $k \geq 2$, for each $v \in E_\infty$, if we let $v(g)$ denote the polynomial $\sum_i v(b_i) x^i$ for $g(x) = \sum_i b_i x^i \in E[x]$, we obtain the inequality

$$-\min_{\substack{0 \leq i \leq k+1 \\ i \not\equiv k \ (2)}} v(g_{k+1})(\beta_{i,v}^{(k)}) < v(a_{m-k-1}) < -\max_{\substack{0 \leq i \leq k+1 \\ i \equiv k \ (2)}} v(g_{k+1})(\beta_{i,v}^{(k)});$$

here, $\beta_{1,v}^{(k)}, \ldots, \beta_{k,v}^{(k)}$ denote the roots of $v(f_k(x))$, and $\beta_{0,v}^{(k)}, \beta_{k+1,v}^{(k)}$ are computed in an analogous way using Lagrange multipliers. In this situation, we have a_{m-k-1} contained in an honest rectangular box, and the results of the previous subsection apply directly.

For each polynomial which satisfies these bounds, we perform similar tests to discard polynomials as in §2.3. One has the option of working always relative to the ground field or immediately computing the corresponding absolute field; in practice, for the small base fields under consideration, these approaches seem to be comparable, with a slight advantage to working with the absolute field.

3.2 Conclusion and Timing

Putting together the primitive and imprimitive fields computed in §§2–3, we have proven Theorem 2. In Table 2 in §2.4, we list some timing details arising from the computation. Note that in high degrees (presumably because we enumerate an exponentially large space) we recover all imprimitive fields already during the search for primitive fields.

4 Tables of Totally Real Fields

In Table 3, we count the number of totally real fields F with root discriminant $\delta_F \leq 14$ by degree, and separate out the primitive and imprimitive fields. We also list the minimal discriminant and root discriminant for $n \leq 9$. The polynomial

$$x^{10} - 11x^8 - 3x^7 + 37x^6 + 14x^5 - 48x^4 - 22x^3 + 20x^2 + 12x + 1$$

with $d_F = 443952558373 = 61^2 397^2 757$ and $\delta_F \approx 14.613$ is the dectic totally real field with smallest discriminant that we found—the corresponding number field (though not this polynomial) already appears in the tables of Klüners-Malle [17] and is a quadratic extension of the second smallest real quintic field, of discriminant 24217. It is reasonable to conjecture that this is indeed the smallest such field.

Table 3. Totally real fields F with $\delta_F \leq 14$

$n = [F : \mathbb{Q}]$	$\#NF(n, 14)$	Primitive F	Imprimitive F	Minimal d_F	Minimal δ_F
2	59	59	0	5	2.236
3	86	86	0	49	3.659
4	277	117	160	725	5.189
5	170	170	0	14641	6.809
6	263	104	159	300125	8.182
7	301	301	0	20134393	11.051
8	62	19	43	282300416	11.385
9	11	6	5	9685993193	12.869
10	0	0	0	443952558373?	14.613?
Total	1229	862	367	-	-

In Tables 4–5, we list the octic and nonic fields F with $\delta_F \leq 14$. For each field, we specify a maximal subfield E by its discriminant and degree—when more than one such subfield exists, we choose the one with smallest discriminant.

Table 4. Octic totally real fields F with $\delta_F \leq 14$

d_F	f	$[E:\mathbb{Q}]$	d_E
282300416	$x^8 - 4x^7 + 14x^5 - 8x^4 - 12x^3 + 7x^2 + 2x - 1$	4	2624
309593125	$x^8 - 4x^7 - x^6 + 17x^5 - 5x^4 - 23x^3 + 6x^2 + 9x - 1$	4	725
324000000	$x^8 - 7x^6 + 14x^4 - 8x^2 + 1$	4	1125
410338673	$x^8 - x^7 - 7x^6 + 6x^5 + 15x^4 - 10x^3 - 10x^2 + 4x + 1$	4	4913
432640000	$x^8 - 2x^7 - 7x^6 + 16x^5 + 4x^4 - 18x^3 + 2x^2 + 4x - 1$	4	1600
442050625	$x^8 - 2x^7 - 12x^6 + 26x^5 + 17x^4 - 36x^3 - 5x^2 + 11x - 1$	4	725
456768125	$x^8 - 2x^7 - 7x^6 + 11x^5 + 14x^4 - 18x^3 - 8x^2 + 9x - 1$	4	725
483345053	$x^8 - x^7 - 7x^6 + 4x^5 + 15x^4 - 3x^3 - 9x^2 + 1$	1	1
494613125	$x^8 - x^7 - 7x^6 + 4x^5 + 13x^4 - 4x^3 - 7x^2 + x + 1$	4	725
582918125	$x^8 - 2x^7 - 6x^6 + 9x^5 + 11x^4 - 9x^3 - 6x^2 + 2x + 1$	4	725
656505625	$x^8 - 3x^7 - 4x^6 + 13x^5 + 5x^4 - 13x^3 - 4x^2 + 3x + 1$	4	725
661518125	$x^8 - x^7 - 7x^6 + 5x^5 + 15x^4 - 7x^3 - 10x^2 + 2x + 1$	2	5
707295133	$x^8 - 8x^6 - 2x^5 + 19x^4 + 7x^3 - 13x^2 - 4x + 1$	1	1
733968125	$x^8 - 2x^7 - 6x^6 + 10x^5 + 11x^4 - 11x^3 - 7x^2 + 2x + 1$	2	5
740605625	$x^8 - x^7 - 9x^6 + 8x^5 + 21x^4 - 12x^3 - 14x^2 + 4x + 1$	4	725
803680625	$x^8 - 2x^7 - 9x^6 + 12x^5 + 22x^4 - 24x^3 - 14x^2 + 14x - 1$	4	725
852038125	$x^8 - 10x^6 - 5x^5 + 17x^4 + 5x^3 - 10x^2 + 1$	4	725
877268125	$x^8 - 3x^7 - 6x^6 + 20x^5 + 5x^4 - 25x^3 - x^2 + 7x + 1$	4	725
898293125	$x^8 - x^7 - 9x^6 + 10x^5 + 15x^4 - 10x^3 - 9x^2 + x + 1$	4	725
1000118125	$x^8 - 3x^7 - 4x^6 + 14x^5 + 5x^4 - 19x^3 - x^2 + 7x - 1$	2	5
1024000000	$x^8 - 8x^6 + 19x^4 - 12x^2 + 1$	4	1600
1032588125	$x^8 - 9x^6 - 2x^5 + 23x^4 + 9x^3 - 17x^2 - 9x - 1$	2	5
1064390625	$x^8 - 13x^6 + 44x^4 - 17x^2 + 1$	4	725
1077044573	$x^8 - x^7 - 8x^6 + 8x^5 + 16x^4 - 17x^3 - 2x^2 + 5x - 1$	1	1
1095205625	$x^8 - 3x^7 - 5x^6 + 18x^5 + 2x^4 - 23x^3 + 2x^2 + 8x - 1$	2	5
1098290293	$x^8 - 3x^7 - 4x^6 + 16x^5 + x^4 - 23x^3 + 7x^2 + 5x - 1$	1	1
1104338125	$x^8 - 2x^7 - 8x^6 + 15x^5 + 17x^4 - 31x^3 - 9x^2 + 17x - 1$	4	725
1114390153	$x^8 - 8x^6 - 2x^5 + 16x^4 + 3x^3 - 10x^2 + 1$	1	1
1121463125	$x^8 - 3x^7 - 4x^6 + 15x^5 + 2x^4 - 18x^3 + 5x + 1$	2	5
1136700613	$x^8 - x^7 - 7x^6 + 4x^5 + 14x^4 - 4x^3 - 8x^2 + x + 1$	1	1
1142440000	$x^8 - 3x^7 - 5x^6 + 15x^5 + 8x^4 - 15x^3 - 5x^2 + 4x + 1$	4	4225
1152784549	$x^8 - 4x^7 - x^6 + 15x^5 - 3x^4 - 16x^3 + 4x^2 + 4x - 1$	4	1957
1153988125	$x^8 - 2x^7 - 7x^6 + 11x^5 + 12x^4 - 16x^3 - 5x^2 + 6x - 1$	4	2525
1166547493	$x^8 - x^7 - 7x^6 + 6x^5 + 14x^4 - 9x^3 - 9x^2 + 3x + 1$	1	1
1183423341	$x^8 - x^7 - 8x^6 + 9x^5 + 17x^4 - 20x^3 - 8x^2 + 10x - 1$	4	1957
1202043125	$x^8 - 3x^7 - 4x^6 + 16x^5 - 21x^3 + 9x^2 + 2x - 1$	2	5
1225026133	$x^8 - 3x^7 - 4x^6 + 18x^5 - 6x^4 - 17x^3 + 9x^2 + 2x - 1$	1	1
1243893125	$x^8 - x^7 - 8x^6 + 3x^5 + 18x^4 - x^3 - 12x^2 - 2x + 1$	2	5
1255718125	$x^8 - 2x^7 - 8x^6 + 19x^5 + 10x^4 - 41x^3 + 13x^2 + 10x - 1$	4	725
1261609229	$x^8 - 2x^7 - 6x^6 + 12x^5 + 9x^4 - 19x^3 - x^2 + 6x - 1$	1	1
1292203125	$x^8 - 4x^7 - x^6 + 17x^5 - 6x^4 - 21x^3 + 6x^2 + 8x + 1$	4	1125
1299600812	$x^8 - 2x^7 - 6x^6 + 10x^5 + 12x^4 - 13x^3 - 8x^2 + 3x + 1$	1	1
1317743125	$x^8 - x^7 - 8x^6 + 7x^5 + 19x^4 - 14x^3 - 12x^2 + 8x - 1$	2	5
1318279381	$x^8 - x^7 - 7x^6 + 5x^5 + 14x^4 - 6x^3 - 9x^2 + x + 1$	1	1
1326417388	$x^8 - 2x^7 - 6x^6 + 10x^5 + 12x^4 - 13x^3 - 9x^2 + 4x + 2$	4	2777
1348097653	$x^8 - 2x^7 - 6x^6 + 11x^5 + 11x^4 - 17x^3 - 6x^2 + 6x + 1$	1	1
1358954496	$x^8 - 8x^6 + 20x^4 - 16x^2 + 1$	4	2048
1359341129	$x^8 - 8x^6 - x^5 + 18x^4 + 2x^3 - 12x^2 - x + 2$	1	1
1377663125	$x^8 - 12x^6 + 33x^4 - 5x^3 - 22x^2 + 5x + 1$	4	725
1381875749	$x^8 - 3x^7 - 4x^6 + 14x^5 + 4x^4 - 18x^3 + x^2 + 5x - 1$	1	1
1391339501	$x^8 - 3x^7 - 4x^6 + 15x^5 + 4x^4 - 22x^3 + 9x - 1$	1	1
1405817381	$x^8 - 9x^6 - x^5 + 20x^4 + 6x^3 - 12x^2 - 7x - 1$	1	1
1410504129	$x^8 - 9x^6 - x^5 + 22x^4 + x^3 - 15x^2 - x + 1$	4	3981
1410894053	$x^8 - 2x^7 - 6x^6 + 9x^5 + 12x^4 - 11x^3 - 8x^2 + 3x + 1$	1	1
1413480448	$x^8 - 4x^7 - 2x^6 + 16x^5 - x^4 - 16x^3 + 2x^2 + 4x - 1$	4	2048
1424875717	$x^8 - x^7 - 7x^6 + 5x^5 + 15x^4 - 6x^3 - 10x^2 + x + 1$	1	1
1442599461	$x^8 - 3x^7 - 4x^6 + 15x^5 + 4x^4 - 21x^3 - 2x^2 + 8x + 1$	4	7053
1449693125	$x^8 - x^7 - 9x^6 + 10x^5 + 20x^4 - 20x^3 - 14x^2 + 11x + 1$	2	5
1459172469	$x^8 - 4x^7 - x^6 + 17x^5 - 6x^4 - 21x^3 + 8x^2 + 6x - 1$	4	1957
1460018125	$x^8 - 3x^7 - 5x^6 + 13x^5 + 11x^4 - 14x^3 - 10x^2 + x + 1$	4	2525
1462785589	$x^8 - 2x^7 - 6x^6 + 11x^5 + 10x^4 - 17x^3 - 3x^2 + 6x - 1$	1	1
1472275625	$x^8 - 3x^7 - 6x^6 + 19x^5 + 13x^4 - 35x^3 - 12x^2 + 13x - 1$	4	725

Table 5. Nonic totally real fields F with $\delta_F \leq 14$

d_F	f	$[E:\mathbb{Q}]$	d_E
9685993193	$x^9 - 9x^7 + 24x^5 - 2x^4 - 20x^3 + 3x^2 + 5x - 1$	1	1
11779563529	$x^9 - 9x^7 - 2x^6 + 22x^5 + 5x^4 - 17x^3 - 4x^2 + 4x + 1$	1	1
16240385609	$x^9 - x^8 - 9x^7 + 4x^6 + 26x^5 - 2x^4 - 25x^3 - x^2 + 7x + 1$	3	49
16440305941	$x^9 - 2x^8 - 9x^7 + 11x^6 + 28x^5 - 18x^4 - 34x^3 + 8x^2 + 13x + 1$	3	229
16898785417	$x^9 - 2x^8 - 7x^7 + 11x^6 + 18x^5 - 17x^4 - 19x^3 + 6x^2 + 7x + 1$	1	1
16983563041	$x^9 - x^8 - 8x^7 + 7x^6 + 21x^5 - 15x^4 - 20x^3 + 10x^2 + 5x - 1$	3	361
17515230173	$x^9 - 4x^8 - 3x^7 + 29x^6 - 26x^5 - 24x^4 + 34x^3 - 2x^2 - 5x + 1$	3	49
18625670317	$x^9 - 9x^7 - x^6 + 23x^5 + 4x^4 - 19x^3 - 3x^2 + 4x + 1$	1	1
18756753353	$x^9 - 3x^8 - 4x^7 + 15x^6 + 4x^5 - 22x^4 - x^3 + 10x^2 - 1$	1	1
19936446593	$x^9 - 3x^8 - 5x^7 + 17x^6 + 7x^5 - 30x^4 - x^3 + 16x^2 - 2x - 1$	3	49
20370652633	$x^9 - 2x^8 - 8x^7 + 12x^6 + 15x^5 - 17x^4 - 8x^3 + 8x^2 + x - 1$	1	1

References

1. Aguirre, J., Bilbao, M., Peral, J.C.: The trace of totally positive algebraic integers. Math. Comp. 75(253), 385–393 (2006)
2. Belabas, K.: A fast algorithm to compute cubic fields. Math. Comp. 66(219), 1213–1237 (1997)
3. Bhargava, M.: Gauss composition and generalizations. In: Fieker, C., Kohel, D.R. (eds.) ANTS 2002. LNCS, vol. 2369, pp. 1–8. Springer, Heidelberg (2002)
4. Cohen, H.: Advanced Topics in Computational Number Theory. In: Graduate Texts in Mathematics, vol. 193, Springer, New York (2000)
5. Cohen, H., Diaz y Diaz, F.: A polynomial reduction algorithm. Sém. Théor. Nombres Bordeaux 3(2), 351–360 (1991)
6. Cohen, H., Diaz y Diaz, F., Olivier, M.: A table of totally complex number fields of small discriminants. In: Buhler, J.P. (ed.) ANTS 1998. LNCS, vol. 1423, pp. 381–391. Springer, Heidelberg (1998)
7. Cohen, H., Diaz y Diaz, F., Olivier, M.: Constructing complete tables of quartic fields using Kummer theory. Math. Comp. 72(242), 941–951 (2003)
8. Cohn, H., Elkies, N.: New upper bounds on sphere packings I. Ann. Math. 157, 689–714 (2003)
9. Conway, J.H.,, Sloane, N.J.A.: Sphere packings, lattices and groups. In: Grund. der Math. Wissenschaften, 3rd edn., vol. 290, Springer, New York (1999)
10. De Loera, J., Hemmecke, R., Tauzer, J., Yoshia, R.: Effective lattice point counting in rational convex polytopes. J. Symbolic Comput. 38(4), 1273–1302 (2004)
11. De Loera, J.: LattE: Lattice point Enumeration (2007), http://www.math.ucdavis.edu/~latte/
12. Ellenberg, J.S., Venkatesh, A.: The number of extensions of a number field with fixed degree and bounded discriminant. Ann. of Math. 163(2), 723–741 (2006)
13. Fincke, U., Pohst, M.: Improved methods for calculating vectors of short length in a lattice, including a complexity analysis. Math. Comp. 44, 170, 463–471 (1985)
14. Hajir, F., Maire, C.: Tamely ramified towers and discriminant bounds for number fields. Compositio Math. 128, 35–53 (2001)
15. Hajir, F., Maire, C.: Tamely ramified towers and discriminant bounds for number fields. II. J. Symbolic Comput. 33, 415–423 (2002)
16. Number field tables, ftp://megrez.math.u-bordeaux.fr/pub/numberfields/
17. Klüners, J., Malle, G.: A database for number fields, http://www.math.uni-duesseldorf.de/~klueners/minimum/minimum.html

18. Klüners, J., Malle, G.: A database for field extensions of the rationals. LMS J. Comput. Math. 4, 82–196 (2001)
19. Kreuzer, M., Skarke, H.: PALP: A Package for Analyzing Lattice Polytopes (2006), http://hep.itp.tuwien.ac.at/~kreuzer/CY/CYpalp.html
20. Lenstra, A.K., Lenstra, H.W., Lovász, L.: Factoring polynomials with rational coefficients. Math. Ann. 261, 515–534 (1982)
21. Long, D.D., Maclachlan, C., Reid, A.W.: Arithmetic Fuchsian groups of genus zero. Pure Appl. Math. Q. 2, 569–599 (2006)
22. Malle, G.: The totally real primitive number fields of discriminant at most 10^9. In: Hess, F., Pauli, S., Pohst, M. (eds.) ANTS 2006. LNCS, vol. 4076, pp. 114–123. Springer, Heidelberg (2006)
23. Martin, J.: Improved bounds for discriminants of number fields (submitted)
24. Martinet, J.: Petits discriminants des corps de nombres. In: Journées Arithmétiques (Exeter, 1980). London Math. Soc. Lecture Note Ser., vol. 56, pp. 151–193. Cambridge Univ. Press, Cambridge (1982)
25. Martinet, J.: Tours de corps de classes et estimations de discriminants. Invent. Math. 44, 65–73 (1978)
26. Martinet, J.: Methodes geometriques dans la recherche des petitis discriminants. In: Sem. Théor. des Nombres (Paris 1983–84), pp. 147–179. Birkhäuser, Boston (1985)
27. Odlyzko, A.M.: Bounds for discriminants and related estimates for class numbers, regulators and zeros of zeta functions: a survey of recent results. Sém. Théor. Nombres Bordeaux 2(2), 119–141 (1990)
28. The PARI Group: PARI/GP (version 2.3.2), Bordeaux (2006), http://pari.math.u-bordeaux.fr/
29. Pohst, M.: On the computation of number fields of small discriminants including the minimum discriminants of sixth degree fields. J. Number Theory 14, 99–117 (1982)
30. Roblot, X.-F.: Totally real fields with small root discriminant, http://math.univ-lyon1.fr/~roblot/tables.html
31. Stein, W.: SAGE Mathematics Software (version 2.8.12). The SAGE Group (2007), http://www.sagemath.org/
32. Smyth, C.J.: The mean values of totally real algebraic integers. Math. Comp. 42, 663–681 (1984)
33. Takeuchi, K.: Totally real algebraic number fields of degree 9 with small discriminant. Saitama Math. J. 17, 63–85 (1999)
34. Verschelde, J.: Algorithm 795: PHCpack: A general-purpose solver for polynomial systems by homotopy continuation. ACM Transactions on Mathematical Software 25, 251–276 (1999)
35. Voight, J.: Totally real number fields, http://www.cems.uvm.edu/~voight/nf-tables/

Computing Hilbert Class Polynomials

Juliana Belding[1], Reinier Bröker[2], Andreas Enge[3], and Kristin Lauter[2]

[1] Dept. of Mathematics, University of Maryland, College Park, MD 20742, USA
jbelding@math.umd.edu
[2] Microsoft Research, One Microsoft Way, Redmond, WA 98052, USA
{reinierb,klauter}@microsoft.com
[3] INRIA Futurs & Laboratoire d'Informatique (CNRS/UMR 7161)
École polytechnique, 91128 Palaiseau cedex, France
enge@lix.polytechnique.fr

Abstract. We present and analyze two algorithms for computing the Hilbert class polynomial H_D. The first is a p-adic lifting algorithm for inert primes p in the order of discriminant $D < 0$. The second is an improved Chinese remainder algorithm which uses the class group action on CM-curves over finite fields. Our run time analysis gives tighter bounds for the complexity of all known algorithms for computing H_D, and we show that all methods have comparable run times.

1 Introduction

For an imaginary quadratic order $\mathcal{O} = \mathcal{O}_D$ of discriminant $D < 0$, the j-invariant of the complex elliptic curve $\mathbf{C}/\mathcal{O}$ is an algebraic integer. Its minimal polynomial $H_D \in \mathbf{Z}[X]$ is called the *Hilbert class polynomial*. It defines the ring class field $K_{\mathcal{O}}$ corresponding to $\mathcal{O}$, and within the context of explicit class field theory, it is natural to ask for an algorithm to *explicitly compute H_D*.

Algorithms to compute H_D are also interesting for elliptic curve primality proving [2] and for cryptographic purposes [6]; for instance, pairing-based cryptosystems using ordinary curves rely on complex multiplication techniques to generate the curves. The classical approach to compute H_D is to approximate the values $j(\tau_{\mathfrak{a}}) \in \mathbf{C}$ of the complex analytic j-function at points $\tau_{\mathfrak{a}}$ in the upper half plane corresponding to the ideal classes $\mathfrak{a}$ for the order $\mathcal{O}$. The polynomial H_D may be recovered by rounding the coefficients of $\prod_{\mathfrak{a} \in \mathrm{Cl}(\mathcal{O})} (X - j(\tau_{\mathfrak{a}})) \in \mathbf{C}[X]$ to the nearest integer. It is shown in [9] that an optimized version of that algorithm has a complexity that is essentially linear in the output size.

Alternatively one can compute H_D using a p-adic lifting algorithm [7,3]. Here, the prime p splits completely in $K_{\mathcal{O}}$ and is therefore relatively large: it satisfies the lower bound $p \geq |D|/4$. In this paper we give a p-adic algorithm for *inert* primes p. Such primes are typically much smaller than totally split primes, and under GRH there exists an inert prime of size only $O((\log |D|)^2)$. The complex multiplication theory underlying all methods is more intricate for inert primes p, as the roots of $H_D \in \mathbf{F}_{p^2}[X]$ are now j-invariants of *supersingular* elliptic curves. In Section 2 we explain how to define the canonical lift of a supersingular elliptic curve, and in Section 4 we describe a method to explicitly compute this lift.

A.J. van der Poorten and A. Stein (Eds.): ANTS-VIII 2008, LNCS 5011, pp. 282–295, 2008.
© Springer-Verlag Berlin Heidelberg 2008

In another direction, it was suggested in [1] to compute H_D modulo several totally split primes p and then combine the information modulo p using the Chinese remainder theorem to compute $H_D \in \mathbf{Z}[X]$. The first version of this algorithm was quite impractical, and in Section 3 we improve this 'multi-prime approach' in two different ways. We show how to incorporate inert primes, and we improve the original approach for totally split primes using the class group action on CM-curves. We analyze the run time of the new algorithm in Section 5 in terms of the logarithmic height of H_D, its degree, the largest prime needed to generate the class group of $\mathcal{O}$ and the discriminant D. Our tight bounds on the first two quantities from Lemmata 1 and 2 apply to all methods to compute H_D. For the multi-prime approach, we derive the following result.

Theorem 1. *The algorithm presented in Section 3 computes, for a discriminant $D < 0$, the Hilbert class polynomial H_D. If GRH holds true, the algorithm has an expected run time $O\left(|D|(\log|D|)^{7+o(1)}\right)$. Under heuristic assumptions, the complexity becomes $O\left(|D|(\log|D|)^{3+o(1)}\right)$.*

We conclude by giving examples of the presented algorithms in Section 6.

2   Complex Multiplication in Characteristic p

Throughout this section, $D < -4$ is any discriminant, and we write $\mathcal{O}$ for the imaginary quadratic order of discriminant D. Let $E/K_{\mathcal{O}}$ be an elliptic curve with endomorphism ring isomorphic to $\mathcal{O}$. As $\mathcal{O}$ has rank 2 as a $\mathbf{Z}$-algebra, there are *two* isomorphisms $\varphi : \mathrm{End}(E) \xrightarrow{\sim} \mathcal{O}$. We always assume we have chosen the *normalized* isomorphism, i.e., for all $y \in \mathcal{O}$ we have $\varphi(y)^*\omega = y\omega$ for all invariant differentials ω. For ease of notation, we write E for such a 'normalized elliptic curve,' the isomorphism φ being understood.

For a field F, let $\mathrm{Ell}_D(F)$ be the set of isomorphism classes of elliptic curves over F with endomorphism ring $\mathcal{O}$. The ideal group of $\mathcal{O}$ acts on $\mathrm{Ell}_D(K_{\mathcal{O}})$ via

$$j(E) \mapsto j(E)^{\mathfrak{a}} = j(E/E[\mathfrak{a}]),$$

where $E[\mathfrak{a}]$ is the group of $\mathfrak{a}$-torsion points, i.e., the points that are annihilated by all $\alpha \in \mathfrak{a} \subset \mathcal{O} = \mathrm{End}(E)$. As principal ideals act trivially, this action factors through the class group $\mathrm{Cl}(\mathcal{O})$. The $\mathrm{Cl}(\mathcal{O})$-action is transitive and free, and $\mathrm{Ell}_D(K_{\mathcal{O}})$ is a principal homogeneous $\mathrm{Cl}(\mathcal{O})$-space.

Let p be a prime that splits completely in the ring class field $K_{\mathcal{O}}$. We can embed $K_{\mathcal{O}}$ in the p-adic field $\mathbf{Q}_p$, and the reduction map $\mathbf{Z}_p \to \mathbf{F}_p$ induces a bijection $\mathrm{Ell}_D(\mathbf{Q}_p) \to \mathrm{Ell}_D(\mathbf{F}_p)$. The $\mathrm{Cl}(\mathcal{O})$-action respects reduction modulo p, and the set $\mathrm{Ell}_D(\mathbf{F}_p)$ is a $\mathrm{Cl}(\mathcal{O})$-torsor, just like in characteristic zero. This observation is of key importance for the improved 'multi-prime' approach explained in Section 3.

We now consider a prime p that is *inert* in $\mathcal{O}$, fixed for the remainder of this section. As the principal prime $(p) \subset \mathcal{O}$ splits completely in $K_{\mathcal{O}}$, all primes of $K_{\mathcal{O}}$ lying over p have residue class degree 2. We view $K_{\mathcal{O}}$ as a subfield of the unramified degree 2 extension L of $\mathbf{Q}_p$. It is a classical result, see [8] or [15, Th. 13.12], that for $[E] \in \mathrm{Ell}_D(L)$, the reduction E_p is *supersingular*. It can be defined over the finite field $\mathbf{F}_{p^2}$, and its endomorphism ring is a maximal order in the unique quaternion algebra $\mathcal{A}_{p,\infty}$ which is ramified at p and ∞. The reduction map $\mathbf{Z}_L \to \mathbf{F}_{p^2}$ also induces an embedding $f : \mathcal{O} \hookrightarrow \mathrm{End}(E_p)$. This embedding is not surjective, as it is in the totally split case, since $\mathrm{End}(E_p)$ has rank 4 as a $\mathbf{Z}$-algebra, and $\mathcal{O}$ has rank 2.

We let $\mathrm{Emb}_D(\mathbf{F}_{p^2})$ be the set of isomorphism classes of pairs (E_p, f) with $E_p/\mathbf{F}_{p^2}$ a supersingular elliptic curve and $f : \mathcal{O} \hookrightarrow \mathrm{End}(E_p)$ an embedding. Here, (E_p, f) and (E'_p, f') are isomorphic if there exists an isomorphism $h : E_p \xrightarrow{\sim} E'_p$ of elliptic curves with $h^{-1} f'(\alpha) h = f(\alpha)$ for all $\alpha \in \mathcal{O}$. As an analogue of picking the normalized isomorphism $\mathcal{O} \xrightarrow{\sim} \mathrm{End}(E)$ in characteristic zero, we now identify (E_p, f) and (E'_p, f') if f equals the complex conjugate of f'.

Theorem 2. *Let $D < -4$ be a discriminant. If p is inert in $\mathcal{O} = \mathcal{O}_D$, the reduction map $\pi : \mathrm{Ell}_D(L) \to \mathrm{Emb}_D(\mathbf{F}_{p^2})$ is a bijection. Here, L is the unramified extension of $\mathbf{Q}_p$ of degree 2.*

Proof. By the Deuring lifting theorem, see [8] or [15, Th. 13.14], we can lift an element of $\mathrm{Emb}_D(\mathbf{F}_{p^2})$ to an element of $\mathrm{Ell}_D(L)$. Hence, the map is surjective.

Suppose that we have $\pi(E) = \pi(E')$. As E and E' both have endomorphism ring $\mathcal{O}$, they are isogenous. We let $\varphi_{\mathfrak{a}} : E \to E^{\mathfrak{a}} = E'$ be an isogeny. Writing $\mathcal{O} = \mathbf{Z}[\tau]$, we get

$$f' = f^{\mathfrak{a}} : \tau \mapsto \overline{\varphi}_{\mathfrak{a}} f(\tau) \widehat{\overline{\varphi}}_{\mathfrak{a}} \otimes (\deg \overline{\varphi}_{\mathfrak{a}})^{-1} \in \mathrm{End}(E_p) \otimes \mathbf{Q}.$$

The map $\overline{\varphi}_{\mathfrak{a}}$ commutes with $f(\tau)$ and is thus contained in $S = f(\mathrm{End}(E)) \otimes \mathbf{Q}$.

Write $\mathcal{O}' = S \cap \mathrm{End}(E_p)$, and let m be the index $[\mathcal{O}' : f(\mathrm{End}(E))]$. For any $\delta \in \mathcal{O}'$, there exists $\gamma \in \mathrm{End}(E)$ with $m\delta = f(\gamma)$. As $f(\gamma)$ annihilates the m-torsion $E_p[m]$, γ annihilates $E[m]$, thus it is a multiple of m inside $\mathrm{End}(E)$. We derive that δ is contained in $f(\mathrm{End}(E))$, and $\mathcal{O}' = f(\mathrm{End}(E))$. Hence, $\varphi_{\mathfrak{a}}$ is an endomorphism of E, and E and $E^{\mathfrak{a}}$ are isomorphic. $\square$

The *canonical lift* $\widetilde{E}$ of a pair $(E_p, f) \in \mathrm{Emb}_D(\mathbf{F}_{p^2})$ is defined as the inverse $\pi^{-1}(E_p, f) \in \mathrm{Ell}_D(L)$. This generalizes the notion of a canonical lift for ordinary elliptic curves, and the main step of the p-adic algorithm described in Section 4 is to compute $\widetilde{E}$: its j-invariant is a zero of the Hilbert class polynomial $H_D \in L[X]$.

The reduction map $\mathrm{Ell}_D(L) \to \mathrm{Emb}_D(\mathbf{F}_{p^2})$ induces a transitive and free action of the class group on the set $\mathrm{Emb}_D(\mathbf{F}_{p^2})$. For an $\mathcal{O}$-ideal $\mathfrak{a}$, let $\varphi_{\mathfrak{a}} : E \to E^{\mathfrak{a}}$ be the isogeny of CM-curves with kernel $E[\mathfrak{a}]$. Writing $\mathcal{O} = \mathbf{Z}[\tau]$, let $\beta \in \mathrm{End}(E)$ be the image of τ under the normalized isomorphism $\mathcal{O} \xrightarrow{\sim} \mathrm{End}(E)$. The normalized isomorphism for $E^{\mathfrak{a}}$ is now given by

$$\tau \mapsto \varphi_{\mathfrak{a}} \beta \widehat{\varphi}_{\mathfrak{a}} \otimes (\deg \varphi_{\mathfrak{a}})^{-1}.$$

We have $E_p^{\mathfrak{a}} = (E^{\mathfrak{a}})_p$ and $f^{\mathfrak{a}}$ is the composition $\mathcal{O} \xrightarrow{\sim} \mathrm{End}(E^{\mathfrak{a}}) \hookrightarrow \mathrm{End}(E_p^{\mathfrak{a}})$. Note that principal ideals indeed act trivially: $\varphi_{\mathfrak{a}}$ is an endomorphism in this case and, as $\mathrm{End}(E)$ is commutative, we have $f = f^{\mathfrak{a}}$.

To explicitly compute this action, we fix one supersingular curve $E_p/\mathbf{F}_{p^2}$ and an isomorphism $i_{E_p} : \mathcal{A}_{p,\infty} \xrightarrow{\sim} \text{End}(E_p) \otimes \mathbf{Q}$ and view the embedding f as an injective map $f : \mathcal{O} \hookrightarrow \mathcal{A}_{p,\infty}$. Let $R = i_{E_p}^{-1}(\text{End}(E_p))$ be the maximal order of $\mathcal{A}_{p,\infty}$ corresponding to E_p. For $\mathfrak{a}$ an ideal of $\mathcal{O}$, we compute the curve $E_p^{\mathfrak{a}} = \overline{\varphi}_{\mathfrak{a}}(E_p)$ and choose an auxiliary isogeny $\varphi_{\mathfrak{b}} : E_p \to E_p^{\mathfrak{a}}$. This induces an isomorphism $g_{\mathfrak{b}} : \mathcal{A}_{p,\infty} \xrightarrow{\sim} \text{End}(E_p^{\mathfrak{a}}) \otimes \mathbf{Q}$ given by

$$\alpha \mapsto \varphi_{\mathfrak{b}} i_{E_p}(\alpha) \widehat{\varphi}_{\mathfrak{b}} \otimes (\deg \varphi_{\mathfrak{b}})^{-1}.$$

The left R-ideals $Rf(\mathfrak{a})$ and $\mathfrak{b}$ are left-isomorphic by [22, Th. 3.11] and thus we can find $x \in \mathcal{A}_{p,\infty}$ with $Rf(\mathfrak{a}) = \mathfrak{b}x$. As $y = f(\tau)$ is an element of $Rf(\mathfrak{a})$, we get the embedding $\tau \mapsto xyx^{-1}$ into the right order $R_{\mathfrak{b}}$ of $\mathfrak{b}$. By construction, the induced embedding $f^{\mathfrak{a}} : \mathcal{O} \hookrightarrow \text{End}(E_p^{\mathfrak{a}})$ is precisely

$$f^{\mathfrak{a}}(\tau) = g_{\mathfrak{b}}(xyx^{-1}) \in \text{End}(E_p^{\mathfrak{a}}),$$

and this is independent of the choice of $\mathfrak{b}$. For example, if $E_p^{\mathfrak{a}} = E_p$, then choosing $\varphi_{\mathfrak{b}}$ as the identity, we find x with $Rf(\mathfrak{a}) = Rx$ to get the embedding $f^{\mathfrak{a}} : \tau \mapsto i_{E_p}(xyx^{-1}) \in \text{End}(E_p)$.

3   The Multi-prime Approach

This section is devoted to a precise description of the new algorithm for computing the Hilbert class polynomial $H_D \in \mathbf{Z}[X]$ via the Chinese remainder theorem.

Algorithm 1
INPUT: an imaginary quadratic discriminant D
OUTPUT: the Hilbert class polynomial $H_D \in \mathbf{Z}[X]$

0. Let $(A_i, B_i, C_i)_{i=1}^{h(D)}$ be the set of primitive reduced binary quadratic forms of discriminant $B_i^2 - 4A_iC_i = D$ representing the class group $\text{Cl}(\mathcal{O})$. Compute

$$n = \left\lceil \log_2 \left(2.48\, h(D) + \pi\sqrt{|D|} \sum_{i=1}^{h(D)} \frac{1}{A_i} \right) \right\rceil + 1, \tag{1}$$

   which by [9] is an upper bound on the number of bits in the largest coefficient of H_D.
1. Choose a set $\mathcal{P}$ of primes p such that $N = \prod_{p \in \mathcal{P}} p \geq 2^n$ and each p is either inert in $\mathcal{O}$ or totally split in $K_{\mathcal{O}}$.
2. For all $p \in \mathcal{P}$, depending on whether p is split or inert in $\mathcal{O}$, compute $H_D \bmod p$ using either Algorithm 2 or 3.
3. Compute $H_D \bmod N$ by the Chinese remainder theorem, and return its representative in $\mathbf{Z}[X]$ with coefficients in $\left(-\frac{N}{2}, \frac{N}{2}\right)$.

The choice of $\mathcal{P}$ in Step 1 leaves some room for different flavors of the algorithm. Since Step 2 is exponential in $\log p$, the primes should be chosen as small as possible. The simplest case is to only use split primes, to be analyzed in Section 5. As the run time of Step 2 is worse for inert primes than for split primes, we view the use of inert primes as a practical improvement.

3.1 Split Primes

A prime p splits completely in $K_{\mathcal{O}}$ if and only if the equation $4p = u^2 - v^2 D$ has a solution in integers u, v. For any prime p, we can efficiently test if such a solution exists using an algorithm due to Cornacchia. In practice, we generate primes satisfying this relation by varying u and v and testing if $(u^2 - v^2 D)/4$ is prime.

Algorithm 2

INPUT: an imaginary quadratic discriminant D and a prime p that splits completely in $K_{\mathcal{O}}$

OUTPUT: $H_D \bmod p$

1. Find a curve E over $\mathbf{F}_p$ with endomorphism ring $\mathcal{O}$. Set $j = j(E)$.
2. Compute the Galois conjugates $j^{\mathfrak{a}}$ for $\mathfrak{a} \in \mathrm{Cl}(\mathcal{O})$.
3. Return $H_D \bmod p = \prod_{\mathfrak{a} \in \mathrm{Cl}(\mathcal{O})} (X - j^{\mathfrak{a}})$.

Note: The main difference between this algorithm and the one proposed in [1] is that the latter determines *all* curves with endomorphism ring $\mathcal{O}$ via exhaustive search, while we search for one and obtain the others via the action of $\mathrm{Cl}(\mathcal{O})$ on the set $\mathrm{Ell}_D(\mathbf{F}_p)$.

Step 1 can be implemented by picking j-invariants at random until one with the desired endomorphism ring is found. With $4p = u^2 - v^2 D$, a necessary condition is that the curve E or its quadratic twist E' has $p + 1 - u$ points. In the case that D is fundamental and $v = 1$, this condition is also sufficient. To test if one of our curves E has the right cardinality, we pick a random point $P \in E(\mathbf{F}_p)$ and check if $(p + 1 - u)P = 0$ or $(p + 1 + u)P = 0$ holds. If neither of them does, E does not have endomorphism ring $\mathcal{O}$. If E survives this test, we select a few random points on both E and E' and compute the orders of these points *assuming* they divide $p + 1 \pm u$. If the curve E indeed has $p + 1 \pm u$ points, we quickly find points $P \in E(\mathbf{F}_p)$, $P' \in E'(\mathbf{F}_p)$ of maximal order, since we have $E(\mathbf{F}_p) \cong \mathbf{Z}/n_1\mathbf{Z} \times \mathbf{Z}/n_2\mathbf{Z}$ with $n_1 \mid n_2$ and a fraction $\varphi(n_2)/n_2$ of the points have maximal order. For P and P' of maximal order and $p > 457$, either the order of P or the order of P' is at least $4\sqrt{p}$, by [19, Theorem 3.1], due to J.-F. Mestre. As the Hasse interval has length $4\sqrt{p}$, this then *proves* that E has $p + 1 \pm u$ points.

Let $\Delta = \frac{D}{f^2}$ be the fundamental discriminant associated to D. For $f \neq 1$ or $v \neq 1$ (which happens necessarily for $D \equiv 1 \bmod 8$), the curves with $p + 1 \pm u$ points admit any order $\mathcal{O}_{g^2 \Delta}$ such that $g \mid fv$ as their endomorphism rings. In this case, one possible strategy is to use Kohel's algorithm described in [13, Th. 24] to compute g, until a curve with $g = f$ is found. This variant is easiest to analyze and enough to prove Theorem 1.

In practice, one would rather keep a curve that satisifes $f \mid g$, since by the class number formula $g = vf$ with overwhelming probability. As v and thus $\frac{fv}{g}$ is small, it is then possible to use another algorithm due to Kohel and analyzed in detail by Fouquet–Morain [13,11] to quickly apply an isogeny of degree $\frac{fv}{g}$ leading to a curve with endomorphism ring $\mathcal{O}$.

Concerning Step 2, let $\mathrm{Cl}(\mathcal{O}) = \bigoplus \langle \mathfrak{l}_i \rangle$ be a decomposition of the class group into a direct product of cyclic groups generated by invertible degree 1 prime ideals $\mathfrak{l}_i$ of order h_i and norm ℓ_i not dividing pv. The $j^{\mathfrak{a}}$ may then be obtained successively by computing the Galois action of the $\mathfrak{l}_i$ on j-invariants of curves with endomorphism ring $\mathcal{O}$ over $\mathbf{F}_p$, otherwise said, by computing ℓ_i-isogenous curves: $h_1 - 1$ successive applications of $\mathfrak{l}_1$ yield $j^{\mathfrak{l}_1}, \ldots, j^{\mathfrak{l}_1^{h_1-1}}$; to each of them, $\mathfrak{l}_2$ is applied $h_2 - 1$ times, and so forth.

To explicitly compute the action of $\mathfrak{l} = \mathfrak{l}_i$, we let $\Phi_\ell(X, Y) \in \mathbf{Z}[X]$ be the classical modular polynomial. It is a model for the modular curve $Y_0(\ell)$ parametrizing elliptic curves together with an ℓ-isogeny, and it satisfies $\Phi_\ell(j(z), j(\ell z)) = 0$ for the modular function $j(z)$. If $j_0 \in \mathbf{F}_p$ is the j-invariant of some curve with endomorphism ring $\mathcal{O}$, then all the roots in $\mathbf{F}_p$ of $\Phi_\ell(X, j_0)$ are j-invariants of curves with endomorphism ring $\mathcal{O}$ by [13, Prop. 23]. If $\mathfrak{l}$ is unramified, there are two roots, $j_0^{\mathfrak{l}}$ and $j_0^{\mathfrak{l}^{-1}}$. For ramified $\mathfrak{l}$, we find only one root $j_0^{\mathfrak{l}} = j_0^{\mathfrak{l}^{-1}}$. So Step 2 is reduced to determining roots of univariate polynomials over $\mathbf{F}_p$.

3.2 Inert Primes

Algorithm 3
INPUT: an imaginary quadratic discriminant D and a prime p that is inert in $\mathcal{O}$
OUTPUT: $H_D \bmod p$

1. Compute the list of supersingular j-invariants over $\mathbf{F}_{p^2}$ together with their endomorphism rings inside the quaternion algebra $\mathcal{A}_{p,\infty}$.
2. Compute an optimal embedding $f : \mathcal{O} \hookrightarrow \mathcal{A}_{p,\infty}$ and let R be a maximal order that contains $f(\mathcal{O})$.
3. Select a curve $E/\mathbf{F}_{p^2}$ in the list with $\mathrm{End}(E) \cong R$, and let j be its j-invariant.
4. Compute the Galois conjugates $j^{\mathfrak{a}}$ for $\mathfrak{a} \in \mathrm{Cl}(\mathcal{O})$.
5. Return $H_D \bmod p = \prod_{\mathfrak{a} \in \mathrm{Cl}(\mathcal{O})}(X - j^{\mathfrak{a}})$.

As the number of supersingular j-invariants grows roughly like $(p - 1)/12$, this algorithm is only feasible for *small* primes. For the explicit computation, we use an algorithm due to Cerviño [4] to compile our list. The list gives a bijection between the set of $\mathrm{Gal}(\mathbf{F}_{p^2}/\mathbf{F}_p)$-conjugacy classes of supersingular j-invariants and the set of maximal orders in $\mathcal{A}_{p,\infty}$.

In Step 2 we compute an element $y \in \mathcal{A}_{p,\infty}$ satisfying the same minimal polynomial as a generator τ of $\mathcal{O}$. For non-fundamental discriminants we need to ensure that the embedding is optimal, i.e., does not extend to an embedding of the maximal overorder of $\mathcal{O}$ into $\mathcal{A}_{p,\infty}$. Using standard algorithms for quaternion algebras, Step 2 poses no practical problems. To compute the action of an ideal $\mathfrak{a}$ in Step 4, we note that the right order R' of the left R-ideal $Rf(\mathfrak{a})$ is isomorphic to the endomorphism ring $\mathrm{End}(E')$ of a curve E' with $j(E') = j^{\mathfrak{a}}$ by [22, Prop. 3.9]. The order R' is isomorphic to a unique order in the list, and we get a conjugacy class of supersingular j-invariants. Since roots of $H_D \bmod p$ which are not in $\mathbf{F}_p$ come in conjugate pairs, this allows us to compute all the Galois conjugates $j^{\mathfrak{a}}$.

4   Computing the Canonical Lift of a Supersingular Curve

In this section we explain how to compute the Hilbert class polynomial H_D of a discriminant $D < -4$ using a p-adic lifting technique for an inert prime $p \equiv 1 \bmod 12$. Our approach is based on the outline described in [7]. The condition $p \equiv 1 \bmod 12$ ensures that the j-values $0, 1728 \in \mathbf{F}_p$ are not roots of $H_D \in \mathbf{F}_p[X]$. The case where one of these two values is a root of $H_D \in \mathbf{F}_p[X]$ is more technical due to the extra automorphisms of the curve, and will be explained in detail in the first author's PhD thesis.

Under GRH, we can take p to be *small*. Indeed, our condition amounts to prescribing a Frobenius symbol in the degree 8 extension $\mathbf{Q}(\zeta_{12}, \sqrt{D})/\mathbf{Q}$, and by effective Chebotarev [14] we may take p to be of size $O((\log |D|)^2)$.

The first step of the algorithm is the same as for Algorithm 3 in Section 3: we compute a pair $(j(E_p), f_0) \in \mathrm{Emb}_D(\mathbf{F}_{p^2})$. The main step of the algorithm is to compute to sufficient p-adic precision the canonical lift $\widetilde{E}_p$ of this pair, defined in Section 2 as the inverse under the bijection π of Theorem 2.

For an arbitrary element $\eta \in \mathrm{Emb}_D(\mathbf{F}_{p^2})$, let

$$X_D(\eta) = \{(j(E), f) \mid j(E) \in \mathbf{C}_p, (j(E) \bmod p, f) = \eta\}$$

be a 'disc' of pairs lying over η. Here, $\mathbf{C}_p$ is the completion of an algebraic closure of $\mathbf{Q}_p$. The disc $X_D(\eta)$ contains the points of $\mathrm{Ell}_D(L)$ that reduce modulo p to the j-invariant corresponding to η.

These discs are similar to the discs used for the *split* case in [7,3]. The main difference is that now we need to keep track of the embedding as well. We can adapt the key idea of [7] to construct a p-adic analytic map from the set of discs to itself that has the CM-points as fixed points in the following way. Let $\mathfrak{a}$ be an $\mathcal{O}$-ideal of norm N that is coprime to p. We define a map

$$\rho_{\mathfrak{a}} : \bigcup_{\eta} X_D(\eta) \to \bigcup_{\eta} X_D(\eta)$$

as follows. For $(j(E), f) \in X_D(\eta)$, the ideal $f(\mathfrak{a}) \subset \mathrm{End}(E_p)$ defines a subgroup $E_p[f(\mathfrak{a})] \subset E_p[N]$ which lifts canonically to a subgroup $E[\mathfrak{a}] \subset E[N]$. We define $\rho_{\mathfrak{a}}((j(E), f)) = (j(E/E[\mathfrak{a}]), f^{\mathfrak{a}})$, where $f^{\mathfrak{a}}$ is as in Section 2. If the map f is clear, we also denote by $\rho_{\mathfrak{a}}$ the induced map on the j-invariants.

For principal ideals $\mathfrak{a} = (\alpha)$, the map $\rho_{\mathfrak{a}} = \rho_{\alpha}$ stabilizes every disc. Furthermore, as $\widetilde{E}_p[(\alpha)]$ determines an endomorphism of $\widetilde{E}_p$, the map ρ_{α} *fixes* the canonical lift $j(\widetilde{E}_p)$. As $j(E_p)$ does not equal $0, 1728 \in \mathbf{F}_p$, the map ρ_{α} is p-adic analytic by [3, Theorem 4.2].

Writing $\alpha = a + b\tau$, the derivative of ρ_{α} in a CM-point $j(\widetilde{E})$ equals $\alpha/\overline{\alpha} \in \mathbf{Z}_L$ by [3, Lemma 4.3]. For $p \nmid a, b$ this is a p-adic unit and we can use a modified version of Newton's method to converge to $j(\widetilde{E})$ starting from a random lift $(j_1, f_0) \in X_D(\eta)$ of the chosen point $\eta = (j(E_p), f_0) \in \mathbf{F}_{p^2}$. Indeed, the sequence

$$j_{k+1} = j_k - \frac{\rho_{\alpha}((j_k, f_0)) - j_k}{\alpha/\overline{\alpha} - 1} \tag{2}$$

converges quadratically to $j(\widetilde{E})$. The run time of the resulting algorithm to compute $j(\widetilde{E}) \in L$ up to the necessary precision depends heavily on the choice of α. We find a suitable α by sieving in the set $\{a + b\tau \mid a, b \in \mathbf{Z}, \gcd(a, b) = 1, a, b \neq 0 \bmod p\}$. We refer to the example in Section 6.3 for the explicit computation of the map ρ_α.

Once the canonical lift has been computed, the computation of the Galois conjugates is easier. To compute the Galois conjugate $j(\widetilde{E}_p)^{\mathfrak{l}}$ of an ideal $\mathfrak{l}$ of prime norm $\ell \neq p$, we first compute the value $j(E_p)^{\mathfrak{l}} \in \mathbf{F}_{p^2}$ as in Algorithm 3 in Section 3. We then compute all roots of the ℓ-th modular polynomial $\Phi_\ell(j(\widetilde{E}_p), X) \in L[X]$ that reduce to $j(E_p)^{\mathfrak{l}}$. If there is only one such root, we are done: this is the Galois conjugate we are after. In general, if $m \geq 1$ is the p-adic precision required to distinguish the roots, we compute the value $\rho_{\mathfrak{l}}((j(\widetilde{E}_p), f_0))$ to $m + 1$ p-adic digits precision to decide which root of the modular polynomial is the Galois conjugate. After computing all conjugates, we expand the product $\prod_{\mathfrak{a} \in \mathrm{Cl}(\mathcal{O})} \left(X - j(\widetilde{E}_p)^{\mathfrak{a}} \right) \in \mathbf{Z}_L[X]$ and recognize the coefficients as integers.

5   Complexity Analysis

This section is devoted to the run time analysis of Algorithm 1 and the proof of Theorem 1. To allow for an easier comparison with other methods to compute H_D, the analysis is carried out with respect to all relevant variables: the discriminant D, the class number $h(D)$, the logarithmic height n of the class polynomial and the largest prime generator $\ell(D)$ of the class group, before deriving a coarser bound depending only on D.

5.1   Some Number Theoretic Bounds

For the sake of brevity, we write llog for $\log\log$ and lllog for $\log\log\log$.

The bound given in Algorithm 1 on n, the bit size of the largest coefficient of the class polynomial, depends essentially on two quantities: the class number $h(D)$ of $\mathcal{O}$ and the sum $\sum_{[A,B,C]} \frac{1}{A}$, taken over a system of primitive reduced quadratic forms representing the class group $\mathrm{Cl}(\mathcal{O})$.

Lemma 1. *We have* $h(D) = O(|D|^{1/2} \log |D|)$. *If GRH holds true, we have* $h(D) = O(|D|^{1/2} \mathrm{llog} |D|)$.

Proof. By the analytic class number formula, we have to bound the value of the Dirichlet L-series $L(s, \chi_D)$ associated to D at $s = 1$. The unconditional bound follows directly from [20], the conditional bound follows from [16].   □

Lemma 2. *We have* $\sum_{[A,B,C]} \frac{1}{A} = O((\log |D|)^2)$. *If GRH holds true, we have* $\sum_{[A,B,C]} \frac{1}{A} = O(\log |D| \, \mathrm{llog} |D|)$.

Proof. The bound $\sum_{[A,B,C]} \frac{1}{A} = O((\log |D|)^2)$ is proved in [18] with precise constants in [9]; the argument below will give a different proof of this fact.

By counting the solutions of $B^2 \equiv D \bmod 4A$ for varying A and using the Chinese remainder theorem, we obtain

$$\sum_{[A,B,C]} \frac{1}{A} \leq \sum_{A \leq \sqrt{|D|}} \frac{\prod_{p|A}\left(1 + \left(\frac{D}{p}\right)\right)}{A}.$$

The Euler product expansion bounds this by $\prod_{p \leq \sqrt{|D|}} \left(1 + \frac{1}{p}\right)\left(1 + \frac{\left(\frac{D}{p}\right)}{p}\right)$. By Mertens theorem, this is at most $c \log|D| \prod_{p \leq \sqrt{|D|}} \frac{1}{1-\left(\frac{D}{p}\right)/p}$ for some constant $c > 0$. This last product is essentially the value of the Dirichlet L-series $L(1, \chi_D)$ and the same remarks as in Lemma 1 apply. $\qquad\square$

Lemma 3. *If GRH holds true, the primes needed for Algorithm 1 are bounded by* $O\left(h(D)\max(h(D)(\log|D|)^4, n)\right)$.

Proof. Let $k(D)$ be the required number of splitting primes. We have $k(D) \in O\left(\frac{n}{\log|D|}\right)$, since each prime has at least $\log_2|D|$ bits.

Let $\pi_1(x, K_{\mathcal{O}}/\mathbf{Q})$ be the number of primes up to $x \in \mathbf{R}_{>0}$ that split completely in $K_{\mathcal{O}}/\mathbf{Q}$. By [14, Th. 1.1] there is an effectively computable constant $c \in \mathbf{R}_{>0}$, independent of D, such that

$$\left|\pi_1(x, K_{\mathcal{O}}/\mathbf{Q}) - \frac{\mathrm{Li}(x)}{2h(D)}\right| \leq c\left(\frac{x^{1/2}\log(|D|^{h(D)}x^{2h(D)})}{2h(D)} + \log(|D|^{h(D)})\right), \quad (3)$$

where we have used the bound $\mathrm{disc}(K_{\mathcal{O}}/\mathbf{Q}) \leq |D|^{h(D)}$ proven in [3, Lemma 3.1]. It suffices to find an $x \in \mathbf{R}_{>0}$ for which $k(D) - \mathrm{Li}(x)/(2h(D))$ is larger than the right hand side of (3). Using the estimate $\mathrm{Li}(x) \sim x/\log x$, we see that the choice $x = O\left(\max(h(D)^2\log^4|D|, h(D)n)\right)$ works. $\qquad\square$

5.2 Complexity of Algorithm 2

Let us fix some notation and briefly recall the complexities of the asymptotically fastest algorithms for basic arithmetic. Let $M(\log p) \in O(\log p \, \mathrm{llog}\, p \, \mathrm{lllog}\, p)$ be the time for a multiplication in $\mathbf{F}_p$ and $M_X(\ell, \log p) \in O(\ell \log \ell \, M(\log p))$ the time for multiplying two polynomials over $\mathbf{F}_p$ of degree ℓ.

As the final complexity will be exponential in $\log p$, we need not worry about the detailed complexity of polynomial or subexponential steps. Writing $4p = u^2 - v^2 D$ takes polynomial time by the Cornacchia and Tonelli–Shanks algorithms [5, Sec 1.5]. By Lemma 3, we may assume that v is polynomial in $\log|D|$.

Concerning Step 2, we expect to check $O(p/h(D))$ curves until finding one with endomorphism ring $\mathcal{O}$. To test if a curve has the desired cardinality, we need to compute the orders of $O(\mathrm{llog}\, p)$ points, and each order computation takes time $O\left((\log p)^2\, M(\log p)\right)$. Among the curves with the right cardinality, a fraction of $\frac{h(D)}{H(v^2D)}$, where $H(v^2D)$ is the Kronecker class number, has the desired endomorphism ring. So we expect to apply Kohel's algorithm with run time $O(p^{1/3+o(1)})$ an expected $\frac{H(v^2 D)}{h(D)} \in O(v \operatorname{llog} v)$ times. As $p^{1/3}$ is dominated by $p/h(D)$ of order about $p^{1/2}$, Step 2 takes time altogether

$$O\left(\frac{p}{h(D)}(\log p)^2 M(\log p) \operatorname{llog} p\right). \tag{4}$$

Heuristically, we only check if some random points are annihilated by $p+1 \pm u$ and do not compute their actual orders. The $(\log p)^2$ in (4) then becomes $\log p$.

In Step 3, the decomposition of the class group into a product of cyclic groups takes subexponential time. Furthermore, since all involved primes ℓ_i are of size $O((\log |D|)^2)$ under GRH, the time needed to compute the modular polynomials is negligible. Step 3 is thus dominated by $O(h(D))$ evaluations of reduced modular polynomials and by the computation of their roots.

Once $\Phi_\ell \bmod p$ is computed, it can be evaluated in time $O(\ell^2 M(\log p))$. Finding its roots is dominated by the computation of X^p modulo the specialized polynomial of degree $\ell + 1$, which takes time $O(\log p \, M_X(\ell, \log p))$. Letting $\ell(D)$ denote the largest prime needed to generate the class group, Step 3 takes time

$$O\left(h(D)\ell(D) M(\log p)(\ell(D) + \operatorname{llog} |D| \log p)\right). \tag{5}$$

Under GRH, $\ell(D) \in O((\log |D|)^2)$, and heuristically, $\ell(D) \in O\left((\log |D|)^{1+\varepsilon}\right)$.

By organizing the multiplications of polynomials in a tree of height $O(\log h)$, Step 4 takes $O(\log h(D) M_X(h(D), \log p))$, which is dominated by Step 3. We conclude that the total complexity of Algorithm 2 is dominated by Steps 2 and 3 and given by the sum of (4) and (5).

5.3   Proof of Theorem 1

We assume that $\mathcal{P} = \{p_1, p_2, \ldots\}$ is chosen as the set of the smallest primes p that split into principal ideals of $\mathcal{O}$. Notice that $\log p, \log h(D) \in O(\log |D|)$, so that we may express all logarithmic quantities with respect to D.

The dominant part of the algorithm are the $O(n/\log |D|)$ invocations of Algorithm 2 in Step 2. Specializing (4) and (5), using the bound on the largest prime of Lemma 3 and assuming that $\ell(D) \in \Omega(\log |D| \operatorname{llog} |D|)$, this takes time

$$O\left(n \, M(\log |D|) \left(h(D)\frac{\ell(D)^2}{\log |D|} + \log |D| \operatorname{llog}|D| \max\left(h(D)(\log |D|)^4, n\right)\right)\right). \tag{6}$$

Finally, the fast Chinese remainder algorithm takes $O(M(\log N) \operatorname{llog} N)$ by [12, Th. 10.25], so that Step 3 can be carried out in $O(h(D) M(n) \log |D|)$, which is also dominated by Step 2. Plugging the bounds of Lemmata 1 and 2 into (6) proves the rigorous part of Theorem 1.

For the heuristic result, we note that Lemma 3 overestimates the size of the primes, since it gives a very high bound already for the *first* split prime. Heuristically, one would rather expect that all primes are of size $O(nh)$. Combined with the heuristic improvements to (4) and (5), we find the run time

$$O\left(n \, M(\log |D|) \left(n + h(D)\frac{\ell(D)^2}{\log |D|}\right)\right). \qquad \square$$

5.4 Comparison

The bounds under GRH of Lemmata 1 and 2 also yield a tighter analysis for other algorithms computing H_D. By [9, Th. 1], the run time of the complex analytic algorithm turns out to be $O(|D|(\log|D|)^3(\operatorname{llog}|D|)^3)$, which is essentially the same as the heuristic bound of Theorem 1.

The run time of the p-adic algorithm becomes $O(|D|(\log|D|)^{6+o(1)})$. A heuristic run time analysis of this algorithm has not been undertaken, but it seems likely that again $O(|D|(\log|D|)^{3+o(1)})$ would be reached.

6 Examples and Practical Considerations

6.1 Inert Primes

For very small primes there is a unique supersingular j-invariant in characteristic p. For example, for $D \equiv 5 \bmod 8$, the prime $p = 2$ is inert in $\mathcal{O}_D$ and we immediately have $H_D \bmod 2 = X^{h(D)}$.

More work needs to be done if there is more than one supersingular j-invariant in $\mathbf{F}_{p^2}$, as illustrated by computing $H_{-71} \bmod 53$. The ideal $\mathfrak{a} = (2, 3 + \tau)$ generates the order 7 class group of $\mathcal{O} = \mathbf{Z}[\tau]$. The quaternion algebra $\mathcal{A}_{p,\infty}$ has a basis $\{1, i, j, k\}$ with $i^2 = -2, j^2 = -35, ij = k$, and the maximal order R with basis $\{1, i, 1/4(2 - i - k), -1/2(1 + i + j)\}$ is isomorphic to the endomorphism ring of the curve with j-invariant 50. We compute the embedding $f : \tau \mapsto y = 1/2 - 3/2i + 1/2j \in R$, where y satisfies $y^2 - y + 18 = 0$. Calculating the right orders of the left R-ideals $Rf(\mathfrak{a}^i)$ for $i = 1, \ldots, 7$, we get a sequence of orders corresponding to the j-invariants $28 \pm 9\sqrt{2}, 46, 0, 46, 28 \pm 9\sqrt{2}, 50, 50$ and compute $H_{-71} \bmod 53 = X(X - 46)^2(X - 50)^2(X^2 + 50X + 39)$.

6.2 Totally Split Primes

For $D = -71$, the smallest totally split prime is $p = 107 = \frac{12^2 + 4 \cdot 71}{4}$. Any curve over $\mathbf{F}_p$ with endomorphism ring $\mathcal{O}$ is isomorphic to a curve with $m = p + 1 \pm 12 = 96$ or 120 points. By trying randomly chosen j-invariants, we find that $E : Y^2 = X^3 + X + 35$ has 96 points. We either have $\operatorname{End}(E) = \mathcal{O}_D$ or $\operatorname{End}(E) = \mathcal{O}_{4D}$. In this simple case there is no need to apply Kohel's algorithm. Indeed, $\operatorname{End}(E)$ equals $\mathcal{O}_D$ if and only if the complete 2-torsion is $\mathbf{F}_p$-rational. The curve E has only the point $P = (18, 0)$ as rational 2-torsion point, and therefore has endomorphism ring $\mathcal{O}_{4D}$. The 2-isogenous curve $E' = E/\langle P \rangle$ given by $Y^2 = X^3 + 58X + 59$ of j-invariant 19 has endomorphism ring $\mathcal{O}_D$.

The smallest odd prime generating the class group is $\ell = 3$. The third modular polynomial $\Phi_\ell(X, Y)$ has the two roots $46, 63$ when evaluated in $X = j(E') = 19 \in \mathbf{F}_p$. Both values are roots of $H_D \bmod p$. We successively find the other Galois conjugates $64, 77, 30, 57$ using the modular polynomial Φ_ℓ and expand

$$H_{-71} \bmod 107 = X^7 + 72X^6 + 93X^5 + 73X^4 + 46X^3 + 29X^2 + 30X + 19.$$

6.3 Inert Lifting

We illustrate the algorithm of Section 4 by computing H_D for $D = -56$.

The prime $p = 37$ is inert in $\mathcal{O} = \mathcal{O}_D$. The supersingular j-invariants in characteristic p are $8, 3 \pm 14\sqrt{-2}$. We fix a curve $E = E_p$ with j-invariant 8. We take the basis $\{1, i, j, k\}$ with $i^2 = -2, j^2 = j - 5, ij = k$ of the quaternion algebra $\mathcal{A}_{p,\infty}$. This basis is also a $\mathbf{Z}$-basis for a maximal order $R \subset \mathcal{A}_{p,\infty}$ that is isomorphic to the endomorphism ring $\mathrm{End}(E_p)$.

Writing $\mathcal{O}_D = \mathbf{Z}[\tau]$, we compute an element $y = [0, 1, 1, -1] \in R$ satisfying $y^2 + 56 = 0$. This determines the embedding $f = f_0$ and we need to lift the pair (E, f) to its canonical lift. As element α for the 'Newton map' ρ_α, we use a generator of $\mathfrak{a}^4$ where $\mathfrak{a} = (3, 1 + \tau)$ is a prime lying over 3.

To find the kernel $E[f(\mathfrak{a})]$ we check which 3-torsion points $P \in E[3]$ are killed by $f(1 + \tau) \in \mathrm{End}(E)$. We find $P = 18 \pm 9\sqrt{-2}$, and use Vélu's formulas to find $E^\mathfrak{a} \cong E$ of j-invariant 8. As E and $E^\mathfrak{a}$ are isomorphic, it is easy to compute $f^\mathfrak{a}$. We compute a left-generator $x = [1, 1, 0, 0] \in R$ of the left R-ideal $Rf(\mathfrak{a})$ to find $f^\mathfrak{a}(\tau) = xy/x = [-1, 0, 1, 1] \in R$.

Next, we compute the $\mathfrak{a}$-action on the pair $(E^\mathfrak{a}, f^\mathfrak{a}) = (E, f^\mathfrak{a})$. We find that $P = 19 \pm 12\sqrt{a}$ is annihilated by $f^\mathfrak{a}(1 + \tau) \in \mathrm{End}(E)$. The curve $E^{\mathfrak{a}^2}$ of j-invariant $3 - 14\sqrt{-2}$ is not isomorphic to E. We pick a 2-isogeny $\varphi_\mathfrak{b} : E^\mathfrak{a} \to E^{\mathfrak{a}^2}$ with kernel $\langle 19 + 23\sqrt{-2}\rangle$. The ideal $\mathfrak{b}$ has basis $\{2, i + j, 2j, k\}$ and is left-isomorphic to $Rf^\mathfrak{a}(\mathfrak{a})$ via left-multiplication by $x' = [-1, 1/2, 1/2, -1/2] \in R$. We get $f^{\mathfrak{a}^2}(\tau) = x'y/x' = [0, 1, 1, -1] \in R_\mathfrak{b}$ and we use the map $g_\mathfrak{b}$ from Section 2 to view this as an embedding into $\mathrm{End}(E^{\mathfrak{a}^2})$.

The action of $\mathfrak{a}^3$ and $\mathfrak{a}^4$ is computed in the same way. We find a cycle of 3-isogenies

$$(E, f) \to (E^\mathfrak{a} = E, f^\mathfrak{a}) \to (E^{\mathfrak{a}^2}, f^{\mathfrak{a}^2}) \to (E^{\mathfrak{a}^3}, f^{\mathfrak{a}^3}) \to (E^{\mathfrak{a}^4}, f^{\mathfrak{a}^4}) = (E, f)$$

where each element of the cycle corresponds uniquely to a root of H_D. We have now also computed $H_D \bmod p = (X - 8)^2(X^2 - 6X - 6)$.

As a lift of E we choose the curve defined by $Y^2 = X^3 + 210X + 420$ over the unramified extension L of degree 2 of $\mathbf{Q}_p$. We lift the cycle of isogenies over $\mathbf{F}_{p^2}$ to L in 2 p-adic digits precision using Hensel's lemma, and update according to the Newton formula (2) to find $j(\widetilde{E}) = -66 + 148\sqrt{-2} + O(p^2)$. Next we work with 4 p-adic digits precision, lift the cycle of isogenies and update the j-invariant as before. In this example, it suffices to work with 16 p-adic digits precision to recover $H_D \in \mathbf{Z}[X]$.

Since we used a generator of an ideal generating the class group, we get the Galois conjugates of $j(\widetilde{E})$ as a byproduct of our computation. In the end we expand the polynomial $H_{-56} = \prod_{\mathfrak{a}\in\mathrm{Cl}(\mathcal{O})}(X - j(\widetilde{E})^\mathfrak{a}) \in \mathbf{Z}[X]$ which has coefficients with up to 23 decimal digits.

6.4   Chinese Remainder Theorem

As remarked in Section 5.4, the heuristic run time of Theorem 1 is comparable to the expected run times of both the complex analytic and the p-adic approaches from [9] and [7,3]. To see if the CRT-approach is comparable *in practice* as well, we computed an example with a reasonably sized discriminant $D = -108708$, the first discriminant with class number 100.

The *a posteriori* height of H_D is 5874 bits, and we fix a target precision of $n = 5943$. The smallest totally split prime is 27241. If only such primes are used, the largest one is 956929 for a total of 324 primes. Note that these primes are indeed of size roughly $|D|$, in agreement with Lemma 3. We have partially implemented the search for a suitable curve: for each $4p = u^2 - v^2 D$ we look for the first j-invariant such that for a random point P on an associated curve, $(p+1)P$ and uP have the same X-coordinate. This allows us to treat the curve and its quadratic twist simultaneously. The largest occurring value of v is 5. Altogether, 487237 curves need to be checked for the target cardinality.

On an Athlon-64 2.2 GHz computer, this step takes roughly 18.5 seconds. As comparison, the third authors' complex analytic implementation takes 0.3 seconds on the same machine. To speed up the multi-prime approach, we incorporated some inert primes. Out of the 168 primes less than 1000, there are 85 primes that are inert in $\mathcal{O}$. For many of them, the computation of H_D mod p is trivial. Together, these primes contribute 707 bits and we only need 288 totally split primes, the largest one being 802597. The required 381073 curve cardinalities are tested in 14.2 seconds.

One needs to be careful when drawing conclusions from only few examples, but the difference between 14.2 and 0.3 seconds suggests that the implicit constants in the O-symbol are worse for the CRT-approach.

6.5   Class Invariants

For many applications, we are mostly interested in a generating polynomial for the ring class field $K_{\mathcal{O}}$. As the Hilbert class polynomial has very large coefficients, it is then better to use 'smaller functions' than the j-function to save a constant factor in the size of the polynomials. We refer to [17,21] for the theory of such *class invariants*.

There are theoretical obstructions to incorporating class invariants into Algorithm 1. Indeed, if a modular function f has the property that there are class invariants $f(\tau_1)$ and $f(\tau_2)$ with different minimal polynomials, we cannot use the CRT-approach. This phenomenon occurs for instance for the double eta quotients described in [10]. For the discriminant D in Section 6.4, we can use the double eta quotient of level $3 \cdot 109$ to improve the 0.3 seconds of the complex analytic approach. For CRT, we need to consider less favourable class invariants.

Acknowledgement

We thank Dan Bernstein, François Morain and Larry Washington for helpful discussions.

References

1. Agashe, A., Lauter, K., Venkatesan, R.: Constructing elliptic curves with a known number of points over a prime field. In: van der Poorten, A.J., Stein, A. (eds.) High Primes and Misdemeanours: Lectures in Honour of the 60th Birthday of H C Williams. Fields Inst. Commun., vol. 41, pp. 1–17 (2004)

2. Atkin, A.O.L., Morain, F.: Elliptic curves and primality proving. Math. Comp. 61(203), 29–68 (1993)
3. Bröker, R.: A p-adic algorithm to compute the Hilbert class polynomial. Math. Comp. (to appear)
4. Cerviño, J.M.: Supersingular elliptic curves and maximal quaternionic orders. In: Math. Institut G-A-Univ. Göttingen, pp. 53–60 (2004)
5. Cohen, H.: A Course in Computational Algebraic Number Theory. In: Graduate Texts in Mathematics, vol. 138, Springer, Heidelberg (1993)
6. Cohen, H., Frey, G., Avanzi, R., Doche, C., Lange, T., Nguyen, K., Vercauteren, F.: Handbook of Elliptic and Hyperelliptic Curve Cryptography. In: Discrete Mathematics and its Applications, Chapman & Hall/CRC (2005)
7. Couveignes, J.-M., Henocq, T.: Action of modular correspondences around CM points. In: Fieker, C., Kohel, D.R. (eds.) ANTS 2002. LNCS, vol. 2369, pp. 234–243. Springer, Heidelberg (2002)
8. Deuring, M.: Die Typen der Multiplikatorenringe elliptischer Funktionenkörper. Abh. Math. Sem. Univ. Hamburg 14, 197–272 (1941)
9. Enge, A.: The complexity of class polynomial computation via floating point approximations. HAL-INRIA 1040 = arXiv:cs/0601104, INRIA (2006), http://hal.inria.fr/inria-00001040
10. Enge, A., Schertz, R.: Constructing elliptic curves over finite fields using double eta-quotients. J. Théor. Nombres Bordeaux 16, 555–568 (2004)
11. Fouquet, M., Morain, F.: Isogeny volcanoes and the SEA algorithm. In: Fieker, C., Kohel, D.R. (eds.) ANTS 2002. LNCS, vol. 2369, pp. 276–291. Springer, Heidelberg (2002)
12. von zur Gathen, J., Gerhard, J.: Modern Computer Algebra. Cambridge University Press, Cambridge (1999)
13. Kohel, D.: Endomorphism Rings of Elliptic Curves over Finite Fields. PhD thesis, University of California at Berkeley (1996)
14. Lagarias, J.C., Odlyzko, A.M.: Effective versions of the Chebotarev density theorem. In: Fröhlich, A. (ed.) Algebraic Number Fields (L-functions and Galois properties), pp. 409–464. Academic Press, London (1977)
15. Lang, S.: Elliptic Functions. In: GTM 112, 2nd edn., Springer, New York (1987)
16. Littlewood, J.E.: On the class-number of the corpus $P(\sqrt{-k})$. Proc. London Math. Soc. 27, 358–372 (1928)
17. Schertz, R.: Weber's class invariants revisited. J. Théor. Nombres Bordeaux 14(1), 325–343 (2002)
18. Schoof, R.: The exponents of the groups of points on the reductions of an elliptic curve. In: van der Geer, G., Oort, F., Steenbrink, J. (eds.) Arithmetic Algebraic Geometry, pp. 325–335. Birkhäuser, Basel (1991)
19. Schoof, R.: Counting points on elliptic curves over finite fields. J. Théor. Nombres Bordeaux 7, 219–254 (1995)
20. Schur, I.: Einige Bemerkungen zu der vorstehenden Arbeit des Herrn G. Pólya: Über die Verteilung der quadratischen Reste und Nichtreste. Nachr. Kön. Ges. Wiss. Göttingen, Math.-Phys. Kl, pp. 30–36 (1918)
21. Stevenhagen, P.: Hilbert's 12th problem, complex multiplication and Shimura reciprocity. In: Miyake, K. (ed.) Class Field Theory—its Centenary and Prospect, pp. 161–176. Amer. Math. Soc. (2001)
22. Waterhouse, W.C.: Abelian varieties over finite fields. Ann. Sci. École Norm. Sup. (4) 2, 521–560 (1969)

Computing Zeta Functions in Families of $C_{a,b}$ Curves Using Deformation

Wouter Castryck[2], Hendrik Hubrechts[1,*], and Frederik Vercauteren[2,*]

[1] Department of Mathematics, University of Leuven,
Celestijnenlaan 200B, B-3001 Leuven-Heverlee, Belgium
hendrik.hubrechts@wis.kuleuven.be
[2] Department of Electrical Engineering, University of Leuven
Kasteelpark Arenberg 10, B-3001 Leuven-Heverlee, Belgium
firstname.lastname@esat.kuleuven.be

Abstract. We apply deformation theory to compute zeta functions in a family of $C_{a,b}$ curves over a finite field of small characteristic. The method combines Denef and Vercauteren's extension of Kedlaya's algorithm to $C_{a,b}$ curves with Hubrechts' recent work on point counting on hyperelliptic curves using deformation. As a result, it is now possible to generate $C_{a,b}$ curves suitable for use in cryptography in a matter of minutes.

1 Introduction

The development of algorithms that compute the Hasse-Weil zeta function of a curve over a finite field has witnessed several revolutions in the past 20 years, partly motivated by applications in cryptography. The first was the Schoof-Elkies-Atkin algorithm [18] to compute the number of points on an elliptic curve over a finite field. Although this algorithm readily generalises to higher genus, it is not really practical except in the genus 2 case for moderately sized finite fields [7]. The second revolution was the canonical lift approach introduced by Satoh [17] and reinterpreted by Mestre [15] using the AGM. Extensions and improvements of this algorithm (an overview is given in [2]) resulted in very efficient point counting methods for ordinary elliptic and hyperelliptic curves over finite fields of small characteristic. The third revolution was the p-adic cohomological approach introduced by Kedlaya [10] and Lauder and Wan [12]. Although the resulting algorithms are polynomial time for fixed characteristic, they are only practical for hyperelliptic curves. Finally, the fourth revolution consists of two components, deformation and fibration, and was introduced by Lauder [13,14] to compute the zeta function of higher dimensional hypersurfaces.

Despite the efforts of many researchers, the ultimate goal of having a set of algorithms that can handle any given curve of genus g over any finite field $\mathbb{F}_q$ where q^g is limited to having several hundred bits, is still far off. In fact, up to the time of writing, only the case of elliptic curves (both in large and small

---

* Postdoctoral Fellow of the Research Foundation – Flanders (FWO).

A.J. van der Poorten and A. Stein (Eds.): ANTS-VIII 2008, LNCS 5011, pp. 296–311, 2008.

characteristic) and the case of hyperelliptic and superelliptic [6] curves in small characteristic have a satisfactory solution.

Although tiny steps towards tackling the large characteristic case have been made [4], handling all curves over finite fields of small characteristic looks much more feasible. In the latter case, there has been partial progress to include $C_{a,b}$ [3] and non-degenerate curves [1], but these algorithms are not sufficiently practical. Although the approach is similar to Kedlaya's algorithm for hyperelliptic curves, these algorithms use a different Frobenius lifting technique, which makes them slow.

The goal of this paper is to remedy this situation by taking a totally different approach based on deformation theory. Although this theory was primarily introduced for high dimensional hypersurfaces, Hubrechts [9,8] showed it to be efficient in the hyperelliptic case.

The advantage of using deformation for the broader classes of $C_{a,b}$, non-degenerate or even more general curves is twofold: firstly, it avoids the explicit computation of the Frobenius lift that makes the algorithms in [3] and [1] slow and secondly, the core of the algorithms, i.e. solving a p-adic differential equation, is always the same. Only the computation of the so-called connection matrix differs for each class of curves, but is in itself a much easier problem than developing an efficient differential reduction method as needed in Kedlaya's approach.

In this paper we present a detailed version of this method for $C_{a,b}$ curves, which should readily extend to non-degenerate curves. Our algorithm is used in two applications: firstly, given a random $C_{a,b}$ curve over a finite field $\mathbb{F}_q$, compute its zeta function and secondly, given a finite field $\mathbb{F}_q$, generate $C_{a,b}$ curves whose Jacobian has nearly prime order for use in cryptography. The speed-up over known techniques for the second application is remarkable: after a precomputation, computing the zeta function of each member of a family with a Jacobian of 160-bit order only takes a few seconds. As a result, generating cryptographically useful $C_{a,b}$ curves now is feasible in a matter of minutes.

The remainder of this paper is organised as follows: Section 2 reviews p-adic cohomology and deformation for general curves and Section 3 covers the necessary background on $C_{a,b}$ curves. Section 4 studies relative Monsky-Washnitzer cohomology for a family of $C_{a,b}$ curves, resulting in a practical algorithm described and analysed in Section 5. Finally, Section 6 reports on a preliminary Magma implementation of this algorithm.

2 p-Adic Cohomology and Deformation

Throughout this section, the survey paper on Monsky-Washnitzer cohomology by van der Put [16] and the AWS 2007 lecture notes by Kedlaya [11, Chapter 3] are implicit references.

2.1   Zeta Functions and Cohomology

Let $\mathbb{F}_q$ be a finite field of characteristic p with q elements. The zeta function of a polynomial $\overline{C}(x,y) \in \mathbb{F}_q[x,y]$ defining a non-singular affine curve is determined by the action of the Frobenius endomorphism $\overline{\mathcal{F}}_q : \overline{A} \to \overline{A} : a \mapsto a^q$ on a certain cohomology space $H^1_{MW}(\overline{A}/\mathbb{Q}_q)$. Here $\overline{A}$ denotes $\mathbb{F}_q[x,y]/(\overline{C}(x,y))$ and $H^1_{MW}(\overline{A}/\mathbb{Q}_q)$ is constructed as follows. Let $\mathbb{Q}_q$ be an unramified extension of the field of p-adic numbers $\mathbb{Q}_p$, with valuation ring $\mathbb{Z}_q$ and residue field $\mathbb{F}_q$. Let $C(x,y) \in \mathbb{Z}_q[x,y]$ be such that it reduces to $\overline{C}(x,y)$ mod p and consider the $\mathbb{Z}_q$-algebra

$$A^\dagger = \frac{\mathbb{Z}_q\langle x,y\rangle^\dagger}{(C(x,y))}$$

where $\mathbb{Z}_q\langle x,y\rangle^\dagger$ is the *weak completion* of $\mathbb{Z}_q[x,y]$. It consists of power series $\sum a_{i,j}x^iy^j \in \mathbb{Z}_q[[x,y]]$ for which there is a $\rho \in \,]0,1[$ such that $|a_{i,j}|_p/\rho^{i+j} \to 0$ as $i+j \to \infty$. The idea behind this convergence condition is that $\mathbb{Z}_q\langle x,y\rangle^\dagger$ should be closed under integration. Let $D^1(A^\dagger)$ be the universal module of differentials on $A^\dagger$ over $\mathbb{Z}_q$ and let $d : A^\dagger \to D^1(A^\dagger)$ be the usual exterior derivation. Then $H^1_{MW}(\overline{A}/\mathbb{Q}_q)$ is defined as

$$\frac{D^1(A^\dagger)}{d(A^\dagger)} \otimes_{\mathbb{Z}_q} \mathbb{Q}_q,$$

which turns out to be the right object for the following theorem to hold.

Theorem 1 (Monsky, Washnitzer). *There exists a $\mathbb{Z}_q$-algebra endomorphism $\mathcal{F}_q : A^\dagger \to A^\dagger$ that lifts $\overline{\mathcal{F}}_q$ in the sense that $\overline{\mathcal{F}}_q \circ \pi = \pi \circ \mathcal{F}_q$, where $\pi : A^\dagger \to \overline{A}$ is reduction mod p. For any such lift, the induced map $\mathcal{F}^*_q : D^1(A^\dagger) \to D^1(A^\dagger)$ is well-defined modulo $d(A^\dagger)$ and acts on $H^1_{MW}(\overline{A}/\mathbb{Q}_q)$ as an invertible $\mathbb{Q}_q$-vector space morphism, which does not depend on the choice of $\mathcal{F}_q$. Moreover, the zeta function of $\overline{C}$ is given by*

$$Z_{\overline{C}}(T) = \frac{\det\left(\mathbb{I} - q\mathcal{F}^{*\,-1}_q \middle| H^1_{MW}(\overline{A}/\mathbb{Q}_q)\right)}{1-qT}.$$

2.2   Relative Cohomology

Let $\overline{C}(x,y,t) \in \mathbb{F}_q[t][x,y]$ define a family of smooth curves over an open dense subset Spec $\overline{S}$ of the affine t-line. Thus $\overline{S} = \mathbb{F}_q[t,\overline{r}(t)^{-1}]$ for some nonzero $\overline{r}(t) \in \mathbb{F}_q[t]$. Write $\overline{A} = \overline{S}[x,y]/(\overline{C}(x,y,t))$ and, for every $\overline{t}_0 \in \mathbb{F}_q$ where $\overline{r}(t)$ does not vanish, write $\overline{A}_{\overline{t}_0}$ for $\overline{A}/(t-\overline{t}_0)$, the coordinate ring of the fiber at $\overline{t}_0$. Then the aim of relative cohomology is to describe how the action of Frobenius on $H^1_{MW}(\overline{A}_{\overline{t}_0}/\mathbb{Q}_q)$ alters as $\overline{t}_0$ varies. Let $C(x,y,t) \in \mathbb{Z}_q[t][x,y]$ and $r(t) \in \mathbb{Z}_q[t]$ be such that they reduce mod p to $\overline{C}(x,y,t)$ and $\overline{r}(t)$ respectively. Define $S^\dagger = \mathbb{Z}_q\langle t,r(t)^{-1}\rangle^\dagger = \mathbb{Z}_q\langle t,z\rangle^\dagger/(zr(t)-1)$ along with the $S^\dagger$-module

$$A^\dagger = \frac{\mathbb{Z}_q\langle t,r(t)^{-1},x,y\rangle^\dagger}{(C(x,y,t))}$$

(the weak completion being realised as in the bivariate case). Note that there is a well-defined p-adic valuation on $S^\dagger$ and $A^\dagger$. Let $D^1_t(A^\dagger)$ be the universal module of differentials on $A^\dagger$ over $S^\dagger$ and let $d_t : A^\dagger \to D^1_t(A^\dagger)$ be the corresponding

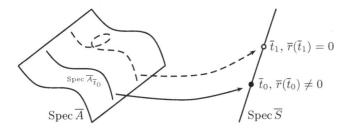

exterior derivation. Thus in all this, t is left constant. Write $S^\dagger_{\mathbb{Q}_q} = S^\dagger \otimes_{\mathbb{Z}_q} \mathbb{Q}_q$. Then our object of interest is the $S^\dagger_{\mathbb{Q}_q}$-module

$$H^1_{MW}(\overline{A}/S^\dagger_{\mathbb{Q}_q}) = \frac{D^1_t(A^\dagger)}{d_t(A^\dagger)} \otimes_{\mathbb{Z}_q} \mathbb{Q}_q.$$

As above, one can show that there exists a $\mathbb{Z}_q$-algebra endomorphism $\mathcal{F}_q$ on $A^\dagger$ that lifts the Frobenius action $\overline{\mathcal{F}}_q$ on $\overline{A}$. Moreover, one can realise that $\mathcal{F}_q(t) = t^q$ (we will illustrate this in Section 4 in our specific families of $C_{a,b}$ curves). The induced map $\mathcal{F}^*_q$ on $H^1_{MW}(\overline{A}/S^\dagger_{\mathbb{Q}_q})$ is well-defined, though in general it is not an $S^\dagger_{\mathbb{Q}_q}$-module endomorphism.

Let $\bar{t}_0 \in \mathbb{F}_q$ be a non-zero of $\bar{r}(t)$ and let $\hat{t}_0 \in \mathbb{Z}_q$ be its Teichmüller lift, i.e. the unique root of $X^q - X \in \mathbb{Z}_q[X]$ that reduces to $\bar{t}_0$ mod p. Then one sees that $H^1_{MW}(\overline{A}_{\bar{t}_0}/\mathbb{Q}_q)$ can be identified with $H^1_{MW}(\overline{A}/S^\dagger_{\mathbb{Q}_q})/(t - \hat{t}_0)$, and that $\mathcal{F}^*_q$ induces a well-defined map on $H^1_{MW}(\overline{A}_{\bar{t}_0}/\mathbb{Q}_q)$ which exactly matches with the Frobenius action described in Theorem 1.

In summary, the action of Frobenius on a single fiber can be obtained from the relative Frobenius action by substituting for t a suitable Teichmüller representative. So one could think of the relative Frobenius action as an interpolation of the Frobenius actions on all fibers in the family.

2.3   The Gauss-Manin Connection

In addition to the notation from above, we introduce $D^1(A^\dagger)$, denoting the module of differentials on $A^\dagger$ over $\mathbb{Z}_q$ (so t is no longer left constant), and $D^2(A^\dagger) = \bigwedge^2 D^1(A^\dagger)$, denoting the corresponding module of 2-forms. Let d be the usual exterior derivation, both on $A^\dagger$ and $D^1(A^\dagger)$, giving rise to the Monsky-Washnitzer complex

$$0 \to A^\dagger \overset{d}{\to} D^1(A^\dagger) \overset{d}{\to} D^2(A^\dagger) \to 0$$

of the surface $\operatorname{Spec}\overline{A}$ over $\mathbb{F}_q$. Note that we have a natural surjective morphism $D^1(A^\dagger) \to D^1_t(A^\dagger) : dt \mapsto 0$, thus we can identify $D^1_t(A^\dagger)$ with $D^1(A^\dagger)/(dt)$.

Definition 1. *The* Gauss-Manin connection

$$\nabla : H^1_{MW}(\overline{A}/S^\dagger_{\mathbb{Q}_q}) \to H^1_{MW}(\overline{A}/S^\dagger_{\mathbb{Q}_q}) : \omega \mapsto \nabla(\omega)$$

is constructed as follows. For large enough $e \in \mathbb{N}$, let $p^e \omega$ be represented by a 1-form $\widetilde{\omega} \in D^1(A^\dagger)$ and take its exterior derivative $d(\widetilde{\omega}) \in D^2(A^\dagger)$, which one can always write as $\widetilde{\varphi} \wedge dt$ for some $\widetilde{\varphi} \in D^1(A^\dagger)$. Reduce $\widetilde{\varphi}$ modulo (dt) to end up in $D^1_t(A^\dagger)$. Then reduce modulo $d_t(A^\dagger)$ and tensor with p^{-e}, so that one ends up in $H^1_{MW}(\overline{A}/S^\dagger_{\mathbb{Q}_q})$: this is $\nabla(\omega)$.

We leave it to the reader to show that the above is well-defined, i.e. $\nabla(\omega)$ does not depend on the choice of e, $\widetilde{\omega}$ and $\widetilde{\varphi}$. Remark that the above construction does not result in a geometric connection in the usual sense of the word, in which case ∇ should take values in $H^1_{MW}(\overline{A}/S^\dagger_{\mathbb{Q}_q}) \otimes D^1(S^\dagger_{\mathbb{Q}_q})$. But for our purposes, we prefer to think of the Gauss-Manin connection as mapping $H^1_{MW}(\overline{A}/S^\dagger_{\mathbb{Q}_q})$ into itself. Then the following observation is the key towards deformation theory.

Theorem 2. *One has* $\nabla \circ \mathcal{F}^*_q = qt^{q-1} \circ \mathcal{F}^*_q \circ \nabla$, *where* qt^{q-1} *denotes the corresponding multiplication map on* $H^1_{MW}(\overline{A}/S^\dagger_{\mathbb{Q}_q})$.

Proof. (sketch only) This follows from the commutativity of the diagram of $\mathbb{Z}_q$-module morphisms

$$
\begin{array}{ccc}
D^1(A^\dagger) & \xrightarrow{\ d\ } & D^2(A^\dagger) \\
\downarrow \mathcal{F}^*_q & & \downarrow \mathcal{F}^*_q \\
D^1(A^\dagger) & \xrightarrow{\ d\ } & D^2(A^\dagger).
\end{array}
$$

2.4   Deformation

Suppose that $H^1_{MW}(\overline{A}/S^\dagger_{\mathbb{Q}_q})$ is finitely generated and free over $S^\dagger_{\mathbb{Q}_q}$, having a basis that for any $\bar{t}_0 \in \mathbb{F}_q$ for which $\bar{r}(\bar{t}_0) \neq 0$, reduces mod $(t - \hat{t}_0)$ to a basis of $H^1_{MW}(\overline{A}_{\bar{t}_0}/\mathbb{Q}_q)$. Here, $\hat{t}_0$ is the Teichmüller lift of $\bar{t}_0$. In Section 4 we will prove this assumption for our concrete families of $C_{a,b}$ curves.

Let $s_1, \ldots, s_d$ be an $S^\dagger_{\mathbb{Q}_q}$-basis of $H^1_{MW}(\overline{A}/S^\dagger_{\mathbb{Q}_q})$ and let $F = (F_{i,j})$, $G = (G_{i,j})$ be $(d \times d)$-matrices with entries in $S^\dagger_{\mathbb{Q}_q}$ such that

$$
\mathcal{F}^*_q(s_j) = \sum_{i=1}^{d} F_{i,j} s_i, \qquad \nabla(s_j) = \sum_{i=1}^{d} G_{i,j} s_i
$$

for $j = 1, \ldots, d$. Then the quasi-commutativity of the Gauss-Manin connection with the Frobenius action gives rise to a first-order differential equation

$$
G \cdot F - \frac{d}{dt} F = qt^{q-1} \cdot F \cdot G(t^q).
$$

This allows one to compute F from an initial value. Typically, this is the matrix of Frobenius acting on $H^1_{MW}(\overline{A}_{\bar{t}_0}/\mathbb{Q}_q) = H^1_{MW}(\overline{A}/S^\dagger_{\mathbb{Q}_q})/(t - \hat{t}_0)$ for some 'easy' fiber $\mathrm{Spec}\,\overline{A}_{\bar{t}_0}$. When expressed with respect to the basis $s_1(\hat{t}_0), \ldots, s_d(\hat{t}_0)$ of $H^1_{MW}(\overline{A}_{\bar{t}_0}/\mathbb{Q}_q)$, this exactly matches with $F(\hat{t}_0)$. If $n := \log_p q$ is large, a substantial speed-up in the algorithms can be achieved by working with the matrix

F_p of the pth power Frobenius $\overline{\mathcal{F}}_p : \overline{A} \to \overline{A}$ suitably acting on $H^1_{MW}(\overline{A}/S^\dagger_{\mathbb{Q}_q})$, and then reconstructing F as $F_p^{\sigma^{n-1}} \cdot F_p^{\sigma^{n-2}} \cdots F_p^{\sigma} \cdot F_p$. Here $\sigma : S^\dagger_{\mathbb{Q}_q} \to S^\dagger_{\mathbb{Q}_q}$ maps t to t^p, acts on $\mathbb{Q}_q$ by Frobenius substitution and extends by linearity and continuity. Furthermore, F_p can be computed from an initial $F_p(\hat{t}_0)$ as the solution to the differential equation

$$G \cdot F_p - \frac{d}{dt}F_p = pt^{p-1} \cdot F_p \cdot G^\sigma(t^p). \tag{1}$$

3 Generalities on $C_{a,b}$ Curves

3.1 Definition and First Properties

Let a and b be coprime integers ≥ 2 and let k be any field. An algebraic curve C/k is said to be $C_{a,b}$ if it admits a non-singular affine 'Weierstrass model'

$$C(x,y) = y^a + c_{b,0}x^b + \sum_{ai+bj<ab} c_{i,j}x^iy^j \in k[x,y] \qquad (c_{b,0} \neq 0). \tag{2}$$

Such a model has a unique, generally singular point at infinity. One can prove that this point is dominated by a single place P on the non-singular model, and the pole divisors of x and y are aP and bP respectively. Since a and b are coprime, this allows us to determine the pole divisor of any function $f(x,y)$ in the affine coordinate ring $A = k[x,y]/(C)$. Indeed, using $C(x,y) = 0$ one can write

$$f(x,y) = \sum_{j=0}^{a-1} \sum_{i=0}^{\deg_x f} f_{i,j}x^iy^j,$$

in which no two monomials have the same pole order at P. Hence $-\text{ord}_P(f) = \max\{ai + bj \,|\, i = 0, \ldots, \deg_x f; j = 0, \ldots, a-1, f_{i,j} \neq 0\}$, and the Weierstrass semigroup

$$\{-\text{ord}_P(f) \,|\, f \in k(C) \setminus \{0\}\} \subset \mathbb{N}$$

of P equals $a\mathbb{N}+b\mathbb{N}$. From the Riemann-Roch theorem it follows that the geometric genus of C equals $g = (a-1)(b-1)/2$. Hyperelliptic curves of genus g having a rational Weierstrass point are $C_{2,2g+1}$, and are therefore special instances of $C_{a,b}$ curves.

Let $\Delta \subset \mathbb{R}^2$ be the convex hull of $(0,0)$, $(b,0)$ and $(0,a)$. It contains (and generically equals) the Newton polytope of $C(x,y)$. Then the following property is a key feature of $C_{a,b}$ curves. One can copy the proof of [3, Lemma 1], replacing $\mathbb{F}_q$ and $\mathbb{Z}_q$ with k and R respectively.

Lemma 1 (Effective Nullstellensatz). *Let R be a discrete valuation ring or a field with maximal ideal $\mathfrak{m}$. Let $\overline{C}(x,y) \in \frac{R}{\mathfrak{m}}[x,y]$ define a $C_{a,b}$ curve. Let $C(x,y) \in R[x,y]$ be such that it reduces to $\overline{C}(x,y)$ mod $\mathfrak{m}$ and such that it is again supported on Δ. Then there exist polynomials $\alpha, \beta, \gamma \in R[x,y]$ that are supported on 2Δ, such that $1 = \alpha C + \beta C_x + \gamma C_y$. In particular, if $C(x,y)$ was chosen to be monic in y, it defines a $C_{a,b}$ curve over the fraction field of R.*

Here C_x (resp. C_y) denotes $\partial C/\partial x$ (resp. $\partial C/\partial y$).

3.2   Cohomology

Write $A = k[x, y]/(C)$ and suppose first that $\operatorname{char}(k) = 0$. Then in [3], it is shown that

$$\{x^r y^s dx \mid r = 0, \ldots, b - 2; \; s = 1, \ldots, a - 1\} \tag{3}$$

is a basis for the k-vector space $H^1_{DR}(A/k) = D^1(A)/d(A)$. The proof moreover gives an explicit procedure to express a differential form $\omega \in D^1(A)$ in terms of this basis: using $C(x, y) = 0$ and the exactness of forms of the type $d(x^r y^s)$, one immediately sees that $H^1_{DR}(A/k)$ is generated by $x^r y^s dx$ for $0 < s < a$. These generators are totally ordered by $-\operatorname{ord}_P$ and as long as $r \geq b - 1$, each of them can be rewritten in terms of forms $x^r y^s dx$ having strictly smaller pole order. This is because

$$\omega_{r,s} = x^{r-(b-1)} y^s dC - d\left(x^{r-(b-1)} \left(\frac{a}{a+s} y^{a+s} + \sum_{ai+bj<ab} \frac{jc_{i,j}}{s+j} x^i y^{s+j} \right) \right)$$

is exact, and after expanding and reducing mod $C(x, y)$ one can check that its pole order is determined by the term $\lambda x^r y^s dx$, where

$$\lambda = \left(b + (r - b + 1) \frac{a}{a+s} \right) c_{b,0} \neq 0.$$

Therefore, subtracting $\omega_{r,s}/\lambda$ from $x^r y^s dx$ reduces the pole order. Continuing in this way will reduce everything onto the basis.

Next, suppose that $k = \mathbb{F}_q$. Let $\widetilde{C}(x, y) \in \mathbb{Z}_q[x, y]$ be monic in y and supported on Δ, such that it reduces to $C(x, y) \bmod p$. Let $\widetilde{A}^\dagger$ be the weak completion of $\widetilde{A} = \mathbb{Z}_q[x, y]/(\widetilde{C})$. Then from [3, Lemma 4], it follows that the canonical map

$$H^1_{DR}(\widetilde{A}/\mathbb{Q}_q) = \frac{D^1(\widetilde{A})}{d(\widetilde{A})} \otimes \mathbb{Q}_q \longrightarrow H^1_{MW}(A/\mathbb{Q}_q) = \frac{D^1(\widetilde{A}^\dagger)}{d(\widetilde{A}^\dagger)} \otimes \mathbb{Q}_q$$

is an isomorphism, more precisely the above reduction process converges. Since $\widetilde{C}$ is $C_{a,b}$ by Lemma 1, the set given in (3) is a $\mathbb{Q}_q$-basis for $H^1_{MW}(A/\mathbb{Q}_q)$.

3.3   Families of $C_{a,b}$ Curves

Let k be any field. Let $C(x, y, t) \in k[t][x, y]$ be supported on Δ and suppose that the coefficient of y^a is 1. Let $c_{b,0}(t) \in k[t]$ be the coefficient of x^b. Denote the monic polynomial generating the $k[t]$-ideal $(C, C_x, C_y) \cap k[t]$ by $f(t)$ and let $\mathfrak{r}(t) = c_{b,0}(t) f(t)$. One can then check that for any $t_0 \in \overline{k}$, $C(x, y, t_0) \in \overline{k}[x, y]$ defines a $C_{a,b}$ curve (in Weierstrass form) if and only if $\mathfrak{r}(t_0) \neq 0$. We will say that $C(x, y, t)$ defines a (one-dimensional) family of $C_{a,b}$ curves if $\mathfrak{r}(t) \neq 0$; the polynomial $\mathfrak{r}(t)$ will be referred to as the resultant of the family. Any $C(x, y, t)$ defining a family of $C_{a,b}$ curves gives rise to a flat family of smooth curves

$$\operatorname{Spec} \frac{k[t][x, y]}{(C)} \longrightarrow \operatorname{Spec} k[t, \mathfrak{r}(t)^{-1}].$$

A condition equivalent to $\mathfrak{r}(t) \neq 0$ is: $C(x, y, t)$ defines a $C_{a,b}$ curve over the function field $k(t)$. Indeed, consider the system of equations $C = C_x = C_y = z\mathfrak{r}(t) - 1$, where z is a new variable. It has no solutions over $\overline{k}$, and therefore there are polynomials $\alpha, \beta, \gamma, \delta \in k[x, y, z, t]$ for which

$$1 = \alpha C + \beta C_x + \gamma C_y + \delta(z\mathfrak{r}(t) - 1).$$

If $\mathfrak{r}(t) \neq 0$, we can replace z by $1/\mathfrak{r}(t)$, to get an expansion

$$1 = \alpha' C + \beta' C_x + \gamma' C_y, \quad \alpha', \beta', \gamma' \in k(t)[x, y]. \tag{4}$$

Together with $c_{b,0}(t) \neq 0$ this implies that $C(x, y, t)$ indeed defines a $C_{a,b}$ curve over $k(t)$. Conversely, for any expansion (4), $f(t)$ must divide the least common multiple of the denominators appearing in α', β' and γ' and can therefore not be zero. Together with $c_{b,0}(t) \neq 0$ this gives $\mathfrak{r}(t) \neq 0$.

The above observation allows us to bound the degree of the resultant.

Lemma 2. *Let $C(x, y, t)$ define a family of $C_{a,b}$ curves and let $\mathfrak{r}(t) \in k[t]$ be its resultant. Then $\deg \mathfrak{r}(t) \leq (9g + 6(a + b) - 1) \deg_t C$.*

Proof. Let $\alpha, \beta, \gamma \in k(t)[x, y]$ be as in the effective Nullstellensatz (Lemma 1) applied over $k(t)$. The coefficients of α, β and γ can be obtained by solving a system of $\#(3\Delta \cap \mathbb{Z}^2)$ equations in $3\#(2\Delta \cap \mathbb{Z}^2)$ unknowns. Both numbers are bounded by $9g + 6(a + b) - 2$. Now by Cramer's theorem, the denominators of α, β and γ can be chosen to be the determinant $d(t)$ of some fixed minor matrix of our system; since $d(t)$ contains $f(t)$ as a factor, $\deg f(t)$ is clearly bounded by $(9g + 6(a + b) - 2) \deg_t C$. Together with $\deg c_{b,0}(t) \leq \deg_t C$, this gives the desired result.

Lemma 3. *Let R be a discrete valuation ring with maximal ideal $\mathfrak{m}$ and suppose that $\overline{C}(x, y, t) \in (R/\mathfrak{m})[t][x, y]$ defines a family of $C_{a,b}$ curves. Let $C(x, y, t) \in R[t][x, y]$ be supported on Δ, such that it reduces to $\overline{C}(x, y, t)$ mod $\mathfrak{m}$ and such that the coefficient of y^a is 1. Then $C(x, y, t)$ defines a family of $C_{a,b}$ curves (over the fraction field K of R).*

Proof. This follows from Lemma 1, when applied over the discrete valuation ring $R[t]_{\mathfrak{m} R[t]}$, i.e. the subring of $K(t)$ consisting of rational functions that can be written as a quotient of two integral polynomials whose denominator does not reduce to zero modulo $\mathfrak{m}$.

4   Relative MW Cohomology of a Family of $C_{a,b}$ Curves

Let $\overline{C}(x, y, t) \in \mathbb{F}_q[t][x, y]$ define a family of $C_{a,b}$ curves. Let $C(x, y, t) \in \mathbb{Z}_q[t][x, y]$ lift $\overline{C}(x, y, t)$ such that it is monic in y and again supported on Δ. By Lemma 3, $C(x, y, t)$ defines a family of $C_{a,b}$ curves over $\mathbb{Q}_q$.

Instead of the resultant $\mathfrak{r}(t)$ of $C(x, y, t)$, we will work with a possibly larger polynomial $r(t) = c_{b,0}(t)d(t)$, where $d(t)$ is obtained as in the proof of Lemma 2 by linear algebra over the discrete valuation ring $\mathbb{Z}_q[t]_{p\mathbb{Z}_q[t]}$ (see also the proof of Lemma 3). In particular, $d(t)$ has p-adic valuation 0 and there exists a completely integral Nullstellensatz expansion

$$r(t) = \alpha C + \beta C_x + \gamma C_y \tag{5}$$

where α, β and γ are supported (in x and y) on 2Δ and where $\deg r(t)$, $\deg_t \alpha$, $\deg_t \beta$ and $\deg_t \gamma$ are bounded by $(9g + 6(a + b) - 1)\tau$, with $\tau := \deg_t C(x, y, t)$.

Let $\bar{r}(t)$ be the reduction modulo p. Since (5) is integral, it follows that $\overline{C}(x, y, t)$ defines a family of smooth curves over $\operatorname{Spec} \mathbb{F}_q[t, \bar{r}(t)^{-1}]$, so the theory explained in Section 2 applies. We inherit the notation introduced there, where for simplicity we drop the lower indices from d_t and D_t. Below, we give a basis for $H^1_{MW}(\overline{A}/S^\dagger_{\mathbb{Q}_q})$ and discuss the action of Frobenius on it. We will intensively make use of [1] and [3], so the proof-verifying reader should take these references at hand. The following lemma is easily proved.

Lemma 4. *Let $f(t, z) \in S^\dagger_{\mathbb{Q}_q}$ have p-adic valuation ν. There is only a finite number of Teichmüller elements $\hat{t}_0$ in $\overline{\mathbb{Z}}_q$ for which both $r(\hat{t}_0) \neq 0$ and the p-adic valuation of $f(\hat{t}_0, r(\hat{t}_0)^{-1})$ is $> \nu$.*

Lemma 5. *Let $r, s \in \mathbb{N}$ with $0 \leq s < a$. Then in $D^1(A^\dagger)$, $x^r y^s dx$ can be rewritten as*

$$\sum_{j=1}^{a-1} \sum_{i=0}^{b-2} \alpha_{i,j}(t, z) x^i y^j dx + d \left(\sum_{j=0}^{a-1} \sum_{i=0}^{r+b+1} \beta_{i,j}(t, z) x^i y^j \right),$$

where

1. $\alpha_{i,j}$ *and* $\beta_{i,j}$ *are polynomial expressions of degree* $\leq (ar+b)(9g+7a+6b-1)\tau$ *in* t, *and of degree* $\leq ar + b$ *in* z;
2. $p^m \alpha_{i,j}$ *and* $p^m \beta_{i,j}$ *are integral, with* $m = \lfloor \log_p((r + 1)a + sb) \rfloor + 4(a - 1)b\lfloor \log_p(2a - 1) \rfloor$.

Proof. One can follow the procedure described in Section 3.2. The factor $1/c_{b,0}(t)$ that is introduced in each reduction step can be rewritten as $\frac{r(t)}{c_{b,0}(t)}z$, which is a polynomial expression of degree at most $(9g+6(a+b)-2)\tau$ in t and of degree 1 in z. The $\alpha_{i,j}(t, z)$ are obtained by subsequently *(i)* expanding $x^r y^s dx - w_{r,s}/\lambda(t)$ and *(ii)* consecutively substituting $y^a - C(x, y, t)$ for y^a until only monomial forms of the type $x^i y^j dx$ with $j < a$ remain, so that one can start over again. The corresponding operations to compute the $\beta_{i,j}(t, z)$ are *(i)* computing

$$x^{r-(b-1)} \left(\frac{a}{a + s} y^{a+s} + \sum_{ai+bj<ab} \frac{j c_{i,j}(t)}{s + j} x^i y^{s+j} \right)$$

and *(ii)* substituting $y^a - C(x, y, t)$ for y^a until only monomial forms of the type $x^i y^j$ with $j < a$ remain. Since there are at most $ar+bs-a(b-2)-b(a-1) < ar+b$ reduction steps, the degree bounds follow.

The $r+b+1$ bound on the degree in x in the $d(\ldots)$-part follows from the fact that all terms that are introduced have pole order $\leq (r-(b-1))a + (2a-1)b$.

The bound on the p-adic valuations follows from the above lemma, together with [3, Lemma 4].

Corollary 1. $\{x^r y^s dx \mid r = 0, \ldots, b-2; \ s = 1, \ldots, a-1\}$ is an $S_{\mathbb{Q}_q}^\dagger$-module basis of $H^1_{MW}(\overline{A}/S_{\mathbb{Q}_q}^\dagger)$.

Proof. Linear independence follows from the corresponding statement in the absolute case. To see that it is a generating set, note that the above lemma implies the convergence of the reduction process described in Section 3.2.

Next, we determine the action of the pth power Frobenius on this basis. To this end, we construct a $\mathbb{Z}_p$-algebra endomorphism $\mathcal{F}_p : A^\dagger \to A^\dagger$ (along with explicit bounds on its rate of convergence) that lifts $\overline{A} \to \overline{A} : \overline{a} \mapsto \overline{a}^p$. The concrete aim is to find polynomials $\delta_x, \delta_y \in \mathbb{Z}_q[x, y, t, z]$ and overconvergent series $W, Z \in p\mathbb{Z}_q\langle x, y, t, z\rangle^\dagger$ such that

$$\mathcal{F}_p : \begin{cases} x \mapsto x^p(1 + \delta_x W) \\ y \mapsto y^p(1 + \delta_y W) \\ t \mapsto t^p \\ z \mapsto z^p + Z \end{cases}$$

(acting on $\mathbb{Z}_q$ by Frobenius substitution σ) extends by linearity and continuity to a well-defined map $A^\dagger \to A^\dagger$, i.e. modulo the relations $C(x, y, t) = 0$ and $r(t)z - 1 = 0$. Using Newton iteration, the latter relation allows one to determine Z, which should satisfy $r^\sigma(t^p)(z^p + Z) - 1 = 0$. As we are not interested in its rate of convergence, we move on to the determination of W.

The former relation implies that W should satisfy

$$H(W) := C^\sigma\left(x^p(1 + \delta_x W), y^p(1 + \delta_y W), t^p\right) = 0$$

over $A^\dagger$. We try to find δ_x and δ_y such that this equation can be solved using Newton iteration, starting from the approximate solution $W = 0$. From (5) it follows that

$$1 = z\alpha C + z\beta C_x + z\gamma C_y - (r(t)z - 1),$$

so we can take $\delta_x = z^p \beta^p$ and $\delta_y = z^p \gamma^p$: indeed, then $H(W) = 0$ satisfies the initial conditions for Newton iteration over $A^\dagger$. To find a unique representative however, we will instead solve

$$\tilde{H}(W) := H(W) - C^p + (r^p z^p - 1 - z^p \alpha^p C^p)W = 0,$$

for which these conditions are satisfied over the base ring $\mathbb{Z}_q\langle x, y, t, z\rangle^\dagger$.

If we expand $\tilde{H}(W) = \sum h_k W^k$, one verifies that the polynomials $h_k \in \mathbb{Z}_q[x, y][t][z]$ are supported on

$$(2k+1)p\Delta \times (\chi k + \tau)p[0, 1] \times kp[0, 1] \qquad \subset \mathbb{R}^4$$

where $\chi = \max\{\deg_t \alpha, \deg_t \beta, \deg_t \gamma\} \le (9g + 6(a + b) - 1)\tau$. This is contained in $(k + 1)p\Delta_{t,z}$, where $\Delta_{t,z} = 2\Delta \times [0, \chi] \times [0, 1]$. Proceeding as in [1], we finally find that $1 + \delta_x W \bmod p^N, 1 + \delta_y W \bmod p^N$ are supported on $5p(N + 1)\Delta_{t,z}$, for any $N \in \mathbb{N}$.

Lemma 6. *Let $F_p(t, z)$ be a matrix of the induced action of $\mathcal{F}_p$ on $H^1_{MW}(\overline{A}/S^\dagger_{\mathbb{Q}_q})$, with respect to the basis $\{x^r y^s dx \mid r = 0, \ldots, b - 2;\ s = 1, \ldots, a - 1\}$. Then for any $N \in \mathbb{N}$ we can represent any entry of $F_p(t, z)$ modulo p^N as a polynomial of degree $\le 7p(a+b)(ab+1)(N+\theta+1)\kappa\tau$ in t and of degree $\le 7p(a+b)(ab+1)(N+\theta+1)$ in z. Here $\kappa = 9g + 7a + 6b - 1$ and θ is the smallest positive integer satisfying $\theta \ge \log_p(8pab(a + b)(N + \theta + 1)) + 4ab \log_p(2a)$.*

Proof. Write $\mu = p(1 + 5(a + b - 2)(N + \theta + 1))$. Then one can check that the differential form $x^i y^j dx$ is mapped to an expression $x^{p-1} f dx$, where f is supported modulo $p^{N+\theta}$ on $\mu\Delta_{t,z}$ (use that $(i, j, 0, 0) \in \Delta_{t,z}$ and that $i + j + 1 \le a + b - 2$). Rewrite the polynomial $x^{p-1} f \bmod p^{N+\theta}$ as $\sum_{j=0}^{a-1} \sum_{i=0}^{r} f_{i,j}(t, z) x^i y^j$ by subsequently substituting $y^a - C(x, y, t)$ for y^a. Since there are less than $a\mu$ substitution steps, this adds at most $a\tau\mu$ to the degree in t. Therefore $\deg_t f_{i,j} \le \chi\mu + a\tau\mu = \kappa\tau\mu$ and $\deg_z f_{i,j} \le \mu$. By pole order arguments, one finds $r \le p - 1 + b\mu$. Following Lemma 5, this reduces further to

$$x^{p-1} f dx \equiv \sum_{j=1}^{a-1} \sum_{i=0}^{b-2} f'_{i,j}(t, z) x^i y^j dx,$$

where the congruence is valid modulo p^N since the valuations of the denominators introduced during reduction are bounded by

$$\lfloor \log_p((p + b\mu)a + (a - 1)b) \rfloor + 4(a - 1)b \lfloor \log_p(2a - 1) \rfloor \le \theta.$$

Moreover, $\deg_z f'_{i,j} \le a(p-1) + (ab+1)\mu + b$ and $\deg_t f'_{i,j} \le (a(p-1) + (ab+1)\mu + b)\kappa\tau$. One can verify that $a(p - 1) + (ab + 1)\mu + b \le 7p(a + b)(ab + 1)(N + \theta + 1)$.

5   Our Deformation Algorithm

We follow the strategy explained in Section 2.4, applied to families of the type considered in Section 4. In this, we aim for two applications: *(i)* computing the zeta function of a given $C_{a,b}$ curve over a given finite field $\mathbb{F}_q$, and *(ii)* generating $C_{a,b}$ curves having an (almost) prime order Jacobian over a given finite field $\mathbb{F}_q$ for given a and b.

As for *(i)*, let $\overline{C}_1(x, y) \in \mathbb{F}_q[x, y]$ be the $C_{a,b}$ curve of interest and let $\overline{C}_0(x, y)$ be a $C_{a,b}$ curve defined over the prime subfield $\mathbb{F}_p$. E.g. one can take $\overline{C}_0(x, y) = y^a - x^b + \varphi(x, y)$ where $\varphi(x, y) = 1$ if $p \nmid a, b$, and $\varphi(x, y) = y$ resp. $\varphi(x, y) = x$ if $p \mid a$ resp. $p \mid b$. Then our family of interest is $\overline{C}(x, y, t) = t\overline{C}_1(x, y) + (1 - t)\overline{C}_0(x, y)$ and the goal is to compute $F_p(1)$ from $F_p(0)$ by solving equation (1).

In *(ii)*, we take a 'random' family $\overline{C}(x, y, t) \in \mathbb{F}_p[t][x, y]$ and compute $F_p(t)$ from $F_p(0)$ by solving equation (1). Afterwards, we substitute various Teichmüller elements $\hat{t}_0 \in \mathbb{Z}_q$ until we find a curve with an (almost) prime order Jacobian. We remark that some special families are unsuited for this application, such as the supersingular family $y^2 = x^3 + tx$ with $t \in \mathbb{F}_q$ and $q \equiv 3 \bmod 4$.

First we compute the polynomial $r(t)$ as explained in the beginning of Section 4. Note that $\bar{r}(t)$ contains the actual resultant as an in general non-trivial factor, so it may accidentally happen that e.g. $\bar{r}(0) = 0$ or $\bar{r}(1) = 0$. We will assume that this is *not* the case, i.e. all fibers of interest correspond to non-roots of $\bar{r}(t)$.

Before describing the main steps of the algorithm we define several constants. As before, $\tau = \deg_t C(x, y, t)$, and we define $\rho := \deg r(t)$, so that $\rho = \mathcal{O}(g\tau)$. We will (see [3]) have to compute both $F_p(t)$ and $F_p(1)$ modulo p^m with

$$m := \left\lceil \log_p \left(2 \binom{2g}{g} q^{g/2} \right) \right\rceil + (g+1)ng \log_p a.$$

Let $\alpha := (2g - 1)g \log_p a + g$ and $\gamma := 2g^2 \log_p a + g$, and choose θ and κ as in Lemma 6, where the accuracy N is now equal to m. Now we define $M := 7p(a + b)(ab + 1)(m + \theta + 1)$ and $\ell := \kappa\tau M + \rho M + 1$. The matrices $F_p(0)$ and G will be computed with p-adic accuracy $\varepsilon := m + (5\gamma + 1)\lceil \log_p \ell \rceil + 12\alpha$ and all computations are modulo t^ℓ.

5.1   Step I: Computing $F_p(0)$

In all instances, we can reduce to the case where the 0-fiber $\overline{C}_0(x, y)$ is defined over the prime subfield $\mathbb{F}_p$. Computing the Frobenius matrix of such a curve is of course easier, but note that we need the Frobenius matrix $F_p(0)$ up to a much higher precision than required for computing the zeta function of $\overline{C}_0(x, y)$.

Currently, we use two very basic methods. The first method consists of computing $F_p(0)$ using the $C_{a,b}$-algorithm described in [3]. For the basic forms of the 0-fiber suggested above, the action of Frobenius has much nicer properties than in the general case, thereby circumventing the problem we originally set out to solve. The second method is more efficient and relies on the extension of Kedlaya's algorithm to superelliptic curves [6]. Note that all basic forms suggested in application *(i)* above fall in the category of superelliptic curves, so the algorithm of [6] applies. However, the basis used in [6] is different from ours, so we need to apply a basis transformation obtained by reducing our basis onto the basis of [6] using the reduction procedure given there.

5.2   Step II: Computing G

To compute the Gauss-Manin connection ∇, we simply apply Definition 1. For each basis differential $x^i y^j dx$ with $i = 0, \ldots, b-2$ and $j = 1, \ldots, a-1$, we rewrite $d(x^i y^j dx)$ as $\varphi_{i,j} \wedge dt$ to obtain $\nabla(x^i y^j dx) = \varphi_{i,j}$.

Define $\beta' = \beta/r(t)$ and $\gamma' = \gamma/r(t)$ with β, γ as in Equation (5), i.e. $1 \equiv \beta' C_x + \gamma' C_y \bmod C(t)$, then a short computation shows that

$$d(x^i y^j dx) = x^i jy^{j-1}(\beta' C_x + \gamma' C_y)dy \wedge dx = x^i jy^{j-1}(\gamma' dx - \beta' dy)C_t \wedge dt.$$

So all that remains to do is to apply the reduction formulae given in Section 3.2 to $x^i j y^{j-1}(\gamma' dx - \beta' dy)C_t$, the result of which gives a column of G.

Note that $x^i j y^{j-1}(\gamma' dx - \beta' dy)C_t$ can be rewritten as $h_{i,j} dx$ where $h_{i,j}$ is supported (in x and y) on 4Δ. So the pole order is at most $4ab$ and we can write $h_{i,j}$ in terms of $x^k y^\ell$ with $0 \leq \ell < a$ and $0 \leq k \leq 4b$. From Lemma 5 it follows that the entries of G are of degree $\leq (4ab + b)(9g + 7a + 6b - 1)\tau$ in t and of degree $\leq 4ab + b$ in z. The p-adic valuations of the denominators are $\tilde{\mathcal{O}}(g)$.

5.3   Step III: Solving the Differential Equation

We first reformulate the differential equation in a way that ensures that the coefficients as well as the solution modulo p^m of the equation are all polynomials, rather than just rational functions or power series in t. From Lemma 6 above, it follows that $K(t) := r(t)^M \cdot F_p(t) \mod p^m$ has polynomial entries of degree less than ℓ. Let $d_G(t) \in \mathbb{Z}_q[t]$ be a factor of some power of $r(t)$ such that $d_G(t)G(t)$ consists of polynomials. As follows from the end of Section 5.2, we can take $\deg d_G(t) = \mathcal{O}(g^2\tau)$. Rewriting equation (1) using $K(t)$ gives

$$r(t)\frac{dK(t)}{dt} - \left(M\frac{dr(t)}{dt} + r(t)G(t)\right) \cdot K(t) + K(t) \cdot \left(pt^{p-1}r(t)G^\sigma(t^p)\right) = 0. \quad (6)$$

After multiplying this equation with $d_G(t)d_G^\sigma(t^p)$ we find an equation of the form $A\frac{dK}{dt}B + AKX + YKB$, where $A(t) := r(t)d_G(t)$, $B(t) := d_G^\sigma(t^p)$,

$$X(t) := pt^{p-1}d_G^\sigma(t^p)G^\sigma(t^p) \quad \text{and} \quad Y(t) := -Md_G(t)\frac{dr(t)}{dt} - r(t)d_G(t)G(t),$$

all consisting of polynomials of degree bounded by $\mathcal{O}(g^2\tau)$. In [8, Theorem 2], it is explained how to solve this equation for $K(t)$ with precision (p^m, t^ℓ) respectively $K(1) \mod p^m$, given that the initial precision p^ε is large enough. From [3] it follows that $\mathrm{ord}_p(K(t)) \geq -g \log_p a$, and as shown in [9, Lemma 18] this implies that $\mathrm{ord}_p(K^{-1}(t)) \geq -\alpha$. Let $C(t) = \sum_i C_i t^i$ and $D(t) = \sum_i D_i t^i$ be matrices in $\mathbb{Q}_q[[t]]^{2g \times 2g}$ that satisfy $A\frac{dC}{dt} + YC = 0$, $C(0) = \mathbb{I}$, and $\frac{dD}{dt}B + DX = 0$, $D(0) = \mathbb{I}$ respectively. Denote with C_i' and D_i' the respective coefficients of t^i in $C(t)^{-1}$ and $D(t)^{-1}$. As shown in [9, Proposition 20] for $C(t)$ and $C(t)^{-1}$ and in [8, Section 3.2] for $D(t)$ and $D(t)^{-1}$ we then have that

$$\mathrm{ord}_p(C_i), \mathrm{ord}_p(C_i'), \mathrm{ord}_p(D_i), \mathrm{ord}_p(D_i') \geq -\gamma \cdot \lceil \log_p(i+1) \rceil - 2\alpha.$$

These properties, together with straightforward estimates on the valuation of A, A^{-1}, B, B^{-1}, X and Y, guarantee that working modulo p^ε suffices for finding the correct result modulo p^m as proved in [8, Theorem 2].

For application *(ii)* we now have to compute the Teichmüller lift $\hat{t}_0$ and compute $F_p(\hat{t}_0) = K(\hat{t}_0)r(\hat{t}_0)^{-M}$. Application *(i)* requires us only to compute $F_p(1) = K(1)r(1)^{-M}$. As final steps the calculation of the qth power Frobenius and the characteristic polynomial of Frobenius are needed, but for this we can refer to Steps 9 and 10 of the algorithm in [1]. The loss of precision in these steps is easily seen to be at most $(g+1)ng\log_p a$, where $n = \log_p q$, so that working modulo p^m guarantees correctness of the zeta function modulo p to the power $\left\lceil \log_p\left(2\binom{2g}{g}q^{g/2}\right)\right\rceil$. The latter precision allows us to determine the zeta function correctly, as follows from the Weil conjectures, see [3, Section 4].

5.4 Complexity Analysis

We will throughout suppose that asymptotically fast arithmetic is used [5]. We see that $m = \mathcal{O}(g^2 n \log a)$ and $\varepsilon = \tilde{\mathcal{O}}(g^2 n)$. From the analysis in [3] it is clear that Step I requires both time and space $\tilde{\mathcal{O}}(g^6 n^2)$. For Step II and application (ii) we need to reduce $2g$ basis elements (each one requiring $\mathcal{O}(g)$ steps) and the objects have size $\mathcal{O}(g\tau\varepsilon)$, whence working over the prime field requires time $\mathcal{O}(g^5 n\tau)$. In the situation of application (i) this step needs time $\mathcal{O}(g^5 n^2\tau)$.

Next we need an estimate on

$$\zeta := \max\{\deg A + \deg B, \deg A + \deg X + 1, \deg Y + \deg B + 1\}.$$

From the estimates in Step II we see that $\zeta = \mathcal{O}(g^2\tau)$. Now Theorem 2 from [8] shows that the computation of $F_p(t)$ requires time $\tilde{\mathcal{O}}(\ell\zeta g^\omega\varepsilon) = \tilde{\mathcal{O}}(g^{9+\omega}n^2\tau^2)$ (with ω as an exponent for matrix multiplication, e.g. $\omega = 2.376$ [5]) and space $\mathcal{O}(\ell g^2\varepsilon) = \tilde{\mathcal{O}}(g^9 n^2\tau)$. Note that in the estimates in [8] we have to take $n = 1$ as we are working over the field $\mathbb{Q}_p$.

For application (i) the time requirements are $\tilde{\mathcal{O}}(\ell\zeta g^\omega n\varepsilon) = \tilde{\mathcal{O}}(g^{9+\omega}n^3\tau^2)$ and we need $\mathcal{O}(\zeta g^2 n\varepsilon) = \tilde{\mathcal{O}}(g^6 n^2\tau)$ space. Finally for the computation of the matrix of the qth power Frobenius and the zeta function we can follow [1], needing $\tilde{\mathcal{O}}((n+g)n^2 g^3)$ time and $\mathcal{O}(n^2 g^3)$ space. Taking the maximum over all these steps gives the following result for the respective applications:

(i) time $\tilde{\mathcal{O}}(g^{9+\omega}n^3\tau^2)$ and space $\tilde{\mathcal{O}}(g^9 n^2\tau)$,
(ii) time $\tilde{\mathcal{O}}(g^{9+\omega}n^3\tau^2)$ and space $\tilde{\mathcal{O}}(g^6 n^2\tau)$.

The complexity in g seems bad but in all concrete examples, multiplication with $p^{\mathcal{O}(\log g)}$ suffices to make the matrices of the pth as well as the qth power Frobenius integral. If we take this into account, we can remove at least a factor g^2. Moreover, the implementation results below show that the algorithm performs quite well for relatively high genera.

We note that for the second application, where we compute zeta functions within families defined over the prime field, it is possible to achieve a time complexity of $\tilde{\mathcal{O}}(n^{2.667})$ (where g is fixed) by computing a suitable defining polynomial for $\mathbb{Q}_q$. For more details we refer to Section 6.3 of [9].

6 Preliminary Implementation Results

In this section, we briefly report on some experiments with application (ii), i.e. with families defined over prime fields, using the computer algebra system Magma V2.13-14 running on a Pentium IV 2.4 GHz. From a cryptographic viewpoint, the goal is a curve whose Jacobian order has a prime factor $> 2^{160}$. We can achieve this by trying many curves over a suitable field and verifying whether this condition holds. A consequence is that if we fix a family and vary the parameter in a field $\mathbb{F}_q$, we can consider Steps I, II and the computation of $K(t)$ in Step III as precomputation. The results of our experiments are given in Table 1. For Step I, we used the algorithm described in [6] in the column 'G.-G.', and the algorithm presented in [3] in the column 'D.-V.'. The column 'Precomp.' accounts for the precomputations other than $F_p(0)$, and 't/c' gives the time required for each curve after these precomputations.

Table 1. Running times (in seconds) and memory usage to compute the zeta function of a fiber in a family over a prime field

Equation $C(X,Y,t)$	$\mathbb{F}_{p^n}$	g	G.-G.	D.-V.	Precomp.	t/c	Memory
$Y^3 + X^4 + X + X^3 + t(XY^2 + 1)$	2^{59}	3	6.83	46.38	379	12.31	59 MB
$Y^3 + X^5 + X^2 + t + 1$	2^{43}	4	11.81	261.37	27	6.94	43 MB
$Y^4 + X^5 + Y + t(XY + 1)$	2^{27}	6	10.96	10.45	1080	57.52	126 MB
$Y^3 + X^9 + 1 + tX^4Y$	2^{20}	8	2.29	37.56	25	7.6	60 MB
$Y^3 - X^4 + Y + tXY$	3^{37}	3	2.86	4.77	10	10.63	46 MB
$Y^3 + (t+2)X^5 + (t+1)Y + t$	3^{29}	4	4.30	6.22	11	2.42	28 MB
$Y^4 - X^5 + tXY + tY - 1$	3^{21}	6	1.64	21.83	876	77.30	102 MB
$Y^3 - X^4 + tX^2 + t - 1$	5^{23}	3	0.75	7.82	4.5	1.27	69 MB
$Y^4 - X^5 - X - t(X + Y)$	5^{12}	6	9.76	7.14	4260	77.21	290 MB

These results have to be compared with [3], where, for curves comparable to the first line in this table, each curve required 5000 to 7000 seconds of computing time (albeit on a somewhat slower AMD XP 1700+) and 130 to 147 MB of memory.

Acknowledgements

The authors would like to thank an anonymous referee for his/her detailed verification of the article and useful suggestions.

References

1. Castryck, W., Denef, J., Vercauteren, F.: Computing zeta functions of nondegenerate curves. IMRP Int. Math. Res. Pap. 57, Art. ID 72017 (2006)
2. Cohen, H., Frey, G., Avanzi, R., Doche, C., Lange, T., Nguyen, K., Vercauteren, F.: Handbook of Elliptic and Hyperelliptic Curve Cryptography. In: Discrete Mathematics and its Applications, Chapman & Hall/CRC (2005)
3. Denef, J., Vercauteren, F.: Counting points on C_{ab} curves using Monsky-Washnitzer cohomology. Finite Fields Appl. 12(1), 78–102 (2006)

4. Edixhoven, B., Couveignes, J.M., de Jong, R., Merkl, F., Bosman, J.: On the computation of coefficients of a modular form (2006), http://arxiv.org/abs/math/0605244
5. von zur Gathen, J., Gerhard, J.: Modern Computer Algebra. Cambridge University Press, New York (1999)
6. Gaudry, P., Gürel, N.: An extension of Kedlaya's point-counting algorithm to super-elliptic curves. In: Boyd, C. (ed.) ASIACRYPT 2001. LNCS, vol. 2248, pp. 480–494. Springer, Heidelberg (2001)
7. Gaudry, P., Schost, É.: Construction of secure random curves of genus 2 over prime fields. In: Cachin, C., Camenisch, J.L. (eds.) EUROCRYPT 2004. LNCS, vol. 3027, pp. 239–256. Springer, Heidelberg (2004)
8. Hubrechts, H.: Memory efficient hyperelliptic curve point counting (preprint, 2006), http://arxiv.org/abs/math/0609032
9. Hubrechts, H.: Point counting in families of hyperelliptic curves. In: Foundations of Computational Mathematics (to appear)
10. Kedlaya, K.S.: Counting points on hyperelliptic curves using Monsky-Washnitzer cohomology. J. Ramanujan Math. Soc. 16(4), 323–338 (2001)
11. Kedlaya, K.S.: p-Adic Cohomology: From Theory to Practice. Arizona Winter School 2007 Lecture Notes (2007)
12. Lauder, A.G.B., Wan, D.: Counting points on varieties over finite fields of small characteristic. In: Buhler, J.P., Stevenhagen, P. (eds.) Algorithmic Number Theory: Lattices, Number Fields, Curves and Cryptography, vol. 44, Mathematical Sciences Research Institute Publications (to appear, 2007)
13. Lauder, A.G.B.: Deformation theory and the computation of zeta functions. Proc. London Math. Soc. (3) 88(3), 565–602 (2004)
14. Lauder, A.G.B.: A recursive method for computing zeta functions of varieties. LMS J. Comput. Math. 9, 222–269 (2006)
15. Mestre, J.F.: Lettre adressée à Gaudry et Harley (December 2000), http://www.math.jussieu.fr/~mestre/
16. van der Put, M.: The cohomology of Monsky and Washnitzer. In: Mém. Soc. Math. France (N.S.), vol. 23(4), pp. 33–59 (1986); Introductions aux cohomologies p-adiques (Luminy, 1984)
17. Satoh, T.: The canonical lift of an ordinary elliptic curve over a finite field and its point counting. J. Ramanujan Math. Soc. 15(4), 247–270 (2000)
18. Schoof, R.: Counting points on elliptic curves over finite fields. J. Théor. Nombres Bordeaux 7(1), 219–254 (1995); Les Dix-huitièmes Journées Arithmétiques (Bordeaux, 1993)

Computing *L*-Series of Hyperelliptic Curves

Kiran S. Kedlaya* and Andrew V. Sutherland

Department of Mathematics
Massachusetts Institute of Technology
77 Massachusetts Avenue
Cambridge, MA 02139
{kedlaya,drew}@math.mit.edu

Abstract. We discuss the computation of coefficients of the *L*-series associated to a hyperelliptic curve over $\mathbb{Q}$ of genus at most 3, using point counting, generic group algorithms, and *p*-adic methods.

1 Introduction

For C a smooth projective curve of genus g defined over $\mathbb{Q}$, the *L*-function $L(C, s)$ is conjecturally (and provably for $g = 1$) an entire function containing much arithmetic information about C. Most notably, according to the conjecture of Birch and Swinnerton-Dyer, the order of vanishing of $L(C, s)$ at $s = 1$ equals the rank of the group $J(C/\mathbb{Q})$ of rational points on the Jacobian of C.

It is thus natural to ask to what extent we are able to compute with the *L*-function. This splits into two subproblems:

1. For appropriate N, compute the first N coefficients of the Dirichlet series expansion $L(C, s) = \prod_p L_p(p^{-s})^{-1} = \sum_{n=1}^{\infty} c_n n^{-s}$.
2. From the Dirichlet series, compute $L(C, s)$ at various values of s to suitable numerical accuracy. (The Dirichlet series converges for Real$(s) > 3/2$.)

In this paper, we address problem 1 for hyperelliptic curves of genus $g \leq 3$ with a distinguished rational Weierstrass point. This includes in particular the case of elliptic curves, and indeed we have something new to say in this case; we can handle significantly larger coefficient ranges than other existing implementations. We say nothing about problem 2; we refer instead to [5].

Our methods combine efficient point enumeration with generic group algorithms as discussed in the second author's PhD thesis [22]. For $g > 2$, we also apply *p*-adic cohomological methods, as introduced by the first author [11] and refined by Harvey [8]. Since what we need is adequately described in these papers, we focus our presentation on the point counting and generic group techniques and use an existing *p*-adic cohomological implementation provided by Harvey. (The asymptotically superior Schoof-Pila method [15, 14] only becomes practically better far beyond the ranges we can hope to handle.)

* Kedlaya was supported by NSF CAREER grant DMS-0545904 and a Sloan Research Fellowship.

A.J. van der Poorten and A. Stein (Eds.): ANTS-VIII 2008, LNCS 5011, pp. 312–326, 2008.
© Springer-Verlag Berlin Heidelberg 2008

As a sample application, we compare statistics for Frobenius eigenvalues of particular curves to theoretical predictions. These include the Sato-Tate conjecture for $g = 1$, and appropriate analogues in the Katz-Sarnak framework for $g > 1$; for the latter, we find little prior numerical evidence in the literature.

2 The Problem

Let C be a smooth projective curve over $\mathbb{Q}$ of genus g . We wish to determine the polynomial $L_p(T)$ appearing in $L(C, s) = \prod L_p(p^{-s})^{-1}$, for $p \leq N$. We consider only p for which C is defined and nonsingular over $\mathbb{F}_p$ (almost all of them), referring to [16, 4] in the case of bad reduction. The polynomial $L_q(T)$ appears as the numerator of the local zeta function

$$Z(C/\mathbb{F}_q; T) = \exp\left(\sum_{k=1}^{\infty} N_k T^k / k\right) = \frac{L_q(T)}{(1-T)(1-qT)}, \qquad (1)$$

where N_k counts the points on C over $\mathbb{F}_{q^k}$. Here q is any prime power, however we are primarily concerned with $q = p$ an odd prime. The rationality of $Z(C/\mathbb{F}_q; T)$ is part of the well known theorem of Weil [24], which also requires

$$L_q(T) = \sum_{i=0}^{2g} a_j T^j \qquad (2)$$

to have integer coefficients satisfying $a_0 = 1$ and $a_{2g-j} = p^{g-j} a_j$, for $0 \leq j < g$. To determine $L_q(T)$, it suffices to compute $a_1, \ldots, a_g$.

For reasons of computational efficiency we restrict ourselves to curves which may be described by an affine equation of the form $y^2 = f(x)$, where $f(x)$ is a monic polynomial of degree $d = 2g+1$ (hyperelliptic curves with a distinguished rational Weierstrass point). We denote by $J(C/\mathbb{F}_q)$ the group of $\mathbb{F}_q$-rational points on the Jacobian variety of C over $\mathbb{F}_q$ (the *Jacobian* of C over $\mathbb{F}_q$), and use $J(\tilde{C}/\mathbb{F}_q)$ to denote the Jacobian of the quadratic twist of C over $\mathbb{F}_q$.

We consider three approaches to determining $L_p(T)$ for $g \leq 3$:

1. **Point counting:** Compute $N_1, \ldots, N_g$ of (1) by enumerating the points on C over $\mathbb{F}_p, \mathbb{F}_{p^2}, \ldots, \mathbb{F}_{p^g}$. The coefficients $a_1, \ldots, a_g$ can then be readily derived from (1) [3, p. 135]. This requires $O(p^g)$ field operations.
2. **Group computation:** Use generic algorithms to compute $L_p(1) = \#J(C/\mathbb{F}_p)$, and, for $g > 1$, compute $L_p(-1) = \#J(\tilde{C}/\mathbb{F}_p)$. Then use $L_p(1)$ and $L_p(-1)$ to determine $L_p(T)$ [21, Lemma 4]. This involves a total of $O(p^{(2g-1)/4})$ group operations.
3. **p-adic methods:** Apply extensions of Kedlaya's algorithm [11, 8] to compute (modulo p) the characteristic polynomial $\chi(T) = T^{-2g} L_p(T)$ of the Frobenius endomorphism on $J(C/\mathbb{F}_p)$, then use generic algorithms to compute the exact coefficients of $L_p(T)$. The asymptotic complexity is $\tilde{O}(p^{1/2})$.[1]

---

[1] For fixed $g \geq 4$, one works modulo $p^{\lfloor g/2 - 1 \rfloor}$ to obtain the same complexity.

Computing the coefficients of $L_p(T)$ for all $p \leq N$ necessarily requires time and space exponential in $\lg N$, since the output contains $\Theta(N)$ bits. In practice, we are limited to N of moderate size: on the order of 2^{40} in genus 1, 2^{28} in genus 2, and 2^{26} in genus 3 (larger in parallel computations). We expect to compute $L_p(T)$ for a large number of relatively small values of p. Constant factors will have considerable impact, however we first consider the asymptotic situation.

The $O(p^g)$ complexity of point counting makes it an impractical method to compute $a_1, \ldots, a_g$ unless p is very small. However, point counting over $\mathbb{F}_p$ is an efficient way to compute $a_1 = N_1 - p - 1$ for a reasonably large range of p when $g > 1$, requiring only $O(p)$ field operations. Knowledge of a_1 aids the computation of $\#J(C/\mathbb{F}_p)$, reducing the complexity of the baby-steps giant-steps search to $O(p^{1/4})$ in genus 2 and $O(p)$ in genus 3. The optimal strategy then varies according to genus and range of p:

Genus 1. The $O(p^{1/4})$ complexity of generic group computation makes it the compelling choice, easily outperforming point counting for $p > 2^{10}$.

Genus 2. There are three alternatives: (i) $O(p)$ field operations followed by $O(p^{1/4})$ group operations, (ii) $O(p^{3/4})$ group operations, or (iii) an $\tilde{O}(p^{1/2})$ p-adic computation. We find the range in which (iii) becomes optimal to be past the feasible values of N.

Genus 3. The choice is between (i) $O(p)$ field operations followed by $O(p)$ group operations and (ii) an $\tilde{O}(p^{1/2})$ p-adic computation followed by $O(p^{1/4})$ group operations. Here the p-adic algorithm plays the major role once $p > 2^{15}$.

3   Point Counting

Counting points on C over $\mathbb{F}_p$ plays a key role in our strategy for genus 2 and 3 curves. Moreover, it is a useful tool in its own right. If one wishes to study the distribution of $\#J(C/\mathbb{F}_p) = L_p(1)$, or to simply estimate $L_p(p^{-s})$, the value a_1 may be all that is required.

Given C in the form $y^2 = f(x)$, the simplest approach is to build a table of the quadratic residues in $\mathbb{F}_p$ (typically stored as a bit-vector), then evaluate $f(x)$ for all $x \in \mathbb{F}_p$. If $f(x) = 0$, there is a single point on the curve, and otherwise either two points (if $f(x)$ is a residue) or none. Additionally, we add a single point at infinity (recall that f has odd degree). A not-too-naïve implementation computes the table of quadratic residues by squaring half the field elements, then uses d field multiplications and d field additions for each evaluation of $f(x)$, where d is the degree of f. A better approach uses finite differences, requiring only d field additions (subtractions) to compute each $f(x)$.

Let $f(x) = \sum f_j x^j$ be a degree d polynomial over a commutative ring R. Fix a nonzero $\delta \in R$ and define the linear operator $\boldsymbol{\Delta}$ on $R[x]$ by

$$(\boldsymbol{\Delta} f)(x) = f(x + \delta) - f(x). \tag{3}$$

For any $x_0 \in R[x]$, given $f(x_0)$, we may enumerate the values $f(x_0 + n\delta)$ via

$$f(x_0 + (n+1)\delta) = f(x_0 + n\delta) + \boldsymbol{\Delta} f(x_0 + n\delta). \tag{4}$$

To enumerate $f(x_0 + n\delta)$ it suffices to enumerate $\boldsymbol{\Delta} f(x_0 + n\delta)$, which we also do via (4), replacing f with $\boldsymbol{\Delta} f$. Since $\boldsymbol{\Delta}^{d+1} f = 0$, each step requires only d additions in R, starting from the initial values $\boldsymbol{\Delta}^k f(x_0)$ for $0 \leq k \leq d$.

When $R = \mathbb{F}_p$, this process enumerates $f(x)$ over the entire field and we simply set $\delta = 1$ and $x_0 = 0$. As subtraction modulo p is typically faster than addition, instead of (4) we use

$$f(x_0 + (n + 1)\delta) = f(x_0 + n\delta) - (-\boldsymbol{\Delta} f)(x_0 + n\delta). \tag{5}$$

The necessary initial values are then $(-1)^k \boldsymbol{\Delta} f(0)$.

Algorithm 1 (Point Counting over $\mathbb{F}_p$). *Given a polynomial $f(x)$ over $\mathbb{F}_p$ of odd degree d and a vector M identifying nonzero quadratic residues in $\mathbb{F}_p$:*

1. Set $t_k \leftarrow (-1)^k \boldsymbol{\Delta}^k f(0)$, for $0 \leq k \leq d$, and set $N \leftarrow 1$.

2. For i from 1 to p:
   (a) If $t_0 = 0$, set $N \leftarrow N + 1$, and if $M[t_0]$, set $N \leftarrow N + 2$.
   (b) Set $t_0 \leftarrow t_0 - t_1$, $t_1 \leftarrow t_1 - t_0$, ..., and $t_{d-1} \leftarrow t_{d-1} - t_d$.

Output N.

The computation $t_k = t_k - t_{k+1}$ is performed using integer subtraction, adding p if the result is negative. The map M is computed by enumerating the polynomial $f(x) = x^2$ for x from 1 to $(p-1)/2$ and setting $M[f(x)] = 1$, using a total of p subtractions (and no multiplications).

The size of M may be cut in half by only storing residues less than $p/2$. One then uses $M[\min(t_0, p - t_0)]$, inverting $M[p - t_0]$ when $p \equiv 3 \bmod 4$. This slows down the algorithm, but is worth doing if M exceeds the size of cache memory.

It remains only to compute $\boldsymbol{\Delta}^k f(0)$. We find that

$$\boldsymbol{\Delta}^k f(0) = \sum_j k! \left\{ \begin{matrix} j \\ k \end{matrix} \right\} f_j = \sum_j T_{j,k} f_j, \tag{6}$$

where the bracketed coefficient denotes a Stirling number of the second kind. The triangle of values $T_{j,k}$ is represented by sequence A019538 in the OEIS [17]. Since (6) does not depend on p, it is computed just once for each $k \leq d$.

In the process of enumerating $f(x)$, we can also enumerate $f(x) + g(x)$ with $e + 1$ additional field subtractions, where e is the degree of $g(x)$. The case where $g(x)$ is a small constant is particularly efficient, since nearby entries in M are used. The last two columns in Table 1 show the amortized cost per point of applying this approach to the curves $y^2 = f(x)$, $f(x) + 1$, ..., $f(x) + 31$.

4   Group Computations

The performance of generic group algorithms is typically determined by two quantities: the time required to perform a group operation, and the number of operations performed. We briefly mention two techniques that reduce the former, then consider the latter in more detail.

Table 1. Point counting $y^2 = f(x)$ over $\mathbb{F}_p$ (CPU nanoseconds/point)

$p \approx$	Polynomial Evaluation		Finite Differences		Finite Differences $\times 32$	
	Genus 2	Genus 3	Genus 2	Genus 3	Genus 2	Genus 3
2^{16}	195.1	257.2	6.1	7.8	1.1	1.1
2^{17}	196.3	262.6	6.0	6.9	1.1	1.1
2^{18}	192.4	259.8	6.0	6.8	1.1	1.1
2^{19}	186.3	251.1	6.0	6.8	1.1	1.1
2^{20}	187.3	244.1	7.2	8.0	1.1	1.3
2^{21}	172.3	240.8	8.8	9.4	1.2	1.3
2^{22}	197.9	233.9	12.1	13.4	1.2	1.3
2^{23}	229.2	285.8	12.8	14.6	2.6	2.7
2^{24}	258.1	331.8	41.2	44.0	3.5	4.7
2^{25}	304.8	350.4	53.6	55.7	4.8	4.9
2^{26}	308.0	366.9	65.4	67.8	4.8	4.6
2^{27}	318.4	376.8	70.5	73.1	4.9	5.0
2^{28}	332.2	387.8	74.6	76.5	5.1	5.2

The middle rows of Table 1 show the transition of M from $L2$ cache to general memory. The top section of the table is the most relevant for the algorithms considered here, as asymptotically superior methods are used for larger p.

4.1   Faster Black Boxes

The performance of the underlying finite field operations used to implement the group law on the Jacobian can be substantially improved using a Montgomery representation to perform arithmetic modulo p [13]. Another optimization due to Montgomery that is especially useful for the algorithms considered here is the simultaneous inversion of field elements (see [3, Alg. 11.15]).[2] With an affine representation of the Jacobian each group operation requires a field inversion, but uses fewer multiplications than alternative representations. To ameliorate the high cost of field inversions, we then modify our algorithms to perform group operations "in parallel".

In the baby-steps giant-steps algorithm, for example, we fix a small constant n, compute n "babies" β, β^2, ..., β^n, then march them in parallel using steps of size n (the giant steps are handled similarly). In each parallel step we execute n group operations to the point where a field inverse is required, perform all the field inversions together for a cost of $3n - 3$ multiplications and one inversion, then use the results to complete the group operations. Exponentiation can also benefit from parallelization, albeit to a lesser extent.

These two optimizations are most effective when applied in combination, as may be seen in Table 2.

---

[2] This algorithm can be applied to any group.

Table 2. Black box performance (CPU nanoseconds/group operation)

g	p	Standard			Montgomery		
		$\times 1$	$\times 10$	$\times 100$	$\times 1$	$\times 10$	$\times 100$
1	$2^{20} + 7$	501	245	215	239	89	69
1	$2^{25} + 35$	592	255	216	286	93	69
1	$2^{30} + 3$	683	264	217	333	98	69
2	$2^{20} + 7$	1178	933	902	362	216	196
2	$2^{25} + 35$	1269	942	900	409	220	197
2	$2^{30} + 3$	1357	949	902	455	225	196
3	$2^{20} + 7$	2804	2556	2526	642	498	478
3	$2^{25} + 35$	2896	2562	2528	690	502	476
3	$2^{30} + 3$	2986	2574	2526	736	506	478

The heading $\times n$ indicates n group operations performed "in parallel". All times are for a single thread of execution.

4.2   Generic Order Computations

Our approach to computing $\#J(C/\mathbb{F}_q) = L_q(1)$ is based on a generic algorithm to compute the structure of an arbitrary abelian group [22]. We are aided both by absolute bounds on $L_q(1)$ derived from the Weil conjectures (theorems), as well as predictions regarding its distribution within these bounds based on a generalized form of the Sato-Tate conjecture (proven for most genus 1 curves over $\mathbb{Q}$ in [6]). We first consider the general algorithm.

We assume we have a black box for an abelian group G (written multiplicatively) that can generate uniformly random group elements. For Jacobians, these can be obtained via decompression techniques [3, 14.1-2].[3] We also suppose we are given bounds M_0 and M_1 such that $M_0 \le |G| \le M_1$.

The first (typically only) step is to compute the group exponent, $\lambda(G)$, the least common multiple of the orders of all the elements of G. This is accomplished by initially setting $E = 1$, and for a random $\alpha \in G$, computing the order of $\beta = \alpha^E$ using a baby-steps giant-steps search on the interval $[M_0/E, M_1/E]$. We then update $E \leftarrow |\beta|E$ and repeat the process until either (1) there is only one multiple of E in the interval $[M_0, M_1]$, or (2) we have generated c random elements, where c is a confidence parameter. In the former case we must have $|G| = E$, and in the latter case $E = \lambda(G)$, with probability greater than $1 - 2^{2-c}$ [22, Proposition 8.3]. For large Jacobians, (1) almost always applies, however for the relatively small groups considered here, (2) arises more often, particularly when $g > 1$. Fortunately, this does not present undue difficulty.

Proposition 2. *Given $\lambda(G)$ and M_0 such that $M_0 \le |G| < 2M_0$, the value of $|G|$ can be computed using $O(|G|^{1/4})$ group operations.*

---

[3] This becomes costly when $g > 2$, where we use the simpler approach of [3, p. 307].

Proof (sketch). The bounds on $|G|$ imply that it is enough to know the order of all but one of the p-Sylow subgroups of G (the p dividing $|G|$ are obtained from $\lambda(G)$). Following Algorithm 9.1 of [22], we use $\lambda(G)$ to compute the order of each p-Sylow subgroup $H \subseteq G$ using $O(|H|^{1/2})$ group operations; however, we abandon the computation for any p-Sylow subgroup that proves to be larger than $\sqrt{|G|}$. This can happen at most once, and the remaining successful computations uniquely determine $|G|$ within the interval $[M_0, 2M_0)$. $\square$

From the Weil interval (see (8) in section 4.4) we find that $M_1 < 2M_0$ for all $q > 300$ and $g \leq 3$. Proposition 2 implies that group structure computations will not impact the complexity of our task. Indeed, computing $\#J(C/\mathbb{F}_q)$ is almost always dominated by the first computation of $|\beta|$.

   Given $\beta \in G$ and the knowledge that the interval $[M_0, M_1]$ contains an integer M for which $\beta^M = 1_G$, a baby-steps giant-steps search may be used to find such an M. This is not necessarily the order of β, it is a multiple of it. We can then factor M and compute $|\beta|$ using $\tilde{O}(\lg M)$ group operations [22, Ch. 7]. The time to factor M is negligible in genus 2 and 3 (compared to the group computations), and in genus 1 we note that if a sieve is used to enumerate the primes up to N, the factorization of every M in the interval $[M_0, M_1]$ can be obtained at essentially no additional cost, using $O(\sqrt{N})$ bytes of memory.

   An alternative approach avoids the computation of $|\beta|$ from M by attempting to prove that M is the only multiple of $|\beta|$ in the interval. Write $[M_0, M_1]$ as $[C - R, C + R]$, and suppose the search to find $M = C \pm r$ has shown $\beta^n \neq 1_G$ for all n in $(C - r, C + r)$. If M is not the only multiple of $|\beta|$ in $[C - R, C + R]$, then $|\beta|$ is a divisor of M satisfying $2r \leq |\beta| \leq R + r$. In particular, if P is the largest prime factor of M and $P > R + r$ and $M/P < 2r$, then M must be unique. When $R = O(M^{1/2})$ this happens fairly often (about half the time). When it does not happen, one can avoid an $\tilde{O}(\lg M)$ order computation at the cost of $O(R^{1/2})$ group operations by searching the remainder of the interval *on the opposite side* of M. This is only worthwhile when R is quite small, but can be helpful in genus 1.[4]

4.3   Optimized Baby-Steps Giant-Steps in the Jacobian — Part I

The Mumford representation of $J(C/\mathbb{F}_q)$ uniquely represents a reduced divisor of the curve $y^2 = f(x)$ by a pair of polynomials (u, v). The polynomial u is monic, with degree at most g, and divides $v^2 - f$ [3, p. 307]. The inverse of (u, v) is simply $(u, -v)$, which makes two facts immediate:

1. The cost of group inversions is effectively zero.
2. The element (u, v) has order 2 if and only if $v = 0$ and u divides f.

Fact 1 allows us to apply the usual optimization for fast inverses [2, p. 250], reducing the number of group operations by a factor of $\sqrt{2}$ (we no longer count inversions). Fact 2 gives us a bijection between the 2-torsion subgroup of $J(C/\mathbb{F}_q)$

---

[4] These ideas were sparked by a conversation with Mark Watkins, who also credits Geoff Bailey.

and polynomials dividing f of degree at most g (exactly half the polynomials dividing f). If k counts the irreducible polynomials in the unique factorization of f, then the 2-rank of $J(C/\mathbb{F}_q)$ is $k-1$ and 2^{k-1} divides $\#J(C/\mathbb{F}_q)$.[5]

When $k > 1$, we start with $E = 2^{k-1}$ in our computation of $\lambda(G)$ above, reducing the number of group operations by a factor of $2^{(k-1)/2}$. Otherwise, we know $\#J(C/\mathbb{F}_q)$ is odd and can reduce the number of group operations by a factor of $\sqrt{2}$. The total expected benefit of fast inversions and knowledge of 2-rank is at least a factor of 2.10 in genus 1, 2.31 in genus 2, and 2.48 in genus 3.

4.4   Optimized Baby-Steps Giant-Steps in the Jacobian — Part II

We come now to the most interesting class of optimizations, those based on the distribution of $\#J(C/\mathbb{F}_q)$. The Riemann hypothesis for curves (proven by Weil) states that $L_q(T)$ has roots lying on a circle of radius $q^{-1/2}$ about the origin of the complex plane. As $L_q(T)$ is a real polynomial of even degree with $L_q(0) = 1$, these roots may be grouped into conjugate pairs.

Definition 3. *A* unitary symplectic *polynomial $p(z)$ is a real polynomial of even degree with roots $\alpha_1, ...\alpha_g, \bar{\alpha}_1, ...\bar{\alpha}_g$ all on the unit circle.*

The unitary symplectic polynomials are precisely those arising as the characteristic polynomial of a unitary symplectic matrix. The Riemann hypothesis for curves implies that $p(z) = L_q(zq^{-1/2})$ is a unitary symplectic polynomial. The coefficients of $p(z) = \sum a_j z^j$ may be bounded by

$$|a_j| \leq \binom{2g}{j}. \tag{7}$$

The corresponding bounds on the coefficients of $L_q(T)$ constrain the value of $L_q(1) = \#J(C/\mathbb{F}_q)$, yielding the Weil interval

$$(\sqrt{q} - 1)^{2g} \leq \#J(C/\mathbb{F}_q) \leq (\sqrt{q} + 1)^{2g}. \tag{8}$$

For the a_j with j odd, the well known bounds in (7) are tight, however for even j they are not. We are particularly interested in the coefficient a_2.

Proposition 4. *Let $p(z) = \sum a_j z^j$ be a unitary symplectic polynomial of degree $2g$. For fixed a_1, a_2 is bounded by an interval of radius at most g. In fact*

$$a_2 \leq g + \left(\frac{g-1}{2g}\right) a_1^2; \tag{9}$$

$$a_2 \geq -g + 2 + \left(a_1^2 - \delta^2\right)/2. \tag{10}$$

The value $\delta \leq 2$ is the distance from a_1 to the nearest integer congruent to 0 mod 4 *(when g is odd), or* 2 mod 4 *(when g is even).*

---

[5] Computing k requires only a distinct-degree factorization of f, see [2, Alg. 3.4.3].

Proof. Define $\beta_j = \alpha_j + \bar{\alpha}_j$ for $1 \le j \le g$, where the α_j are the roots of $p(z)$. Then $a_1 = \sum \beta_j$ and $a_2 = g + (a_1^2 - t_2)/2$, where $t_2 = \sum \beta_j^2$. For fixed a_1, t_2 is minimized by $\beta_j = a_1/g$, yielding (9), and t_2 is maximized by $\beta_j = \pm 2$ for $j < g$ and $\beta_g = \delta$, yielding (10) (note that $|\beta_j| \le 2$). The proposition follows.   □

We have as a corollary, independent of a_1, the bound $a_2 \ge -g$, and for g odd, $a_2 \ge 2 - g$. In genus 2, the proposition reduces to Lemma 1 of [12], however we are especially interested in the genus 3 case, where our estimate of a_2 will determine the leading constant factor in the time to compute $\#J(C/\mathbb{F}_q)$. In genus 3, Proposition 4 constrains a_2 to an interval of radius 3 once a_1 is known, whereas (7) would give a radius of 15.

Having bounded the interval as tightly as possible, we consider the search within. We suppose we are seeking the value of a random variable X with some distribution over $[M_0, M_1]$. We assume that we start from an initial estimate M and search outward in both directions using a standard baby-steps giant-steps search with all baby steps taken first (see [19] for a more general analysis). Ignoring the boundaries, the cost of the search is

$$c = s + 2|X - M|/s \tag{11}$$

group operations. As our cost function is linear in $|X - M|$, we minimize the mean absolute error in our estimate by setting M to the median value of X and $s = \sqrt{2E}$, where E is the expectation of $|X - M|$. This holds for any distribution on X, we simply need the median value of X and its expected distance from it.

If we consider $p(z) = L_q(zq^{-1/2})$ as a "random" unitary symplectic polynomial, a natural distribution for $p(z)$ can be derived from the Haar measure on the compact Lie group $USp(2g)$ (the group of $2g \times 2g$ matrices over $\mathbb{C}$ that are both unitary and symplectic). Each $p(z)$ corresponds to a conjugacy class of matrices with $p(z)$ as their characteristic polynomial. Let the eigenvalues of a random matrix in $USp(2g)$ be $e^{\pm i\theta_1}, \ldots, e^{\pm i\theta_g}$, with $\theta_j \in [0, \pi)$. The joint probability density function on the θ_j given by the Haar measure on $USp(2g)$ is

$$\mu(USp(2g)) = \frac{1}{g!} \left(\prod_{j<k} (2\cos\theta_j - 2\cos\theta_k) \right)^2 \prod_j \frac{2}{\pi} \sin^2\theta_j d\theta_j. \tag{12}$$

This distribution is derived from the Weyl integration formula [25, p. 218] and can be found in [10, p. 107]. For $g = 1$, this simplifies to $(2/\pi)\sin^2\theta d\theta$, which corresponds to the Sato-Tate distribution. We may apply (12) to compute various statistical properties of random unitary symplectic polynomials. The coefficient a_1 is simply the negative sum of the eigenvalues,

$$a_1 = -\sum_{j=1}^{g} 2\cos\theta_j, \tag{13}$$

and we find that the median (and expectation) of a_1 is 0. In genus 1, the expected distance of a_1 from its median is

$$\mathbf{E}\left[|a_1|\right] = \frac{2}{\pi} \int_0^{\pi} |2\cos\theta| \sin^2\theta d\theta = \frac{8}{3\pi}. \tag{14}$$

The value $8/(3\pi) \approx 0.8488$ is not much smaller than 1, which corresponds to a uniform distribution, so the potential benefit is small in genus 1. In genus 2, however, the expected distance of a_1 from its median is $4096/(625\pi^2) \approx 0.7905$, versus an expected distance of 2 for the uniform distribution. The corresponding values for genus 3 are ≈ 0.7985 and 3.

Given the value of a_1 we can take this approach further, computing the median and expected distance for a_2 conditioned on a_1. Applying (12), we precompute a table of median and expected distance values for a_2 for various ranges of a_1. In genus 3, we find that the largest expected distance for a_2 given a_1 is about 0.66, much smaller than the value 7.5 for a uniform distribution of a_2 over the interval given by (7).

Of course such optimizations are effective only when the polynomials $L_p(T)$ for a particular curve and relatively small values of p actually correspond to (apparently) random unitary symplectic polynomials. For $g > 1$, it is not known whether this occurs at all, even as $p \to \infty$.[6] In genus 1, while the Sato-Tate conjecture is now largely proven over $\mathbb{Q}$ [6], the convergence rate remains the subject of conjecture. Indeed, the investigation of such questions was one motivation for undertaking these computations. It is only natural to ask whether our assumptions are met.

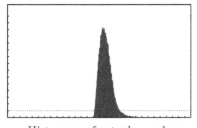   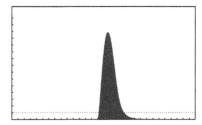

Histogram of actual a_2 values   Distribution of a_2 given by (12)

The figure on the left is a histogram of a_2 coefficient values obtained by computing $L_p(T)$ for $p \le 2^{24}$ for an arbitrarily chosen genus 3 curve (see Table 6). The figure on the right is the distribution of a_2 predicted by the Haar measure on $USp(2g)$, obtained by numerically integrating

$$a_2 = g + \prod_{j<k} 4\cos\theta_j \cos\theta_k \tag{15}$$

over the distribution in (12). The dotted lines show the height of the uniform distribution. Similarly matching graphs are found for the other coefficients.

This remarkable degree of convergence is typical for a randomly chosen curve. We should note, however, that the generalized form of the Sato-Tate conjecture considered here applies only to curves whose Jacobian over $\mathbb{Q}$ has a trivial endomorphism ring (isomorphic to $\mathbb{Z}$), so there are exceptional cases. In genus 1 these are curves with complex multiplication. In higher genera, other exceptional cases occur, such as the genus 2 QM-curves considered in [9].

---

[6] Results are known for certain universal families of curves, e.g. [10, Thm. 10.8.2].

5   Results

To compare different methods for computing $L_p(T)$ and to assess the feasible range of L-series computations, we conducted extensive performance tests. Our test platform consisted of eight networked PCs, each equipped with a 2.5GHz AMD Athlon processor running a 64-bit Linux operating system. The point-counting and generic group algorithms were implemented using the techniques described in this paper, and we incorporated David Harvey's source code for the p-adic computations (the algorithm of [8], including recent improvements described in [7]). All code was compiled with the GNU C/C++ compiler using the options "-O2 -m64 -mtune=k8" [18].

In genus 1 there are several existing implementations of the computation contemplated here: given an elliptic curve defined over $\mathbb{Q}$, determine the coefficient a_1 of $L_p(T) = pT^2 + a_1T + 1$ for all $p \leq N$. We were able to compare our implementation with two software packages specifically optimized for this purpose: Magma [1], and the PARI library [23] as incorporated in SAGE [20]. The range of N we could use in this comparison was necessarily limited; results for larger N may be found in Table 5.

Before undertaking similar computations in genus 2 and 3, we first determined the appropriate algorithm to use for various ranges of p using Table 4. Each row gives timings for the algorithms considered here, averaged over a small sample of primes of similar size.

Table 3. L-series computations in genus 1 (CPU seconds)

N	PARI	Magma	smalljac
2^{16}	0.26	0.29	0.07
2^{17}	0.55	0.59	0.15
2^{18}	1.17	1.24	0.30
2^{19}	2.51	2.53	0.62
2^{20}	5.46	5.26	1.29
2^{21}	11.67	11.09	2.65
2^{22}	25.46	23.31	5.53
2^{23}	55.50	49.22	11.56
2^{24}	123.02	104.50	24.31
2^{25}	266.40	222.56	51.60
2^{26}	598.16	476.74	110.29
2^{27}	1367.46	1017.55	233.94
2^{28}	3152.91	2159.87	498.46
2^{29}	7317.01	4646.24	1065.28
2^{30}	17167.29	10141.28	2292.74

Each row lists CPU times for a single thread of execution to compute the coefficient a_1 of $L_p(T)$ for all $p \leq N$, using the elliptic curve $y^2 = x^3 + 314159x + 271828$. In SAGE, the function aplist(N) performs this computation via the PARI function $ellap(N)$. The corresponding function in Magma is $TracesOfFrobenius(N)$. The column labeled "smalljac" list times for our implementation.

Table 4. $L_p(T)$ computations (CPU milliseconds)

$p \approx 2^k$	Genus 2 – $L_p(T)$			Genus 3 – $L_p(T)$		Genus 3 – a_1
	pts/grp	group	p-adic	pts/grp	p-adic/grp	points
2^{14}	**0.22**	0.55	4	**10**	15	**0.12**
2^{15}	**0.34**	0.88	6	**21**	23	**0.23**
2^{16}	**0.56**	1.33	8	43	**31**	**0.45**
2^{17}	**0.98**	2.21	11	82	**40**	**0.89**
2^{18}	**1.82**	3.42	17		**51**	**1.78**
2^{19}	**3.44**	5.87	27		**67**	**3.57**
2^{20}	**7.98**	10.1	40		**97**	**8.48**
2^{21}	18.9	**17.9**	66		**148**	**19.7**
2^{22}	52	**35**	104		**212**	**56**
2^{23}		**54**	176		**355**	**123**
2^{24}		**104**	288		**577**	738
2^{25}		**173**	494		**995**	1870
2^{26}		**306**	871		**1753**	4550
2^{27}		**505**	1532		**3070**	9800

Random curves of the appropriate genus were generated with coefficients uniformly distributed over $[1, 2^k)$. The polynomial $L_p(T)$ was then computed for 100 primes $\approx 2^k$, with the average CPU time listed. Columns labeled "pts/grp" compute a_1 by point counting over $\mathbb{F}_p$, followed by a group computation to obtain $L_p(T)$. The column "p-adic/grp" computes $L_p(T)$ mod p, then applies a group computation to get $L_p(T)$. The rightmost column computes just the coefficient a_1, via point counting over $\mathbb{F}_p$.

The task of computing *L*-series coefficients is well-suited to parallel computation. We implemented a simple distributed program which partitions the range $[1, N]$ into subintervals $I_1, I_2, \ldots, I_m$, distributes the task of computing $L_p(T)$ for $p \in I_m$ to n CPUs on a network, then collects and collates the results. This is useful even on a single computer whose microprocessor may have two or more cores. On our 8 node test platform we had 16 CPUs available for computation. Tables 5 and 6 lists elapsed times for *L*-series computations in single and 8-node configurations.

For practical reasons, we limited the duration of any single test. Larger computations could be undertaken with additional time and/or computing resources, without requiring software modifications. As they stand, the results extend to values of N substantially larger than any we could find in the literature.

Source code for the software can be freely obtained under a GNU General Public License (GPL) and is expected to be incorporated into SAGE. It is a pleasure to thank William Stein for access to the SAGE computational resources at the University of Washington, and especially David Harvey for providing the code used for the p-adic computations.

Table 5. L-series computations in genus 1 (elapsed times)

	Genus 1			Genus 1	
N	$\times 1$	$\times 8$	N	$\times 1$	$\times 8$
2^{21}	1.5	0.5	2^{30}	20:43	2:41
2^{22}	3.1	0.7	2^{31}	45:13	5:52
2^{23}	6.3	1.1	2^{32}	1:45:45	13:12
2^{24}	13.3	2.0	2^{33}	4:24:50	32:51
2^{25}	28.2	4.2	2^{34}	10:16:11	1:16:18
2^{26}	59.2	8.1	2^{35}	23:15:58	2:52:47
2^{27}	126.2	16.6	2^{36}		6:29:46
2^{28}	271.3	35.1	2^{37}		14:44:33
2^{29}	578.0	74.5	2^{38}		33:11:08

For the elliptic curve $y^2 = x^3 + 314159x + 271828$, the coefficients of $L_p(T)$ were computed for all $p \leq N$. Columns labeled $\times n$ list total elapsed times (seconds or hh:mm:ss) for a computation performed on n nodes (two cores per node), including communication overhead and time spent collating responses.

Table 6. L-series computations in genus 2 and 3 (elapsed times)

	Genus 2		Genus 3		Genus 3 - a_1 only	
N	$\times 1$	$\times 8$	$\times 1$	$\times 8$	$\times 1$	$\times 8$
2^{16}	1	< 1	43	13	1	< 1
2^{17}	4	2	1:49	18	5	1
2^{18}	12	3	4:42	41	11	2
2^{19}	40	7	12:43	1:47	41	6
2^{20}	2:32	24	36:14	4:52	2:41	21
2^{21}	10:46	1:38	1:45:36	13:40	11:33	1:27
2^{22}	40:20	5:38	5:23:31	41:07	53:26	6:38
2^{23}	2:23:56	19:04	16:38:11	2:05:40	4:33:26	33:00
2^{24}	8:00:09	1:16:47		6:28:25	38:51:07	4:42:43
2^{25}	26:51:27	3:24:40		20:35:16		20:35:16
2^{26}		11:07:28				
2^{27}		36:48:52				

The coefficients of $L_p(T)$ were computed for the genus 2 and 3 hyperelliptic curves

$$y^2 = x^5 + 31419x^3 + 271828x^2 + 1644934x + 57721566;$$
$$y^2 = x^7 + 314159x^5 + 271828x^4 + 1644934x^3 + 57721566x^2 + 1618034x + 141421,$$

for all $p \leq N$ where the curves had good reduction. Columns labeled $\times n$ list total elapsed wall times (hh:mm:ss) for a computation performed on n nodes, including all overhead. The last two columns give times to compute just the coefficient a_1.

References

[1] Cannon, J.J., Bosma, W. (eds.): Handbook of Magma functions, 2.14 ed. (2007), http://magma.maths.usyd.edu.au/magma/htmlhelp/MAGMA.htm

[2] Cohen, H.: A Course in Computational Algebraic Number Theory. Graduate Texts in Mathematics, vol. 138. Springer, Heidelberg (1993)

[3] Cohen, H., et al. (eds.): Handbook of Elliptic and Hyperelliptic Curve Cryptography. Chapman and Hall, Boca Raton (2006)

[4] Deninger, C., Scholl, A.J.: The Beilinson conjectures. In: *L*-functions and Arithmetic (Durham 1989) London Math. Soc. Lecture Note Series, vol. 153, pp. 173–209. Cambridge University Press, Cambridge (1991)

[5] Dokchitser, T.: Computing special values of motivic *L*-functions. Experimental Math. 13, 137–149 (2004)

[6] Harris, M., Shepherd-Barron, N., Taylor, R.: A family of Calabi-Yau varieties and potential automorphy May 2006 (preprint)

[7] Harvey, D.: Faster polynomial multiplication via multipoint Kronecker substitution (preprint, 2007), http://arxiv.org/abs/0712.4046v1

[8] Harvey, D.: Kedlaya's algorithm in larger characteristic. Int. Math. Res. Notices (2007)

[9] Hashimoto, K.-I., Tsunogai, H.: On the Sato-Tate conjecture for QM-curves of genus two. Math. Comp. 68, 1649–1662 (1999)

[10] Katz, N.M., Sarnak, P.: Random Matrices, Frobenius Eigenvalues, and Monodromy. American Mathematical Society (1999)

[11] Kedlaya, K.: Counting points on hyperelliptic curves using Monsky-Washnitzer cohomology. J. Ramanujan Math. Soc. 16, 332–338 (2001)

[12] Matsuo, K., Chao, J., Tsujii, S.: An improved baby step giant step algorithm for point counting of hyperelliptic curves over finite fields. In: Fieker, C., Kohel, D.R. (eds.) ANTS 2002. LNCS, vol. 2369, pp. 461–474. Springer, Heidelberg (2002)

[13] Montgomery, P.L.: Modular multiplication without trial division. Math. Comp. 44, 519–521 (1985)

[14] Pila, J.: Frobenius maps of abelian varieties and finding roots of unity in finite fields. Math. Comp. 55, 745–763 (1990)

[15] Schoof, R.: Counting points on elliptic curves over finite fields. J. Théor. Nombres Bordeaux 7, 219–254 (1995)

[16] Silverman, J.: Advanced topics in the arithmetic of elliptic curves. Springer, Heidelberg (1999)

[17] Sloane, N.J.A.: The on-line encyclopedia of integer sequences (2007), http://www.research.att.com/~njas/sequences/

[18] Stallman, R., et al.: GNU compiler collection 4.1.2 (February 2007), http://gcc.gnu.org/index.html

[19] Stein, A., Teske, E.: Optimized baby step-giant step methods. J. Ramanujan Math. Soc. 20(1), 1–32 (2005)

[20] Stein, W., Joyner, D.: SAGE: System for Algebra and Geometry Experimentation. Communications in Computer Algebra (SIGSAM Bulletin) (2005), version 2.8.5 (September 2007), http://sage.sourceforge.net/

[21] Sutherland, A.V.: A generic approach to searching for Jacobians. Math. Comp. (to appear), http://arxiv.org/abs/0708.3168v1

[22] Sutherland, A.V.: Order Computations in Generic Groups. PhD thesis, M.I.T. (2007), http://groups.csail.mit.edu/cis/theses/sutherland-phd.pdf
[23] The PARI Group: Bordeaux PARI/GP, version 2.3.2 (2007), http://pari.math.u-bordeaux.fr/
[24] Weil, A.: Numbers of solutions of equations in finite fields. Bull. AMS 55, 497–508 (1949)
[25] Weyl, H.: Classical groups, 2nd edn. Princeton University Press, Princeton (1946)

Point Counting on Singular Hypersurfaces

Remke Kloosterman

Institut für Algebraische Geometrie, Leibniz Universität Hannover
Welfengarten 1, D-30167 Hannover Germany

1 Introduction

Let $q = p^r$ be a prime power. Let $\overline{F} \in \mathbf{F}_q[X_0, \ldots, X_{n+1}]$ be a homogenous polynomial of degree d. Let $\overline{V} \subset \mathbf{P}^{n+1}$ be the hypersurface defined by $\overline{F} = 0$. A natural question to ask is how to determine $\#\overline{V}(\mathbf{F}_q)$.

Recently, several algorithms were presented that calculate $\#\overline{V}(\mathbf{F}_q)$ if $\overline{V}$ is a smooth hypersurface. We would like to investigate whether these algorithms extend to singular hypersurfaces.

In the case $n = 1$ (curves) there are many special algorithms to determine $\#\overline{V}(\mathbf{F}_q)$. For the sake of simplicity we leave these out of consideration, and we focus on the case $n > 1$. To our knowledge, there exist the following types of algorithms to determine $\#\overline{V}(\mathbf{F}_q)$ for a smooth hypersurface of degree d:

- A direct method by Abbott, Kedlaya and Roe [1].
- A deformation method by Lauder [8] and a slightly different one by Gerkmann [3].
- A recursive method by Lauder [9].

In this paper we identify an obstruction to extend the deformation method to singular varieties; for singular $\overline{V}$ the deformation method might give an output different from $\#\overline{V}(\mathbf{F}_q)$. Since the recursive method is based on the deformation method we expect that a similar obstruction plays a role there. Therefore we leave that method out of consideration.

Theorem 1. *There exist hypersurfaces $\overline{V} \subset \mathbf{P}_{\mathbf{F}_q}^{n+1}$ such that*

1. *$H^i_{\mathrm{rig}}(\overline{V}, \mathbf{Q}_q) \cong H^i_{\mathrm{rig}}(\mathbf{P}^{n+1}, \mathbf{Q}_q)$ for $i \neq n, 2n + 2$.*
2. *Lauder's Deformation algorithm and Gerkmann's Deformation algorithm terminate, but the output of the algorithm differs from $\#V(\mathbf{F}_q)$.*
3. *a modification of Abbott-Kedlaya-Roe's algorithm gives $\#V(\mathbf{F}_q)$.*

How one needs to modify Abbott–Kedlaya–Roe is explained in 2.4. We illustrate this theorem by giving two explicit examples of hypersurfaces for satisfying 1–3 of Theorem 1. Due to space restrictions we will not describe the precise class of hypersurfaces for which Theorem 1 holds, we intend to come back to this issue in [7].

Unfortunately, in the smooth case the algorithm of Abbott–Kedlaya–Roe is expected to have worse complexity than the Lauder–Gerkmann type of algorithm. This latter algorithm requires $(pd^n \log(q))^{O}(1)$ bit operations (for a discussion

A.J. van der Poorten and A. Stein (Eds.): ANTS-VIII 2008, LNCS 5011, pp. 327–341, 2008.

see [8]). Abbott, Kedlaya and Roe did not include an analysis of the complexity of their algorithm.

We will use a variant of Abbott–Kedlaya–Roe where we replace the Frobenius operator Frob_q^* with the so-called ψ-operator. This ψ-operator is a left inverse to Frob_q^*. In the smooth case the replacement of Frob_q^* by ψ allows one to do the computation with slightly less precision, hence improves the running time of the algorithm.

However, in the case of a singular hypersurface the choice for ψ is essential, since the original version of Abbott–Kedlaya–Roe will encounter the problem of 'exploding coefficients' if applied to a singular hypersurface: Abbott–Kedlaya–Roe relate the trace of Frob_q^* on a certain $\mathbf{Q}_q$-vector space W with $\#\overline{V}(\mathbf{F}_q)$. If $\overline{V}$ is singular ψ on this vector space W might have eigenvalues with small p-adic absolute value, hence Frob_q^* might have eigenvalues with very large p-adic absolute value. The eigenvalues of ψ with small p-adic absolute value should be ignored if one wants to calculate $\#\overline{V}(\mathbf{F}_q)$.

If $H^i_{\mathrm{rig}}(\overline{V}, \mathbf{Q}_q) \not\cong H^i_{\mathrm{rig}}(\mathbf{P}^{n+1}, \mathbf{Q}_q)$ for some i with $n + 1 \leq i \leq 2n + 2$ then it is easy to see that none of [1,3,8] can work. This follows from Obstruction 4 (PD-Failure). An approach to resolve this PD-Failure will be given in the paper [7]. In the sequel we will assume that the hypersurfaces under consideration do not have this obstruction.

The organization of this paper is as follows. In Section 2 we describe the deformation methods of Lauder and of Gerkmann, and the method of Abbott–Kedlaya–Roe. We indicate which results from algebraic geometry are used. Some of these results hold only for smooth varieties, whereas many other results hold only for certain classes of singular varieties.

In the case of the deformation method we describe an obstruction that is very hard to resolve. In the case of the direct method we indicate how one can bypass the obstructions for a certain class of varieties. The main difference between our method and that of [1] is that we use Dwork's left-inverse ψ of a lift of Frobenius instead of the lift itself.

In Section 3 we study the surface $X^2 + Y^2 + Z^2 = 0$ in $\mathbf{P}^3$. This is a cone over a conic, i.e. a quadric with an A_1 singularity. This is the prototype of an example for which in principle [3,8] cannot work, while [1] does work.

2   A Short Description of the Algorithms under Consideration

Notation 1. *Let p be a prime number, $q = p^r$ a power of p. Let $\mathbf{F}_q$ be the finite field with q elements. Denote the ring of Witt vectors of $\mathbf{F}_q$ by $\mathbf{Z}_q$, its maximal ideal by π, and its fraction field by $\mathbf{Q}_q$. Equivalently, the field $\mathbf{Q}_q$ is the unique unramified extension of degree r of $\mathbf{Q}_p$.*

We proceed by giving a short summary of the ideas used in [1,3,8].

In all three papers the authors prefer to calculate $\#\overline{U}(\mathbf{F}_q)$, where $\overline{U} = \mathbf{P}^{n+1} \setminus \overline{V}$ is the complement of $\overline{V}$, instead of calculating $\#\overline{V}(\mathbf{F}_q)$. The main advantage is that $\overline{U}$ is a smooth *affine* variety.

The idea now is to use cohomology. Denote by $H^i(\overline{U}, \mathbf{Q}_q)$ the i-th Monsky–Washnitzer, rigid or Dwork cohomology of $\overline{U}$. (In our case, all these groups are isomorphic as vector spaces with Frobenius action.) We can use the Lefschetz trace formula, which reads as

$$\sum_{i=0}^{n+1} q^i - \#\overline{V}(\mathbf{F}_q) = \#\overline{U}(\mathbf{F}_q) = \sum(-1)^i \operatorname{trace}\left((q^{n+1}\operatorname{Frob}_q^{*-1}) \mid H^i(\overline{U}, \mathbf{Q}_q)\right).$$

The use of $q^{n+1}\operatorname{Frob}_q^{*-1}$ rather than Frob_q^* is due to the fact that the usual Lefschetz trace formula holds for (rigid) cohomology $H_c^\bullet(\overline{U}, \mathbf{Q}_q)$ with compact support, which is Poincaré dual to $H^{2n+2-\bullet}(\overline{U}, \mathbf{Q}_q)$.

We can simplify the Lefschetz trace formula by:

Proposition 2 (Lefschetz hyperplane theorem). *Suppose $\overline{V}$ is smooth then*

- $H^i(\overline{U}, \mathbf{Q}_q) = 0$ *if $i \neq 0, n+1$ and*
- $H^0(\overline{U}, \mathbf{Q}_q)$ *is one-dimensional and Frobenius acts as the identity.*

From this lemma it follows that it suffices to determine the eigenvalues of Frob_q^* on $H^{n+1}(\overline{U}, \mathbf{Q}_q)$. All methods under consideration calculate the action of Frobenius on $H^{n+1}(\overline{U}, \mathbf{Q}_q)$.

Remark 3. Actually this Proposition is a combination of Lefschetz hyperplane theorem with Poincaré duality on $H^\bullet(\overline{V}, \mathbf{Q}_q)$. If $\overline{V}$ is singular then Poincaré duality might not hold. In that case one can show that $H^i(\overline{U}, \mathbf{Q}_q) = 0$ for $i > n+1$.

Here is the first obstruction to extending these algorithms to singular varieties that occurs:

Obstruction 4 (PD-Failure). *If $\overline{V}$ is a singular hypersurface then Proposition 2 might fail for i such that $n - \dim \overline{V}_{\text{sing}} \leq i \leq n$. If this happens, one needs a separate algorithm to calculate the Frobenius action on $H^i(\overline{U}, \mathbf{Q}_q)$ for these i.*

For a strategy to resolve PD-Failure in some cases we refer to [7]. We give two examples of varieties which have PD-failure:

Example 5. Suppose $\overline{V}$ is a hypersurface with two irreducible components. Then $H^{2n}(\overline{V}, \mathbf{Q}_q)$ is two-dimensional. A standard argument using Gysin long exact sequence and Poincaré duality yields that $H^1(\overline{U}, \mathbf{Q}_q)$ is 1-dimensional.

Example 6. Let $V : x_0^5 + x_1^5 + x_2^5 + x_3^5 + x_4^5 - 5x_0x_1x_2x_3x_4 = 0$ in $\mathbf{P}^4$. Then V is an irreducible surface with 125 ordinary double points. If $p \neq 2, 5$ then $H^4(\overline{V}, \mathbf{Q}_q)$ is 25-dimensional [10] and using a similar standard argument as in the previous example we obtain that $H^3(\overline{U}, \mathbf{Q}_q)$ is 24-dimensional.

At this stage the methods under consideration diverge. We start to consider them separately.

2.1 Deformation Method, Smooth Case

Consider the family

$$\overline{V}_{\overline{\lambda}} : (1 - \overline{\lambda}) \left(\sum_{i=0}^{n+1} X_i^d \right) + \overline{\lambda} \, \overline{F} = 0.$$

Then $\overline{V}_0$ is the diagonal hypersurface of degree d and $\overline{V}_1 = \overline{V}$. Let $\overline{U}_\lambda$ denote the corresponding family of complements. Let V_λ be a family of hypersurfaces lifting $\overline{V}_{\overline{\lambda}}$ to $\mathbf{Z}_q$, i.e. a family given by $F_\lambda \in \mathbf{Z}_q[X_0, \ldots, X_{n+1}]$ such that $F_\lambda \equiv \overline{F}_{\overline{\lambda}} \bmod \pi$ for all $\lambda \in \mathbf{Z}_q$, where $\overline{\lambda} \equiv \lambda \bmod \pi$.

The deformation method is built around the following diagram (cf. [5]):

$$
\begin{array}{ccc}
H^{n+1}(U_{\lambda^q}, \mathbf{Q}_q) & \xrightarrow{\mathrm{Frob}^*_{q,\lambda}} & H^{n+1}(U_\lambda, \mathbf{Q}_q) \\
\scriptstyle A(\lambda^q) \big\downarrow & & \scriptstyle A(\lambda) \big\downarrow \\
H^{n+1}(U_0, \mathbf{Q}_q) & \xrightarrow{\mathrm{Frob}^*_{q,0}} & H^{n+1}(U_0, \mathbf{Q}_q).
\end{array}
$$

It is relatively easy to calculate the Frobenius action on $H^{n+1}(\overline{U}_0, \mathbf{Q}_q)$ and we leave this aside. The operator $A(\lambda)$ is the unique solution to the p-adic Picard–Fuchs equation associated with the family V_λ, such that $A(0)$ is the identity. Equivalently, one can express $A(\lambda)$ in terms of the Gauß–Manin connection of the local system $H^{n+1}(V_\lambda, \mathbf{Q}_q)$.

To calculate $\mathrm{Frob}^*_{q,1} : H^{n+1}(U_1, \mathbf{Q}_q) \to H^{n+1}(U_1, \mathbf{Q}_q)$ it suffices to calculate

$$\lim_{\lambda \to 1} A(\lambda)^{-1} \, \mathrm{Frob}^*_{q,0} \, A(\lambda^q).$$

It should be remarked that the operator $A(\lambda)$ itself does not converge on the p-adic unit disc.

The methods of Gerkmann and Lauder consist of an efficient calculation of the solution of the Picard–Fuchs equation.

2.2 Deformation Method, Singular Case

We describe which of the above ideas differ in the case that V_1 is singular.

We start with some (false) heuristics. One expects that the dimension of H^n to drop; thus $\dim H^n(\overline{V}_1, \mathbf{Q}_q) < \dim H^n(\overline{V}_0, \mathbf{Q}_q)$ and $\dim H^{n+1}(U_1, \mathbf{Q}_q) < \dim H^{n+1}(U_0, \mathbf{Q}_q)$. However,

$$Fr(\lambda) := \lim_{\mu \to \lambda} A(\mu)^{-1} \, \mathrm{Frob}^*_{q,0} \, A(\mu^q)$$

defines for $\lambda = 1$ an operator on a vector space W of dimension equal to the dimension of $H^n(U_0, \mathbf{Q}_q)$.

At the same time one expects that the singularities of the Picard–Fuchs equation are related to the singularities in the family V_λ, so Fr might have singularities at $\lambda = 1$. This suggests that $Fr^{-1}(1) = \lim_{\lambda \to 1} Fr^{-1}(\lambda)$ has a kernel K, that $W = W_1 \oplus K$, such that Fr^{-1} respects this decomposition and $\dim W_1 = \dim H^{n+1}(U_1, \mathbf{Q}_q)$. When this happens then it would be likely that $W_1 \cong H^{n+1}(U_1, \mathbf{Q}_q)$ as vector space with Frobenius action, and the trace of Frob^{*-1} on $H^{n+1}(U_1, \mathbf{Q}_q)$ would equal the trace of Frob^{*-1} on W.

Unfortunately, this does not happen very often: one can construct examples such that the Picard–Fuchs equation is 'less' singular than the drop in the dimension of H^{n+1} predicts, i.e. $\dim W_1 > \dim H^{n+1}(U_1, \mathbf{Q}_q)$. This is due to the fact that the family V_λ over the punctured disc $\{\lambda : 0 <| \lambda - 1 |< 1\}$, considered as a family of abstract varieties, can be completed in different ways. Since the Picard–Fuchs equation depends only on the family V_λ considered in a neighborhood of $\lambda = 0$ all these families have the same Picard–Fuchs equation and therefore the same operator $A(\lambda)$. However, the dimension of $H^n(V_1, \mathbf{Q}_q)$ depends on how one completes the family V_λ. The number of points $\#\overline{V}_1(\mathbf{F}_q)$ depends also on the way one completes the family $\overline{V}_\lambda$. So the main obstruction to extend the deformation algorithm is:

Obstruction 7. *If $\overline{V}$ is singular then the deformation algorithm might calculate $\#\overline{V}'(\mathbf{F}_p)$ for a variety $\overline{V}'$ different from $\overline{V}$.*

Remark 8. If $n = 1$ it is quite predictable how $\overline{V}$ and $\overline{V}'$ are related; one expects $\overline{V}'$ to be the stable reduction of $\overline{V}$. For $n > 1$ the variety $\overline{V}'$ is related to $\overline{V}$, but (in general) it seems quite unclear just how. The variety $\overline{V}'$ might be 'the' stable reduction of $\overline{V}$ (if one can find an appropriate moduli problem) or if, for example, $\overline{V}$ is a surface with isolated $A - D - E$ singularities then $\overline{V}'$ might be the resolution of singularities of $\overline{V}$. To extend the deformation algorithm to singular varieties, one should first start by studying the relation between $\overline{V}$ and $\overline{V}'$.

2.3   Direct Method, Smooth Case

The idea used in [1] is easier to explain. Suppose for the moment that $\overline{V}$ is a smooth hypersurface. Then $\#\overline{V}(\mathbf{F}_q)$ can be calculated by determining the action of Frobenius on the rigid cohomology group

$$H_{\mathrm{rig}}^{n+1}(\overline{U}, \mathbf{Q}_q).$$

Fix a lift V of $\overline{V}$ to $\mathbf{Z}_q$, let U be the complement of V. A theorem of Baldassarri–Chiarellotto [2] states that

$$H_{\mathrm{rig}}^{n+1}(\overline{U}, \mathbf{Q}_q) \cong H_{\mathrm{dR}}^{n+1}(U, \mathbf{Q}_q).$$

Due to work of Griffiths [4], the latter group $H_{\mathrm{dR}}^{n+1}(U, \mathbf{Q}_q)$ is very well understood:

Let $\Omega := \prod_i X_i \sum_j (-1)^j \frac{dX_0}{X_0} \wedge \cdots \wedge \widehat{\frac{dX_j}{X_j}} \wedge \cdots \wedge \frac{dX_n}{X_n}$. Let $F = 0$ be an equation defining V, such that $\overline{F} \equiv F \mod \pi$. Then $H_{dR}^{n+1}(U, \mathbf{Q}_q)$ consists of

$$\Omega^{n+1}(U) := \left\{ \frac{G}{F^t} \Omega : t \in \mathbf{Z}, t > 0, \deg(G) = t \deg(F) - n - 1 \right\}$$

modulo the following relations

$$\frac{(t-1)GF_{X_i}}{F^t} \Omega = \frac{G_{X_i}}{F^{t-1}} \Omega \tag{1}$$

where X_i is a coordinate on $\mathbf{P}$ and the subscript X_i means the partial derivative with respect to X_i. In particular, one can show that H_{dR}^{n+1} can be generated by forms with $t \leq n + 1$. Let $\{\omega_j\}$ be a basis of $H_{dR}^{n+1}(U, \mathbf{Q}_q)$ (which in turn is a basis for $H_{rig}^{n+1}(\overline{U}, \mathbf{Q}_q)$).

Let $\overline{A}$ be the coordinate ring of $\overline{U}$. Let A be the coordinate ring of U. Fix a representation

$$A = \mathbf{Q}_q[Y_0, \ldots, Y_m]/(G_1, \ldots, G_k).$$

Definition 9. *Set*

$$A^{\dagger} = \frac{\{H \in \mathbf{Q}_q[[Y_0, \ldots, Y_m]] : \text{the radius of convergence of } H \text{ is at least } r > 1\}}{(G_1, \ldots, G_k)}.$$

Then $A^{\dagger}$ is called an overconvergent completion *(or weak completion) of A.*

An overconvergent completion depends on the representation of A. However, the results mentioned below are independent of the chosen representation of A.

Fix a lift of Frobenius $\mathrm{Frob}_q^* : A^{\dagger} \to A^{\dagger}$. To calculate the Frobenius action on $H_{rig}^{n+1}(\overline{U}, \mathbf{Q}_q)$ we need to express

$$\mathrm{Frob}_q^*(\omega_j) = \left(\sum_{i=0}^{\infty} \frac{G_i}{F^i} \right) \Omega \tag{2}$$

in terms of the basis $\{\omega_i\}$.

For our purposes it suffices to know the characteristic polynomial of Frobenius up to a certain p-adic precision. For this reason we can truncate the series (2) after N steps, where N can be computed in terms of p, n and d. This truncated series gives a class in H_{dR}^{n+1}. We can use the expression (1) to reduce the pole order, and hence to write $\mathrm{Frob}_q^*(\omega_j)$ in the form $\sum_i a_{i,j}\omega_i$. This suffices to calculate the characteristic polynomial of Frobenius.

2.4   Direct Method, Singular Case

If $\overline{V}$ is singular then several of the above ideas fail to work. It turns out that a combination of these obstructions yields an outline for an algorithm that works for singular varieties.

The following three steps fail in the singular case:

1. First of all, the comparison theorem of Baldassarri–Chiarellotto does not hold. Instead one only has a natural map

$$H_{\mathrm{dR}}^{n+1}(U, \mathbf{Q}_q) \to H_{\mathrm{rig}}^{n+1}(\overline{U}, \mathbf{Q}_q).$$

One of the problems here is that the dimension of the left hand side depends on the choice of the lift U, whereas the dimension of the right hand side is independent of the dimension of the lift, so there is no hope that an arbitrary choice of a lift will work.

2. To reduce expression (2) one needs to be able to write polynomials G of large degree as a combination $\sum H_i F_{X_i}$. This is possible, since the Jacobian ring of F

$$R = \mathbf{Q}_q[X_0, \ldots, X_{n-1}]/(F_{X_0}, \ldots, F_{X_{n+1}})$$

is a finite dimensional $\mathbf{Q}_q$-vector space, provided that F is smooth. If F is singular then R is infinite dimensional.

3. If one chooses the lift F of $\overline{F}$ such that $F = 0$ is smooth, then the reduction of

$$\lim_{N \to \infty} \left(\sum_{i=0}^{N} \frac{G_i}{F^i} \right) \Omega$$

might diverge.

The following remark gives an algo-geometric explanation for these phenomena.

Remark 10. The second point is the most fundamental obstruction. One can filter Ω_U^k, the k-form on U, by the order of the pole along V. The filtered complex $\Omega_U^\bullet$ yields a spectral sequence $E_k^{i,j}$ abutting to $H_{\mathrm{dR}}^{i+j}(U, \mathbf{Q}_q)$. The relations (1) describe $E_2^{i,n+1-j}$: Let R be the Jacobian ring of F. Since the Jacobian ideal is homogenous we can grade elements of R by their degree. Then $\oplus_i E_2^{i,n+i-1} = \oplus_p R_{id-n-2}$.

If V is smooth then this spectral sequence degenerates at E_2, hence this suffices to calculate $H_{\mathrm{dR}}^{n+1}(U, \mathbf{Q}_q)$. If V is singular then this spectral sequence *cannot* degenerate at E_2 but degenerates at a higher step. One could try to adjust the algorithm [1] by trying to take an 'equisingular' lift, and try to identify the extra relations one needs to obtain $H^{n+1}(U, \mathbf{Q}_q)$ as a quotient of Ω_U^n. Unfortunately, such a lift might not exist and it is not clear at all which relations one needs to add, except for a few cases.

We give a procedure to determine the kernel of $H_{\mathrm{dR}}^{n+1}(U, \mathbf{Q}_q) \to H_{\mathrm{rig}}^{n+1}(\overline{U}, \mathbf{Q}_q)$ under some restrictions on the singularities of $\overline{V}$. In practice (e.,g., the case of a surface with $A - D - E$ singularities in sufficiently large characteristic) it turns out that this kernel has the same size as the difference between dim $H_{\mathrm{dR}}^{n+1}(U, \mathbf{Q}_q)$ and dim $H_{\mathrm{rig}}^{n+1}(\overline{U}, \mathbf{Q}_q)$.

For simplicity, let us assume we have a sequence $F_k \in \mathbf{Z}_q[X_0, \ldots, X_{n+1}]$, such that

- $F_k \equiv F_{k-1} \bmod \pi^{k-1}$,
- the singular locus of $F_k \bmod \pi^k$ coincides with a lift of the singular locus of $\overline{F}$,
- $F_k \bmod p^{k+1}$ is smooth.

That is, we have a series of polynomials F_k, defining smooth hypersurfaces, but lifting the singular locus modulo π^k. In general such a sequence of polynomials might not exist.

Since the Jacobian ideal of F_k is finite-dimensional we can try to mimic [1], that is we make a power series expansion $\mathrm{Frob}_q(\omega_j)$, truncate this after N steps, and try to reduce this form in $H_{\mathrm{dR}}^{n+1}(U_k, \mathbf{Q}_q)$.

It turns out that if $N \to \infty$ or $k \to \infty$ then the p-adic absolute value of some of the coefficient of the reduction tend to increase. This is due to the fact that the Jacobian ideal of $\overline{F}$ is infinite-dimensional:

Example 11. Suppose we have a form

$$\frac{G}{F_k^t} \Omega.$$

After dividing or multiplying by π we may assume that $G \in \mathbf{Z}_q[X_0, \ldots, X_n]$ and $G \not\equiv 0 \bmod \pi$.

In order to reduce the pole order we need to write G as $\sum H_i F_{k,X_i}$. Let P be a lift of a point in the singular locus. Suppose G is general, thus $G(P) \not\equiv 0 \bmod \pi$. Now,

$$G(P) \equiv \sum H_i(P) F_{k,X_i}(P) \equiv 0 \bmod \pi^k.$$

Hence some of the coefficients in H_i need to have negative p-adic valuation. In practice this means that the after each reduction step the p-adic valuation of the coefficient decreases rapidly.

Since $\mathrm{Frob}_p^*(\omega_j) = \sum \frac{G_t}{F_k^t} \Omega$ is an overconvergent power series one has that the p-adic valuation of the coefficient of G_t increases when t increases. However, the minimum of the valuation of the coefficients of G_t is around t/p. This turns out to be insufficient to compensate for the high power of p in the denominator obtained by reducing the pole order. In the next section we give an example where the inverse of Frobenius has an eigenvalue with very small p-adic absolute value, hence Frobenius has an eigenvalue with large absolute value.

Next, the main idea is to consider the action of Frob_q^{*-1}. We could do this by considering $\mathrm{Frob}_q^*(\omega)$ and truncating at pole order N and then inverting the obtained operator. This operator has several eigenvalues with small q-adic absolute value, that is, very positive q-adic valuation. At the same time we know that the eigenvalues of $q^{n+1} \mathrm{Frob}_q^{*-1}$ on $H_{\mathrm{rig}}^{n+1}(\overline{U}, \mathbf{Q}_q)$ are algebraic integers with complex absolute value at most q^{n+1}. In particular, the q-adic valuation of such an eigenvalue is between 0 and $n+1$. Therefore, all eigenvalues that have q-adic valuation bigger than $n+1$ cannot be eigenvalues of Frobenius on $H_{\mathrm{rig}}^{n+1}(\overline{U}, \mathbf{Q}_q)$, hence the corresponding eigenvectors lie in the kernel of $H_{\mathrm{dR}}^{n+1}(U_k, \mathbf{Q}_q) \to H_{\mathrm{rig}}^{n+1}(\overline{U}, \mathbf{Q}_q)$.

This idea seems to be very hard to use in practice, since by inverting the approximation of the operator Frob_q^* one encounters severe problems in obtaining the necessary p-adic precision.

Instead we study a left-inverse of Frob_q^*:

Notation 12. *Let* $\psi : A^\dagger \to A^\dagger$, *be the* $\mathbf{Q}_q$-*linear operator defined by*

$$\psi\left(\prod X_i^{a_i}\right) = \begin{cases} \prod X_i^{a_i/q} & \text{if } a_i \equiv 0 \bmod q \text{ for all } i, \\ 0 & \text{otherwise.} \end{cases}$$

and $\psi(\Omega/\prod X_i) = \Omega/(p^{n+1}\prod X_i)$.

Since Frob_q^* on $H_{\mathrm{rig}}^{n+1}(\overline{U}, \mathbf{Q}_q)$ is invertible and $\psi \circ \mathrm{Frob}_q^*$ is the identity, one has that ψ^* on $H_{\mathrm{rig}}^{n+1}(\overline{U}, \mathbf{Q}_q)$ is the inverse of Frob_q^*.

Remark 13. This operator ψ^* behaves much better than Frob_q^*. Assume for simplicity that $n < q$. We need only consider forms with pole order $t \leq q$

$$\psi\left(\frac{G}{F^t}\Omega\right) = \psi\left(\frac{F^{q-t}G\prod X_k}{F^q}\frac{\Omega}{\prod X_k}\right) \tag{3}$$

$$= \psi\left(\sum_i \frac{F^{q-t}G\prod X_k \Delta^i}{F(X_0^q,\ldots,X_{n+1}^q)^{i+1}}\frac{\Omega}{\prod X_k}\right) \tag{4}$$

$$= \left(\sum_i \frac{\psi(F^{q-t}G\prod X_k \Delta^i)}{F(X_0,\ldots,X_{n+1})^{i+1}}\right)\frac{\Omega}{p^{n+1}\prod X_k} \tag{5}$$

with $\Delta = F(X_0^q,\ldots,X_{n+1}^q) - F(X_0,\ldots,X_{n+1})^q$.

Abbott–Kedlaya–Roe reduce the form

$$\mathrm{Frob}_q^*\left(\frac{G}{F^t}\Omega\right) = \sum_i \binom{t+i-1}{i}(-\Delta)^i\frac{\mathrm{Frob}^*(G\prod X_k)}{F^{qi+t}}p^{n+1}\frac{\Omega}{\prod X_k}. \tag{6}$$

Very roughly the convergence of power series in (5) is q times faster than in (6). If we reduce the pole order in (5) then the valuation of Δ^i is sufficiently high to compensate for the high power of π one gets in the denominator by reducing.

We would like to remark that in the case of a *smooth* hypersurface one can also use ψ rather than Frob_q^*. By using ψ one can lower the necessary pole order roughly by a factor q.

3   Examples

We apply the above observations to one particular example.

Let q be an odd prime power. In this section we consider the surface S_1 : $X^2 + Y^2 + Z^2 = 0$ in $\mathbf{P}_{\mathbf{F}_q}^3$. The surface S_1 is a cone over a conic in $\mathbf{P}^2$. This implies that S_1 has an A_1-singularity at $P := [1:0:0:0]$. Let $\tilde{S}_1$ be the blow-up of S_1 at P. Then $\tilde{S}_1$ is a ruled surface over $\mathbf{P}^1$. In particular it has the following Betti numbers:

$$h^0(\tilde{S}_1) = h^4(\tilde{S}_1) = 1, h^2(\tilde{S}_1) = 2$$

and all other Betti numbers vanish. From this it follows that $h^2(S_1) = 1$ and $h^i(S_1) = h^i(\tilde{S}_1)$ for $i \neq 2$.

One can easily see that $\#V(\mathbf{F}_q) = q^2 + q + 1$. We will show that a slight modification of Lauder's (or Gerkmann's) method yield the output $q^2 + 2q + 1$, whereas a slight modification of Abbott–Kedlaya–Roe gives the correct answer $\#V(\mathbf{F}_q) = q^2 + q + 1$.

3.1   Deformation Method

Consider the family

$$\overline{V}_\lambda : (1 - \lambda)W^2 + X^2 + Y^2 + Z^2 = 0.$$

Let $\overline{U}_\lambda$ be the complement of $\overline{V}_\lambda$. The methods of Lauder and Gerkmann require to calculate the Frobenius action on $H^3(U_0, \mathbf{Q}_q)$. It is easy to see that Frobenius acts as multiplication by p on this one-dimensional vector space.

Secondly, one defines an operator $A(\lambda) : H^3(U_\lambda) \to H^3(U_0)$. For this we need the following definition:

Definition 1. *Let r, s be non-negative integers, let $\alpha_i \in \mathbf{Q}_q$, for $i \in \{1, 2, \ldots, r\}$, let $\beta_j \in \mathbf{Q}_q \setminus \mathbf{Z}_{<0}$ for $j \in \{1, 2, \ldots, s\}$. We define the (generalized) hypergeometric function*

$$_rF_s\left(\begin{matrix} \alpha_1\ \alpha_2\ \cdots\ \alpha_r \\ \beta_1\ \beta_2\ \cdots\ \beta_s \end{matrix}; z\right)$$

to be

$$\sum_{k=0}^{\infty} b_j z^j,$$

with $b_0 = 1$, and

$$\frac{b_{j+1}}{b_j} = \frac{(j + \alpha_1) \ldots (j + \alpha_r)}{(j + \beta_1) \ldots (j + \beta_s)(j + 1)},$$

for all positive integers j.

Using the methods presented in [6, Section 5] one can calculate $A(\lambda)$. This yields that

$$A(\lambda) = {}_1F_0\left(\frac{1}{2}; \lambda\right),$$

hence the composition $A(\lambda)^{-1} \operatorname{Frob}_{q,0} A(\lambda^q)$ equals

$$Fr(\lambda) = q\frac{{}_1F_0\left(\frac{1}{2}; -\lambda^q\right)}{{}_1F_0\left(\frac{1}{2}; -\lambda\right)} = q\frac{(1 + \lambda)^{1/2}}{(1 + \lambda^q)^{1/2}}.$$

Now, $q^2 Fr(\lambda)^{-2} = \frac{(1+\lambda^q)}{(1+\lambda)} = \sum_{i=0}^{q-1}(-\lambda)^i$. Hence if λ is the Teichmüller lift of $\bar\lambda \in \mathbf{F}_q^*$ (the unique lift such that $\lambda^q = \lambda$), then $Fr(\lambda)^2 = q$. Slightly more involved is the following equality:

$$\frac{(1+\lambda)^{1/2}}{(1+\lambda^q)^{1/2}} = \chi(\bar\lambda)$$

where $\chi : \mathbf{F}_q^* \to \{\pm 1\}$ is the unique non-trivial quartic character of $\mathbf{F}_q^*$, in other words $\chi(\bar\lambda) = 1$ if and only if $\bar\lambda$ is a square in $\mathbf{F}_q$.

Lauder's and Gerkmann's algorithm would give $\#\overline{U}_\lambda(\mathbf{F}_q) = q^3 - \chi(\bar\lambda)q$, whence

$$\#\overline{V}_{\bar\lambda}(\mathbf{F}_q) = \begin{cases} q^2 + 2q + 1 & \text{if } \bar\lambda \text{ is a square modulo } q \text{ or } \bar\lambda = 0 \\ q^2 + 1 & \text{if } \bar\lambda \text{ is a not square modulo } q. \end{cases}$$

It is clear that this answer is wrong if $\bar\lambda = 1$, and correct if $\bar\lambda \neq 1$. The following remark gives an algo-geometric explanation for this phenomena.

Remark 2. As remarked in the previous section, the deformation method might give wrong answers because one can complete the family $V_\lambda, 0 <| \lambda - 1 |< 1$ in a non-unique way. We construct now the family Y_λ such that $V_\lambda = Y_\lambda$ for $\lambda \neq 1$ and the deformation method calculates the zeta function of Y_1.

It is known that over an algebraically closed field one can construct a family of vector bundles $\mathcal{V}_\lambda$ on $\mathbf{P}^1$ such that

$$\mathcal{V}_\lambda = \begin{cases} \mathcal{O} \oplus \mathcal{O} & \text{if } \lambda \neq 1 \\ \mathcal{O}(-1) \oplus \mathcal{O}(1) & \text{if } \lambda = 1. \end{cases}$$

This yields a family of projective bundles $Y_\lambda := \mathbf{P}(\mathcal{V}_\lambda)$. For $\lambda \neq 1$ we have that $\mathbf{P}(\mathcal{V}_\lambda) \cong \mathbf{P}^1 \times \mathbf{P}^1$, whereas for $\lambda = 1$ we have that $\mathbf{P}(\mathcal{V}_\lambda)$ is isomorphic to the Hirzebruch surface F_2.

We can map this family in to $\mathbf{P}^2$ by fixing a degree 2 line bundle $\mathcal{L}_\lambda$ on Y_λ. On $\mathbf{P}^1 \times \mathbf{P}^1$, let f_1 be a fiber of the first projection, f_2 be a fiber of the second projection, then $\mathcal{L}_\lambda := \mathcal{O}(f_1 + f_2)$ has degree 2 and $\mathcal{L}_\lambda$ is ample. Actually, the family of line bundles $\mathcal{L}_\lambda$ for $\lambda \neq 1$ is a line bundle on the 3-dimensional variety $\cup_{\lambda,\lambda \neq 1} Y_\lambda$. We can extend $\mathcal{L}$ to all of $\cup_\lambda Y_\lambda$: On $Y_1 \cong F_2$ there is only one ruling, let f be a fiber of this ruling, let z be the exceptional section, that is, the self-intersection (z, z) equals -2 and $(z, f) = 1$. Then $\mathcal{L} \mid_{Y_1} = \mathcal{O}(2f + z)$. This line bundle is of degree 2, but not ample, since $(2f + z, z) = 0$. If we use $\mathcal{L}$ to map the family Y_λ in $\mathbf{P}^3$ then we obtain a family of surfaces V_λ in $\mathbf{P}^3$ such that $V_\lambda \cong Y_\lambda$ for $\lambda \neq 1$ and Y_1 is a resolution of singularities of V_1. I.e., the map $Y_1 \to V_1$ contracts z.

The deformation method calculates $\#Y_1(\mathbf{F}_q)$ rather than $\#V_1(\mathbf{F}_q)$.

3.2 Direct Method

For the direct method we only need to consider $F = X^2 + Y^2 + Z^2 = 0$. To simplify the exposition, assume that $q = p$ a prime number.

Let $F_k := X^2 + Y^2 + Z^2 + p^k W^2$. Then $F_k = 0$ defines a smooth hypersurface, such that its reduction modulo p^k is singular. The cohomology group $H^{n+1}_{\mathrm{dR}}(U_k, \mathbf{Q}_p)$ is one-dimensional and it is generated by

$$\frac{1}{F_k^2}\Omega.$$

From (5) it follows that

$$\psi\left(\frac{1}{F_k^2}\Omega\right) = \sum_i \frac{\psi(XYZWF_k^{p-2}\Delta^i)}{F_k^{i+1}} \frac{\Omega}{p^3 XYZW}$$

If we truncate this expression at pole order N we get

$$\sum_{j=0}^{N-1}(-1)^j \left(\sum_{i=j}^{N-1}\binom{i}{j}\right) \frac{\psi(XYZWF_k^{(j+1)p-2})}{F_k^{j+1}} \frac{\Omega}{p^3 XYZW}.$$

From the definition of ψ it follows that we only have to consider monomials in $XYZWF_k^{(j+1)p-2}$ such that all the exponents are divisible by p. This observation combined by writing out $XYZWF_k^{(j+1)p-2}$ yields:

Lemma 3. Set $T_j = \{(t_1, t_2, t_3, t_4) : t_1, t_2, t_3, t_4 \geq 0, \sum t_i = j - 1\}$. For t_1, t_2, t_3, t_4 in T_j set

$$B(t_1, t_2, t_3, t_4) := \binom{(j+1)p - 2}{\frac{p-1}{2} + t_1 p \quad \frac{p-1}{2} + t_2 p \quad \frac{p-1}{2} + t_3 p \quad \frac{p-1}{2} + t_4 p}.$$

Then $\psi(XYZWF_k^{(j+1)p-2})$ equals

$$\sum_{(t_1, t_2, t_3, t_4) \in T_j} B(t_1, t_2, t_3, t_4) p^{k(\frac{p-1}{2} + t_4 p)} X^{1+2t_1} Y^{1+2t_2} Z^{1+2t_3} W^{1+2t_4}.$$

Denote by $(a)_m$ the Pochhammer symbol $a(a+1)\ldots(a+m-1)$. Successively applying (1) yields the following result:

Lemma 4. The reduction of

$$\frac{X^{2t_1} Y^{2t_2} Z^{2t_3} W^{2t_4}}{F_k^{t_1+t_2+t_3+t_4+2}} \Omega$$

in $H^{n+1}_{\mathrm{dR}}(U, \mathbf{Q}_q)$ equals

$$\frac{(1/2)_{t_1}(1/2)_{t_2}(1/2)_{t_3}(1/2)_{t_4}}{(t_1 + t_2 + t_3 + t_4 + 1)! p^{kt_4}} \frac{1}{F_k^2} \Omega.$$

Combing the above Lemmas yields:

Lemma 5. *For $j > 0$ the reduction of*

$$\frac{\psi(XYZWF^{(j+1)p-2})}{F^{j+1}}\frac{\Omega}{p^3XYZW}$$

in $H^{n+1}_{dR}(U, \mathbf{Q}_q)$ equals

$$\left(\sum_{(t_1,t_2,t_3,t_4)\in T_j} B(t_1,t_2,t_3,t_4)p^{\frac{1+2t_4}{2}k(p-1)}\frac{(1/2)_{t_1}(1/2)_{t_2}(1/2)_{t_3}(1/2)_{t_4}}{(t_1+t_2+t_3+t_4+1)!}\right)\frac{1}{F_k^2}\frac{\Omega}{p^3}.$$

Lemma 6. *The quantity*

$$\gamma = B(t_1,t_2,t_3,t_4)\frac{(1/2)_{t_1}(1/2)_{t_2}(1/2)_{t_3}(1/2)_{t_4}}{(t_1+t_2+t_3+t_4+1)!}$$

is a p-adic integer.

Proof. Let $\alpha = (p-1)/2$. Note that $B_0 := B(t_1,t_2,t_3,t_4)$ equals

$$\binom{(t_1+t_2+t_3+t_4)p+4\alpha}{t_1p+\alpha}\binom{(t_2+t_3+t_4)p+3\alpha}{t_2p+\alpha}\binom{(t_3+t_4)p+2\alpha}{t_3+\alpha}.$$

It is well-known that the p-adic valuation of $\binom{m}{i}$ equals the number of carries $c(i, m-i)$ (in base p) if one sums i and $m-i$. Hence $v(B_0)$ equals

$$c(t_1p+\alpha, (t_2+t_3+t_4)p+3\alpha)+c(t_2p+\alpha, (t_3+t_4)p+2\alpha)+c(t_3p+\alpha, t_4p+\alpha).$$

We want to compare the valuation of B_0 with the valuation of $\frac{(t_1+t_2+t_3+t_4+1)!}{t_1!t_2!t_3!t_4!}$. Let B_1 denote the latter quantity. One has that B_1 equals

$$\binom{t_1+t_2+t_3+t_4+1}{t_1}\binom{t_2+t_3+t_4+1}{t_2}\binom{t_3+t_4}{t_3}(t_3+t_4+1),$$

whence its valuation $v(B_1)$ equals

$$c(t_1, t_2+t_3+t_4+1)+c(t_2, t_3+t_4+1)+c(t_3, t_4)+v(t_3+t_4+1).$$

It is easy to see that

$$c(t_1p+\alpha, (t_2+t_3+t_4)+3\alpha) = c(t_1, t_2+t_3+t_4+1) \text{ and } c(t_3p+\alpha, t_4p+\alpha) = c(t_3, t_4).$$

Let $m := v(t_3+t_4+1)$. Since $t_3+t_4 \equiv 1 \bmod p$ we can write

$$t_3+t_4 = (p-1)+(p-1)p+\cdots+(p-1)p^{m-1}+\beta_m,$$

with $\beta_m \equiv 0 \bmod p^{m-1}$. Since $\alpha \not\equiv 0 \bmod p$ we get

$$c(t_2p+\alpha, (t_3+t_4+1)p-1) = m+c(t_2p, \beta_m+p^m) = m+c(t_2, t_3+t_4+1),$$

whence $v(B_0) = v(B_1)$.

Since $(1/2)_{t_j}$ is the product of the first t_j odd number divided by 2^{t_j}, we get that $v((1/2)_{t_j}) \geq v(t_j!)$ and

$$v(\gamma) \geq v\left(\frac{B_0}{B_1}\right) = 0,$$

which shows that γ is a p-adic integer.   □

Combining these lemmas shows that the reduction ω_N of $p^3\psi\left(\frac{1}{F_k^2}\Omega\right)$ truncated after N steps satisfies $\omega_N \equiv 0 \bmod p^{k(p-1)/2}$, provided that $N > 1$. The eigenvalues of ψp^3 on $H^3_{\mathrm{rig}}(\overline{U}, \mathbf{Q}_q)$ are algebraic integers with complex absolute value at most p^3. Take k such that $k(p-1) \geq 8$ then $\frac{1}{F_k}\Omega$ lies in the kernel of $H^3_{\mathrm{dR}}(U, \mathbf{Q}_q) \to H^3_{\mathrm{rig}}(\overline{U}, \mathbf{Q}_q)$, and the latter group vanishes.

If k is chosen large enough, then (modified) Abbott–Kedlaya–Roe does not see the eigenvalue corresponding to ω_N hence its output is $p^2 + p + 1$, which is the correct number of points.

3.3 Another Example

We did some computer experiments with the cubic surface S defined by

$$W^3 + X^3 + Y^3 + Z^3 + \overline{3}WX^2$$

in $\mathbf{F}_5$. This cubic surface has a D_4 singularity.

For the same reason as above, Gerkmann's and Lauder's algorithm (with sufficiently high precision) yield the number of points of $\tilde{S}$, the resolution of singularities of S.

We applied the modified algorithm of Abbott–Kedlaya–Roe (with ψ rather than Frob_q^*), where we took the naive lift $W^3 + X^3 + Y^3 + Z^3 + 3WX^2$. Truncating at $N = 3$ revealed that $p^3\psi$ has eigenvalues $p, -p$ and four eigenvalues with valuation at least 2, two of which are only defined over a degree 2 extension of $\mathbf{Q}_p$. One can show that for a surface with A-D-E-singularities in 'large' characteristic (where large depends on the type of singularity) the eigenvalues of Frobenius on $H^3_{\mathrm{rig}}(\overline{U}, \mathbf{Q}_q)$ have complex absolute value p, (thus, the Riemann hypothesis holds for such surfaces). Hence the eigenvectors corresponding to eigenvalues with p-adic valuation at least 2 generate the kernel $H^3_{\mathrm{dR}}(U, \mathbf{Q}_q) \to H^3_{\mathrm{rig}}(\overline{U}, \mathbf{Q}_q)$. This yields that the zeta function $Z(S, t)$ equals $\left((1-t)(1-5t)(1+5t)(1-5^2t)\right)^{-1}$ and that $\#S(\mathbf{F}_5) = 5^2 + 1 = 26$, which is correct.

References

1. Abbott, T.G., Kedlaya, K., Roe, D.: Bounding Picard numbers of surfaces using p-adic cohomology. In: Arithmetic, Geometry and Coding Theory (AGCT 2005), Societé Mathématique de France (to appear, 2007)
2. Baldassarri, F., Chiarellotto, B.: Algebraic versus rigid cohomology with logarithmic coefficients. In: Barsotti Symposium in Algebraic Geometry (Abano Terme, 1991), Perspect. Math., vol. 15, pp. 11–50. Academic Press, San Diego (1994)

3. Gerkmann, R.: Relative rigid cohomology and deformation of hypersurfaces. Intern. Math. Research Papers (to appear, 2007)
4. Griffiths, P.A.: On the periods of certain rational integrals I, II. Ann. of Math. 90(2), 460–495 (1969); ibid. 90(2), 496–541 (1969)
5. Katz, N.M.: On the differential equations satisfied by period matrices. Inst. Hautes Études Sci. Publ. Math. 35, 223–258 (1968)
6. Kloosterman, R.: The zeta-function of monomial deformations of Fermat hypersurfaces. Algebra Number Theory 1, 421–450 (2007)
7. Kloosterman, R.: An algorithm for point counting on singular hypersurfaces (in preparation)
8. Lauder, A.G.B.: Counting solutions to equations in many variables over finite fields. Found. Comput. Math. 4, 221–267 (2004)
9. Lauder, A.G.B.: A recursive method for computing zeta functions of varieties. LMS J. Comput. Math. 9, 222–269 (2006)
10. Schoen, C.: Algebraic cycles on certain desingularized nodal hypersurfaces. Math. Ann. 270, 17–27 (1985)

Efficient Hyperelliptic Arithmetic Using Balanced Representation for Divisors

Steven D. Galbraith[1], Michael Harrison[2], and David J. Mireles Morales[1]

[1] Mathematics Department
Royal Holloway, University of London
{steven.galbraith,d.mireles-morales}@rhul.ac.uk
[2] School of Mathematics and Statistics
University of Sydney
mch@maths.usyd.edu.au

Abstract. We discuss arithmetic in the Jacobian of a hyperelliptic curve C of genus g. The traditional approach is to fix a point $P_\infty \in C$ and represent divisor classes in the form $E - d(P_\infty)$ where E is effective and $0 \leq d \leq g$. We propose a different representation which is balanced at infinity. The resulting arithmetic is more efficient than previous approaches when there are 2 points at infinity.

1 Introduction

The study of efficient addition algorithms for divisors on genus 2 curves has come to a point where cryptography based on these curves provides an alternative to its well-established elliptic curve counterpart. The most commonly used case is when the curve has 1 point at infinity and addition corresponds to Cantor's ideal composition and reduction algorithm in [2]. Explicit formulae have been given by Lange in [8] and a comprehensive account of the different addition algorithms can be found in [3].

It is then only natural to extend this work to hyperelliptic curves with 2 points at infinity since curves with a rational Weierstrass point are rare among all hyperelliptic curves. Further motivation is given by pairing based cryptography, since Galbraith, Pujolas, Ritzenthaler and Smith gave in [5] an explicit construction of a pairing-friendly genus 2 curve C which typically cannot be given a model with 1 point at infinity. It is an interesting question to determine how efficiently pairings can be implemented for these curves.

Scheidler, Stein and Williams [11] gave algorithms to compute in the so-called infrastructure of a function field (also see [7]). Their approach included composition and reduction algorithms used by Cantor, as well as an algorithm that had no analogue in his theory, known as a "baby-step". The relationship between the infrastructure and divisor class groups was studied by Paulus and Rück [9]. It is well-known that arithmetic on curves with two points at infinity is slower than the simpler case of one point at infinity (our methods do not change this).

A.J. van der Poorten and A. Stein (Eds.): ANTS-VIII 2008, LNCS 5011, pp. 342–356, 2008.
© Springer-Verlag Berlin Heidelberg 2008

In this article we view the Cantor and infrastructure algorithms as operations on the Mumford representation of affine effective semi-reduced divisors, rather than as operations on the Jacobian of a curve. This simple change of perspective suggests a representation of elements in the Jacobian of C which is more "balanced" at infinity. We therefore show that arithmetic in the Jacobian may be performed more efficiently than done by [4,6,9,10]. In the case of genus 2 curves, all explicit addition formulae presented so far [4] can be used with our representation, giving improved results (see Table 1).

We interpret the algorithms developed for the infrastructure, in particular the baby-step, from our new perspective. This gives, in our opinion, a simpler explanation of them. In particular, we do not need to discuss continued fraction expansions. Note however that we only discuss the application of these ideas to arithmetic in the Jacobian, rather than computation in the infrastructure itself. We observe that computing inverses of elements using an unbalanced representation is non-trivial, whereas with our representation it is easy. Previous literature (e.g., [4]) has suggested that the baby step has no analogue for curves with one point at infinity; however we explain that one can develop a fast baby step operation in all settings.

We would like to point out that the group law for hyperelliptic curves with 2 rational points at infinity for the computer algebra system Magma [1], implemented by the second author, follows the approach described in this article. It was first released in Magma V2.12, in July 2005.

2   Divisor Class Groups of Hyperelliptic Curves

In this paper we consider a genus g hyperelliptic curve C defined over a field K given by a non-singular planar model

$$y^2 + h(x)y = F(x) = \sum_{i=0}^{2g+2} F_i x^i,$$

where $h(x), F(x) \in K[x]$ satisfy $\deg(F) \leq 2g+2$ and $\deg(h) \leq g+1$. If $P = (x, y)$ is a point on C, the point $(x, -h(x) - y)$ also lies on C, we will call this point the *hyperelliptic conjugate* of P and we will denote it by $\overline{P}$.

If $F_{2g+2} = 0$ then C will have one K-rational point at infinity, in this case we say that this is an *imaginary* model for C. If $F_{2g+2} \neq 0$ then C will have two points at infinity, possibly defined over a quadratic extension of K, in this case we say that C is represented by a *real* model. If the curve C has a K-rational point we can always move it to the line at infinity so that the points at infinity of the curve are K-rational.

Let C be an algebraic curve defined over a field K. All divisors considered in this article will be K-rational unless otherwise stated. Denote by $\mathrm{Div}^0(C)$ the group of degree zero K-rational divisors on X. Two divisors D_0 and D_1 are *linearly equivalent*, denoted $D_0 \equiv D_1$, if there is a function f such that

$$\operatorname{div}(f) = D_1 - D_0 \,,$$

where $\operatorname{div}(f)$ is the divisor of f.

Definition 1. *The divisor class group of C is the group of K-rational divisor classes modulo linear equivalence. We will denote it as $\operatorname{Cl}(C)$. The class of a divisor D in $\operatorname{Cl}(C)$ will be denoted by $[D]$. We define $\operatorname{Cl}^0(C)$ as the degree zero subgroup of $\operatorname{Cl}(C)$.*

Definition 2. *We say that an effective divisor $D = \sum_i P_i$ is semi-reduced if $i \neq j$ implies $P_i \neq \overline{P}_j$. We say that a divisor D on a curve of genus g is reduced if it is semi-reduced, and has degree $d \leq g$. Throughout this article we will denote the degree of a divisor D_i as d_i.*

There is a standard way to represent an effective affine semi-reduced divisor D_0 on a hyperelliptic curve C: Mumford's representation. In this case we will represent our divisor using a pair of polynomials $u(x), v(x) \in K[x]$, where $u(x)$ is a polynomial of degree d_0 whose roots are the X-coordinates of the points in D_0 (with the appropriate multiplicity) and u divides $F - hv - v^2$. This last condition implies that if x_i is a root of u, the linear polynomial $v(x_i)$ gives the Y-coordinate of the corresponding point in D_0. Because of this last condition, D_0 must be a semi-reduced divisor. We will denote the divisor associated to the pair of polynomials $u(x)$ and $v(x)$ as $\operatorname{div}[u, v]$. Notice that Mumford's representation can be used to describe any effective affine semi-reduced divisor. Describing elements of $\operatorname{Cl}^0(C)$ is a more delicate matter.

To describe elements of $\operatorname{Cl}^0(C)$ we will need a degree g effective divisor D_∞. Throughout this article, unless otherwise stated, this divisor will be as below.

Definition 3.   – *If C has a unique point at infinity ∞, then $D_\infty = g\infty$.*
- *If g is even and C has two points at infinity ∞^+ and ∞^- then $D_\infty = \frac{g}{2}(\infty^+ + \infty^-)$.*
- *If g is odd and C has two points at infinity, then $D_\infty = \frac{g+1}{2}\infty^+ + \frac{g-1}{2}\infty^-$. In this case we will further assume that ∞^+ and ∞^- are K-rational points.*

Proposition 1. *Let D_∞ be a K-rational degree g divisor, and let $D \in \operatorname{Div}^0(C)$ be a K-rational divisor on the curve C. Then $[D]$ has a unique representative in $\operatorname{Cl}^0(C)$ of the form $[D_0 - D_\infty]$, where D_0 is an effective K-rational divisor of degree g whose affine part is reduced.*

Proof. The case $D_\infty = g\infty^+$ is Proposition 4.1 of [9]. Now let D_∞ be any degree g divisor. If D is a representative of a class in $\operatorname{Cl}^0(C)$, using Proposition 4.1 in [9] we know that $D + (D_\infty - g\infty^+) \equiv D_1 - g\infty^+$, where D_1 is an effective degree g divisor with affine reduced part. This implies that $D \equiv D_1 - D_\infty$ and proves existence.

To prove uniqueness, suppose that D_1 and D_2 are two effective degree g divisors with affine reduced support, and $D_1 - D_\infty \equiv D_2 - D_\infty$. Adding $D_\infty - g\infty^+$ to both sides gives $D_1 - g\infty^+ \equiv D_2 - g\infty^+$. Proposition 4.1 from [9] implies that $D_1 = D_2$. $\qquad\square$

A small problem from a computational point of view is that this proposition does not guarantee that the supports of D_0 and D_∞ are disjoint, and indeed, in some cases they will have points in common which should be "cancelled out". However, divisors of the form $D_0 - D_\infty$ with D_0 and D_∞ having disjoint support are generic, so it is enough to describe their arithmetic for many applications. In this article we will give a complete addition algorithm for hyperelliptic curves, that becomes very efficient in the generic case.

If the curve C has two different points at infinity ∞^+ and ∞^-, it is possible to prove that the function y/x^{g+1} is well defined and not zero at each of ∞^+ and ∞^-. One can further prove that

$$\frac{y}{x^{g+1}}(\infty^+) \neq \frac{y}{x^{g+1}}(\infty^-),$$

so if we define

$$a_+ = (y/x^{g+1})(\infty^+), \quad a_- = (y/x^{g+1})(\infty^-),$$

it follows that $a_+ \neq a_-$. Hence, for $p(x)$ a polynomial of the form $p(x) = (a_+x^{g+1} + \sum_{0 \leq i \leq g} b_i x^i)$, the function $y - p(x)$ will have valuation strictly larger than $-(g+1)$ at ∞^+ and valuation $-(g+1)$ at ∞^-.

Definition 4. *In the notation of the previous paragraph, among all degree $g+1$ polynomials with leading coefficient a_+, there is a unique polynomial in $\overline{K}[x]$ for which the valuation of the function at ∞^+ is maximal; we will denote this polynomial by H^+. Define the polynomial H^- analogously.*

If $C(x, y)$ is the equation of the curve, then $H^+(x)$ and $H^-(x)$ are the polynomials with leading coefficient a_+ and a_- such that $C(x, H^\pm(x))$ has minimal degree. Their coefficients can thus be found recursively. The polynomials $H^\pm(x)$ are just a technical tool to specify a point at infinity, similar to the choice of sign when computing the square root of a complex number. Note that the polynomials $H^\pm$ are defined over K if and only if the points ∞^+ and ∞^- are K-rational.

Definition 5. *Given two divisors D_1 and D_2, we will denote the set of pairs of integers ω^+, ω^- such that*

$$D_1 \equiv D_2 + \omega^+\infty^+ + \omega^-\infty^-,$$

as $\omega(D_1, D_2)$. We say that the numbers ω^+ and ω^- are counterweights for D_1 and D_2 if $(\omega^+, \omega^-) \in \omega(D_1, D_2)$.

The set $\omega(D_1, D_2)$ may be empty. If $[\infty^+ - \infty^-]$ is a torsion point on $\mathrm{Cl}^0(C)$, and the set $\omega(D_1, D_2)$ is not empty, then it is infinite; however this will not affect our algorithms. Given two divisors D_1 and D_2, calculating the values of the counterweights relating them is a difficult problem. When these values are needed in our algorithms, there will be a simple way to calculate them.

3   Operations on the Mumford Representation

In this section we recall some well-known algorithms due to Cantor [2] for computing with divisor classes of hyperelliptic curves. We will analyse them as operations on the Mumford representation of an affine semi-reduced divisor. Our main contribution is to give a geometric interpretation of these algorithms.

Algorithm 1. Composition

INPUT: Semi-reduced affine divisors $D_1 = \text{div}[u_1, v_1]$ and $D_2 = \text{div}[u_2, v_2]$.
OUTPUT: A semi-reduced affine divisor $D_3 = \text{div}[u_3, v_3]$ and a pair (ω^+, ω^-), such that $(\omega^+, \omega^-) \in \omega(D_1 + D_2, D_3)$.
1: Compute s (monic), $f_1, f_2, f_3 \in K[x]$ such that

$$s = \gcd(u_1, u_2, v_1 + v_2 + h) = f_1 u_1 + f_2 u_2 + f_3(v_1 + v_2 + h).$$

2: Set $u_3 := u_1 u_2 / s^2$ and $v_3 := (f_1 u_1 v_2 + f_2 u_2 v_1 + f_3(v_1 v_2 + F)) / s \mod u_3$
3: **return** $\text{div}[u_3, v_3]$ and $(\deg(s), \deg(s))$.

The result D_3 of Algorithm 1 will be denoted $D_3, (\omega^+, \omega^-) = \text{comp}(D_1, D_2)$. The divisor of the function s from Algorithm 1 is

$$\text{div}(s) = D_1 + D_2 - D_3 - \frac{d_1 + d_2 - d_3}{2}(\infty^+ + \infty^-), \tag{1}$$

which proves that

$$(\omega^+, \omega^-) \in \omega(D_1 + D_2, D_3).$$

Algorithm 1 is also known as *divisor composition*.

Given an affine semi-reduced divisor D_0, of degree $d_0 \geq g + 2$, Algorithm 2 finds another affine semi-reduced divisor D_1 with smaller degree d_1, and a pair of integers (ω^+, ω^-) such that

$$(\omega^+, \omega^-) \in \omega(D_0, D_1) \tag{2}$$

Algorithm 2 is known as *divisor reduction*.

The result D_1 of Algorithm 2 will be denoted as $D_1, (\omega^+, \omega^-) = \text{red}(D_0)$. The geometric interpretation of Algorithm 2 is very simple: given the effective affine divisor $D_0 = \text{div}[u_0, v_0]$, we know (by definition of the Mumford representation) that the divisor of zeros D_z of the function $y - v_0(x)$ has (in the notation of Algorithm 2) $D_z = D_0 + \overline{D}_1$, and if $\deg(u_0) \geq g + 2$, then the degree of D_z satisfies $\deg(D_z) < 2\deg(D_0)$, hence $\deg(D_1) < \deg(D_0)$, and if the leading term of v_0 is different to that of $H^\pm$ we have

$$\text{div}\left(\frac{y - v_0(x)}{u_0}\right) = \overline{D}_0 - \overline{D}_1 - \frac{d_0 - d_1}{2}(\infty^+ + \infty^-). \tag{3}$$

It follows that

$$D_0 - D_1 \equiv \frac{d_0 - d_1}{2}(\infty^+ + \infty^-).$$

Algorithm 2. Reduction

INPUT: A semi-reduced affine divisor $D_0 = \mathrm{div}[u_0, v_0]$, with $d_0 \geq g + 2$.

OUTPUT: A semi-reduced affine divisor $D_1 = \mathrm{div}[u_1, v_1]$ and a pair (ω^+, ω^-), such that $d_1 < d_0$ and Equation (2) holds.

1: Set $u_1 := (v_0^2 + h v_0 - F)/u_0$ made monic.
2: Let $v_1 := (-v_0 - h) \mod u_1$.
3: **if** the leading term of v_0 is $a_+ x^{g+1}$ (in the notation of Definition 4) **then**
4:     Let $(\omega^+, \omega^-) := (d_0 - g - 1, g + 1 - d_1)$.
5: **else if** the leading term of v_0 is $a_- x^{g+1}$ **then**
6:     Let $(\omega^+, \omega^-) := (g + 1 - d_1, d_0 - g - 1)$.
7: **else**
8:     Let $(\omega^+, \omega^-) := (\frac{d_0 - d_1}{2}, \frac{d_0 - d_1}{2})$.
9: **end if**
10: **return** $\mathrm{div}[u_1, v_1], (\omega^+, \omega^-)$.

A similar analysis when the leading coefficient of v_0 coincides with that of $H^\pm$ shows that if $D_1, (\omega^+, \omega^-) = \mathrm{red}(D_0)$, then we always have $(\omega^+, \omega^-) \in \omega(D_0, D_1)$.

If C is an imaginary model of a curve with point at infinity ∞, this relation degenerates into

$$D_0 - D_1 \equiv (d_0 - d_1)\infty. \qquad (4)$$

In this case, if D_0 is a divisor of degree $d_0 = g + 1$, Algorithm 2 will produce a divisor of degree $d_1 < d_0$, satisfying Equation (4).

Algorithm 3. Composition at Infinity and Reduction

INPUT: A semi-reduced affine divisor $D_0 = \mathrm{div}[u_0, v_0]$ of degree $d_0 \leq g + 1$.

OUTPUT: A reduced affine divisor $D_1 = \mathrm{div}[u_1, v_1]$ and a pair of integers (ω^+, ω^-) such that $(\omega^+, \omega^-) \in \omega(D_0, D_1)$.

1: $v_1' := H^\pm + (v_0 - H^\pm \mod u_0)$,
2: $u_1 := (v_1'^2 + h v_1' - F)/u_0$ made monic.
3: $v_1 := -h - v_1' \mod u_1$.
4: **if** H^+ was used **then**
5:     Let $(\omega^+, \omega^-) := (d_0 - g - 1, g + 1 - d_1)$.
6: **else if** H^- was used **then**
7:     Let $(\omega^+, \omega^-) := (g + 1 - d_1, d_0 - g - 1)$.
8: **end if**
9: **return** $\mathrm{div}[u_1, v_1], (\omega^+, \omega^-)$.

Algorithm 3 is only defined for affine semi-reduced divisors on curves given by a real model. If it were applied on a divisor of degree at least $g + 2$, Algorithm 3 would coincide with Algorithm 2. When applied on a divisor D_0 degree at most $g + 1$, Algorithm 3 can be interpreted as composing the divisor D_0 with some divisor at infinity, followed by Algorithm 2. The polynomial v_1' in this algorithm is the equivalent to polynomial v_3 in Algorithm 1. The result D_1 of this algorithm will be denoted as $D_1, (\omega^+, \omega^-) = \text{red}_\infty(D_0)$. Formally, the action of this algorithm is given by the following.

Proposition 2. *Given an effective semi-reduced divisor with affine support D_0, with Mumford representation $\text{div}[u_0, v_0]$ and degree $d_0 \leq g+1$. If $D_1, (\omega^+, \omega^-) = \text{red}_\infty(D_0)$, then*

$$(\omega^+, \omega^-) \in \omega(D_0, D_1).$$

Proof. We will only prove this when the algorithm is applied using H^+. Notice that the polynomial $v_1'(x)$ has the property that the function $f = y - v_1'(x)$ has all the points in D_0 in its divisor of zeros.

The $(g+1) - d_0$ highest degree coefficients of $v_1'(x)$ coincide with those of $H^+(x)$, so the function

$$(v_1'(x))^2 + h v_1'(x) - F(x),$$

which finds the affine support of f, has degree at most $g + d_0$, and it follows that the affine support of f has at most $g + d_0$ points.

We know that the function $y - v_1'(x)$ will have valuation $-(g+1)$ at ∞^-. The divisor of f is then:

$$\text{div}(f) = D_0 + D_2 - (d_0 + d_2 - (g+1))\infty^+ - (g+1)\infty^- \tag{5}$$

If we denote by D_1 the hyperelliptic conjugate of D_2, we know that

$$\text{div}(u_1) = D_2 + D_1 - d_2(\infty^+ + \infty^-)$$

which together with Equation (5) implies

$$\frac{y - v_1'(x)}{u_1} = D_0 - D_1 - (d_0 - (g+1))\infty^+ - (g+1-d_2)\infty^- \tag{6}$$

which trivially becomes

$$D_0 \equiv D_1 + (d_0 - (g+1))\infty^+ + (g+1-d_1)\infty^-. \tag{7}$$

The proposition follows at once. □

Remark 1. When dealing with explicit computations, the divisors D_0 and D_1 will very often have degree g, in which case we can re-write Equation (7) as

$$D_0 + (\infty^+ - \infty^-) \equiv D_1.$$

Choosing any degree g base divisor D_∞ to represent the points on the class group of C, this equation tells us that

$$(D_0 - D_\infty) + (\infty^+ - \infty^-) \equiv (D_1 - D_\infty),$$

in other words, Algorithm 3 is nothing but addition of $\infty^+ - \infty^-$; this turns out to be such a simple operation because the divisor composition is elementary and can easily be incorporated in the divisor reduction process, which is itself very simple.

We would like to emphasize that Algorithm 3 is independent of the choice of base divisor, so one has the freedom to choose a divisor D_∞ optimal in each specific case.

Remark 2. We have just seen that Algorithm 3 generically corresponds to addition of $\infty^+ - \infty^-$, however, it has long been claimed that this operation[1] has no analogue in the imaginary curve case. Using the previous remark, we propose the following.

Let $C : y^2 = G(x)$, where $\deg(G(x)) = 2g + 1$, be a non-singular imaginary model for a hyperelliptic curve of genus g. Take a point $P = (x_P, y_P)$ on C. Given an effective affine divisor $D = \text{div}[u_0, v_0]$ on C, where $\deg(v_0) < \deg(u_0)$, define a P-baby step on D as follows:

$$a = (y_P - v_0(x_P))/u_0(x_P)$$
$$\tilde{v}_1(x) = au_0(x) + v_0(x)$$
$$u_1(x) = \frac{(\tilde{v}_1)^2 - G(x)}{(x - x_P)u_0(x)}$$
$$v_1(x) = -\tilde{v}_1 \mod u_1(x)$$

The result of applying a P-baby step on the divisor D_0 is, generically, a divisor D_1 such that $D_0 + ([P] - \infty) = D_1$. This algorithm will fail when P is in the support of D_0. Doing some precomputations and using an appropriate implementation, this operation should be as efficient as a baby step. A good choice of P (for instance, having a very small x_P, or even $x_P = 0$) could have a big impact on the efficiency of this algorithm.

The following technical lemma will be used in the next section to prove that our proposed addition algorithm finishes. It can be safely ignored by readers interested only in the computational aspects of the paper.

Lemma 1. *Let D_0 be an effective divisor of degree $d_0 = 2g$ and D_1 be an effective affine divisor of degree $d_1 \le g$. If $(\omega_1^+, \omega_1^-) \in w(D_0, D_1)$,*

$$D_2, (\omega_r^+, \omega_r^-) = \text{red}_\infty(D_1) \quad \text{(using } H^+ \text{),}$$

and we denote $(\omega_2^+, \omega_2^-) = (\omega_1^+ + \omega_r^+, \omega_1^- + \omega_r^-)$, then $(\omega_2^+, \omega_2^-) \in w(D_0, D_2)$ and

$$\omega_1^+ - \omega_1^- > \omega_2^+ - \omega_2^-.$$

If $\omega_1^- < (g-1)/2$ then $\omega_2^+ \le g/2$.

Proof. From the hypotheses we know that $\omega_1^+ + \omega_1^- = 2g - d_1$. Proposition 2 says that

$$(\omega_r^+, \omega_r^-) = (d_1 - (g + 1), g + 1 - d_2), \tag{8}$$

---

[1] Some authors call it a "baby step", see Section 4.1.

this implies that

$$\omega_1^+ - \omega_1^- = \omega_2^+ - \omega_2^- + (2g + 2 - d_0 - d_1),$$

which proves the first assertion. Equation (8) together with $\omega_1^+ = 2g - d_1 - \omega_1^-$ implies

$$\begin{aligned} \omega_2^+ &= \omega_1^+ + d_1 - g - 1 \\ &= (2g - d_1 - \omega_1^-) + d_1 - g - 1 \\ &= g - 1 - \omega_1^- \end{aligned}$$

by hypothesis $\omega_1^- < (g-1)/2$, so that $\omega_2^+ > (g-1)/2$, and since ω_2^+ is an integer, the result follows.   □

Remark 3. Previous authors have used the notation "baby steps" and "giant steps". We explain these using our notation. Given two divisors $D_1 = \mathrm{div}[u_1, v_1]$ and $D_2 = \mathrm{div}[u_2, v_2]$ on C, a *"giant step"* on D_1 and D_2 is the result of computing $D_3 = \mathrm{comp}(D_1, D_2)$ and succesively applying reduction steps (using a red_∞ reduction) on the result until the degree of $\mathrm{red}_\infty^i(D_3)$ is at most g. "Baby steps" are only defined on reduced affine effective divisors, and the result of a "baby step" on a reduced divisor D is the divisor $\mathrm{red}_\infty(D)$.

In [6], an algorithm is given to efficiently compute a giant step. It can then be used in any arithmetic application that requires such an operation, regardless of the representation of divisors in $\mathrm{Cl}^0(C)$ being used.

4   Addition on Real Models

Throughout this section C will denote a genus g hyperelliptic curve defined over a field K, given by the equation

$$C : y^2 + h(x)y = F(x),$$

where $F(x)$ is a degree $2g + 2$ polynomial. If $\mathrm{char}(K) \neq 2$, then we will further assume that $h = 0$. If $\mathrm{char}(K) = 2$, then h will be monic and $\deg(h) = g + 1$.

We will also assume that the divisor D_∞ from Definition 3 is K-rational. This condition holds automatically for even g. For odd values of g one needs to further assume that the leading coefficient of F is a square in K if $\mathrm{char}(K) \neq 2$ or that the leading coefficient of F is of the form $\omega^2 + \omega$ if $\mathrm{char}(K) = 2$.

Every element $[a_0]$ of $\mathrm{Cl}^0(C)$ has a unique representative of the form $a_0 = D_0 - D_\infty$, where D_0 is a degree g effective divisor with reduced affine part. Any effective, degree g divisor D_0 can be uniquely written as $D_0 = D_0' + n_0\infty^+ + m_0\infty^-$, where D_0' is the affine support of D_0, and $n_0, m_0 \in \mathbb{Z}_{\leq 0}$; in this case we will denote the divisor $D_0 - D_\infty$ as $\mathrm{div}([u_0, v_0], n_0)$, where $\mathrm{div}[u_0, v_0] = D_0'$ is the Mumford representation of D_0'. This representation of a divisor is unique.

We would like to remark that in the notation we have just described for divisors, we always have $\deg v_0 < \deg u_0$ and n_0 is an integer such that $0 \leq n_0 \leq$

$g - \deg(u_0)$. The implementation used in Magma represents elements of the class group as $\langle u, v', d \rangle$, where $\mathrm{div}[u, v' \bmod u]$ is the Mumford representation of an affine reduced divisor and d is an even integer such that $\deg(u) \leq d \leq g+1$. We do not have enough space to describe this notation, for which we refer the reader to the Magma documentation. The element represented in Magma as $\langle u, v', d \rangle$ corresponds in our notation to the divisor $\mathrm{div}([u, v' \bmod u], n)$, where n is an integer given by:

$$n = \lceil \tfrac{g-d}{2} \rceil, \qquad \text{if } d = \deg(u) \text{ or } \deg(v' - H^-) \leq g.$$
$$n = \lceil \tfrac{g-(-1)^g d}{2} \rceil - \deg(u), \qquad \text{otherwise.}$$

The representation used in Magma is sub-optimal for cryptographic applications since it can have $\deg(v') \geq \deg(u)$.

Given two divisors $a_1 = \mathrm{div}([u_1, v_1], n_1)$ and $a_2 = \mathrm{div}([u_2, v_2], n_2)$ of $\mathrm{Cl}^0(C)$, we want to find $a_3 = \mathrm{div}([u_3, v_3], n_3)$ such that

$$[a_1] + [a_2] = [a_3].$$

To fix notation, let

$$a_i = \mathrm{div}[u_i, v_i] + n_i \infty^+ + m_i \infty^- - D_\infty,$$
$$\tilde{D}_i = \mathrm{div}[u_i, v_i] + n_i \infty^+ + m_i \infty^-,$$
$$D_i = \mathrm{div}[u_i, v_i]$$

for $i \in 1, 2$.

Algorithm 4. Divisor Addition

INPUT: Divisors $a_i = \mathrm{div}([u_i, v_i], n_i)$ for $i \in \{1, 2\}$.
OUTPUT: $a_3 = \mathrm{div}([u_3, v_3], n_3)$, $[a_3] = [a_1] + [a_2]$.
1: Set $(\omega^+, \omega^-) := (n_1 + n_2, m_1 + m_2)$.
2: Let $D, (a, b) := \mathrm{comp}(D_1, D_2)$. Update $(\omega^+, \omega^-) := (\omega^+ + a, \omega^- + b)$.
3: **while** $\deg(D) > g + 1$ **do**
4:   $D, (a, b) := \mathrm{red}(D)$. Update $(\omega^+, \omega^-) := (\omega^+ + a, \omega^- + b)$.
5: **end while**
6: **while** $\omega^+ < g/2$ or $\omega^- < (g-1)/2$ **do**
7:   $D, (a, b) := \mathrm{red}_\infty(D)$. Update $(\omega^+, \omega^-) := (\omega^+ + a, \omega^- + b)$.
8:   Use H^+ in red_∞ if $\omega^+ > \omega^-$, else use H^-.
9: **end while**
10: Let $E := D + \omega^+ \infty^+ + \omega^- \infty^- - D_\infty$.
11: Now E is an effective degree g divisor. Write $E = D + n_3 \infty^+ + m_3 \infty^-$, where D is an effective affine divisor.
12: **return** $\mathrm{div}(D, n_3)$.

Some comments are in order. Throughout the algorithm we always have that $(\omega^+, \omega^-) \in \omega(\tilde{D}_1 + \tilde{D}_2, D)$. We have mentioned that if $\deg(D) \geq g + 2$ then $\deg(\mathrm{red}(D)) < \deg(D)$, so step 3 always finishes. Lemma 1 proves that step 4, and hence the algorithm, always finish.

Cantor's addition algorithm for curves given by an imaginary model (see [2]) can be seen as a degenerate case of our algorithm. We can think of Algorithm 4 as: 1. Divisor composition; 2. Reduction steps until the degree is at most $g + 1$; 3. Use red_∞ to balance the divisor at infinity. Since imaginary models have a unique point at infinity, to perform divisor addition it suffices to compute the composition and reduction steps, making the balancing step redundant. In the following section we will argue that our divisor D_∞ is the correct choice to have an algorithm analogous to that of Cantor.

If C has even genus, the points ∞^+ and ∞^- are not K-rational and the divisors a_1 and a_2 are K-rational, by a simple rationality argument the counterweights will always be equal, hence the addition algorithm will get a divisor D with equal counterweights such that $\deg(D) \leq g$ in step 3. Algorithm 4 will then finish and step 4 will not be necesary. In this case the (non K-rational) polynomials $H^\pm$ will not be used and no red_∞ step will be computed.

This last observation suggests that, given a hyperelliptic curve C with even genus, one should move two non K-rational points to infinity and get an addition law completely analogous to Cantor's algorithm. This trivial trick could greatly simplify the arithmetic on C.

One key operation in an efficiently computable group is element inversion. Algorithm 5 describes this operation in $\text{Cl}^0(C)$.

Algorithm 5. Divisor Inversion

INPUT: A divisor $a_1 = \text{div}([u_1, v_1], n_1)$.
OUTPUT: A divisor $a_2 = \text{div}([u_2, v_2], n_2)$ such that $[a_1] = -[a_2]$.
1: **if** g is even **then**
2:     **return** $\text{div}([u_1, (-h - v_1 \mod u_1)], g - \deg(u_1) - n_1)$.
3: **else if** g is odd and $n_1 > 0$ **then**
4:     **return** $\text{div}([u_1, (-h - v_1 \mod u_1)], g - m_1 - \deg(u_1) + 1)$.
5: **else**
6:     Let $D_1 = \text{red}_\infty(\text{div}[u_1, -h - v_1])$.
7:     **return** $\text{div}(D_1, 0)$.
8: **end if**

Given the geometric analysis that we have made of the addition algorithm, computing pairings on the class group of an arbitrary hyperelliptic curve can be done following Miller's algorithm. There is not enough space in this paper to give a complete description of an algorithm to compute pairings, but Miller's functions can be calculated from Equations (1),(3) and (6).

4.1   Other Proposals

Previous proposals for addition algorithms on hyperelliptic curves given by a real model use $D_\infty = g\infty^+$ instead of the divisor D_∞ we used in the previous section [9,10]. In particular, this implies that the points ∞^+ and ∞^- need to be K-rational.

A simple modification of Algorithm 4 can be used to add divisors in $\mathrm{Cl}^0(C)$ using $D_\infty = g\infty^+$ as base divisor. All one needs to do is change the finishing condition in step 4 from $((\omega^+ < g/2)$ or $(\omega^- < (g-1)/2))$ to $(\omega^+ < g)$. Indeed, one can verify that using Algorithm 4 with a modified terminating condition coincides with the addition algorithms presented in [9,10].

We will now compare the two proposals for addition algorithms on $\mathrm{Cl}^0(C)$. Since the performance of the algorithms, specially for cryptographic applications, will depend exclusively on its behaviour when adding generic divisors, we will restrict our analysis to this case.

Assume for a moment that the curve C has even genus g, and that D_1 and D_2 are two effective affine divisors of degree g. Generically, the result D_3 of applying succesive reductions to $\mathrm{comp}(D_1, D_2)$ until the degree is at most $g+1$ is a divisor D_3 of degree g. If this is the case, we have

$$D_1 + D_2 \equiv D_3 + (g/2)(\infty^+ + \infty^-), \tag{9}$$

Notice that the counterweights between $D_1 + D_2$ and D_3 are equal, this is a consequence of Equation (4). Using Equation (9) with $D_\infty = (g/2)(\infty^+ + \infty^-)$, we get

$$D_1 - D_\infty + D_2 - D_\infty \equiv D_3 - D_\infty,$$

which means that we have found the result of adding $D_1 - D_\infty$ and $D_2 - D_\infty$, and no "composition at infinity and reduction" steps were necessary.

If instead we work with a divisor at infinity $D'_\infty = g\infty^+$, Equation (9) becomes

$$D_1 - D'_\infty + D_2 - D'_\infty = D_3 - D'_\infty - (g/2)(\infty^+ - \infty^-),$$

so typically one will need $g/2$ extra red_∞ steps to find D_4 such that

$$D_4 - D'_\infty = (D_1 - D'_\infty) + (D_2 - D'_\infty),$$

it is not difficult to see that the need for the red_∞ steps is related to the fact that the valuations of D'_∞ at the two points at infinity are so different.

Now consider a curve C of odd genus g, and let again D_1 and D_2 be degree g affine divisors. Typically, the result after step 2 in Algorithm 4 on the divisors D_1 and D_2 will be a divisor D_3 of degree $g+1$ such that

$$D_1 + D_2 \equiv D_3 + \frac{g-1}{2}(\infty^+ + \infty^-). \tag{10}$$

Again, the counterweights between $D_1 + D_2$ and D_3 are equal as a consequence of Equation (4), and if we now compute $D_4 = \mathrm{red}_\infty(D_3)$, then generically

$$D_3 \equiv D_4 + \infty^-,$$

which together with Equation (10) gives us

$$D_1 + D_2 \equiv D_4 + \frac{g+1}{2}\infty^+ + \frac{g-1}{2}\infty^-. \tag{11}$$

Using our base divisor $D_\infty = (g+1)/2\infty^+ + (g-1)/2\infty^-$, we get

$$D_1 - D_\infty + D_2 - D_\infty \equiv D_4 - D_\infty,$$

and only one red_∞ step was needed. Notice that in this case the addition algorithm consists of composition, a series of standard reduction steps, and the last step is a single application of red_∞.

Using the base divisor $D'_\infty = g\infty^+$, Equation (10) becomes

$$D_1 - D'_\infty + D_2 - D'_\infty \equiv D_3 - D'_\infty - (g-1)/2(\infty^+ - \infty^-),$$

so one will typically need $(g-1)/2$ extra steps to find D_4 such that

$$D_4 - D'_\infty = (D_1 - D'_\infty) + (D_2 - D'_\infty).$$

Again, the need for the red_∞ steps stems from the difference in the valuations of D_∞ at both points at infinity.

We have seen that using a "balanced" divisor at infinity, generically the number of red_∞ steps needed to compute the addition of two divisor classes in $\mathrm{Cl}^0(C)$ is 0 when g is even and 1 when g is odd; whereas when using a non-balanced divisor, the number of red_∞ steps needed to compute the addition of two divisors is generically $g/2$ for even g and $(g-1)/2$ for odd g.

In order to compare the two proposals for arithmetic in $\mathrm{Cl}^0(C)$, we must also consider the computation of inverses, a fundamental operation in a computable group which has, surprisingly, been ignored in the literature. Besides its trivial use to invert divisors, this operation is fundamental to achieve fast divisor multiplication through signed representations.

We will just analyse inversion in the generic case. To do this let D be a degree g affine effective divisor on C. Assume for a moment that g is even. The inverse of the divisor $P = D - (g/2)(\infty^+ + \infty^-)$ is the divisor $\overline{D} - (g/2)(\infty^+ + \infty^-)$, whereas if we now assume that g is odd, the divisor

$$\left(D - \frac{g+1}{2}\infty^+ - \frac{g-1}{2}\infty^-\right) + \left(\overline{D} - \frac{g-1}{2}\infty^+ - \frac{g+1}{2}\infty^-\right)$$

is principal, which means that $\overline{D} - (g-1)/2\infty^+ - (g+1)/2\infty^-$ is the inverse of P, and in order to fix the divisor at infinity, using Proposition 2 it is easy to see that generically only one application of Algorithm 3 will suffice. In other words, using the "balanced" representation at infinity, 0 or 1 applications of Algorithm 3 will be needed, depending on the parity of g.

We now analyze the computation of inverses using $D'_\infty = g\infty^+$ as base divisor. Clearly, the divisor

$$(D - g\infty^+) + (\overline{D} - g\infty^-)$$

is principal, so we need to find an appropriate representative of the divisor class $[\overline{D} - g\infty^-]$. Again, this can be done through g applications of Algorithm 3, as can be easily seen using Proposition 2.

Table 1. Operation counts for genus 2 arithmetic using formulae of [4]

	Imaginary	Balanced	Non-balanced
Addition	1I, 2S, 22M [8]	1I, 2S, 26M	2I, 4S, 30M
Doubling	1I, 5S, 22M [8]	1I, 4S, 28M	2I, 6S, 32M
Inversion	0	0	2I, 4S, 8M

It is now clear that computing the inverse of a divisor class is easier when the divisor at infinity is as balanced as possible, supporting our claim that a "balanced" representation is a closer analogue to that of Cantor for imaginary models, where the inverse of a divisor is its hyperelliptic conjugate, just as in our case when the genus of C is even.

Table 1 gives the cost of addition and doubling in a genus 2 curve using the explicit formulae for Algorithms 1, 2 and 3 presented in [4]. If $S = M$ and $I = 4M$ then balanced representations give a saving of around 15% for addition and 13% for doubling (if $I = 30M$ the savings become 62% and 58% respectively). The extra operations in the non-balanced case come from an additional application of Algorithm 3 in each case.

5   Conclusion

We have given an explicit geometric interpretation of Algorithm 3, which made it clear that all the composition and reduction algorithms presented in this paper (all of which have been known for a long time) really act on semi-reduced affine divisors rather than on elements of $\mathrm{Cl}^0(C)$; that is to say, they can be seen as acting on the Mumford representation of a divisor. Having made this simple observation, a number of interesting consequences follow. One such observation is that in order to get simple arithmetic operations one needs to find an optimal base divisor D_∞, and we have argued that in cryptography-related applications the optimal choice is a balanced divisor D_∞. When the genus of the curve is even, if the points at infinity are non-rational (which can always be achieved), using a balanced base divisor yields an algorithm identical to that of Cantor, where the rationality takes care of the counterweights; this is impossible to achieve with non-balanced divisors.

The question of finding explicit addition formulae for curves in real representation using our proposed divisor already has an answer: since generic addition formulas have been given for Algorithms 1, 2 and 3 in a genus 2 real curve [4], we can use these formulas to calculate an addition law on $\mathrm{Cl}^0(C)$ by just changing the divisor at infinity one is working with. All the explicit addition formulae presented so far (specially for $g = 2$) that we have knowledge of (including those of [4,6]) first compute the composition of the two affine divisors in the summands, then find the divisor with degree at most $g + 1$ which is the result of successively applying reduction steps, and finally give an explicit form of Algorithm 3. Hence, it is possible to use these formulae to compute divisor addition using our proposal with no alterations.

Acknowledgments

We would like to thank Mike Jacobson and the anonymous referees for their helpful comments. The first author is supported by EPSRC Grant EP/D069904/1. The third author thanks CONACyT for its financial support.

References

1. Bosma, W., Cannon, J., Playoust, C.: The Magma algebra system. I. The user language. J. Symbolic Comput. 24(3-4), 235–265 (1997)
2. Cantor, D.G.: Computing in the Jacobian of a hyperelliptic curve. Math. Comp. 48, 95–101 (1987)
3. Cohen, H., Frey, G., Avanzi, R., Doche, C., Lange, T., Nguyen, K., Vercauteren, F.: Handbook of elliptic and hyperelliptic curve cryptography. Discrete Mathematics and its Applications (Boca Raton). Chapman & Hall/CRC, Boca Raton (2006)
4. Erickson, S., Jacobson, M.J., Shang, N., Shen, S., Stein, A.: Explicit formulas for real hyperelliptic curves of genus 2 in affine representation. In: Carlet, C., Sunar, B. (eds.) WAIFI 2007. LNCS, vol. 4547, pp. 202–218. Springer, Heidelberg (2007)
5. Galbraith, S.D., Pujolas, J., Ritzenthaler, C., Smith, B.: Distortion maps for genus two curves
6. Jacobson, M., Scheidler, R., Stein, A.: Fast arithmetic on hyperelliptic curves via continued fraction expansions. In: Shaska, T., Huffman, W., Joyner, D., Ustimenko, V. (eds.) Advances in Coding Theory and Cryptography. Series on Coding Theory and Cryptology, vol. 3, pp. 201–244. World Scientific Publishing, Singapore (2007)
7. Jacobson, M.J., Scheidler, R., Stein, A.: Cryptographic protocols on real hyperelliptic curves. Adv. Math. Commun. 1(2), 197–221 (2007)
8. Lange, T.: Formulae for arithmetic on genus 2 hyperelliptic curves. Appl. Algebra Engrg. Comm. Comput. 15(5), 295–328 (2005)
9. Paulus, S., Rück, H.-G.: Real and imaginary quadratic representations of hyperelliptic function fields. Math. Comp. 68(227), 1233–1241 (1999)
10. Paulus, S., Stein, A.: Comparing real and imaginary arithmetics for divisor class groups of hyperelliptic curves. In: Buhler, J.P. (ed.) ANTS 1998. LNCS, vol. 1423, pp. 576–591. Springer, Heidelberg (1998)
11. Scheidler, R., Stein, A., Williams, H.C.: Key exchange in real quadratic congruence function fields. Designs, Codes and Cryptography 7, 153–174 (1996)

Tabulation of Cubic Function Fields with Imaginary and Unusual Hessian

Pieter Rozenhart and Renate Scheidler

Department of Mathematics and Statistics, University of Calgary,
2500 University Drive NW, Calgary, Alberta, Canada, T2N 1N4
{pieter,rscheidl}@math.ucalgary.ca

Abstract. We give a general method for tabulating all cubic function fields over $\mathbb{F}_q(t)$ whose discriminant D has odd degree, or even degree such that the leading coefficient of $-3D$ is a non-square in $\mathbb{F}_q^*$, up to a given bound on $|D| = q^{\deg(D)}$. The main theoretical ingredient is a generalization of a theorem of Davenport and Heilbronn to cubic function fields. We present numerical data for cubic function fields over $\mathbb{F}_5$ and over $\mathbb{F}_7$ with $\deg(D) \leq 7$ and $\deg(D)$ odd in both cases.

1 Introduction

In 1997, Belabas [2] presented an algorithm for tabulating all non-isomorphic cubic number fields of discriminant D with $|D| \leq X$ for any $X > 0$. The results make use of the reduction theory for binary cubic forms with integral coefficients. A theorem of Davenport and Heilbronn [8] states that there is a discriminant-preserving bijection between $\mathbb{Q}$-isomorphism classes of cubic number fields of discriminant D and a certain explicitly characterizable set $\mathcal{U}$ of equivalence classes of primitive irreducible integral binary cubic forms of the same discriminant D. Using this one-to-one correspondence, one can enumerate all cubic number fields of discriminant D with $|D| \leq X$ by computing the unique reduced representative $f(x, y)$ of every equivalence class in $\mathcal{U}$ of discriminant D with $|D| \leq X$. The corresponding field is then obtained by simply adjoining a root of the irreducible cubic $f(x, 1)$ to $\mathbb{Q}$. Belabas' algorithm is essentially linear in X, and performs quite well in practice.

In this paper, we give an extension of the above approach to function fields. That is, we present a method for tabulating all cubic function fields over a fixed finite field up to a given upper bound on the degree of the discriminant, using the theory for binary cubic forms with coefficients in $\mathbb{F}_q[t]$, where $\mathbb{F}_q$ is a finite field with $\operatorname{char}(\mathbb{F}_q) \neq 2, 3$. While some of the ideas of [2] translate essentially directly from number fields to function fields, there are in fact a number of obstructions to a straightforward adaptation of Belabas' algorithm [2] to the function field setting. Firstly, there is a very simple connection between the signatures of cubic and quadratic number fields of the same discriminant D, which are simply characterized as real or complex/imaginary according to whether $D > 0$ or $D < 0$. In cubic function fields, this connection is far more complicated and in some

A.J. van der Poorten and A. Stein (Eds.): ANTS-VIII 2008, LNCS 5011, pp. 357–370, 2008.
© Springer-Verlag Berlin Heidelberg 2008

cases no longer exists, due to the increased level of flexibility in how the place at infinity of $\mathbb{F}_q(t)$ splits in the cubic extension. Secondly, the case of unusual quadratic function fields, where the place at infinity is inert, has no number field analogue. Thirdly, the extensions of the degree map on $\mathbb{F}_q(t)$ to any function field are non-Archimedean valuations, i.e. satisfy the strong triangle inequality $|a + b| \leq \max\{|a|, |b|\}$, whereas the absolute value on any number field is Archimedean, satisfying the ordinary triangle inequality $|a + b| \leq |a| + |b|$. This results in somewhat different bounds on the coefficients of the binary cubic forms that the function field version of the tabulation algorithm uses for its search.

Our main tool is the function field analogue of the Davenport-Heilbronn theorem [8] mentioned above (see [10,13]). We also make use of the association of any binary cubic form f of discriminant D over $\mathbb{F}_q[t]$ to its Hessian H_f which is a binary quadratic form over $\mathbb{F}_q[t]$ of discriminant $-3D$. Under certain conditions, this association can be exploited to develop a reduction theory for binary cubic forms over $\mathbb{F}_q[t]$ that is analogous to the reduction theory for integral binary cubic forms. Suppose that $\deg(D)$ is odd, i.e. H_f is an imaginary binary quadratic form, or that $\deg(D)$ is even and the leading coefficient of $-3D$ is a non-square in $\mathbb{F}_q^*$, i.e. H_f is an unusual binary quadratic form. We will establish that under these conditions, the equivalence class of f contains a unique reduced form, i.e. a binary cubic form that satisfies certain normalization conditions and has an associated Hessian that is a reduced binary quadratic form. Thus, equivalence classes of binary cubic forms can be efficiently identified via their unique representatives. This result no longer holds when H_f is a real binary quadratic form, i.e. $\deg(D)$ is even and the leading coefficient of $-3D$ is a square in $\mathbb{F}_q^*$. In this case, the equivalence class of f contains many — in fact, generally exponentially many — reduced forms, and a different reduction theory needs to be developed. This is the subject of future research.

Our tabulation procedure proceeds analogously to the number field scenario. The function field analogue of the Davenport-Heilbronn theorem states that there is again a discriminant-preserving bijection between $\mathbb{F}_q(t)$-isomorphism classes of cubic function fields of discriminant $D \in \mathbb{F}_q[t]$ and a certain set $\mathcal{U}$ of primitive irreducible binary cubic forms over $\mathbb{F}_q[t]$ of discriminant D. Hence, in order to list all $\mathbb{F}_q(t)$-isomorphism classes of cubic function fields up to an upper bound X on $|D|$, it suffices to enumerate the unique reduced representatives of all equivalence classes of binary cubic forms of discriminant D for all $D \in \mathbb{F}_q[t]$ with $|D| = q^{\deg(D)} \leq X$. Bounds on the coefficients of such a reduced form show that there are only finitely many candidates for any reduced form of a fixed discriminant. These bounds can then be employed in nested loops to test whether each form found lies in $\mathcal{U}$. As mentioned earlier, the coefficient bounds obtained for function fields are different from those used by Belabas for number fields, due to the fact that the degree valuation is non-Archimedean.

This paper is organized as follows. After a brief overview of binary quadratic and cubic forms over $\mathbb{F}_q[t]$ in Section 2, the reduction theory for imaginary and unusual binary cubic forms is developed in Sections 3 and 4, respectively. We present the Davenport-Heilbronn theorem for function fields and an explicit characterization of the set $\mathcal{U}$ in Section 6. Bounds on the coefficients of a reduced binary cubic form are derived in Section 5. Finally, we present the tabulation algorithm as well as numerical results in Section 7.

2   Binary Quadratic and Cubic Forms over $\mathbb{F}_q[t]$

For a general introduction to algebraic function fields, we refer the reader to Rosen [9] or Stichtenoth [12]. Let $\mathbb{F}_q$ be a finite field of characteristic at least 5, and set $\mathbb{F}_q^* = \mathbb{F}_q \backslash \{0\}$. Denote by $\mathbb{F}_q[t]$ and $\mathbb{F}_q(t)$ the ring of polynomials and the field of rational functions in the variable t over $\mathbb{F}_q$, respectively. For any non-zero $H \in \mathbb{F}_q[t]$ of degree $n = \deg(H)$, we let $|H| = q^n = q^{\deg(H)}$, and denote by $\mathrm{sgn}(H)$ the leading coefficient of H. For $H = 0$, we set $|H| = 0$. This absolute value extends in the obvious way to $\mathbb{F}_q(t)$. Note that in contrast to the absolute value on the rationals, the absolute value on $\mathbb{F}_q(t)$ is non-Archimedean.

Any non-zero $r \in \mathbb{F}_q(t)$ can be written as $r = a_n t^n + a_{n-1} t^{n-1} + \cdots + a_0 + a_{-1} t^{-1} + \cdots$ with $n \in \mathbb{Z}$ and $a_i \in \mathbb{F}_q$ for $i \leq n$. We set $\lfloor r \rfloor = a_n t^n + \cdots + a_1 t + a_0$ to be the polynomial part of r; note that $\lfloor r \rfloor = 0$ if $n < 0$. We also set $\lfloor 0 \rfloor = 0$. The function $\lfloor r \rfloor$ is analogous to the floor function for integers.

We give a brief overview of binary quadratic and cubic forms with coefficients in $\mathbb{F}_q[t]$; their reduction theory will be developed in Sections 3 and 4 respectively. Much of this material is completely analogous to the theory for binary cubic forms over the integers.

A *binary quadratic form over* $\mathbb{F}_q[t]$ is a homogeneous quadratic polynomial in two variables with coefficients in $\mathbb{F}_q[t]$. If $H(x, y) = Px^2 + Qxy + Ry^2$ is a binary quadratic form over $\mathbb{F}_q[t]$, then we write $H = (P, Q, R)$ for brevity. The *discriminant* of H is the polynomial $\mathrm{disc}(H) = Q^2 - 4PR \in \mathbb{F}_q[t]$. H is said to be *imaginary* if $\deg(\mathrm{disc}(H))$ is odd, *unusual* if $\deg(\mathrm{disc}(H))$ is even and $\mathrm{sgn}(\mathrm{disc}(H))$ is a non-square in $\mathbb{F}_q^*$, and *real* if $\deg(\mathrm{disc}(H))$ is even and $\mathrm{sgn}(\mathrm{disc}(H))$ is a square in $\mathbb{F}_q^*$.

A *binary cubic form over* $\mathbb{F}_q[t]$ is a homogeneous cubic polynomial in two variables with coefficients in $\mathbb{F}_q[t]$. If $f(x, y) = ax^3 + bx^2 y + cxy^2 + dy^3$ is a binary cubic form over $\mathbb{F}_q[t]$, then we write $f = (a, b, c, d)$ for brevity. The *discriminant* of $f = (a, b, c, d)$ is the polynomial

$$\mathrm{disc}(f) = 18abcd + b^2 c^2 - 4ac^3 - 4b^3 d - 27a^2 d^2 \in \mathbb{F}_q[t] \,.$$

For the remainder of this paper, we assume that all binary cubic forms $f = (a, b, c, d)$ are *primitive*, i.e. $\gcd(a, b, c, d) = 1$.

Definition 2.1. *Let F be a binary quadratic or cubic form over $\mathbb{F}_q[t]$. If*

$$M = \begin{pmatrix} \alpha & \beta \\ \gamma & \delta \end{pmatrix},$$

is a 2×2 matrix with entries in $\mathbb{F}_q[t]$, then the action of M on F is defined by $F \circ M = f(\alpha x + \beta y, \gamma x + \delta y)$.

We obtain an equivalence relation from this action by restricting to matrices $M \in GL_2(\mathbb{F}_q[t])$, the group of 2×2 matrices over $\mathbb{F}_q[t]$ whose determinant lies in $\mathbb{F}_q^*$. That is, two binary quadratic or cubic forms F and G over $\mathbb{F}_q[t]$ are said to be *equivalent* if

$$\mu F(\alpha x + \beta y, \gamma x + \delta y) = G(x, y)$$

for some $\mu \in \mathbb{F}_q^*$ and $\alpha, \beta, \gamma, \delta \in \mathbb{F}_q[t]$ with $\alpha\delta - \beta\gamma \in \mathbb{F}_q^*$. Up to associates, equivalent binary forms have the same discriminant. Furthermore, the action of the group $GL_2(\mathbb{F}_q[t])$ on binary forms over $\mathbb{F}_q[t]$ preserves irreducibility over $\mathbb{F}_q(t)$.

As in the case of integral binary cubic forms, any binary cubic form $f = (a, b, c, d)$ over $\mathbb{F}_q[t]$ is closely associated with its *Hessian*

$$H_f(x, y) = -\frac{1}{4} \begin{vmatrix} \dfrac{\partial^2 f}{\partial x \partial x} & \dfrac{\partial^2 f}{\partial x \partial y} \\ \dfrac{\partial^2 f}{\partial y \partial x} & \dfrac{\partial^2 f}{\partial y \partial y} \end{vmatrix} = (P, Q, R),$$

where $P = b^2 - 3ac$, $Q = bc - 9ad$, and $R = c^2 - 3bd$. Note that H_f is a binary quadratic form over $\mathbb{F}_q[t]$. The Hessian has a number of useful properties, which are easily verified by direct computation:

Proposition 2.1. *Let* $f = (a, b, c, d)$ *be a binary cubic form over* $\mathbb{F}_q[t]$ *with Hessian* $H_f = (P, Q, R)$. *Then the following are satisfied.*

1. $H_{f \circ M} = (\det M)^2 (H_f \circ M)$ *for any* $M \in GL_2(\mathbb{F}_q[t])$.
2. $\operatorname{disc}(H_f) = -3 \operatorname{disc}(f)$.

A binary cubic form f over $\mathbb{F}_q[t]$ is said to be *imaginary, unusual,* or *real* according to whether its Hessian H_f is an imaginary, unusual, or real binary quadratic form. By Proposition 2.1, f is imaginary if $\operatorname{disc}(f)$ has odd degree, unusual if $\operatorname{disc}(f)$ has even degree and $-3 \operatorname{sgn}(\operatorname{disc}(f))$ is a non-square in $\mathbb{F}_q^*$, and real if $\operatorname{disc}(f)$ has even degree and $-3 \operatorname{sgn}(\operatorname{disc}(f))$ is a square in $\mathbb{F}_q^*$.

For the tabulation of cubic function fields, it will be important to represent equivalence classes of binary cubic forms over $\mathbb{F}_q[t]$ via a unique and efficiently identifiable representative. This can be accomplished via reduction. As in the case of integral forms, reduction of cubic forms is accomplished via reduction of their associated binary quadratic forms. Specifically, in the imaginary and unusual cases, a binary cubic form over $\mathbb{F}_q[t]$ is declared to be reduced essentially if its associated Hessian is reduced and certain normalization conditions are satisfied.

3   Reduction Theory of Imaginary Binary Cubic Forms

We begin with an overview of the reduction theory for imaginary binary quadratic forms over $\mathbb{F}_q[t]$ which can be found in Artin [1]. We then use this theory to develop a reduction theory for imaginary binary cubic forms via their associated Hessians. This theory is quite similar to its counterpart for integral binary forms.

In the case of unusual binary cubic forms, we will proceed in an analogous fashion to the approach for imaginary forms; this is done in Section 4.

An imaginary binary quadratic form $H = (P, Q, R)$ of discriminant $D = \mathrm{disc}(H)$ is said to be *reduced* if $|Q| < |P| \leq |D|^{1/2}$, $\mathrm{sgn}(P) = 1$, and either $Q = 0$ or $\mathrm{sgn}(Q) \in S$, where $S \subset \mathbb{F}_q$ is a set such that if $a \in S$, then $-a \notin S$ and $|S| = (q - 1)/2$. Such a set can always be found. One such choice is as follows: order the non-zero elements of $\mathbb{F}_q$ lexicographically and let S consist of the first $(q - 1)/2$ elements. If $q = p$ is a prime, this is simply the set $\{1, 2, ..., (p - 1)/2\}$. Note that since $\deg(D)$ is odd, the exponent in $\sqrt{|D|} = q^{\deg(D)/2}$ is a half integer, so the second inequality is in fact equivalent to the strict inequality $|P| < \sqrt{|D|}$. Note also that in contrast to integral binary quadratic forms, the only matrices $M \in GL_2(\mathbb{F}_q[t])$ whose action on H leaves H unchanged are the identity matrix, its negative and $\pm \begin{pmatrix} 1 & 0 \\ 0 & -1 \end{pmatrix}$ when $Q = 0$ (see [1]).

The algorithm for reducing a binary quadratic form over $\mathbb{F}_q[t]$ is almost the same as for integral imaginary binary quadratic forms. If $H = (P, Q, R)$ with $|Q| \geq |P|$, then compute $s = \lfloor -Q/2P \rfloor$ and apply the matrix

$$T = \begin{pmatrix} 1 & s \\ 0 & 1 \end{pmatrix} \in GL_2(\mathbb{F}_q[t])$$

to H to obtain a new form $H_1(x, y) = H(x + sy, y) = (P_1, Q_1, R_1)$ equivalent to H. Now the inequality $|Q_1| < |P_1|$ is satisfied. If $|P_1| > |D|^{1/2}$, then apply the matrix

$$S = \begin{pmatrix} 0 & -1 \\ 1 & 0 \end{pmatrix} \in GL_2(\mathbb{F}_q[t])$$

to H_1 to obtain the equivalent form $H_2(x, y) = H_1(-y, x) = (P_2, Q_2, R_2)$ with $(P_2, Q_2, R_2) = (R_1, -Q_1, P_1)$. If as a result of this last transformation, the condition $|Q_2| < |P_2|$ is not satisfied, then we repeat this procedure from the beginning with $H = H_2 = (P_2, Q_2, R_2)$. Since P_i, Q_i, R_i are polynomials in $\mathbb{F}_q[t]$, the process must eventually terminate after a finite number of steps, as we reduce the degree of P_i at each step.

Now suppose $H_j = (P_j, Q_j, R_j)$ satisfies $|Q_j| < |P_j| \leq \sqrt{|D|}$ for some j. To obtain the condition $\mathrm{sgn}(P_j) = 1$, we apply

$$N = \begin{pmatrix} 1 & 0 \\ 0 & \nu \end{pmatrix} \in GL_2(\mathbb{F}_q[t])$$

to H_j, where $\nu \in \mathbb{F}_q^*$ is chosen appropriately. It follows from the above reduction procedure that every imaginary binary quadratic form over $\mathbb{F}_q[t]$ is equivalent to a unique reduced quadratic form, see [1].

Now let $f = (a, b, c, d)$ be an imaginary binary cubic form over $\mathbb{F}_q[t]$ of discriminant $D = \mathrm{disc}(f)$ with (imaginary) Hessian $H_f = (P, Q, R)$. Then f is said to be *reduced* if H_f is reduced, $\mathrm{sgn}(a) = 1$ and if $Q = 0$, then $\mathrm{sgn}(d) \in S$, where $S \subset \mathbb{F}_q$ as described above. Equivalently, by Proposition 2.1, f is reduced if

$$|Q| < |P| \leq |D|^{1/2}, \quad \mathrm{sgn}(P) = 1, \quad \mathrm{sgn}(a) = 1, \quad \mathrm{sgn}(Q) \in S \text{ or } \mathrm{sgn}(d) \in S,$$

depending on whether or not $Q = 0$. Analogous to [2], one can deduce that any two equivalent reduced imaginary forms are equal, so equivalence classes of such forms can be efficiently identified by their unique reduced representative.

Theorem 3.1

1. *Every equivalence class of imaginary binary cubic forms over $\mathbb{F}_q[t]$ has a unique reduced representative.*
2. *Every imaginary binary cubic form over $\mathbb{F}_q[t]$ is equivalent to a unique reduced binary cubic form.*

4   Reduction Theory of Unusual Binary Cubic Forms

As in the previous section, we first outline reduction for unusual binary quadratic forms over $\mathbb{F}_q[t]$ and then apply this theory to unusual binary cubic forms over $\mathbb{F}_q[t]$. Both the reduction theory and the algorithm for the unusual case are almost identical to that of imaginary forms, with one crucial difference: the analogous definition of reducedness does not lead to a unique reduced representative in each equivalence class, but instead to $q+1$ equivalent reduced forms. To achieve uniqueness, a distinguished representative among these $q + 1$ equivalent forms will need to be identified.

Again, the reduction theory for unusual binary quadratic forms over $\mathbb{F}_q[t]$ goes back to Artin [1]. An unusual binary quadratic form $H = (P, Q, R)$ of discriminant $D = \text{disc}(H)$ is said to be *reduced* if $|Q| < |P| \le \sqrt{|D|}$, $\text{sgn}(P) = 1$ and either $Q = 0$ or $\text{sgn}(Q) \in S$ where $S \subset \mathbb{F}_q$ is a set such that if $a \in S$, then $-a \notin S$ and $|S| = (q - 1)/2$, as for imaginary quadratic forms. At first glance, this definition looks exactly like the definition of a reduced imaginary binary quadratic form. However, the crucial difference is that here, the exponent in $\sqrt{|D|} = q^{\deg(D)/2}$ is an integer, whereas in the imaginary scenario, it was a half integer. So here, equality $|P| = \sqrt{|D|}$ can in fact be achieved. The algorithm for reducing an unusual binary quadratic form is the same as for imaginary binary quadratic forms, so every unusual binary quadratic form is equivalent to a reduced form.

Unusual reduced binary quadratic forms $H = (P, Q, R)$ with $|P| < \sqrt{|D|}$ behave exactly like reduced imaginary binary quadratic forms. However, if $H = (P, Q, R)$ is an unusual reduced binary quadratic form with $|P| = \sqrt{|D|}$, then so is $H_\alpha = (P_\alpha, Q_\alpha, R_\alpha)$ for all $\alpha \in \mathbb{F}_q$, where

$$H_\alpha = H \circ \left(\frac{1}{\mu_\alpha} \begin{pmatrix} \alpha & \text{sgn}(D) \\ 4 & \alpha \end{pmatrix} \right) = H \left(\frac{\alpha}{\mu_\alpha} x + \frac{\text{sgn}(D)}{\mu_\alpha} y, \frac{4}{\mu_\alpha} x + \frac{\alpha}{\mu_\alpha} y \right),$$

with $\alpha \in \mathbb{F}_q$ and $\mu_\alpha = \alpha^2 - 4\,\text{sgn}(D)$. Note that $\mu_\alpha \ne 0$ for all $\alpha \in \mathbb{F}_q$, since $\text{sgn}(D)$ is a non-square in $\mathbb{F}_q^*$. Hence, we have a family of $q + 1$ equivalent reduced unusual binary quadratic forms when $|P| = \sqrt{|D|}$. These $q + 1$ forms can be sorted according to lexicographical order in $\mathbb{F}_q[t]$ of their x^2-coefficients. To identify a unique representative in the class of H, one selects the form

$H' = (P', Q', R') \in \{H, H_\alpha\}_{\alpha \in \mathbb{F}_q}$ so that P' is minimal in terms of lexicographical order in $\mathbb{F}_q[t]$ amongst $\{P, P_\alpha\}_{\alpha \in \mathbb{F}_q}$. We call the form H' *distinguished*. Thus, to find such a representative, it is necessary to execute the reduction algorithm described in Section 3 and then computing the q forms H_α, $\alpha \in \mathbb{F}_q$. This is slower than reduction for imaginary binary quadratic forms, especially for large values of q.

Now let $f = (a, b, c, d)$ be an unusual binary cubic form over $\mathbb{F}_q[t]$ of discriminant $D = \mathrm{disc}(f)$ with (unusual) Hessian $H_f = (P, Q, R)$. Then f is said to be *reduced* if H_f is reduced, H_f is distinguished if $|P| = \sqrt{|D|}$, $\mathrm{sgn}(a) = 1$ and either $Q = 0$ or $\mathrm{sgn}(Q) \in S$, where $S \subset \mathbb{F}_q$ is a set such that if $a \in S$, then $-a \notin S$ and $|S| = (q-1)/2$. Equivalently, by Proposition 2.1, f is reduced if

$$|Q| < |P| \le |D|^{1/2}, \quad \mathrm{sgn}(P) = 1, \quad \mathrm{sgn}(a) = 1 \,, \quad \mathrm{sgn}(Q) \in S \text{ or if } Q = 0 \text{ then}$$
$$\mathrm{sgn}(d) \in S, \text{ where } S \text{ is as described above,}$$

if $|P| = \sqrt{|D|}$, then P is lexicographically minimal in the set $\{\tilde{P} \mid \tilde{H} = (\tilde{P}, \tilde{Q}, \tilde{R}) \text{ is a reduced form equivalent to } H\}$.

Analogous to the imaginary case, we again obtain

Theorem 4.1

1. *Every equivalence class of unusual binary cubic forms over $\mathbb{F}_q[t]$ has a unique reduced representative.*
2. *Every unusual binary cubic form over $\mathbb{F}_q[t]$ is equivalent to a unique reduced binary cubic form.*

5   Bounds on Reduced Binary Cubic Forms

For our tabulation algorithm, we will need to search over all candidates for reduced imaginary or unusual binary cubic forms $f = (a, b, c, d)$ of discriminant D where $|D|$ is bounded above by some given bound X. It then remains to test via Algorithm 6.4 whether such a reduced form lies in the Davenport-Heilbronn set $\mathcal{U}$ defined in Section 6 below. If this is the case, then the reduced form corresponds to a triple of $\mathbb{F}_q(t)$-isomorphic cubic function fields.

In order to establish that this set of candidates for reduced forms of discriminant D of absolute value at most X is in fact finite, and to ensure that the search procedure is as efficient as possible, we develop good bounds on the absolute values of the coefficients a, b, c, d of an imaginary or unusual reduced binary cubic form in terms of the absolute value of D. The following inequality appears in Cremona [5] and is easily verified by straightforward computation.

Lemma 5.1. $f = (a, b, c, d)$ be a binary cubic form over $\mathbb{F}_q[t]$ of discriminant D and Hessian $H_f = (P, Q, R)$, where we recall that $P = b^2 - 3ac$. Set $U = 2b^3 + 27a^2d - 9abc$. Then $4P^3 = U^2 + 27a^2D$.

The above identity can be used to establish degree bounds on the coefficients of an imaginary or unusual reduced binary cubic form over $\mathbb{F}_q[t]$ in terms of the degree of its discriminant.

Proposition 5.2. *Let $f = (a, b, c, d)$ be an imaginary or unusual binary cubic form over $\mathbb{F}_q[t]$ of discriminant D, and set $P = b^2 - 3ac$ and $U = 2b^3 + 27a^2d - 9abc$. Then $|U|^2 \le |P|^3$.*

Proof. By Lemma 5.1, we have

$$4P^3 = U^2 + 27a^2D = U^2 - (-3D)(9a)^2. \tag{5.1}$$

Now $|P|^3 < |U|^2$ if and only if the leading terms of the polynomials U^2 and $(-3D)(9a)^2$ in the right hand side of (5.1) cancel, which is the case if and only if $\deg(U^2) = \deg((-3D)(9a)^2)$ and $\operatorname{sgn}(U^2) = \operatorname{sgn}((-3D)(9a)^2)$. The first of these two equalities implies that $\deg(D)$ is even, and the second one forces $\operatorname{sgn}(-3D)$ to be a square in $\mathbb{F}_q^*$, which would imply that H_f is a real binary quadratic form, a contradiction.

We can now derive our desired degree bounds for imaginary or unusual reduced binary cubic forms.

Corollary 5.3. *Let $f = (a, b, c, d)$ be a reduced imaginary or unusual binary cubic form over $\mathbb{F}_q[t]$ of discriminant D. Then*

$$|a|, |b| \le |D|^{1/4}, \quad |c| \le |D|^{1/2}/|a|, \quad |d| \le \max\{|bc|/|a|, |b|^2/|a|q, |c|/q\}.$$

Proof. Let $H_f = (P, Q, R)$ be the Hessian of f. Then $P = b^2 - 3ac$ and $|Q| < |P| \le \sqrt{|D|}$. Set $U = 2b^3 + 27a^2d - 9abc$. Then $4P^3 = U^2 + 27a^2D$ by Lemma 5.1, and $|U|^2 \le |P|^3$ by Proposition 5.2. It follows that

$$|a^2D| = |4P^3 - U^2| \le \max\{|P|^3, |U|^2\} \le |P|^3 \le |D|^{3/2},$$

and hence $|a| \le |D|^{1/4}$.

A straightforward computation shows that $U = 2bP - 3aQ$. Hence,

$$|bP| = |U + 3aQ| \le \max\{|U|, |aQ|\} \le \max\{|P|^{3/2}, |a||P|\},$$

so $|b| \le \max\{|P|^{1/2}, |a|\} \le |D|^{1/4}$.

To obtain the upper bound for c, we observe that $3ac = b^2 - P$, so

$$|ac| \le \max\{|b|^2, |P|\} \le |D|^{1/2},$$

and hence $|c| \le |D|^{1/2}/|a|$. Finally, $Q = bc - 9ad$, $P = b^2 - 3ac$, and $|Q| \le |P|/q$ imply

$$|d| = |bc - Q|/|a| \le \max\{|bc/a|, |Q|/|a|\} \le \max\{|bc/a|, |P|/|a|q\}$$
$$= \max\{|bc/a|, |b^2 - 3ac|/|a|q\} \le \max\{|bc|/|a|, |b|^2/|a|q, |c|/q\}.$$

This concludes the proof.

The bounds for a and b are essentially of the same order of magnitude as the corresponding bounds for integral imaginary binary cubic forms. However, the bounds for c and d are different.

Corollary 5.4. *For any fixed discriminant D in $\mathbb{F}_q[t]$, there are only finitely many imaginary and unusual reduced binary cubic forms over $\mathbb{F}_q[t]$ of discriminant D.*

6   The Davenport–Heilbronn Theorem

Recall that the Davenport-Heilbronn theorem [8] states that there is a discriminant-preserving bijection from a certain set $\mathcal{U}$ of equivalence classes of integral binary cubic forms of discriminant D to the set of $\mathbb{Q}$-isomorphism classes of cubic fields of the same discriminant D. Therefore, if one can compute the unique reduced representative f of any class of forms in $\mathcal{U}$ of discriminant D with $|D| < X$, then this leads to a list of minimal polynomials $f(x, 1)$ for all cubic fields of discriminant D with $|D| \leq X$.

The situation for cubic function fields is completely analogous. We now describe the Davenport-Heilbronn set $\mathcal{U}$ for function fields, state the function field version of the Davenport-Heilbronn theorem, and provide a fast algorithm for testing membership in $\mathcal{U}$ that is in fact more efficient than its counterpart for integral forms.

For brevity, we let $[f]$ denote the equivalence class of any primitive binary cubic form f over $\mathbb{F}_q[t]$. Fix any irreducible polynomial $p \in \mathbb{F}_q[t]$. We define $\mathcal{V}_p$ to be the set of all equivalence classes $[f]$ of binary cubic forms such that $p^2 \nmid \operatorname{disc}(f)$. In other words, if $\operatorname{disc}(f) = i^2 \Delta$ where Δ is squarefree, then $f \in \mathcal{V}_p$ if and only if $p \nmid i$. Hence, $f \in \bigcap_p \mathcal{V}_p$ if and only if $\operatorname{disc}(f)$ is squarefree.

Now let $\mathcal{U}_p$ be the set of equivalence classes $[f]$ of binary cubic forms over $\mathbb{F}_q[t]$ such that

- either $[f] \in \mathcal{V}_p$, or
- $f(x, y) \equiv \lambda(\delta x - \gamma y)^3 \pmod{p}$ for some $\lambda \in \mathbb{F}_q[t]/(p)^*$, $\gamma, \delta \in \mathbb{F}_q[t]/(p)$, $x, y \in \mathbb{F}_q[t]/(p)$ not both zero, and in addition, $f(\gamma, \delta) \not\equiv 0 \pmod{p^2}$.

For brevity, we summarize the condition $f(x, y) \equiv \lambda(\delta x - \gamma y)^3 \pmod{p(t)}$ for some $\gamma, \delta \in \mathbb{F}_q[t]/(p)$ and $\lambda \in \mathbb{F}_q[t]/(p)^*$ with the notation $(f, p) = (1^3)$ as was done in [7,8].

Finally, we set $\mathcal{U} = \bigcap_p \mathcal{U}_p$; this is the set under consideration in the Davenport-Heilbronn theorem for function fields. The version given below appears in [10]. A more general version of this theorem for Dedekind domains appears in Taniguchi [13].

Theorem 6.1. *Let q be a prime power with $\gcd(q, 6) = 1$. Then there exists a discriminant-preserving bijection between $\mathbb{F}_q(t)$-isomorphism classes of cubic function fields and classes of binary cubic forms over $\mathbb{F}_q[t]$ belonging to $\mathcal{U}$.*

In order to to convert Theorem 6.1 into an algorithm, we require a fast method for testing membership in the set $\mathcal{U}$. This is aided by the following efficiently testable conditions:

Proposition 6.2. *Let $f = (a, b, c, d)$ be a binary cubic form over $\mathbb{F}_q[t]$ with Hessian $H_f = (P, Q, R)$. Let $p \in \mathbb{F}_q[t]$ be irreducible. Then the following hold:*

1. *$(f, p) = (1^3)$ if and only if $p \mid \gcd(P, Q, R)$.*
2. *If $(f, p) = (1^3)$ then $f \in \mathcal{U}_p$ if and only if $p^3 \nmid \operatorname{disc}(f)$.*

In addition, classes in $\mathcal{U}$ contain only irreducible forms; this result can be found for integral cubic forms in [4] and is completely analogous for forms over $\mathbb{F}_q[t]$. In other words, by Theorem 6.1, if $[f] \in \mathcal{U}$, then $f(x, 1)$ is the minimal polynomial of a cubic function field over $\mathbb{F}_q(t)$. This useful fact eliminates the necessity for a potentially costly irreducibility test when testing membership in $\mathcal{U}$.

Theorem 6.3. *Any binary cubic form whose equivalence class belongs to $\mathcal{U}$ is irreducible.*

Using Proposition 6.2, we can now formulate an algorithm for testing membership in $\mathcal{U}$. This algorithm will be used in our tabulation routines for cubic function fields.

Algorithm 6.4
Input: A binary cubic form $f = (a, b, c, d)$ over $\mathbb{F}_q[t]$.
Output: true if $[f] \in \mathcal{U}$, false otherwise.
Algorithm:

1. *If f is not primitive, return false.*
2. *Put $P := b^2 - 3ac$, $Q := bc - 9ad$, $R := c^2 - 3bd$, $H_f := (P, Q, R)$, $\ell_H := \gcd(P, Q, R)$, $D := Q^2 - 4PR$ (so that $D = -3\operatorname{disc}(f)$).*
3. *If ℓ_H is not squarefree, return false.*
4. *Put $s := D/(\ell_H)^2$. If $\gcd(s, \ell_H) \neq 1$, return false*
5. *If s is squarefree, return true. Otherwise return false.*

Proposition 6.5. *Algorithm 6.4 is correct.*

Proof. Step 1 is correct, as $\mathcal{U}$ only contains classes of primitive forms by definition. If $p^2 \mid \ell_H$, then $p^4 \mid D$. If $p \mid \ell_H$ and $p \mid s$, then $p^3 \mid D$. In both cases, it follows that $p^3 \mid \operatorname{disc}(f)$, so $[f] \notin \mathcal{U}_p$, and hence $[f] \notin \mathcal{U}$, by part 2 of Proposition 6.2. This proves the correctness of steps 3 and 4.

Assume now that f passes steps 1-4, so $p^2 \nmid \ell_H$ and $p \mid \gcd(s, \ell_H)$ for some irreducible polynomial $p \in \mathbb{F}_q[t]$. Then s is not squarefree if and only if there exists an irreducible polynomial $z \in \mathbb{F}_q[t]$ with $z^2 \mid s$ and hence $z \nmid \ell_H$. By part 1 of Proposition 6.2, this rules out $(f, z) = (1^3)$. On the other hand, we also have $z^2 \mid \operatorname{disc}(f)$, so $f \notin V_z$, and hence $f \notin \mathcal{U}_z$, by steps 3 and 4 above. Thus, s is squarefree if and only if $[f] \in \mathcal{U}_p$ for all p, or equivalently, $[f] \in \mathcal{U}$, proving the validity of step 5.

Note that steps 3 and 5 of Algorithm 6.4 require tests for whether a polynomial $F \in \mathbb{F}_q[t]$ is squarefree. This can be accomplished very efficiently with a simple gcd computation, namely by checking whether $\gcd(F, F') = 1$, where F' denotes the formal derivative of F with respect to t. This is in contrast to the integral case, where squarefree testing of integers is generally difficult; in fact, squarefree factorization of integers is just as difficult as complete factorization. Hence, the membership test for $\mathcal{U}$ is more efficient than its counterpart for integral forms.

7   Tabulation Algorithm and Numerical Results

We now describe the tabulation algorithms for cubic function fields corresponding to imaginary and unusual reduced binary cubic forms over $\mathbb{F}_q[t]$; that is, cubic extensions of $\mathbb{F}_q(t)$ of discriminant D where $\deg(D)$ is odd, or $\deg(D)$ is even and $\text{sgn}(-3D)$ is a non-square in $\mathbb{F}_q^*$, respectively.

The idea of both algorithms is as follows. Input a prime power q coprime to 6 and a bound $X \in \mathbb{N}$. The first algorithm outputs minimal polynomials for all $\mathbb{F}_q(t)$-isomorphism classes of cubic extension of $\mathbb{F}_q(t)$ of discriminant D such that $\deg(D)$ is odd and $|D| \leq X$. For the second algorithm, the output is analogous, except that all the discriminants D satisfy $\deg(D)$ even, $\text{sgn}(-3D)$ is a non-square in $\mathbb{F}_q^*$, and again $|D| \leq X$. Both algorithms search through all coefficient 4-tuples (a, b, c, d) that satisfy the degree bounds of Corollary 5.3 with $|D|$ replaced by X such that the form $f = (a, b, c, d)$ satisfies the following conditions:

1. f is reduced;
2. f is imaginary, respectively, unusual;
3. f belongs to an equivalence class in $\mathcal{U}$;
4. f has a discriminant D whose degree is bounded above by X.

If f passes all these tests, the algorithms outputs $f(x, 1)$ which by Theorem 6.1 is the minimal polynomial of a triple of $\mathbb{F}_q(t)$-isomorphic cubic function fields of discriminant D.

Algorithm 7.1
Input: A prime power q not divisible by 2 or 3, and a positive integer X.
Output: Minimal polynomials for all $\mathbb{F}_q(t)$-isomorphism classes of cubic function fields of discriminant D with $\deg(D)$ odd and $|D| \leq X$.
Algorithm:
for $|a| \leq X^{1/4}$
  for $|b| \leq X^{1/4}$
    for $|c| \leq X^{1/2}/|a|$
      for $|d| \leq \max\{|bc|/|a|, |b|^2/|a|q, |c|/q\}$
        Set $f := (a, b, c, d)$;
        compute $D = \text{disc}(f)$;
        if $\deg(D)$ is odd AND $|D| \leq X$ AND $[f] \in \mathcal{U}$ AND f is reduced
        then output $f(x, 1)$.

Each loop of the form "for $|f| \leq M$" runs through all polynomials $f \in \mathbb{F}_q[t]$ with $\deg(f) = 0, 1, \ldots, \log_q(M)$. The algorithm for unusual forms (Algorithm 7.2) is completely analogous, except that the test of whether or not f is reduced in Algorithm 7.2 is more involved. Recall that if $H_f = (P, Q, R)$ is the Hessian of f and $|P| = \sqrt{|D|}$, then this test requires the computation and sorting of $q + 1$ reduced binary quadratic forms equivalent to H_f. This makes Algorithm 7.2 a good deal slower than Algorithm 7.1.

Algorithm 7.2

Input: A prime power q not divisible by 2 or 3, and a positive integer X.
Output: Minimal polynomials for all $\mathbb{F}_q(t)$-isomorphism classes of cubic function
fields of discriminant D with $\deg(D)$ is even, $\mathrm{sgn}(-3D)$ is a non-square in $\mathbb{F}_q^*$,
and $|D| \leq X$.
Algorithm:
```
for |a| ≤ X^{1/4}
  for |b| ≤ X^{1/4}
    for |c| ≤ X^{1/2}/|a|
      for |d| ≤ max{|bc|/|a|, |b|²/|a|q, |c|/q}
        Set f := (a, b, c, d);
        compute D = disc(f);
        if deg(D) is even AND sgn(−3D) is not a square in F_q AND
        |D| ≤ X AND [f] ∈ U AND f is reduced
          then output f(x, 1).
```

The algorithms presented here have some of the same advantages as Belabas'
algorithm [2]. In particular, there is no need to check for irreducibility of binary
cubic forms lying in $\mathcal{U}$, no need to factor the discriminant, and no need to keep
all fields found so far in memory. Our algorithm has the additional advantage
that there is no overhead computation needed for using a sieve to compute numbers that are not squarefree, since by the remarks following Algorithm 6.4, we
need only perform a gcd computation of a polynomial and its formal derivative.
There is an additional bottleneck for Algorithm 7.2, namely the computation
of additional Hessians and subsequently finding the smallest one in terms of
lexicographical ordering in $\mathbb{F}_q[t]$.

The following tables present the results of our computations for cubic function
fields with imaginary Hessian for $q = 5, 7$ for various degrees. In the interests
of space, we only include our computational results on imaginary forms. We
implemented the tabulation algorithm using the C++ programming language
coupled with the number theory library NTL [11]. The lists of cubic function
fields were computed on a 3 GHz Pentium 4 machine running Linux with 1 GB
of RAM.

Table 1. Cubic Function Fields over $\mathbb{F}_5$ with imaginary Hessian

Degree bound X	# of fields	Elapsed time
3	50	0.06 seconds
5	2050	53.09 sec
7	33290	24 min 21.36 sec

In [2], Belabas derived essentially the same bounds on the coefficients a and
b as ours, i.e. $O(X^{1/4})$. However, his bounds on c and d are different and were
obtained using analytic methods that do not seem to have an obvious analogue
in function fields. Using the bounds of Corollary 5.3, it is possible to show that
$O(X^{5/4})$ forms need to be checked. Belabas obtained a quasi-linear complexity

Table 2. Cubic Function Fields over $\mathbb{F}_7$ with imaginary Hessian

Degree bound X	# of fields	Elapsed time
3	147	0.52 seconds
5	12495	29 min 53.22 sec
7	365421	1 day, 3 hours, 45 min 58.78 sec

for his algorithm for tabulating cubic number fields, using the fact that the number of reduced binary cubic forms of discriminant up to $|X|$ is $O(|X|)$, see Theorem 3.7 of [4]. For function fields, we have no such asymptotic available, but we conjecture an analogous complexity of $O(X)$; this is a subject of future research.

8   Conclusions and Future Work

This paper presented a method for computing all cubic function fields with imaginary and unusual Hessian. We computed all cubic function fields with imaginary Hessian up to $|D| \leq q^7$ for $q = 5, 7$.

An immediate question is how to obtain a more exact complexity analysis of Algorithms 7.1 and 7.2; in particular whether the bound of $O(X^{5/4})$ on the number of forms searched can be improved to $O(|X|^{1+\epsilon})$, as in the case of Belabas' algorithm. In addition, a method for finding a distinguished representative in each class of reduced unusual cubic forms that is more efficient than brute force exhaustive search would significantly improve the performance of Algorithm 7.2.

We intend to extend our computations to function fields whose associated binary cubic form is unusual, and to larger values of q and $\deg(D)$. We also hope to derive an algorithm analogous to Algorithms 7.1 and 7.2 for cubic function fields where the associated binary cubic form is real. It is unclear how to develop a reduction theory for binary cubic forms with real Hessian that guarantees a unique reduced cubic form in each equivalence class. Achieving this goal via the Hessian of the cubic form is impossible, since this Hessian is a real binary quadratic form. It well-known that the number of real reduced binary quadratic forms in each equivalence class of discriminant D is of order $\sqrt{|D|}$, i.e. exponential in the size of the discriminant.

In addition, we plan to apply our methods to the task of finding quadratic function fields with large 3-rank, in a similar way to Belabas' method [3] for number fields.

Finally, recall that a cubic function field can have 5 different signatures at infinity, whereas a cubic number field can only have 2 (three real roots or one real root and two non-real complex roots, according to whether the discriminant is positive or negative). For some of the possible signatures of a cubic function field of a given discriminant, it is unclear how they relate to the signature of the quadratic function field of the same discriminant. For cubic fields that are not totally ramified at infinity, it is possible to establish the connection between the cubic and the quadratic signature through the Hilbert class field. If the place at infinity is totally ramified, the situation is unclear. It would also be interesting to analyze density results like those of [6] according to the signature of a cubic function field or of the underlying quadratic field. Such density results are the subject of future investigation.

References

1. Artin, E.: Quadratische Körper im Gebiete der höheren Kongruenzen I. Math. Zeitschrift 19, 153–206 (1924)
2. Belabas, K.: A fast algorithm to compute cubic fields. Math. Comp. 66(219), 1213–1237 (1997)
3. Belabas, K.: On quadratic fields with large 3-rank. Math. Comp. 73(248), 2061–2074 (2004)
4. Cohen, H.: Advanced Topics in Computational Number Theory. Springer, New York (2000)
5. Cremona, J.E.: Reduction of binary cubic and quartic forms. LMS J. Comput. Math. 2, 62–92 (1999)
6. Datskovsky, B., Wright, D.J.: Density of discriminants of cubic extensions. J. reine angew. Math. 386, 116–138 (1988)
7. Davenport, H., Heilbronn, H.: On the density of discriminants of cubic fields I. Bull. London Math. Soc. 1, 345–348 (1969)
8. Davenport, H., Heilbronn, H.: On the density of discriminants of cubic fields II. Proc. Royal Soc. London A 322, 405–420 (1971)
9. Rosen, M.: Number Theory in Function Fields. Springer, New York (2002)
10. Rozenhart, P.: Fast Tabulation of Cubic Function Fields. PhD Thesis, University of Calgary (in progress)
11. Shoup, V.: NTL: A Library for Doing Number Theory. Software (2001), http://www.shoup.net/ntl
12. Stichtenoth, H.: Algebraic Function Fields and Codes. Springer, New York (1993)
13. Taniguchi, T.: Distributions of discriminants of cubic algebras (preprint, 2006), http://arxiv.org/abs/math.NT/0606109

Computing Hilbert Modular Forms over Fields with Nontrivial Class Group

Lassina Dembélé and Steve Donnelly

Institut für Experimentelle Mathematik, Ellernstrasse 29, 45326 Essen, Germany
lassina.dembele@uni-due.de
School of Mathematics and Statistics F07, University of Sydney
NSW 2006, Sydney, Australia
donnelly@maths.usyd.edu.au

Abstract. We exhibit an algorithm for the computation of Hilbert modular forms over an arbitrary totally real number field of even degree, extending results of the first author. We present some new instances of the conjectural Eichler-Shimura construction for totally real number fields over the fields $\mathbb{Q}(\sqrt{10})$ and $\mathbb{Q}(\sqrt{85})$ and their Hilbert class fields, and in particular some new examples of modular abelian varieties with everywhere good reduction over those fields.

Introduction

Let F be a totally real number field of even degree. Let B be the quaternion algebra over F which is ramified at all infinite places and no finite places. The Jacquet-Langlands correspondence ([10, Chap. XVI] and [9]), establishes isomorphisms of Hecke modules between spaces of Hilbert modular forms over F and certain spaces of automorphic forms on B. The latter objects are combinatorial by nature and can be computed by using the theory of Brandt matrices. In [4] and [5], the first author presented an algorithm which adopts an alternative approach to the theory of Brandt matrices that is computationally more efficient than the classical one. Both papers considered only fields with narrow class number one.

In this paper we present a general algorithm that is practical for a large range of fields and levels. This opens the possibility of experimenting systematically, especially over fields with nontrivial class group. One technical difficulty arising from nontrivial class groups is that ideals in B are no longer free $\mathcal{O}_F$-modules. This is now handled smoothly in the package for quaternion algebras over number fields contained in the Magma computational algebra system [2] (version 2.14). Our computations rely heavily on this package, in which algorithms from [23] and [14] are implemented.

There are not many explicit examples in the literature of Hilbert modular forms in the nontrivial class group case. Okada [17] provides several examples of systems of Hecke eigenvalues of level 1 and parallel weight 2 on the quadratic

A.J. van der Poorten and A. Stein (Eds.): ANTS-VIII 2008, LNCS 5011, pp. 371–386, 2008.
© Springer-Verlag Berlin Heidelberg 2008

fields $\mathbb{Q}(\sqrt{257})$ and $\mathbb{Q}(\sqrt{401})$, computed using explicit trace formulae. One drawback with this method is that it computes the characteristic polynomials of the Hecke operators rather than the matrices themselves, and it seems difficult to recover the eigenforms from this. Also, it would not be easy to use the trace formula as the basis of an algorithm for arbitrary totally real number fields, levels and weights.

In the last few years, there has been tremendous progress towards the Langlands correspondence for $\mathbf{GL}_2/\mathbb{Q}$, culminating in the recent proof of the Serre conjecture for $\bmod\, p$ Galois representations by Khare and Wintenberger [13], and Kisin [12] et al, which in turn led to a proof of the Shimura-Taniyama-Weil conjecture for abelian varieties of $\mathbf{GL}_2$-type over $\mathbb{Q}$. We hope that our algorithm, which we implemented in Magma, will be helpful in gaining more insight as to the natural generalizations of those conjectures to the totally real case, as well as the Birch and Swinnerton-Dyer conjecture. In fact, such a project is currently under way in Dembélé, Diamond and Roberts [7] in which we use a $\bmod\, \mathfrak{p}$ version of this algorithm to investigate the Serre conjecture for some totally real number fields. See also Schein [18] for another such application.

The paper is organized as follows. Section 1 contains the necessary theoretical background. In section 2 we state the general algorithm, and describe some improvements to its implementation. Section 3 provides some numerical data over the real quadratic fields $\mathbb{Q}(\sqrt{10})$ and $\mathbb{Q}(\sqrt{85})$ and their Hilbert class fields. We also revisit the results in [17]. In section 4 we use our data to give new examples of the Eichler-Shimura construction over totally real number fields.

1   Theoretical Background

In this section, we given an explicit presentation of Hilbert modular forms as Hecke modules. By the Jacquet-Langlands correspondence, it is equivalent to give an explicit presentation of certain spaces of automorphic forms on a quaternion algebra B, which are in turn given in terms of automorphic forms on quaternion orders. A good reference for the material on Hilbert modular forms is [22]. For the theory of Brandt matrices, we refer to [8], and also to [5] for the adelic framework used here.

Let F be a totally real number field of even degree g. Let v_i, $i = 1, \ldots, g$, be all the real embeddings of F. For every $a \in F$, we let $a_i = v_i(a)$ be the image of a under v_i. We let $\mathcal{O}_F$ be the ring of integers of F, and fix an integral ideal $\mathfrak{N}$ of F. We let B be the quaternion algebra over F ramified at all infinite places and no finite places. We choose a maximal order R of B. Let K be a finite extension of F contained in $\mathbb{C}$ which splits B. We choose an isomorphism $B \otimes_F K \cong \mathbf{M}_2(K)^g$, and let $j : B^\times \hookrightarrow \mathbf{GL}_2(\mathbb{C})^g$ be the resulting embedding. For each prime $\mathfrak{p}$ in $\mathcal{O}_F$, we choose a local isomorphism $B_\mathfrak{p} \cong \mathbf{M}_2(F_\mathfrak{p})$ which sends $R_\mathfrak{p}$ to $\mathbf{M}_2(\mathcal{O}_{F,\mathfrak{p}})$. Combining these local isomorphisms, one obtains an isomorphism $\hat{B} \cong \mathbf{M}_2(\hat{F})$ under which $\hat{R}$ goes to $\mathbf{M}_2(\hat{\mathcal{O}}_F)$, where $\hat{F}$ and $\hat{\mathcal{O}}_F$ are the finite adeles of F and $\mathcal{O}_F$ respectively. We define the compact open subgroup $U_0(\mathfrak{N})$ of $\hat{R}^\times$ by

$$U_0(\mathfrak{N}) := \left\{ \begin{pmatrix} a & b \\ c & d \end{pmatrix} \in \mathbf{GL}_2(\hat{\mathcal{O}}_F) : \ c \equiv 0 (\mathrm{mod}\ \mathfrak{N}) \right\}.$$

Let $\mathrm{Cl}(R)$ denote a complete set of representatives of all the right ideal classes of R; it is in bijection with the double coset space $B^\times \backslash \hat{B}^\times / \hat{R}^\times$. Let S be a finite set of primes of $\mathcal{O}_F$ that generate the narrow class group $\mathrm{Cl}^+(F)$ and such that $\mathfrak{q}$ is coprime with $\mathfrak{N}$ for any $\mathfrak{q} \in S$. Applying the strong approximation theorem, we choose the representatives $\mathfrak{a} \in \mathrm{Cl}(R)$ such that the primes dividing $\mathrm{nr}(\mathfrak{a})$ belong to S. For any $\mathfrak{a} \in \mathrm{Cl}(R)$, we let $R_\mathfrak{a}$ be the left (maximal) order of $\mathfrak{a}$. Then there are well-defined surjective reduction maps $\hat{R}_\mathfrak{a}^\times \to \mathbf{GL}_2(\mathcal{O}_F/\mathfrak{N})$ that all differ by conjugation in $\mathbf{GL}_2(\mathcal{O}_F/\mathfrak{N})$. From this, we obtain a transitive action of each $\hat{R}_\mathfrak{a}^\times$ on $\mathbf{P}^1(\mathcal{O}_F/\mathfrak{N})$.

Let $\underline{k} \in \mathbb{Z}^g$ be a vector such that $k_i \geq 2$ and $k_i \equiv k_j (\mathrm{mod}\ 2)$ for all $i, j = 1, \ldots, g$. Set $\underline{t} = (1, \ldots, 1)$ and $\underline{m} = \underline{k} - 2\underline{t}$, then choose $\underline{n} \in \mathbb{Z}^g$ such that each $n_i \geq 0$, $n_i = 0$ for some i, and $\underline{m} + 2\underline{n} = \mu\underline{t}$ for some $\mu \in \mathbb{Z}_{\geq 0}$. Let $L_{\underline{k}}$ be the representation of $\mathbf{GL}_2(\mathbb{C})^g$ given by

$$L_{\underline{k}} := \bigotimes_{i=1}^{g} \det{}^{n_i} \otimes \mathrm{Sym}^{m_i}(\mathbb{C}^2).$$

We then obtain a representation of $B^\times$ by composing with $j : B^\times \hookrightarrow \mathbf{GL}_2(\mathbb{C})^g$.

The space of automorphic forms of level $\mathfrak{N}$ and weight $\underline{k}$ on B is defined as

$$M_{\underline{k}}^B(\mathfrak{N}) := \left\{ f : \hat{B}^\times / U_0(\mathfrak{N}) \to L_{\underline{k}} : f|_{\underline{k}} \gamma = f \text{ for all } \gamma \in B^\times \right\},$$

where $f|_{\underline{k}} \gamma(x) := f(\gamma x)\gamma$.

By the Jacquet-Langlands correspondence [10, Chap. XVI], there is an isomorphism of Hecke modules between $M_{\underline{k}}^B(\mathfrak{N})$ and $M_{\underline{k}}(\mathfrak{N})$, the space of Hilbert modular forms of weight $\underline{k}$ and level $\mathfrak{N}$ over F. On the other hand, we will now describe $M_{\underline{k}}^B(\mathfrak{N})$ in terms of automorphic forms on maximal orders of B.

The space of automorphic forms of level $\mathfrak{N}$ and weight $\underline{k}$ on the order $R_\mathfrak{a}$ is defined as

$$M_{\underline{k}}^{R_\mathfrak{a}}(\mathfrak{N}) := \left\{ f : \mathbf{P}^1(\mathcal{O}_F/\mathfrak{N}) \to L_{\underline{k}} : f|_{\underline{k}} \gamma = f \text{ for all } \gamma \in \Gamma_\mathfrak{a} \right\},$$

where $\Gamma_\mathfrak{a} = R_\mathfrak{a}^\times / \mathcal{O}_F^\times$ is a finite arithmetic group. For each $\mathfrak{a}, \mathfrak{b} \in \mathrm{Cl}(R)$ and any prime $\mathfrak{p}$ in $\mathcal{O}_F$, put

$$\Theta^{(S)}(\mathfrak{p}; \mathfrak{a}, \mathfrak{b}) := R_\mathfrak{a}^\times \backslash \left\{ u \in \mathfrak{a}\mathfrak{b}^{-1} : \ \frac{(\mathrm{nr}(u))}{\mathrm{nr}(\mathfrak{a})\mathrm{nr}(\mathfrak{b})^{-1}} = \mathfrak{p} \right\},$$

where $R_\mathfrak{a}^\times$ acts by multiplication on the left. We define the linear map

$$T_{\mathfrak{a}, \mathfrak{b}}(\mathfrak{p}) : \ M_{\underline{k}}^{R_\mathfrak{b}}(\mathfrak{N}) \to M_{\underline{k}}^{R_\mathfrak{a}}(\mathfrak{N})$$

$$f \mapsto \sum_{u \in \Theta^{(S)}(\mathfrak{p}; \mathfrak{a}, \mathfrak{b})} f|_{\underline{k}} u.$$

The following result, relating the spaces $M_{\underline{k}}^B(\mathfrak{N})$ and $M_{\underline{k}}^{R_\mathfrak{a}}(\mathfrak{N})$, was proved by the first author in [5], without restriction on the class group.

Proposition 1 ([5, Theorem 2]). *There is an isomorphism of Hecke modules*

$$M_{\underline{k}}^B(\mathfrak{N}) \longrightarrow \bigoplus_{\mathfrak{a} \in \mathrm{Cl}(R)} M_{\underline{k}}^{R_\mathfrak{a}}(\mathfrak{N}),$$

where the action of the Hecke operator $T(\mathfrak{p})$ on the right is given by the collection of linear maps $(T_{\mathfrak{a},\mathfrak{b}}(\mathfrak{p}))$ for all $\mathfrak{a}, \mathfrak{b} \in \mathrm{Cl}(R)$.

Remark 1. Proposition 1 may also be deduced from [6, Theorem 1] as a special case.

We now describe the action of the class group $\mathrm{Cl}(F)$ on $M_{\underline{k}}^B(\mathfrak{N})$. Note that $\mathrm{Cl}(F)$ acts on the set $\mathrm{Cl}(R)$ via ideal multiplication, with the class $[\mathfrak{m}] \in \mathrm{Cl}(F)$ sending $[\mathfrak{a}] \mapsto [\mathfrak{m}\mathfrak{a}]$. We then let $\mathrm{Cl}(F)$ act on $M_{\underline{k}}^B(\mathfrak{N})$ by permuting the direct summands: the class $[\mathfrak{m}] \in \mathrm{Cl}(F)$ sends an element $(f_\mathfrak{a})_\mathfrak{a} \in \bigoplus M_{\underline{k}}^{R_\mathfrak{a}}(\mathfrak{N})$ to $(f_{\mathfrak{m}\mathfrak{a}})_\mathfrak{a}$.

For each character χ of the abelian group $\mathrm{Cl}(F)$, let $M_{\underline{k}}^B(\mathfrak{N}, \chi)$ denote the χ-equivariant subspace $\{f \in M_{\underline{k}}^B(\mathfrak{N}) : \mathfrak{m} \cdot f = \chi(\mathfrak{m})f\}$. One then has the decomposition

$$M_{\underline{k}}^B(\mathfrak{N}) = \bigoplus_\chi M_{\underline{k}}^B(\mathfrak{N}, \chi).$$

2   Algorithmic Issues

Our algorithm for computing Brandt matrices using the adelic framework has already been discussed in the case of real quadratic fields in [4, sec. 2] and [5, sec. 6]. Here, we give an outline of the algorithm for any totally real number field F of even degree and any weight and level. We then discuss new optimisations to some of the key steps.

We keep the notation of section 1. Our goal is to compute the space $M_{\underline{k}}^B(\mathfrak{N})$ as a Hecke module, meaning we determine its dimension, and matrices representing the Hecke operators $T(\mathfrak{p})$ for primes $\mathfrak{p}$ with $\mathrm{N}\mathfrak{p} \le b$, for a given bound b (which must be chosen at the outset). When b is large enough, this data enables us to compute the Hecke constituents, thus the eigenforms. The precomputation stage is independent of the level and weight. Algorithms for steps $(2), (3)$ and (4) of the precomputation are given in [23].

Precomputation. The input is a field F as above, and a bound b.

1. Find a set of prime ideals S not dividing $\mathfrak{N}$ that generate $\mathrm{Cl}^+(F)$.
2. Find a presentation of the quaternion algebra B/F ramified at precisely the infinite places, and compute a maximal order R of B.
3. Compute a complete set $\mathrm{Cl}(R)$ of representatives $\mathfrak{a}$ for the right ideal classes of R such that the primes dividing $\mathrm{nr}(\mathfrak{a})$ belong to S.

4. For each representative $\mathfrak{a} \in \mathrm{Cl}(R)$, compute its left order $R_\mathfrak{a}$, and compute the unit group $\Gamma_\mathfrak{a} = R_\mathfrak{a}^\times / \mathcal{O}_F^\times$.
5. Compute the sets $\Theta^{(S)}(\mathfrak{p}; \mathfrak{a}, \mathfrak{b})$, for all primes $\mathfrak{p}$ with $N\mathfrak{p} \le b$ and all $\mathfrak{a}, \mathfrak{b} \in \mathrm{Cl}(R)$. (See Section 2.1 for details.)

Algorithm. The input consists of F and b together with the precomputed data, and also $\mathfrak{N}$ and $\underline{k}$. The output consists of a matrix $T(\mathfrak{p})$ for each prime $\mathfrak{p}$ with $N\mathfrak{p} \le b$ (and possibly additional primes), and also the Hecke constituents.

1. Compute splitting isomorphisms $R_\mathfrak{p}^\times \cong \mathbf{GL}_2(\mathcal{O}_{F,\mathfrak{p}})$, for each prime $\mathfrak{p} \mid N$.
2. For each $\mathfrak{a} \in \mathrm{Cl}(R)$, compute $M_{\underline{k}}^{R_\mathfrak{a}}(\mathfrak{N})$ as a module of coinvariants

$$M_{\underline{k}}^{R_\mathfrak{a}}(\mathfrak{N}) = K[\mathbf{P}^1(\mathcal{O}_F/\mathfrak{N})] \otimes L_{\underline{k}}/\langle x - \gamma x, \gamma \in \Gamma_\mathfrak{a} \rangle.$$

3. Combine the results of step (2), forming the direct sum

$$M_{\underline{k}}^B(\mathfrak{N}) = \bigoplus_{\mathfrak{a} \in \mathrm{Cl}(R)} M_{\underline{k}}^{R_\mathfrak{a}}(\mathfrak{N}).$$

4. For each prime $\mathfrak{p}$ with $N\mathfrak{p} \le b$, compute the families of linear maps $(T_{\mathfrak{a}, \mathfrak{b}}(\mathfrak{p}))$. (These determine the Hecke operator $T(\mathfrak{p})$ as a block matrix.)
5. Find a common basis of eigenvectors of $M_{\underline{k}}^B(\mathfrak{N})$ for the $T(\mathfrak{p})$.
6. If Step (5) does not completely diagonalize $M_{\underline{k}}^B(\mathfrak{N})$, increase b and extend the precomputation, obtaining $\Theta^{(S)}(\mathfrak{p}; \mathfrak{a}, \mathfrak{b})$ for $N\mathfrak{p} \le b$. Then return to Step (4).

Remark 2. In practice, it is extremely rare that one resorts to Step (6) since very few Hecke operators $T(\mathfrak{p})$ are required to diagonalize the space $M_{\underline{k}}^B(\mathfrak{N})$. In the cases we tested, which included levels with norm as large as 5000, we never needed more than 10 primes.

The steps in the main algorithm involve only local computations and linear algebra. The expensive steps in the process all occur in the precomputation; these involve lattice enumeration and are discussed below. For a given field F, if one wishes to compute forms of all levels up to some large bound, it is practical to simply take the primes in S to be larger than that bound, so the precomputation need only be done once.

2.1   Computing $\Theta^{(S)}(\mathfrak{p}; \mathfrak{a}, \mathfrak{b})$

Lemma 2. *The correspondence $u \leftrightarrow u^{-1}\mathfrak{a}$ gives a bijection between $\Theta^{(S)}(\mathfrak{p}; \mathfrak{a}, \mathfrak{b})$ and the set of fractional right R-ideals $\mathfrak{c} \supset \mathfrak{b}$ such that $\mathrm{nr}(\mathfrak{b}) = \mathrm{nr}(\mathfrak{c})\mathfrak{p}$ and $\mathfrak{c} \cong \mathfrak{a}$ as right R-ideals.*

Proof. The fractional right R-ideals $\mathfrak{c}$ isomorphic to $\mathfrak{a}$ are the ideals $u^{-1}\mathfrak{a}$ for $u \in B^\times$. Note that $u^{-1}\mathfrak{a} = v^{-1}\mathfrak{a}$ if and only if $v \in R_\mathfrak{a}^\times u$. It is clear that $u^{-1}\mathfrak{a}$ contains $\mathfrak{b}$ if and only if $u \in \mathfrak{a}\mathfrak{b}^{-1}$, and that $\mathrm{nr}(\mathfrak{b}) = \mathrm{nr}(u^{-1}\mathfrak{a})\mathfrak{p}$ if and only if $\mathrm{nr}(u)\mathcal{O}_F = \mathrm{nr}(\mathfrak{a})\mathrm{nr}(\mathfrak{b})^{-1}\mathfrak{p}$.

Algorithm. This computes $\Theta^{(S)}(\mathfrak{p}; \mathfrak{a}, \mathfrak{b})$ for all $\mathfrak{a} \in \mathrm{Cl}(R)$, where $\mathfrak{p}$ and $\mathfrak{b}$ are fixed.

1. Compute the fractional right R-ideals $\mathfrak{c} \supset \mathfrak{b}$ with $\mathrm{nr}(\mathfrak{b}) = \mathrm{nr}(\mathfrak{c})\mathfrak{p}$.
2. For each such $\mathfrak{c}$, compute the representative $\mathfrak{a} \in \mathrm{Cl}(R)$ and some $u \in B$ such that $\mathfrak{c} = u^{-1}\mathfrak{a}$. Append u to $\Theta^{(S)}(\mathfrak{p}; \mathfrak{a}, \mathfrak{b})$.

Remark 3. In step (1), the number of ideals $\mathfrak{c}$ obtained is $N\mathfrak{p} + 1$. Thus for each $\mathfrak{p}$ and $\mathfrak{b}$

$$\sum_{\mathfrak{a} \in \mathrm{Cl}(R)} \#\Theta^{(S)}(\mathfrak{p}; \mathfrak{a}, \mathfrak{b}) = N\mathfrak{p} + 1,$$

however this fact is not used in the algorithm.

Step (1) is a local computation; the ideals are obtained by pulling back local ideals under a splitting homomorphism $R_{\mathfrak{p}} \cong \mathbf{M}_2(F_{\mathfrak{p}})$. Step (2) is the standard problem of isomorphism testing for right ideals; we discuss below an improvement to the standard algorithm for this, such that the complexity of each isomorphism test will not depend on $\mathfrak{p}$.

2.2 Lattice-Based Algorithms for Definite Quaternion Algebras

In this section, we let B be any definite quaternion algebra over a totally real number field F, and let R be an order of B. Two basic computational problems are:

1. to find an isomorphism between given right R-ideals $\mathfrak{a}$ and $\mathfrak{b}$, and
2. to compute the unit group of R (modulo the unit group of $\mathcal{O}_F$).

The standard approach to both problems (as in [23]) reduces them to finitely many instances of the following problem.

Quaternionic Norm Equation. Let $L \cong \mathbb{Z}^d$ be a lattice contained in B (not necessarily of full rank). Given a totally positive element $\alpha \in F$, compute all $x \in L$ with $\mathrm{nr}(x) = \alpha$.

In the context of isomorphism testing, L is the fractional ideal $\mathfrak{a}\mathfrak{b}^{-1}$ and α is some generator of $\mathrm{nr}(\mathfrak{a}\mathfrak{b}^{-1})$. (One can show that it suffices to consider a finite set of candidates for α.) In the context of computing units, L is R (or occasionally an $\mathcal{O}_F$-submodule of R), and α is some unit of $\mathcal{O}_F$.

One may solve this by considering the positive definite quadratic form on L given by $\mathrm{Tr}(\mathrm{nr}(x))$. (Note that its values are positive since $\mathrm{nr}(x)$ is a totally positive element of F, for all $0 \neq x \in B$.) One captures all $x \in L$ with $\mathrm{nr}(x) = \alpha$ by enumerating all x for which the quadratic form takes value $\mathrm{Tr}(\alpha)$ (using the standard Fincke-Pohst algorithm for enumeration). The drawback is that $\mathrm{Tr}(\alpha)$ might not be particularly small in relation to the determinant of the lattice (even when α is a unit), in which case the lattice enumeration can be very time-consuming.

We now present a variation which avoids this bottleneck. For any nonzero $c \in F$, one may instead consider the lattice $cL \subset B$, again under the positive definite quadratic form given by $\mathrm{Tr}(\mathrm{nr}(x))$. One captures all $x \in L$ with $\mathrm{nr}(x) = \alpha$ by enumerating all $y \in cL$ with $\mathrm{Tr}(\mathrm{nr}(y)) = \mathrm{Tr}(c^2\alpha)$ and taking $x = y/c$. In the special case that $c \in \mathbb{Q}$, this merely rescales the enumeration problem. However, we will see that $c \in F$ may be chosen so that, in the applications (1) and (2) above, one only needs to find relatively short vectors in the lattice.

Let $g = \deg(F)$ and $d = \dim(L)$. Note that $\det(cL) = |\mathrm{N}(c)|^{d/g} \det(L)$. Heuristically, as c varies, the complexity of the enumeration process will be roughly proportional to the number of lattice elements with length up to the desired length, and this is asymptotically equal to

$$\frac{\mathrm{Tr}(c^2\alpha)^{d/2}}{\det(cL)} = \frac{\mathrm{Tr}(c^2\alpha)^{d/2}}{|\mathrm{N}(c)|^{d/g}\det(L)} = \frac{\mathrm{Tr}(c^2\alpha)^{d/2}}{\mathrm{N}(c^2\alpha)^{d/2g}}\frac{\mathrm{N}(\alpha)^{d/2g}}{\det(L)}.$$

Given that α is totally positive, $\mathrm{Tr}(c^2\alpha)/\mathrm{N}(c^2\alpha)^{1/g}$ cannot be less than g, and is close to g when all the real embeddings of $c^2\alpha$ lie close together. It is straightforward to find $c \in \mathcal{O}_F$ with this property, as follows.

Algorithm. Given some totally positive $\alpha \in F$, and some $\epsilon > 0$, this returns $c \in \mathcal{O}_F$ such that $\mathrm{Tr}(c^2\alpha)/\mathrm{N}(c^2\alpha)^{1/g} < g + \epsilon$.

1. Fix a $\mathbb{Z}$-module basis $\mathfrak{bas}(\mathcal{O}_F)$ of $\mathcal{O}_F$.
2. Initialize $C := 100$.
3. Calculate $r_i := C/\sqrt{v_i(\alpha)}$ (note that the real embeddings of α are positive).
4. Represent the vector (r_i) in terms of the basis $\mathfrak{bas}(\mathcal{O}_F)$, then round the coordinates to integers, thus obtaining an element $c \in \mathcal{O}_F$.
5. If c does not have the desired property, multiply C by 100 and return to step (3).

Proof. Fix α and let $C \to \infty$, regarding $r_i \in \mathbb{R}$ and $c \in \mathcal{O}_F$ as functions of C. Since we use a fixed basis of $\mathcal{O}_F$, $v_i(c) - r_i$ is bounded by a constant independent of C. Therefore as $C \to \infty$, $v_i(c^2\alpha) = r_i^2\alpha_i + O(C) = C^2 + O(C)$. This implies that for any i and j, the ratio $v_i(c^2\alpha)/v_j(c^2\alpha) \to 1$ as $C \to \infty$, and the lemma follows.

The complexity of the enumeration thus depends on the ratio $\mathrm{N}(\alpha)^{d/2g}/\det(L)$. In both the applications above, this ratio is small: in computing units, α is a unit, and in isomorphism testing, α generates the fractional ideal $\mathrm{nr}(L)$ where $L = \mathfrak{a}\mathfrak{b}^{-1}$.

3   Examples of Hilbert Modular Forms

In this section we give some examples of Hilbert modular forms computed using our algorithm, which we have implemented in Magma (and which will be available in a future version of Magma).

3.1 The Quadratic Field $\mathbb{Q}(\sqrt{85})$

Let $F = \mathbb{Q}(\sqrt{85})$. The class number of F is the same as its narrow class number: $h_F = h_F^+ = 2$. The maximal order in F is $\mathcal{O}_F = \mathbb{Z}[\omega_{85}]$, where $\omega_{85} = \frac{1+\sqrt{85}}{2}$. Let B be the Hamilton quaternion algebra over F. As an F-algebra, B is generated by i, j subject to the relations $i^2 = j^2 = (ij)^2 = -1$. Since the prime 2 is inert in F, the algebra B is ramified only at the two infinite places. Using Magma, we find that the class number of B is 8. The Hecke module of Hilbert modular forms of level 1 and weight $(2, 2)$ over F is therefore an 8-dimensional $\mathbb{Q}$-space, and it can be diagonalized by using the Hecke operator T_2. There are two Eisenstein series and two Galois conjugacy classes of newforms. The eigenvalues of the Hecke operators for the first few primes are given in Table 1 (only one eigenform in each Galois conjugacy class of newforms is listed). Each newform is given by a column, and we use the following labeling. For a quadratic field F, we label each form by a roman letter preceded by the discriminant of F. For the Hilbert class field of F, everything is just preceded by an H. For example, 85A is the first newform of level 1 over $\mathbb{Q}(\sqrt{85})$, and H85A is the first newform of level 1 over the Hilbert class field of $\mathbb{Q}(\sqrt{85})$.

The Hilbert class field of $\mathbb{Q}(\sqrt{85})$ is $H := \mathbb{Q}(\sqrt{5}, \sqrt{17}) = \mathbb{Q}(\alpha)$, where the minimal polynomial of α is $x^4 - 4x^3 - 5x^2 + 18x - 1$. The narrow class number of H is 1, and $B \otimes_F H$ (the quaternion algebra over H ramified at the four infinite places) has class number 4. Thus the space of Hilbert modular forms of level 1 and weight $(2,2)$ is 4-dimensional. The eigenvalues of the Hecke action for the first few primes are listed in Table 1. There is one Eisenstein series and two classes of newforms. Elements of $\mathcal{O}_H$ are expressed in terms of the integral basis

$$1, \quad \tfrac{1}{6}(\alpha^3 - 3\alpha^2 - 5\alpha + 10), \quad \tfrac{1}{6}(-\alpha^3 + 3\alpha^2 + 11\alpha - 10), \quad \tfrac{1}{6}(-\alpha^3 + 14\alpha + 5),$$

which we use to write generators of the ideals in the table.

Table 1. Hilbert modular forms of level 1 and parallel weight 2 over $\mathbb{Q}(\sqrt{85})$ and its Hilbert class field H. The minimal polynomial of β (resp. β') is $x^4 - 6x^2 + 2$ (resp. $x^2 + 6x + 2$).

$N(\mathfrak{p})$	$\mathfrak{p}$	EIS1	EIS2	85A	85B	$N(\mathfrak{p})$	$\mathfrak{p}$	EIS	H85A	H85B
3	$(3, 2\omega_{85})$	4	-4	$2\sqrt{-1}$	β	4	$[1, -1, 0, 1]$	5	1	$3 + \beta'$
3	$(3, 4 + 2\omega_{85})$	4	-4	$-2\sqrt{-1}$	β	4	$[0, 2, -1, 1]$	5	1	$3 + \beta'$
4	(2)	5	5	1	$-\beta^3 + 3$	9	$[0, 1, -1, 0]$	10	2	β'
5	$(5, -1 + 2\omega_{85})$	6	-6	0	$-\beta^3 + 4\beta$	9	$[1, -1, -1, 0]$	10	2	β'
7	$(7, 2\omega_{85})$	8	-8	$-2\sqrt{-1}$	$\beta^3 - 5\beta$	19	$[0, 1, 0, -1]$	20	-4	2
7	$(7, 12 + 2\omega_{85})$	8	-8	$2\sqrt{-1}$	$\beta^3 - 5\beta$	19	$[-1, 2, 0, 1]$	20	-4	2
17	$(17, -1 + 2\omega_{85})$	18	-18	0	$2\beta^3 - 14\beta$	19	$[1, -1, -1, 1]$	20	-4	2
19	$(19, 2 + 2\omega_{85})$	20	20	-4	2	19	$[-1, 2, -1, 1]$	20	-4	2

We also computed some spaces over $\mathbb{Q}(\sqrt{85})$ with nontrivial level. The dimensions of the spaces with prime level of norm less than 100 are given in Table 2. (It suffices to consider just one prime in each pair of conjugate primes, and for the precomputation we took $S = \{(3, -1 + \omega_{85})\}$.) For example, for level $\mathfrak{N} = (5, \sqrt{85})$, $M_2(\mathfrak{N})$ has dimension 20, and the Hecke operator $T_{\mathfrak{p}}$ with $\mathfrak{p} = (7, 2\omega_{85})$ acting on $M_2(\mathfrak{N})$ has characteristic polynomial

$$(x - 8)(x + 8)(x^2 + 4)^2(x^4 - 10x^2 + 18)^2(x^6 + 28x^4 + 104x^2 + 100).$$

Comparing this with the space $M_2(1)$ of level 1, on which $T_{\mathfrak{p}}$ has characteristic polynomial

$$(x - 8)(x + 8)(x^2 + 4)(x^4 - 10x^2 + 18),$$

one sees that the Hecke action on the subspace of newforms $M_2(\mathfrak{N})$ is irreducible, and the cuspidal oldform space embeds in $M_2(\mathfrak{N})$ under two degeneracy maps (as expected).

Table 2. Dimensions of spaces of Hilbert modular forms over $\mathbb{Q}(\sqrt{85})$ with weight $(2, 2)$ and prime level of norm less than 100

$N(\mathfrak{N})$	dim $M_2(\mathfrak{N})$	dim $S_2(\mathfrak{N})$	dim $S_2^{\mathrm{new}}(\mathfrak{N})$
3	16	14	8
4	24	22	16
5	20	18	12
7	32	30	24
17	56	54	48
19	68	66	60
23	72	70	64
37	124	122	116
59	180	178	172
73	232	230	224
89	272	270	264
97	304	302	296

3.2   The Quadratic Field $\mathbb{Q}(\sqrt{10})$

Let $F = \mathbb{Q}(\sqrt{10})$. The Hilbert class field of F is $H := \mathbb{Q}(\sqrt{2}, \sqrt{5}) = \mathbb{Q}(\alpha)$, where the minimal polynomial of α is $x^4 - 2x^3 - 5x^2 + 6x - 1$. The narrow class number of H is 1. We computed the space of Hilbert modular forms of level 1 and weight $(2, 2)$ over F and H, and the Hecke eigenvalues for the first few primes are listed in Table 3 (only one eigenform in each Galois conjugacy class of newforms is listed). Elements of $\mathcal{O}_H$ are expressed in terms of the integral basis

$1, \quad \frac{1}{3}(2\alpha^3 - 3\alpha^2 - 10\alpha + 7), \quad \frac{1}{3}(-2\alpha^3 + 3\alpha^2 + 13\alpha - 7), \quad \frac{1}{3}(-\alpha^3 + 3\alpha^2 + 5\alpha - 8).$

Table 3. Hilbert modular forms of level 1 and parallel weight 2 over $\mathbb{Q}(\sqrt{10})$ and its Hilbert class field

N($\mathfrak{p}$)	$\mathfrak{p}$	EIS1	EIS2	40A
2	$(2, \omega_{40})$	-3	3	$-\sqrt{2}$
3	$(3, \omega_{40} + 4)$	-4	4	$\sqrt{2}$
3	$(3, \omega_{40} + 2)$	-4	4	$\sqrt{2}$
5	$(5, \omega_{40})$	-6	6	$-2\sqrt{2}$
13	$(13, \omega_{40} + 6)$	-14	14	0
13	$(13, \omega_{40} + 7)$	-14	14	0
31	$(31, \omega_{40} + 14)$	32	32	4
31	$(31, \omega_{40} + 17)$	32	32	4

N($\mathfrak{p}$)	$\mathfrak{p}$	EIS	H40A
4	$[0, 0, 1, 0]$	5	-2
9	$[1, 1, -1, 0]$	10	-4
9	$[0, 1, -1, 1]$	10	-4
25	$[1, -2, 0, 0]$	26	-2
31	$[1, 1, 1, -1]$	32	4
31	$[1, -1, -1, -1]$	32	4
31	$[1, 1, -1, 1]$	32	4
31	$[-3, 2, -1, 0]$	32	4

3.3 Revisiting the Examples by Okada

Okada [17] computes systems of Hecke eigenvalues of Hilbert newforms of level 1 and weight $(2,2)$ over the fields $\mathbb{Q}(\sqrt{257})$ and $\mathbb{Q}(\sqrt{401})$ by using explicit trace formulas. We now compare that data with results obtained using our algorithm.

First, let $F = \mathbb{Q}(\sqrt{257})$, which has $h_F = h_F^+ = 3$. Using our algorithm, we obtain that $\dim M_2(1) = 39$ and $\dim S_2(1) = 36$. The forms that are base change come from the space of classical modular forms $S_2(257, (\frac{257}{\cdot}))$ which has dimension 20. Thus the dimension of the subspace of Hilbert newforms that are not base change is $36 - 20/2 = 26$. For each character $\chi : \mathrm{Cl}^+(F) \to \mathbb{C}^\times$, let $S_2(1, \chi)$ be the subspace of $S_2(1)$ corresponding to χ. The space computed in [17] is the 12-dimensional subspace $S_2(1, \mathbf{1})$, where $\mathbf{1}$ is the trivial character. Furthermore, since $h_F^+ = 3$ is odd, $S_2(1, \mathbf{1})$ maps isomorphically onto $S_2(1, \chi)$ by twisting, for all χ. Hence $\dim S_2(1) = 3 \dim S_2(1, \mathbf{1})$. In Table 4, we list all the eigenforms of level 1 and weight $(2,2)$ whose fields of coefficients have degree at most 4. There are two additional newforms 257E and 257F whose fields of coefficients are given respectively by the polynomials:

$$f = x^9 + x^8 - 14x^7 - 10x^6 + 66x^5 + 25x^4 - 114x^3 - x^2 + 39x - 9$$
$$g = x^{18} - x^{17} + 15x^{16} - 6x^{15} + 140x^{14} - 33x^{13} + 771x^{12} + 75x^{11} + 2969x^{10}$$
$$+ 559x^9 + 7056x^8 + 2982x^7 + 10627x^6 + 2430x^5 + 4672x^4 + 2091x^3$$
$$+ 1512x^2 + 351x + 81.$$

The forms 257A and 257E are base change from $S_2(257, (\frac{257}{\cdot}))$, and 257B is the form discussed in [17].

Next, let $F = \mathbb{Q}(\sqrt{401})$, in which case $h_F = h_F^+ = 5$. Our algorithm gives the dimensions $\dim M_2(1) = 125$ and $\dim S_2(1) = 120$. The forms that are base change come from the space of classical modular forms $S_2(401, (\frac{401}{\cdot}))$, which has dimension 32. Thus the dimension of the subspace of newforms that are not base change is $120 - 32/2 = 104$.

Table 4. Hilbert modular forms of level 1 and weight $(2,2)$ over $\mathbb{Q}(\sqrt{257})$. The minimal polynomial of β is $x^4 + x^3 + 4x^2 - 3x + 9$.

$N(\mathfrak{p})$	$\mathfrak{p}$	EIS1	257A	257B	257C	EIS2
2	$(2, \omega_{257})$	3	-1	$\frac{1+\sqrt{13}}{2}$	$\frac{1+\sqrt{-3}}{2}$	$\frac{-3+3\sqrt{-3}}{2}$
2	$(2, 1 - \omega_{257})$	3	-1	$\frac{1-\sqrt{13}}{2}$	$\frac{1-\sqrt{-3}}{2}$	$\frac{-3-3\sqrt{-3}}{2}$
9	(3)	10	4	-4	4	10
11	$(11, 4 + \omega_{257})$	12	0	1	0	$-6 + 6\sqrt{-3}$
11	$(11, 5 - \omega_{257})$	12	0	1	0	$-6 - 6\sqrt{-3}$
13	$(13, 9 + \omega_{257})$	14	2	$\sqrt{13}$	$-1 + \sqrt{-3}$	$-7 - 7\sqrt{-3}$
13	$(13, 10 - \omega_{257})$	14	2	$-\sqrt{13}$	$-1 - \sqrt{-3}$	$-7 + 7\sqrt{-3}$
17	$(17, 11 + \omega_{257})$	18	4	$4 + \sqrt{13}$	$-2 - 2\sqrt{-3}$	$-9 + 9\sqrt{-3}$
17	$(17, 12 - \omega_{257})$	18	4	$4 - \sqrt{13}$	$-2 + 2\sqrt{-3}$	$-9 - 9\sqrt{-3}$

$N(\mathfrak{p})$	$\mathfrak{p}$	257D
2	$(2, \omega_{257})$	β
2	$(2, 1 - \omega_{257})$	$(\beta^3 + \beta^2 + 4\beta - 3)/3$
9	(3)	-4
11	$(11, 4 + \omega_{257})$	$(-\beta^3 - 4\beta^2 - 4\beta - 9)/12$
11	$(11, 5 - \omega_{257})$	$(\beta^3 + 4\beta^2 + 4\beta - 3)/12$
13	$(13, 9 + \omega_{257})$	$(-7\beta^3 - 4\beta^2 - 28\beta + 21)/12$
13	$(13, 10 - \omega_{257})$	$(-\beta^3 - 4\beta^2 - 28\beta - 9)/12$
17	$(17, 11 + \omega_{257})$	$(-\beta^3 - 4\beta^2 + 4\beta - 9)/4$
17	$(17, 12 - \omega_{257})$	$(11\beta^3 + 20\beta^2 + 44\beta - 33)/12$

4   Examples of the Eichler-Shimura Construction

In the study of Hilbert modular forms, the following conjecture is important and wide open. We refer to Shimura [19] or Knapp [15] for the classical case, and to Oda [16], Zhang [24] and Blasius [1] for the number field case.

Conjecture 3 (Eichler-Shimura). *Let f be a Hilbert newform of level $\mathfrak{N}$ and parallel weight 2 over a totally real field F. Let K_f be the number field generated by the Fourier coefficients of f. Then there exists an abelian variety A_f defined over F, with good reduction outside of $\mathfrak{N}$, such that $K_f \hookrightarrow \mathrm{End}(A_f) \otimes \mathbb{Q}$ and*

$$L(A_f, s) = \prod_{\sigma \in \mathrm{Gal}(K_f/\mathbb{Q})} L(f^\sigma, s),$$

where f^σ is obtained by letting σ act on the Fourier coefficients of f.

In the classical setting, namely when $F = \mathbb{Q}$, this is a theorem known as the Eichler-Shimura construction. In general, many cases of the conjecture are also known. In those cases the abelian variety A_f is often constructed as a quotient of the Jacobian of some Shimura curve of level $\mathfrak{N}$. See, for example, Zhang [24] and references therein. In the case when $[F : \mathbb{Q}]$ is even, the level of such a Shimura curve must contain at least one finite prime, which means its Jacobian must have at least one prime of bad reduction. So when A_f has everywhere good reduction, such a parametrization is simply not available. In this section, we provide new examples of such A_f. We note that similar examples have already been discussed in Socrates and Whitehouse [21].

Remark 4. We refer back to the final paragraph of section 3.1. The characteristic polynomials given there, viewed in terms of Conjecture 3, indicate that the newsubspace of $M_2(\mathfrak{N})$ corresponds to a simple abelian variety of dimension 6.

4.1   The Quadratic Field $\mathbb{Q}(\sqrt{85})$

Keeping the notation of subsection 3.1, let E/H be the elliptic curve with the following coefficients:

	a_1	a_2	a_3	a_4	a_6
$E:$	$[1,0,0,1]$	$[0,-1,0,-1]$	$[0,1,1,0]$	$[-5,-6,-1,0]$	$[-8,-7,-3,2]$

It is a global minimal model which has everywhere good reduction. Hence, the restriction of scalars $A = \mathrm{Res}_{H/F}(E)$ is an abelian surface over F also with everywhere good reduction.

Remark 5. The j-invariant of E is $64047678245 - 12534349815\omega_{85} \in F$, and in fact E is H-isomorphic to its conjugate under $\mathrm{Gal}(H/F)$. Therefore A is isomorphic to $E \times E$ over H. Let E' denote one of the other two conjugates with respect to the Galois group $\mathrm{Gal}(H/\mathbb{Q})$, which have j-invariant $51513328430 + 12534349815\omega_{85}$; there is an isogeny of degree 2 from E to E'. The restriction of scalars $\mathrm{Res}_{H/F}(E')$ over F is isomorphic to $E' \times E'$ over H, and is therefore isogenous to A.

To establish the modularity of E and A, we will apply the following result of Skinner and Wiles. Here we state the *nearly ordinary* assumption (Condition (iv)) in a slightly different way.

Theorem 4 ([20, Theorem A]). *Let F be a totally real abelian extension of $\mathbb{Q}$. Suppose that $p \geq 3$ is prime, and let $\rho : \mathrm{Gal}(\overline{F}/F) \longrightarrow \mathbf{GL}_2(\overline{\mathbb{Q}}_p)$ be a continuous, absolutely irreducible and totally odd representation unramified away from a finite set of places of F. Suppose that the reduction of ρ is of the form $\bar{\rho}^{ss} = \chi_1 \oplus \chi_2$, where χ_1 and χ_2 are characters, and suppose that:*

(i) *the splitting field $F(\chi_1/\chi_2)$ of χ_1/χ_2 is abelian over $\mathbb{Q}$,*

(ii) *$(\chi_1/\chi_2)|_{D_v} \neq 1$ for each $v \mid p$,*

(iii) *$\rho|_{I_v} \cong \begin{pmatrix} \psi\epsilon_p^{k-1} & * \\ 0 & 1 \end{pmatrix}$ for each prime $v \mid p$,*

(iv) *$\det \rho = \psi\epsilon_p^{k-1}$, with $k \geq 2$ an integer, ψ a character of finite order, and ϵ_p the p-adic cyclotomic character.*

Then ρ comes from a Hilbert modular from.

Proposition 5

(a) *The elliptic curves E is modular and corresponds to Table 1's form* H85A.
(b) *The abelian surface A is modular and corresponds to the form* 85A *in Table 1.*

Proof. (a) Let $\rho_{E,3}$ be the 3-adic representation attached to E, and $\bar{\rho}_{E,3}$ the corresponding residual representation. Also, let $\mathfrak{p} \subset \mathcal{O}_H$ be any prime above 3. Using Magma, we compute the torsion subgroup $E(H)_{tors} \cong \mathbb{Z}/2 \oplus \mathbb{Z}/2$, and the trace of Frobenius $a_{\mathfrak{p}}(E) = 2$. The latter implies that the representation $\rho_{E,3}$ is ordinary at $\mathfrak{p}$. By direct calculation, we find that $j(E)$ is the image of a H-rational point on the modular curve $X_0(3)$:

$$j(E) = \frac{(\tau + 27)(\tau + 3)^3}{\tau}, \text{ where } \tau = [2166, 527, -527, 1054].$$

This implies that E has a Galois-stable subgroup of order 3, so the representation $\bar{\rho}_{E,3}$ is reducible. Since it is ordinary, there exist characters χ, χ' unramified away from $\mathfrak{p} \mid 3$, with χ unramified at $\mathfrak{p}$, such that $\bar{\rho}_{E,3}^{ss} = \chi \oplus \chi'$ and $\chi\chi' = \epsilon_3$ is the mod 3 cyclotomic character. The field $H(\chi/\chi')$ is clearly abelian. Therefore the representation $\rho_{E,3}$ satisfies the conditions of Skinner and Wiles, and E is modular. Comparing traces of Frobenius with the eigenvalues given in Table 1, we see that the corresponding form is H85A.

(b) Let f be the base change from F to H of the newform 85A in Table 1. Since the Hilbert class field extension H/F is totally unramified, the form f has level 1 and trivial character. By comparing the Fourier coefficients at the split primes above 19, we see that $f = $ H85A in Table 2. The result then follows from properties of restriction of scalars and base change.

Remark 6. To find E, we reasoned as follows. The eigenvalues of H85A in Table 1 suggest that the curve corresponding to it admits a 2-isogeny. This curve must have good reduction everywhere, and so must its conjugates; if these are also modular, then they share the same L-series and are therefore isogenous to each other. This would mean the curve comes from an H-rational point on $X_0(2)$ whose j-invariant is integral. Using a parametrisation of $X_0(2)$, we searched for such points. We would like to thank Noam Elkies for suggesting this approach. (Note that it would be extremely arduous to find E by computing all elliptic curves over H with trivial conductor, via the general algorithm described in Cremona and Lingham [3].)

Remark 7. If we assume Conjecture 3, then there exists a modular abelian surface A over H with real multiplication by $\mathbb{Q}(\sqrt{7})$ which corresponds to the form H85B in Table 1. The restriction of scalars of A from H to F is a modular abelian fourfold with real multiplication by $\mathbb{Q}(\beta)$ which corresponds to the form 85B in Table 1.

4.2 The Quadratic Field $\mathbb{Q}(\sqrt{10})$

Keeping the notation of subsection 3.2, let E/H be the elliptic curve with the following coefficients:

a_1	a_2	a_3	a_4	a_6
$E : [0,0,1,0]$	$[1,0,1,-1]$	$[0,1,0,0]$	$-[15,44,21,26]$	$-[91,123,48,97]$

This is a global minimal model with everywhere good reduction over H. In contrast with the previous example, the four Galois conjugates have distinct j-invariants. The restriction of scalars $A = \operatorname{Res}_{H/F}(E)$ is an abelian surface over F with everywhere good reduction.

Proposition 6. *The elliptic curve E/H and the abelian surface A/F are modular; E corresponds to H40A in Table 3, and A corresponds to 40A in Table 3.*

Proof. Let $\rho_{E,3}$ be the 3-adic representation attached to E, and $\bar{\rho}_{E,3}$ its reduction modulo 3. Then $\bar{\rho}_{E,3}$ is reducible since

$$j(E) = \frac{(\tau + 27)(\tau + 3)^3}{\tau}, \text{ where } \tau = [5, 52, -18, -26].$$

As before, it is easy to see that $\rho_{E,3}$ satisfies the conditions of Skinner and Wiles. So E is modular, and hence A is also modular. Comparing traces of Frobenius with Fourier coefficients, it is easy to see which forms in the tables they correspond to.

Alternatively, we could consider the 7-adic representation $\rho_{E,7}$. Its reduction mod 7 is reducible since the point $([16, 23, 9, 18] : [-157, -268, -119, -184] : [1, 0, 0, 0])$ is an H-rational point of order 7 on E. Furthermore, for any prime $\mathfrak{p} \mid 7$, we have $a_{\mathfrak{p}}(E) = 8$, and it is easy to see that $\rho_{E,7}$ satisfies the conditions of Skinner and Wiles.

Remark 8. It was shown by Kagawa [11, Theorem 3.2] that there is no elliptic curve with everywhere good reduction over $\mathbb{Q}(\sqrt{10})$. Our results show that if we assume modularity in addition, there is only one such simple abelian variety: an abelian surface with real multiplication by $\mathbb{Z}[\sqrt{2}]$.

Remark 9. To find E, we were again assisted by the eigenvalues of the corresponding form H40A in Table 3, which suggest that E has an H-rational point of order 14. The modular curve $X_0(14)/\mathbb{Q}$ is an elliptic curve (14A1 in Cremona's table), which (using Magma) was found to have rank 1 over H and also rank 1 over $\mathbb{Q}(\sqrt{10})$; this enabled us to obtain a point of infinite order simply by finding a $\mathbb{Q}$-rational point on the quadratic twist by $\sqrt{10}$. We considered curves corresponding to points of small height in $X_0(14)(H)$, and twists of these curves, until we found one with good reduction everywhere.

Remark 10. Although we have restricted the discussion in this paper to fields of even degree, the algorithm can clearly be used over fields of odd degree as well. In that case, the ramification $Ram(B)$ of the quaternion algebra B must contains some finite primes, and we only obtain the newforms whose corresponding automorphic representations are special or supercuspidal at the primes in $Ram(B)$.

Acknowledgements

This project was started when the first author was a PIMS postdoctoral fellow at the University of Calgary, and parts of it were written during his visit to the University of Sydney in August 2007. He would like to thank both PIMS and the University of Calgary for their financial support and the Department of Mathematics and Statistics of the University of Sydney for its hospitality. In particular, he would like to thank Anne and John Cannon for their invitation to visit the Magma group. He would also like to thank Clifton Cunningham for his constant support and encouragement in the early stage of the project. Finally, the authors would like to thank Fred Diamond, Noam Elkies and Haruzo Hida for helpful email exchanges.

References

1. Blasius, D.: Elliptic curves, Hilbert modular forms, and the Hodge conjecture. In: Hida, Ramakrishnan, Shahidi (eds.) Contributions to Automorphic forms, Geometry, and Number Theory, pp. 83–103. Johns Hopkins Univ. Press, Baltimore (2004)
2. Bosma, W., Cannon, J., Playoust, C.: The Magma algebra system. I. The user language. J. Symbolic Comput. 24(3-4), 235–265 (1997)
3. Cremona, J., Lingham, M.: Finding all elliptic curves with good reduction outside a given set of primes. Experimental Math. (to appear)
4. Dembélé, L.: Explicit computations of Hilbert modular forms on $\mathbb{Q}(\sqrt{5})$. Experimental Math. 14, 457–466 (2005)
5. Dembélé, L.: Quaternionic M-symbols, Brandt matrices and Hilbert modular forms. Math. Comp. 76, 1039–1057 (2007)
6. Dembélé, L.: On the computation of algebraic modular forms (submitted)
7. Dembélé, L., Diamond, F., Roberts, D.: Examples and numerical evidence for the Serre conjecture over totally real number fields (in preparation)
8. Eichler, M.: On theta functions of real algebraic number fields. Acta Arith. 33, 269–292 (1977)
9. Gelbart, S.: Automorphic forms on adele groups. In: Annals of Maths. Studies, vol. 83, Princeton Univ. Press, Princeton (1975)
10. Jacquet, H., Langlands, R.P.: Automorphic forms on GL(2). Lectures Notes in Math, vol. 114. Springer, Berlin, New York (1970)
11. Kagawa, T.: Elliptic curves with everywhere good reduction over real quadratic fields. Ph. D Thesis, Waseda University (1998)
12. Kisin, M.: Modularity of 2-adic Barsotti-Tate representations (preprint), http://www.math.uchicago.edu/~kisin/preprints.html
13. Khare, C., Wintenberger, J.-P.: On Serre's conjecture for 2-dimensional mod preresentations of the absolute Galois group of the rationals. Annals of Mathematics (to appear), http://www.math.utah.edu/~shekhar/serre.pdf
14. Kirschmer, M.: Konstruktive Idealtheorie in Quaternionenalgebren. Diplom Thesis, Universität Ulm (2005)
15. Knapp, A.W.: Elliptic Curves. Mathematical Notes, vol. 40. Princeton University Press, Princeton (1992)
16. Oda, T.: Periods of Hilbert Modular Surfaces. Progress in Mathematics, vol. 19. Birkhäuser, Boston, Mass. (1982)

17. Okada, K.: Hecke eigenvalues for real quadratic fields. Experiment. Math. 11, 407–426 (2002)
18. Schein, M.: Weights in Serre's conjecture for Hilbert modular forms: the ramified case. Israel Journal of Mathematics (to appear),
    http://www.math.huji.ac.il/~mschein/wt5rev.pdf
19. Shimura, G.: Introduction to the Arithmetic Theory of Automorphic Functions. Kanô Memorial Lectures, No. 1. Publications of the Mathematical Society of Japan, No. 11. Iwanami Shoten, Publishers, Tokyo; Princeton University Press, Princeton (1971)
20. Skinner, C.M., Wiles, A.J.: Residually reducible representations and modular forms. Inst. Hautes Études Sci. Publ. Math. (89), 5–126 (1999)
21. Socrates, J., Whitehouse, D.: Unramified Hilbert modular forms, with examples relating to elliptic curves. Pacific J. Math. 219, 333–364 (2005)
22. Taylor, R.: On Galois representations associated to Hilbert modular forms. Invent. Math. 98, 265–280 (1989)
23. Voight, J.: Quadratic forms and quaternion algebras: Algorithms and arithmetic. PhD thesis, University of California, Berkeley (2005)
24. Zhang, S.: Heights of Heegner points on Shimura curves. Ann. of Math. 153(2), 27–147 (2001)

Hecke Operators and Hilbert Modular Forms

Paul E. Gunnells and Dan Yasaki

University of Massachusetts Amherst, Amherst, MA 01003, USA

Abstract. Let F be a real quadratic field with ring of integers $\mathcal{O}$ and with class number 1. Let Γ be a congruence subgroup of $\mathrm{GL}_2(\mathcal{O})$. We describe a technique to compute the action of the Hecke operators on the cohomology $H^3(\Gamma; \mathbb{C})$. For F real quadratic this cohomology group contains the cuspidal cohomology corresponding to cuspidal Hilbert modular forms of parallel weight 2. Hence this technique gives a way to compute the Hecke action on these Hilbert modular forms.

1 Introduction

1.1 Modular Symbols

Let $\mathbf{G}$ be a reductive algebraic group defined over $\mathbb{Q}$, and let $\Gamma \subset \mathbf{G}(\mathbb{Q})$ be an arithmetic subgroup. Let $Y = \Gamma \backslash X$ be the locally symmetric space attached to $G = \mathbf{G}(\mathbb{R})$ and Γ, where X is the global symmetric space, and let $\mathcal{M}$ be a local system on Y attached to a rational finite-dimensional complex representation of Γ. The cohomology $H^*(Y; \mathcal{M})$ plays an important role in number theory, through its connection with automorphic forms and (mostly conjectural) relationship to representations of the absolute Galois group $\mathrm{Gal}(\overline{\mathbb{Q}}/\mathbb{Q})$ (cf. [18,5,9,28]). This relationship is revealed in part through the action of the *Hecke operators* on the cohomology spaces. Hecke operators are endomorphisms induced from a family of correspondences associated to the pair $(\Gamma, \mathbf{G}(\mathbb{Q}))$; the arithmetic nature of the cohomology is contained in the eigenvalues of these linear maps.

For $\Gamma \subset \mathrm{SL}_n(\mathbb{Z})$, *modular symbols* provide a concrete method to compute the Hecke eigenvalues in $H^\nu(Y; \mathcal{M})$, where $\nu = n(n-1)/2$ is the top nonvanishing degree [25,10].[1] Using modular symbols many people have studied the arithmetic significance of this cohomology group, especially for $n = 2$ and 3 [14,27,5,9,8,28]; these are the only two values of n for which $H^\nu(Y; \mathcal{M})$ can contain cuspidal cohomology classes, in other words cohomology classes coming from cuspidal automorphic forms on $\mathrm{GL}(n)$. Another setting where automorphic cohomology has been profitably studied using modular symbols is that of $\Gamma \subset \mathrm{SL}_2(\mathcal{O})$, where $\mathcal{O}$ is the ring of integers in a complex quadratic field [13,15,12,24]. In this case Y is a three-dimensional hyperbolic orbifold; modular symbols allow investigation of $H^2(Y; \mathcal{M})$, which again contains cuspidal cohomology classes.

---

[1] Here and throughout the paper by modular symbol we mean *minimal modular symbol* in the sense of [2], in contrast to [3].

A.J. van der Poorten and A. Stein (Eds.): ANTS-VIII 2008, LNCS 5011, pp. 387–401, 2008.
© Springer-Verlag Berlin Heidelberg 2008

Now let F be a real quadratic field with ring of integers $\mathcal{O}$, and let $\mathbf{G}$ be the $\mathbb{Q}$-group $\mathrm{Res}_{F/\mathbb{Q}}(\mathrm{GL}_2)$. Let $\Gamma \subseteq \mathbf{G}(\mathbb{Q})$ be a congruence subgroup. In this case we have $X \simeq \mathfrak{H} \times \mathfrak{H} \times \mathbb{R}$, where $\mathfrak{H}$ is the upper halfplane (§2.1). The locally symmetric space Y is topologically a circle bundle over a Hilbert modular surface, possibly with orbifold singularities if Γ has torsion. The cuspidal cohomology of Y is built from cuspidal Hilbert modular forms. Hence an algorithm to compute the Hecke eigenvalues on the cuspidal cohomology gives a topological technique to compute the Hecke eigenvalues of such forms. But in this case there is a big difference from the setting above: the top degree cohomology occurs in degree $\nu = 4$, but the cuspidal cohomology appears in degrees $2, 3$.[2] Thus modular symbols cannot "see" the cuspidal Hilbert modular forms, and cannot directly be used to compute the Hecke eigenvalues.

1.2 Results

In this article we discuss a technique, based on constructions in [19], that in practice allows one to compute the Hecke action on the cohomology space $H^3(Y; \mathbb{C})$. Moreover it is easy to modify our technique to compute with other local systems; all the geometric complexity occurs for trivial coefficients. Here we must stress the phrase *in practice*, since we cannot prove that our technique will actually work. Nevertheless, the ideas in [19] have been successfully used in practice [7, 6], and the modifications presented here have been extensively tested for $F = \mathbb{Q}(\sqrt{2}), \mathbb{Q}(\sqrt{3})$.

The basic idea is the following. We first identify a finite topological model for $H^3(Y; \mathbb{C})$, the *Voronoĭ reduced* cocycles. This uses a generalization of Voronoĭ's reduction theory for positive definite quadratic forms [22, 1], which constructs a Γ-equivariant tessellation of X (§2.1). The Hecke operators do not act directly on this model, and to accommodate the Hecke translates of reduced cocycles we work with a larger model for the cohomology, the (infinite-dimensional) space $S_1(\Gamma)$ of 1-*sharblies* modulo Γ (§2.2). The space $S_1(\Gamma)$ is part of a homological complex $S_*(\Gamma)$ with Hecke action that naturally computes the cohomology of Y. Any Voronoĭ reduced cocycle in H^3 gives rise to a 1-sharbly cycle, which allows us to identify a finite dimensional subspace $S_1^{\mathrm{red}}(\Gamma) \subset S_1(\Gamma)$.

The main construction is then to take a general 1-sharbly cycle ξ and to modify it by subtracting an appropriate coboundary to obtain a homologous cycle ξ' that is closer to being Voronoĭ reduced (§3). By iterating this process, we eventually obtain a cycle that lies in our finite-dimensional subspace $S_1^{\mathrm{red}}(\Gamma)$. Unfortunately, we are unable to prove that at each step the output cycle ξ' is better than the input cycle ξ, in other words that it is somehow "more reduced." However, in practice this always works.

---

[2] The reader is probably more familiar with the case of $\mathbf{G}' = \mathrm{Res}_{F/\mathbb{Q}}\mathrm{SL}_2$. In this case the locally symmetric space is a Hilbert modular surface, and the cuspidal Hilbert modular forms contribute to H^2. Our symmetric space is slightly larger since the real rank of $\mathbf{G}$ is larger than that of $\mathbf{G}'$. However, regardless of whether one studies the Hilbert modular surface or our GL_2 symmetric space, the cusp forms contribute to the cohomology in degree one below the top nonvanishing degree.

The passage from ξ to ξ' is based on ideas found in [19], which describes an algorithm to compute the Hecke action on H^5 of congruence subgroups of $\mathrm{SL}_4(\mathbb{Z})$. The common feature that this case has with that of subgroups of $\mathrm{GL}_2(\mathcal{O})$ is that the cuspidal cohomology appears in the degree one less than the highest. This means that from our point of view the two cases are geometrically very similar. There are some complications, however, coming from the presence of non-torsion units in $\mathcal{O}$, complications leading to new phenomena requiring ideas not found in [19]. This is discussed in §4. We conclude the article by exhibiting the reduction of a 1-sharbly to a sum of Voronoï reduced 1-sharblies where the base field is $\mathbb{Q}(\sqrt{2})$ (§5).

We remark that there is another case sharing these same geometric features, namely that of subgroups of $\mathrm{GL}_2(\mathcal{O}_K)$, where K is a complex quartic field. We are currently applying the algorithm in joint work with F. Hajir and D. Ramakrishnan for $K = \mathbb{Q}(\zeta_5)$ to compute the cohomology of congruence subgroups of $\mathrm{GL}_2(\mathcal{O}_K)$ and to investigate the connections between automorphic cohomology and elliptic curves over K. Details of these cohomology computations, including some special features of the field K, will appear in [21]; the present paper focuses on the Hilbert modular case.

Finally, we remark that there is a rather different method to compute the Hecke action on Hilbert modular forms using the Jacquet–Langlands correspondence. For details we refer to work of L. Dembélé [17,16]. However, the Jacquet–Langlands technique works only with the complex cohomology of subgroups of $\mathrm{GL}_2(\mathcal{O})$, whereas our method in principle allows one to compute with torsion classes in the cohomology.

2   Background

Let F be a real quadratic field with class number 1. Let $\mathcal{O} \subset F$ denote the ring of integers. Let $\mathbf{G}$ be the $\mathbb{Q}$-group $\mathrm{Res}_{F/\mathbb{Q}}(\mathrm{GL}_2)$ and let $G = \mathbf{G}(\mathbb{R})$ the corresponding group of real points. Let $K \subset G$ be a maximal compact subgroup, and let A_G be the identity component of the maximal $\mathbb{Q}$-split torus in the center of G. Then the symmetric space associated to G is $X = G/KA_G$. Let $\Gamma \subseteq \mathrm{GL}_2(\mathcal{O})$ be a finite index subgroup.

In §2.1 we present an explicit model of X in terms of positive-definite binary quadratic forms over F and construct a $\mathrm{GL}_2(\mathcal{O})$-equivariant tessellation of X following [22,1]. Section 2.2 recalls the sharbly complex [23,11,19].

2.1   The Voronoï Polyhedron

Let ι_1, ι_2 be the two real embeddings of F into $\mathbb{R}$. These maps give an isomorphism $F \otimes_{\mathbb{Q}} \mathbb{R} \simeq \mathbb{R}^2$, and more generally, an isomorphism

$$G \xrightarrow{\sim} \mathrm{GL}_2(\mathbb{R}) \times \mathrm{GL}_2(\mathbb{R}). \tag{1}$$

When the meaning is clear from the context, we use ι_1, ι_2 to denote all such induced maps. In particular, (1) is the map

$$g \longmapsto (\iota_1(g), \iota_2(g)). \tag{2}$$

Under this identification, A_G corresponds to $\{(rI, rI) \mid r > 0\}$, where I is the 2×2 identity matrix.

Let C be the cone of real positive definite binary quadratic forms, viewed as a subset of V, the $\mathbb{R}$-vector space of 2×2 real symmetric matrices. The usual action of $\mathrm{GL}_2(\mathbb{R})$ on C is given by

$$(g \cdot \phi)(v) = \phi(^tgv), \quad \text{where } g \in \mathrm{GL}_2(\mathbb{R}) \text{ and } \phi \in C. \tag{3}$$

Equivalently, if A_ϕ is the symmetric matrix representing ϕ, then $g \cdot \phi = gA_\phi{}^tg$. In particular a coset $gO(2) \in \mathrm{GL}_2(\mathbb{R})/O(2)$ can be viewed as the positive definite quadratic form associated to the symmetric matrix $g\,^tg$.

Let $\mathcal{C} = C \times C$. Then (2) and (3) define an action of G on $\mathcal{C}$. Specifically, $g \cdot (\phi_1, \phi_2) = (\alpha_1, \alpha_2)$, where α_i is represented by $\iota_i(g)A_{\phi_i}\iota_i(^tg)$. Let ϕ_0 denote the quadratic form represented by the identity matrix. Then the stabilizer in G of (ϕ_0, ϕ_0) is a maximal compact subgroup K. The group A_G acts on $\mathcal{C}$ by positive real homotheties, and we have

$$X = \mathcal{C}/\mathbb{R}_{>0} = (C \times C)/\mathbb{R}_{>0} \simeq \mathfrak{H} \times \mathfrak{H} \times \mathbb{R},$$

where $\mathfrak{H}$ is the upper halfplane.

Let $\bar{\mathcal{C}}$ denote the closure of $\mathcal{C}$ in $V \times V$. Each vector $w \in \mathbb{R}^2$ gives a rank 1 positive semi-definite form $w\,^tw$ (here w is regarded as a column vector). Combined with ι_1 and ι_2, we get a map $L : \mathcal{O}^2 \to \bar{\mathcal{C}}$ given by

$$L(v) = \left(\iota_1(v) \cdot {}^t(\iota_1(v)), \iota_2(v) \cdot {}^t(\iota_2(v))\right). \tag{4}$$

Let $R(v)$ be the ray $\mathbb{R}_{>0} \cdot L(v) \subset \bar{\mathcal{C}}$. Note that

$$L(cv) = (\iota_1(c)^2 L_1(v), \iota_2(c)^2 L_2(v))$$

so that if $c \in \mathbb{Q}$, then $L(cv) \in R(v)$, and in particular $L(-v) = L(v)$. The set of *rational boundary components* $\mathcal{C}_1$ of $\mathcal{C}$ is the set of rays of the form $R(v)$, $v \in F^2$ [1]. These are the rays in $\bar{\mathcal{C}}$ that correspond to the usual cusps of the Hilbert modular variety.

Let $\Lambda \subset V \times V$ be the lattice

$$\Lambda = \left\{ (\iota_1(A), \iota_2(A)) \,\Big|\, A = \begin{bmatrix} a & c \\ c & b \end{bmatrix}, \quad a, b, c \in \mathcal{O} \right\}.$$

Then $\mathrm{GL}_2(\mathcal{O})$ preserves Λ.

Definition 1. *The* Voronoï polyhedron Π *is the closed convex hull in* $\bar{\mathcal{C}}$ *of the points* $\mathcal{C}_1 \cap \Lambda \smallsetminus \{0\}$.

Since F has class number 1, one can show that any vertex of Π has the form $L(v)$ for $v \in \mathcal{O}^2$. We say that $v \in \mathcal{O}^2$ is *primitive* if $L(v)$ is a vertex of Π. Note that v is primitive only if $L(v)$ is primitive in the usual sense as a lattice point in Λ.

By construction $GL_2(\mathcal{O})$ acts on Π. By taking the cones on the faces of Π, one obtains a Γ-*admissible decomposition* of $\mathcal{C}$ for $\Gamma = GL_2(\mathcal{O})$ [1]. Essentially this means that the cones form a fan in $\bar{\mathcal{C}}$ and that there are finitely many cones modulo the action of $GL_2(\mathcal{O})$. Since the action of $GL_2(\mathcal{O})$ commutes with the homotheties, this decomposition descends to a $GL_2(\mathcal{O})$-equivariant tessellation of X.[3]

We call this decomposition the *Voronoĭ decomposition*. We call the cones defined by the faces of Π *Voronoĭ cones*, and we refer to the cones corresponding to the facets of Π as *top cones*. The sets $\sigma \cap \mathcal{C}$, as σ ranges over all top cones, cover $\mathcal{C}$. Given a point $\phi \in \mathcal{C}$, there is a finite algorithm that computes which Voronoĭ cone contains ϕ [20].

For some explicit examples of the Voronoĭ decomposition over real quadratic fields, we refer to [26] (see also §5).

2.2   The Sharbly Complex

Let S_k, $k \geq 0$, be the Γ-module A_k/C_k, where A_k is the set of formal $\mathbb{Z}$-linear sums of symbols $[v] = [v_1, \cdots, v_{k+2}]$, where each v_i is in F^2, and C_k is the submodule generated by

1. $[v_{\sigma(1)}, \cdots, v_{\sigma(k+2)}] - \text{sgn}(\sigma)[v_1, \cdots, v_{k+2}]$,
2. $[v, v_2, \cdots, v_{k+2}] - [w, v_2, \cdots v_{k+2}]$ if $R(v) = R(w)$, and
3. $[v]$, if v is *degenerate*, i.e., if $v_1, \cdots, v_{k+2}$ are contained in a hyperplane.

We define a boundary map $\partial \colon S_{k+1} \to S_k$ by

$$\partial[v_1, \cdots, v_{k+2}] = \sum_{i=1}^{k+2} (-1)^i [v_1, \cdots, \hat{v}_i, \cdots, v_{k+2}]. \qquad (5)$$

This makes S_* into a homological complex, called the *sharbly complex* [4].

The basis elements $\mathbf{u} = [v_1, \cdots, v_{k+2}]$ are called k-*sharblies*. Notice that in our class number 1 setting, using the relations in C_k one can always find a representative for $\mathbf{u}$ with the v_i primitive. In particular, one can always arrange that each $L(v_i)$ is a vertex of Π. When such a representative is chosen, the v_i are unique up to multiplication by ± 1. In this case the v_i—or by abuse of notation the $L(v_i)$—are called the *spanning vectors* for $\mathbf{u}$.

Definition 2. *A sharbly is* Voronoĭ reduced *if its spanning vectors are a subset of the vertices of a Voronoĭ cone.*

The geometric meaning of this notion is the following. Each sharbly $\mathbf{u}$ with spanning vectors v_i determines a closed cone $\sigma(\mathbf{u})$ in $\bar{\mathcal{C}}$, by taking the cone generated by the points $L(v_i)$. Then $\mathbf{u}$ is reduced if and only if $\sigma(\mathbf{u})$ is contained in some Voronoĭ cone. It is clear that there are finitely many Voronoĭ reduced sharblies modulo Γ.

---

[3] If one applies this construction to $F = \mathbb{Q}$, one obtains the Farey tessellation of $\mathfrak{H}$, with tiles given by the $SL_2(\mathbb{Z})$-orbit of the ideal geodesic triangle with vertices at $0, 1, \infty$.

Using determinants, we can define a notion of size for 0-sharblies:

Definition 3. *Given a* 0-*sharbly* **v***, the size* Size(**v**) *of* **v** *is given by the absolute value of the norm determinant of the* 2 × 2 *matrix formed by spanning vectors for* **v***.*

By construction Size takes values in $\mathbb{Z}_{>0}$. We remark that the size of a 0-sharbly **v** is related to whether or not **v** is Voronoĭ reduced, but that in general there exist Voronoĭ reduced 0-sharblies with size > 1.

The boundary map (5) commutes with the action of Γ, and we let $S_*(\Gamma)$ be the homological complex of coinvariants. Note that $S_*(\Gamma)$ is infinitely generated as a $\mathbb{Z}\Gamma$-module. One can show

$$H_k((S_* \otimes \mathbb{C})(\Gamma)) \xrightarrow{\sim} H^{4-k}(\Gamma; \mathbb{C}) \tag{6}$$

(cf. [4]), with a similar result holding for cohomology with nontrivial coefficients. Moreover, there is a natural action of the Hecke operators on $S_*(\Gamma)$ (cf. [19]). Thus to compute with $H^3(\Gamma; \mathbb{C})$, which will realize cuspidal Hilbert modular forms over F of weight $(2,2)$, we work with 1-sharbly cycles. We note that the Voronoĭ reduced sharblies form a *finitely generated* subcomplex of $S_*(\Gamma)$ that also computes the cohomology of Γ as in (6). This is our finite model for the cohomology of Γ.

3 The Reduction Algorithm

3.1 The Strategy

The general idea behind our algorithm is simple. To compute the action of a Hecke operator on the space of 1-sharbly cycles, it suffices to describe an algorithm that writes a general 1-sharbly cycle as a sum of Voronoĭ reduced 1-sharblies. Now any basis 1-sharbly **u** contains three sub-0-sharblies (the *edges* of **u**), and the Voronoĭ reduced 1-sharblies tend to have edges of small size. Thus our first goal is to systematically replace all the 1-sharblies in a cycle with edges of large size with 1-sharblies having smaller size edges. This uses a variation of the classical modular symbol algorithm, although no continued fractions are involved. Eventually we produce a sum of 1-sharblies with all edges Voronoĭ reduced. However, having all three edges Voronoĭ reduced is (unfortunately) not a sufficient condition for a 1-sharbly to be Voronoĭ reduced.[4] Thus a different approach must be taken for such 1-sharblies to finally make the cycle Voronoĭ reduced. This is discussed further in §4.

3.2 Lifts

We begin by describing one technique to encode a 1-sharbly cycle using some mild extra data, namely that of a choice of *lifts* for its edges:

---

[4] This is quite different from what happens with classical modular symbols, and reflects the infinite units in $\mathcal{O}$.

Definition 4 ([19]). *A* 2×2 *matrix M with coefficients of F with columns A_1, A_2 is said to be a* lift *of a* 0-*sharbly* $[u, v]$ *if* $\{R(A_1), R(A_2)\} = \{R(u), R(v)\}$.

The idea behind the use of lifts is the following. Suppose a linear combination of 1-sharblies $\xi = \sum a(\mathbf{u})\mathbf{u} \in S_1$ becomes a cycle in $S_1(\Gamma)$. Then its boundary must vanish modulo Γ. In the following algorithm, we attempt to pass from ξ to a "more reduced" sharbly ξ' by modifying the edges of each $\mathbf{u}$ in the support of ξ. To guarantee that ξ' is a cycle modulo Γ, we must make various choices in the course of the reduction Γ-equivariantly across the boundary of ξ. This can be done by first choosing 2×2 integral matrices for each sub-0-sharbly of ξ. We refer to [19] for more details and discussion. For the present exposition, we merely remark that we always view a 1-sharbly $\mathbf{u} = [v_1, v_2, v_3]$ as a triangle with vertices labelled by the v_i and with a given (fixed) choice of lifts for each edge (Figure 1). If two edges $\mathbf{v}, \mathbf{v}'$ satisfy $\gamma \cdot \mathbf{v} = \mathbf{v}'$, then we choose the corresponding lifts to satisfy $\gamma M = M'$. The point is that we can then work individually with 1-sharblies enriched with lifts; we don't have to know explicitly the matrices in Γ that glue the 1-sharblies into a cycle modulo Γ.

We emphasize that the lift matrices for any given 1-sharbly in the support of ξ are essentially forced on us by the requirement that ξ be a cycle modulo Γ. There is almost no flexibility in choosing them. Such matrices form an essential part of the input data for our algorithm.

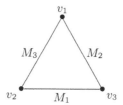

Fig. 1. A 1-sharbly with lifts

3.3 Reducing Points

Definition 5. *Let* $\mathbf{v}$ *be a* 0-*sharbly with spanning vectors* $\{x, y\}$. *Assume* $\mathbf{v}$ *is not Voronoï reduced. Then* $u \in \mathcal{O}^2 \smallsetminus \{0\}$ *is a* reducing point *for* $\mathbf{v}$ *if the following hold:*

1. $R(u) \neq R(x), R(y)$.
2. $L(u)$ *is a vertex of the unique Voronoï cone* σ *(not necessarily top-dimensional) containing the ray* $R(x + y)$.
3. *If* $x = ty$ *for some* $t \in F^\times$, *then* $R(u)$ *lies in the cone spanned by* $R(x)$ *and* $R(y)$.
4. *Of the vertices of* σ, *the point* u *minimizes the sum of the sizes of the* 0-*sharblies* $[x, u]$ *and* $[u, y]$.

Given a non-Voronoĭ reduced 0-sharbly $\mathbf{v} = [x, y]$ and a reducing point u, we apply the relation

$$[x, y] = [x, u] + [u, y] \tag{7}$$

in the hopes that the two new 0-sharblies created are closer to being Voronoĭ reduced. Note that choosing u uses the geometry of the Voronoĭ decomposition instead of (a variation of) the continued fraction algorithms of [25, 14, 10]. Unfortunately we cannot guarantee that the new 0-sharblies on the right of (7) are better than $\mathbf{v}$, but this is true in practice.

3.4 Γ-Invariance

The reduction algorithm proceeds by picking reducing points for non-Voronoĭ reduced edges. We want to make sure that this is done Γ-equivariantly; in other words that if two edges $\mathbf{v}$, $\mathbf{v}'$ satisfy $\gamma \cdot \mathbf{v} = \mathbf{v}'$, then if we choose u for $\mathbf{v}$ we want to make sure that we choose γu for $\mathbf{v}'$.

We achieve this by making sure that the choice of reducing point for $\mathbf{v}$ only depends on the lift matrix M that labels $\mathbf{v}$. The matrix is first put into *normal form*, which is a unique representative M_0 of the coset $\mathrm{GL}_2(\mathcal{O})\backslash M$. This is an analogue of Hermite normal form that incorporates the action of the units of $\mathcal{O}$. There is a unique 0-sharbly associated to M_0. We choose a reducing point u for this 0-sharbly and translate it back to obtain a reducing point for $\mathbf{v}$. Note that u need not be unique. However we can always make sure that the same u is chosen any time a given normal form M_0 is encountered, for instance by choosing representatives of the Voronoĭ cones modulo $\mathrm{GL}_2(\mathcal{O})$ and then fixing an ordering of their vertices.

We now describe how M_0 is constructed from M. Let Ω_* be a fundamental domain for the action of $(\mathcal{O}^\times, \cdot)$ on $F^\times$. For $t \in \mathcal{O}$, let $\Omega_+(t)$ be a fundamental domain for the action of $(t\mathcal{O}, +)$ on F.

Definition 6. *A nonzero matrix $M \in \mathrm{Mat}_2(F)$ is in* normal form *if M has one of the following forms:*

1. $\begin{bmatrix} 0 & b \\ 0 & 0 \end{bmatrix}$, *where $b \in \Omega_*$.*

2. $\begin{bmatrix} a & b \\ 0 & 0 \end{bmatrix}$, *where $a \in \Omega_*$ and $b \in F$.*

3. $\begin{bmatrix} a & b \\ 0 & d \end{bmatrix}$, *where $a, d \in \Omega_*$ and $b \in \Omega_+(d)$.*

It is easy to check that the normal form for M is uniquely determined in the coset $\mathrm{GL}_2(\mathcal{O}) \cdot M$.

To explicitly put $M = \begin{bmatrix} a & b \\ c & d \end{bmatrix}$ in normal form, the first step is to find $\gamma \in \mathrm{GL}_2(\mathcal{O})$ such that $\gamma \cdot M$ is upper triangular. Such a γ can be found after finitely many computations as follows. Let $N \colon F \to \mathbb{R}$ be defined by $N(\alpha) = |\mathrm{Norm}_{F/\mathbb{Q}}(\alpha)|$. If

$$0 < N(c) < N(a),$$

then let $\alpha \in \mathcal{O}$ be an element of smallest distance from a/c. Let

$$\gamma' = \begin{bmatrix} 0 & 1 \\ 1 & 0 \end{bmatrix} \begin{bmatrix} 1 & -\alpha \\ 0 & 1 \end{bmatrix}.$$

Then $\gamma' \in \mathrm{GL}_2(\mathcal{O})$ and $\gamma'M = \begin{bmatrix} a' & b' \\ c' & d' \end{bmatrix}$ with $N(c') < N(c)$ and $N(a') < N(a)$. Repeating this procedure will yield the desired result.

After a reducing point is selected for $\mathbf{v}$ and the relation (7) is applied, we must choose lifts for the 0-sharblies on the right of (7). This we do as follows:

Definition 7. *Let $[v_1, v_2]$ be a non-reduced 0-sharbly with lift matrix M and reducing point u. Then the inherited lift $\hat{M}_i$ for $[v_i, u]$ is the matrix obtained from M by keeping the column corresponding to v_i and replacing the other column by u.*

3.5 The Algorithm

Let $T = [v_1, v_2, v_3]$ be a non-degenerate sharbly. Let M_i be the lifts of the edges of T as shown in Figure 1. The method of subdividing the interior depends on the number of edges that are Voronoï reduced. After each subdivision, lift data is attached using inherited lifts for the exterior edges. The lift for each interior edge can be chosen arbitrarily as long as the same choice is made for the edge to which it is glued. We note that steps (I), (II), and (III.1) already appear in [19], but (III.2) and (IV) are new subdivisions needed to deal with the complications of the units of $\mathcal{O}$.

(I) Three Non-reduced Edges. If none of the edges are Voronoï reduced, then we subdivide each edge by choosing reducing points u_1, u_2, and u_3. In addition, form three additional edges $[u_1, u_2]$, $[u_2, u_3]$, and $[u_3, u_1]$. We then replace T by the four 1-sharblies

$$[v_1, v_2, v_3] \longmapsto [v_1, u_3, u_2] + [u_3, v_2, u_1] + [u_2, u_1, v_3] + [u_1, u_2, u_3]. \qquad (8)$$

(II) Two Non-reduced Edges. If only one edge is Voronoï reduced, then we subdivide the other two edges by choosing reducing points u_1 and u_3. We form two additional edges $[u_1, u_3]$ and ℓ, where ℓ is taken to be either $[v_1, u_1]$ or $[v_3, u_3]$, whichever has smaller size. More precisely:

1. If $\mathrm{Size}([v_1, u_1]) \leq \mathrm{Size}([u_3, v_3])$, then we form two additional edges $[u_1, u_3]$ and $[v_1, u_1]$, and replace T by the three 1-sharblies

$$[v_1, v_2, v_3] \longmapsto [v_1, u_3, u_1] + [u_3, v_2, u_1] + [v_1, u_1, v_3]. \qquad (9)$$

2. Otherwise, we form two additional edges $[u_1, u_3]$ and $[v_3, u_3]$, and replace T by the three 1-sharblies

$$[v_1, v_2, v_3] \longmapsto [v_1, u_3, v_3] + [u_3, v_2, u_1] + [u_3, u_1, v_3]. \qquad (10)$$

(III) One Non-reduced Edge. If two edges are Voronoï reduced, then we subdivide the other edge by choosing a reducing point u_1. The next step depends on the configuration of $\{v_1, v_2, v_3, u_1\}$.

1. If $[v_2, u_1]$ or $[u_1, v_3]$ is not Voronoï reduced or $v_2 = tv_1$ for some $t \in F^\times$, then we form one additional edge $[v_1, u_1]$ and replace T by the two 1-sharblies

$$[v_1, v_2, v_3] \longmapsto [v_1, v_2, u_1] + [v_1, u_1, v_3]. \tag{11}$$

2. Otherwise, a central point w is chosen. The central point w is chosen from the vertices of the top cone containing the barycenter of $[v_1, v_2, v_3, w]$ so that it maximizes the number of Voronoï reduced edges in the set

$$S = \{[v_1, w], [v_2, w], [v_3, w], [u_1, w]\}.$$

We do not allow v_1, v_2 or v_3 to be chosen as a central point. We form four additional edges $[v_1, w], [v_2, w], [u_1, w]$, and $[v_3, w]$ and replace T by the four 1-sharblies

$$[v_1, v_2, v_3] \longmapsto [v_1, v_2, w] + [w, v_2, u_1] + [w, u_1, v_3] + [w, v_3, v_1]. \tag{12}$$

(IV) All Edges Voronoï Reduced. If all three edges are Voronoï reduced, but T is not Voronoï reduced, then a central point w is chosen. The central point w is chosen from the vertices of the top cone containing the barycenter of $[v_1, v_2, v_3]$ so that it maximizes the sum $\#E + \#P$, where E is the set of Voronoï reduced edges in $\{[v_1, w], [v_2, w], [v_3, w]\}$ and P is the set of Voronoï reduced triangles in $\{[v_1, v_2, w], [v_2, v_3, w], [v_3, v_1, w]\}$. We do not allow v_1, v_2 or v_3 to be chosen as a central point. We form three additional edges $[v_1, w], [v_2, w]$, and $[v_3, w]$ and replace T by the three 1-sharblies

$$[v_1, v_2, v_3] \longmapsto [v_1, v_2, w] + [w, v_2, v_3] + [w, v_3, v_1]. \tag{13}$$

4   Comments

First, the transformations (8)–(13) do not follow from the relations in the sharbly complex. Rather they only make sense in the complex of coinvariants when applied to an entire 1-sharbly cycle ξ that has been locally encoded by lifts for the edges, and where the reducing points have been chosen Γ-equivariantly. More discussion of this point, as well as pictures illustrating some of the transformations, can be found in [19, §4.5].

Next, we emphasize that the reducing point u of Definition 5 works in practice to shrink the size of a 0-sharbly $\mathbf{v}$, but we have no proof that it will do so. The difficulty is that Definition 5 chooses u using the geometry of the Voronoï polyhedron Π and not the size of $\mathbf{v}$ directly. Moreover, our experience with examples shows that this use of the structure of Π is essential to reduce the original 1-sharbly cycle (cf. §5.2).

As mentioned in §3.1, case (IV) is necessary: there are 1-sharblies T with all three edges Voronoï reduced, yet T is itself not Voronoï reduced. An example is given in the next section. The point is that in $\bar{C}$ the points $L(v)$ and $L(\varepsilon v)$ are different if ε is not a torsion unit, but after passing to the Hilbert modular surface $L(v)$ and $L(\varepsilon v)$ define the same cusp. This means one can take a geodesic triangle Δ in the Hilbert modular surface with vertices at three cusps that by any measure should be considered reduced, and can lift Δ to a 3-cone in the GL_2-symmetric space that is far from being Voronoï reduced.

Finally, the reduction algorithm can be viewed as a two stage process. When a 1-sharbly T has 2 or 3 non-reduced edges or 1 non-reduced edge and satisfies the criteria for case 1, then in some sense T is "far" from being Voronoï reduced. One tries to replace T by a sum of 1-sharblies that are more reduced in that the edges have smaller size. However, this process will not terminate in Voronoï reduced sharblies. In particular, if T is "close" to being Voronoï reduced, then one must use the geometry of the Voronoï cones more heavily. This is why we need the extra central point w in (III.2) and (IV).

For instance, suppose $T = [v_1, v_2, v_3]$ is a 1-sharbly with 1 non-reduced edge such that the criteria for (III.2) are satisfied when the reducing point is chosen. One can view choosing the central point and doing the additional subdivision as first moving the bad edge to the interior of the triangle, where the choices of reducing points no longer need to be Γ-invariant. The additional freedom allows one to make a better choice. Indeed, without the central point chosen wisely, this does lead to some problems. In particular, there are examples where $[v_1, u_1]$ is not Voronoï reduced, and the choice of the reducing point for this edge is v_2, leading to a repeating behavior.

5   The Field $F = \mathbb{Q}(\sqrt{2})$

5.1   The Voronoï Polyhedron

Let $F = \mathbb{Q}(\sqrt{2})$ and let $\varepsilon = 1 + \sqrt{2}$, a fundamental unit of norm -1. Computations of H. Ong [26, Theorem 4.1.1] with positive definite binary quadratic forms over F allow us to describe the Voronoï polyhedron Π and thus the Voronoï decomposition of $\mathcal{C}$:

Proposition 1 ([26, Theorem 4.1.1]). *Modulo the action of* $GL_2(\mathcal{O})$, *there are two inequivalent top Voronoï cones. The corresponding facets of* Π *have 6 and 12 vertices, respectively.*

We fix once and for all representative 6-dimensional cones A_0 and A_1. To describe these cones, we give sets of points $S \subset \mathcal{O}^2$ such that the points $\{L(v) \mid v \in S\}$ are the vertices of the corresponding face of Π. Let e_1, e_2 be the canonical basis of $\mathcal{O}^2$. Then we can take A_0 to correspond to the 6 points

$$e_1, e_2, e_1 - e_2, \bar{\varepsilon}e_1, \bar{\varepsilon}e_2, \bar{\varepsilon}(e_1 - e_2),$$

and A_0 to correspond to the 12 points

$$e_1, e_2, \bar{\varepsilon}e_1, \bar{\varepsilon}e_2, e_1 - e_2, e_1 + \bar{\varepsilon}e_2, e_2 + \bar{\varepsilon}e_1, \bar{\varepsilon}(e_1 + e_2), \alpha, \beta, \bar{\varepsilon}\alpha, \bar{\varepsilon}\beta,$$

where $\alpha = e_1 - \sqrt{2}e_2$, $\beta = e_2 - \sqrt{2}e_1$. Since A_1 is not a simplicial cone, there exist basis sharblies that are Voronoĭ reduced but do not correspond to Voronoĭ cones.

Now we consider cones of lower dimension. Modulo $GL_2(\mathcal{O})$, every 2-dimensional Voronoĭ cone either lies in $\bar{C} \setminus C$ or is equivalent to the cone corresponding to $\{e_1, e_2\}$. The $GL_2(\mathcal{O})$-orbits of 3-dimensional Voronoĭ cones are represented by $\{e_1, e_2\} \cup U$, where U ranges over

$$\{e_1 - e_2\}, \{\bar{\varepsilon}e_1\}, \{\bar{\varepsilon}(e_1 - e_2)\}, \{e_1 - \sqrt{2}e_2, e_2 - \sqrt{2}e_1\}, \{e_1 + \bar{\varepsilon}e_2\}.$$

Note that all but one of the 3-cones are simplicial.

5.2   Reducing 1-Sharblies

Now we consider reducing a 1-sharbly T. Let us represent T by a 2×3 matrix whose columns are the spanning vectors of T. We take T to be

$$T = \begin{bmatrix} \sqrt{2} + 3 & 4\sqrt{2} + 4 & 3\sqrt{2} - 4 \\ \sqrt{2} & 5\sqrt{2} - 1 & -3\sqrt{2} - 5 \end{bmatrix},$$

and we choose arbitrary initial lifts for the edges of T. This data is typical of what one encounters when trying to reduce a 1-sharbly cycle modulo Γ.

The input 1-sharbly T has 3 non-reduced edges with edge sizes given by the vector $[5299, 529, 199]$. The first pass of the algorithm follows (I) and splits all 3 edges, replacing T by the sum $S_1 + S_2 + S_3 + S_4$, where

$$S_1 = \begin{bmatrix} \sqrt{2} + 3 & -\sqrt{2} - 1 & 1 \\ \sqrt{2} & -\sqrt{2} & 0 \end{bmatrix}, \quad S_2 = \begin{bmatrix} 4\sqrt{2} + 4 & 0 & -\sqrt{2} - 1 \\ 5\sqrt{2} - 1 & -\sqrt{2} - 1 & -\sqrt{2} \end{bmatrix},$$

$$S_3 = \begin{bmatrix} 3\sqrt{2} - 4 & 1 & 0 \\ -3\sqrt{2} - 5 & 0 & -\sqrt{2} - 1 \end{bmatrix}, \quad S_4 = \begin{bmatrix} 0 & 1 & -\sqrt{2} - 1 \\ -\sqrt{2} - 1 & 0 & -\sqrt{2} \end{bmatrix}.$$

We compute that $\mathrm{Size}(S_1) = [2, 2, 8]$, $\mathrm{Size}(S_2) = [1, 1, 16]$, $\mathrm{Size}(S_3) = [1, 2, 7]$, and $\mathrm{Size}(S_4) = [2, 1, 1]$. Notice that the algorithm replaces T by a sum of sharblies with edges of significantly smaller size. This kind of performance is typical, and looks similar to the performance of the usual continued fraction algorithm over $\mathbb{Z}$. Note also that S_4, which is the 1-sharbly spanned by the three reducing points of the edges T, also has edges of very small size. This reflects our use of Definition 5 to choose the reducing points; choosing them without using the geometry of Π often leads to bad performance in the construction of this 1-sharbly.

Now S_4 has 3 Voronoĭ reduced edges, but is itself not Voronoĭ reduced. The algorithm follows (IV), replaces S_4 by $R_1 + R_2 + R_3$, and now each R_i is Voronoĭ reduced.

The remaining 1-sharblies S_1, S_2, and S_3 have only 1 non-reduced edge. They are almost reduced in the sense that they satisfy the criteria for (III.2). The algorithm replaces S_1 by $O_1 + O_2 + O_3 + O_4$, where O_1 and O_2 are degenerate and O_3 and O_4 are Voronoï reduced. The 1-sharbly S_2 is replaced by a $P_1 + P_2 + P_3 + P_4$, and each P_i is Voronoï reduced. S_3 is replaced by $Q_1 + Q_2 + Q_3 + Q_4$, where Q_1 and Q_2 are degenerate, Q_3 is Voronoï reduced, and Q_4 is not Voronoï reduced. This 1-sharbly is given by

$$Q_4 = \begin{bmatrix} -\sqrt{2}+1 & 0 & 3\sqrt{2}-4 \\ 2\sqrt{2}+3 & -\sqrt{2}-1 & -3\sqrt{2}-5 \end{bmatrix}$$

and has 3 Voronoï reduced edges. Once again the algorithm is in case (IV), and replaces Q_4 by a sum $N_1 + N_2 + N_3$ of Voronoï reduced sharblies.

To summarize, the final output of the reduction algorithm applied to T is a sum

$$N_1 + N_2 + N_3 + O_3 + O_4 + P_1 + P_2 + P_3 + P_4 + Q_3 + R_1 + R_2 + R_3, \quad \text{where}$$

$$N_1 = \begin{bmatrix} -\sqrt{2}+1 & 0 & 0 \\ 2\sqrt{2}+3 & -\sqrt{2}-1 & -2\sqrt{2}-3 \end{bmatrix},$$

$$N_2 = \begin{bmatrix} 0 & 3\sqrt{2}-4 & 0 \\ -\sqrt{2}-1 & -3\sqrt{2}-5 & -2\sqrt{2}-3 \end{bmatrix},$$

$$N_3 = \begin{bmatrix} 3\sqrt{2}-4 & -\sqrt{2}+1 & 0 \\ -3\sqrt{2}-5 & 2\sqrt{2}+3 & -2\sqrt{2}-3 \end{bmatrix}, \quad O_3 = \begin{bmatrix} -\sqrt{2}-1 & -\sqrt{2}-1 & 1 \\ -1 & -\sqrt{2} & 0 \end{bmatrix},$$

$$O_4 = \begin{bmatrix} -\sqrt{2}-1 & 1 & \sqrt{2}+3 \\ -1 & 0 & \sqrt{2} \end{bmatrix}, \quad P_1 = \begin{bmatrix} 2\sqrt{2}+3 & 4\sqrt{2}+4 & 1 \\ \sqrt{2}+2 & 5\sqrt{2}-1 & -2\sqrt{2}+2 \end{bmatrix},$$

$$P_2 = \begin{bmatrix} 2\sqrt{2}+3 & 1 & 0 \\ \sqrt{2}+2 & -2\sqrt{2}+2 & -\sqrt{2}-1 \end{bmatrix}, \quad P_3 = \begin{bmatrix} 2\sqrt{2}+3 & 0 & -\sqrt{2}-1 \\ \sqrt{2}+2 & -\sqrt{2}-1 & -\sqrt{2} \end{bmatrix},$$

$$P_4 = \begin{bmatrix} 2\sqrt{2}+3 & -\sqrt{2}-1 & 4\sqrt{2}+4 \\ \sqrt{2}+2 & -\sqrt{2} & 5\sqrt{2}-1 \end{bmatrix}, \quad Q_3 = \begin{bmatrix} -\sqrt{2}+1 & 1 & 0 \\ 2\sqrt{2}+3 & 0 & -\sqrt{2}-1 \end{bmatrix},$$

$$R_1 = \begin{bmatrix} -\sqrt{2}+1 & 0 & 0 \\ 2\sqrt{2}+3 & -\sqrt{2}-1 & -2\sqrt{2}-3 \end{bmatrix},$$

$$R_2 = \begin{bmatrix} 0 & 3\sqrt{2}-4 & 0 \\ -\sqrt{2}-1 & -3\sqrt{2}-5 & -2\sqrt{2}-3 \end{bmatrix}, \quad \text{and}$$

$$R_3 = \begin{bmatrix} 3\sqrt{2}-4 & -\sqrt{2}+1 & 0 \\ -3\sqrt{2}-5 & 2\sqrt{2}+3 & -2\sqrt{2}-3 \end{bmatrix},$$

and each of the above is Voronoï reduced. Some of these 1-sharblies correspond to Voronoï cones and some don't. In particular, one can check that the spanning vectors for P_3, P_4, R_1, and N_1 do form Voronoï cones, and all others don't. However, the spanning vectors of O_3 and O_4 almost do, in the sense that they are subsets of 3-dimensional Voronoï cones with four vertices.

References

1. Ash, A.: Deformation retracts with lowest possible dimension of arithmetic quotients of self-adjoint homogeneous cones. Math. Ann. 225, 69–76 (1977)
2. Ash, A.: A note on minimal modular symbols. Proc. Amer. Math. Soc. 96(3), 394–396 (1986)
3. Ash, A.: Nonminimal modular symbols for GL(n). Invent. Math. 91(3), 483–491 (1988)
4. Ash, A.: Unstable cohomology of $SL(n, \mathcal{O})$. J. Algebra 167(2), 330–342 (1994)
5. Ash, A., Grayson, D., Green, P.: Computations of cuspidal cohomology of congruence subgroups of $SL_3(\mathbf{Z})$. J. Number Theory 19, 412–436 (1984)
6. Ash, A., Gunnells, P.E., McConnell, M.: Cohomology of congruence subgroups of $SL_4(\mathbf{Z})$ II. J. Number Theory (submitted)
7. Ash, A., Gunnells, P.E., McConnell, M.: Cohomology of congruence subgroups of $SL_4(\mathbf{Z})$. J. Number Theory 94, 181–212 (2002)
8. Ash, A., McConnell, M.: Experimental indications of three-dimensional Galois representations from the cohomology of $SL(3, \mathbf{Z})$. Experiment. Math. 1(3), 209–223 (1992)
9. Ash, A., Pinch, R., Taylor, R.: An $\widehat{A_4}$ extension of $\mathbf{Q}$ attached to a non-selfdual automorphic form on $GL(3)$. Math. Ann. 291, 753–766 (1991)
10. Ash, A., Rudolph, L.: The modular symbol and continued fractions in higher dimensions. Invent. Math. 55, 241–250 (1979)
11. Ash, A.: Unstable cohomology of SL($n, \mathcal{O}$). J. Algebra 167(2), 330–342 (1994)
12. Bygott, J.: Modular forms and modular symbols over imaginary quadratic fields. PhD thesis, Exeter (1999)
13. Cremona, J.E.: Hyperbolic tessellations, modular symbols, and elliptic curves over complex quadratic fields. Compositio Math. 51(3), 275–324 (1984)
14. Cremona, J.E.: Algorithms for modular elliptic curves, 2nd edn. Cambridge University Press, Cambridge (1997)
15. Cremona, J.E., Whitley, E.: Periods of cusp forms and elliptic curves over imaginary quadratic fields. Math. Comp. 62(205), 407–429 (1994)
16. Dembélé, L.: Explicit computations of Hilbert modular forms on $\mathbf{Q}(\sqrt{5})$. Experiment. Math. 14(4), 457–466 (2005)
17. Dembélé, L.: Quaternionic Manin symbols, Brandt matrices, and Hilbert modular forms. Math. Comp. 76, 1039–1057 (2007)
18. Franke, J.: Harmonic analysis in weighted L_2-spaces. Ann. Sci. École Norm. Sup. 31(4), 181–279 (1998)
19. Gunnells, P.E.: Computing Hecke eigenvalues below the cohomological dimension. Experiment. Math. 9(3), 351–367 (2000)
20. Gunnells, P.E.: Modular symbols for $\mathbf{Q}$-rank one groups and Voronoïreduction. J. Number Theory 75(2), 198–219 (1999)
21. Gunnells, P.E., Yasaki, D.: Computing Hecke operators on modular forms over real quadratic and complex quartic fields (in preparation)
22. Koecher, M.: Beiträge zu einer Reduktionstheorie in Positivitätsbereichen I. Math. Ann. 141, 384–432 (1960)
23. Lee, R., Szczarba, R.H.: On the homology and cohomology of congruence subgroups. Invent. Math. 33(1), 15–53 (1976)
24. Lingham, M.: Modular Forms and Elliptic Curves over Imaginary Quadratic Fields. Ph.D. thesis, Nottingham (2005)

25. Manin, Y.-I.: Parabolic points and zeta-functions of modular curves. Math. USSR Izvestija 6(1), 19–63 (1972)
26. Ong, H.E.: Perfect quadratic forms over real-quadratic number fields. Geom. Dedicata. 20(1), 51–77 (1986)
27. Stein, W.: Modular forms, a computational approach. In: Graduate Studies in Mathematics, vol. 79, American Mathematical Society, Providence (2007); With an appendix by Gunnells, P.E.
28. van Geemen, B., van der Kallen, W., Top, J., Verberkmoes, A.: Hecke eigenforms in the cohomology of congruence subgroups of $SL(3, \mathbf{Z})$. Experiment. Math. 6(2), 163–174 (1997)

A Birthday Paradox for Markov Chains, with an Optimal Bound for Collision in the Pollard Rho Algorithm for Discrete Logarithm

Jeong Han Kim[1,*], Ravi Montenegro[2], Yuval Peres[3,**], and Prasad Tetali[4,***]

[1] Department of Mathematics, Yonsei University, Seoul, 120-749 Korea
jehkim@yonsei.ac.kr
[2] Department of Mathematical Sciences, University of Massachusetts at Lowell,
Lowell, MA 01854
ravi_montenegro@uml.edu
[3] Microsoft Research, Redmond and University of California, Berkeley, CA 94720
peres@microsoft.com
[4] School of Mathematics and School of Computer Science,
Georgia Institute of Technology, Atlanta, GA 30332
tetali@math.gatech.edu

Abstract. We show a Birthday Paradox for self-intersections of Markov chains with uniform stationary distribution. As an application, we analyze Pollard's Rho algorithm for finding the discrete logarithm in a cyclic group G and find that, if the partition in the algorithm is given by a random oracle, then with high probability a collision occurs in $\Theta(\sqrt{|G|})$ steps. This is the first proof of the correct bound which does not assume that every step of the algorithm produces an i.i.d. sample from G.

Keywords: Birthday Paradox, Pollard Rho, Discrete Logarithm, self intersection, collision time.

1 Introduction

The Birthday Paradox states that if $C\sqrt{N}$ items are sampled uniformly at random, with replacement, from a set of N items, then for large C, with high probability some item will be chosen twice. This can be interpreted as a statement that with high probability, a Markov chain on the complete graph K_N with transitions $P(i,j) = 1/N$ will intersect its past in $C\sqrt{N}$ steps; we refer to such a self-intersection as a *collision*, and say the *"collision time"* is $O(\sqrt{N})$. In [7], this was generalized: for a general Markov chain, the collision time was bounded by $O(\sqrt{N}\,T_s(1/2))$, where $T_s(\epsilon) = \min\{n : \forall u, v, \ P^n(u,v) \geq (1-\epsilon)\pi(v)\}$ measures the time required for the n-step distribution to assign every state a suitable

* Research supported by the Korea Science and Engineering Foundation (KOSEF) grant funded by the Korea government(MOST) (No. R16-2007-075-01000-0).
** Research supported in part by NSF grant DMS-0605166.
*** Research supported in part by NSF grants DMS 0401239, 0701043.

A.J. van der Poorten and A. Stein (Eds.): ANTS-VIII 2008, LNCS 5011, pp. 402–415, 2008.

multiple of its stationary probability. In [5], the bound on collision time was improved to $O(\sqrt{N}\,T_s(1/2))$.

The motivation of [7,5] was to study the collision time for a Markov chain involved in Pollard's Rho algorithm for finding the discrete logarithm on a cyclic group G of prime order $N = |G| \neq 2$. For this walk $T_s(1/2) = \Omega(\log N)$ and so the results of [7,5] are insufficient to show the widely believed $\Theta(\sqrt{N})$ collision time for this walk. In this paper we improve upon these bounds and show that if a finite ergodic Markov chain has uniform stationary distribution over N states, then $O(\sqrt{N})$ steps suffice for a collision to occur, as long as the relative-pointwise distance (L_∞ of the densities of the current and the stationary distribution) drops steadily *early* in the random walk; it turns out that the precise mixing time is largely, although not entirely, unimportant. See Theorem 4 for a precise statement. This is then applied to the Rho walk to give the first proof of collision in $\Theta(\sqrt{N})$ steps.

We note here that it is also well known (see e.g. [1], Section 4.1) that a sample of length L from a Markov chain is roughly equivalent to $L\lambda$ samples from the stationary measure (of the Markov chain) for the purpose of sampling, where λ is the spectral gap of the chain. This yields another estimate on collision time for a Markov chain, which is also of a multiplicative nature (namely, $\sqrt{N}$ *times* a function of the mixing time) as in [7,5]. A main point of the present work is to establish sufficient criteria under which the collision time has an *additive* bound: $C\sqrt{N}$ plus an estimate on the mixing time. While the Rho algorithm provided the main motivation for the present work, we find the more general Birthday paradox result to be of independent interest, and as such expect to have other applications in the future.

A bit of detail about the Pollard Rho algorithm is in order. The classical discrete logarithm problem on a cyclic group deals with computing the exponents, given the generator of the group; more precisely, given a generator g of a cyclic group G and an element $h = g^x$, one would like to compute x efficiently. Due to its presumed computational difficulty, the problem figures prominently in various cryptosystems, including the Diffie-Hellman key exchange, El Gamal system, and elliptic curve cryptosystems. About 30 years ago, J.M. Pollard suggested algorithms to help solve both factoring large integers [10] and the discrete logarithm problem [11]. While the algorithms are of much interest in computational number theory and cryptography, there has been little work on rigorous analysis. We refer the reader to [7] and other existing literature (e.g., [15,2]) for further cryptographic and number-theoretical motivation for the discrete logarithm problem.

A standard variant of the classical Pollard Rho algorithm for finding discrete logarithms can be described using a Markov chain on a cyclic group G. While there has been no rigorous proof of rapid mixing of this Markov chain of order $O(\log^c |G|)$ until recently, Miller-Venkatesan [7] gave a proof of mixing of order $O(\log^3 |G|)$ steps and collision time of $O(\sqrt{|G|} \log^3 |G|)$, and Kim et al. [5] showed mixing of order $O(\log |G| \log\log |G|)$ and collision time of $O(\sqrt{|G|} \log |G| \log\log |G|)$. In this paper we give the first proof of the correct

$\Theta(\sqrt{|G|})$ collision time. By recent results of Miller-Venkatesan [8] this collision will be non-degenerate with probability $1 - o(1)$ for almost every prime order $|G|$, if the start point of the algorithm is chosen at random or if there is no collision in the first $O(\log |G| \, \log \log |G|)$ steps.

The paper proceeds as follows. Section 2 contains some preliminaries; primarily an introduction to the Pollard Rho Algorithm, and a simple multiplicative bound on the collision time in terms of the mixing time. The more general Birthday Paradox for Markov chains with uniform stationary distribution is shown in Section 3. In Section 4 we bound the appropriate constants for the Rho walk and show the optimal collision time. We finish in Section 5 with a few comments on the sharpness of our result.

2   Preliminaries

Our intent in generalizing the Birthday Paradox was to bound the collision time of the Pollard Rho algorithm for Discrete Logarithm. As such, we briefly introduce the algorithm here. Throughout the analysis in the following sections, we assume that the size $N = |G|$ of the cyclic group on which the random walk is performed is odd. Indeed there is a standard reduction – see [12] for a very readable account and also a classical reference [9] – justifying the fact that it suffices to study the discrete logarithm problem on cyclic groups of *prime* order.

Suppose g is a generator of G, that is $G = \{g^i\}_{i=0}^{N-1}$. Given $h \in G$, the discrete logarithm problem asks us to find x such that $g^x = h$. Pollard suggested an algorithm on $\mathbb{Z}_N^\times$ based on a random walk and the Birthday Paradox. A common extension of his idea to groups of prime order is to start with a partition of G into sets S_1, S_2, S_3 of roughly equal sizes, and define an iterating function $F : G \to G$ by $F(y) = gy$ if $y \in S_1$, $F(y) = hy = g^x y$ if $y \in S_2$, and $F(y) = y^2$ if $y \in S_3$. Then consider the walk $y_{i+1} = F(y_i)$. If this walk passes through the same state twice, say $g^{a+xb} = g^{\alpha+x\beta}$, then $g^{a-\alpha} = g^{x(\beta-b)}$ and so $a - \alpha \equiv x(\beta - b) \mod N$ and $x \equiv (a - \alpha)(\beta - b)^{-1} \mod N$, which determines x as long as $(\beta - b, N) = 1$. Hence, if we define a *collision* to be the event that the walk passes over the same group element twice, then the first time there is a collision it might be possible to determine the discrete logarithm.

To estimate the running time until a collision, one heuristic is to treat F as if it outputs uniformly random group elements. By the Birthday Paradox if $O(\sqrt{|G|})$ group elements are chosen uniformly at random, then there is a high probability that two of these are the same. We analyze instead the actual Markov chain in which it is assumed only that each $y \in G$ is assigned independently and at random to a partition S_1, S_2 or S_3. In this case, although the iterating function F described earlier is deterministic, because the partition of G was randomly chosen then the walk is equivalent to a Markov chain (i.e. a random walk), at least until the walk visits a previously visited state and a collision occurs. The problem is then one of considering a walk on the exponent of g, that is a walk P on the cycle $\mathbb{Z}_N$ with transitions $P(u, u+1) = P(u, u+x) = P(u, 2u) = 1/3$.

Remark 1. By assuming each $y \in G$ is assigned independently and at random to a partition we have eliminated one of the key features of the Pollard Rho algorithm, space efficiency. However, if the partitions are given by a hash function $f : (G, N) \rightarrow \{1, 2, 3\}$ which is sufficiently pseudo-random then we might expect behavior similar to the model with random partitions.

Remark 2. While we are studying the time until a collision occurs, there is no guarantee that the first collision will be non-degenerate. If the first collision is degenerate then so also will be all collisions, as the algorithm becomes deterministic after the first collision.

A simple multiplicative bound on collision time was obtained in [5] which relates $T_s(1/2)$ to the time until a collision occurs for any Markov chain P with uniform distribution on G as the stationary distribution.

Proposition 3. *With the above definitions, a collision occurs after*

$$1 + T_s(1/2) + 2\sqrt{2c|G|T_s(1/2)}$$

steps, with probability at least $1 - e^{-c}$, for any $c > 0$.

Obtaining a more refined additive bound on collision time will be the focus of the next section. While the proof can be seen as another application of the well-known second moment method, it turns out that bounding the second moment of the number of collisions *before* the mixing time is somewhat subtle. To handle this, we use an idea from [6], who in turn credit their line of calculation to [4].

3 Collision Time

Consider a finite ergodic Markov chain P with uniform stationary distribution (i.e. doubly stochastic), state space Ω of cardinality $N = |\Omega|$, and let $X_0, X_1, \cdots$ denote a particular instance of the walk. In this section we determine the number of steps of the walk required to have a high probability that a "collision" has occurred, i.e. a self-intersection $X_i = X_j$ for some $i \neq j$.

First, some notation. Fix some $T \geq 0$. Define

$$S = \sum_{i=0}^{\beta\sqrt{N}} \sum_{j=i+2T}^{\beta\sqrt{N}+2T} 1_{\{X_i = X_j\}}$$

to be the number of times the walk intersects itself in $\beta\sqrt{N} + 2T$ steps, where i and j are at least $2T$ steps apart. Also, for $u, v \in \Omega$, let

$$G_T(u, v) = \sum_{i=0}^{T} P^i(u, v)$$

be the expected number of times a walk beginning at u hits state v in T steps. Finally, let

$$A_T = \max_u \sum_v G_T^2(u, v) \qquad \text{and} \qquad A_T^* = \max_u \sum_v G_T^2(v, u).$$

To see the connection between these and the collision time, observe that

$$\sum_v G_T^2(u, v) = \sum_v \left(\sum_{i=0}^T \sum_{j=0}^T P^i(u, v) P^j(u, v) \right)$$

$$= \sum_{i=0}^T \sum_{j=0}^T \sum_v P^i(u, v) P^j(u, v)$$

$$= \sum_{i=0}^T \sum_{j=0}^T \mathbb{P}_{u,u}(X_i = Y_j)$$

$$= \sum_{i=0}^T \sum_{j=0}^T E\left(1_{\{X_i = Y_j\}} \right) = E \sum_{i,j=0}^T 1_{\{X_i = Y_j\}},$$

where $\{X_i\}, \{Y_j\}$ are i.i.d. copies of the chain, both having started at u at time 0. Hence A_T is the maximal expected number of collisions of two T-step i.i.d. walks of P starting at the same state u, while A_T^* is the same for P^*.

The main result of this section is the following.

Theorem 4 (Birthday Paradox for Markov chains). *Consider a finite ergodic Markov chain with uniform stationary distribution on a state space of N vertices. Let T be such that $\frac{m}{N} \leq P^T(u, v) \leq \frac{M}{N}$ for some $m \leq 1 \leq M$ and every u, v. After*

$$4c \left(\frac{M}{m} \right)^2 \left(\sqrt{\frac{2N}{M} \max\{A_T, A_T^*\}} + T \right)$$

steps a collision occurs with probability at least $1 - e^{-c}$, for any $c \geq 0$.

Proof. First recall the standard second moment bound: using Cauchy-Schwarz,

$$E[S] = E[S 1_{\{S>0\}}] \leq E[S^2]^{1/2} E[1_{\{S>0\}}]^{1/2}$$

and hence $\Pr[S > 0] \geq E[S]^2 / E[S^2]$. By Lemma 6, if $\beta = 2\sqrt{2 \max\{A_T, A_T^*\}/M}$ then

$$\Pr[S > 0] \geq \frac{m^2/M^2}{1 + \frac{8 \max\{A_T, A_T^*\}}{M\beta^2}} \geq \frac{m^2}{2M^2}, \tag{1}$$

independent of the starting point. Hence the probability that there is no collision after $k(\beta\sqrt{N} + 2T)$ steps is at most $(1 - m^2/2M^2)^k \leq e^{-km^2/2M^2}$. Taking $k = 2cM^2/m^2$ completes the proof.

Remark 5. Observe that if $A_T, A_T^*, m, M = \Theta(1)$ and $T = O(\sqrt{N})$ then the collision time is $O(\sqrt{N})$, as in the standard Birthday Paradox. By Lemma 7, it will suffice that P^T be sufficiently close to uniform after $T = o(\sqrt{N})$ steps, and that $P^j(u, v) = o(T^{-2}) + d^j$ for all u, v, for $j \leq T$ and some $d < 1$.

When applied to the standard Birthday Paradox equation (1) with $T = 1$ is $2/\sqrt{\ln 2} \approx 2.4$ times the correct number of steps required to reach probability $1/2$. In the final section of the paper, we present an example to illustrate the need for the pre-mixing term A_T in Theorem 4. A slight strengthening of Theorem 4 is also shown there, at the cost of a somewhat less intuitive bound.

The proof of Theorem 4 relied largely on the following:

Lemma 6. *Under the conditions of Theorem 4,*

$$E[S] \geq \frac{m}{N}\binom{\beta\sqrt{N}+2}{2}, \quad E[S^2] \leq \frac{M^2}{N^2}\binom{\beta\sqrt{N}+2}{2}^2\left(1 + \frac{8\max\{A_T, A_T^*\}}{M\beta^2}\right).$$

Proof. We will repeatedly use the relation that there are $\binom{\beta\sqrt{N}+2}{2}$ choices for i, j appearing in the summation for S, i.e. $0 \leq i$ and $i + 2T \leq j \leq \beta\sqrt{N} + 2T$.

Now to the proof. The expectation $E[S]$ satisfies

$$E[S] = E\sum_{i=0}^{\beta\sqrt{N}}\sum_{j=i+2T}^{\beta\sqrt{N}+2T} 1_{\{X_i = X_j\}} = \sum_{i=0}^{\beta\sqrt{N}}\sum_{j=i+2T}^{\beta\sqrt{N}+2T} E[1_{\{X_i = X_j\}}] \geq \binom{\beta\sqrt{N}+2}{2}\frac{m}{N}$$

because if $j \geq i + T$ then

$$Pr(X_j = X_i) = \sum_u Pr(X_i = u)P^{j-i}(u, u) \geq \sum_u Pr(X_i = u)\frac{m}{N} = \frac{m}{N}.$$

Similarly, $Pr(X_j = X_i) \leq \frac{M}{N}$ when $j \geq i + T$.

Now for $E[S^2]$. Note that

$$E[S^2] = E\left(\sum_{i=0}^{\beta\sqrt{N}}\sum_{j=i+2T}^{\beta\sqrt{N}+2T} 1_{\{X_i = X_j\}}\right)\left(\sum_{k=0}^{\beta\sqrt{N}}\sum_{l=k+2T}^{\beta\sqrt{N}+2T} 1_{\{X_k = X_l\}}\right)$$

$$= \sum_{i=0}^{\beta\sqrt{N}}\sum_{k=0}^{\beta\sqrt{N}}\sum_{j=i+2T}^{\beta\sqrt{N}+2T}\sum_{l=k+2T}^{\beta\sqrt{N}+2T} Prob(X_i = X_j, X_k = X_l).$$

To evaluate this quadruple sum we break it into 3 cases.

Case 1: Suppose $|j - l| \geq T$. Without loss, assume $l \geq j$, so in particular $l \geq \max\{i, j, k\} + T$. Then

$$Prob(X_i = X_j, X_k = X_l) = Prob(X_i = X_j)\,Prob(X_l = X_k \mid X_i = X_j)$$
$$\leq Prob(X_i = X_j)\max_{u,v} Prob(X_l = v \mid X_{\max\{i,j,k\}} = u)$$

$$\leq Prob(X_i = X_j)\frac{M}{N} \leq \left(\frac{M}{N}\right)^2.$$

The first inequality is because $\{X_t\}$ is a Markov chain and so given X_i, X_j, X_k the walk at any time $t \geq \max\{i, j, k\}$ depends only on the state $X_{\max\{i,j,k\}}$.

Case 2: Suppose $|i - k| \geq T$ and $|j - l| < T$. Without loss, assume $i \leq k$. If $j \leq l$ then

$$Prob(X_i = X_j, X_k = X_l) = \sum_{u,v} Prob(X_i = u) \, P^{k-i}(u,v) P^{j-k}(v,u) P^{l-j}(u,v)$$

$$\leq \sum_u Prob(X_i = u) \frac{M}{N} \frac{M}{N} \sum_v P^{l-j}(u,v) = \left(\frac{M}{N}\right)^2$$

because $k \geq i + T$, $j \geq k + T$, and $\sum_v P^t(u,v) = 1$ for any t because P and hence also P^t is a stochastic matrix. If, instead, $l < j$ then essentially the same argument works, but with $\sum_v P^t(v,u) = 1$ because P and hence also P^t is doubly-stochastic.

Case 3: Finally, consider those terms with $|j - l| < T$ and $|i - k| < T$. Without loss, assume $i \leq k$. If $l \leq j$ then

$$Prob(X_i = X_j, X_k = X_l) = \sum_{u,v} Prob(X_i = u) P^{k-i}(u,v) P^{l-k}(v,v) P^{j-l}(v,u)$$

$$\leq \sum_u Prob(X_i = u) \sum_v P^{k-i}(u,v) \frac{M}{N} P^{j-l}(v,u).$$

The sum over $i \leq k < i + T$ and $l \leq j < l + T$ is upper bounded as follows:

$$\sum_{i=0}^{\beta\sqrt{N}} \sum_{k=i}^{i+T} \sum_{l=k+2T}^{\beta\sqrt{N}+2T} \sum_{j=l}^{l+T} Prob(X_i = X_j, X_k = X_l) \qquad (2)$$

$$\leq \frac{M}{N} \sum_{i=0}^{\beta\sqrt{N}} \sum_{l=i+2T}^{\beta\sqrt{N}+2T} \max_u \sum_v \sum_{k \in [i, i+T)} P^{k-i}(u,v) \sum_{j \in [l, l+T)} P^{j-l}(v,u) \qquad (3)$$

$$\leq \frac{M}{N} \sum_{i=0}^{\beta\sqrt{N}} \sum_{l=i+2T}^{\beta\sqrt{N}+2T} \max_u \sum_v G_T(u,v) G_T(v,u)$$

$$\leq \frac{M}{N} \sum_{i=0}^{\beta\sqrt{N}} \sum_{l=i+2T}^{\beta\sqrt{N}+2T} \max_u \sqrt{\sum_v G_T^2(u,v) \sum_v G_T^2(v,u)}$$

$$\leq \frac{M}{N} \left(\frac{\beta\sqrt{N}+2}{2}\right) \sqrt{A_T A_T^*} \,.$$

The case when $j < l$ gives the same bound, but with the observation that $j \geq k + T$ and with A_T instead of $\sqrt{A_T A_T^*}$.

Putting together these various cases we get that

$$E[S^2]$$

$$\leq \left(\frac{\beta\sqrt{N}+2}{2}\right)^2 \left(\frac{M}{N}\right)^2 + 2\left(\frac{\beta\sqrt{N}+2}{2}\right) \frac{M}{N} A_T + 2\left(\frac{\beta\sqrt{N}+2}{2}\right) \frac{M}{N} \sqrt{A_T A_T^*}$$

The $\left(\frac{\beta\sqrt{N}+2}{2}\right)^2$ term is the total number of values of i, j, k, l appearing in the sum for $E[S^2]$, and hence also an upper bound on the number of values in Cases 1 and 2. Along with the relation $\left(\frac{\beta\sqrt{N}+2}{2}\right) \geq \frac{\beta^2 N}{2}$ this simplifies to complete the proof.

To upper bound A_T and A_T^* it suffices to show that the maximum probability of being at a vertex decreases quickly.

Lemma 7. *If a finite ergodic Markov chain has uniform stationary distribution then*

$$A_T, A_T^* \leq 2 \sum_{j=0}^{T} (j+1) \max_{u,v} P^j(u,v).$$

Proof. If u is such that equality occurs in the definition of A_T, then

$$A_T = \sum_v G_T^2(u,v) = \sum_{i=0}^{T} \sum_{j=0}^{T} \sum_v P^i(u,v) P^j(u,v)$$

$$\leq 2 \sum_{j=0}^{T} \sum_{i=0}^{j} \max_y P^j(u,y) \sum_v P^i(u,v)$$

$$\leq 2 \sum_{j=0}^{T} (j+1) \max_y P^j(u,y).$$

The same bound holds for A_T^*, which plays the role of A_T for the reversed chain, because the upper bound just shown is the same for the chain and its reversal.

In particular, if $P^j(u,v) \leq c + d^j$ for every $u, v \in \Omega$ and some $c, d \in [0,1)$ then

$$\sum_{j=0}^{T} (j+1)(c + d^j) \leq (1 + o(1))\frac{cT^2}{2} + \frac{1}{(1-d)^2},$$

and so if $P^j(u,v) \leq o(T^{-2}) + d^j$ for every $u, v \in \Omega$ then $A_T, A_T^* = \frac{2+o(1)}{(1-d)^2}$.

4   Convergence of the Rho Walk

Let us now turn our attention to the Pollard Rho walk for discrete logarithm. To apply the collision time result we will first show that $\max_{u,v \in \mathbb{Z}_N} P^s(u,v)$ decreases quickly in s so that Lemma 7 may be used. We then find T such that $P^T(u,v) \approx 1/N$ for every $u, v \in \mathbb{Z}_N$. However, instead of studying the Rho walk directly, most of the work will instead involve a "block walk" in which only a certain subset of the states visited by the Rho walk are considered.

Definition 8. *Let us refer to the three types of moves that the Pollard Rho random walk makes, namely* $(u, u+1), (u, u+x)$*, and* $(u, 2u)$*, as moves of Type 1, Type 2, and Type 3, respectively. In general, let the random walk be denoted by* $Y_0, Y_1, Y_2, \ldots$*, with* Y_t *indicating the position of the walk (modulo N) at time* $t \geq 0$*. Let* T_1 *be the first time that the walk makes a move of Type 3. Let* $b_1 = Y_{T_1 - 1} - Y_{T_0}$ *(i.e., the ground covered, modulo N, only using consecutive moves of Types 1 and 2.) More generally, let* T_i *be the first time, since* T_{i-1}*, that a move of Type 3 happens and set* $b_i = Y_{T_i - 1} - Y_{T_{i-1}}$*. Then the block walk* B *is the walk* $X_s = Y_{T_s} = 2^s Y_{T_0} + 2 \sum_{i=1}^s 2^{s-i} b_i$*. Also, for* $\delta \in [0, 1]$ *the* $(1 + \delta)$*-block walk has transition matrix* $B_{1+\delta} = (1 - \delta)B + \delta B^2$*.*

By combining our Birthday Paradox for Markov chains with several lemmas to be shown in this section we obtain the main result of the paper:

Theorem 9. *For every choice of starting state, the expected number of steps required for the Pollard Rho algorithm for discrete logarithm on a group G to have a collision is at most*

$$(1 + o(1)) \, 12\sqrt{19} \, \sqrt{|G|} < (1 + o(1)) \, 52.5 \, \sqrt{|G|}.$$

Proof. We work with Theorem 14, shown in the Concluding Remarks, because this gives a somewhat sharper bound. Alternatively, Theorem 4 and Lemma 7 can be applied nearly identically to get the slightly weaker $(1 + o(1))72\sqrt{|G|}$.

First consider steps of the $(1 + \delta)$-block walk with $\delta = 1/\log_2 N$. Note that $B_{1+\delta}^s(u, v) \leq \max_{k \in [s, 2s]} B^k(u, v)$, and so Lemma 10 implies that $B_{1+\delta}^s(u, v) \leq \frac{3/2}{\sqrt{N}} + (\frac{2}{3})^s$, for $s \geq 0$, and for all u, v. Hence, by equation (5), if $T = o(\sqrt[4]{N})$ then $1 + \sum_{j=1}^{2T} 3j \, P^j(u, v) \leq 19 + o(1)$. By Lemma 12, after $T = 500(\log_2^4 N) = o(\sqrt[4]{N})$ steps, we have $M \leq 1 + 1/N^2$ and $m \geq 1 - 1/N^2$. Plugging this into Theorem 14, a collision fails to occur in

$$k \left(2\sqrt{\left(1 + \sum_{j=1}^{2T} 3j \max_{u,v} P^j(u, v)\right) \frac{N}{M} + 2T} \right) = (1 + o(1)) \, 2\sqrt{19} \, k\sqrt{N}$$

steps with probability at most $(1 - \delta)^k$ where $\delta = m^2/2M^2 = (1 - o(1))/2$. By Chebyshev's Inequality this requires $(1 + o(1))^2 \, 2\sqrt{19} \, k\sqrt{N}$ steps of the Block walk with probability $1 - o(1)$, and so in $(1 + o(1)) \, 2\sqrt{19} \, k\sqrt{N}$ steps of the Block walk there is a collision with probability $\frac{1 - o(1)}{2}$.

Now let us return to the Rho walk. Recall that T_i denotes the number of Rho steps required for i block steps. The difference $T_{i+1} - T_i$ is an i.i.d. random variable with the same distribution as $T_1 - T_0$. Hence, if $i \geq j$ then $E[T_i - T_j] = (i - j) E[T_1 - T_0] = 3(i - j)$. In particular, if we let $r = (1 + o(1)) \, 2\sqrt{19 \, N}$, let R denote the number of Rho steps before a collision, and let B denote the number of block steps before a collision, then

$$E[R] \leq \sum_{k=0}^{\infty} Pr[B > kr]\, E[T_{(k+1)r} - T_{kr} \mid B > kr]$$

$$= \sum_{k=0}^{\infty} Pr[B > kr]\, E[T_{(k+1)r} - T_{kr}]$$

$$\leq \sum_{k=0}^{\infty} \left(\frac{1 + o(1)}{2}\right)^k 3r = (1 + o(1))\, 12\sqrt{19}\,\sqrt{N}\,.$$

Now to the first lemma required for the collision bound, a proof that $B^s(u, v)$ decreases quickly for the block walk:

Lemma 10. *If $s \leq \lfloor \log_2 N \rfloor$ then for every $u, v \in \mathbb{Z}_N$ the block walk satisfies*

$$B^s(u, v) \leq (2/3)^s\,.$$

If $s > \lfloor \log_2 N \rfloor$ then $B^s(u, v) \leq \dfrac{3/2}{N^{1 - \log_2 3}} \leq \dfrac{3/2}{\sqrt{N}}$.

Proof. We start with a weaker, but somewhat more intuitive, proof of a bound on $B^s(u, v)$ and then improve it to obtain the result of the lemma. The key idea here will be to separate out a portion of the Markov chain which is tree-like with some large depth L, namely the moves induced solely by $b_i = 0$ and $b_i = 1$ moves. Because of the high depth of the tree, the walk spreads out for the first L steps, and hence the probability of being at a vertex also decreases quickly.

Let $S = \{i \in [1 \ldots s] : b_i \in \{0, 1\}\}$ and $z = \sum_{i \notin S} 2^{s-i} b_i$. Then $Y_{T_s} = 2^s Y_{T_0} + 2z + 2\sum_{i \in S} 2^{s-i} b_i$. Hence, choosing $Y_{T_0} = u$, $Y_{T_s} = v$, we may write

$$B^s(u, v)$$

$$= \sum_S Prob(S) \sum_{z \in \mathbb{Z}_N} Prob(z \mid S)\, Prob\left(\sum_{i \in S} 2^{s-i} b_i = v/2 - 2^{s-1} u - z \mid z, S\right)$$

$$\leq \sum_S Prob(S) \max_{w \in \mathbb{Z}_N} Prob\left(\sum_{i \in S} 2^{s-i} b_i = w \mid S\right),$$

and so for a fixed choice of S, we can ignore what happens on S^c.

Each $w \in [0 \ldots N-1]$ has a unique binary expansion, and so if $s \leq \lfloor \log_2 N \rfloor$ then modulo N each w can still be written in at most one way as an s bit string. For the block walk, $Prob(b_i = 0) \geq 1/3$ and $Prob(b_i = 1) \geq 1/9$, and so $\max\{Prob(b_i = 0 \mid i \in S), Prob(b_i = 1 \mid i \in S)\} \leq \frac{8}{9}$. It follows that

$$\max_{w \in \mathbb{Z}_N} Prob\left(\sum_{i \in S} 2^{s-i} b_i = w \mid S\right) \leq (8/9)^{|S|}, \tag{4}$$

using independence of the b_i's. Hence,

$$\mathsf{B}^s(u,v) \leq \sum_S Prob(S)\,(8/9)^{|S|} = \sum_{r=0}^{s} Prob(|S| = r)\,(8/9)^r$$

$$\leq \sum_{r=0}^{s}\binom{s}{r}\left(\frac{4}{9}\right)^r\left(1 - \frac{4}{9}\right)^{s-r}\left(\frac{8}{9}\right)^r = \left(\frac{4}{9}\frac{8}{9} + \frac{5}{9}\right)^s = \left(\frac{77}{81}\right)^s .$$

The second inequality was because $(8/9)^{|S|}$ is decreasing in $|S|$ and so underestimating $|S|$ by assuming $Prob(i \in S) = 4/9$ will only increase the upper bound on $\mathsf{B}^s(u,v)$.

In order to improve on this, we will shortly re-define S (namely, events $\{i \in S\}, \{i \notin S\}$) and auxiliary variables c_i, using the steps of the Rho walk. Also note that the block walk is induced by a Rho walk, so we may assume that the b_i were constructed by a series of steps of the Rho walk. With probability $1/4$ set $i \in S$ and $c_i = 0$, otherwise if the first step is of Type 1 then set $i \in S$ and $c_i = 1$, while if the first step is of Type 3 then put $i \notin S$ and $c_i = 0$, and finally if the first step is of Type 2, then again repeat the above decision making process, using the subsequent steps of the walk. Note that the above construction can be summarized as consisting of one of four equally likely outcomes (at each time), where the last three outcomes depend on the type of the step that the Rho walk takes; indeed each of these three outcomes happens with probability $\frac{3}{4} \times \frac{1}{3} = 1/4$; finally, a Type 2 step forces us to reiterate the four-way decision making process.

Then $Pr(i \in S) = \sum_{l=0}^{\infty}(1/4)^l\,(1/2) = 2/3$. Also observe that $Pr(c_i = 0|i \in S) = Pr(c_i = 1|i \in S)$, and that $Pr(b_i - c_i = x \mid i \in S,\ c_i = 0) = Pr(b_i - c_i = x \mid i \in S,\ c_i = 1)$. Hence the steps done earlier (leading to the weaker bound) carry through with $z = \sum_i 2^{s-i}(b_i - c_i)$ and with $\sum_{i \in S} 2^{s-i}b_i$ replaced by $\sum_{i \in S} 2^{s-i}c_i$. In (4) replace $(8/9)^{|S|}$ by $(1/2)^{|S|}$, and in showing the final upper bound on $\mathsf{B}^s(u,v)$ replace $4/9$ by $2/3$. This leads to the bound $\mathsf{B}^s(u,v) \leq (2/3)^s$.

Finally, when $s > \lfloor \log_2 N \rfloor$, simply apply the preceding argument to $S' = S \cap [1 \ldots \lfloor \log_2 N \rfloor]$. Alternately, note that when $s \geq \lfloor \log_2 N \rfloor$ then $\mathsf{B}^s(u,v) \leq \max_w \mathsf{B}^{\lfloor \log_2 N \rfloor}(u,w)$, for every doubly-stochastic Markov chain B.

In order to use the Birthday Paradox on the Rho walk it suffices to show a mixing time bound of $T = O(\sqrt[4]{N})$ (to guarantee that $A_T, A_T^* = O(1)$). The first such bound was shown by Miller and Venkatesan [7] using characters and quadratic forms, albeit for the Rho walk rather than the Block walk; other sufficiently strong bounds are shown in [5] using canonical paths or Fourier analysis. The argument given here is chosen for brevity alone.

Perhaps the most widely used approach to bounding mixing times is the method of canonical paths. Canonical path methods [14] can be used to lower bound the spectral gap of a Markov kernel P in terms of paths involving edges of P. Fill [3] showed a bound on the mixing time in terms of the smallest singular value of P, or equivalently the spectral gap of PP^*, where the time-reversed walk is $\mathsf{P}^*(v,u) = \frac{\pi(u)\mathsf{P}(u,v)}{\pi(v)} = \mathsf{P}(u,v)$, when the stationary distribution π is uniform. By combining these two methods we obtain a bound on mixing time in terms of even length paths alternating between edges of P and P^*.

Theorem 11. *Consider a finite Markov chain* P *on state space* Ω *with stationary distribution* π, *and set* $\pi_* = \min_{v \in \Omega} \pi(v)$. *For every* $u, v \in \Omega$, $u \neq v$, *define a path* γ_{uv} *from* u *to* v *along edges of* PP^*, *and let*

$$A = A(\Gamma) = \max_{x \neq y : \mathsf{PP}^*(x,y) \neq 0} \frac{1}{\pi(x)\mathsf{PP}^*(x,y)} \sum_{a \neq b : (x,y) \in \gamma_{ab}} \pi(a)\pi(b)|\gamma_{ab}|.$$

Then, for every $u, v \in \Omega$,

$$\left| \frac{\mathsf{P}^T(u,v)}{\pi(v)} - 1 \right| \leq \epsilon \quad if \quad T \geq 2A \log \frac{1}{\epsilon \pi_*}.$$

To apply this we need only construct paths for the $(1 + \delta)$-block walk:

Lemma 12. *If* $T \geq \frac{486}{\delta(1-\delta)} \lceil \log_2 N \rceil^3$ *then* $\forall u, v \in \mathbb{Z}_N$: $\left| \frac{\mathsf{B}_{1+\delta}^T(u,v)}{\pi(v)} - 1 \right| \leq \frac{1}{N^2}$.

Proof. We will construct paths and apply Theorem 11 with $\epsilon = \pi_* = 1/N$. If $u \in \mathbb{Z}_N$ then

$$\mathsf{B}_{1+\delta}\mathsf{B}_{1+\delta}^*(u, 2u+1) \geq \mathsf{B}_{1+\delta}(u, 4u+2)\mathsf{B}_{1+\delta}^*(4u+2, 2u+1) \geq \frac{\delta}{27} \frac{1-\delta}{3} = \frac{\delta(1-\delta)}{81},$$

and likewise $\mathsf{B}_{1+\delta}\mathsf{B}_{1+\delta}^*(u, 2u) \geq \mathsf{B}_{1+\delta}(u, 4u)\mathsf{B}_{1+\delta}^*(4u, 2u) \geq \frac{\delta}{9} \frac{1-\delta}{3} \geq \frac{\delta(1-\delta)}{81}$.

To construct a path from u to v, set $n = \lceil \log_2 N \rceil$ and $x = (v - 2^n u) \mod N$. Then x has a unique n-bit binary expansion $x = x_0 x_1 \cdots x_{n-2} x_{n-1}$. To describe the path let $u_0 = u$ and inductively define $u_{i+1} = 2u_i + x_i$. Then $u_n \equiv 2^n u + x \equiv v \mod N$ and $|\gamma_{uv}| = n$.

It remains to count the number of paths through each edge. Fix edge (a, b) with $b \equiv 2a \mod N$ or $b \equiv 2a + 1 \mod N$. There are 2^{i-1} potential values of u, and 2^{n-i} potential values of v, such that (a, b) is the i-th edge of path γ_{uv}, and there are n potential values for i, for a total of at most $n\,2^{n-1} \leq n\,N$ paths passing through edge (a, b).

5   Concluding Remarks

As promised in Section 3, we now present an example that illustrates the need for the pre-mixing term A_T in Theorem 4.

Example 13. Consider the random walk on $\mathbb{Z}_N$ which transitions from $u \to u+1$ with probability $1 - 1/\sqrt{N}$, and with probability $1/\sqrt{N}$ transitions $u \to v$ for a uniformly random choice of v.

Heuristically the walk proceeds as $u \to u+1$ for $\approx \sqrt{N}$ steps, then randomizes, then proceeds as $u \to u + 1$ for another $\sqrt{N}$ steps. This effectively splits the state space into $\sqrt{N}$ blocks of size about $\sqrt{N}$ each, so by the standard Birthday Paradox it should require about $\sqrt{N^{1/2}}$ of these randomizations before a collision will occur. In short, about $N^{3/4}$ steps in total.

To see the need for the pre-mixing term, observe that $T_s \approx \sqrt{N} \log 2$ while if $T = T_\infty \approx \sqrt{N} \log(2(N-1))$ then we may take $m = 1/2$ and $M = 3/2$ in Theorem 4. So, whether T_s or T_∞ are considered, it will be insufficient to take $O(T+\sqrt{N})$ steps. However, the number A_T of collisions between two independent copies of this walk is about $\sqrt{N}$, since once a randomization step occurs then the two independent walks are unlikely to collide anytime soon. Our collision time bound says that $O(N^{3/4})$ steps will suffice, which is the correct bound.

A proper analysis shows that $\frac{1-o(1)}{\sqrt{2}} N^{3/4}$ steps are necessary to have a collision with probability $1/2$. Conversely, when $T = \sqrt{N} \log^2 N$ then $m = 1 - o(1)$, $M = 1 + o(1)$ and $A_T, A_T^* \leq \frac{1+o(1)}{2} \sqrt{N}$, so by equation (1), $(2+o(1))N^{3/4}$ steps are sufficient to have a collision with probability at least $1/2$. Our upper bound is thus off by at most a factor of $2\sqrt{2} \approx 2.8$.

Also, the slight sharpening that was used to derive our improved bound for the Pollard Rho walk:

Theorem 14 (Improved Birthday Paradox). *Under the conditions of Theorem 4, after*

$$2c\left(\sqrt{\left(1 + \sum_{j=1}^{2T} 3j \max_{u,v} P^j(u,v)\right)\frac{N}{M}} + T\right)$$

steps a collision occurs with probability at least $1 - \left(1 - \frac{m^2}{2M^2}\right)^c$, *independent of the starting state.*

Proof. We give only the steps that differ from before. First, in equation (3), note that the triple sum after $\max_u$ can be re-written as

$$\sum_{\alpha \in [0,T)} \sum_{\beta \in [0,T)} \sum_v P^\alpha(u,v)P^\beta(v,u) \leq \sum_{\gamma=0}^{2(T-1)} (\gamma+1)P^\gamma(u,u)$$

and so (2) reduces to $\frac{M}{N}\binom{\beta\sqrt{N}+2}{2} \max_u \sum_{\gamma=0}^{2(T-1)} (\gamma+1)P^\gamma(u,u)$.

When $i < k$ and $j < l$ proceed similarly, finishing as in Lemma 7 to obtain

$$\frac{M}{N}\binom{\beta\sqrt{N}+2}{2} \sum_{\alpha=1}^{T-1}\sum_{\beta=1}^{T-1}\sum_v P^\alpha(u,v)P^\beta(u,v)$$

$$\leq \frac{M}{N}\binom{\beta\sqrt{N}+2}{2} \sum_{\gamma=1}^{T-1} (2\gamma-1) \max_v P^\gamma(u,v).$$

Adding these two expressions gives an expression of at most

$$\frac{M}{N}\binom{\beta\sqrt{N}+2}{2}\left(1 + \sum_{\gamma=1}^{2T} 3\gamma \max_v P^\gamma(u,v)\right).$$

The remaining two cases add to the same bound, so a $4\max\{A_T, A_T^*\}$ in the original theorem is replaced by $2\left(1 + \max_u \sum_{\gamma=1}^{2T} 3\gamma \max_v P^\gamma(u,v)\right)$.

To simplify the improved bound, note that if $\max_{u,v} P^j(u,v) \le c + d^j$ then

$$1 + \sum_{j=1}^{2T} 3j \max_{u,v} P^j(u,v) \le 1 + \frac{3d}{(1-d)^2} + 3cT(2T+1). \tag{5}$$

Acknowledgment

The authors thank S. Kijima, S. Miller, R. Venkatesan and D. Wilson for several helpful discussions and for the pointers to E. Teske's work on discrete logarithms.

References

1. Aldous, D., Fill, J.: Reversible Markov Chains and Random walks on Graphs (in preparation), http://www.stat.berkeley.edu/~aldous
2. Crandall, R., Pomerance, C.: Prime Numbers: a Computational Perspective, 2nd edn. Springer, Heidelberg (2005)
3. Fill, J.: Eigenvalue bounds on convergence to stationarity for nonreversible Markov chains, with an application to the exclusion process. The Annals of Applied Probability 1, 62–87 (1991)
4. Le Gall, J.F., Rosen, J.: The range of stable random walks. Ann. Probab. 19, 650–705 (1991)
5. Kim, J.-H., Montenegro, R., Tetali, P.: Near optimal bounds for collision in Pollard Rho for discrete log. In: Proc. 48th Annual Symposium on Foundations of Computer Science (FOCS 2007) (2007)
6. Lyons, R., Peres, Y., Schramm, O.: Markov chain intersections and the loop-erased walk. Ann. Inst. H. Poincaré Probab. Statist. 39(5), 779–791 (2003)
7. Miller, S., Venkatesan, R.: Spectral analysis of Pollard Rho collisions. In: Hess, F., Pauli, S., Pohst, M. (eds.) ANTS 2006. LNCS, vol. 4076, pp. 573–581. Springer, Heidelberg (2006)
8. Miller, S., Venkatesan, R.: Personal communications (2007)
9. Pohlig, S., Hellman, M.: An improved algorithm for computing logarithms over $GF(p)$ and its cryptographic significance. IEEE Trans. Information Theory 24, 106–110 (1978)
10. Pollard, J.M.: A Monte Carlo method for factorization. BIT Nord. Tid. f. Inf. 15, 331–334 (1975)
11. Pollard, J.M.: Monte Carlo methods for index computation (mod p). Math. Comp. 32(143), 918–924 (1978)
12. Pomerance, C.: Elementary thoughts on discrete logarithms. In: Buhler, J.P., Stevenhagen, P. (eds.) Algorithmic Number Theory: Lattices, Number Fields, Curves and Cryptography, vol. 44, Mathematical Sciences Research Institute Publications (to appear, 2007), http://www.math.dartmouth.edu/~carlp
13. Shoup, V.: Lower bounds for discrete logarithms and related problems. In: Fumy, W. (ed.) EUROCRYPT 1997. LNCS, vol. 1233, pp. 256–266. Springer, Heidelberg (1997)
14. Sinclair, A.: Improved bounds for mixing rates of Markov chains and multicommodity flow. Combinatorics, Probability and Computing 1(4), 351–370 (1992)
15. Teske, E.: Square-root algorithms for the discrete logarithm problem (a survey). In: Public Key Cryptography and Computational Number Theory, pp. 283–301. Walter de Gruyter (2001)

An Improved Multi-set Algorithm for the Dense Subset Sum Problem

Andrew Shallue

University of Calgary
Calgary AB T2N 1N4 Canada
ashallue@math.ucalgary.ca

Abstract. Given sets $L_1, \ldots, L_k$ of elements from $\mathbb{Z}/m\mathbb{Z}$, the k-set birthday problem is to find an element from each list such that their sum is 0 modulo m. We give a new analysis of the algorithm in [16], proving that it returns a solution with high probability. By the work of Lyubashevsky [10], we get as an immediate corollary an improved algorithm for the random modular subset sum problem. Assuming the modulus $m = 2^{n^\epsilon}$ for $\epsilon < 1$, this problem is now solvable using time and space

$$\widetilde{O}(2^{\frac{n^\epsilon}{(1-\epsilon)\log n}}).$$

Let $a_1, a_2, \ldots, a_n, t \in \mathbb{Z}/m\mathbb{Z}$ be given. The modular subset sum problem is to find a subset of the a_i that sum to t in $\mathbb{Z}/m\mathbb{Z}$, i.e. to find $x_i \in \{0, 1\}$ such that

$$\sum_{i=1}^{n} a_i x_i = t \mod m . \tag{1}$$

The corresponding decision problem is to determine whether or not there exists $x = (x_1, x_2, \ldots, x_n)$ that satisfies (1).

A subset sum problem is called random if n, m, and t are all fixed parameters but the a_i are drawn uniformly at random from $\mathbb{Z}/m\mathbb{Z}$. In addition to being interesting in their own right, random subset sum problems accurately model problems that arise naturally in number theory and combinatorics. We will use the shorthand MSS for modular subset sum and RMSS for random modular subset sum.

A useful way of classifying subset sum problems is by density.

Definition 1. *The* density *of a MSS instance is* $\frac{n}{\log_2 m}$. *Problems with density less than one are called* sparse, *while those with density greater than one are called* dense.

Now let sets $L_1, \ldots, L_k$ of elements of $\mathbb{Z}/m\mathbb{Z}$ be given. The k-set birthday problem is to find $b_i \in L_i$ such that $b_1 + \cdots + b_k = 0 \mod m$. We will assume that the elements of the L_i are uniformly generated and independent.

A.J. van der Poorten and A. Stein (Eds.): ANTS-VIII 2008, LNCS 5011, pp. 416–429, 2008.

In this paper all logarithms will have base 2. Since the main algorithm has exponential complexity, we will often use "Soft-Oh" notation (see [5] for a definition) to highlight the main term and will assume "grade-school" arithmetic for simplicity.

This work was part of the author's dissertation research. Contact the author for further details or for proofs omitted from this paper.

Thanks to Eric Bach, Matt Darnall, Tom Kurtz, and Dieter van Melkebeek for thoughtful discussions that proved essential, to NSF award CCF-8635355 and the William F. Vilas Trust Estate for monetary support, and to the referees for helpful comments.

1   Previous Work and Results

The subset sum problem is of great practical and theoretical interest. Its decision version was proven NP-complete by R. Karp in his seminal 1972 paper on reductions among combinatorial problems [8]. It has seen application in the creation of public key cryptosystems [3], and is a vital tool in discovering Carmichael numbers [6].

Trivial algorithms for MSS include brute force enumeration at $O(2^n)$ time and constant space, basic time-space tradeoff at $O(2^{n/2})$ time and space, and dynamic programming at $O(n \cdot m)$ time and space. Schroeppel and Shamir [14] were first to discover a nontrivial method for solving the subset sum problem. Their algorithm takes time $O(2^{n/2})$ and space $O(2^{n/4})$, using a technique of decomposition that is reflected in this paper.

Despite its status as an NP-complete problem, many cases are quite tractable. If m is polynomial in n (giving a very dense instance), the problem is solvable in polynomial time using dynamic programming. More sophisticated methods can improve the running time, for example [1] achieved a running time of $O(n^{7/4}/\log^{3/4} n)$ for instances with $m \log m = \Theta(n^2)$. In [4], the range of problems solvable in polynomial time was extended to cases with $m = 2^{O(\log n)^2}$.

For sparse instances, the current favored technique is that of lattice basis reduction. If we have density $d < 0.64$, then Lagarias and Odlyzko [9] proved that almost all (as n goes to infinity) subset sum problems reduce to the shortest vector problem in polynomial time. The density bound was improved to 0.98 in [2]. Note that this work was on the integer subset sum problem, and so the definition of density used was different from the one in this paper. It was also proven in [9] that if $m = \Omega(2^{n^2})$, almost all subset sum problems are solvable in polynomial time using lattice basis reduction.

The inspiration for the present paper is the work of Lyubashevsky [10], who gave a rigorous analysis of Wagner's algorithm [16] for the k-set birthday problem over $\mathbb{Z}/m\mathbb{Z}$, proving that a solution is output with high probability. However, in order to preserve the independence and uniformity of set elements at all levels of the algorithm, the complexity of the algorithm was weakened to $\widetilde{O}(km^{2/\log k})$ time and space from Wagner's proposed complexity of $\widetilde{O}(km^{1/\log k})$ time and space. Lyubashevsky also leveraged the k-set birthday problem into a new algorithm for the random subset sum problem. This algorithm uses $\widetilde{O}(km^{2/\log k})$ time and space, though by assuming $m = 2^{n^\epsilon}$, $\epsilon < 1$ and choosing $k = \frac{1}{2}n^{1-\epsilon}$ this becomes $\widetilde{O}(2^{\frac{2n^\epsilon}{(1-\epsilon)\log n}})$.

This paper extends this research by providing a rigorous analysis of Wagner's original algorithm.

Theorem 2. *Let sets $L_1, \ldots, L_k$ each contain $\alpha m^{1/\log k}$ independent and uniformly generated elements from $\mathbb{Z}/m\mathbb{Z}$. We make the technical assumptions that $\alpha > \max\{1024, k\}$ and that $\log m > 7(\log \alpha)(\log k)$. Then Wagner's algorithm for the k-set birthday problem has complexity $\widetilde{O}(k\alpha \cdot m^{1/\log k})$ time and space and finds a solution with probability greater than $1 - m^{1/\log k}e^{-\Omega(\alpha)}$.*

The most novel part of this result is that an exponentially small failure probability is achieved despite the fact that the elements at higher levels of the algorithm are neither independent nor uniform. The key tool that makes this possible is the theory of martingales.

This theorem has profound implications for cryptographic applications that use Wagner's k-set birthday algorithm. Though this analysis has only been done for the case of $\mathbb{Z}/m\mathbb{Z}$, it is anticipated that the techniques developed will work for other algebraic objects where the k-set birthday algorithm is applied, most notably the case of bit strings with the bitwise exclusive-or operation. This will provide justification for the use of the k-set birthday algorithm in cryptography.

Following [10], we get the following new result for RMSS as a corollary. This gives the fastest known algorithm for dense problems of asymptotic density smaller than $n/(\log n)^2$.

Theorem 3. *Let $m = 2^{n^\epsilon}$, $\epsilon < 1$, and assume that $n^\epsilon = \Omega((\log n)^2)$. Then there is a randomized algorithm that runs using time and space $\widetilde{O}(2^{\frac{n^\epsilon}{(1-\epsilon)\log n}})$ and finds a solution to RMSS with probability greater than $1 - 2^{-\Omega(n^\epsilon)}$.*

Here the probability of success is over the random bits of the algorithm and also over the random choice of inputs.

Note that by choosing $n^\epsilon = O((\log n)^2)$ the running time becomes polynomial, though not as small of a polynomial as that in [4].

The algorithm works just as well on problems of large enough constant density. Let $m = 2^{cn/k}$ for $c < \log k/(\log k + 4)$, giving problems of density greater than $k(1+\frac{4}{\log k})$. Then the randomized algorithm runs using time and space $\widetilde{O}(m^{1/\log k})$ and finds a solution to RMSS with probability greater than $1 - 2^{-\Omega(n)}$. The constant in the exponent of the success probability depends on c and on k in such a way that the probability of success increases with increasing density.

2   Outline

The outline of this paper is as follows. In Section 3 we present Wagner's algorithm for the k-set birthday problem. In Section 4 we discuss what probability distribution the elements of the lists in the algorithm have, and show that it is close to uniform. We show that the elements are close to independent in Section 5 and then give our new analysis of the k-set birthday algorithm in Section 6. The final section applies the k-set brithday algorithm to the RMSS problem.

3   The k-Set Birthday Algorithm

We next provide a description of Wagner's k-set algorithm from [16]. A key subroutine is Algorithm ListMerge, which takes as input two lists L_1 and L_2 of integers in the interval $[-\frac{R}{2}, \frac{R}{2})$ and outputs elements $b + c \in [-\frac{Rp}{2}, \frac{Rp}{2})$ where $b \in L_1$, $c \in L_2$. Here $p < 1$ is a parameter set at the beginning. This subroutine is implemented by sorting L_1, L_2 and then for all $b \in L_1$, searching for $c \in L_2$ in the interval $[-b - \frac{Rp}{2}, -b + \frac{Rp}{2})$. Assuming that $|L_1| = |L_2| = N$, the cost of sorting is $O(N \log N)$ and the cost of N searches is $O(N \log N)$, giving a resource usage of $O(N \log N)$ time and space.

Algorithm 1 (ListMerge)
  Input: two lists L_1, L_2 of integers in the interval $[-\frac{R}{2}, \frac{R}{2})$, parameter $p < 1$
  Output: list L_{12} of integers $b + c \in [-\frac{Rp}{2}, \frac{Rp}{2})$ where $b \in L_1$, $c \in L_2$

1. sort L_1, L_2
2. **for** $b \in L_1$ **do:**
3.   pick random $c \in L_2$ from those in interval $[-b - \frac{Rp}{2}, -b + \frac{Rp}{2})$
4. $L_{12} \leftarrow L_{12} \cup \{b + c\}$
5. output L_{12}

Note that at most one $b + c$ is taken as output for each $b \in L_1$, so the output list again has at most N elements.

   For the k-set birthday problem we will choose $p = m^{-1/\log k}$, and assume that the initial k sets are populated with α/p elements of $\mathbb{Z}/m\mathbb{Z}$ chosen uniformly and independently at random. Treating the elements of the lists as integers in the interval $[-\frac{m}{2}, \frac{m}{2})$, we apply ListMerge to pairs of lists in a binary tree fashion, so that after $\log k$ levels we are left with a single list of integers in the interval $[-\frac{mp^{\log k}}{2}, \frac{mp^{\log k}}{2}) = [-\frac{1}{2}, \frac{1}{2})$. Having kept track of how each element is composed of elements from level 0, we have solved the problem (assuming the final list is nonempty) since we have found $s_1, \ldots, s_k$ such that $s_1 + \cdots + s_k = 0 \mod m$.

   The resource usage of the algorithm is dominated by that of ListMerge applied to $2k$ lists of size at most $\alpha/p = \alpha \cdot m^{1/\log k}$, and so the k-set birthday problem is solvable using $\widetilde{O}(k\alpha \cdot m^{1/\log k})$ time and space. This proves the complexity claim of Theorem 2. However, proving that the algorithm outputs a solution with reasonable probability is much more difficult. The elements of the lists at levels greater than 0 are not uniformly generated over $\mathbb{Z}/m\mathbb{Z}$, nor are they independent. In the sections that follow we will analyze the distributions that arise, finishing the proof of Theorem 2.

4   Symmetric Unimodal Distributions

We choose parameters $p = m^{-1/\log k}$ and $\alpha > \max\{1024, k\}$. We also make the weak technical assumption from Theorem 2 that $\log m > 7(\log \alpha)(\log k)$. In most of the lemmas that follow, simplification requires an assumption that p is small. Note that the condition on $\log m$ implies $p < \frac{1}{128\alpha k^5}$. This is sufficient for the results that follow, though each has a weaker condition on p if allowed.

Our strategy over the next several sections is to prove by induction that the output list of Algorithm 1 has at least α/p elements. At level 0, the initial lists have α/p elements which are independent and uniformly generated. Our inductive hypothesis is that at level $\lambda - 1$ the remaining lists have α/p elements which are close to uniform and close to independent (to be defined precisely later). In this section we analyze the distributions of the elements at level λ. For this we need the following definitions.

Definition 4. *Let independent random variables X and Y have distributions F and G with probability mass functions f and g.*

1. *The distribution F is* symmetric *about the origin if $f(-x) = f(x)$ for all x.*
2. *The distribution F is* unimodal *at a if f is nondecreasing on $(-\infty, a]$ and nonincreasing on $[a, \infty)$.*
3. *The* convolution *of F and G, denoted $F * G$, is defined by*

$$f * g(s) = \sum_{x} f(x)g(s - x)$$

where the sum is over the probability space (for ease of notation, this is extended to $(-\infty, \infty)$).

From now on we take symmetric, unimodal to mean symmetric and unimodal about the origin. We will also use symmetric, unimodal to refer to the corresponding mass function of a distribution. The following are standard facts from probability theory.

Proposition 5. *Let X and Y be independent random variables with discrete distributions F and G.*

1. *The random variable $S = X + Y$ has distribution $F * G$.*
2. *If X and Y are symmetric, $F * G$ is symmetric.*
3. *If X and Y are symmetric and unimodal, $F * G$ is unimodal.*

Proof. 1. and 2. follow directly from the definitions. The proof of 3. is more technical. For a continuous version that is nicely written, see [13].   □

Fix the following notation. At level λ of the algorithm (note that $\lambda < \log k$), we have lists L_1 and L_2 of integers in the interval $[-\frac{mp^\lambda}{2}, \frac{mp^\lambda}{2})$ with $|L_1| = |L_2| = N = \alpha/p$. Let b_i be the elements of L_1 and c_i the elements of L_2. Let I be the interval $[-\frac{mp^{\lambda+1}}{2}, \frac{mp^{\lambda+1}}{2})$ and I_b the interval $[-\frac{mp^{\lambda+1}}{2} - b, \frac{mp^{\lambda+1}}{2} - b)$ for $b \in L_1$.

Let D_λ be the distribution of the elements of L_1 and L_2 with probability mass function f_λ. The support of f_λ is $[-\frac{mp^\lambda}{2}, \frac{mp^\lambda}{2})$, so that in particular $f_{\log k}$ is supported on $[-\frac{1}{2}, \frac{1}{2})$. Since elements at level λ are sums of elements from level $\lambda - 1$ that fall in the restricted interval $[-\frac{mp^\lambda}{2}, \frac{mp^\lambda}{2})$, we see that D_λ is the convolution of two copies of $D_{\lambda-1}$ with the tails thrown out and the remainder normalized to make a new probability distribution. Symbolically this looks like

$$f_\lambda(x) = \frac{1}{\sum_{a=-mp^\lambda/2}^{mp^\lambda/2} f_{\lambda-1} * f_{\lambda-1}(a)} \cdot f_{\lambda-1} * f_{\lambda-1}(x)$$

where x ranges over the support of f_λ. Here summing over an interval will always mean summing over the integers in the interval.

Elements from different lists are independent, so we conclude from Proposition 5 that at all levels, the distributions D_λ are symmetric unimodal.

A very surprising and useful fact is that f_λ is always close to a uniform distribution. The following lemma supports this claim by bounding the largest difference between f_λ and the uniform distribution on the support of f_λ. Intuitively, $6^\lambda p$ is small since $\lambda < \log k$ and p is small. Note that $\log m > 7(\log \alpha)(\log k)$ implies the required condition that $p \leq 1/(24k^3)$.

Lemma 6. Let U be the uniform distribution on $[-\frac{mp^\lambda}{2}, \frac{mp^\lambda}{2})$, and assume that $p \leq 1/(24k^3)$. Then for all $x \in [-\frac{mp^\lambda}{2}, \frac{mp^\lambda}{2})$,

$$|f_\lambda(x) - U(x)| \leq \frac{6^\lambda p}{mp^\lambda} .$$

Consider that if two uniform distributions are convolved the result is a triangle distribution. While far from uniform, if we only consider part of the distribution above a small interval centered at the origin the result is much closer to uniform. Carefully bounding the highest and lowest points while using induction on λ gives the proof, the details of which may be found in [15, Sect. 5.1].

The next result allows us to bound the expected number of elements in the output of Algorithm 1.

Proposition 7. Assume notation for level λ, and that $p \leq 1/(2k^3)$. Let $b \in L_1$ be a random variable and let $\ell_i = b + c_i$ for $c_i \in L_2$. Assume that $|L_2| = \alpha/p$. Then the expected number of ℓ_i in I is at least $\alpha/8$.

Proof. The work is in bounding $\Pr[\ell_i \in I] = \sum_b \Pr[b] \Pr[c_i \in I_b]$. Assume first that the level is not $\log k - 1$.

The number of integers in I_b is at least $\lfloor \frac{mp^{\lambda+1}}{2} \rfloor$, this lower bound corresponding to the case when $b = \pm\frac{mp^\lambda}{2}$. Using Lemma 6, we conclude that

$$\Pr[c_i \in I_b] \geq \frac{1 - 6^\lambda p}{mp^\lambda} \left(\frac{mp^{\lambda+1}}{2} - 1 \right) \geq \frac{1}{2mp^\lambda} \left(\frac{mp^{\lambda+1}}{2} - 1 \right) \tag{2}$$

$$= \frac{p}{4} - \frac{1}{2mp^\lambda} \geq \frac{p}{8} \tag{3}$$

where for (2), $p \leq 1/(2k^3) \leq 1/(2 \cdot 6^\lambda)$ by assumption and for (3) we assume $mp^{\lambda+1} \geq 4$ (satisfied since $\lambda + 1 < \log k$).

We conclude that $\Pr[\ell_i \in I] \geq p/8$ for all i and hence that the expected number of ℓ_i in I is at least $\alpha/8$.

If the level is $\log k - 1$, then $mp^\lambda = 1/p$ and $mp^{\lambda+1} = 1$. Thus I_b contains exactly one integer unless $b = \pm mp^\lambda/2$. Since the distribution is symmetric unimodal, these values of b are the least likely and so

$$\Pr[\ell_i \in I] \geq \sum_{b \neq \pm 1/(2p)} \Pr[b] \left(\frac{1 - 6^\lambda p}{mp^\lambda} \right) \geq \frac{1}{2} \cdot \frac{p}{2}$$

again using the assumption that $p \leq 1/(2 \cdot 6^\lambda)$. $\square$

We conclude from this lemma that if L_1 and L_2 have α/p elements, the output of Algorithm ListMerge is expected to again have at least α/p elements. Our task in the next two sections is to prove that the number of elements is close to the expected value with high probability.

5   Bounding Dependency

Since we have uniform bounds for most distributions, we often suppress the value a random variable takes in expressing a probability. For example, $\Pr[\ell]$ means the probability that a random variable ℓ takes some unspecified value in its interval of support.

In the last section we showed that the distributions which arise in the k-set birthday algorithm are close to uniform, which allowed us to bound the expected size of the output of Algorithm 1. In this section we analyze what dependencies arise among list elements.

The first observation is that they are not independent. Consider the following example using the notation for combining lists L_1 and L_2 at level λ, where X_i is a Bernoulli random variable taking value 1 if $\ell_i \in I$ and 0 otherwise. If $\ell_1 = b_1 + c_1$, $\ell_2 = b_2 + c_1$, $\ell_3 = b_1 + c_2$, and $\ell_4 = b_2 + c_2$, then $\ell_4 = \ell_2 + \ell_3 - \ell_1$. Thus the random variable X_4 is functionally dependent upon X_1, X_2, X_3. Avoiding similar examples is the inspiration for the following definition.

Definition 8. *Organize the elements of $L_1 + L_2$ at level λ into a table, where if $\ell = b + c$ it appears in the row corresponding to b and column corresponding to c. Then $\ell_1, \ldots, \ell_j$ are called* row distinct *if they each appear in a distinct row.*

To motivate the next lemma, suppose that the distributions of the elements of L_1 and L_2 (at level 0) are uniform over $\mathbb{Z}/m\mathbb{Z}$, and that sums are taken over $\mathbb{Z}/m\mathbb{Z}$. Then if ℓ_1 shares column c with ℓ_2,

$$\Pr[\ell_1, \ell_2] = \sum_{z \in \mathbb{Z}/m\mathbb{Z}} \Pr[c = z] \Pr[b_1 = \ell_1 - z] \Pr[b_2 = \ell_2 - z] = \frac{1}{m^2}$$

while if L_1 and L_2 share neither row nor column they are also independent.

This extends easily to larger numbers of ℓ_i, proving that in this simple situation row distinct implies independent. At higher levels the sums start dropping terms due to exceeding interval bounds. However, since we are interested in the dependence only among those ℓ_i in the restricted interval, the number of terms lost is small. Combining these ideas along with Lemma 6 and induction on λ yields the following technical result. The proof may be found in [15, Sect. 5.5].

Lemma 9. *Let the current level of the algorithm be λ, and let X' be the event $X_1 = 1 \wedge \cdots \wedge X_{r-1} = 1$. Assume that $\ell_1, \ldots, \ell_r$ are row distinct. Then*

$$\frac{(1-p)^{4^\lambda}(1 - 3 \cdot 6^\lambda p)^{4^{\lambda-1}}}{(1 + 4 \cdot 6^\lambda p)^{4^{\lambda-1}}} \leq \frac{\Pr[\ell_1, \ldots, \ell_r \mid X_r = 1, X']}{\Pr[\ell_r \mid X_r = 1] \Pr[\ell_1, \ldots, \ell_{r-1} \mid X']} \quad and$$

$$\frac{\Pr[\ell_1, \ldots, \ell_r \mid X_r = 1, X']}{\Pr[\ell_r \mid X_r = 1] \Pr[\ell_1, \ldots, \ell_{r-1} \mid X']} \leq \frac{(1 + 4 \cdot 6^\lambda p)^{4^{\lambda-1}}}{(1-p)^{4^\lambda}(1 - 3 \cdot 6^\lambda p)^{4^{\lambda-1}}}$$

unless both numerator and denominator are 0.

Using power series and the assumption that $p \leq \frac{1}{864k^5} \leq \frac{1}{864 \cdot 24^\lambda}$ gives conceptually simpler bounds of $1 \pm 2 \cdot 24^\lambda p$. Also note that the distribution of b or c from level $\lambda + 1$ is the same as that of ℓ_i given $X_i = 1$ from level λ, so we can rewrite the statement of Lemma 9 for level $\lambda + 1$ as

$$1 - 2 \cdot 24^\lambda p \leq \frac{\Pr[c_1, \ldots, c_r]}{\Pr[c_1] \Pr[c_2, \ldots, c_r]} \leq 1 + 2 \cdot 24^\lambda p . \tag{4}$$

This bound on the dependence will allow us to prove in the next section that Algorithm 1 outputs $N = \frac{\alpha}{p}$ row distinct elements with high probability.

6   Correctness Proof

Recall our induction hypothesis that the lists at level $\lambda - 1$ have α/p elements (one from each row, making them row distinct), and that these elements are close to uniform in the sense of Lemma 6 and close to independent in the sense of Lemma 9. Lemmas 6 and 9 are true at all levels, so to finish the induction it is enough to prove that every row contains an element in the restricted interval I. In fact we apply a tail bound to show that the probability is low that the number of row elements in the restricted interval strays too far from the expected value of $\alpha/8$.

Recall our previous notation for level λ of the k-set birthday algorithm. We are interested in proving that at least one ℓ_j per row is in the restricted interval $I = [-\frac{mp^{\lambda+1}}{2}, \frac{mp^{\lambda+1}}{2})$. Towards this end we fix $b \in L_1$ and relabel indices so that $\ell_i = b + c_i$ for $1 \leq i \leq N$. Let $I_b = [-\frac{mp^{\lambda+1}}{2} - b, \frac{mp^{\lambda+1}}{2} - b)$, and redefine the random variable X_i to take value 1 if $c_i \in I_b$ and 0 otherwise.

The next set of notation follows a survey paper by McDiarmid [12, Sect. 3.2] that covers numerous concentration inequalities and their applications to problems in combinatorics and computer science.

Let $f(\boldsymbol{X})$ be a bounded real valued function on $X_1, \ldots, X_N$, which for our purposes will be $S = \sum_{i=1}^{N} X_i$. Let B denote the event that $X_i = x_i$ for $i = 1, \ldots, j-1$ where x_i is either 0 or 1. For $x = 0, 1$ let

$$K_j(x) = \mathrm{E}[f(\boldsymbol{X}) \mid B, X_j = x] - \mathrm{E}[f(\boldsymbol{X}) \mid B] .$$

Define $dev(x_1, \ldots, x_{j-1})$ to be $\sup\{|K_j(0)|, |K_j(1)|\}$, while $ran(x_1, \ldots, x_{j-1})$ is defined to be $|K_j(0) - K_j(1)|$.

Let the *sum of squared ranges* be $R^2(\boldsymbol{x}) = \sum_{j=1}^{N} ran(x_1, \ldots, x_{j-1})^2$ and let $\hat{r}^2$, the *maximum sum of squared ranges*, be the supremum of $R^2(\boldsymbol{x})$ over all choices of $\boldsymbol{x} = (x_1, \ldots, x_N)$. Let $maxdev$ be the maximum of $dev(x_1, \ldots, x_{j-1})$ over all choices of j and all choices of x_i.

The context of all this notation is the theory of martingales. By the Doob construction, $Y_j = \mathrm{E}[f(\boldsymbol{X}) \mid X_1, \ldots, X_j]$, $1 \leq j \leq N$ forms a martingale sequence. The standard tail bound for martingales is the Azuma-Hoeffding theorem, but in our case this is not tight enough to be meaningful. The theorem we will use instead is the following martingale version of Bernstein's inequality, proven by McDiarmid [12, Sect. 3.2].

Theorem 10. *Let $X_1, \ldots, X_N$ be a family of random variables with X_i taking values in $\{0,1\}$, and let f be a bounded real-valued function defined on $\{0,1\}^N$. Let μ denote the mean of $f(\boldsymbol{X})$, let b denote the maximum deviation maxdev, and let $\hat{r}^2$ denote the maximum sum of squared ranges. Suppose that X_i takes two values with the smaller probability being $p < \frac{1}{2}$. Then for any $t \geq 0$,*

$$\Pr[|f(\boldsymbol{X}) - \mu| \geq t] \leq 2 \exp\left(-\frac{t^2}{2p\hat{r}^2(1 + bt/(3p\hat{r}^2))}\right) .$$

In our application, μ is $\alpha/8$ and thus we choose $t = \alpha/16$. We will prove that $bt/(3p\hat{r}^2)$ is small, which means that $\hat{r}^2$ needs to be not much bigger than α/p (the value it would take if the X_i are independent) in order for the bound to be meaningful. The next lemma is crucial for finding good bounds on $\hat{r}^2$ and $maxdev$.

Lemma 11. *Use the notation for level λ, with X_i being the indicator event for $c_i \in I_b$. Assume that $c_1, \ldots, c_N$ are row distinct when treated as ℓ_i from level $\lambda - 1$, and independent if $\lambda = 0$. Then for any $i > j$,*

$$|\mathrm{E}[X_i \mid X_1, \ldots, X_j] - \mathrm{E}[X_i]| \leq 4 \cdot 24^\lambda p^2$$

assuming that $p \leq 1/(864k^5)$.

Proof. We will reduce the result to finding a uniform bound for $|\Pr[c_i \mid X_1, \ldots, X_j] - \Pr[c_i]|$. If the level λ is 0, then $c_1, \ldots, c_j$, c_i are all fully independent and hence $|\Pr[c_i \mid X_1, \ldots, X_j] - \Pr[c_i]| = 0$. In the general case

$$\Pr[c_i \mid X_1, \ldots, X_j] = \frac{\Pr[c_i \wedge X_1 \wedge \cdots \wedge X_j]}{\Pr[X_1 \wedge \cdots \wedge X_j]}$$

$$= \frac{\sum_{d_1} \cdots \sum_{d_j} \Pr[c_i \wedge c_1 = d_1 \wedge \cdots \wedge c_j = d_j]}{\sum_{d_1} \cdots \sum_{d_j} \Pr[c_1 = d_1 \wedge \cdots \wedge c_j = d_j]}$$

$$\leq (1 + 2 \cdot 24^{\lambda-1} p) \Pr[c_i]$$

where each sum in the numerator and denominator ranges over $d \in I_b$ if $X = 1$ and $d \notin I_b$ if $X = 0$. For the last step we have used (4) to break off the $\Pr[c_i]$ term, after which the rest of the terms cancel. A similar argument using the other inequality in (4) gives a lower bound for $\Pr[c_i \mid X_1, \ldots, X_j]$ of $(1 - 2 \cdot 24^{\lambda-1} p) \Pr[c_i]$. We now have

$$|\Pr[c_i \mid X_1, \ldots, X_j] - \Pr[c_i]| \leq |\Pr[c_i](1 \pm 2 \cdot 24^\lambda p) - \Pr[c_i]|$$

$$= \Pr[c_i] \cdot 2 \cdot 24^\lambda p$$

$$\leq 2 \cdot 24^\lambda p \left(\frac{1}{mp^\lambda} + \frac{6^\lambda p}{mp^\lambda} \right)$$

using Lemma 6 to bound $\Pr[c_i]$, and conclude that

$$|\mathrm{E}[X_i \mid X_1, \ldots, X_j] - \mathrm{E}[X_i]| \leq \sum_{c_i \in I_b} |\Pr[c_i \mid X_1, \ldots, X_j] - \Pr[c_i]|$$

$$\leq mp^{\lambda+1} \cdot 2 \cdot 24^\lambda p \left(\frac{1}{mp^\lambda} + \frac{6^\lambda p}{mp^\lambda} \right)$$

$$\leq 4 \cdot 24^\lambda p^2$$

by our assumption on p.   □

With ingredients in hand, we next present the correctness proof for Algorithm 1. Note that the result requires and preserves the property that each list contains a row distinct sublist. This is prevented from circular reasoning by the fact that list elements at level 0 are independent.

Lemma 12 (k-set ListMerge). *Use the notation for level λ. Let A be the following event: for every $b \in L_1$, there exists $c \in L_2$ such that $b + c \in [-\frac{mp^{\lambda+1}}{2}, \frac{mp^{\lambda+1}}{2})$. Then*

$$\Pr[A] \geq 1 - (\alpha/p) e^{-\alpha/1024}$$

assuming that $p \leq 1/(128\alpha k^5)$.

Proof. First consider one row of the table. Fix $b \in L_1$, and let X_i be indicator variables for $c_i \in I_b$, $1 \leq i \leq N$. Then by Proposition 7 we have $\mathrm{E}[X_i] \geq p/8$ and hence that $\mathrm{E}[S] \geq \alpha/8$.

Our main goal now is to find upper bounds for *maxdev* and $\hat{r}^2$.

Consider $dev(x_1, \ldots, x_{j-1})$ for any $j \leq N$, any choice of $x_1, \ldots, x_{j-1}$, and $x_j = 0$ or 1. Note that

$$|K_j(x_j)| = |E[S \mid X_1, \ldots, X_j] - E[S \mid X_1, \ldots, X_{j-1}]|$$

$$\leq \sum_{i=1}^{N} |E[X_i \mid X_1, \ldots, X_j] - E[X_i \mid X_1, \ldots, X_{j-1}]| \; .$$

If $i < j$ the corresponding term is 0 since its value has already been fixed. If $i = j$ the term can be at most 1 since that is the range of X_i. If $i > j$ then we apply Lemma 11 to see that

$$|E[X_i \mid X_1, \ldots, X_j] - E[X_i \mid X_1, \ldots, X_{j-1}]|$$
$$= |E[X_i \mid X_1, \ldots, X_j] - E[X_i] + E[X_i] - E[X_i \mid X_1, \ldots, X_{j-1}]|$$
$$\leq 8 \cdot 24^\lambda p^2$$

Thus $maxdev$ is no more than $1 + \frac{\alpha}{p} \cdot 8 \cdot 24^\lambda p^2 = 1 + 8 \cdot 24^\lambda \alpha p$. By definition

$$ran(x_1, \ldots, x_{j-1}) = |K_j(0) - K_j(1)|$$
$$= |E[S \mid B, X_j = 1] - E[S \mid B, X_j = 0]|$$
$$\leq \sum_{i=1}^{N} |E[X_i \mid B, X_j = 1] - E[X_i \mid B, X_j = 0]| \; .$$

Following the same reasoning as we did for $maxdev$, if $i < j$ the corresponding term is 0, if $i = j$ the corresponding term is at most 1 since X_i is an indicator variable, while if $i > j$ the term is at most $8 \cdot 24^\lambda p^2$ by Lemma 11.

So $ran(x_1, \ldots, x_{j-1})$ has a uniform upper bound of $1 + 8 \cdot 24^\lambda \alpha p$ and thus $\hat{r}^2 \leq \frac{\alpha}{p}(1 + 8 \cdot 24^\lambda \alpha p)^2 \leq \frac{\alpha}{p} + 32 \cdot 24^\lambda \alpha^2$, assuming that $p \leq \frac{1}{4\alpha 24^\lambda}$.

We now conclude from Theorem 10 that

$$\Pr\left[S \leq \frac{\alpha}{8} - \frac{\alpha}{16}\right] \leq \Pr\left[S \leq \mu - \frac{\alpha}{16}\right]$$

$$\leq \exp\left(-\frac{\alpha^2/256}{2(\alpha + 32 \cdot 24^\lambda \alpha^2 p) + \frac{2}{3}\frac{\alpha}{16}(1 + 8 \cdot 24^\lambda \alpha p)}\right)$$

$$\leq \exp\left(-\frac{\alpha^2/256}{2\alpha + \frac{\alpha}{2} + \frac{2}{3}\alpha + \frac{\alpha}{2}}\right) \leq e^{-\alpha/1024}$$

assuming that $p \leq 1/(128\alpha 24^\lambda)$.

Finally, by using the union bound the probability that some row fails to have at least $\alpha/16$ elements fall in I_b is smaller than $(\alpha/p)e^{-\alpha/1024}$ and the bound on the probability of event A follows.   □

The proof of Theorem 2 now follows quite easily.

Proof (Theorem 2)

Lemma 12 completed our proof by induction on the level that all lists have α/p elements with high probability. An application of Algorithm Listmerge on lists of size α/p will again result in a list of size α/p with probability at least $1 - (\alpha/p)e^{-\alpha/1024}$.

The k-set birthday algorithm successfully finds a solution as long as this occurs for all $2k$ applications of Algorithm Listmerge. Thus the algorithm succeeds with probability at least

$$(1 - (\alpha/p)e^{-\alpha/1024})^{2k} > 1 - 2k(\alpha/p)e^{-\alpha/1024} = 1 - m^{1/\log k}e^{-\Omega(\alpha)} .$$

As for the complexity, it is dominated by storing, sorting, and searching through $2k$ lists of size $\alpha m^{1/\log k}$, giving a time and space bound of $\widetilde{O}(k\alpha \cdot m^{1/\log k})$.

$\square$

7   Application to RMSS

In this section we show how to use the multi-set birthday problem to solve dense instances of RMSS. This work can be found in [10] and [11]; we include it here for completeness.

Consider the following random variable $Z_{\boldsymbol{a}}$ taking values on $\mathbb{Z}/m\mathbb{Z}$, where $\boldsymbol{a} = (a_1, \ldots, a_n)$ with $a_i \in \mathbb{Z}/m\mathbb{Z}$. Let $\boldsymbol{x} = (x_1, \ldots, x_n)$ be an n-bit vector, where each element is drawn uniformly and independently from $\{0, 1\}$. Then we define

$$Z_{\boldsymbol{a}} := \sum_{i=1}^{n} x_i a_i \quad \mod m . \tag{5}$$

In the case where $\boldsymbol{a}$ is fixed and understood from context (say where it is the input of an RMSS instance) we will suppress the $\boldsymbol{a}$ in the notation. Note that for fixed $\boldsymbol{a}$ and varying $\boldsymbol{x}$, the collection $\{Z_{\boldsymbol{a}}(\boldsymbol{x})\}$ is a collection of independent random variables .

Our goal is to show that the distribution of $Z_{\boldsymbol{a}}$ is close to uniform, thus it is vital that we formalize what we mean by "close."

Definition 13. *Let X and Y be random variables taking values in a probability space A. The statistical distance between X and Y, denoted $\Delta(X, Y)$, is*

$$\Delta(X, Y) = \frac{1}{2} \sum_{a \in A} |Pr[X = a] - Pr[Y = a]| .$$

The next proposition states that for most choices of $\boldsymbol{a}$, $Z_{\boldsymbol{a}}$ is exponentially close to uniform. The proof involves showing that $\{Z_{\boldsymbol{a}} : \{0, 1\}^n \to \mathbb{Z}/m\mathbb{Z}\}_{\boldsymbol{a} \in (\mathbb{Z}/m\mathbb{Z})^n}$ is a universal (and hence almost universal) family of hash functions, and then applying the leftover hash lemma. Here we encode elements of $\mathbb{Z}/m\mathbb{Z}$ as bit strings of length cn, $c < 1$.

Proposition 14 (Impagliazzo, Naor [7]). *Let $m = 2^{cn}$ with $c < 1$. Then the probability over all choices of input vector $\boldsymbol{a} = (a_1, \ldots, a_n)$ that $\Delta(Z_{\boldsymbol{a}}, U) <$ $2^{-\frac{(1-c)n}{4}}$ is greater than $1 - 2^{-\frac{(1-c)n}{4}}$.*

Definition 15. *We call the multiset $\boldsymbol{a} = (a_1, \ldots, a_n)$ well-distributed if it is one of the good choices from Proposition 14, i.e.*

$$\Delta(Z_{\boldsymbol{a}}, U) < 2^{-\frac{(1-c)n}{4}} \ .$$

So if the a_i are chosen uniformly at random, we lose very little by assuming that $Z_{\boldsymbol{a}}$ is uniform. This is the only place we use the fact that our subset sum problem is random, and it is possible to apply the k-set algorithm to MSS instances with the additional assumption that $\boldsymbol{a}$ is well-distributed. This might be preferable if a constructive criterion could be found for $\boldsymbol{a}$ being well-distributed, but for now that remains an open problem. Note that a necessary condition for being well-distributed is that the a_i contain no common factor. If $\gcd(a_1, \ldots, a_n, m) \neq 1$ then $Z_{\boldsymbol{a}}$ is only nonzero on a subgroup of $\mathbb{Z}/m\mathbb{Z}$, and thus is far from uniform.

Just as in [10], our subset sum algorithm is as follows. Choose parameters k and α. Break up the a_i into k sets, and generate k lists where each list contains $\alpha m^{1/\log k}$ random subset sums of that portion of the a_i. Apply the k-set birthday algorithm to find a solution.

Proof (Theorem 3)
For the analysis we make parameter choices of $\alpha = n$ and $k = \frac{1}{2}n^{1-\epsilon}$. Our assumption that $n^\epsilon = \Omega((\log n)^2)$ satisfies the requirement of Theorem 2 that $\log m > 7(\log \alpha)(\log k)$.

The probability of success is greater than the probability that all k subsets of $\boldsymbol{a}$ are well-distributed, times the probability that the algorithm succeeds given that all subsets are well-distributed. By applying Proposition 14, the probability that one of the subsets is well-distributed is greater than $1 - 2^{-\frac{(1-c)n}{4k}}$, where $c = kn^{\epsilon-1}$ since $2^{n^\epsilon} = 2^{cn/k}$. Thus the probability that all are well-distributed is greater than

$$\left(1 - 2^{-\frac{(1-.5)n}{4k}}\right)^k > 1 - \frac{1}{2}n^{1-\epsilon} \cdot 2^{-\frac{n^\epsilon}{4}} \geq 1 - 2^{-\Omega(n^\epsilon)} \ .$$

In addition, the distance between these distributions and uniform ones is less than $2^{-\frac{n^\epsilon}{4}}$.

Now, assume that all elements from all initial lists are drawn independently from uniform distributions. Then the probability that the k-set birthday algorithm succeeds is at least

$$1 - n^{1-\epsilon} \cdot n2^{\frac{n^\epsilon}{(1-\epsilon)\log n}} e^{-n/1024} \geq 1 - n^{2-\epsilon}2^{-n\left(\frac{1}{1024} - \frac{1}{(1-\epsilon)n^{1-\epsilon}\log n}\right)} \geq 1 - 2^{-\Omega(n)} \ .$$

Accounting for the fact that the elements of the initial lists are only close to uniform, the probability of the birthday algorithm succeeding is reduced by $2^{-\Omega(n^\epsilon)}$ (see [11]).

Thus the probability of success of the multi-set subset sum algorithm is greater than

$$\left(1 - 2^{-\Omega(n^{\epsilon})}\right)\left(1 - 2^{-\Omega(n)} - 2^{-\Omega(n^{\epsilon})}\right) \geq 1 - 2^{-\Omega(n^{\epsilon})} \ .$$

The complexity is dominated by the complexity of the k-set birthday algorithm. Thus the algorithm takes $\widetilde{O}(k\alpha \cdot m^{1/\log k}) = \widetilde{O}(2^{\frac{n^{\epsilon}}{(1-\epsilon)\log n}})$ time and space.

$\square$

References

1. Chaimovich, M.: New algorithm for dense subset-sum problem. Astérisque 258, 363–373 (1999)
2. Coster, M.J., Joux, A., LaMacchia, B.A., Odlyzko, A.M., Schnorr, C.P., Stern, J.: Improved low–density subset sum algorithms. Comput. Complexity 2(2), 111–128 (1992)
3. Diffie, W., Hellman, M.E.: New directions in cryptography. IEEE Trans. Information Theory IT-22(6), 644–654 (1976)
4. Flaxman, A., Przydatek, B.: Solving medium-density subset sum problems in expected polynomial time. In: Diekert, V., Durand, B. (eds.) STACS 2005. LNCS, vol. 3404, pp. 305–314. Springer, Heidelberg (2005)
5. von zur Gathen, J., Gerhard, J.: Modern Computer Algebra, 2nd edn. Cambridge University Press, Cambridge (2003)
6. Howe, E.W.: Higher-order Carmichael numbers. Math. Comp. 69(232), 1711–1719 (2000)
7. Impagliazzo, R., Naor, M.: Efficient cryptographic schemes provably as secure as subset sum. J. of Cryptology 9(4), 199–216 (1996)
8. Karp, R.M.: Reducibility among combinatorial problems. In: Complexity of Computer Computations, pp. 85–103. Plenum Press, New York (1972)
9. Lagarias, J., Odlyzko, A.: Solving low-density subset sum problems. JACM: Journal of the ACM 32(1), 229–246 (1985)
10. Lyubashevsky, V.: The parity problem in the presence of noise, decoding random linear codes, and the subset sum problem. In: Chekuri, C., Jansen, K., Rolim, J.D.P., Trevisan, L. (eds.) APPROX 2005 and RANDOM 2005. LNCS, vol. 3624, pp. 378–389. Springer, Heidelberg (2005)
11. Lyubashevsky, V.: On random high density subset sums. Electronic Colloquium on Computational Complexity (ECCC) 12 (2005), http://eccc.hpi-web.de/eccc-reports/2005/TR05-007/index.html
12. McDiarmid, C.: Concentration. In: Probabilistic Methods for Algorithmic Discrete Mathematics. Algorithms Combin., vol. 16, pp. 195–248. Springer, Berlin (1998)
13. Purkayastha, S.: Simple proofs of two results on convolutions of unimodal distributions. Statist. Prob. Lett. 39(2), 97–100 (1998)
14. Schroeppel, R., Shamir, A.: A $T = O(2^{n/2})$, $S = O(2^{n/4})$ algorithm for certain NP-complete problems. SIAM J. Comput. 10(3), 456–464 (1981)
15. Shallue, A.: Two Number-Theoretic Problems that Illustrate the Power and Limitations of Randomness. PhD thesis, University of Wisconsin–Madison (2007)
16. Wagner, D.: A generalized birthday problem (extended abstract). In: Yung, M. (ed.) CRYPTO 2002. LNCS, vol. 2442, pp. 288–303. Springer, Heidelberg (2002)

On the Diophantine Equation $x^2 + 2^\alpha 5^\beta 13^\gamma = y^n$

Edray Goins[1], Florian Luca[2], and Alain Togbé[3,*]

[1] Department of Mathematics, Purdue University
150 North University Street, West Lafayette IN 47907 USA
egoins@purdue.edu
[2] Instituto de Matemáticas UNAM, Campus Morelia
Apartado Postal 27-3 (Xangari), C.P. 58089, Morelia, Michoacán, Mexico
fluca@matmor.unam.mx
[3] Department of Mathematics, Purdue University
North Central, 1401 S, U.S. 421, Westville IN 46391 USA
atogbe@pnc.edu

Abstract. In this paper, we find all the solutions of the Diophantine equation $x^2 + 2^\alpha 5^\beta 13^\gamma = y^n$ in nonnegative integers $x, y, \alpha, \beta, \gamma, n \geq 3$ with x and y coprime. In fact, for $n = 3, 4, 6, 8, 12$, we transform the above equation into several elliptic equations written in cubic or quartic models for which we determine all their $\{2, 5, 13\}$-integer points. For $n \geq 5$, we apply a method that uses primitive divisors of Lucas sequences. Again we are able to obtain several elliptic equations written in cubic models for which we find all their $\{2, 5, 13\}$-integer points. All the computations are done with MAGMA [12].

Keywords: Exponential equations, Diophantine equation, Computer solution of Diophantine equations.

1 Introduction

The Diophantine equation

$$x^2 + C = y^n, \quad x \geq 1, \quad y \geq 1, \quad n \geq 3 \tag{1}$$

in integers x, y, n once C is given has a rich history. In 1850, Lebesgue [18] proved that the above equation has no solutions when $C = 1$. In 1965, Chao Ko [15] proved that the only solution of the above equation with $C = -1$ is $x = 3$, $y = 2$. J. H. E. Cohn [14] solved the above equation for several values of the parameter C in the range $1 \leq C \leq 100$. A couple of the remaining values of C in the above range were covered by Mignotte and De Weger in [23], and the remaining ones in the recent paper [13]. In [26], all solutions of the similar looking equation $x^2 + C = 2y^n$, where $n \geq 2$, x and y are coprime, and $C = B^2$ with $B \in \{3, 4, \ldots, 501\}$ were found.

---

* The first and third authors were respectively partially supported by Purdue University and by Purdue University North Central. The second author was partially supported by Grant SEP-CONACyT 46755.

A.J. van der Poorten and A. Stein (Eds.): ANTS-VIII 2008, LNCS 5011, pp. 430–442, 2008.

Recently, several authors become interested in the case when only the prime factors of C are specified. For example, the case when $C = p^k$ with a fixed prime number p was dealt with in [5] and [17] for $p = 2$, in [7], [6] and [19] for $p = 3$, and in [8] for $p = 5$ and k odd. Partial results for a general prime p appear in [10] and [16]. All the solutions when $C = 2^a 3^b$ were found in [20], and when $C = p^a q^b$ where $p, q \in \{2, 5, 13\}$, were found in the sequence of papers [4], [21] and [22]. For an analysis of the case $C = 2^\alpha 3^\beta 5^\gamma 7^\delta$, see [24]. See also [9], [25], as well as the recent survey [3] for further results on this type of equations.

In this note, we consider the equation

$$x^2 + 2^\alpha 5^\beta 13^\gamma = y^n, \ x \geq 1, \ y \geq 1, \ \gcd(x, y) = 1,$$
$$n \geq 3, \ \alpha \geq 0, \ \beta \geq 0, \ \gamma \geq 0. \quad (2)$$

We have the following result.

Theorem 1. *The equation* (2) *has no solution except for*

$n = 3$	*the solutions given in Table 1;*
$n = 4$	*the solutions given in Table 2;*
$n = 5$	$(x, y, \alpha, \beta, \gamma) = (401, 11, 1, 3, 0);$
$n = 6$	$(x, y, \alpha, \beta, \gamma) \in \{(25, 3, 3, 0, 1), (23, 3, 3, 2, 0),$
	$(333, 7, 3, 1, 2), (521, 9, 5, 4, 1)\};$
$n = 7$	$(x, y, \alpha, \beta, \gamma) = (43, 3, 1, 0, 2);$
$n = 8$	$(x, y, \alpha, \beta, \gamma) \in \{(79, 3, 6, 1, 0), (49, 3, 6, 1, 1)\};$
$n = 12$	$(x, y, \alpha, \beta, \gamma) = (521, 3, 5, 4, 1).$

One can deduce from the above result the following corollary.

Corollary 1. *The equation*

$$x^2 + 13^c = y^n, \ x \geq 1, \ y \geq 1, \ \gcd(x, y) = 1, \ n \geq 3, \ c > 0 \quad (3)$$

has only the solution $(x, y, c, n) = (70, 17, 1, 3).$

For the proof, we apply the method used in [4]. In Section 2, we treat the case $n = 3$. In this case, we transform equation (2) into several elliptic equations written in cubic models for which we need to determine all their $\{2, 5, 13\}$-integer points. We use the same method in Section 3 to determine the solutions of (2) for $n = 4$. However, in this case, we use quartic models of elliptic curves. In the last section, we study the equation for $n \geq 5$ and $n \neq 6, 8, 12$. The method here uses primitive divisors of Lucas sequences. All the computations are done with MAGMA [12]. Our results from the last section contain some results already obtained in the literature as well as some new results.

Table 1. Solutions for $n = 3$

α_1	β_1	γ_1	z	α	β	γ	x	y
0	0	1	1	0	0	1	70	17
0	2	2	1	0	2	2	142	29
0	2	2	2	6	2	2	98233	2129
1	0	0	1	1	0	0	5	3
1	0	0	$2 \cdot 5$	7	6	0	383	129
1	0	1	1	1	0	1	1	3
1	0	1	1	1	0	1	207	35
1	0	1	2	7	0	1	57	17
1	0	1	2	7	0	1	18719	705
1	0	1	5	1	6	1	8553	419
1	0	1	2^4	25	0	1	15735	881
1	2	3	1	1	2	3	151	51
1	2	5	2^2	13	2	5	1075281	10721
1	3	2	2	7	3	2	3114983	21329
1	4	0	1	1	4	0	9	11
1	4	2	1	1	4	2	9823	459
1	4	2	5^2	1	16	2	46679827	130659
2	0	0	1	2	0	0	11	5
2	0	2	1	2	0	2	27045	901
2	0	2	2	8	0	2	6183	337
2	2	4	$2^2 \cdot 13$	14	2	10	137411503	422369
2	4	1	1	2	4	1	441	61
3	0	1	1	3	0	1	25	9
3	0	1	$2 \cdot 5^2$	9	12	1	1071407	14049
3	0	3	2	9	0	3	181	105
3	1	0	$2 \cdot 13$	9	1	6	83149	2681
3	1	2	1	3	1	2	333	49
3	2	0	2	9	2	0	17771	681
3	2	0	1	3	2	0	23	9
3	2	1	2	9	2	1	109513	2289
3	4	3	2^2	15	4	3	11706059	51561
4	0	2	1	4	0	2	47	17
4	4	0	13	4	4	6	1397349	12601
5	1	2	1	5	1	2	3017	209
5	2	0	1	5	2	0	261	41
5	2	1	2	11	2	1	1217	129
5	2	3	1	5	2	3	103251	2201
5	4	1	1	5	4	1	521	81

Table 2. Solutions for $n = 4$

α_1	β_1	γ_1	z	α	β	γ	x	y
0	1	0	2	4	1	0	1	3
0	1	1	1	0	1	1	4	3
0	1	1	2^3	12	1	1	959	33
1	0	0	2	5	0	0	7	3
1	0	1	$2 \cdot 5$	5	4	1	521	27
1	2	1	2	5	2	1	2599	51
2	1	0	2	6	1	0	79	9
2	1	1	2	6	1	1	49	9
2	1	1	2^2	10	1	1	16639	129
3	2	1	2	7	2	1	391	21

2   The Case $n = 3$, 6, or 12

Lemma 1. *When $n = 3$, then the only solutions to equation (2) are given in Table 1; when $n = 6$, the only solutions are*

$$(25, 3, 3, 0, 1), (23, 3, 3, 2, 0), (333, 7, 3, 1, 2), (521, 9, 5, 4, 1);$$

when $n = 12$, the only solution is $(521, 3, 5, 4, 1)$.

Proof. Equation (2) can be rewritten as

$$\left(\frac{x}{z^3}\right)^2 + A = \left(\frac{y}{z^2}\right)^3, \tag{4}$$

where A is cube-free and defined implicitly by $2^\alpha 5^\beta 13^\gamma = Az^6$. One can see that $A = 2^{\alpha_1} 5^{\beta_1} 13^{\gamma_1}$ with $\alpha_1, \beta_1, \gamma_1 \in \{0, 1, 2, 3, 4, 5\}$. We thus get

$$V^2 = U^3 - 2^{\alpha_1} 5^{\beta_1} 13^{\gamma_1}, \tag{5}$$

with $U = y/z^2$, $V = x/z^3$ and $\alpha_1, \beta_1, \gamma_1 \in \{0, 1, 2, 3, 4, 5\}$. We need to determine all the $\{2, 5, 13\}$-integral points on the above 216 elliptic curves. Recall that if S is a finite set of prime numbers, then an S-integer is rational number a/b with coprime integers a and b, where the prime factors of b are in S. We use MAGMA [12] to determine all the $\{2, 5, 13\}$-integer points on the above elliptic curves. Here are a few remarks about the computations:

1. We avoid the solutions with $UV = 0$ because they yield to $xy = 0$.
2. We don't consider solutions such that the numerators of U and V are not coprime.
3. If U and V are integers then $z = 1$, therefore $\alpha_1 = \alpha$ and $\beta_1 = \beta$.
4. If U and V are rational numbers which are not integers, then z is determined by the denominators of U and V. The numerators of these rational numbers give x and y. Then α and β are computed knowing that $2^\alpha 5^\beta 13^\gamma = Az^6$.

Therefore, we first determine $(U, V, \alpha_1, \beta_1, \gamma_1)$ and then we use the relations

$$U = \frac{y}{z^2}, \quad V = \frac{x}{z^3}, \quad 2^\alpha 5^\beta 13^\gamma = Az^6,$$

to find the solutions $(x, y, \alpha, \beta, \gamma)$ listed in Table 1.

For $n = 6$, equation

$$x^2 + 2^\alpha 5^\beta 13^\gamma = y^6 \tag{6}$$

becomes equation

$$x^2 + 2^\alpha 5^\beta 13^\gamma = \left(y^2\right)^3. \tag{7}$$

We look in the list of solutions of equation Table 1 and observe that the only solutions in Table 1 whose y is a perfect square are

$$(25, 9, 3, 0, 1), (23, 9, 3, 2, 0), (333, 49, 3, 1, 2), (521, 81, 5, 4, 1).$$

Therefore, the only solutions to equation (2) for $n = 6$ are

$$(25, 3, 3, 0, 1), (23, 3, 3, 2, 0), (333, 7, 3, 1, 2), (521, 9, 5, 4, 1).$$

In the same way, one can see that the value of y above which is a perfect square is $y = 4$ for the solution $(521, 9, 5, 4, 1)$, therefore the only solution with $n = 12$ is $(521, 9, 5, 4, 1)$. This completes the proof of Lemma 2.1.

3   The Case $n = 4$ or 8

Here, we have the following result.

Lemma 2. *If $n = 4$, then the only solutions to equation (2) are given in Table 2. If $n = 8$, then the only solutions to equation (2) are $(79, 3, 6, 1, 0), (49, 3, 6, 1, 1)$.*

Proof. Equation (2) can be written as

$$\left(\frac{x}{z^2}\right)^2 + A = \left(\frac{y}{z}\right)^4, \tag{8}$$

where A is fourth-power free and defined implicitly by $2^\alpha 5^\beta 13^\gamma = Az^4$. One can see that $A = 2^{\alpha_1} 5^{\beta_1} 13^{\gamma_1}$ with $\alpha_1, \beta_1, \gamma_1 \in \{0, 1, 2, 3\}$. Hence, the problem consists in determining the $\{2, 5, 13\}$-integer points on the totality of the 64 elliptic curves

$$V^2 = U^4 - 2^{\alpha_1} 5^{\beta_1} 13^{\gamma_1}, \tag{9}$$

with $U = y/z$, $V = x/z^2$ and $\alpha_1, \beta_1, \gamma_1 \in \{0, 1, 2, 3\}$. Here, we use again MAGMA [12] to determine the $\{2, 5, 13\}$-integer points on the above elliptic curves. As in Section 2, we first find $(U, V, \alpha_1, \beta_1, \gamma_1)$, and then using the co-primality conditions on x and y and the definition of U and V, we determine all the corresponding solutions $(x, y, \alpha, \beta, \gamma)$ listed in Table 2.

Looking in the list of solutions of equation Table 2, we observe that the only solutions whose values for y are perfect squares are $(79, 9, 6, 1, 0), (49, 9, 6, 1, 1)$. Thus, $(79, 3, 6, 1, 0), (49, 3, 6, 1, 1)$ are the only solutions to equation (2) with $n = 8$. One can notice that we can also recover the known solution for $n = 12$ from Table 2 also. This concludes the proof of Lemma 3.1.

If (x, y, α, β, n) is a solution of the Diophantine equation (2) and d is any proper divisor of n, then $(x, y^d, \alpha, \beta, n/d)$ is also a solution of the same equation. Since $n \geq 3$ and we have already dealt with the case $n = 3$ and 4, it follows that it suffices to look at the solutions n for which $p \mid n$ for some odd prime $p \geq 5$. In this case, we may certainly replace n by p, and thus assume for the rest of the paper that n is an odd prime.

4   The Case $n \geq 5$ and Prime

Lemma 3. *The Diophantine equation* (2) *has no solution with* $n \geq 5$ *prime except for*

$$
\begin{aligned}
n = 5 \qquad & (x, y, \alpha, \beta, \gamma) = (401, 11, 1, 3, 0) \\
n = 7 \qquad & (x, y, \alpha, \beta, \gamma) = (43, 3, 1, 0, 2).
\end{aligned}
\qquad ;
$$

Proof. We change n to p to emphasize that it is a prime number. We write the Diophantine equation (2) as $x^2 + dz^2 = y^p$, where $d = 1, 2, 5, 10, 13, 26, 65, 130$ according to the parities of the exponents α, β, and γ. Here, $z = 2^a 5^b 13^c$ for some nonnegative integers a, b and c. Let $\mathbb{K} = \mathbb{Q}[i\sqrt{d}]$. We factor the above equation in $\mathbb{K}$ getting

$$
\left(x + i\sqrt{d}\, z\right)\left(x - i\sqrt{d}\, z\right) = y^p. \tag{10}
$$

Since $\gcd(x, y) = 1$, if dz^2 is even, we get that y is odd. If $dz^2 = 5^\beta 13^\gamma$ is odd, we get that $dz^2 \equiv 1 \pmod 4$. Thus, x cannot be odd otherwise $x^2 + dz^2 \equiv 2 \pmod 4$ cannot be a perfect power. So, x is even when dz^2 is odd, therefore y is odd in this case also. Since y is odd, a standard argument shows that the ideals generated by $x + i\sqrt{d}z$ and $x - i\sqrt{d}z$ are coprime in $\mathbb{K}$. Hence, the ideal $x + i\sqrt{d}z$ is a pth power of some ideal in $\mathcal{O}_\mathbb{K}$. The class number of $\mathbb{K}$ belongs to $\{1, 2, 4, 6, 8\}$. In particular, it is coprime to p. Thus, by a standard argument, it follows that $x + i\sqrt{d}z$ is associated to a pth power in $\mathcal{O}_\mathbb{K}$. Since the group of units in $\mathbb{K}$ is of order 2 or 4 coprime to p, it follows that we may assume that

$$
x + i\sqrt{d}z = \eta^p \tag{11}
$$

holds with some algebraic integer $\eta \in \mathcal{O}_\mathbb{K}$. Finally, since the discriminant of $\mathbb{K}$ is $-4d$, it follows that $\{1, i\sqrt{d}\}$ is a base for $\mathcal{O}_\mathbb{K}$. In conclusion, we can write $\eta = u + i\sqrt{d}v$. Conjugating equation (11) and subtracting the two relations, we get

$$
2i\sqrt{d}\, 2^a\, 5^b\, 13^c = \eta^p - \bar{\eta}^p. \tag{12}
$$

The right hand side of the above equation is a multiple of $2i\sqrt{d}v = \eta - \bar{\eta}$. We deduct that $v \mid 2^a\, 5^b\, 13^c$, and that

$$
\frac{2^a\, 5^b\, 13^c}{v} = \frac{\eta^p - \bar{\eta}^p}{\eta - \bar{\eta}} \in \mathbb{Z}. \tag{13}
$$

Let $\{L_m\}_{m\geq0}$ be the sequence of general term $L_m = (\eta^m - \bar{\eta}^m)/(\eta - \bar{\eta})$, for all $m \geq 0$. This is called a *Lucas sequence* and it consists of integers. Its discriminant is $(\eta - \bar{\eta})^2 = -4dv^2$. For any nonzero integer k, we write $P(k)$ for the largest prime factor of k. Equation (13) leads to the conclusion that

$$P(L_p) = P\left(\frac{2^a 5^b 13^c}{v}\right). \tag{14}$$

A prime factor q of L_m is called *primitive* if $p \nmid L_k$ for any $0 < k < m$ and $q \nmid (\eta - \bar{\eta})^2$. When q exists, we have that $q \equiv \pm1 \pmod{m}$, where the sign coincides with $\left(\frac{-4d}{q}\right)$. Here, and in what follows, $\left(\frac{a}{q}\right)$ stands for the Legendre symbol of a with respect to the odd prime q. Recall that a particular instance of the Primitive Divisor Theorem for Lucas sequences implies that, if $p \geq 5$, then L_p has a *primitive* prime factor except for finitely many pairs $(\eta, \bar{\eta})$ and all of them appear in Table 1 in [11] (see also [1]). These exceptional Lucas numbers are called *defective*.

For $p = 5$, we look again in Table 1 in [11]. Of the seven possible values, only the possibility $(u, d, v) = (2, 10, 2)$ leads to a number $\eta = 2 + 2i\sqrt{10} \in \mathbb{Q}[i\sqrt{d}]$ with a value of d in the set $\{1, 2, 5, 10, 13, 26, 65, 130\}$, which gives the solution with $p = 5$.

Aside from the above mentioned possibility, we get that L_p must have a primitive divisor q. Clearly, $q \in \{2, 5, 13\}$ and $q \equiv \pm1 \pmod{p}$, where $p \geq 5$. Hence, the only possibility is $q = 13$, and we conclude that $p \mid 12, 14$. The only possibility is $p = 7$, and since $13 \equiv -1 \pmod{7}$, we must have that $\left(\frac{-4d}{13}\right) = -1$. Since $d \in \{1, 2, 5, 10, 13, 26, 65, 130\}$, we conclude that $d \in \{2, 5, 10\}$.

4.1   The Case $d = 2$

Using equation (12) with $p = 7$, we obtain

$$v\left(7u^6 - 70u^4v^2 + 84u^2v^2 - 8v^6\right) = 2^a\, 5^b\, 13^c. \tag{15}$$

Since u and v are coprime, we have the possibilities

$$v = \pm 2^a 5^b 13^c;\ v = \pm 5^b 13^c;\ v = \pm 2^a 13^c;\ v = \pm 13^c; \\ v = \pm 2^a 5^b;\ v = \pm 5^b;\ v = \pm 2^a;\ v = \pm 1. \tag{16}$$

The first four cases lead to the conclusion that $P(L_p) = P(2^a 5^b 13^c/v) \leq 5$, which is impossible, so we look at the last four possibilities.

Case 1: $v = \pm 2^a 5^b$.
In this case, the Diophantine equation (15) gives

$$7u^6 - 70u^4v^2 + 84u^2v^2 - 8v^6 = \pm 13^c. \tag{17}$$

Dividing both sides of the above equation by v^6, we obtain the following elliptic equations

$$7X^3 - 70X^2 + 84X - 8 = D_1Y^2; \tag{18}$$

where
$$X = \frac{u^2}{v^2}, \quad Y = \frac{13^{c_1}}{v^3}, \quad c_1 = \lfloor c/2 \rfloor, \quad D_1 = \pm 1, \pm 13.$$

• In the case $D_1 = \pm 1$ (changing X to $-X$ when $D_1 = -1$), we have to find the $\{2,5\}$-integer points on the elliptic curve

$$7X^3 + \varepsilon 70 X^2 + 84 X + \varepsilon 8 = Y^2, \qquad \varepsilon \in \{-1, 1\}; \qquad (19)$$

We multiply both sides of (19) by 7^2 to obtain

$$U^3 + \varepsilon 70 U^2 + 588 U + \varepsilon 392 = V^2, \qquad (20)$$

where $(U, V) = (\varepsilon 7X, 7Y)$ are $\{2,5\}$-integer points on the above elliptic curve. We use MAGMA [12] to determine all the $\{2,5\}$-integer points on the above elliptic curves. We find only the points $(U, V) = (14, 56)$, $(7, 91)$ corresponding to $\varepsilon = 1$. This gives us $(X, Y) = (1, 13)$, then $a = 0$, $b = 2$, $u = v = 1$. This leads to the solution $(43, 3, 1, 0, 2)$ of equation (2).
• When $D = \pm 13$, we multiply both sides of equation (19) by $7^2 \, 13^3$ and obtain the elliptic curves

$$U^3 + \varepsilon 910 U^2 + 99372 U + \varepsilon 861224 = V^2, \qquad \varepsilon \in \{-1, 1\}; \qquad (21)$$

where
$$U = \varepsilon 91 X, \quad V = 1183 Y,$$

for which we need again all their $\{2,5\}$-integer points. We obtain a totality of nine solutions for (U, V).

Case 2: $v = \pm 5^b$.
In this case, the Diophantine equation (15) becomes

$$7u^6 - 70u^4 v^2 + 84 u^2 v^2 - 8 v^6 = \pm 2^a \, 13^c. \qquad (22)$$

Dividing both sides of the above equation by v^6, we obtain the following elliptic equations

$$7X^3 - 70X^2 + 84X - 8 = D_1 Y^2, \qquad (23)$$

where
$$X = \frac{u^2}{v^2}, \quad Y = \frac{2^{a_1} \, 13^{c_1}}{v^3}, \quad a_1 = \lfloor a/2 \rfloor, \quad c_1 = \lfloor c/2 \rfloor, \quad D_1 = \pm 1, \pm 2, \pm 13, \pm 26.$$

• In the case $D_1 = \pm 1$, we obtain again equation (20) and we know the result.
• In the case $D_1 = \pm 2$ (changing X to $-X$ when $D_1 = -1$), we have to find the $\{2,5\}$-integer points on the elliptic curves

$$7X^3 + \varepsilon 70 X^2 + 84 X + \varepsilon 8 = \pm 2 Y^2, \qquad \varepsilon \in \{-1, 1\}. \qquad (24)$$

We now multiply both sides of equation (24) by $2^3 \, 7^2$ and obtain

$$U^3 + \varepsilon 140 U^2 + 2352 U + \varepsilon 3136 = V^2, \qquad (25)$$

where $(U, V) = (\varepsilon 14X, 28Y)$ is a $\{5\}$-integer points on the above elliptic curve. We use MAGMA [12] to determine the $\{5\}$-integer points on the above elliptic curves. We find thirteen solutions in (U, V).

- In the case $D_1 = \pm 13$, we arrive at equation (21).
- When $D = \pm 26$, we multiply both sides of equation (23) by $7^2\, 26^3$ and obtain the elliptic curves

$$U^3 + \varepsilon 1820 U^2 + 397488 U + \varepsilon 6889792 = V^2, \qquad \varepsilon \in \{-1, 1\}, \qquad (26)$$

where
$$U = \varepsilon 182 X, \quad V = 4732 Y,$$

for which we need again its $\{5\}$-integer points. In the same way as before, we find a total of twelve solutions in (U, V).

Case 3: $v = \pm 2^a$.
In this case, the Diophantine equation (15) is

$$7u^6 - 70u^4 v^2 + 84u^2 v^2 - 8v^6 = \pm 5^b\, 13^c. \qquad (27)$$

Dividing both sides of the above equation by v^6, we obtain the following elliptic equations
$$7X^3 - 70X^2 + 84X - 8 = D_1 Y^2, \qquad (28)$$

where

$$X = \frac{u^2}{v^2}, \; Y = \frac{5^{b_1}\, 13^{c_1}}{v^3}, \; b_1 = \lfloor b/2 \rfloor, \; c_1 = \lfloor c/2 \rfloor, \; D_1 = \pm 1, \, \pm 5, \, \pm 13, \, \pm 65.$$

- In the case $D_1 = \pm 1$, we obtain again equation (20).

- In the case $D_1 = \pm 5$ (changing X to $-X$ when $D_1 = -1$), we have to find the $\{2\}$-integer points on the elliptic curve

$$7X^3 + \varepsilon 70X^2 + 84X + \varepsilon 8 = \pm 5 Y^2, \qquad \varepsilon \in \{-1, 1\}. \qquad (29)$$

We multiply both sides of equation (29) by $5^3\, 7^2$ and obtain

$$U^3 + \varepsilon 350 U^2 + 14700 U + \varepsilon 49000 = V^2, \qquad (30)$$

where $(U, V) = (\varepsilon 35X, 175Y)$ is a $\{2\}$-integer points on the above elliptic curve. We use MAGMA [12] to determine the $\{2\}$-integer points on the above elliptic curve. We find only $(U, V) = (54, 344)$.

- In the case $D_1 = \pm 13$, we arrive at equation (21).

- When $D = \pm 65$, we multiply both sides of equation (28) by $65^3\, 7^2$ and obtain the elliptic curve

$$U^3 + \varepsilon 4550 U^2 + 2484300 U + \varepsilon 107653000 = V^2, \qquad \varepsilon \in \{-1, 1\}, \qquad (31)$$

where
$$U = \varepsilon 455 X, \quad V = 29575 Y,$$

for which we need again all its $\{2\}$-integer points. In the same way, we find a totality of nine solutions of which the only convenient one is $(U, V) = (1001, 34307)$.

Case 4: $v = \pm 1$.

Here, we obtain the following Thue-Mahler equation

$$7u^6 - 70u^4 + 84u^2 - 8 = 2^a \, 5^b \, 13^c. \qquad (32)$$

By the same method, we can rewrite the above equation as

$$7X^3 - 70X^2 + 84X - 8 = D_1 Y^2, \qquad (33)$$

where

$$X = u^2, \quad Y = 2^{a_1} 5^{b_1} 13^{c_1}, \quad a_1 = \lfloor a/2 \rfloor, \quad b_1 = \lfloor b/2 \rfloor, \quad c_1 = \lfloor c/2 \rfloor,$$

and

$$D_1 \in \pm \{1, 2, 5, 10, 13, 26, 65, 130\}.$$

We will study the cases $D_1 = \pm 10, \ \pm 130$, because all the other cases have been studied (except that now we need only the integer points on these curves which have already been computed).

• When $D_1 = \pm 10$, we then multiply both sides of equation (33) by $7^2 \, 10^3$ and get the two elliptic curves

$$U^3 + \varepsilon 700 U^2 + 58800 U + \varepsilon 392000 = V^2, \qquad \varepsilon \in \{-1, 1\}, \qquad (34)$$

where $U = \varepsilon 70 X$, $V = 700Y$, and we need their integer points. Here also we use MAGMA [12] to find two integer points but none leads to a solution.

• Finally, for the case $D = \pm 130$, we multiply both sides of equation (33) by $7^2 \, 130^3$ to obtain

$$U^3 + \varepsilon 9100 U^2 + 9937200 U + \varepsilon 861224000 = V^2, \qquad \varepsilon \in \{-1, 1\}, \qquad (35)$$

where $U = \varepsilon 910 X$, $V = 118300Y$, whose integer points we need to compute. We find two solutions (U, V).

4.2   The Case $d = 5$

In this case, equation (12) with $p = 7$ is

$$v \left(7u^6 - 175u^4 v^2 + 525u^2 v^4 - 125v^6 \right) = 2^a \, 5^b \, 13^c. \qquad (36)$$

Since u and v are coprime, we have the possibilities

$$v = \pm 2^a 5^b 13^c; \ v = \pm 5^b 13^c; \ v = \pm 2^a 13^c; \ v = \pm 13^c; \\ v = \pm 2^a 5^b; \ v = \pm 5^b; \ v = \pm 2^a; \ v = \pm 1. \qquad (37)$$

The first four cases lead to the conclusion that $P(L_p) = P(2^a 5^b 13^c/v) \leq 5$, which is impossible, so we look at the last four possibilities. Then we use the same method as in subsection 4.1. In fact, each v considered is used to simplify the equation (36). After dividing the simplified expression obtained by v^6, we get an equation of the form $f(X) = D_1 Y^2$, where X and Y depend on u, v, powers of $2, 5, 13$, and f is a third degree polynomial in X with integer coefficients. When it is necessary, for each D_1, we multiply the equation $f(X) = D_1 Y^2$ by an appropriate product of powers of $7, 2, 5, 13$ to obtain an elliptic equation $g(U) = V^2$. Then we use MAGMA to determine all the S-integer points on the resulting elliptic curve. We find no solution. The values of v, D_1, and S are given by Table 3.

Table 3. The case $d = 5$

Case	v	D_1	S
1	$\pm 2^a\, 5^b$	$\pm 1, \pm 13$	$\{2, 5\}$
2	$\pm 5^b$	$\pm 1, \pm 2, \pm 13,\ \pm 26$	$\{5\}$
3	$\pm 2^a$	$\pm 1, \pm 5, \pm 13,\ \pm 65$	$\{2\}$
4	± 1	$\pm 1, \pm 2, \pm 5,\ \pm 10, \pm 13,\ \pm 26, \pm 65, \pm 130$	$\{1\}$

4.3   The Case $d = 10$

In this case, equation (12) with $p = 7$ is

$$v\left(7u^6 - 350u^4 v^2 + 2100u^2 v^4 - 1000v^6\right) = 2^a\, 5^b\, 13^c. \tag{38}$$

Since u and v are coprime, we have the possibilities

$$v = \pm 2^a\, 5^b\, 13^c;\ v = \pm 5^b\, 13^c;\ v = \pm 2^a\, 13^c;\ v = \pm 13^c; \\ v = \pm 2^a\, 5^b;\ v = \pm 5^b;\ v = \pm 2^a;\ v = \pm 1. \tag{39}$$

The first four cases lead to the conclusion that $P(L_p) = P(2^a\, 5^b\, 13^c / v) \le 5$, which is impossible, so we look at the last four possibilities. Then we use the same method as in subsection 4.1. In fact, we use each v considered to simplify the equation (38). Then we divide both sides of the simplified expression obtained by v^6 to get an equation of the form $f(X) = D_1 Y^2$, where f is a cubic polynomial in X, and X, Y depend on u, v, and powers of $2, 5, 13$. When it is necessary, for each D_1, we multiply the equation $f(X) = D_1 Y^2$ by an appropriate product of powers of $7, 2, 5, 13$ to obtain an elliptic equation of the form $g(U) = V^2$. Finally, we find all the S-integer points on the elliptic curve using MAGMA. No solution is obtained. The values of v, D_1, and S are contained in Table 4.

Let us specify that although we have obtained two identical tables, we did not always get the same elliptic curves. Thus, one cannot draw a quick conclusion about the cases $d = 5$ and $d = 10$ and a full investigation of each of these two cases is necessary.

For each point (U, V) found on any of the above curves, we determine the corresponding x and y and none of these cases lead to a solution to the equation (2) except for the case of the equation (20) which gives one solution. This completes the proof of Theorem 1.

Table 4. The case $d = 10$

$Case$	v	D_1	S
1	$\pm 2^a\, 5^b$	$\pm 1, \pm 13$	$\{2,5\}$
2	$\pm 5^b$	$\pm 1, \pm 2, \pm 13, \pm 26$	$\{5\}$
3	$\pm 2^a$	$\pm 1, \pm 5, \pm 13, \pm 65$	$\{2\}$
4	± 1	$\pm 1, \pm 2, \pm 5, \pm 10, \pm 13, \pm 26, \pm 65, \pm 130$	$\{1\}$

Acknowledgement

We thank the three referees for a careful reading of the manuscript and for useful suggestions, which, in particular, helped us reduce the length of an earlier draft. The first author was partially supported by Purdue University. The second author was partially supported by Grant SEP-CONACyT 46755. The third author was partially supported by Purdue University North Central.

References

1. Abouzaid, M.: Les nombres de Lucas et Lehmer sans diviseur primitif. J. Théor. Nombres Bordeaux 18, 299–313 (2006)
2. Abu Muriefah, F.S.: On the diophantine equation $x^2 + 5^{2k} = y^n$. Demonstratio Mathematica 319(2), 285–289 (2006)
3. Abu Muriefah, F.S., Bugeaud, Y.: The Diophantine equation $x^2 + c = y^n$: a brief overview. Rev. Colombiana Mat. 40, 31–37 (2006)
4. Abu Muriefah, F.S., Luca, F., Togbé, A.: On the equation $x^2 + 5^a \cdot 13^b = y^n$. Glasgow J. Math. 50, 143–161 (2008)
5. Arif, S.A., Abu Muriefah, F.S.: On the Diophantine equation $x^2 + 2^k = y^n$, Internat. J. Math. Math. Sci. 20, 299–304 (1997)
6. Arif, S.A., Abu Muriefah, F.S.: On a Diophantine equation. Bull. Austral. Math. Soc. 57, 189–198 (1998)
7. Arif, S.A., Abu Muriefah, F.S.: The Diophantine equation $x^2 + 3^m = y^n$. Internat. J. Math. Math. Sci. 21, 619–620 (1998)
8. Arif, S.A., Abu Muriefah, F.S.: The Diophantine equation $x^2 + 5^{2k+1} = y^n$. Indian J. Pure Appl. Math. 30, 229–231 (1999)
9. Arif, S.A., Abu Muriefah, F.S.: The Diophantine equation $x^2 + q^{2k} = y^n$. Arab. J. Sci. Eng. Sect. A Sci. 26, 53–62 (2001)
10. Arif, S.A., Abu Muriefah, F.S.: On the Diophantine equation $x^2 + q^{2k+1} = y^n$. J. Number Theory 95, 95–100 (2002)
11. Bilu, Y., Hanrot, G., Voutier, P.M.: Existence of primitive divisors of Lucas and Lehmer numbers. With an appendix by Mignotte. M. J. reine angew. Math. 539, 75–122 (2001)
12. Bosma, W., Cannon, J., Playoust, C.: The Magma algebra system. I. The user language. J. Symbolic Comput. 24, 235–265 (1997)
13. Bugeaud, Y., Mignotte, M., Siksek, S.: Classical and modular approaches to exponential Diophantine equations. II. The Lebesgue-Nagell equation. Compositio Math. 142, 31–62 (2006)

14. Cohn, J.H.E.: The Diophantine equation $x^2 + c = y^n$. Acta Arith. 65, 367–381 (1993)
15. Ko, C.: On the Diophantine equation $x^2 = y^n + 1$, $xy \neq 0$. Sci. Sinica 14, 457–460 (1965)
16. Le, M.: An exponential Diophantine equation. Bull. Austral. Math. Soc. 64, 99–105 (2001)
17. Le, M.: On Cohn's conjecture concerning the Diophantine equation $x^2 + 2^m = y^n$. Arch. Math. (Basel) 78, 26–35 (2002)
18. Lebesgue, V.A.: Sur l'impossibilité en nombres entiers de l'équation $x^m = y^2 + 1$ Nouv. Annal. des Math. 9, 178–181 (1850)
19. Luca, F.: On a Diophantine Equation. Bull. Austral. Math. Soc. 61, 241–246 (2000)
20. Luca, F.: On the equation $x^2 + 2^a \cdot 3^b = y^n$. Int. J. Math. Math. Sci. 29, 239–244 (2002)
21. Luca, F., Togbé, A.: On the equation $x^2 + 2^a \cdot 5^b = y^n$. Int. J. Number Theory (to appear)
22. Luca, F., Togbé, A.: On the equation $x^2 + 2^a \cdot 13^b = y^n$ (preprint, 2007)
23. Mignotte, M., de Weger, B.M.M.: On the Diophantine equations $x^2 + 74 = y^5$ and $x^2 + 86 = y^5$. Glasgow Math. J. 38, 77–85 (1996)
24. Pink, I.: On the diophantine equation $x^2 + 2^\alpha \cdot 3^\beta \cdot 5^\gamma \cdot 7^\delta = y^n$. Publ. Math. Debrecen 70, 149–166 (2006)
25. Tengely, S.: On the Diophantine equation $x^2 + q^{2m} = 2y^p$. Acta Arith. 127, 71–86 (2007)
26. Tengely, S.: On the Diophantine equation $x^2 + a^2 = 2y^p$. Indag. Math. (N.S.) 15, 291–304 (2004)

Non-vanishing of Dirichlet L-functions at the Central Point

Sami Omar

Faculty of Sciences of Tunis
Mathematics Department 2092 Campus Universitaire, Tunis. Tunisia
sami.omar@fst.rnu.tn

Abstract. This paper deals with the matter of the non-vanishing of Dirichlet L-functions at the central point for all primitive characters χ. More precisely, S. Chowla conjectured that $L(\frac{1}{2}, \chi) \neq 0$, but this remains still unproved. We first give an efficient algorithm to compute the order n_χ of zero of $L(s, \chi)$ at $s = \frac{1}{2}$. This enables us to efficiently compute n_χ for L-functions with very large conductor near 10^{16}. Then, we prove that $L(\frac{1}{2}, \chi) \neq 0$ for all real characters χ of modulus less than 10^{10}. Finally we give some estimates for n_χ and the lowest zero of $L(s, \chi)$ on the critical line in terms of the conductor q.

1   Introduction

In 1859 Riemann published his only paper in number theory, a short eight-page note which introduced the use of complex analysis into the subject of the prime number theory. In the course of this paper, he conjectured that all non-trivial zeros of the Riemann zeta function lie on the line $\mathcal{R}e(s) = \frac{1}{2}$. This conjecture is now known as the "Riemann Hypothesis". Further, it is expected that this conjecture would also hold for most L-functions used in number theory which share some basic analytic properties, in particular meromorphic continuation, an Euler product and a functional equation of a certain type. Beyond the classical Riemann zeta function, one may mention the example of the Dirichlet L-functions. In the latter case, it is believed that there are no $\mathbb{Q}$-linear relations among the positive ordinates of the zeros. Therefore, it is expected that $L(1/2, \chi) \neq 0$ for all primitive characters χ. This appears to have been first conjectured by S. D. Chowla [6] when χ is a quadratic character. In connection with this conjecture, we mention the work of R. Balasubramanian and V. K. Murty [1] in which they showed that for any fixed s in the critical strip a positive small portion of the $L(s, \chi)$ do not vanish as χ ranges over all characters to a sufficiently large prime modulus. More recently, H. Iwaniec and P. Sarnak [11] proved that this portion is at least one third. However, assuming the Riemann Hypothesis, it is shown in [16] that this portion is at least one half by using the Weil explicit formulas for suitable test functions. Further much numerical evidence for Chowla's conjecture has been accumulated; these calculations use the approximate formula of Bateman/Grosswald [3] and Chowla/Selberg [7] to obtain the best previous

A.J. van der Poorten and A. Stein (Eds.): ANTS-VIII 2008, LNCS 5011, pp. 443–453, 2008.
© Springer-Verlag Berlin Heidelberg 2008

results for non-vanishing of $L(1/2, \chi)$. More precisely, it is shown numerically by M. Watkins [25] that $L(s, \chi)$ has no positive real zeros for real odd characters of modulus d up to 3×10^8, extending the previous record of 8×10^5 due to Low and Purdy. We also remark that the best non-vanishing result for real even characters has been obtained by C. Kok Seng [12] for modulus d up to 2×10^5 extending the previous result of 986 due to J. B. Rosser. Some theoretical progress towards these non-vanishing questions can be seen in the work of B. Conrey and of K. Soundararajan [24], [8]. In this paper, we show how a careful use of Weil's explicit formula enables us to compute efficiently the order n_χ of zero of $L(s, \chi)$ at $s = \frac{1}{2}$ for primitive real odd or even characters with no restrictions on the modulus $d \leq 10^{16}$.

Therefore, we particularly checked the conjecture $L(\frac{1}{2}, \chi) \neq 0$ for real characters of large modulus $d \leq 10^{10}$. It should be mentioned that one possible method to do so is to compute explicit values of L-functions at the central point by writing the approximate formula of the functional equation of $L(s, \chi)$ (see p. 98 in the book [10]). Such algorithms are usually based on writing $L(s, \chi)$ as a series in incomplete Gamma functions associated to the inverse Mellin transform of Γ factors of $L(s, \chi)$. By doing that, these algorithms become very slow for large conductors and require about $O(\sqrt{q})$ terms to compute; see [13] and [22]. However the explicit formulas used here are clean to implement and quickly provide sharp estimates for low zeros; see [5] and [19] for computation of the lowest zero of zeta functions with conductors of magnitude 10^{28}. The arguments used in [19] allow us to give faster algorithms when we assume the Generalized Riemann Hypothesis (GRH). The computations have been done using the PARI-GP package version 2.3.2 [4]. Finally, we improve, under GRH, the Siegel bounds [23] for the lowest zero of Dirichlet L-functions in terms of the conductor q.

2   Functional Equations

Let χ be a primitive Dirichlet character of conductor q. The Dirichlet L-function attached to this character is defined by

$$L(s, \chi) = \sum_{n=1}^{+\infty} \frac{\chi(n)}{n^s}, \quad (\mathcal{R}e(s) > 1).$$

For the trivial character $\chi = 1$, $L(s, \chi)$ is the Riemann zeta function. It is well known [9] that if $\chi \neq 1$ then $L(s, \chi)$ can be extended to an entire function in the whole complex plane and satisfies the functional equation

$$\Lambda(s, \chi) = W_\chi \overline{\Lambda(1 - \bar{s}, \chi)},$$

where

$$\Lambda(s, \chi) = \left(\frac{q}{\pi}\right)^{\frac{s}{2}} \Gamma\left(\frac{s + \delta}{2}\right) L(s, \chi),$$

$$\delta = \begin{cases} 0 & \text{if } \chi(-1) = 1 \\ 1 & \text{if } \chi(-1) = -1, \end{cases}$$

and

$$W_\chi = \frac{\tau(\chi)}{\sqrt{q}\, i^\delta} \,,$$

where $\tau(\chi)$ is the Gauss sum

$$\tau(\chi) = \sum_{m=1}^{q} \chi(m) e^{2\pi i m/q} \,.$$

Note that the quadratic twists of $\zeta(s)$ are the particular Dirichlet L-functions with $\chi(n) = \chi_d(n) = (\frac{d}{n})$, where $(\frac{d}{n})$ is the Kronecker symbol. Then the functional equation of the completed Dirichlet L-function is

$$\Lambda(1 - s, \chi_d) = \Lambda(s, \chi_d) \,.$$

3   An Explicit Formula

In this section, we give an explicit formula to compute efficiently the order n_χ of the zero of $L(s, \chi)$ at $s = \frac{1}{2}$. For that purpose, we use Weil's explicit formula first given by Weil [26], and reformulated by K. Barner [2] in an easier and more manageable way for computations. One can adapt this formula to $L(s, \chi)$ and then evaluate the sum on the zeros of the Dirichlet L-function $L(s, \chi)$ in the explicit formula.

Theorem 1. *Consider functions $F : \mathbb{R} \to \mathbb{R}$ which satisfy $F(0) = 1$ together with the following conditions:*

(A) *F is even, continuous and continuously differentiable everywhere except at a finite number of points a_i, where $F(x)$ and $F'(x)$ have only a discontinuity of the first kind, such that $F(a_i) = \frac{1}{2}(F(a_i + 0) + F(a_i - 0))$.*
(B) *There exists a number $b > 0$ such that $F(x)$ and $F'(x)$ are $O(e^{-(\frac{1}{2}+b)|x|})$ as $|x| \to \infty$.*

Then the Mellin transform of F:

$$\Phi(s) = \int_{-\infty}^{+\infty} F(x) e^{(s-\frac{1}{2})x} dx$$

is holomorphic in every vertical strip $-a \le \sigma \le 1 + a$ where $0 < a < b$, $a < 1$, and the sum $\sum \Phi(\rho)$ running over the non trivial zeros $\rho = \beta + i\gamma$ of $L(s, \chi)$ with $|\gamma| < T$ tends to a limit as T tends to infinity. This limit is given by the formula:

$$\sum_\rho \Phi(\rho) = \ln\left(\frac{q}{\pi}\right) - I_\delta(F) - 2 \sum_{p, m \ge 1} \mathcal{R}e(\chi^m(p)) F(m \ln(p)) \frac{\ln(p)}{p^{m/2}} \,,$$

where

$$I_\delta(F) = \int_0^{+\infty} \left(\frac{F(x/2) e^{-(\frac{1}{4}+\frac{\delta}{2})x}}{1 - e^{-x}} - \frac{e^{-x}}{x} \right) dx \,;$$

δ is defined in §2 above.

The last integral can be also expressed as follows:

$$I_\delta(F) = \int_0^\infty \frac{F(x/2) - 1}{1 - e^{-x}} e^{-(\frac{1}{4} + \frac{1}{2}\delta)} \, dx + \int_0^\infty \left(\frac{e^{-(\frac{1}{4} + \frac{1}{2}\delta)}}{1 - e^{-x}} - \frac{e^{-x}}{x} \right) dx \, ,$$

where the second integral is equal to $\gamma + 3\ln 2 + \frac{1}{2}\pi$ for $\delta = 0$, and is equal to $\gamma + 3\ln 2 - \frac{1}{2}\pi$ for $\delta = 1$.

4   Efficient Computation of n_χ

4.1   Conditional Bounds

Now we assume the Generalized Riemann Hypothesis (GRH) for $L(s, \chi)$ which asserts that all the non-trivial zeros of $L(s, \chi)$ lie on the critical line $\Re(s) = \frac{1}{2}$. We rewrite Theorem 1 for Serre's choice $F_y(x) = e^{-yx^2}$ ($y > 0$). The Mellin transform $\Phi(s)$ of F_y is

$$\Phi_y(s) = \sqrt{\frac{\pi}{y}} e^{(s - \frac{1}{2})^2 / 4y}$$

and the Fourier transform $\widehat{F}_y$ of F_y is

$$\widehat{F}_y(t) = \sqrt{\frac{\pi}{y}} e^{-t^2 / 4y}.$$

If we assume the Generalized Riemann Hypothesis (GRH) for $L(s, \chi)$, we can write $\Phi_y(\rho_k) = \widehat{F}_y(t)$ where $\rho_k = \frac{1}{2} + i\gamma_k$. We denote by γ_k the imaginary part of the kth zero of the Dirichlet L-function $L(s, \chi)$, and n_k its multiplicity. Thus we have

$$\ldots < \gamma_{-3} < \gamma_{-2} < \gamma_{-1} < 0 < \gamma_1 < \gamma_2 < \gamma_3 < \ldots \, .$$

We set

$$S(y) = n_\chi + \sum_{k \neq 0} n_k e^{-\gamma_k^2 / 4y} \, .$$

By the explicit formulas, we have the identity

$$S(y) = \sqrt{\frac{y}{\pi}} \left(\ln(\frac{q}{\pi}) - I_\delta(F_y) - 2 \sum_{\mathfrak{p}, m} \frac{\ln(p)}{p^{m/2}} \mathcal{R}e(\chi^m(p)) e^{-y(m \ln(p))^2} \right).$$

In the following proposition, we give an upper bound of n_χ.

Proposition 2. *Assuming GRH, we have for all $y > 0$*

$$n_\chi \leq S(y)$$

and

$$\lim_{y \to 0} S(y) = n_\chi \, .$$

One should notice that the advantage of Serre's choice in Weil's explicit formula is that the series $S(y)$ converges rapidly to n_χ when $y \to 0$. In practice, one should find a non-negative value y so that we have $n_\chi \le S(y) < 1$ and so $n_\chi = 0$. Thus we can numerically check Chowla's conjecture by the following result.

Corollary 3. *Under GRH, $L(\frac{1}{2}, \chi) \neq 0$ holds if and only if there exists $y > 0$ such that $S(y) < 1$.*

It is obvious that if there exists $y > 0$ such that $S(y) < 1$ then $n_\chi \le S(y) < 1$. Thus $L(\frac{1}{2}, \chi) \neq 0$. Conversely, if $L(\frac{1}{2}, \chi) \neq 0$ then $n_\chi = 0$. Since

$$\lim_{y \to 0} S(y) = n_\chi = 0,$$

then for sufficiently many small positive values y, we have $S(y) < 1$.

4.2 Unconditional Bounds

The unconditional bounds of n_χ are less good than the (GRH) ones in Proposition 2 because of the requirement that $\Re\,\Phi(s) \ge 0$ on the whole critical strip. By using an argument of Odlyzko [17], this last condition holds when we take in Theorem 1 the function $G_y(x) = \frac{F_y(x)}{\cosh(x/2)}$ with $F_y(x) = e^{-yx^2}$ $(y > 0)$. Actually, on both lines $\sigma = 0, 1$, $\Re\,\Phi(\sigma + it) = \widehat{F_y}(t) \ge 0$. Since $\Re\,\Phi(s)$ is harmonic, then it is positive inside the whole critical strip. Let us define

$$T(y) = n_\chi + \frac{1}{2 \int_0^{+\infty} \frac{e^{-yx^2}}{\cosh(x/2)}\, dx} \sum_{\rho \neq \frac{1}{2}} \Re\,\Phi(\rho).$$

By the explicit formulas, we see that $T(y)$ is given by

$$\frac{1}{2 \int_0^{+\infty} \frac{e^{-yx^2}}{\cosh(x/2)}\, dx} \left(\ln(\tfrac{q}{\pi}) - I_\delta(G_y) - 4 \sum_{p,m} \frac{\ln(p)}{1 + p^m} \Re(\chi^m(p)) e^{-y(m \ln(p))^2} \right).$$

Thus we obtain the following bound for n_χ.

Proposition 4. *For all $y > 0$, we have*

$$n_\chi \le T(y) \quad and \quad \lim_{y \to 0} T(y) = n_\chi.$$

Using the same idea as in Corollary 3, we also obtain the following similar result.

Corollary 5. *$L(\frac{1}{2}, \chi) \neq 0$ holds if and only if there exists $y > 0$ such that $T(y) < 1$.*

4.3 Numerical Evidence for the Chowla Conjecture

To compute $S(y)$ and $T(y)$, we begin by computing the integrals $I_\delta(F_y)$ and $I_\delta(G_y)$ to a high enough precision; then we compute the series over primes in the explicit formula by computing $\mathcal{R}e(\chi^m(p))$ for each prime number p less than some large enough p_0. The series over primes $v_\infty(y)$ in the Weil explicit formula is truncated to

$$v_{p_0}(y) = \sum_{p \leq p_0} \ln(p) \sum_{\substack{m \\ m \ln(p) \leq \text{cons}}} \mathcal{R}e(\chi^m(p)) \frac{e^{-y(m\ln(p))^2}}{D_m(p)},$$

where

$$D_m(p) = \begin{cases} p^{m/2} & \text{under GRH} \\ p^m + 1 & \text{otherwise} \end{cases}$$

and cons $= \sqrt{c\ln(10)/y}$.

The condition $m\ln(p) < \text{cons}$ means that we don't take into account the terms of the series less than 10^{-c}. In practice we take $c = 30$ and p_0 less than 10^6 for conductors $q \approx 10^{16}$. Actually the experimental value of $S(y)$ is $\tilde{S}(y) \geq S(y)$ and so $n_\chi \leq \tilde{S}(y)$. By a simple use of the prime number theorem, the main error term of these computations is derived from the following estimate.

Proposition 6. *If we take* cons $= \infty$, *then we have*

$$|v_\infty(y) - v_{p_0}(y)| \ll_y \frac{\sqrt{p_0}}{\ln(p_0)} e^{-y\ln(p_0)^2}.$$

It should be noted that when the conductor q is large, the computation of $S(y)$ and $T(y)$ is slower; this is essentially due to the low lying zeros of the Dirichlet L-function $L(s, \chi)$. Actually, when the low zeros of $L(s, \chi)$ distinct from $\frac{1}{2}$ are close to the real axis, one has to compute the series $S(y)$ and $T(y)$ for small positive values of y in order to be able to bound $S(y)$ and $T(y)$ above by 1 (note Corollaries 3 and 5). An intuitive approach to the low lying zeros and the order of vanishing of $L(s, \chi)$ at $s = \frac{1}{2}$ is given in section 6.

The following table gives the maximum of values of $S(y_0)$ and $T(y)$ in the intervals $10^k \leq d < 10^{k+1}$ where $1 \leq k \leq 9$ and the real characters associated to those maximum values.

q	y_0	$\max_\chi S(y_0)$	y	$\max_\chi T(y)$	q_0	p_0	n_χ	time
$10 \leq q < 10^2$	0.3	0.50410	0.3	0.67812	48 (odd)	100	0	50 m
$10^2 \leq q < 10^3$	0.2	0.46543	0.2	0.57037	768 (even)	10^3	0	14 h
$10^3 \leq q < 10^4$	0.11	0.41720	0.11	0.52140	3596 (even)	4×10^3	0	4 d
$10^4 \leq q < 10^5$	0.09	0.34512	0.09	0.37528	15736 (odd)	10^4	0	6 d
$10^5 \leq q < 10^6$	0.08	0.28643	0.07	0.31726	176717 (odd)	6×10^4	0	9 d
$10^6 \leq q < 10^7$	0.07	0.13242	0.07	0.09642	1447681 (odd)	10^5	0	14 d
$10^7 \leq q < 10^8$	0.05	0.07830	0.05	0.08347	18476264 (odd)	5×10^5	0	20 d
$10^8 \leq q < 10^9$	0.04	0.05176	0.04	0.06941	154862795 (even)	10^6	0	50 d
$10^9 \leq q < 10^{10}$	0.01	0.25871	0.01	0.35762	1710534545 (even)	5×10^6	0	180 d

The complexity of the method can be seen as the number of primes less than p_0 needed to compute the sum $v_{p_0}(y_0)$ so that $S(y_0) < 1$ for a suitable positive value y_0. According to the table, The latter value of y_0 is determined by considering conductors close to 10^k ($2 \leq k \leq 10$). Then, these values of y_0 are considered for all conductors between 10^{k-1} and 10^k. Actually, when the conductor is larger in that range, the parameter y_0 decreases slightly so we do not need many more terms to compute $S(y_0)$. We should mention that computations of n_χ by this technique overcome different problems of other previous methods which distinguish the odd and even character cases. Indeed, the complexity of the algorithm depends only on the size of the conductor.

5 Upper Bounds for n_χ

The following theorem gives upper bounds for n_χ in terms of the conductor q.

Theorem 7. *Under GRH, we have:*

$$n_\chi \ll \frac{\ln(q)}{\ln \ln(q)}, \tag{1}$$

Unconditionally, we have the following estimate

$$n_\chi < \ln(q). \tag{2}$$

Proof. To prove Theorem 7, we follow the method of Mestre [14] proved in the case of modular forms. We first need an estimate for the sum over primes in Theorem 1. Let F be a function of support contained in $[-1, 1]$ satisfying the hypotheses of Theorem 1 and let $F_T(x) = F(x/T)$. By using the prime number theorem, one can prove the following estimate:

Lemma 8. *The sum over primes in Theorem 1 is bounded by the inequality*

$$\left| \sum_{p,m} \frac{\ln(p)}{p^{m/2}} \mathcal{R}e(\chi^m(p)) F_T(m \ln(p)) \right| \leq C_0 \, e^{T/2},$$

where C_0 is a non-negative constant.

We also need the following result.

Lemma 9. *We define F by*

$$F(x) = \begin{cases} 1 - |x| & \text{if } |x| \leq 1 \\ 0 & \text{otherwise.} \end{cases}$$

Then F satisfies the hypotheses of Theorem 1 and $\widehat{F}(u) = (2\sin(\frac{1}{2}u)/u)^2$.

Thus, writing $F_T(x) = F(x/T)$, we obtain $\widehat{F}_T(u) = T\widehat{F}(Tu)$. Applying Weil's explicit formula to F_T and using Lemma 8, yields the estimate

$$n_\chi T \leq \ln(\frac{q}{\pi}) + C_0\, e^{T/2} - I_\delta(F_T)\,.$$

Because $I_\delta(F_T)$ is bounded as T tends to $+\infty$, we see that replacing T by $2\ln\ln(q)$ provides

$$n_\chi \ll \frac{\ln(q)}{\ln\ln(q)}\,,$$

and so (1) holds.

To prove the estimate (2) we define the function H_T with compact support by $H_T(x) = F_T(x)/\cosh(\frac{1}{2}x)$.

The argument used in paragraph 4.2 yields that the Mellin transform Φ_T of H_T satisfies $\Re\,\Phi_T(s) \geq 0$ in the critical strip. Thus, when we apply Theorem 1 to H_T we obtain the inequality

$$n_\chi\,\Phi_T(\tfrac{1}{2}) \leq \ln(\frac{q}{\pi}) - I_\delta(H_T) - 2\sum_{p,m}\frac{\ln(p)}{p^{m/2}}\mathcal{R}e(\chi^m(p))H_T(m\ln(p))\,. \tag{3}$$

Since H_T is a decreasing function on $[0, +\infty[$ we have the following result.

Lemma 10. *We have the inequality*

$$\left|\sum_{p,m}\frac{\ln(p)}{p^{m/2}}\mathcal{R}e(\chi^m(p))H_T(m\ln(p))\right| \leq \sum_{p^m \leq e^T}\frac{\ln(p)}{p^{m/2}}H_T(m\ln(p))\,.$$

Thus, by using (3), we deduce the inequality:

$$n_\chi\,\Phi_T(\frac{1}{2}) \leq \ln(q) - \ln(\pi) - I_\delta(H_T) + 2\sum_{p^m \leq e^T}\frac{\ln(p)}{p^{m/2}}H_T(m\ln(p))\,.$$

If we now put $T = \ln(3)$ we obtain for any $\delta \in \{0, 1\}$ the estimate

$$1.072\,n_\chi < \ln(q)\,,$$

and we deduce that

$$n_\chi < \ln(q)\,.$$

6   An Upper Bound for the Lowest Zero of $L(s, \chi)$

Let $N(T, \chi)$ be the number of zeros of $L(s, \chi)$ in the rectangle $0 < \sigma < 1$, $0 < |t| < T$. It is a basic fact of the standard theory of L-functions that $N(T, \chi)$ has an asymptotic as $T \to +\infty$ (see more details in [9])

$$N(T, \chi) = \frac{T}{\pi}\ln(\frac{T}{2\pi e}) + O(\ln(qT))\,.$$

Hence, intuitively at least, one can expect that the number of non-trivial zeros of $L(s, \chi)$ with imaginary part less than $1/\ln(q)$ is on "average" absolutely bounded. If one can reach the limit of the resolution provided by harmonic analysis, and justify this intuitive argument, it will be possible to deduce that n_χ is bounded by an absolute positive constant (which is roughly close to Chowla's conjecture). However, as seen in section 5, the best conditional estimate for n_χ is $\frac{\ln(q)}{\ln \ln(q)}$ which is clearly not bounded as $q \to +\infty$. To understand better this problem, Siegel studied the analogy between the behaviour of the Riemann zeta function for variable $s = \sigma + it$ and $t \to +\infty$, and that of $L(s, \chi)$ for variable χ and $q \to +\infty$ [23]. He proved that the lowest zero of $L(s, \chi)$ is essentially bounded by $C/\ln \ln \ln(q)$, where C is an effective positive constant. Next, we give a conditional improvement of the upper bound for the lowest zero $\rho_\chi = \frac{1}{2} + i \gamma_\chi$ of $L(s, \chi)$ distinct from $\frac{1}{2}$ (i.e. $|\gamma_\chi| = \min(\gamma_1, -\gamma_{-1})$). For this purpose, we apply Theorem 1 to suitable functions with compact support. If we assume GRH, then one can prove more precise estimates on γ_χ. Such improvements have been also considered in [18] and [20] for Dedekind zeta functions as an application of the positivity technique in the explicit formula.

Theorem 11. *Under GRH, we have*

$$|\gamma_\chi| \ll \frac{1}{\ln \ln(q)} \, . \tag{4}$$

Proof. To prove the estimate (4), we use another even function G with compact support defined in the following lemma [21].

Lemma 12. *Let*

$$G(x) = \begin{cases} (1-x)\cos(\pi x) + \frac{3}{\pi}\sin(\pi x) & \text{if } x \in [0, 1] \\ 0 & \text{otherwise} \, . \end{cases}$$

Then G satisfies the hypotheses of Theorem 1 and we have

$$\widehat{G}(u) = \left(2 - \frac{u^2}{\pi^2}\right)\left[\frac{2\pi}{\pi^2 - u^2}\cos(u/2)\right]^2 \, .$$

We now apply once more Weil's explicit formula to $G_T(x) = G(x/T)$ and we replace T by $\sqrt{2}\,\pi/|\gamma_\chi|$. We obtain the estimate

$$\frac{8}{\pi^2} n_\chi T \geq \ln(\tfrac{q}{\pi}) - I_\delta(G_T) - 2\sum_{\mathfrak{p}, m} \frac{\ln(p)}{p^{m/2}}\mathcal{R}e(\chi^m(p))G_T(m\ln(p)).$$

Using Lemma 8, the above estimate (a) on n_χ and the fact that the integral $I_\delta(G_T)$ is bounded as T tends to $+\infty$, we deduce the following inequality for some positive constants A and B

$$\frac{\ln(q)}{\ln \ln(q)}A\,T + Be^{T/2} \geq \ln(q) \, ,$$

so that

$$T \geq \min(\frac{1}{2A}, 1 - \frac{\ln(2B)}{\ln\ln(q)}) \ln\ln(q) .$$

Thus for sufficiently large q we have

$$T \gg \ln\ln(q),$$

and so

$$|\gamma_\chi| \ll \frac{1}{\ln\ln(q)} .$$

Corollary 13. *If we assume GRH, we have*

$$\lim_{q \to +\infty} \rho_\chi = \frac{1}{2} .$$

The above corollary shows more particularly that the lowest zero of $L(s, \chi)$ is lower than the first zero of the Riemann zeta function with respect to their imaginary parts (i.e. $|\gamma_\chi| \leq 14.13472$) for sufficiently large q. More recently, S.D. Miller showed [15] that this assumption holds for arbitrary q.

Acknowledgments

I would like to thank the American Institute of Mathematics (AIM) in Palo Alto California for their support (NSF Grant DMS0111966) and for the excellent conditions where part of this article was completed during the workshop "L-Functions and Modular Forms" in August 2007. I thank the referee and Stéphane Louboutin for their comments on the manuscript.

References

1. Balasubramanian, R., Murty, V.K.: Zeros of Dirichlet L-functions. Ann. Scient. Ecole Norm. Sup. 25, 567–615 (1992)
2. Barner, K.: On A. Weil's explicit formula. J. reine angew. Math. 323, 139–152 (1981)
3. Bateman, P.T., Grosswald, E.: On Epstein's zeta function. Acta Arith. 9, 365–373 (1964)
4. Batut, C., Belabas, K., Bernardi, D., Cohen, H., Olivier, M.: User's Guide to PARI/GP, version 2.3.2, Bordeaux (2007), http://pari.math.u-bordeaux.fr/
5. Booker, A.R.: Artin's conjecture, Turing's method, and the Riemann hypothesis. Experiment. Math. 15, 385–407 (2006)
6. Chowla, S.D.: The Riemann Hypothesis and Hilbert's tenth problem. Gordon and Breach Science Publishers, New York, London, Paris (1965)
7. Chowla, S.D., Selberg, A.: On Epstein's zeta function. J. reine angew. Math. 227, 86–110 (1967)
8. Conrey, B., Soundararajan, K.: Real zeros of quadratic Dirichlet L-functions. Invent. Math. 150, 1–44 (2002)

9. Davenport, H.: Multiplicative Number Theory. Graduate Texts in Math., vol. 74. Springer, Heidelberg (1980)
10. Iwaniec, H., Kowalski, E.: Analytic Number Theory. American Mathematical Society Colloquium Publications. vol. 53 American Mathematical Society, Providence, RI (2004)
11. Iwaniec, H., Sarnak, P.: Dirichlet L-functions at the central point. In: Number Theory in Progress, vol. 2, pp. 941–952. de Gruyter, Berlin (1999)
12. Kok Seng, X.: Real zeros of Dedekind zeta functions of real quadratic fields. Math. Comp. 74, 1457–1470 (2005)
13. Lagarias, J.C., Odlyzko, A.M.: On computing Artin L-functions in the critical strip. Math. Comp. 33, 1081–1095 (1979)
14. Mestre, J.-F.: Formules explicites et minorations de conducteurs de variétés algébriques. Compositio. Math. 58, 209–232 (1986)
15. Miller, S.D.: The highest lowest zero and other applications of positivity. Duke Math. J. 112, 83–116 (2002)
16. Murty, M.R., Murty, V.K.: Non-vanishing of L-functions and Applications. In: Progress in Mathematics, vol. 157, Birkhäuser Verlag, Basel (1997)
17. Odlyzko, A.M.: Bounds for discriminants and related estimates for class numbers, regulators and zeroes of zeta functions: a survey of recent results. Séminaire de Théorie des Nombres, Bordeaux 2, 119–141 (1990)
18. Omar, S.: Majoration du premier zéro de la fonction zêta de Dedekind. Acta Arith. 95, 61–65 (2000)
19. Omar, S.: Localization of the first zero of the Dedekind zeta function. Math. Comp. 70, 1607–1616 (2001)
20. Omar, S.: Note on the low zeros contribution to the Weil explicit formula for minimal discriminants. LMS J. Comput. Math. 5, 1–6 (2002)
21. Poitou, G.: Sur les petits discriminants. Séminaire Delange-Pisot-Poitou, 18e année, n 6 (1976/77)
22. Rumely, R.: Numerical computations concerning the ERH. Math. Comp. 61, 415–440 (1993)
23. Siegel, C.L.: On the zeros of the Dirichlet L-functions. Ann. of Math. 46, 409–422 (1945)
24. Soundararajan, K.: Non-vanishing of quadratic Dirichlet L-functions at $s = \frac{1}{2}$. Ann. of Math. 152, 447–488 (2000)
25. Watkins, M.: Real zeros of real odd Dirichlet L-functions. Math. Comp. 73(245), 415–423 (2004)
26. Weil, A.: Sur les formules explicites de la théorie des nombres. Izv. Akad. Nauk SSSR Ser. Mat. 36, 3–18 (1972); Reprinted in: Oeuvres Scientifiques, vol. 3, pp. 249–264. Springer, Heidelberg (1979)

Author Index

Lecture Notes in Computer Science

Sublibrary 1: Theoretical Computer Science and General Issues

For information about Vols. 1– 4666
please contact your bookseller or Springer